PLASTICS PROCESS ENGINEERING

PLASTICS PROCESS ENGINEERING

JAMES L. THRONE
Naperville, Illinois

MARCEL DEKKER, INC.   New York and Basel

Library of Congress Cataloging in Publication Data

Throne, James L. [Date]
  Plastics process engineering.

  Includes bibliographical references and index.
  1. Polymers and polymerization. 2. Plastics.
I. Title.
TP156. P6T44   668. 4'1   78-9883
ISBN 0-8247-6700-4

MARCEL DEKKER, INC.
270 Madison Avenue, New York, New York 10016

Current printing (last digit):
10 9 8 7 6 5 4 3 2

PRINTED IN THE UNITED STATES OF AMERICA

To my loving, devoted, and understanding wife, Jean, whose faith in me has been constant throughout my otherwise variable, nomadic career

Foreword

There has long been a need for a definitive publication on the engineering aspects of plastics processing. The Society of Plastics Engineers is pleased to sponsor and endorse this new volume, Plastics Process Engineering, as the publication which serves this long-standing need. This book represents yet another contribution by the publisher and the Society to literature concerning the emerging technology of plastics.

Through the medium of its Technical Volumes Committee, SPE has been involved in sponsoring books of this nature since 1956. During this period it has been the catalyst for publication of at least one book annually in some vital segment of plastics not previously covered in the literature. The committee has surveyed and determined needs for specific titles, located authors and editors, considered broad content, and, most importantly, reviewed final manuscrips to ensure accuracy of technical material.

It is this same technical competence that is the Society's hallmark in its other activities—education programs, conferences, periodicals, and meetings. Its reservoir of 20,000 practicing plastics engineers is its greatest resource, a resource which has made it the largest organization of its type in plastics worldwide.

Again, it is with particular pleasure that the Society sponsors this much needed volume authored by a member who has long been active in SPE.

Robert D. Forger
Executive Director
The Society of Plastics Engineers, Inc.

"Halfway through my Ph.D. treatise on Ebonite and some joker comes along inventing plastics…"

Preface

In the closing years of the nineteenth century, consumer demand for consistent quality in petroleum distillates forced many organic chemists out of the laboratory and into design of industrial processing equipment. Similarly, mechanical engineers with rudimentary training in chemistry were pressed into chemical plant operation and maintenance. From this interface came the new engineer, the chemical engineer. A similar demand for consistent quality in metals brought physicists and civil engineers together to form the foundations for the metallurgical engineering profession. Today the consumer demand for articles of plastics are bringing together the chemical engineer and the organic chemist into what will be called plastics process engineering.

The tailoring and engineering of macromolecules is in contrast to the refining and cracking technology of petrochemicals and to the admixturing and solution technology of metals. Creation and processing of the intimately entangled molecules made of thousands of simple units offer a challenge and a frustration to any engineer. Many processes that have been developed to convert these materials into useful products depend on the ability of these molecules to seemingly defy natural laws. Many engineers, upon first encountering such anomalies, abandon hope for a logical understanding and resort to correlations and hunches. Others become enamored with the cuteness of the material property behavior, and propose extreme theoretical hypotheses which cannot be substantiated except for the purest laboratory materials. Few are equipped with sufficient understanding of the materials and the processes to predict their behavior.

It is the intent of this work to demonstrate to the engineer that sound engineering principles, accepted and used daily in other fields of engineering, can be applied with good success in plastics engineering. No attempt is made, however, to hide the differences in processibility of macromolecules and the simpler materials. Likewise, the absence of sound engineering reasons for a particular processing phenomenon or the lack of suitable engineering data upon which to construct a reason are not concealed. It has been my intent, however, to move away from the rheological and morphological foundations that are found in other texts on plastics processing. This does not mean that these areas are not important. Nor does it mean that these areas are not applicable. It has been my experience, however, that most engineers are not grounded in rheology and morphology, whereas they have good backgrounds in the macroscopic systems of heat transfer, fluid flow, reactor design, thermodynamics, and economics. Enhancement and rethinking of these principles will frequently enable a good engineer to solve a large portion of the problems he faces in plastics engineering. Understanding rheology and morphology in an effort to improve processes and products is essential, but it appears that an understanding of the process from an engineering viewpoint is necessary first. This, then, is the main thrust of the book.

The text is written for senior undergraduates and first-year graduate students in either chemical or mechanical engineering. The sections on reaction kinetics do not detail the molecular concepts that are traditional in chemical engineering; the emphasis is instead on stability and control. The basic concepts of heat transfer, most familiar to mechanical engineers, can be understood by the chemical engineer with some diligence. Solution thermodynamics offers sufficiently different fare that neither will have an advantage. Although the book format allows the engineer to move from the formation of the macromolecule through the processing of the plastic, only the sections on fluid flow and heat transfer are referred to extensively throughout the book. Since the field is developing very rapidly, only the major processes are detailed. Certain newer developments, such as thermoplastics structural foam molding and injection blow molding, are referred to only as examples of how present technology has been extended. In nearly every process discussed, the basic principles underlying the process are explored in depth. With some modifications for given materials, these principles should be applicable every time the process is considered.

There are no apologies for the vagaries and difficulty in many of the homework problems. First, many are distilled from actual industrial problems, which are notorious for their vagaries and high levels of difficulty. And second, the few data available are indicative of the present level of sound engineering information in the field. Some care was taken, however, to eliminate trivial questions.

I have tried to draw on those engineering areas that seem to me to be long-lived. In my decade of industrial experience, at duPont, American

Standard, and Beloit Corporation, I have seen these types of problems occur from a resin supplier's point of view, from an end-user's point of view, and from a machinery builder's point of view. Therefore they seem to be valid material for an engineering text. I have also reworked the material into a readable, albeit somewhat simplistic, way so that the underlying principles are somewhat obvious. This is not always the case with industrial problems. I hope that the level of material is sufficiently uniform. However, in my six years as an associate professor at Ohio University and University of Wisconsin-Milwaukee, and as visiting professor at University of Cincinnati, Newark College of Engineering, and Universidade Estadual de Campinas, Brazil, I have learned that what is obvious and transparent to the teacher is frequently obtuse and opaque to the student, no matter how discerning he is. I hope that the text represents a good compromise between the ideal "Run, Jack, Run" version desired by the student and the obfuscated convoluted version sometimes preferred by instructors.

I have also tried to couch the material in a format that is relevant to the practicing engineer. Frequently, material in a given chapter moves from the academic realms, where theory is sound and examples are abundant, to the practical production realm, where a good guess is sometimes better than nothing and the entire operation is confused by overlapping phenomena and the absence of any guidelines. To this end, I have spent many hours compiling an extensive index to help the practicing engineer locate the relevant material, and I have tried to cross-reference the material throughout the book.

I hope that the plastics processing practitioner will not be too hard on me in the area of comprehensiveness. I have not conducted a complete analysis of every area of plastics process engineering. However, I have tried to illustrate fundamentals, but I have not done as thorough a literature review as I would normally do for a research paper. My objective was to select references that would most clearly illustrate the points I was trying to make; and in doing so, other, more relevant references were undoubtedly missed. Furthermore, as with all authors, I admit to a bias in favor of specific ways of looking at problems or regions of analysis. I favor a phenomenological approach, and thus have unduly deemphasized the molecular, statistical, and theoretical approaches. To practitioners in these areas, I apologize and hope that you who are teaching can supplement my material with yours. Your comments and criticisms would be appreciated.

I acknowledge the hundreds of U.S. seniors and graduate students who have suffered through the tensor calculus, the blurred mimeo handouts, and the impossible-to-answer homework problems of earlier versions. I hope that their patience during the early groping and the formative years of the development of this text is rewarded within. Certainly many have become successful plastics process engineers and apparently were not badly crippled or discouraged by the material. I must acknowledge the food process engineering students of Campinas who forced me to rethink the material into book form. The parallels of these fields are quite startling. I acknowledge the patience of Mrs. Patricia Lentz, who typed the original manuscript from the most confusing rough draft. I acknowledge my wife, Jean, to whom the book is dedicated, for her willingness to undertake revision and updating, even when it appeared that we would need to mortgage the homestead to get it published. I acknowledge the excellent graphic artists who helped put the ink onto the vellum in orderly fashion. Certainly, Mike Throne's help in the final stages was most appreciated.

Please note that the original manuscript was completed in first draft on February 1, 1975, with the revision and update in accordance with the reviewer's comments being completed on June 1, 1977. During that period of time, I have acted as an unpaid consultant to Sherwood Technical Service for the purpose of production of this manuscript. My present company is therefore not responsible for any of the statements or comments made in this text.

James L. Throne

Contents

Contents

PLASTICS PROCESS ENGINEERING

1.

Plastics and Their Uses

1.1 Introduction

The most general definition of a plastic is a material that is formable during some step in the manufacturing of a useful article. This definition will allow inclusion of many inorganic materials, such as cement, glass, copper, and cast iron, and is therefore probably not suitable. More specifically, plastics are organic materials comprised of molecules having many thousands of individual units or more. Naturally occurring materials such as cotton, wool, linen, and silk are therefore plastics. Wood is a plastic. Tar and pitch are plastics. Further restrictions limit plastics to those organic materials which are comprised of a limited number of specifically selected or refined molecules having regular orders of chains with thousands of individual units or mers. The requirement of selectivity restricts this definition to man-made or man-altered materials and suggests purposeful and selective control over the type of molecule used to form a given plastic. Furthermore, the number of molecular types in each plastic is limited. This allows the identification of polyethylene as a plastic but not tar or resin. Note that the definition does not restrict the form for the material, however. Therefore, the plastic need be formable for only a short period of time during the transition from a resin to a final product form. Nor does this definition restrict plastics to a specific source of raw material. Thus,

1

plastics based on cellulose, starch, sugars, proteins, and silicon are allowed as plastics, as are hydrocarbon-based plastics. This definition will become more important in the future as alternative sources for plastics are explored.

Man has always sought to imitate nature in the development of utilitarian goods. The fantastic fabrics of today owe allegiance to caveman's use of animal skins in imitation of other animals. The development of naturally occurring fabrics, such as wool, cotton, linen, and silk, was a logical improvement in the quality of the garment and not innovative. The present synthetic fabric market is also a logical extension of man's ability to improve on natural materials in order to achieve certain qualities and consistencies not found in nature. The newest major plastics market, structural foams, is similarly a desire of man to improve on wood. As a result, the primary markets for these materials are those dominated by wood products. DuBois and John [1] attribute the first deliberate molding of plastics materials to Malayan Indians in 1843, where the vegetable protein, gutta-percha, was molded into knife handles. Twenty years before this, however, rubber was being molded under temperature and pressure. Cellulose nitrate was developed in France and Switzerland in the late 1830s by leeching cellulose with nitric acid in the presence of concentrated sulfuric acid. Alfred Critchlow was making rosin-based molded buttons and combs in the late 1840s in Massachusetts. Alfred Parkes exhibited molded knife handles and combs made of precipitated cellulose nitrate at an international exhibition in London in 1862. Nevertheless, the first deliberate pursuit of a synthetic plastic as a replacement material for a naturally occurring material has been attributed to John Hyatt for his invention of camphor-reacted cellulose nitrate in 1870. He responded to a $10,000 invitation to discover a replacement for ivory in billiard balls. He and his brother subsequently patented several methods for compression-molding articles of Celluloid and found a large market in the manufacture of dentures.

Dr. Leo Baekeland found a way of combining phenol and formaldehyde to produce a synthetic substitute for metal and wood. This substitute is called Bakelite and is still used extensively for electrical outlets and boxes. The material was completely moldable and spurred a rapid development of the plastics industry in Germany and the United States prior to World War I. The development of aviation during World War I caused a need for parachutes. Silk was the best naturally occurring material, but owing to the cost, the limited supply, and the deterioration of the material, a substitute was sought. Several companies, including Bayer Fabriken in Germany and E. I. duPont in the United States, developed laboratories dedicated to the search for a substitute fiber. In the late 1920s, Dr. H. C. Carothers of duPont concentrated on the simple building blocks of proteins, the amino acids. His work led to the discovery of the family of nylons, all of which are based on the concept of the simple α-amino acids, and to the discovery of acetals, which are based on formaldehyde. His earliest work, however, led to the discovery of neoprene, as a substitute for natural rubber. This giant did not see his discoveries form the foundation for most of today's worldwide synthetic fiber industry, however, since, feeling that his work was not leading to suitable recognition, he committed suicide in 1937. Neoprene was not a successful substitute for natural rubber, however, since it was too soft. In the 1930s, I. G. Farben in Germany discovered the polymerization techniques to produce butadienes or Perbunan rubbers, and the resulting products enabled the German power to meet demands long after its Asian sources of natural rubber were cut off during World War II.

Otto Röhm found ways of polymerizing acrylates in 1901 and was awarded a doctorate for his work. The need for a tough transparent plastic as a substitute for glass was not apparent until just prior to World War II. High-performance fighter planes required strong materials for windows, and available glasses were not strong enough. As a result, Röhm and Haas in the United States developed the first commercial Plexiglas in 1935. Imperial Chemicals Industry, Ltd. (ICI) concentrated on ways of polymerizing ethylene, obtaining the first commercial high-pressure polyethylene in 1931 and the first laboratory quantities of low-pressure linear polyethylene in 1947. Most of the present olefinic materials, however, are the direct result of Dr. G. F. Natta's researches in organometallic catalysts for polymerizations, beginning in the early 1950s.

It is rather surprising that an industry producing annually more materials by volume than all ferrous metals combined is dependent on inventions that are less than a century old. Many plastics have been discovered, invented, and reduced to production within the last 60 years or so, including polyvinyl chloride (a Russian discovery in 1912); polyvinyl alcohol (discovered in 1924); polystyrene (discovered in Germany in 1929); unsaturated polyester resins, the foundation for the fiberglass-reinforced plastics industry (discovered in 1933); polyurethanes (produced first in 1937); epoxies (invented in Germany in 1939); silicones (produced by Dow Chemical Co. in 1942); Teflon (tetrafluoroethylene) (produced commercially in 1950); polypropylene (discovered in Italy in 1954); polycarbonates (discovered in Germany in 1956); and polyimides (discovered in 1964). These materials form the basis for more than 90% of the plastics consumed worldwide today.

Along with the major materials discoveries have come inventions which enable processing. The compression molding press is of course a simple extension of a hand-operated screw press in common use for more than 2000 years. The injection molding press, however, was invented in Germany in 1921. The screw extruder was invented in 1935, and the blow molding machine, which was to accelerate the transition from glass bottles to plastic bottles, was invented at about the same time in the United States. Rotational molding and thermoforming processes can be traced to rudimentary machines used during World War II. The screw-injection molding machine, most popular in the United States today, was developed in the early 1950s. The structural foam-injection molding machine of today is no more than 10 years old.

In all these examples and many more, plastics have been invented and developed as viable materials primarily when a need for a substitute for more conventional materials was felt. Certainly in times of conflict (World War I, World War II, Korea, Vietnam), shortages of critical metals and fibers force industries to develop plastics as substitutes. The space race of the 1960s spawned many new applications for plastics, including lightweight electrical insulation (polyimides) and platable transparent impact surfaces (polycarbonates). Today, it is difficult to identify a sector of our way of life that is not touched by plastics. Plastics are found in automobiles, in the kitchen, throughout the house, in clothing, in food processing, and in the medical profession, to name a few areas.

Throughout this text, then, the emphasis will be placed on the shaping, forming, or molding of plastics materials. In particular, the processing must put the material into a formed state as efficiently as possible, and this state must remain stable indefinitely under end-use conditions. In most cases, the process begins with a plastic in a solid resin form as powder or pellets, in a liquid resin form to which a catalyst or reactive agent is added, or in a prepared sheet form. In nearly all forms of fabrication, the plastic must be heated to a forming temperature (until it softens or melts), it must be pushed into a final state, and it must be cooled until it retains the desired shape without additional effort. It is the method of heating, forming, and retaining that forms the basis for plastics processing, and it is the understanding of the physical forces that control the processing that is called plastics process engineering.

1.2 What Are Plastics?

Plastics are man-made or man-altered regular chains of a large number of simple molecules. The most common of all plastics, polyethylene, can be used to illustrate this. Ethylene is a simple molecule ($CH_2{=}CH_2$) having a double or unsaturated bond between the two carbons. It is a by-product of the petroleum industry, although it can be obtained by cracking any hydrocarbon source, such as coal. If great pressure (30,000 psi) and high temperature (600°F) are applied, the double bond can be opened to make the molecule reactive. With proper control, reactive molecules can be joined very rapidly, thus forming a long chain of molecules. Ethylene is the monomer or unit; polyethylene is the polymer. As shown in Table 1.2-1, polyethylene is a very long molecule. If the ethylene molecule were magnified 100 million times, it would be 3/8 in. long. A polymer molecule of low-density polyethylene can be 50 ft long on the same scale. Special-purpose polyethylene molecules, known as ultrahigh-molecular-weight polyethylene, have chains on the same scale that would be 2 to 5 miles in length. It is apparent that these chains cannot be found in extended form. Rather, they tangle and are twisted around one another. It should be apparent that

Table 1.2-1

Comparative Molecular Sizes of Various Thermoplastics and Thermosets [2]

Material	End-to-end distance (Å)	Degree of polymerization	Model length, CH_2 groups, 3/8 in. diameter
Phenolics	40.1	8	8 1/2 in.
Melamine	35.4	5	7 1/2 in.
Alkyd (unsaturated polyester)	94.4	19	20 in.
Epoxy (bisphenol A) (e.g., Epon 828, Dow 331)	15.9	1	3 1/3 in.
Epoxy (bisphenol A), medium MW[a]	110.89	6.5	2 ft
Epoxy (bisphenol A), high MW	580.6	34	10 ft
Polyethylene, low density	2,540	1,000	45 ft
Polyethylene, ultrahigh MW	907,000	36,000	3 miles

[a]MW = molecular weight.

this entangling, even in a melt state, should cause the viscosity of the polymer to be much higher than that of the monomer. Thus polymers are liquids characterized by viscosities that are 10^5 to 10^{25} times greater than their monomers. This is duly recognized in the processing of the materials, as will be seen. Plastics that are characterized by long, entangled, but identifiable chains are called thermoplastics. Roughly two-thirds of all plastics used worldwide are thermoplastics.

Thermosetting plastics are characterized by short bulky molecules that are on the average only 10 to 20 monomer units long. As shown in Table 1.2-1, on the 100,000,000:1 scale, phenol-formaldehyde is about 9 in. long and nearly 1 in. in diameter. In this state, the resin is plastic and formable. However, if the resin is heated under pressure, reaction will take place in the middle of the molecule as well as on the reactive ends. This additional reaction causes the molecule to cross-link into a permanent three-dimensional structure. As a result, unlike the thermoplastic which can be heated, shaped, and cooled many times, the thermoset will not flow when reheated. It has taken a permanent set. As might be predicted, many reactive sites along a short chain will lead to a high degree of cross-linking and a potentially stiff plastic. Few sites will lead to a potentially soft plastic. Rigidity can also be controlled by selecting the proper plastic in the short backbone and the proper length for the cross-linking agent molecules.

There are more than 20 families of plastics commercially available today. More than 30,000 plastics have been invented and patented. It is possible, then, to combine certain single chains of plastics with chains of another plastic to achieve properties that combine the best of both plastics. These combinations are called copolymers. A good example of how the proportion of one plastic to another can change the final characteristics of the plastic is the copolymer butadiene-styrene. Polystyrene is a clear brittle plastic commonly used for disposable containers. Polybutadiene is a synthetic rubber. A copolymer of 25% polystyrene, 75% butadiene yields SBR rubber, with direct applications to carpeting, padding, seat cushions, and so on. A copolymer of 25% butadiene, 75% styrene yields an impact styrene, with applications in equipment cabinets and appliances. The addition of a small amount of acrylonitrile to impact styrene yields a terpolymer called ABS, with major applications in plumbing systems and telephones. ABS is much tougher than impact styrene and has better surface properties.

Plastics are frequently combined with smaller molecules to aid in processing. Some processing aids or additives help the molten material flow more easily under pressures. Antioxidants keep materials from turning yellow when heated. UV stabilizers screen ultraviolet light to prevent the materials from yellowing and cracking when exposed to sunlight. Colorants are added to the plastic to minimize painting. Fire retardants reduce the fuel value of certain materials. External lubricants aid in removal of the part from the mold. Internal lubricants aid flow of resins in the early stages of processing. Chopped glass strands and mineral wool are added to stiffen an otherwise weak material. Talc, woodflour, and calcium carbonate are used to extend an expensive material. Water can be used as a filler in some thermosets. Nitrogen and volatile fluorocarbons can be used to introduce voids in certain plastics to reduce weight and minimize abnormal shrinkage. Thus, even though much is known about the chemistry of polymerization and processing of pure polymers, little is known about the interaction of additives with the pure polymer. As a result, much ingenuity and application of some basic engineering principles are needed to form final useful products of plastics materials.

Plastics are not cheap substitutes for naturally occurring materials. As shown in Fig. 1.2-1, on a pound-for-pound basis, plastics are among the most expensive raw materials. Even one of the cheapest plastics, general-purpose polystyrene (GPS), costs nearly twice as much as steel and more than 10 times more than glass or clay. Even the monomer, styrene, costs more than steel. Many common plastics raw materials are very expensive, as shown in Fig. 1.2-2. Nylon, for example, costs about

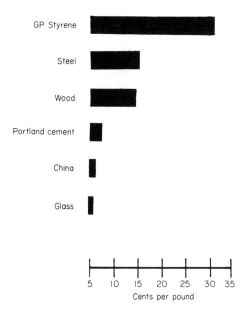

Figure 1.2-1   Comparative raw materials prices.

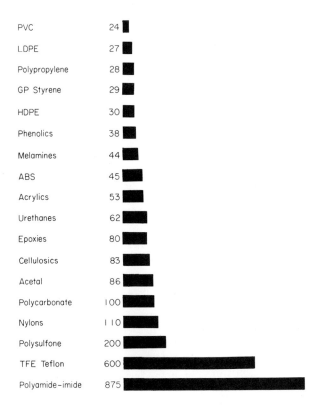

Figure 1.2-2   Relative costs of plastics materials (May 1976).

$1 per pound and Teflon about $6 per pound. With the worldwide intensive effort to restrict the use of petroleum during 1974, the wholesale prices of plastics materials has risen from an all-time low of 80 (based on 1967 = 100) as recently as February 1972 to 175 in October 1974 and 195 in January 1977. During the 20-month period of rapid materials cost inflation, all commodities increased from 115 to 167. In January 1977, the commodity index stood at 185. This is seen in Fig. 1.2-3. This more than 100% increase in plastic raw materials has had a dramatic effect on the cost of the finished plastic goods but has not dramatically affected the demand for the goods. The present economic status of the plastics industry will be considered in detail later.

The principal reason for the tremendous growth in the plastics industry, then, is not the cheapness of the resin. Rather, the key to the utility of plastics is the ease by which they can be converted from the resin to the final part. For example, plastics can normally be converted into useful products at costs that are two to three times the base resin price. China is a very cheap raw material but costs 25 to 30 times the base material costs in conversion to useful materials. Many operations dealing with naturally occurring raw materials are either energy intensive, such as steel or glass, or labor intensive, such as wood. As will be shown, many plastics conversion operations can be run automatically, with a minimum of attention. Most conversion operations use shear heating and hydraulic pressures, both of which are energy-conserving processing techniques. Therefore, in spite of rising resin raw materials prices, the plastics industry should continue to infiltrate the realm of natural materials.

1.2.1 Commodity Plastics

In the United States, nearly two-thirds of the plastics used are polyethylenes, polypropylenes, polystyrenes, or polyvinyl chlorides (PVCs). These materials are all based on the ethylene double bond mentioned above. As shown in Table 1.2-2, the polyethylenes represent more than a quarter of the plastics market. These materials are characterized by their easy processibility and adaptability and by relatively soft, weak mechanical characteristics. They have excellent electrical insulating characteristics and as a result are used extensively as wire sheath. They are excellent moisture barrier materials, as shown in Fig. 1.2-4, and find great application in the building industry. Further, they are nearly chemically inert and thus are extensively used in medical and chemical industries. They cannot be used in boiling water service and cannot be glued. But they have good low-temperature stability and can be heat-sealed and thus find extensive applications as film and packaging wrap. Roughly 16% of the U.S. consumption are PVC-based resins. Polyvinyl chloride in a raw resin state is a hard, bone-like substance that is very difficult to process. Modifiers adapt to PVC very readily, however, allowing a wide range of flexibility to be compounded into the PVC prior to processing. Rigid PVC for piping contains relatively little

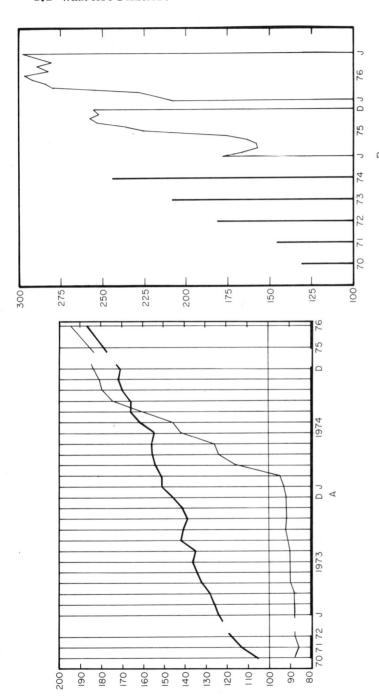

Figure 1.2-3   A: Comparison of plastics and all commodities price indices, 1970 to 1976. Source: Bureau of Labor Statistics, U.S. Department of Labor. Courtesy Modern Plastics Magazine. Wholesale price index, 1967 = 100. Years 1973, 1974 amplified to show rapid rise in plastics price index. Note: heavy line, all commodities; light line, plastics resins and materials. B: Plastics output, 1967 = 100, adjusted. Source: McGraw-Hill Economics Department. Courtesy, Modern Plastics Magazine. Data show effects of U.S. economic recession and recovery.

Table 1.2-2

General Characteristics of Plastics

| | Advantages | Disadvantages | Usage (1000 metric tons) | | | | | % Market (1974) |
			1971	1972	1973	1974		
PE polyethylene	Flexible; easily processed; electrically insulating; water barrier; nearly chemically inert	Soft; low-temperature service; easily scratched	(LDPE) 1974 (HDPE)[a] 864	2372 1026	2691 1248	2769 1275		20.7 9.6
PVC vinyl	Easily processed; can be plasticized to allow rigid-to-soft flexibility; good weathering characteristics	Some properties such as inertness reduced by plasticizers; will sunfade; attacked by hydrocarbons	1571	1975	2151	2180		16.3
PS polystyrene	Clear; easily processed; easily compounded to improve impact resistance; good stain, abrasion resistance	Not outdoor material; fades; brittle unless compounded; loses some resistance to staining; attacked by hydrocarbons	1337	1627	2356	2328		17.4
Polypropylene	Very flexible; very low density; does not stress-crack; excellent electrical resistance; good abrasion resistance; can be electroplated; superior weathering characteristics	Cannot be heat-sealed; difficult to dye; somewhat soft in unmodified state	590	767	1012	1061		8.0

Acetals	Low coefficient of friction; easily processed; high abrasion resistance; solvent resistant	Must be copolymer for dimensional stability; some loss in properties during reprocessing of scrap; not suitable for outdoor applications	24	27	31	30	—
Polycarbonates	High heat resistance; excellent clarity; self-lubricating	Difficult to process owing to water absorption; difficult to color	20	25	47	51	—
Cellulosics	Excellent films; transparent; propionate films very tough	High water absorption; UV sensitive; easily scorched	68	75	77	76	—
Acrylics	Optical clarity; excellent outdoor material	Crazes; attacked by solvents; easily scratched; brittle	185	209	233	243	1.8
Nylons	Low coefficient of friction; nearly chemically inert; easily processed; good impact; fatigue characteristics	Absorbs water; retains set; statically charged	58	67	87	88	—
Polysulfones	High-temperature applications; high strength; self-extinguishing; easily colored	Attacked by hydrocarbons	Small	do[b]	do	do	—
TFE Teflon	Chemically inert; low friction coefficient; high-temperature applications	Cannot be extruded or injection-molded; soft	Small	do	do	do	—

Table 1.2-2 (continued)

	Advantages	Disadvantages	Usage (1000 metric tons)				
			1971	1972	1973	1974	% Market (1974)
Polyimides	600°F applications; very high dielectric strength	Cannot be molded using conventional equipment; expensive	Small	do	do	do	—
Polybenzimidazoles	900°F applications; excellent adhesives; can be glass-reinforced for added strength	In experimental quantities only; cannot be molded on conventional machines	Small	do	do	do	—
Polymethylpentene (TPX)	Good strength at 400°F; good optical quality; lightweight; excellent electrical properties	Available in powder form only; colors must be dry-blended; sun-fades; flammable	Small	do	do	do	—
ABS	Easily tailored properties; excellent impact strength; high gloss retention; can be plated	Fades in outdoor applications; not easily colored	300	388	422	406	3.0
Amino resins	Excellent surface characteristics; excellent strength; solvent resistance; chip resistance	Usually filled with α-cellulose; unfilled not easily colored; not easily injection-molded	21	20	21	20	—

Unsaturated polyesters	Usually used with glass for reinforcing; easily colored; resin available in variety of forms	Flammable; some attack by ultraviolet light	409	496	555	538	4.0
Urethanes	Easily foamed; good bearing surfaces	Difficult to reinforce	414	459	593	622	4.7
Epoxies	Low heat curing; high strength; excellent adhesive; good chemical resistance; easily filled	Viscous liquid; entrains air bubbles; requires separate catalyst	69	78	102	106	—
Diallylphthalate (DAP)	High water resistance; good high-temperature properties; excellent strength	Not easily colored; material not easily processed	Small	do	do	do	—
Phenolics	Good surface finish; low water absorption; excellent flame resistance	Cannot be colored	540	652	624	587	4.4

[a]LDPE = low-density polyethylene; HDPE = high-density polyethylene.
[b]do = ditto.

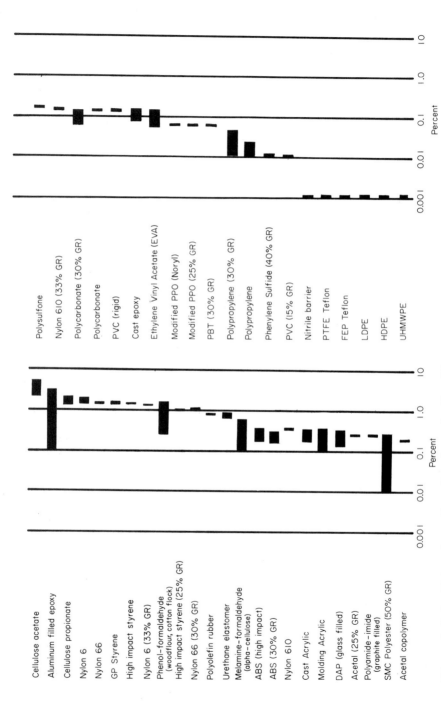

Figure 1.2-4 Water absorption of various plastics [2].

plasticizer. Seating is frequently covered with flexible vinyl which contains a high concentration of plasticizer. The ability to adjust the final product flexibility is somewhat offset by the poor UV and hydrocarbon resistance of the plasticized product.

Until recently, polystyrene exhibited the most rapid growth of the common plastics. Fuel allocations have greatly restricted the supply of styrene, and the resulting price increases have slowed the growth. In a general-purpose state, polystyrene is a clear, somewhat brittle plastic. It is easily processed and has good heat stability. Materials are added to polystyrene to improve its impact resistance and UV characteristics and to minimize hydrocarbon solvent attack. Nevertheless, it has excellent stain resistance, an index of refraction similar to that of crystal, and a very low haze level. As mentioned, polystyrene can be modified by copolymerizing other plastics with it. Butadiene-styrene is an impact styrene or an SBR rubber. Acrylonitrile-butadiene-styrene is a terpolymer with impact characteristics that exceed the best impact styrene. Styrene-acrylonitrile (SAN) is a transparent bluish material that is tougher than acrylic.

Polypropylene is now the most rapidly growing plastic. It is easily processed, is very tough and flexible, and does not stress-crack. It has excellent chemical resistance and superior weatherability and was the first thermoplastic to be electroplated. It is finding wide acceptance as a filled plastic. Polypropylene is the base for the new generation of thermoplastic elastomers and as a result should continue to penetrate such large markets as automotives and housing.

1.2.2 Engineering Plastics

The engineering plastics are normally characterized as materials that have properties superior to the olefinic and styrenic materials. Long-term heat stability, high heat distortion temperature, high tensile and flexural characteristics, good impact strength, and raw material prices about two to three times those of the common plastics characterize most engineering plastics. The most familiar engineering plastic family is nylon. Although familiar to nearly everyone, the entire output represents less than 1% of the entire U.S. plastics consumption. These materials are characterized by low coefficients of friction, good impact and excellent abrasion resistance, and high tolerance to hydrocarbon solvents. The materials, as a family, absorb water, as shown in Fig. 1.2-4, and thus cannot be used where static charge buildup is potentially dangerous. As mentioned, nylons were found to be successful replacement fibers for silk and have continued to play an important role in synthetic fabrics. Acrylics, dominated by polymethyl methacrylate (PMMA), are excellent outdoor plastics. In addition, they have optical clarity nearly equal to and toughness many times that of common glass and can be used for windows, signs, and exterior paints. They are attacked by simple solvents, however, and built-in molding stresses are relieved through crazing. Most recently, acrylics are being copolymerized

with vinyls to increase their strengths and minimize the crazing characteristics. Acetals are formed by polymerizing formaldehyde, the simplest of
the amino acids. The resulting polymer is strong, tough, and self-lubri-
cating and can be used in hot water service without significant deterioration
or dimensional change. Unlike the more complex amino acid family, nylon,
acetals have not been successfully spun into fibers.

Polycarbonates are also relatively new materials that are growing
in applications more rapidly than most engineering resins. These materials
are comprised of very bulky molecules that have little order in the solid
state. As a result, the materials remain amorphous or noncrystalline and
transparent. While not as clear as acrylic, polycarbonate is about three
times tougher than acrylic and thus is finding extensive use in high-impact
transparent applications. The high-impact character of the material at low
temperature has been used extensively by the hand power tool industry with
great success. The all-plastic power tool case has eliminated extensive
electrical grounding precautions. Cellulosics still command about 1% of the
total market, although many of the applications using the properties of cellulosics such as high transparency of cellulose acetate or flexibility of cellulose butyrate have shifted to other thermoplastics. Nevertheless, enormous quantities of cellulose acetate film (cellophane) are made each year
for the packaging industry. Butyrate is still the acknowledged material of
choice between the sheets of glass for automotive safety glass. Rayon, a
xanthated cellulose, finds markets in fabrics, and Japan is working on a
rayon-type cellulose-based synthetic wood-pulp substitute [3]. Technology
in cellulose-based resins is again underway in an effort to find alternative
materials to the petroleum-based resins.

1.2.3 Specialty Plastics

There are hundreds of plastics that form the nucleus of specialty
plastics. Here specific properties make the resin useful, in spite of enormously high resin prices. For example, polysulfones are desirable where
high temperatures are required, such as in an electrical base against which
soldering takes place. Tetrafluoroethylene Teflon is probably the most
commercially important specialty plastic. Of the many types of fluorinated
carbon polymers developed in recent years, TFE Teflon remains the most
useful. It has the lowest coefficient of friction of known plastics materials
and is chemically the most inert. Its high temperature stability has enabled
it to be used in frying pans and greaseless high-temperature bearings.
Because TFE Teflon has a viscosity 10^{12} times that of water at its melting
point, common ways of processing cannot be used. As a result, it is sintered in much the same way brass or nickel is or is flame-sprayed onto a
prepared surface.

Of the other specialty plastics, the polyimides, polyamide-imides,
and polyphenylene sulfide seem likely candidates to become sound engineering resins of the future. The polyimides became important when it was

discovered that they had more than twice the electrical resistivity of conventional extrudable plastics. As a result, wire sheath weight could be reduced significantly. More recently, it has been found that copolymers of polyimides and polyamides (nylons) yield products that are tougher and more dimensionally stable and less water absorbant than nylons. The new class of polyamide-imides will certainly be explored as carefully as the butadiene-styrene copolymers have been. Polyphenylene sulfide (PPS) is an extremely slippery plastic when loaded with molybdenum sulfide and Teflon. As a result, nonlubricated ball bearing races have been used with PPS with great success. Since the price is very high, very careful application must be made in order for the end use to be economically attractive. Nevertheless, PPS seems to have an excellent chance of capturing those areas where the softer Teflon cannot perform well.

1.2.4 Thermosets

Although thermosets were the first plastics materials developed, they occupy only 15% of the market place. Much of the lag in development is attributable to the rudimentary form of processing equipment. Until a few years ago, low-pressure casting and compression molding and its automated improvements were the only ways of processing thermosets. Recent developments of liquid reaction molding, wherein two components of epoxy or urethanes are mixed at high shear and injected under pressure into a mold, and screw-injection molding, wherein a screw-injection molding machine is modified to accept a two-stage phenolic, have drawn much interest. Paralleling the machinery development is a concerted effort to extend the time during which the reactive material remains fluid. Successful development of automated methods of processing reactive materials will most certainly allow thermosets to claim a larger portion of the market. The primary reason for this is the superior properties offered by thermosets. As shown in Fig. 1.2-5, the most common thermoset, phenol-formaldehyde, has a heat distortion temperature greater than most engineering plastics. Glass-reinforced thermoset polyester has a flexural strength that rivals aluminum, and aluminum-filler epoxy has a flexural strength greater than magnesium. Figure 1.2-6 illustrates this. Phenolics command about 40% of the thermoset market with excellent electrical and heat resistance and low water absorption. Amino resins, using melamine in place of phenol, have been used where cosmetic surfaces are required. α-Cellulose is frequently used as a filler, and until the development of polypropylene, melamine-formaldehyde was the only plastic suitable for hot water application. The polyester family is a generic name for materials using maleic acid, phthalic acid, or similar materials as a basic building block. With proper selection of additional materials, a thermoset resin can be formed that can be easily catalyzed with a peroxide. As a result, thermoset polyester in styrene is used as a binder with fiberglass and can be sprayed or layed by hand on forms to form parts at room temperature. Recent

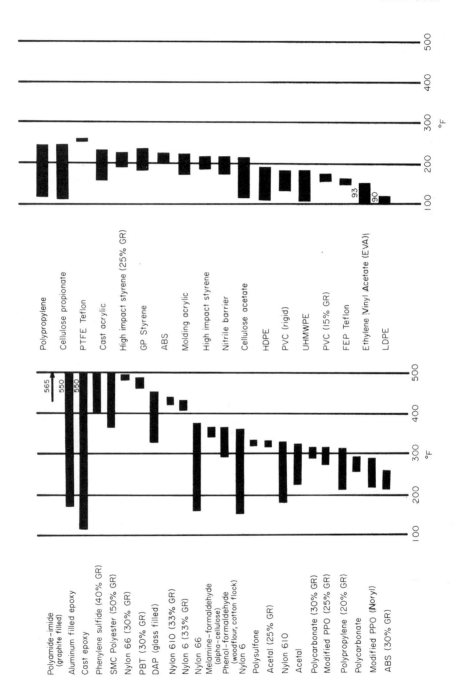

Figure 1.2–5 Heat distortion temperatures for various plastics [1].

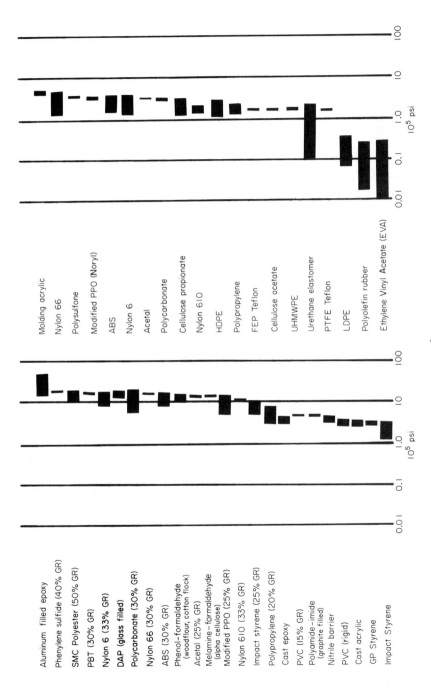

Figure 1.2-6 Flexural modulus of various plastics (10^5 psi) [1].

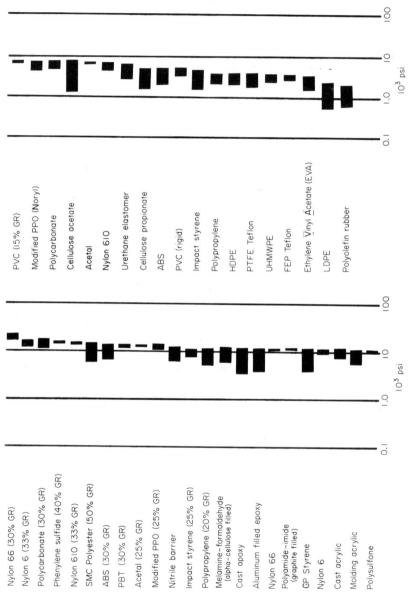

Figure 1.2-7 Tensile strength of various plastics (1000 psi) [1].

developments in catalysts and inhibitors have led to preimpregnated sheet stock that can be formed under heat and pressure in metal stamping machines. Thus, sheet thermoset polyesters are in direct competition with structural foam and sheet metal in the appliance and automotive industries. More than 95% of the pleasure craft made each year have hulls of glass-reinforced polyester. Direct polymerization of phthalic-based materials leads to a family of thermoplastic polyesters. Polyethylene terephthalate was first formed into sheet as Mylar and then into fibers. Recent developments have led to polybutylene terephthalate (PBT) polyesters that can be injection-molded. These materials are extremely tough and will probably be significant engineering materials within a decade.

One form of thermoset polyester that has had great potential without showing major commercial application is water-extended polyester. Water is emulsified into the reactive resin, and the polyester encapsulates it. The cured product is cement-like but can be carved and machined like wood. Unfortunately, water diffusion from the material over long periods leads to weight loss, physical property deterioration, and volume change, and thus the product remains without suitable application.

Epoxies are standard electronic component potting resins. They are transparent or semitransparent when unfilled and have no volatile or corrosive intermediates to interfere with sensitive or delicate embedded specimens. They can be easily filled with glass fibers, sand, and metal powder and form extremely strong bonds. As a result, their tensile strengths are very high, as shown in Fig. 1.2-7. Their excellent bonding characteristics with metals and ceramics make them invaluable as household adhesives. Urethanes represent a family similar to that of polyesters. There are many ways of formulating urethanes, and the resulting products can be extremely flexible and rubbery or very tough. Certain urethanes can be air-cured, and as a result extensive applications of urethane varnishes and sealers have occurred in the marine and housing industries in recent years. Urethanes also accept fluorocarbon refrigerants very well and thus can be foamed either with heat and a catalyst or at room temperature, depending on the formulation. Cold cure urethane foams have very low permanent set characteristics and are used extensively in automotive and commercial seating.

1.3 The General Properties of Plastics

It is generally thought that the properties of plastics are inferior to those of traditional materials. It is certainly the case that no plastic is as strong as steel, as shown in Fig. 1.3-1, and the best elastic modulus of plastic is an order of magnitude below that of the weakest metal, as shown in Fig. 1.3-2. No plastic is as transparent as glass. And none have the low coefficients of thermal expansion possessed by metals, as shown in Fig. 1.3-3. And because

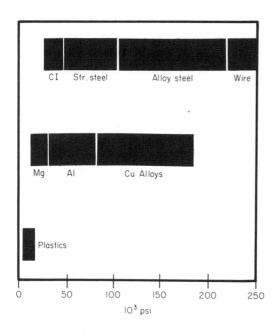

Figure 1.3-1　Room-temperature ultimate tensile strength.

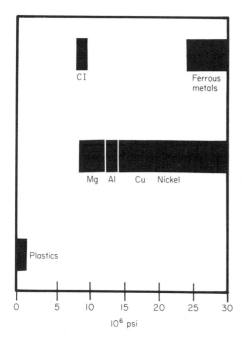

Figure 1.3-2　Elastic modulus.

22

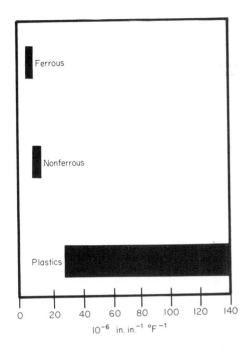

Figure 1.3-3   Coefficient of thermal expansion.

of the organic nature of the plastic materials, none have the useful temperature range of ceramics or metals, as shown in Fig. 1.3-4. No plastic can
match the abrasion resistance of porcelain, as shown in Fig. 1.3-5, and
none can match the fire retardancy of asbestos. Only a few plastics can
meet the electrical resistance of rubber or mica, as shown in Figs. 1.3-6
and 1.3-7, and certainly none can meet the conductivity of copper, as
shown in Fig. 1.3-8. Nevertheless, there are many applications where
these extreme conditions are not needed and plastics can be substituted
readily. For example, in surgical implant applications, the high-temperature performance of stainless steel is not needed, whereas its corrosion
resistance and chemical inertness are. Replacement of the very rigid
metal with a more maleable plastic having equivalent corrosion resistance
and chemical inertness can be accomplished readily. Cast iron has been
replaced with plastic in pump housings not because of operating temperature
constraints but because plastics are not as susceptible to fatigue or corrosion. Synthetic fabrics are known for their wearability and colorfastness,
even though cotton is considerably more absorbant, as shown in Fig. 1.3-9.
Therefore, proper selection of the plastics materials properties to meet
the specific need is required, in the same way a suitable metal is selected

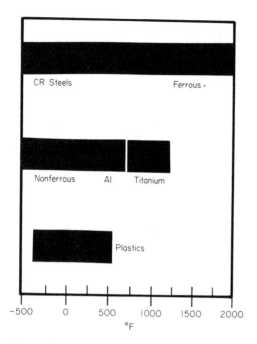

Figure 1.3-4   Useful temperature range.

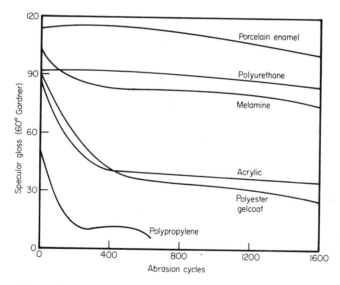

Figure 1.3-5   Specular gloss for various abrasion cycles.

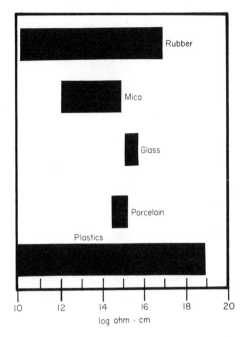

Figure 1.3-6   Electrical resistivity ($\log_{10} \Omega \cdot \text{cm}$).

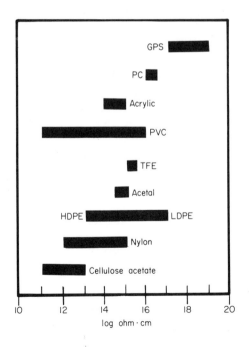

Figure 1.3-7   Electrical resistivity ($\log_{10} \Omega \cdot \text{cm}$).

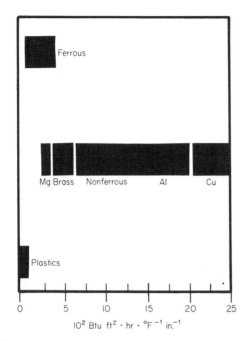

Figure 1.3-8   Thermal conductivity.

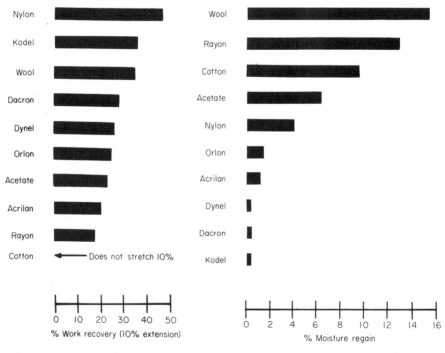

Figure 1.3-9   Work recovery and moisture regain for staple fibers [2].

to meet specific conductive or thermal needs. Table 1.3-1 gives a guide
to the selection of materials to meet specific demands such as elasticity,
abrasion, impact, and so on. Finer breakdowns for specific properties can
be made. For example, in Table 1.3-2, materials are ranked according to
their relative abrasion resistance. As can be seen, urethanes are superior
materials for this application. The loss of gloss with these materials is
also superior to other plastics, as shown in Fig. 1.3-5. In Fig. 1.3-10 it
is shown that the thermal conductivity of polystyrene is considerably lower
than that of most common plastics, and in Fig. 1.3-7 it is shown that cellulose acetate has a lower electrical resistivity than most common plastics.
Polyethylenes have more than 1000 times the moisture barrier potential
than the nylons or acetates, however, and do not exhibit crack propagation
and impact failure. This is in direct contrast to melamine, which exhibits
the lowest resistance to impact failure. This is shown in Fig. 1.3-11.

It is important to note that methods of polymerization can change
the relative properties of a plastic. As mentioned above, the number and
location of unsaturated double bonds in urethanes and polyesters can be
varied to yield a spectrum of physical properties ranging from tough and
flexible to brittle and rigid. Similarly, properties of many thermoplastic
resins can be altered somewhat through molecular-weight distribution and
homogeneity of molecular structure. DuBois and John show this for polyethylene in Table 1.3-3. Other materials have similar material property
changes. It should be noted, however, that the changes are not so dramatic
with homopolymer thermoplastics as with thermosets. Usually, order-of-
magnitude changes are achieved only through co- or terpolymerization with
other plastics having the desired end properties.

Although such plastics as Teflon and polyethylene are nearly inert
to solvent attack, most plastics are attacked by certain solvents. In Table
1.3-4 [4] are given 17 families of plastics and their resistance to 8 families
of common solvents at two temperatures. While the solvent resistance
within a given plastic family may vary somewhat, the tables give a good
qualitative picture. As mentioned, many plastics are deteriorated by UV
light. This is shown in Table 1.3-5, where most families of plastics are
rated in their ability to withstand weather. It should be noted here, however, that accelerated weather testing is not necessarily correlatable with
Florida or Arizona weather exposure tests [1].

It is well known that given sufficient oxygen, all plastics burn. There
have been many testing procedures proposed for evaluation of the rate of
burning of plastics. Nearly every plastics technical program includes a
symposium on plastics flammability. Nevertheless, standards still depend
on the end use of the plastic. Fabric flammability testing programs require
the plastic, as a fiber, to meet one standard, whereas the carpet flammability testing programs may require the plastic, as a carpet filament, to
meet another. As important to the consumer as the burning rate are the
levels of smoke generation and toxic chemicals. Urethanes, long touted for

Table 1.3-1

Materials That Meet Certain Performance Requirements [1]

Property	Thermoplastics	Thermosets
Low temperature	TFE, GBM[a]	DAP
Lost cost	Polystyrene	Phenolic
Low gravity	Polypropylene, methylpentene	Phenolic–nylon
Thermal expansion	Phenoxy–glass, GBM	Epoxy–glass
Volume resistivity	TFE	DAP
Dielectric strength	PVC	DAP
Elasticity	EVA	—
Moisture absorption	Chlorotrifluoroethylene	Alkyd–glass
Steam resistance	Polysulfone	DAP
Flame resistance	TFE–GBM	Melamine
Water immersion	Chlorinated polyether	DAP
Stress–craze resistance	Polypropylene	All
High temperature	TFE–GBM	Silicones
Gasoline resistance	Acetal	Phenolic
Impact	Ultrahigh–molecular–weight PE	Epoxy–glass
Cold flow	Polysulfone	Melamine–glass
Chemical resistance	TFE, FEP, PE, PP	Epoxy
Scratch resistance	Acrylic	Allyl diglycol carbonate
Abrasive wear	Polyurethane	Phenolic–canvas
Colors	Acetate, styrene	Urea, melamine

[a] GBM = glass–bonded mica.

Table 1.3-2

Relative Abrasion Resistance [1]

Material	Index
Polyurethane	3
Polyester film	18
Nylon 11	24
High-density polyethylene	29
PTFE	42
Nitrile rubber	44
Nylon 101	49
Low-density polyethylene	70
High-impact PVC	122
Plasticized PVC	187
Butyl rubber	205
ABS	275
Polystyrene	325
Nylon 6	366

their superior insulating characteristics, are now recognized as potentially dangerous materials insofar as flame spread rating, smoke generation level, and generation of potential toxic chemicals, particularly when openly exposed, are concerned. One test that yields predictable quantitative tests that may be correlatable to actual combustion situations is the limiting oxygen index test. In their development work, Fennimore and Martin [5] showed that if a vertical sample of plastic were ignited at the base and placed in an oxygen-rich environment, a measure of the least amount of oxygen required could be consistently determined. To do this, the sample was ignited in an environment containing nearly pure O_2. Then the ratio of oxygen to nitrogen was changed until the ignited sample was extinguished. The amount of oxygen at that time was called the limiting oxygen index (LOI). As shown in Table 1.3-6, acrylics, polystyrenics, and polyolefins have the lowest LOI and TFE Teflon has the highest. Air contains about 21% oxygen, and thus a material having an LOI of 21 or lower should theoretically burn in air without added combustion. In practice, it appears that a

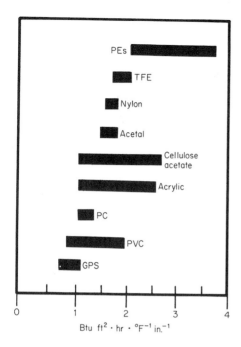

Figure 1.3-10   Thermal conductivity.

material should have an LOI of about 25 to be incombustible in air. Addition of fire retardants such as antimony oxide can raise the LOI of an otherwise combustible material 7 to 10 points, as shown in Table 1.3-6. Unfortunately, addition of these fire-retardant chemicals greatly increases the viscosity and the cost of the resin. Furthermore, recent studies indicate that suppression of open flame may lead to generation of excess quantities of carbon monoxide, a toxic by-product of incomplete combustion. Likewise, intumescent coatings that foam and generate dense smoke during ignition do inhibit flame propagation but produce dangerous quantities of blinding smoke. And, as mentioned, it is not at all clear that the LOI data, however internally consistent, relate to actual conflagration data [6-8].

1.4 The Processing Considerations

In general, thermoplastics have high viscosities in the melt state. At the molding temperatures of 150 to 650°F or so, it is necessary to use high forces to push the materials from one form to another. As a result, the

Table 1.3-3

Effect of Material Structure on Properties for Polyethylenes [1]

As density increases,		As melt index decreases,		As molecular structure becomes more homogeneous,	
Stiffness	Increases	Stiffness	Increases	Tensile strength	Increases
Yield strength	Increases	Tensile strength	Increases	Creep resistance	Increases
Hardness	Increases	Yield strength	Increases	Toughness	Increases
Creep resistance	Increases	Hardness	Increases	Softening temperature	Increases
Toughness	Decreases	Creep resistance	Increases	Stress-crack resistance	Increases
Softening temperature	Increases	Toughness	Increases		
Stress-crack resistance	Decreases	Softening temperature	Increases		
Permeability	Decreases	Stress-crack resistance	Increases		
Gloss	Increases	Permeability	Decreases		
Grease resistance	Increases	Gloss	Decreases		
		Grease resistance	Increases		

Table 1.3-4

Resistance of Various Plastics to Certain Classes of Solvents [4]

Room temperature

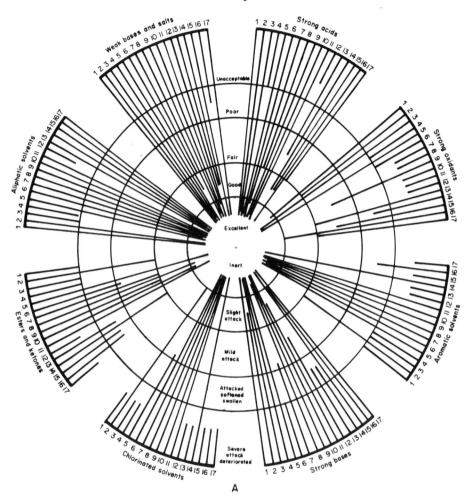

A

Key to Solvent Resistance in Plastics

Plastic no.

1 PVC
2 Polyvinylidene chloride (Saran)
3 Polyethylene
4 Polystyrene
5 Polychlorotrifluoroethylene (Kel-F)

Plastic no.

6 Asbestos-filled phenolic
7 Glass-reinforced thermoset polyester
8 Glass-reinforced amine-cured epoxy

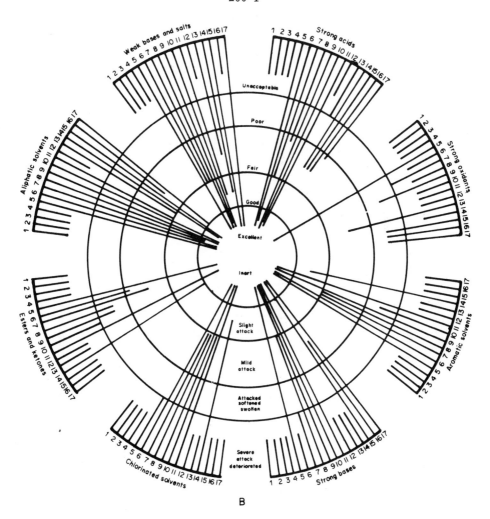

200° F

B

Table 1.3-5

Effect of Weather Exposure on Appearance, Thickness, and Electrical Properties [1]

Material	Surface condition[a]	Thickness change (in.)[a]	Electric strength change (%)[a]
Thermosetting plastics			
Epoxy casting (TETA cured)	Frosted surface (600 hr)	—	+6.7 (3988 hr)
Melamine laminate (G-5)	Loss of gloss (2500 hr)	0 (3188 hr)	-55.1 (3188 hr)
Melamine molding (cellulose filled)	Erosion (2500 hr)	-0.001 (3130 hr)	+7.2 (3130 hr)
Phenolic laminate (XXXP)	Fibers exposed (900 hr)	-0.001 (1611 hr)	+4.7 (1611 hr)
Phenolic molding (woodflour filled)	Filler exposed (1200 hr)	-0.002 (3130 hr)	+32.9 (3130 hr)
Polyester laminate (FR)	Fibers exposed (600 hr)	0 (1597 hr)	+13.6 (1597 hr)
Polyester laminate (GP)	Fibers exposed (600 hr)	0 (1597 hr)	+12.5 (1597 hr)
Silicone laminate (G-7)	None (2500 hr)	0 (4002 hr)	-14.4 (4002 hr)
Urea molding (woodflour filled)	Filler exposed (1600 hr)	0 (3130 hr)	+7.5 (3130 hr)
Thermoplastics			
ABS	Erosion (2500 hr)	-0.002 (3130 hr)	+4.7 (3130 hr)
Acetal	Severe erosion (900 hr)	-0.005 (1611 hr)	+6.1 (1611 hr)
Acetal (stabilized)	Erosion (2500 hr)	-0.003 (3971 hr)	+18.7 (3971 hr)

Acetal (carbon black filled)	+13.0 (3188 hr)	-0.003 (3188 hr)	Erosion (1200 hr)
Acrylic (PMMA)	+3.6 (3988 hr)	-0.001 (3988 hr)	None (2500 hr)
Acrylic (modified, type A)	—	—	None (2500 hr)
Acrylic (modified, type M)	—	—	None (2500 hr)
Cellulose acetate	-13.2 (1580 hr)	0 (1580 hr)	Frosted surface (600 hr)
Cellulose acetate butyrate	-6.7 (3893 hr)	+0.002 (3893 hr)	Erosion & frosting (2500 hr)
Fluorocarbon (PTFE)	-3.6 (4002 hr)	0 (4002 hr)	None (2500 hr)
Fluorocarbon (PCTFE, type G)	+3.4 (4002 hr)	0 (4002 hr)	None (2500 hr)
Fluorocarbon (PCTFE, type H)	—	—	None (2500 hr)
Nylon (type 66)	-3.4 (3988 hr)	-0.001 (3988 hr)	None (2500 hr)
Nylon (66, filled with MoS$_2$)	-6.3 (1611 hr)	0 (1600 hr)	None (1600 hr)
Polycarbonate	+5.0 (2224 hr)	0 (2224 hr)	Frosted surface (900 hr)
Polyethylene (low density)	-17.1 (1611 hr)	+0.001 (1611 hr)	None (1600 hr)
Polyethylene (high density)	+9.4 (1611 hr)	0 (1611 hr)	None (1600 hr)
Polypropylene	+15.1 (1580 hr)	-0.001 (1580 hr)	None (1600 hr)
Polyvinyl chloride (rigid)	-1.8 (3893 hr)	-0.001 (3893 hr)	Erosion (2500 hr)
Polystyrene (general purpose)	+10.0 (1580 hr)	0 (1580 hr)	Frosted surface (1600 hr)

[a]Values based on difference between shielded and exposed portions of panels. Figures in parentheses are length of exposure when observation or measurement was made.

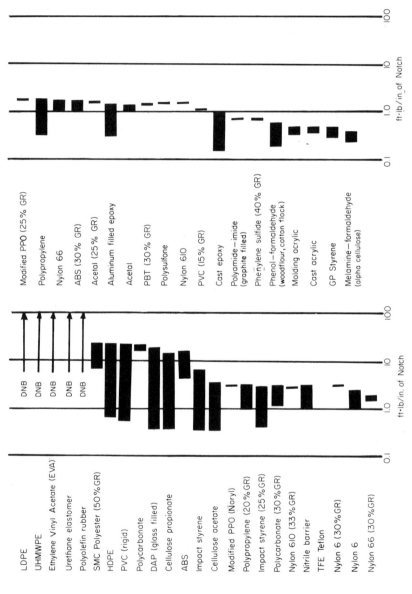

Figure 1.3-11 Izod impact strength of various plastics (DNB = does not break).

Table 1.3-6

Limiting Oxygen Indices for Some Plastics [5]

Material	LOI
Polytetrafluoroethylene (TFE)	95
Chlorinated PVC	60
Phenolic (GR)	55
PVC	52
Melamine	42
Nylon 66 (20% GR)	38–40
Polysulfone	30–33
Polycarbonate (30% GR)	29–33
Phenolic	26–31
ABS-PVC copolymer	27–31
PPO (Noryl)	27–30
Nylon 66	25–30
ABS (FR)	28
Epoxy	28
Unsaturated polyester (FR)	28
Polystyrene (FR)	28
Nylon 6	23–28
Polyethylene (FR)	26
DAP	26
Polycarbonate	25
Polypropylene (FR)	25
Polybutylene terephthalate	23
Polyester (unsaturated)	20–21
Acrylic	19
ABS	18
Polypropylene	18
Polystyrene	18
Polyethylene	17
Acetal (30% GR)	16

machinery used frequently incorporates hydraulic ram systems or heavy-duty high-horsepower electric motors that are geared to deliver very high torques at low revolutions per minute. In principle, however, all processing techniques are quite simple. Some require no heat or pressure. Addition of a catalyst to begin the cross-linking reaction can occur at room temperature. The resulting reactive mixture can be poured into an open mold to form synthetic slate or to pot electrical components. If chopped glass or woven glass mat is used with the resin, hand lay-up or spray up will dispense the reactive resin into or against a mold surface. Again no pressure is required, and curing takes place at room temperature. In either case, however, the glass and polyester must be squeegeed to form strong and intimate mixtures and to expel entrapped air. More recent developments use a hydraulic press and preimpregnated resin to form parts of uniform walls. Matched die molding between heated plates is used to form automotive grille assemblies and large electrical conduit boxes. The normal pressures do not exceed 500 psi of surface. Presses with larger forces can be used to form laminates of melamine, for example, with wood, paper, and aluminum foil. Presses having capacities of 1 ton force/in.^{2} of surface and higher are used to compression-mold thermosets such as phenolics and urea-formaldehydes. The molding powder is placed in a steel mold having a cavity with the internal shape of the part desired. The platens are heated with steam or oil to the molding temperature of the resin, and pressure is applied until the resin cross-links into a permanent shape. Thermoplastics can also be compression-molded, but the platens must have provisions for both heating and cooling so that part removal from the mold can be accomplished. Very heavy slabs of polystyrene and ultrahigh-molecular-weight polyethylene are formed by compression molding also. Many specialty plastics are compression-molded because their molding temperatures are too high to permit processing in conventional injection molding machines. As shown in Table 1.4-1, nearly all plastics can be compression-molded, and very frequently mechanical property evaluations of new plastics are carried out in the laboratory using compression-molded specimens. As a result, nearly every research and development laboratory has at least one fully equipped hand-operated bench-type compression molding press.

A common way of molding thermoplastic resins is injection molding. Here the resin in powder or pellet form is dropped from a hopper into a heated chamber where it is melted or softened. When the plastic has reached the molding temperature, it is pushed with a hydraulic ram through a small nozzle in the heated barrel of the machine and into a cooled mold. The injection mold is made of at least two pieces, held together during injection of the melt with an opposing hydraulic ram or mechanical toggle clamp. The melt squirts through the nozzle and into the cavity machined in the two halves of the mold, where it is quickly cooled to a temperature where it can be removed without distortion. The hydraulic ram is withdrawn to allow more plastic to drop into the heated chamber, and the clamp holding the

Table 1.4-1

Process Considerations

Material	Extrusion	Blow molding	Fiber	Film	Foam	Injection molding	Compression molding	Thermoforming
ABS	X	X	–	–	X	X	X	X
SBR	X	X	–	–	X	X	X	X
EVA	X	X	–	X	X	X	X	–
Melamine	–	–	–	–	–	–	X	–
Nylon	X	X	X	–	X	X	X	X
Phenolic	–	–	–	–	–	–	X	–
PC	X	X	–	–	X	X	X	X
LDPE/HDPE	X	X	–	X	X	X	X	X
PET	–	–	X	X	–	–	–	–
Acetal	X	X	–	–	X	X	X	–
Methyl pentene	X	X	–	–	–	X	X	X
PP	X	X	–	X	X	X	X	X
GPS	X	X	–	–	X	X	X	X
TFE	–	–	–	–	–	–	X	–
Urethanes	–	–	X	X	–	–	–	–
PVC	X	X	–	X	X	X	X	X
Ureas	–	–	–	–	–	–	X	–
PVC/PVA	X	X	–	X	–	X	X	X
Acrylic	X	–	X	–	–	X	X	X

mold halves closed is released, allowing the part to be removed. Sometimes a rotating screw is used to heat and pump the plastic resin into a
reservoir immediately behind the nozzle. When injection occurs, the screw
stops rotating and acts as a ram to push the melt into the mold area.

Injection molding is a cyclic process in that it makes discrete objects.
Extrusion is a process for making continuous objects such as profiles or
film or sheet or pipe. A rotating screw in a heated barrel is used to convey
the plastic resin in bead or pellet form into a region of high shear and heat
where it is melted and softened. From there the melt is pumped through a
die having a shape that is similar to the shape of the article desired. The
profile thus produced is still molten, and it must be cooled either by air or
water quench or by running it over chill rolls. If the die is circular and the
material suitable for film forming, the emerging cylinder of molten plastic
can be simultaneously stretched upward and blown outward, thus forming
very thin films. Almost all the polyethylene film used in packaging and
building is made this way.

Blow molding, as mentioned, is a 40-year-old process that is
presently in direct competition with glass bottle blowing. The machinery
principle is very similar to that for glass blowing in that the molten material is inflated and pressed between mold halves. Some of the features of
injection molding and extrusion are combined in that a cylinder of molten
plastic is extruded in a vertical position. Two mold halves with internal
dimensions having the shape of the bottle to be formed close around the
cylinder or parison, pinching it off at the top and bottom. An air jet in the
neck of the bottle mold then inflates the plastic against the inner surfaces
of the mold halves. Since the mold halves are normally water cooled, the
plastic solidifies quickly into the shape of the mold and a closed container
is formed. The mold then opens, releases the bottle, and is ready for
another parison. Recently, the single market of gallon blow-molded HDPE
bottles was placed at nearly 2 billion per year, in a marketing area where
no plastic was allowed 10 years ago [9].

Many packaging operations require the forming of plastic sheet into
container shapes or blister packs. Thermoforming is normally a continuous
operation with thin sheet stock and a cyclic operation with heavy sheets.
Basically, extruded sheet is placed in an oven until it is soft but not molten.
It is then draped or stretched over a mold having the shape of the desired
part. Frequently a vacuum is drawn beneath the sheet, thus pulling it
tightly to the mold surface, or pressure can be applied above the sheet to
force it onto the mold surface. High-speed automated thermoforming lines
can produce 18,000 containers an hour. At the other end of the size spectrum, mobile home tops as large as 40 ft by 12 ft have been successfully
thermoformed.

Rotational molding is a way of making hollow containers. The earliest rotational molding used a highly plasticized polyvinyl chloride latex
which is liquid at room temperature and cures when heated. The material

is poured into a hollow mold and rotated in a programmed manner in a forced convection oven. As the mold rotates, the liquid plastisol coats the mold inner surface and is cured into this shape. This process is also cyclical in that the operator must open the mold, remove the cured item, and refill the mold with plastisol for the next operation. The technique has been adapted to molding of fine olefinic powders with some good success. Industrial bellows, planters, beach balls, garbage containers, and toys of all kinds are common rotationally molded products.

Much interest has been raised recently in the development of thermoplastic structural foams. In contrast to the foaming of urethanes, where the heat of reaction is sufficient to vaporize the dissolved volatile fluorocarbon refrigerants, and to the foaming of polystyrene, where steam is used to vaporize the dissolved linear hydrocarbon solvents, the foaming of engineering resins into parts having densities 50 to 85% that of the unfoamed plastics requires the addition of raw nitrogen or a thermally unstable chemical to the plastic at molding temperatures and high pressures. To foam the plastic, the pressure must be dropped precipitously as the material enters the molding area. Therefore special provisions are made in the design of injection molding machines that allow for the rapid transfer of the foaming material from the high-pressure region to the low-pressure region. Once the material is transferred to the low-pressure portion of the mold, the dissolved gas is liberated in the form of microscopic bubbles and the foaming continues until the material is cold or fills the mold. Cooling of structural foams is very similar in nature to cooling of conventional injection-molded parts.

As can be seen, there are many ways of converting the raw resin into a usable form. The objective is to know in detail the ways by which a specific resin can be converted most easily and at least cost. To do this, of course, the engineering principles for each of the dominant processes must be understood to a great degree. This, then, is the raison d'etre for this text.

1.5 Growth Projection for Plastics

More things were made of plastics, by volume, in 1976 in the United States than were made of ferrous and nonferrous metals. As shown in Fig. 1.5-1, the plastic growth rate over the last 40 years has been 3 1/2 to 4 times that of the U.S. gross national product (GNP). In spite of the 1974 energy crisis, plastics posted a small gain in tonnage consumed, whereas the GNP showed a substantial real decline. In 1975, the plastics industry suffered its first recession ever, with a nearly 20% drop in U.S. consumption. However, in 1976, plastics usage had recovered to the 1973 to 1974 level. As shown in Fig. 1.5-2, plastics usage is expected to exceed the combined usage of all materials within 70 years. This projection is probably valid

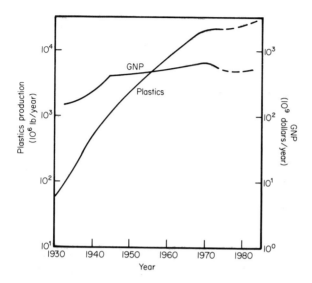

Figure 1.5-1   Comparison of plastics production and U.S. gross national product.

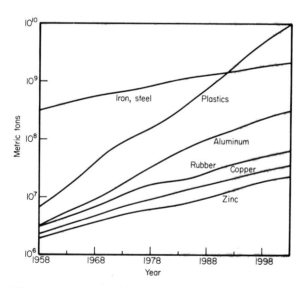

Figure 1.5-2   World consumption of raw materials (metric tons).

even with developing shortages of petroleum, for, as will be discussed below, alternative sources of raw materials for plastics will have been developed by then. As shown in Table 1.5-1, the U.S. consumption in 1971 topped 20 billion lb. In spite of the energy crisis, deepening recession, and shortage of certain raw materials, projections through 1980 of 44.5 billion lb are considered to be reachable and necessary to meet the U.S. and world demand for plastics. As shown in Table 1.5-2, nearly all major markets are showing continued infiltration of plastics. The automotive market, for example, shows a continuing penetration of plastics, from 125 lb/car in the 1972 model year with 11 million cars to 165 lb/car in the 1975 model year with 8 million cars. Part of this market is developing because of increased safety standards that require impact materials in the front end and rear for minimization of low-speed collision damage. Further increases in plastics in vehicles are expected as automakers are forced to reduce car weight to improve gasoline mileage. As shown in Figs. 1.5-3 and 1.5-4 [10], increased air pollution control standards have forced increases in car weights and corresponding reductions in fuel economy. To counter this, lightweight plastic replacements for many major decorative steel components are being actively sought. Thermoplastic structural foam door and trunk lid panels and glass-reinforced polyester sheet molding compound engine bonnets are presently under test. Higher engine compartment temperatures are the major impetus to the shift from rubber components to higher-temperature plastics. Plastics are also being used to reduce the number of parts in assemblies such as the dashboard, horn ring, seat frames, and air

Table 1.5-1

U.S. Consumption of Plastics

Plastics sales (billion lb)	Year
10	1963
17	1968
20	1971
24.2	1972
28.9	1973
29.4	1974
23.9	1975
29.9	1976
44.5 (projected)	1979

Table 1.5-2

Major Markets for Plastics (1000 metric tons)

Market	1971	1972	1973	1974	1975	1976
Appliances	258.61	305.16	418.9	417.3	254.5	320.8
Building and construction	1618.95	1,981.34	2,400.3	2,353.9	1,890.0	2,379.0
Electrical and electronic	543.57	654.65	752.4	755.3	NA[a]	NA
Furniture	384.00	453.16	502.7	493.9	379.0	439.0
Housewares	433.50	545.03	565.0	585.1	NA	NA
Packaging	1987.3	2,362.3	2,536.7	2,671.5	2,524.0	2,928.0
Toys	295.4	323.6	396.8	402.8	NA	NA
Transportation						
Auto	522.33	622.55	577.2	589.5	728.0	815.0
Other	72.44	80.59	81.2	89.4	77.0	88.0
Miscellaneous	2913.81	3,676.62	4,920.8	4,991.3	NA	NA
Total	9030.0	11,005.0	13,152.0	13,350.0	10,672.0	13,353.0

[a]NA = not available.

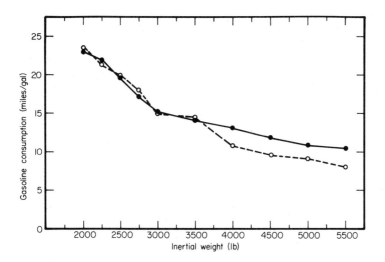

Figure 1.5-3   Comparison of U.S. car weight and fuel consumption; solid line: 1957 to 1967; dashed line: 1974 [10].

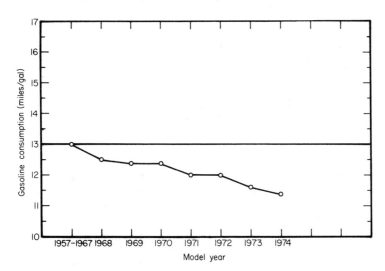

Figure 1.5-4   Average U.S. automobile gasoline usage [10].

conditioning controls. Specialty cars, such as the Bricklin, Phoenix, Lorean, and Bradley, make extensive use of plastics in cosmetic and structural areas. There is little reason, then, to ignore predictions of 200-lb/car plastics consumption by 1980 and an all-plastic car chassis by 1985.

Certainly, similar justifications can be made for infiltration of plastics into the major appliance area, building and housing areas, and so on. It should be apparent from this brief outline, however, that major penetration of plastics as replacement materials for natural materials and steels is continuing. In an effort to understand the mix of plastics materials in the near future, a careful look at the present status of raw materials supply is needed. As will be seen in the next chapter, the majority of plastics, by consumption, are based on the ethylene molecule. With certain modifications, ethylene can be converted to PVC or polystyrene, and it, of course, forms the basis for polyethylene. Nearly all plastics rely on petroleum as a source for the monomer. Projections show that by 1980 nearly 16 billion lb will be needed for polyethylene, 4 billion lb for PVC, and 3 billion lb for polystyrenes. This 23-billion-lb demand is contrasted to the less than 10-billion-lb consumption in 1970 [11]. Natural gas, which has supplied a large portion of the crackable ethane, is in very short supply. Other sources, such as naphtha and gas oil, are more plentiful but yield much lower percentages of crackable ethane. As a result, the basic cost of ethylene has increased from $0.03/lb in 1971 to more than $0.05/lb in 1974 and is expected to increase to $0.14 to $0.16/lb by 1977. This is shown in Fig. 1.5-5 [12]. Other monomer feedstocks, such as propylene and benzene, are also expected to show enormous short-range increases in price. These prices are directly reflected in the price of raw resins. LDPE increased from $0.13/lb in January 1974 to $0.27/lb in October 1974. Similarly, HDPE increased from $0.14 to $0.28/lb, polypropylene from $0.17 to $0.24/lb, and PVC from $0.13 to $0.24/lb. Most engineering resins did not show this dramatic increase, owing to lower demand and increasing technology to maintain pricing. Nevertheless, demands for these resins will increase as the price differential between them and the commodity resins decreases, and pressure will force the prices higher also.

Contrary to the trend of the late 1960s and early 1970s, where the price differential widened and thus important research and development in the properties of new resins were restricted, the present price squeeze is spurring research into newer engineering materials and is encouraging resin suppliers having access to feedstocks to spend more time and money finding improved ways of processing the commodity resins. As an example of this, extensive work is underway to improve the strength and at the same time reduce the weight of the HDPE gallon milk container [13]. The plastic milk container first penetrated the market in 1965 with a 90-g container when the resin price was $0.11/lb. At the present time, the average container weighs 80 g, uses $0.24/lb resin, and has about 50% of the market. Projections indicate that the 65-g bottle is presently feasible, and the market

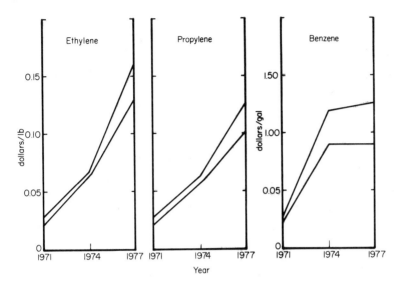

Figure 1.5-5 Projected raw material prices (Source: Union Carbide Corp.) [12].

will be sustained at a resin price of $0.33/lb. Research into very thin wall high-performance containers indicate that with new machinery developments and better tailoring of the plastic, a 45-g container is feasible and thus the market should be sustained even if resin prices were to reach $0.40/lb. At these price levels, however, it would appear that the presently more expensive resins might penetrate the plastics market with a lower-weight container. It is certain that research must be done on proper materials selection in order to be continually competitive in this high-volume market.

The energy crunch is not the only problem facing the plastics industry in the last quarter of the twentieth century. There is great concern about the ecological ramifications of disposing of thousands of tons of plastics. Plastics cannot be removed from trash and garbage using methods that are successful in removing metallic materials. Plastics densities are sufficiently close to conventional materials such as paper and wood to make flotation economically infeasible. They are not metallic and thus cannot be removed magnetically. They are not so easily identifiable as aluminum and glass, and thus recycling centers are of doubtful use. Nevertheless, Smith [14] shows that a very large fraction of materials used in Europe can be recycled if sufficient means for doing so are obtained. As shown in Table 1.5-3, more than 50% of all LDPE and about 20% of the PVC used in disposable articles can be recovered. Obviously, not all this will be recovered, but Smith feels that 35% of all plastics could be recycled if sufficient interest were present. Others feel that 10-20% is a reasonable recovery rate [15].

Table 1.5-3

Recyclable Thermoplastics in the European Common Market—1975 [14]

Material	Product	% of Material consumption	% of Material consumption	Material consumption (metric tons)	Recyclable material (metric tons)
PVC	Bottles	9	19	3,000,000	570,000
	Film and sheet	10			
LDPE	Film	68	74	2,060,000	1,524,000
	Bottles	6			
HDPE	Blow moldings	36	59	1,235,000	481,000
	Film and flat yarn	23			
PP	Film	8	37	540,000	200,000
	Flat yarn	29			
PS	Packaging	39	39	1,120,000	437,000
				7,955,000	3,212,000

Recyclable percentage: $\dfrac{3,212,000}{7,755,000} \times 100 = 40\%$

At present, plastics represent about 3% of the total consumer solids waste, and this is expected to increase to about 5% by 1990. Since many plastics have excellent chemical, bacterial, and UV resistance, landfill methods successful in reducing vegetable and paper waste probably will not yield good results with plastics. Incineration is a viable alternative, however. Recent tests show that if sufficient oxygen is present, incineration of materials such as PVC will not lead to excessive quantities of HCl or SO_2 as has been predicted [15]. It has been found instead that more SO_2 is generated from incineration of paper and more HCl from grass clippings than from an equivalent weight of PVC. As mentioned, all plastics burn, given sufficient oxygen. As shown in Table 1.5-4 [16], only polytetrafluoroethylene has a negative heating value when burned. As shown in Table 1.5-5, polyethylene has nearly the same heating value as natural gas, and styrene has a higher heating value than bituminous coal. This should be the case, since polyethylene is a highly refined hydrocarbon, containing no moisture, air, sulfur, or other noncombustible materials. Incineration is still the

Table 1.5-4

Thermochemical Properties of Common Plastics and Fillers [16]

Composition of subunit	Subunit MW	ΔH_f, 25°C (kcal/g·mol)[a]	$\Delta H_c''$, 25°C (kcal/g·mol)[b]	Heating value (Btu/lb)	Stoichiometric burning			Common name
					Mols O_2/mol compound	Ft³ air/lb 760 torr (25°C)	Theoretical flame temp. (°C/°F)	
C_2H_4	28.05	-12.25	-312.5	20,050	3	200	2120/3850	High-density polyethylene
C_2H_4	28.05	-12.79	-312.0	20,020	3	200	2120/3850	Low-density polyethylene
$C_{2.31}H_{4.62}$	32.04	-14.30	-360.8	20,270	3.46	202	2120/3850	Ethylene–propylene (69:31) random copolymer
C_3H_6	42.08	-18.84	-468.3	20,030	4.5	200	2120/3850	Polypropylene, isotactic or syndiotactic
C_3H_6	42.08	-19.27	-467.8	20,010	4.5	200	2120/3850	Polypropylene, atactic
C_4H_8	56.11	-24.13	-625.4	20,060	6	198	2120/3850	Poly-1-butene, isotactic
C_4H_8	56.11	-21.24	-628.2	20,150	6	198	2130/3870	Polyisobutylene
C_5H_{10}	70.13	-31.07	-780.8	20,040	7.5	198	2120/3850	Poly-1-pentene, isotactic
C_5H_{10}	70.13	-31.64	-780.2	20,030	7.5	198	2120/3850	Poly-3-methyl-1-butene, isotactic

Table 1.5-4 (continued)

Composition of subunit	Subunit MW	ΔH_f°, 25°C (kcal/g·mol)[a]	ΔH_c°, 25°C (kcal/g·mol)[b]	Heating value (Btu/lb)	Stoichiometric burning			Common name
					Mols O_2/mol compound	Ft^3 air/lb 760 torr (25°C)	Theoretical flame temp. (°C/°F)	
C_6H_{12}	84.16	-38.54	-935.7	20,010	9	198	2120/3850	Poly-4-methyl-1-pentene, isotactic
C_4H_6	54.09	+3.13	-584.3	19,440	5.5	189	2220/4020	Poly-1,4-butadiene, atactic
$C_{4.18}H_{6.09}$	56.34	+3.0[c]	-604.2	19,300	5.70	189	2220/4020	Butadiene-styrene (8.58%) copolymer
$C_{4.60}H_{6.29}$	61.59	+3.0[c]	-650.5	19,010	6.17	187	2220/4020	Butadiene-styrene (25.5%) copolymer (GRS rubber)
	53.70	+3.0[c]	-512.6	17,180	4.84	168	2190/3970	Butadiene-acrylonitrile (37%) copolymer
	68.12	-6.0	-737.5	19,490	7	190	2190/3970	Natural rubber (no sulfur)
C_8H_8	104.2	+7.09	-1033	17,850	10	179	2210/4010	Polystyrene, isotactic
C_8H_8	104.2	+8.27	-1034	17,870	10	179	2210/4010	Polystyrene, atactic-crystal
C_9H_{10}	118.2	+7.88	-1196	18,220	11.5	181	2210/4010	Poly-α-methylstyrene
$C_{12}H_8$	152.2	+27.13	-1429	16,900	14.0	183	2230/4040	Polyacenaphthalene
C_2H_3F	46.05	-55.8	-238.8	9,180	2.75	112	1710/3100	Polyvinyl fluoride

$C_2H_2F_2$	Polyvinylidene fluoride	1090/2000	73	2.5	3,940	-140.3	-116.1	64.04
C_2F_4	Polytetrafluoroethylene	d	37	2	-144	+8.01	-196.1	100.0
C_2F_3Cl	Polychlorotrifluoroethylene	320/615	32	2	482	-31.2	-156.9	116.5
C_2H_3Cl	Polyvinyl chloride	1960/3550	82	2.75	7,720	-268.0	-22.6	62.50
$C_2H_2Cl_2$	Polyvinylidene chloride	1840/3340	48	2.5	4,315	-232.4	-24.0	96.95
C_3H_4O	Ethylene–carbon monoxide (1:1) copolymer	2110/3830	117	3.5	11,900	-370.6	-48.2	56.06
C_3O_2	Polycarbon suboxide	2260/3910	55	2	5,940	-224.06	-57.57	68.03
CH_2O	Polyoxymethylene	2050/3750	62	2	7,280	-121.4	-40.93	30.03
C_2H_4O	Polyethylene oxide	2120/3850	106	2.5	11,470	-280.6	-44.13	44.05
C_3H_6O	Polypropylene oxide, 27% isotactic	2100/3810	129	4	13,410	-432.7	-54.41	58.08
C_3H_6O	Polypropylene oxide, 100% atactic	2100/3810	129	4	13,400	-432.3	-54.78	58.08
C_3H_6O	Polyoxytrimethylene	2130/3860	129	4	13,560	-437.6	-49.5	58.08
C_4H_8O	Polytetrahydrofuran	2120/3850	142	5.5	14,790	-592.7	-56.8	72.11
C_2H_4O	Polyvinyl alcohol	1980/3600	106	2.5	10,760	-263.2	-61.53	44.05
$C_3H_4O_2$	Poly-β-propiolactone	2075/3770	78	3	8,330	-33.3	-85.5	72.06
$C_5H_8O_2$	Polymethyl methacrylate	2070/3760	112	6	11,470	-637.7	-105.8	100.1
C_8H_8O	Polyphenylene oxide	2200/3990	148	9.5	14,880	-993.8	-31.88	120.2

Table 1.5-4 (continued)

Composition of subunit	Subunit MW	ΔH_f°, 25°C (kcal/g·mol)[a]	$\Delta H_c''$, 25°C (kcal/g·mol)[b]	Heating value (Btu/lb)	Stoichiometric burning			Common name
					Mols O_2/mol compound	Ft³ air/lb 760 torr (25°C)	Theoretical flame temp. (°C/°F)	
$C_{16}H_{14}O_3$	254.3	-103.26	-1880	13,310	18	132	2190/3980	Polycarbonate
$C_5H_6OCl_2$	155.0	-82.8	-660.7	7,673	6.5	78	1990/3610	Chlorinated polyether
HCN	27.03	+14.0	-150.3	10,010	1.25	86	2410/4370	Polyhydrocyanic acid
$C_2H_3O_2N$	73.05	-19.98	-278.7	6,870	1.75	45	2670/4830	Polynitroethylene
C_3H_3N	53.06	+15.85	-408.6	13,860	3.75	132	1860/3380	Polyacrylonitrile
S	32.06	+3.56	-74.5	4,180	1	58	1740/3160	Polysulfur
$C_3H_6SO_2$	106.1	-95.4	-462.7	7,850	4.5	79	1970/3570	Polypropene sulfone
$C_4H_8SO_2$	120.2	-101.5	-618.9	9,290	6	93	2000/3640	Poly-1-butene sulfone
$C_6H_{12}SO_2$	148.2	-113.7	-913.5	11,310	9	114	2040/3710	Poly-1-hexene sulfone
$C_{15}H_{12}O_2$	224.3	-120.0[c]	-1496	12,000	17	142	1860/3380	Phenol-formaldehyde (1:1)
$C_3H_4N_2O$	84.09	-60.0[c]	-358.8	7,680	3.5	78	1950/3540	Urea-formaldehyde (1:2)

Formula								Name
$C_6H_6N_6$	162.2	-20.0^c	-749.3	8,310	7.5	120	1990/3610	Melamine-formaldehyde (1:3)
$C_{6.3}H_{7.1}NO_{2.1}$	130.3	-100.0^c	-743.1	10,180	7.02	101	2100/3810	Polyurethanes, ester based
$C_{5.77}H_{6.25}O_{1.63}$	101.6	-33.0^c	-723.2	12,810	6.52	120	2250/3910	Unsaturated polyesters (maleic anhydride based; 1:1 styrene)
$C_{11.85}H_{20.37}O_{2.83}N_{0.3}$	212.1	-110.0^c	-1700.0	14,430	15.53	137	2220/4030	Bisphenol A epoxy (3000 MW; 10% diethylenetriamine)
			—	7,590	—	70		Paper
			—	8,520	—	80		Woodflour
C_8H_8	104.2	+24.91	-1050.6	18,160	10	180	—	Styrene
C_6H_{10}	82.15	-9.15	-896.8	19,650	8.5	193	—	Cyclohexane
C_2F_3Cl	116.5	-134.6	-53.5	827	2	32	—	Chlorotrifluoroethylene
C_2H_3Br	107.0	+18.7	-309.3	5,200	2.75	48	—	Vinyl bromide
CH_2O	30.03	-28.2	-134.2	8,040	1	62	—	Formaldehyde
C_2H_4O	44.05	-46.2	-278.5	12,510	2.5	106	—	Acetaldehyde
$C_4H_6O_2$	86.09	-103.6	-477.6	9,985	4.5	98	—	Methacrylic acid
$C_5H_8O_2$	100.1	-92.7	-650.8	11,700	6	112	—	Methyl methacrylate
$C_3H_4O_2$	72.06	-80.54	-338.3	8,450	3	78	—	Acrylic acid
C_3H_6O	56.08	-51.72	-435.4	13,970	4	133	—	Acetone

Table 1.5-4 (continued)

Composition of subunit	Subunit MW	ΔH_f°, 25°C (kcal/g·mol)[a]	$\Delta H_C''$, 25°C (kcal/g·mol)[b]	Heating value (Btu/lb)	Stoichiometric burning			Common name
					Mols O_2/mol compound	Ft^3 air/lb 760 torr (25°C)	Theoretical flame temp. (°C/°F)	
$C_2H_6O_2$	62.07	-93.1	-300.0	8,700	2.5	75	—	Ethylene glycol
$C_4H_6O_2$	86.09	-75.2	-505.9	10,580	4.5	98	—	Vinyl acetate
$C_4H_6O_3$	98.06	-112.4	-332.1	6,100	3.5	67	—	Maleic anhydride
$C_3H_8O_3$	92.10	-159.8	-395.6	7,730	3.5	71	—	Glycerol
$COCl_2$	98.92	-53.3	-40.75	742	0.5	10	—	Phosgene
C_2H_5N	43.07	+30.12	-397.1	16,600	4.25	58	—	Ethylenimine
$C_6H_{15}O_5$	167.2	—	-1011	7,520	—	48	—	Cellulose
	—	—	0	0	—	0	—	Sand
	—	—	0	0	—	0	—	Quartz
	—	—	0	0	—	0	—	Diatomaceous earth (dry)
	—	—	0	0	—	0	—	Kaolin (clay)
	—	—	0	0	—	0	—	Mica

Formula		Mol. wt.	ΔH_f°					Name
$Mg_3Si_2O_5(OH)_4$	—	277.2	+34.4	−234	—	0	—	Asbestos
$CaSiO_3$	—	116.2	+21.3	−331	—	0	—	Calcium metasilicate
$CaCO_3$	—	100.1	+42.5	−736	—	0	—	Calcium carbonate
$BaSO_4$	—	233.4	+145.8	−1,210	—	0	—	Barium sulfate
$CaSO_4$	—	136.2	+119.6	−1,800	—	0	—	Calcium sulfate
MoS_2	—	160.1	+216.4	3,040	—	15	—	Molybdenum sulfide
C	—	12.01	−94.05	14,110	—	156	—	Carbon, graphite
	—	—	0	0	—	0	—	Talc
	—	—	0	0	—	0	—	Glass fiber
$Al_2(OH)_6$	—	156.0	+41.6	−500	—	0	—	Aluminum hydrate
CON_2H_4	−58.7	60.06	−188.2	5,640	3.5	109	—	Urea
$C_3H_6N_6$	−15.35	126.2	−520.3	11,450	10.5	156	—	Melamine

[a] ΔH_f° = heat of formation.
[b] $\Delta H_c^{\ddagger}$ = heat of combustion.
[c] Value estimated by Throne and Griskey [16].
[d] Will not support combustion.

Table 1.5-5

Comparison of Heating Values and Theoretical Flame Temperatures
of Plastics, Fuels, and Other Materials [16]

Material	Heating value (Btu/lb)	Theoretical flame temp. (°C/°F)
Methane	23,590[a]	2012/3654
Butane	21,000[a]	2084/3783
Ethylene	21,400[a]	2250/4082
Propylene	20,750[a]	2180/3956
City gas (MW = 20)	10,300[a]	
Natural gas (MW = 20)	22,700[a]	
No. 1 fuel oil	19,800	
No. 6 fuel oil	18,300	
Bituminous coal (med. volatile, W. Va.)	15,178	
Lignite (Texas)	11,084	
Peat (Minn.)	9,057	
Wood	8,835	
Peat	3,586[b]	
Oil shale	6,300	
Straw	6,700[b]	
Sewage sludge	7,500[b]	
Paper	7,590	
Polyethylene	20,050	2120/3850
Butadiene-styrene copolymer (GRS rubber)	19,010	2220/4020
Polystyrene	17,870	2210/4010
Polyvinyl chloride	7,720	1960/3550
Polycarbonate	13,310	2190/3980
Phenol-formaldehyde resin	12,000	1860/3380

Table 1.5-5 (continued)

Material	Heating value (Btu/lb)	Theoretical flame temp. (°C/°F)
Unsaturated polyesters	12,810	2260/3910
Epoxies	14,430	2220/4030
Glass-reinforced polyester (70% resin)	8,970	
Synthetic marble (30% polyester, 70% calcium carbonate)	3,000	
Melamine laminate (50% melamine-formaldehyde resin, 50% paper)	7,950	
Chipboard (90% woodflour, 10% urea-formaldehyde resin)	8,715	

[a]At 760 torr, 60°F.
[b]As received.

preferred method of disposing of large quantities of homogeneous scrap paper, and frequently the paper is received wet. As shown in Table 1.5-5, dry paper has a heating value when dry of less than 8000 Btu/lb. When wet, this value drops precipitously. Therefore, it is frequently the practice of incinerator operators to mix scrap polyethylene or polystyrene with the wet paper prior to incineration. Given proper incinerator design, there should be no smoke and only a small amount of ash with this practice. Unfortunately, most incinerators are very inefficient. Proper design to maximize energy recovery from scrap or waste materials should lead to an enormous source of quality energy at relatively low costs. It seems far better to convert raw materials into useful articles, use the articles until they are functionless (not obsolete), and then convert them into useful energy with which to produce more useful articles than to simply convert the naturally occurring materials directly into CO_2 and H_2O in portable combustion chambers.

Several studies are underway in an effort to use clean mixed scrap thermoplastics. Paul and coworkers [17] have found that although mixed LDPE and GPS have very poor physical properties, the presence of chlorinated resins such as PVC improves the performance of injection- and compression-molded parts. Fay [18] reported on several methods of handling mixed and contaminated but segregated materials. The Reverzer process using segregated nylon scrap from weaving mills yields a material that is suitable for dunnage stakes. Recent work with thermoplastic structural foam scrap [19] has shown that small amounts of contamination (up to 25% LDPE in GPS or 15% GPS in LDPE) have little effect on the mechanical properties of the final part. Furthermore, very little solvent attack is observed on GPS even with 25% concentration in LDPE. Plastic reclaimed from stripping sheath from copper and aluminum is available at $0.06/lb and is finding immediate use as an inexpensive mulch to improve crop yields. Further work with encapsulation of long-term herbicides and fertilizers in the scrap plastic may yield dramatic yields in wheat, oats, and corn. It would appear then that reuse of high-concentration scrap is feasible and will be an area of increasing interest in the next few years.

As mentioned, plastics have petroleum and natural gas as sources for materials that form the basis for plastics. Stover and Whaley [20] show an average growth of 8.5%/year in the propylene demand. Only 4 billion lb of propylene are recovered from oil, the rest being modified from LPG and gasoline. Seven billion pounds are derived from cracking of ethane, either supplied by natural gas or gas oil. These data support previously mentioned data on ethylene. Skeist [21] has detailed the types of raw materials that might be obtained from similar cracking of coal. He shows (see Table 1.5-6) that synthetic crude oil generated from Utah coal yields nearly 60% naphthenes and 13% paraffins. Reforming the naphthenes could yield a high percentage of trimethyl benzenes, toluene, benzene, and C_8 aromatics. These materials form the basis for nearly all the materials in the styrenic family, phenolics, epoxies, polycarbonate, nylons, and unsaturated and saturated polyesters. Furthermore, the presence of such chemicals as acenaphthene and aromatics of nine carbons or more should lead to the commercialization of new plastics products. For example, 1,2,4,5-tetramethylbenzene is readily obtained in large quantities from coal liquefaction. It can be reacted to produce pyromellitic dianhydride, which in turn is reacted with diamines to make polyimides. Skeist feels that this plastic material, 10 times more plentiful in coal than in gasoline, will fall in price until it is directly competitive with petroleum-based plastics for fibers, films, and coatings.

Synthetic gas can be formed from lignite by treating it at 1900°F in the presence of crushed limestone. The resulting CO_2, CO, and low-molecular-weight hydrocarbons are then methanated to obtain a clean fuel. Extensive investigations to extend the resulting fuel through cracking to ethylene are under way. Smith [14] proposes fixation of atmospheric carbon as a way of calling on the atmospheric reserve of 15×10^{11} tons of carbon dioxide. Until the turn of the century, there was no viable process for the production of ammonia. The Haber-Bosch direct process for catalytically fixing nitrogen in the presence of hydrogen to produce ammonia led to the extensive worldwide consumption of ammoniated products, including refrigerants, plastics, and fertilizers. Certainly the thermodynamics at present do not favor a direct fixation to formaldehyde, but alternative methods,

Table 1.5-6

Chemical Composition of Some Coals and Petroleum[a] [21]

Chemical	Anthracite	Bituminous (medium volability)	Lignite	Petroleum (crude)	Gasoline	Toluene
Carbon[b]	93.7	88.4	72.7	83–87	86	91.3
Hydrogen[c]	2.4	5.0	4.2	11–14	14	8.7
Oxygen	2.4	4.1	21.3	—	—	—
Nitrogen	0.9	1.7	1.2	0.2	—	—
Sulfur	0.6	0.8	0.6	1.0	—	—
Atom ratio: hydrogen to carbon	0.31	0.67	0.69	1.76	1.94	1.14

Composition of a Coed Syncrude Obtained from a Utah Coal

Component	Concentration (%)
Naphthenes	57.8
Polycyclic and thioaromatics	20.8
Paraffins	13.5
Alkyl benzenes	7.9

[a] Exclusive of moisture and 3 to 15% ash.
[b] Carbon fraction is 70% aromatic.
[c] Ratio of aromatic to aliphatic hydrogen is 0.23.

including incorporation of photosynthetic methods, should be explored. Alternative energy sources, including UV, nuclear, and solar radiation, should be considered as ways of altering molecular structure as well as providing valuable energy.

Naturally occurring raw materials have been neglected owing to the availability of low-cost liquid and gaseous hydrocarbon sources. Nevertheless, reduction of complex molecules to potential raw materials should not be overlooked in the future. Production of protein from petroleum on a commercial scale has demonstrated that microorganisms can organize the relatively simple molecules found in liquid materials into complex molecules such as protein. Certainly the anaerobic reduction of complex molecules into CO_2 and H_2O is also well documented. It would seem then that an effort should be made to biologically crack natural polymers such as peptides, proteins, and lignite into small, manageable molecules from which linear structures can be made. And certainly efforts should be made to replace the chemistry of paraffins and olefins with chemistries of sugars and polysaccharides. Attempts to physically combine simple starches with polyolefins with an eye toward biodegradable materials have led instead to the development of a processable polyethylene that seems to have excellent mechanical properties. Additional work on copolymerization of activated starches with other simple plastics will undoubtedly lead to the development of an entirely new generation of plastics based on renewable biological reserves. The primary question is not whether replacement of petroleum and natural gas can be accomplished but whether it can be accomplished before critical shortages of these materials force the world into monetary crises so enormous that necessary research is impossible.

References

1. J. H. DuBois and F. W. John, Plastics, Van Nostrand Reinhold, New York, 1967, Chap. 1.
2. H. R. Simonds, Source Book of the New Plastics, Vol. II, Van Nostrand Reinhold, New York, 1961, p. 21.
3. W. H. Hoge, Plast. Eng., 29(8):51 (1973).
4. Anon., Chem. Eng., 46(23):134 (1968).
5. C. P. Fennimore and F. J. Martin, Chem. Eng., 44(3):141 (1966).
6. J. L. Isaacs, SPE ANTEC, Tech. Pap., 15:251 (1969).
7. A. G. Walker, Europlast. Mon., 72 (Nov. 1973).
8. C. Abbott, Europlast. Mon., 68 (Nov. 1973).
9. Anon., Mod. Plast., 52(1):14 (1975).
10. J. P. DeKany and T. C. Austin, SPE ANTEC Tech. Pap., 20:188 (1974).
11. Anon., Mod. Plast., 49(1):14 (1972).
12. Anon., Mod. Plast., 51(1):10 (1975).
13. Anon., Mod. Plast., 51(1):14 (1975).
14. H. V. Smith, SPE ANTEC, Tech. Pap., 18:490 (1972).
15. C. E. Chastain, SPE ANTEC, Tech. Pap., 18:202 (1972).
16. J. L. Throne and R. G. Griskey, Mod. Plast., 49(11):101 (1972).
17. D. R. Paul, Grant EP 00411, University of Texas, Austin, Texas, 1973.
18. D. M. Fay, SPE ANTEC Tech. Pap., 20:148 (1974).
19. J. L. Throne, Polym. Eng. Sci., 17:682 (1977).

20. R. R. Stover and E. P. Whaley, SPE ANTEC <u>Tech. Pap.</u>, <u>20</u>:146 (1974).
21. I. Skeist, <u>Plast. Eng.</u>, <u>30</u>(2):28 (1974).

2.

Characteristics of Polymer Reactions

2.1 Introduction

As discussed in the first chapter, polymers are composed of synthetic organic molecules of thousands of basic units or mers. They are normally plastic or pliable at some stage in their processing. If they can be heated to a molten state and cooled to a solid state repeatedly, they are thermoplastic, such polyethylene or polystyrene. If, when heated to a plastic state, they undergo reaction so that when cooled the final mass is of a three-dimensional nature, they are thermosetting plastics, such as epoxies or polyurethanes. We should recognize that synthetic organic macromolecules are only one subsection of the larger class of macromolecular plastics. For example, there are nonplastic polymers such as polypeptides, proteins, cellulose, and starches. There are inorganic synthetic polymers such as silicones, silanes, and sulfur. There are nonorganic low-molecular-weight plastics such as concrete, cement, and vitreous clay. And there are organic low-molecular-weight plastics such as asphaltum, paraffin, oils, and alkyd resin paints. The last category is sometimes referred to as the oligomer plastics. Thus, many of the processing techniques discussed here are valid for processing of materials in these categories.

It should be noted that more than 67% of all plastics used in the United States are ethylene derivatives (polyethylene, PVC, and polystyrene), and nearly all plastics are produced from hydrocarbon by-products. There are few commercial plastics that are produced from biological by-products, although this will certainly change as petroleum becomes scarce and expensive.

In this chapter, we shall consider the categories of polymerization of polymers. We shall also consider the various types of chemical reactions that can take place during polymerization. We shall consider the kinetics of these reactions and form the chemical rate equations. And we shall conclude by considering the various mathematical methods for solving these equations. Our objective is to determine the role played by the various materials that participate in the formation of the pure polymer.

2.2 Addition and Condensation Reactions

There are two basic types of polymerization: (1) addition, which occurs when active chains and/or monomer units interact without a by-product, and (2) condensation, which occurs when a Lewis polyacid and a Lewis polybase react to form a chain with the splitting out of a by-product molecule, such as water or CO_2.

2.2.1 Addition Reactions

The vast majority of the plastic materials used today are produced by addition reaction. As shown in Table 2.2-1, the vinyl group of polymers is comprised of typical addition polymers.

To cause these reactions to take place and thus produce a polymer, we must attack the unstable ethylene double bond and open it, thus producing a free-radical or free-valence active molecule. The reaction proceeds much like the linking of hands and is stopped primarily by foreign matter or a terminator molecule. To initiate the reaction, heat and pressure (the packing of highly active molecules) are one choice. Another choice is some source of free radicals, preferably a material with a heat-sensitive double bond, such as a peroxide. Gamma radiation and ultraviolet radiation are other methods for activating the monomer molecule. As regards the use of peroxides, hydrogen peroxide, benzoyl peroxide (BPO), methyl ethyl ketone peroxide (MEKP), and other types of aldehyde and ketone peroxides

Table 2.2-1

Vinyl-Based Monomers

$$\text{Monomer:} \quad \begin{matrix} R_1 & R_3 \\ \overset{\bullet}{C} = \overset{\bullet}{C} \\ \overset{\bullet}{R_2} & \overset{\bullet}{R_4} \end{matrix}$$

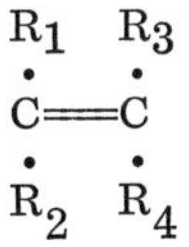

Monomer name	R_1	R_2	R_3	R_4
Ethylene	H	H	H	H
Propylene	H	H	H	CH_3
Butylene	H	H	H	CH_2CH_3
Styrene	H	H	H	ϕ
Methylstyrene	H	H	H	ϕCH_3
Vinyl chloride	H	H	H	Cl
Vinyl dichloride	H	H	Cl	Cl
Vinyl fluoride	H	H	H	F
Vinyl pyrrolidone	H	H	H	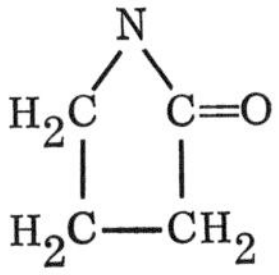
Vinyl acetate	H	H	H	$OOCH_3$
Vinyl alcohol	H	H	H	OH
Butadiene (a double vinyl)	H	H	H	$HC{=}CH_2$
Acrylonitrile	H	H	H	N
Tetrafluoroethylene (TFE Teflon)	F	F	F	F

have been used. During the application of heat the double bond between the two oxygens is opened, producing two free radicals,

$$R-O{=}O-R \xrightarrow{\Delta} R-O\cdot + \cdot O-R \tag{2.2-1}$$

or, in schematic,

$$I_2 \xrightarrow{\Delta} 2I\cdot \tag{2.2-2}$$

We can write the initiation reaction in a kinetic fashion as

$$I + M_1 \xrightarrow{k_i} P_1 \tag{2.2-3}$$

where I is the activated initiator, M_1 is the monomer molecule, and P_1 is the activated monomer molecule. k_i is the initiation reaction rate constant. In this section, we shall restrict ourselves to the initiation, propagation, and termination of a homopolymer, such as polyethylene.

Once we have active species available, the reaction proceeds by adding monomer to the end of the growing chain,

$$\begin{aligned}
M_1 + P_1 &\xrightarrow{k_p} P_2 \\
M_1 + P_2 &\xrightarrow{k_p} P_3 \\
&\cdots \\
M_1 + P_n &\xrightarrow{k_p} P_{n+1} \\
&\cdots
\end{aligned} \tag{2.2-4}$$

where P_n is the activated polymer molecule, with n reacted units, and k_p is the propagation reaction rate constant.

There are many ways of stopping a reaction, either temporarily or permanently. Chemists have identified these primary ways:

1. Spontaneous, either in bulk or by collison with solid surfaces. Here the reaction equation would be

$$P_n \xrightarrow{k_t} M_n \tag{2.2-5}$$

   where M_n represents the "dead" or deactivated polymer.

2. Monomer addition to the end of the growing chain. Here the active site propagates along the growing chain instead of opening the double bond of the monomer. Thus, we have a temporarily stopped reaction, with a molecule having saturated ends but an active

double bond in the interior of the chain. This type of molecule can be reactivated at the double bond site by reaction with another active molecule, thus producing a branched polymer. The termination reaction for the monomer addition is

$$M_1 + P_n \xrightarrow{k_t} M_{n+1} \tag{2.2-6}$$

3. A less likely termination reaction is mutual termination by combination. Here two active, growing polymer molecules combine at the active ends, thus forming a saturated bond and a dead molecule

$$P_n + P_m \xrightarrow{k_t} M_{m+n} \tag{2.2-7}$$

The reason that this is less likely than the first two types of termination reactions is that there are many more monomer molecules available than growing polymer molecules. In addition, the mobility of growing molecules is probably much less than that of the monomer molecules, and thus the probability of this form of termination is very low.

4. Another low-probability termination reaction is mutual termination by disproportionation. Here, instead of forming one dead molecule of two growing molecules, a collision produces two dead molecules:

$$P_n + P_m \xrightarrow{k_t} M_r + M_s \tag{2.2-8}$$

The random scission can produce dead molecules of lengths that are different from those of the growing molecules. Again, the probability of this form of termination is quite low, and as we shall see, Amundson and his coworkers show that there is little effect in the production rate of polymer with or without these last two forms of termination.

2.2.2 Condensation Reactions

Before we consider the kinetics of addition reaction, we should discuss the types of polymers that are produced by condensation. Nearly all condensation reactions split out simple molecules such as H_2O or CO_2. There is no simple listing of these reactions as there is with the vinyl group. Thus, we shall consider several representative examples.

1. A reaction between a dibasic acid and a dialcohol (diol) to pro-
    duce a saturated polyester:

$$
\underset{\text{HO}-\overset{\text{O}}{\overset{\|}{\text{C}}}-\text{R}_1-\overset{\text{O}}{\overset{\|}{\text{C}}}-\text{OH} + \text{H}\,\text{O}-\text{R}_2-\text{OH}}{} \longrightarrow \text{HO}-\overset{\text{O}}{\overset{\|}{\text{C}}}-\text{R}_1-\overset{\text{O}}{\overset{\|}{\text{C}}}-\text{O}-\text{R}_2-\text{OH} + \cdots
$$

(2.2-9)

2. A reaction between two molecules of dihydroxy acid to form a
    polyester:

$$
\text{HO}-\overset{\text{O}}{\overset{\|}{\text{C}}}-\text{R}-\text{OH} + \text{H}\,\text{O}-\overset{\text{O}}{\overset{\|}{\text{C}}}-\text{R}-\text{OH} \longrightarrow \text{HO}-\overset{\text{O}}{\overset{\|}{\text{C}}}-\text{R}-\text{O}-\overset{\text{O}}{\overset{\|}{\text{C}}}-\text{R}-\text{OH} + \cdots
$$

(2.2-10)

3. A reaction between a diol and phosgene to produce a form of
    urethane:

$$
\text{HO}-\text{R}-\text{O}\,\text{H} + \text{Cl}-\overset{\text{O}}{\overset{\|}{\text{C}}}-\text{Cl} \longrightarrow \text{HO}-\text{R}-\text{O}-\overset{\text{O}}{\overset{\|}{\text{C}}}-\text{Cl} + \cdots
$$

(2.2-11)

4. And continuing the above reaction with another molecule of diol
    to produce polycarbonate:

$$
\text{HO}-\text{R}-\text{O}-\overset{\text{O}}{\overset{\|}{\text{C}}}-\text{Cl} + \text{H}\,\text{O}-\text{R}-\text{OH} \longrightarrow \text{HO}-\text{R}-\text{O}-\overset{\text{O}}{\overset{\|}{\text{C}}}-\text{O}-\text{R}-\text{OH} + \cdots
$$

(2.2-12)

5. The classical reaction between a diamine and a dibasic acid to
    produce an amide (known generically as nylon):

$$
\text{H}_2\text{N}-\text{R}_1-\overset{\text{H}}{\overset{|}{\text{N}}}\,\text{H} + \text{HO}-\overset{\text{O}}{\overset{\|}{\text{C}}}-\text{R}_2-\overset{\text{O}}{\overset{\|}{\text{C}}}-\text{OH} \longrightarrow \text{H}_2\text{N}-\text{R}_1-\overset{\text{H}}{\overset{|}{\text{N}}}-\overset{\text{O}}{\overset{\|}{\text{C}}}-\text{R}_2-\overset{\text{O}}{\overset{\|}{\text{C}}}-\text{OH}
$$

(2.2-13)

$R_1 = (CH_2)_6$, hexamethylenediamine; $R_2 = (CH_2)_4$, adipic acid =
                                                              Nylon 66

R_1 = same $R_2 = (CH_2)_8$, sebacic acid =
                                                              Nylon 610

6. The reaction between two molecules of α-amino acid to produce
    an amide:

$$H_2N-R-\overset{O}{\overset{\|}{C}}-OH + H\overset{H}{\underset{|}{N}}-R-\overset{O}{\overset{\|}{C}}-OH \longrightarrow H_2N-R-\overset{O}{\overset{\|}{C}}-\overset{H}{\underset{|}{N}}-R-\overset{O}{\overset{\|}{C}}-OH$$

$$(2.2\text{-}14)$$

R = $(CH_2)_5$, caprolactam = Nylon 6
R = $(CH_2)_{10}$, α-amino undecanoic acid = Nylon 11

7. Reaction between a diisocyanate and a diol to produce a linear urethane:

$$O=C=N-R_1-N=C=O + HO-R_2-OH \longrightarrow O=C=N-R_1-\overset{H}{\underset{|}{N}}-\overset{O}{\overset{\|}{C}}-O-R_2-OH$$

$$(2.2\text{-}15)$$

R_1 = $(CH_2)_6$, hexamethylenediamine;　R_2 = $(CH_2)_2$, ethylene glycol

8. Reaction between a diisocyanate and a triol to produce a thermo-setting urethane:

$$O=C=N-R_1-N=C=O + HO-CH_2$$
$$O=C=N-R_1-N=C=O + HO-CH \longrightarrow \text{three-dimensional thermoset}$$
$$O=C=N-R_1-N=C=O + HO-CH_2$$

$$(2.2\text{-}16)$$

glycerol
R_1 = hemamethylenediamine

9. Reactions between molecules of polyethers or polyoxides:

$$n \; H_2C=O \xrightarrow[\text{acid}]{} HO+CH_2-O+_nH$$

(formaldehyde)　　　　　(polyacetal)

$$(2.2\text{-}17)$$

$$n \; H_2\overset{O}{\overset{\diagup \diagdown}{C-CH_2}} \xrightarrow{-HOH} HO+CH_2-CH_2-O+_nH$$

(ethylene oxide)　　　　　(polyoxyethylene)

10. The formation of the first-stage linear epoxy by reacting bis-phenol A and epichlorohydrin:

(attack point for cross-link)

$$\underset{HO\phi}{\overset{HO\phi}{\diagdown}}\underset{CH_3}{\overset{CH_3}{\diagup}}C + CH_2\overset{O}{\overset{\diagup\diagdown}{-}}CH-CH_2Cl \longrightarrow +O\phi-\underset{CH}{\overset{CH}{\underset{|}{C}}}-\phi O-CH_2-\underset{OH}{\overset{}{\underset{|}{CH}}}-CH_2+_n$$

$$(2.2\text{-}18)$$

11. The classical reaction between a triamine and formaldehyde to form melamine-formaldehyde:

                                                      ⟶ product   (2.2-19)

12. And finally the reaction between a urea and formaldehyde to form urea-formaldehyde (a similar reaction between phenol and formaldehyde was the genesis of the thermoset industry in the 1800s:

$$H_2C=O + H_2N-C-NH_2 + O=CH_2 \longrightarrow \text{urea-formaldehyde}$$

and   (2.2-20)

                                                      dihydroxy diphenyl
                                                      methane (novolac)

Probably the difference in reaction characteristics between addition polymers and condensation polymers is the length of the polymer in process-ready condition. Since most thermoplastics are addition polymers, we find that they gain most of their strength either through crystallization into regular ordered structure or by entanglements of the very long chains. Thus, it is advantageous to have thermoplastic polymer chains of thousands of units. Thermosetting resins, on the other hand, derive their great strength from cross-linking, and thus the process-ready molecule can be short provided it has many attack points for cross-linking. If we could envision an ethylene molecule 1 cm in length (on a scale of about 100,000,000:1), common low-density polyethylene molecules would be on the order of 30 to 50 m long, high-density polyethylene molecules would be 100 to 200 m long, and ultrahigh-molecular-weight polyethylene molecules would be 1000 to 2000 m long. On the same scale, a ready-to-process phenolic molecule might be 20 cm long and a nylon molecule (a thermoplastic polymer) less than 10 m long. As a result, the kinetics for condensation reactions are considerably more complex and less amenable to analysis than those for addition reactions. Probably the best way of thinking about condensation reactions is that they are primarily diffusion-controlled reactions. The very large monomer molecules (and relatively short polymer molecules) offer great resistance for the diffusion of the by-product molecule (H_2O or CO_2). And the movement of these bulky molecules further inhibits the reaction. As a result, there seems to be little relationship between the chemical equations describing the reaction (as given above) and the order of the reaction. This point is discussed in more detail below when we consider the reactor design and again when we consider the processing of thermosetting resins.

2.3 Co- and Terpolymerization Reactions

There are many polymers that are not homopolymers (that is, produced entirely of the same monomer). Impact polystyrene, styrene-acrylonitrile (SAN), ABS, Celcon, and Teflon II are examples of polymers of more than one monomer. There are several categories of co- and terpolymers which should be defined for completeness:

1. Random linear: —A—A—B—B—B—A—B—B—A—A—A—. An example of a random linear copolymer is SAN, which can be written as

$$
\begin{array}{cccc}
\text{H} & \text{H} & \text{H} & \text{H} \\
\text{-----} \; \overset{\textstyle |}{\underset{\textstyle |}{\text{C}}} \text{--} \overset{\textstyle |}{\underset{\textstyle |}{\text{C}}} \text{----------} \overset{\textstyle |}{\underset{\textstyle |}{\text{C}}} \text{--} \overset{\textstyle |}{\underset{\textstyle |}{\text{C}}} \text{-----} \\
\text{H} & \phi & \text{H} & \text{H}
\end{array}
\qquad (2.3\text{-}1)
$$

2. Block linear: —A—A—A—A—A—A—A—B—B—B—B—A—A—A—A—.
   An example of a block linear copolymer is Celcon, which is 95% polyoxymethylene and 5% polyoxyethylene.
3. Branched block or graft:

$$
\begin{array}{c}
\text{—A—A—A—A—A—A—A—A—A—A—} \\
\text{B} \\
\text{B} \\
\text{B} \\
\text{B}
\end{array}
$$

Examples of branched block or graft copolymers are FEP Teflon, where A is TFE Teflon and B is trifluoromonochloroethylene, and vinyl acrylic (which can be a terpolymer of polymethyl methacrylate, polyvinyl acetate, and PVC, as shown):

$$
\begin{array}{c}
\text{—CH}_2\text{—CH—CH}_2\text{—CH—CH}_2\text{—CH—CH}_2\text{—} \\
\;\;\;\;\;\;\;\;\; | \;\;\;\;\;\;\;\;\;\;\;\;\; | \;\;\;\;\;\;\;\;\;\;\;\;\; | \\
\;\;\;\;\;\; \text{COCH}_3 \;\;\;\; \text{COCH}_3 \;\;\;\; \text{COCH}_3 \\
\;\;\;\;\;\;\;\;\; | \;\;\;\;\;\;\;\;\;\;\; || \;\;\;\;\;\;\;\;\;\;\; || \\
\;\;\;\;\;\; \text{O} \;\;\;\;\;\;\;\;\;\; \text{O} \;\;\;\;\;\;\;\;\;\; \text{O} \\
\;\;\;\;\;\;\;\; | \\
\;\;\;\;\;\; \text{CH}_2\text{—CH—O—C—O—CH}_2\text{—CH}_2\text{—Cl} \\
\;\;\;\;\;\;\;\;\;\;\;\;\;\;\;\;\;\;\; | \\
\;\;\;\;\;\;\;\;\;\;\;\;\;\;\;\; \text{CH}_3
\end{array}
\qquad (2.3\text{-}2)
$$

Another polymer that has great commercial use, ABS, is a terpolymer which is made by a random copolymerization of acrylonitrile and styrene which then is grafted to a polybutadiene backbone. On some resins, the graft is made to a copolymer backbone of butadiene and styrene. Again, nearly all copolymers are addition polymers. If both monomers (A and B) are initiated by the same free-radical initiator, we shall have two chemical reactions describing initiation:

$$I + A_1 \xrightarrow{k_{ia}} P_1$$
$$I + B_1 \xrightarrow{k_{ib}} Q_1 \tag{2.3-3}$$

And we shall have four propagation reactions, with active sites on both F and Q adding either A or B monomer:

$$P_n + A_1 \xrightarrow{k_{p_{aa}}} P_{n+1}$$

$$Q_n + A_1 \xrightarrow{k_{p_{ba}}} P_{n+1}$$

$$Q_n + B_1 \xrightarrow{k_{p_{bb}}} Q_{n+1} \tag{2.3-4}$$

$$P_n + B_1 \xrightarrow{k_{p_{ab}}} Q_{n+1}$$

You can see that the notation becomes quite cumbersome in that P always represents the species of monomer A, the last molecule added to the chain, and Q always represents the species of molecule B. Of course, the control of the addition of A to a growing chain that had last added a molecule B and vice versa depends on the reaction rate constants. If $k_{p_{aa}} > k_{p_{ba}}$, A will preferentially add to the growing chain made up primarily of A molecules, and likewise if $k_{p_{bb}} > k_{p_{ab}}$, B will preferentially add to a chain having mostly B molecules. Obviously if these reaction rates are very different, we shall have a reactor containing two polymers, one nearly all A and the other nearly all B. If ($k_{p_{aa}} k_{p_{bb}} / k_{p_{ab}} k_{p_{ba}}$) is on the order of unity, we should get a nearly ideal random copolymerization (—A—B—A—B—A—B—). Even values on the order of 0.4 to 0.5 will yield a nearly random copolymer. Young [1] has compiled reaction rate ratios for nearly all known copolymers.

Just as there are four propagation equations, there are four termination equations for monomer termination, two involving the P_n growing polymer and two involving the Q_n growing polymer. In addition, we can have two spontaneous termination equations, one for each species; combination termination of free radicals involving like species or different species; and disproportionation termination:

$$P_n + A_1 \xrightarrow{k_{t_{aa}}} A_{n+1}$$

$$Q_n + A_1 \xrightarrow{k_{t_{ba}}} A_{n+1}$$

$$P_n + B_1 \xrightarrow{k_{t_{ab}}} B_{n+1}$$

$$Q_n + B_1 \xrightarrow{k_{t_{bb}}} B_{n+1}$$

$$P_n \xrightarrow{k_{t_a}} A_n$$

$$Q_n \xrightarrow{k_{t_b}} B_n$$

$$P_n + P_m \xrightarrow{k_{t_{pp}}} A_{n+m}$$

$$P_n + Q_m \xrightarrow{k_{t_{pq}}} A_{n+m} \quad \text{(or } B_{n+m})$$

$$Q_n + Q_m \xrightarrow{k_{t_{qq}}} B_{n+m}$$

$$P_n + P_m \xrightarrow{k_{d_{pp}}} A_r + B_s$$

$$P_n + Q_m \xrightarrow{k_{d_{pq}}} A_r + B_s$$

$$Q_n + Q_m \xrightarrow{k_{d_{qq}}} A_r + B_s$$

$$(2.3\text{-}5)$$

As you can see from these equations, keeping track of a growing species is very difficult. While some work has been done on the solution of the kinetic equations, most of the effort has been centered around the steady-static kinetic solutions, which we shall discuss in detail below. And as you can imagine, when a third species is added to the process, the equations become even more difficult to analyze or even formulate.

2.4 Degradation Reactions

Since most polymeric reactions take place at high temperatures and all are exothermic, we would expect that as diffusion processes begin to control the growth rate of the molecule various forms of degradation might be initiated that would cause the long chain to be severed. The clearest example of degradation is the effect of hydrolysis on the molecular weight of nylon.

TFE Teflon is also known to "unzip" during polymerization, sometimes with explosive energy. There are many ways of initiating degradation: hydrolysis, pyrolysis (excess localizèd temperature in the presence of oxygen), chemical degradation (such as oxidation), thermal degradation owing to high shear energies or high heats of reaction, photochemical reactions that cause embrittlement (many plastics such as ABS are very sensitive to UV radiation), and gamma and other forms of radiation. The predominant mode of degradation is random chain scission, although chain transfer of unstable free radicals can lead to very rapid degradation. Termination is usually not at the monomer stage but at some stable oligomer species. In TFE Teflon, for example, the stable species is octafluorocyclobutane, $\begin{smallmatrix} CF_2 - CF_2 \\ CF_2 - CF_2 \end{smallmatrix}$, not tetrafluoroethylene, $CF_2{=}CF_2$. Thus the degradation equations are:

$$I + P_n \xrightarrow{\quad k_d \quad} M_i + M_{n-i}$$
$$M_i \xrightarrow{\quad k_{d_2} \quad} M_j + M_{i-j}$$

(2.4-1)

and so on, until a stable species is reached. These equations can be modified to include the effects of hydrolysis on condensation-type reactions as well. Incidentally, I, the initiator, could be UV energy, O_2 in the case of oxidation, and H_2O for hydrolysis. Many chemists consider degradation as a reverse polymerization, including only initiation, depropagation to the monomer, and termination at the stable species. We shall consider briefly some of the arguments concerning degradation within reactors in a later section.

2.5 Molecular-Weight Distributions

Because the rate of reaction and the rate of diffusion of monomer to reactive sites are strongly dependent on temperature, extent of mixing, and microconvection around the growing molecule, most commercial polymers do not have a single molecular weight but consist of molecules with varying chain lengths. A typical molecular-weight spectrum is seen for polystyrene in Fig. 2.5-1. To obtain a measure of the molecular weight of a given polymer, we need to define a statistical molecular weight. If N_i is the number of moles of polymer having i repeat units per weight of sample, then the total number of all species is

$$n = \sum_{i=1}^{\infty} N_i$$

(2.5-1)

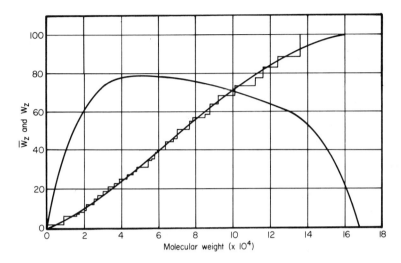

Figure 2.5-1   Molecular-weight distribution for thermally polymerized polystyrene-fractionated data [10]

and the total weight of the sample is

$$w = \sum_{i=1}^{\infty} W_i \tag{2.5-2}$$

where W_i is the weight of polymer having i repeat units. Thus, $W_i = N_i M_i$, where M_i is the molecular weight of polymer having i repeat units. Using this, we can define an average molecular weight based on the number of molecules:

$$\overline{M}_n = w/n = \sum_{i=1}^{\infty} N_i M_i / \sum_{i=1}^{\infty} N_i \tag{2.5-3}$$

And we can define an average molecular weight based on the weight of molecules:

$$\overline{M}_w = (nw)/w = \sum_{i=1}^{\infty} N_i M_i^2 / \sum_{i=1}^{\infty} N_i M_i \tag{2.5-4}$$

In fact, we can define a general molecular weight as

$$\overline{M}_g = \sum_{i=1}^{\infty} N_i M_i^{a+1} \Big/ \sum_{i=1}^{\infty} N_i M_i^a \qquad\qquad (2.5\text{-}5)$$

There are other molecular weights that are measured using various analytical tools, such as gel permeation chromatography, osmometry, ultrafiltration, and solution viscosity. For these molecular weights, $0 < a < 2$. When $a = 0$, we have the number-average molecular weight; when $a = 1$, we have the weight-average molecular weight. The other molecular weights are characterized by the type of test used to measure them, e.g., viscosity-average molecular weight $\overline{M}_v$. Probably the most important information that can be gleaned from knowing the molecular weights is the weight-average to number-average molecular weight ratio, known as the dispersity index:

$$\text{DI} = \overline{M}_w / \overline{M}_n = \boxed{\sum_{i=1}^{\infty} N_i M_i^2 \Big/ \sum_{i=1}^{\infty} N_i M_i} \Big/ \boxed{\sum_{i=1}^{\infty} N_i M_i \Big/ \sum_{i=1}^{\infty} N_i} \qquad (2.5\text{-}6)$$

When the dispersity index is unity, the system is referred to as monodisperse, and the entire batch of polymer has the same molecular weight. Very few commercial polymers have DIs near unity. On the other hand, most proteins and peptides have dispersity indices of unity. According to statistical theory [2], we can show that the most probable distribution is $\overline{M}_w / \overline{M}_n = 2$. Some special polymers have values as low as 1.05 or so, but most commercial resins have dispersity indices of 5 to 20, and some have values as high as 100. Commercially, certain resins, such as polystyrene, exhibit improved physical properties with decreasing dispersity index. However, in other resins, such as PVC, the lower-molecular-weight materials act as a processing aid and lubricate the higher-molecular-weight materials. Thus, no blanket statement about the relationship between molecular-weight distribution and processibility of the final resin can be made.

Another measure of the molecular weight of the polymer is the number-average chain length (or number-average degree of polymerization) $\overline{X}_n$. $\overline{X}_n$ is defined as the original number of monomer units divided by the total number of molecules:

$$\overline{X}_n = N_0 / N \qquad\qquad (2.5\text{-}7)$$

where N_0 is the initial number of monomer molecules and N is the present number of polymer molecules. For a simple condensation reaction, such as caprolactam, $\overline{X}_n = 1/(1 - p)$, where p is the fraction of hydroxyl (or, equivalently, amine) groups reacted. The higher $\overline{X}_n$ becomes, the more complete the reaction is. As expected, there is a weight-average chain length $\overline{X}_w$ also. It is defined as the weight of the polymer divided by the weight of the batch:

$$\overline{X}_w = xn_x/N_0 \tag{2.5-8}$$

where x is the number of monomer units in the polymer chain and xn_x are the number of polymer molecules having x monomer units in the chain. For the same simple condensation reaction, $\overline{X}_w = (1 + p)/(1 - p)$. Thus, the dispersity index can be formed from the ratio of weight-average to number-average chain length:

$$DI = \overline{X}_w/\overline{X}_n = (1 + p) \quad \text{(for the simple condensation} \tag{2.5-9}$$
$$\text{reaction)}$$

It is possible, therefore, to have polymers with dispersity indices of 100 or so and $\overline{X}_n$s of 100 or dispersity indices of 1.02 and $\overline{X}_n$s of 100,000. As a rule of thumb, however, $\overline{M}_w/\overline{M}_n \simeq 5$ to 20 and $\underline{X}_n \simeq 100$ (or 99% reacted). We should note that frequently $\overline{X}_n$ is written as $\overline{DP}_n$, referring of course to the number-average degree of polymerization.

2.6 The Mathematics of Complex Reactions

In any standard text on reactor design [3] the early chapters are given to illustrations of reaction rate equations and possible methods of solution. These are considered without reference to the type of reactor system so that the engineer can gain an insight into the behavior of time- and tempera-ture-dependent reactions. Most of the cases considered are very simple and yield to rather simple, analytical solutions:

$$A \xrightarrow{\text{k}_1} B \xrightarrow{\text{k}_2} C \tag{2.6-1}$$

or

$$A + B \xrightarrow{\text{k}_1} C + D \tag{2.6-2}$$

From a chemistry viewpoint, there is a great distinction between the actual molecular mechanism of the chemical reaction and the "A to B to C" reac-tion. Thus, discussion of molecularity and order is frequently given much consideration. In most instances in chemical engineering reactor design, however, the order of the reaction is simply stated, and the arithmetic is worked out from that point.

Probably the first attempt to fully describe and solve the rate equa-tions for polymers was made by Liu and Amundson in 1961 [4]. Since that time several other methods have been used to solve the set of differential equations that describe polymerization for simple homopolymeric addition reactions (and others as well). In this section, we shall give a brief outline of the several methods available. In the next chapter we shall consider the effects of reactor design, type of polymerization, and presence of catalyst on the kinetics.

For illustrative purposes, consider an isothermal addition reaction, with an initial concentration of monomer of M_1^0 (moles per liter) in the batch reactor. Assume no volume change during reaction, and assume that all the propagation rate constants are the same and independent of molecular weight (e.g., $k_{p_1} = k_{p_2} = \cdots = k_{p_n} = \cdots = k_p$). Further, let us assume that only monomer termination is important. Now the rate of formation of P_1, the activated monomer, is given as

$$\text{rate of formation of } P_1 = k_i M_1 I \tag{2.6-3}$$

where M_i is the monomer concentration and I is the initiator concentration. P_1 is used to produce P_2 and to produce the dead polymer, M_2. Thus, the rate of disappearance of P_1 is

$$\text{rate of disappearance of } P_1 = k_p M_1 P_1 + k_t M_1 P_1 \tag{2.6-4}$$

Thus, the rate expression for P_1 is

$$dP_1/dt = k_i M_1 I - (k_p + k_t)M_1 P_1 \tag{2.6-5}$$

Likewise for P_2, since it is produced from P_1 and is used to produce P_3 and M_3:

$$dP_2/dt = k_p M_1 P_1 - (k_p + k_t)M_1 P_2 \tag{2.6-6}$$

And for the rate of production of a polymer with a chain length of n units,

$$dP_n/dt = k_p M_1 P_{n-1} - (k_p + k_t)M_1 P_n \tag{2.6-7}$$

For the monomer concentration, we see that it is used in every step. Thus,

$$-dM_1/dt = k_i M_1 I + k_p M_1 P_1 + k_p M_1 P_2 + \cdots + k_p M_1 P_n + \cdots \tag{2.6-8}$$

or

$$-dM_1/dt = M_1 \left[k_i I + (k_p + k_t) \sum_{n=1}^{\infty} P_n \right] \tag{2.6-9}$$

For the accumulation of dead polymer (really the desired product), we can write, for the dead polymer with a chain length of m units,

$$dM_m/dt = k_t M_1 P_{m-1} \tag{2.6-10}$$

and for the rate of formation of the initiator,

$$dI/dt = 2k_d I_2 \tag{2.6-11}$$

where k_d is the reaction rate constant for the dissociation of the free-radical source I_2. For other types of termination reactions, the rate equations given above are more complex. For example, for combination termination

$$P_n + P_m \xrightarrow{\quad k_{t_c} \quad} M_{n+m}$$

we see that the rate of production of a polymer with a chain length of n units would become

$$dP_n/dt = k_p M_1 P_n - k_p M_1 P_{n+1} - k_{t_c} P_n \sum_{m=1}^{\infty} P_m \qquad (2.6\text{-}12)$$

and the rate of accumulation of dead polymer would likewise be changed to

$$dM_m/dt = k_{t_c} \sum_{j=1}^{\infty} P_j P_{m-j}/2 \qquad (2.6\text{-}13)$$

(the division by 2 reflects the fact that we are counting each species twice in the summation).

2.6.1 Steady-State Models

One of the earliest forms of solution for the rate equations utilized combination termination. It is argued that one can assume steady state between the rate of generation of free radicals and their rate of dissipation

$$R_i = R_t \qquad (2.6\text{-}14)$$

where R_i is the rate of initiation of free radicals and R_t is their rate of termination. It is further argued that since the rate of initiation k_i is several orders of magnitude greater than the rate of dissociation of the initiator k_d, the rate of initiation of radicals of all types can be represented as

$$R_i = dP/dt = 2k_d I_2 \qquad (2.6\text{-}15)$$

Further, in examining the rate of termination equation for combination termination, we see that the rate of dissipation of radicals of all types can be represented as

$$R_t = -dT/dt = 2k_{t_c} P^2 \qquad (2.6\text{-}16)$$

Here P represents radicals of all types, irrespective of chain length. Now, at steady state, $R_i = R_t$, and thus

$$2k_d I_2 = 2k_{t_c} P^2 \tag{2.6-17}$$

or, rearranging,

$$P = [(k_d/k_{t_c})I_2]^{1/2} \tag{2.6-18}$$

This is the steady-state concentration of chain radicals. You can easily show that this concentration is very low. Let O( ) represent the order of magnitude of the various terms. Now $O(I_2) = 10^{-2}$, $O(k_d) = 10^{-2}$ to 10^{-3}, and $O(k_{t_c}) = 10^7$ to 10^9 (with appropriate units). Thus from Eq. (2.6-18), $O(P) = 10^{-6}$ mol/liter. This argument is used frequently to justify the application of the steady-state model. If we ignore the initiation reaction, we can show that the rate of accumulation of dead polymer of all molecular weights ($\sum_{m=2}^{\infty} M_m$) is approximately equal to the rate of disappearance of monomer. Thus, the rate of propagation can be rewritten as

$$d(\text{polymer})/dt = R_p = k_p M_i P \tag{2.6-19}$$

(Remember that this is an approximate equation.) Now the concentration of radicals is given above in Eq. (2.6-18), and upon substitution we get the rate of accumulation of dead polymer:

$$R_p = d(\text{polymer})/dt = k_p M_1 [(k_d/k_{t_c})I_2]^{1/2} \tag{2.6-20}$$

Thus, we see that the rate of accumulation of polymer varies with the square root of the initiator concentration. Experimentally, it has been found that this simple approach holds for early stages in the polymerization (but not for the initial stages, where the concentration of radicals is building, or the later stages, where transport of reactive material controls the process). We see also that

$$R_p = k_p M_1 (R_i/2k_{t_c})^{1/2} \tag{2.6-21}$$

Solving for the reaction rates,

$$k_p^2/k_{t_c} = 2R_p^2/R_i M^2 \tag{2.6-22}$$

Thus, the larger the ratio of k_p/k_{t_c}, the longer the final chains will be. This, of course, is an obvious conclusion when one considers its implication in terms of the reactions themselves. Frequently, a kinetic chain length v is defined as the ratio of the rate of propagation to the rate of initiation:

$$v = R_p/R_i \tag{2.6-23}$$

But since we have steady state, $R_i = R_t$, and the kinetic chain length can then be written as

$$v = R_p/R_t = k_p^2 M_1^2/2k_t R_p = k_p M_1 (2k_d k_t I_2)^{1/2} \tag{2.6-24}$$

Thus, v represents the average chain length that is developed from one free-radical initiation and is therefore proportional to the number-average degree of polymerization $\overline{DP}_n$ and hence to the number-average molecular weight $\overline{M}_n$. Note for the combination termination reaction that v varies inversely with the square root of initiator concentration and directly with monomer concentration. Thus, we would expect the number-average degree of polymerization $\overline{DP}_n$ and the number-average molecular weight $\overline{M}_n$ to also vary inversely with the square root of initiator concentration and directly with the monomer concentration. Thus, long chains are produced if the initiator concentration is minimized, if the monomer concentration is large, and if the rate of propagation is larger than the combined rate of dissociation of the initiator and the rate of termination. And, as expected, this is common practice, since normally the initiator is much more expensive than the monomer.

The approach outlined can be directly applied to copolymerization, in the following way, following Rodriguez [5]. Recall Eq. (2.3-4), where, if the subscripts are deleted, we can write

$$P + A \xrightarrow{k_{aa}} P$$

$$Q + A \xrightarrow{k_{ba}} P$$

$$Q + B \xrightarrow{k_{bb}} Q \tag{2.6-25}$$

$$P + B \xrightarrow{k_{ab}} Q$$

Again, P and Q are the active species of their respective monomers, A and B. Consider the additional simplification that propagation is the only reaction of importance. As a result, we can make a material balance on the rate of creation of free radical, P, for example, as follows:

$$d(P)/dt = k_{ba}(Q)(A) - k_{ab}(P)(B) = 0 \tag{2.6-26}$$

Again assuming that the concentration of free radicals reaches a steady state after a short induction time. Thus,

$$(P)/(Q) = (k_{ba}/k_{ab})(A)/(B) \tag{2.6-27}$$

Now a similar set of balances can be written for the consumption of monomers A and B:

$$-d(A)/dt = k_{aa}(P)(A) - k_{ba}(Q)(A) \tag{2.6-28}$$

$$-d(B)/dt = k_{bb}(Q)(B) - k_{ab}(P)(B) \tag{2.6-29}$$

In accordance with common practice, reactivity ratios are given as

$$r_a = k_{aa}/k_{ba} \tag{2.6-30}$$

$$r_b = k_{bb}/k_{ab} \tag{2.6-31}$$

Note that these ratios represent the tendency for the monomer to react with itself rather than with the other monomer. To determine whether a reaction will result in a suitable copolymer or simply two mixed homopolymers, it is necessary to invent a fractional conversion rate, defined for A as

$$F_a = d(A)/dt/[d(A)/dt + d(B)/dt] \tag{2.6-32}$$

A similar relationship holds for F_b. Through manipulation, and by definition of the ratio of mole fractions as

$$f_a = (A)/[(A) + (B)] \tag{2.6-33}$$

and

$$f_a/f_b = (A)/(B) \tag{2.6-34}$$

we obtain

$$F_a/F_b = [(r_a f_a/f_b) + 1]/[(r_a f_b/f_a) + 1] \tag{2.6-35}$$

Note that if the copolymerization reactions are negligible (e.g., k_{ab} and k_{ba} are near zero), we have a mixture of homopolymers, since in the limit in Eq. (2.6-35)

$$F_a/F_b = d(A)/d(B) = k_{aa}(A)^2/k_{bb}(B)^2 \tag{2.6-36}$$

On the other hand, if k_{aa} and k_{bb} are small when compared with k_{ab} and k_{ba}, we get the randomly alternating copolymer ($F_a = F_b = 0.5$). From Young's data [1], it appears that for nearly all polymer combinations $r_a r_b$ is less than unity. Alfrey and Young [6] have tried to establish certain parameters that are unique to monomers that enable the prediction of the r ratios. They identify Q and e as parameters that can be related to the reactivity ratio as follows:

$$\ln r_a = \ln(Q_a/Q_b) - e_a(e_a - e_b) \tag{2.6-37}$$

and

$$\ln r_a r_b = -(e_b - e_a)^2 \tag{2.6-38}$$

According to Rodriguez, Q_a is a function of the relative stability of the polymer chain owing to the addition of a molecule of monomer A. e is related to the polarity of the monomer molecule and the resulting free radical when the monomer is added to the end of the growing chain.

For one specific case, as an illustration, consider the values of r_a and r_b to be approximately unity. Thus, $\ln r_a = 0$ and $\ln r_a r_b = 0$. Under these conditions, $e_a = e_b$ and $Q_a = Q_b$. Rodriguez refers to this as the ideal copolymerization. For several commercial systems, ideal copolymerization is difficult to achieve. From Young [1], for styrene monomer, $Q_a = 1.00$ and $e_a = -0.80$; for 1,3-butadiene, $Q_b = 2.39$ and $e_b = -1.05$. These values yield $r_a = 0.322$ and $r_b = 3.31$. Experimentally, $r_a = 0.5$ and $r_b = 1.4$. This copolymerization favors the formation of long butadiene chains with shorter segments of styrene interspersed (e.g., SBR rubber). As another example, vinyl chloride has a Q_a value of 0.044 and $e_a = 0.20$. For this reaction, $r_a = 2.16$ and for 1,3-butadiene, $r_b = 0.075$. Experimentally, $r_a = 8.8$ and $r_b = 0.035$. The very large spread in r values indicates that these two materials may form a mixture of homopolymers. The reader is to be cautioned here, however, that these values should be used only when relative reactivities of combinations of monomers have not been previously determined experimentally.

2.6.2 Difference-Differential Model

Although the steady-state model gives us some information about the relationship of polymer chain length and concentrations of initiator and monomer, the model is not sufficiently flexible to include initial and final stages of polymerization. As mentioned earlier, Liu and Amundson were the first to attempt a solution of the difference-differential equations (2.6-3) through (2.6-11). If we make a simplifying assumption that the initiator free-radical concentration, I, is constant, assume that the monomer is in great excess such that M_1 is constant, and redefine $k_p' = k_p M_1$, $k_i' = k_i M_1 I$, and $k_t' = k_t M_1$ (again assuming monomer termination), our equations become

$$dP_1/dt = k_i' - (k_p' + k_t')P_1 \tag{2.6-39}$$

$$dP_2/dt = k_p' P_1 - (k_p' + k_t')P_2 \tag{2.6-40}$$

$$\cdots$$

$$dP_n/dt = k_p' P_{n-1} - (k_p' + k_t')P_n \tag{2.6-41}$$

$$dM_m/dt = k_t' P_{m-1} \tag{2.6-42}$$

and

$$dM_1/dt = 0 \tag{2.6-43}$$

Note that Eq. (2.6-27) is of the form

$$dy_i/dt = ay_{i-1} + by_i \qquad\qquad (2.6-44)$$

We see further that the solution of equation i requires the solution of equation i - 1, which in turn requires solution of i - 2, and so on. The equation can be directly integrated using a standard Runge-Kutta fourth-order numerical integration scheme, as Liu and Amundson did for 200 sequential equations. The equations are also directly attacked using, for example, a large-capacity analog computer or the continuous system modeling program (CSMP) available on most digital computers. Figures 2.6-1 and 2.6-2 show typical monomer terminated batchwise polymerized free radical and terminated polymer concentrations.

2.6.3 Z Transform Method

Once Liu and Amundson demonstrated that this problem was important to polymerization reactor design, several other attempts to understand the equation solution were presented. Abraham [7] and Kilkson [8] noted that the standard method of solution of backward-difference-differential equations was the Z transform. Briefly, we see that there are n variables $P_i(t)$. If we replace these n variables with a single variable $P(Z,t)$ such that

$$P(Z,t) = 0 + P_1(t)Z^{-1} + P_2(t)Z^{-2} + \cdots + P_i(t)Z^{-i} + \cdots \qquad (2.6-45)$$

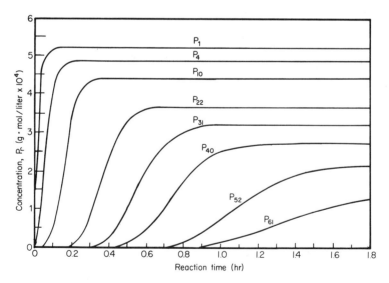

Figure 2.6-1   Batchwise polymerization, addition with monomer termination, active polymer behavior; $(M)_0 = 1.0$; $k_i = 0.03$, $k_p = 6.0$, $k_t = 1.0$ [4].

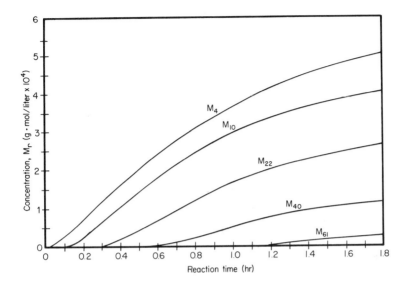

Figure 2.6-2 Batchwise polymerization, addition with monomer termination, dead polymer behavior; $(M)_0 = 1.0$, $k_i = 0.03$, $k_p = 6.0$, $k_t = 1.0$ [4].

and if we then take the partial derivative of $P(Z,t)$ with respect to t, we find that

$$\partial P/\partial t = (dP_1/dt)Z^{-1} + (dP_2/dt)Z^{-2} + \cdots + (dP_i/dt)Z^{-i} + \cdots \quad (2.6\text{-}46)$$

But each term, dP_i/dt, is represented by a differential equation, Eq. (2.6-41). Therefore, the partial differential equation can be written as

$$\partial P/\partial t = k_i'Z^{-1} - (k_p' + k_t')P_1Z^{-1} + k_p'P_1Z^{-2} - (k_p' + k_t')P_2Z^{-2} + \cdots$$

$$+ k_p'P_{i-1}Z^{-i} - (k_p' + k_t')P_iZ^{-i} + \cdots \quad (2.6\text{-}47)$$

Note that the second, fourth, sixth, and the even terms after that are of the form

$$-\sum KP_iZ^{-i} \quad (2.6\text{-}48)$$

According to the definition of $P(Z,t)$, these terms are just $KP(Z,t)$. If we factor Z^{-1} from the first, third, fifth, and subsequent odd terms, we obtain a similar relationship:

$$Z^{-1}(k_i' + k_p' P_1 Z^{-1} + \cdots + k_p' P_{i-1} Z^{-1} + \cdots) \tag{2.6-49}$$

With the exception of the term k_i', the term in the parentheses is again just $k_p' P(Z,t)$. If we now rewrite the partial differential equation for $P(Z,t)$ with this information, we get

$$\partial P/\partial t = k_i' Z^{-1} + k_i' Z^{-1} P(Z,t) - (k_p' + k_t') P(Z,t) \tag{2.6-50}$$

or

$$\partial P/\partial t = k_i' Z^{-1} + [k_p' Z^{-1} - (k_p' + k_t')] P(Z,t) \tag{2.6-51}$$

For illustrative purposes, following Abraham's work, let us ignore the initiation reaction and the termination reaction and just concentrate on the propagation steps. If we do this, we can write the above partial differential equation as

$$\partial P/\partial t = (Z^{-1} - 1) K P(Z,t) \tag{2.6-52}$$

where K is the propagation reaction rate constant k_p'. This partial differential equation can be integrated to obtain

$$P(Z,t) = P(Z,0) \exp[K(Z^{-1} - 1)t] \tag{2.6-53}$$

Now $P(Z,0)$ represents the initial concentration of all growing species at time $t = 0$. If we have only monomer in the reactor, we can assume that $P_1(0) \neq 0$ but that $P_i(0) = 0$ for all values of i greater than 1. Thus, the equation becomes

$$P(Z,t) = P_1(0) \exp[K(Z^{-1} - 1)t] \tag{2.6-54}$$

This equation contains the concentration profiles as functions of time for all species. To obtain, for example, the concentration profile for the chain of i units in length, we simply expand the above equation in negative powers of Z,

$$P(Z,t) = P_1(0) \exp(-Kt)[Z^{-1} + KtZ^{-2} + (Kt)^2 Z^{-3}/2! + \cdots$$

$$+ (Kt)^{i-1} Z^{-i}/(i-1)! + \cdots] \tag{2.6-55}$$

and then pick out the term representing $P_i(t)$ according to the earlier defining equation:

$$P_i(t) = P_1(0) \exp(-Kt)(Kt)^{i-1}/(i-1)! \tag{2.6-56}$$

The term $(Kt)^{i-1}$ represents a time lag. This time lag is necessary when you consider that to produce polymer of i chain length we need polymer of i - 1 chain length and that production of this polymer is also rate dependent. To obtain the concentration of dead polymer of chain length m we write

$$dM_m/dt = k_t' P_{m-1} = k_t'(Kt)^{i-1} P_1(0)\,\exp(-Kt)/(m-1)! \qquad (2.6\text{-}57)$$

or upon integration [where $M_m(0) = 0$], we get

$$M_m^{(t)} = [k_t' K^{m-1} P_1(0)/(m-1)!] \int_0^t t^{m-1}\,\exp(-Kt)\,dt \qquad (2.6\text{-}58)$$

$$= k_t' P_1(0)\gamma(m;t)/(m-1)!\,K \qquad (2.6\text{-}59)$$

where $\gamma(m;t)$ is the incomplete gamma function. See Abramowitz and Stegun [9] for tables of this function. Abraham also shows that from this analysis we can extract various molecular-weight distributions. For example, the number-average chain length is given as

$$\overline{N}_n = \sum_{i=1}^{\infty} i P_i(t) \Big/ \sum_{i=1}^{\infty} P_i(t) \qquad (2.6\text{-}60)$$

$$= \sum_{i=1}^{\infty} i P_i(0)\,\exp(-Kt)(Kt)^{i-1}/(i-1)! \Big/ \sum_{i=1}^{\infty} P_1(0)\,\exp(-Kt)(Kt)^{i-1}/(i-1)! \qquad (2.6\text{-}61)$$

$$= P_1(0)\,\exp(-Kt)(1+Kt)/P_1(0)\,\exp(-Kt) = (1+Kt) \qquad (2.6\text{-}62)$$

Likewise the weight-average chain length is

$$\overline{N}_w = [1 + 3Kt + (Kt)^2]/(1+Kt) \qquad (2.6\text{-}63)$$

and the weight-average to number-average chain length is

$$\overline{N}_w/\overline{N}_n = [1 + 3Kt + (Kt)^2]/[1 + 2Kt + (Kt)^2] \qquad (2.6\text{-}64)$$

Abraham shows that this ratio initially is unity, reaches a peak value of 1.25 at $Kt = 1$, and falls toward 1 as Kt approaches infinity. There are limitations to this method, as Abraham and Kilkson point out. First, the Z transform is a linear transform, and thus linear combinations of rate expressions will produce linear combinations of the transformed variables. If we are to include, for example, combination termination rate expressions, the differential equation for $P(Z,t)$ becomes much more complex. Abraham suggests numerical inversion of the Z transform to produce the individual species equation once the final form for P(Z,t) has been found. Of course, resorting to computation to invert the analytical equation makes the method much less attractive than the straight computer solution.

Kilkson has extended Abraham's Z transform method to include copolymerization and the initiation and termination reactions that were omitted from Abraham's work. The procedure is quite complex, however, and does not provide the flexibility of the numerical methods [8].

2.6.4 Partial Differential Equation Model

Shortly after the publication of the Z transform method, Zeman and Amundson [10] showed that the series of difference-differential equations could be thought of as representing the differencing of a partial differential equation. In their words, "in any mathematical system involving an infinite number of discrete variables an attempt is usually made to approximate the discrete problem by a continuous one."

From a mathematical viewpoint, we can define a "probability density function" based on the definition of the Dirac delta function such that

$$\int_{J_1}^{J_2} f(j) \, dj = \text{fraction of polymers that have chain lengths between } J_1 \text{ and } J_2 \qquad (2.6\text{-}65)$$

If we let $P(j) = P_j$ and $M(j) = M_j$ whenever $j = 1, 2, 3, \ldots, n, \ldots$, we see that $P(j)$ and $M(j)$ are the interpolated mathematical values of P_j and M_j whenever $j = $ integer. According to Zeman and Amundson, we can now use the Euler-Maclaurin summation formula to obtain an expression for $P(j)$ [and $M(j)$]:

$$P_T = \sum_{j=1}^{\infty} P_j = P(j) + P(1)/2 + P(\infty)/2 - (1/12)(dp/dj)_{j=1}$$
$$+ (1/12)(dP/dj)_{j=\infty} + \text{higher-order terms} \qquad (2.6\text{-}66)$$

If we now assume that $P(\infty) = P_j$ approaches zero as j approaches ∞ [$(dP/dj)_{j=\infty} = 0$] and if the degree of polymerization is large [$(dP/dj)_{j=1}$ is the same order of magnitude as that of the higher-order terms (e.g., the slope of P at $j = 1$ is small)], we can rewrite Eq. (2.6-66) as

$$P_T = \sum_{i=1}^{\infty} P_j = \int_1^{\infty} P(j) \, dj + P_1/2 \qquad (2.6\text{-}67)$$

Note further that the Taylor series allows up to expand any "continuous" function around any point. Thus,

$$P(j - 1) = P(j) - (dP/dj)_j + (d^2P/dj^2)_j/2 \mp \dots \qquad (2.6\text{-}68)$$

If we now rewrite our equations for the propagating species P_j and the dead polymer as

$$dP_j/dt = -k'_p(P_j - P_{j-1}) + k'_t P_j \qquad (2.6\text{-}69)$$

and

$$dM_j/dt = k'_t P_{j-1} \qquad (2.6\text{-}70)$$

where we have again assumed M_1 to be constant, we can use the Taylor series expansion to yield first-order partial differential equations:

$$\partial P(j,t)/\partial t = k'_p[\partial P(j,t)/\partial j] - k'_t P(j,t) \qquad (2.6\text{-}71)$$

and

$$\partial M(j,t)/\partial t = k'_t P(j - 1,t) \qquad (2.6\text{-}72)$$

Now the boundary conditions on the first equation are $P(j,0) = 0$ (no active species at time $t = 0$), and $P(1,t) = P_1(t)$, the concentration profile of the activated monomer as a function of time. Using these conditions, the solution to the first equation becomes

$$P(j,t) = P_1[t - (j - 1)/k'_p] \exp[-k'_t(j - 1)/k'_p], \qquad t > (j - 1)/k'_p \quad (2.6\text{-}73)$$

$$P(j,t) = 0, \qquad\qquad\qquad\qquad\qquad\qquad t < (j - 1)/k'_p \quad (2.6\text{-}74)$$

And for $M(j,t)$, we have the boundary condition that $M(j,t) = 0$ unless $j = 1$. For that condition, $M(1,0) = M_1$, the monomer concentration which is assumed to be constant. The solution to the equation for $M(j,t)$ is

$$M(j,t) = k'_t \exp[-k'_t(j - 2)/k'_p] \int_0^t P_1[t - (j - 2)/k'_p] \, dt, \qquad t > (j - 2)/k'_p$$
$$(2.6\text{-}75a)$$

$$M(j,t) = 0, \qquad\qquad\qquad\qquad\qquad\qquad t < (j - 2)/k'_p$$
$$(2.6\text{-}75b)$$

The significance of the fact that the value of $P(j,t) = 0$ until time t exceeds a value $(j - 1)/k'_p$ is very important. A molecule of an arbitrary chain length j cannot be formed in an arbitrarily short period of time. Thus $(j - 1)/k'_p$ represents a measure of the time lag required before $P(j,t)$ can be formed. This approach is most useful when we are trying to determine optimum cycle times in reactors. Zeman and Amundson have illustrated several other types of termination reactions including combination termination (the case discussed for the steady-state model), and they show that in general monomer terminations yield weight distributions that fall abruptly at a given chain length and time, whereas for most other terminations a more gradual decline in weight distribution with chain length at a given time is evident. This effect is seen graphically in Figs. 2.6-3 and 2.6-4 for two types of termination. Nevertheless, the concept of time lag required before appreciable amounts of either growing or dead polymer or arbitrary chain length are produced is invaluable to reactor design, as will be seen in the next chapter.

2.6.5 Moments Method

In 1967, Katz and Saidel [11] presented an important adjunct to the earlier work by considering moments of the size distribution. Using a combination termination reaction, they defined growing polymer chain and dead

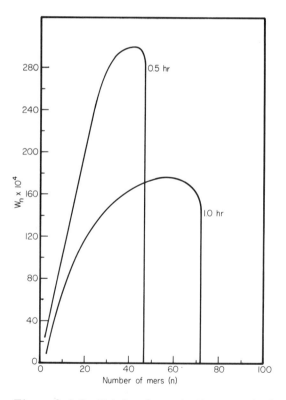

Figure 2.6-3   Batch polymerization, weight distribution for addition reaction, monomer termination [10].

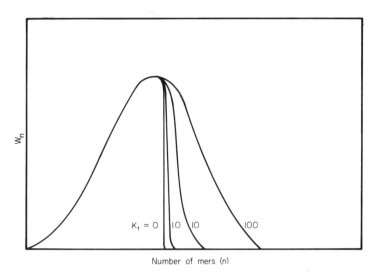

Figure 2.6-4 Batch polymerization, weight distribution for addition reaction, combination termination; termination rate is parameter [10].

polymer chain distributions in a continuous fashion as done by Zeman and Amundson. These functions, $P(j,t)$ and $M(j,t)$, respectively, yield concentrations

$$\int P(j,t)\ dj \quad \text{and} \quad \int M(j,t)\ dj \tag{2.6-76}$$

respectively, at time t. Thus, the equations for initiator concentration $I_2(t)$, monomer concentration $M_1(t)$, and the species $P(j,t)$ and $M(j,t)$ appear as

$$dI_2(t)/dt = -k_d I_2(t) \quad \text{subject to } I_2(0) = I_2^0 \tag{2.6-77}$$

$$dM_1(t)/dt = -k_p M_1(t) \int P(i,t)\ di$$
$$\text{subject to } M_1(0) = M_1^0 \tag{2-6.78}$$

$$\partial P(j,t)/\partial t + k_p M_1(t)[\partial P(j,t)/\partial j] = 2k_d I_2(t)\delta(j) - k_t P(j,t) \int P(i,t)\ di$$
$$\text{subject to } P(j,0) = 0 \tag{2.6-79}$$

and

$$\partial M(J,t)/\partial t = (k_t/2) \int P(j - i,t)P(i,t)\, di$$

subject to $M(j,0) = 0$

(2.6-80)

Here the integrals are taken over all values of i. Thus $\int P(i,t)\, di$ is the total concentration of radicals. Also $\delta(j)$ is the Dirac delta function, and according to Katz and Saidel, this implies that all the primary radicals are formed at size zero. They then introduce the concept of moments for the growing and dead polymer chains:

$$\lambda_k(t) = \int i^k P(i,t)\, di \quad \text{and} \quad \mu_k(t) = \int i^k M(i,t)\, di \qquad (2.6\text{-}81)$$

λ_0 and λ_1 represent the zeroth and first moments of the size distribution of growing polymer and likewise μ_0 and μ_1 the zeroth and first moments for the dead polymer. The zeroth moment here represents the concentration of the growing and dead polymers at any time; the first moment represents the total amount of monomer associated with the growing and dead polymers at any time. Thus, we can show that the number-average degree of polymerization, $\overline{DP}_n$, is the ratio of the sum of the first moments of the dead and growing polymer to the zeroth moments:

$$\overline{DP}_n = (\lambda_1 + \mu_1)/(\lambda_0 + \mu_0) \qquad (2.6\text{-}82)$$

Similarly, we can show that a weight-average degree of polymerization, $\overline{DP}_w$, is the ratio of the sum of the second moments of dead and growing polymer to the first moments. After couching the above equations in terms of the moments of dead and growing chains, Katz and Saidel then solve the initiator equation and an equation for the first moment of the growing polymer chain. The solution for the first moment, λ_0, is

$$\lambda_0 = (4I_2^0/a^2)\phi \left\{ \frac{K_1(\phi) - [K_1(a)/I_1(a)]I_1(\phi)}{K_0(\phi) + [K_1(a)/I_1(a)]I_0(\phi)} \right\} \qquad (2.6\text{-}83)$$

Here $a = (8k_t I_2^0/k_d)^{1/2}$, $\phi(t) = a \exp(-k_d/2t)$, K_0 and K_1 are Bessel functions of the first kind of zero and first order, respectively, and I_0 and I_1 are Bessel functions of the second kind of zero and first order, respectively. These functions are tabulated in Abramowitz and Stegun [9] as well as in other source books. From this, the monomer concentration can be found:

$$m = \left\{ \frac{K_0(\phi) + [K_1(a)/I_1(a)]I_0(\phi)}{K_0(a) + [K_1(a)/I_1(a)]I_0(a)} \right\}^{-b} \qquad (2.6\text{-}84)$$

where $m = M_1(t)/M_1^0$ and $b = k_p/k_t$. Katz and Saidel then show that the other moments are obtained once $I_2(t) = I_2^0(\phi/a)^2$ and m is known:

$$\mu_0 = I_2^0 = I_2 - \lambda_0/2$$

$$\lambda_1 = [M_1^0 b/(1 - b)](m = m^{1/b})$$

$$\mu_1 = [M_1^0/(1 - b)](1 - m) - [M_1^0 b/(1 - b)](1 - m^{1/b})$$

Through the introduction of simplifying expressions for the Bessel functions in Eqs. (2.6-83) and (2.6-84), they find that a satisfactory approximation for m is

$$m = \exp[-b(a - \phi)] = \exp\{-ab[1 - \exp(-k_d/2t)]\} \tag{2.6-85}$$

and

$$\mu_0 + \lambda_0 = I_2^0[1 - (\phi/a)^2] = I_2^0[1 - (\ln m/ab + 1)^2] \tag{2.6-86}$$

$$\lambda_1 + \mu_1 = M_1^0(1 - m) \tag{2.6-87}$$

From this information, we can, of course, get the number-average degree of polymerization:

$$\overline{DP}_n = (M_1^0/I_2^0)(1 - m)/[1 - (\ln m/ab + 1)]^2 \tag{2.6-88}$$

Katz and Saidel show that if I_2 is constant, the zeroth moments become

$$\lambda_0 + \mu_0 = (k_d k_t I_2^0/2k_p^2)^{1/2} \ln m \tag{2.6-89}$$

and the first moments are

$$\lambda_1 + \mu_1 = M_1^0(1 - m) \tag{2.6-90}$$

where m is

$$m = \exp(-k_d k_p t/2k_t) \tag{2.6-91}$$

and the number-average degree of polymerization reduces to

$$\overline{DP}_n = (2k_p^2 I_2^0/k_t k_d)^{1/2}(M_1^0/I_2^0)(1 - m)/\ln (1/m) \tag{2.6-92}$$

Note that this solution is very similar to the steady-state model proposed earlier. If, for example, we were to consider m to be constant [we can show in the equation for $\overline{DP}_n$ that if $m = 1$, the indeterminate equation can be resolved using L'Hopital's rule to yield $\overline{DP}_n = (2k_p^2/k_t k_d)^{1/2}[(M_1^0)^2/I_2]^{1/2}$,

$\overline{DP}_n$ is inversely proportional to the square root of the initiator concentration and directly proportional to the initial monomer concentration. Thus, the Katz-Saidel model is considerably more general than the steady-state model and yet reduces to it when appropriate simplifying assumptions are made. Katz and Saidel extend their analysis to include the case where termination reactions are limited by increasing viscosity (and thus decreasing mobility of the long-chain molecules), the case of termination with disproportionation, and the case of polymerizations with chain transfer either to monomer or to solvent. The important implication here, however, is that this model gives great insight into determination of the degree of polymerization without resorting to computer solution or complex transform model. Nevertheless, the application of the model to condensation polymerization and copolymerization appears to be as limited as the other models discussed in this section.

2.7 Conclusion

You will note the absence of examples to illustrate the methods in Sec. 2.6. Again, our purpose in this chapter was not to develop skills in mathematical manipulation but to illustrate possible ways of solving the most simple cases of polymerization kinetics. These techniques can be extended, with some difficulty, to copolymerization and condensation polymerization. Also, Amundson and his coworkers have examined heterogeneous reaction equations but with major emphasis on reactor design rather than the arithmetic of reaction kinetics. Thus, we shall also consider the design and control of reactors and choose to use these kinetic models as (1) vehicles to the understanding of chain growth and residence times and (2) bases by which we can perhaps analyze more complex chemical kinetics in conjunction with reactor design.

Problems

P2.A Write the reaction equations for a random addition terpolymer, ABC. Consider that the reaction is terminated only with monomer termination and disproportionation termination.

P2.B Write the reaction equations for an addition terpolymer, XYZ, where X and Y are randomly copolymerized in the absence of Z until a given molecular-weight distribution is achieved. At this point, Z is graft-polymerized to Y. Consider that the reaction is terminated only by disproportionation.

P2.C Write the appropriate reaction equations that describe hydrolysis of polycarbonate.

P2.D From fractionation studies, the following weight distribution was found for an aminic condensation homopolymer with a monomer molecular weight of 100 [12]:

Molecular-weight range	Weight fraction of sample
0-500	0.020
501-1500	0.025
1501-2500	0.145
2501-3500	0.165
3501-4500	0.170
4501-5500	0.190
5501-6500	0.150
6501-7500	0.080
7501-8500	0.035
8501-∞	0.020

From these data, determine (1) the weight-average molecular weight, (2) the number-average molecular weight, (3) the dispersity index, (4) the viscosity-average molecular weight (a = 0.7), and (5) the number-average chain length.

P2.E In a particular styrenic polymerization, chemists have determined that the isothermal rates of (100° C) initiator dissociation (k_d), propagation (k_p), and termination (k_{t_c}) are, respectively, 0.01 sec^{-1}, 5 sec^{-1}, and 10^4 liters/mol·sec. The concentration of initiator (I_2) is fixed at 0.001 mol/liter, and the initial monomer concentration (M_1) is 1 mol/liter. Calculate the steady-state concentration of free radicals, P. Calculate the rate of accumulation of dead polymer, the rate of propagation of live polymer, and the kinetic chain length, v. Using the kinetic chain length, calculate the number-average degree of polymerization, $\overline{DP}_n$.

P2.F Obtain a relationship between the kinetic chain length and the initiator and monomer concentrations for a monomer-terminated free-radical reaction, assuming the steady-state model. Determine the kinetic chain length of this system using the numerical values given in P2.E.

P2.G Make a plot of the free-radical concentrations of chain lengths 1, 10, and 100 for $P_1(0) = 1$ mol/liter and K = 0.2 hr^{-1}. Then determine the dead polymer concentrations for chain lengths of 10 and 100 for these conditions, where $k_t^I = 0.1$ hr^{-1}.

P2.H Following the Z transform analysis, show that the total accumulation of dead polymer of chain length m or less is given by

$$\sum_{j=2}^{m} M_j(t) = [k_t^I P_1(0)/K] \sum_{j=2}^{m} \gamma(j;t)/(j-1)!$$

Find a closed solution for the summation of the incomplete gamma function.

P2.I Following the Z transform analysis for the determination of the number-average chain length for a free-radical reaction, obtain the form for the weight-average chain length, $\overline{N}_w$, Eq. (2.6-64). Make a plot of $\overline{N}_w$, $\overline{N}_n$, and $\overline{N}_w/\overline{N}_n$ as a function of time using the data given in P2.G.

P2.J Using the Zeman-Amundson method of Taylor series expansion, calculate the concentrations of free radicals of 1, 10, and 100 chain lengths using the data given in P2.G.

P2.K Using the Katz-Saidel moment method, make a plot of m, the monomer concentration, vs. $2t/k_d$ for ab = 0.25, 1.0, and 5.0. Make a plot of $DP_n/(M_1^0/I_2^0)$ vs. $2t/k_d$ for ab = 1.0.

P2.L Compare and contrast the Katz-Saidel zeroth moment equation for number-average degree of polymerization with the steady-state model discussed in P2.E.

References

1. L. J. Young, J. Polym. Sci., 54:411 (1961).
2. P. J. Flory, Principles of Polymer Chemistry, Cornell University Press, Ithaca, N.Y. 1953, p. 325.
3. O. A. Hougen and K. M. Watson, Chemical Process Principles; Part Three: Kinetics and Catalysis, Wiley, New York, 1974.
4. S.-I. Liu and N. R. Amundson, Rubber Chem. Technol., 34:995-1133 (1961).
5. F. Rodriguez, Principles of Polymer Systems, McGraw-Hill, New York, 1970, Chap. 4.
6. T. Alfrey, Jr., and L. J. Young, in Copolymerization (G. E. Ham, ed.), Wiley-Interscience, New York, 1964, Chap. 4.
7. W. H. Abraham, Ind. Eng. Chem. Fundam., 2:221-224 (1963).
8. H. Kilkson, Ind. Eng. Chem. Fundam., 3:281-293 (1964).
9. M. Abramowitz and I. S. Stegun, eds., Handbook of Mathematical Functions, Dover, New York, 1964, Chap. 6, p. 253.
10. R. Zeman and N. R. Amundson, AIChE J., 9:297-302 (1963).
11. S. Katz and G. N. Saidel, AIChE J., 13:319-326 (1967).
12. E. H. Merz and R. W. Raetz, J. Polym. Sci., 5:587-596 (1950).

3.

Polymerization Reactor Design and Control

3.1 Introduction

It is quite difficult to go into great depth on this subject because of the be-
wildering number of specialized reactors that are in use commercially and
the great difficulty in analyzing the kinetic equations given in Chap. 2. We
have chosen therefore to select the "classical" types of reactors [batch
reactor, continuous stirred tank reactor (CSTR), plug flow reactor, or
tubular reactor] and the "classically simple" homogeneous polymerization
reactions, such as simple addition reaction with either monomer or combination termination steps, to illustrate the methods for designing reactors.
We shall then consider the various categories of heterogeneous reactions
that make up the bulk of today's production (suspension polymerization,
solvent polymerization, emulsion polymerization, heterogeneous catalyza-
tion). We shall then return to the simple reactor-reaction cases to discuss
reactor stability and control.

The serious student of polymerization reactor design and control
should become acquainted with the various types of processes by which addition monomers and polymers are formed. Albright's recent extensive over-
view of this major field is required detailed reading in this regard [1].
Some of the examples used in this chapter are explained in more detail in
Albright's text.

3.2 Traditional Classification of Reactors

In the early days of polymerization reactor design (and even in many small
operations today), the batch reactor was the only way to make high-molecular-weight materials. Certainly, part of the reason for the use of a batch
process rather than a continuous process, such as plug flow or tubular
reactor or a series of CSTRs, was (and is) the long time required for most
polymerization reactions. In contrast to the reaction times of seconds or
minutes for the production of heavy chemicals (such as nitric acid) or petroleum refining products (such as hydrogenation), the reaction times for many
polymers are on the order of hours. This long reaction time is due not only
to the time required to make a polymer of j monomer units once a polymer
of j - 1 units has been produced but also to increasing viscosity with increasing molecular weight. This of course leads to reduced diffusion of
monomer to reaction sites and in the case of condensation reactions to reduced mobility of reaction sites and decreased diffusion rates of the low-
molecular-weight by-product through the viscous polymer syrup. Because
of the uncertainties involved in producing polymeric resins and the difficulties in measuring meaningful physical properties of the resin (such as degree
of polymerization, molecular-weight distributions, monomer concentration,

and activity of initiator), the batch reactor afforded the resin manufacturer
with the ultimate means of control of the final resin properties. If the final
resin product did not meet the specifications, the entire batch could be
scrapped, if necessary, or more likely, blended with another batch that
was off-spec in other ways, thus producing on-spec resin.

On the other hand, the batch process had (and has) some inherent
problems. Scale-up was quite difficult, since, as we shall see, the polymerization reaction generates heat volumetrically and the heat is removed
geometrically. Thus, increasing the size of the reactor was equivalent to
decreasing the control on the reactor temperature. Also, agitation of the
polymer syrup to ensure uniform concentrations throughout the reactor
becomes considerably more difficult as the reactor size increases and as
the syrup becomes more viscous. Sampling techniques for the purpose of
control are much more difficult to design for batch reactors than for CSTRs
or tubular reactors. And, most important, the economics of increasing
capacity were less favorable to the high-labor batch process than the CSTR
or the tubular flow reactor.

Consequently, for large volume production at least, CSTRs and tubular flow reactors are favored over the batch reactor. We should point out,
however, that this does not mean that the batch reactor is being phased out.
On the contrary, as the demand for resins with specific molecular-weight
distributions or certain concentrations of co- and terpolymers increases,
some major suppliers are returning to the batch reactor. The result, of
course, is that these specialized resins cost the resin converter more than
the high-volume multipurpose resin.

Platzer [2] classifies addition polymerization according to the phase
in which the monomer is present during polymerization (see Table 3.2-1).
In most cases, the monomer is present in a liquid phase, but that liquid
phase may not be the phase in which the polymer is growing. Thus, for
example, we can have a polymerization taking place in an oil-water "oil"
phase by diffusion of the monomer from a "water" phase through the oil-water
interface. Thus, we can have emulsion or suspension polymerization. If
the monomer is homogeneous with the polymer, we have a bulk polymerization, and if the polymer is suspended in a solvent during polymerization,
we have solution polymerization. Each of these processes has commercial
importance, and the type of polymerization will, in many cases, dictate the
type of reactor that can be used.

Nevertheless, to understand the interaction of addition polymerization
kinetics and reactor design, we shall examine the various types of reactors
first. After we have a grasp of the parameters needed in reactor design,
we shall then consider the classification of the polymerization according to
bulk, emulsion, suspension, or solution polymerization. A similar argument
can be advanced for condensation or heterogeneous catalyzed reactions.

Table 3.2-1

Classification of Addition Polymerization Processes [2]

Free radical and ionic initiation

I. Monomer in liquid phase

    1. Bulk (mass) polymerization

        a. Polymer soluble in monomer

        b. Polymer insoluble in monomer

        c. Polymer swollen by monomer

        d. Falling drop

    2. a. Solution; polymer soluble in solvent

        b. Solvent/nonsolvent; polymer insoluble in solvent

    3. Suspension; initiator dissolved in monomer

    4. Emulsion; initiator dissolved in dispersing medium

II. Monomer in vapor phase

III. Monomer in solid phase

3.3 The Batch Reactor with Addition Polymerization

Shown in Fig. 3.3-1 is a typical batch reactor, consisting of a kettle of
known volume, an agitator, a "drop-down" valve at the bottom of the kettle,
and some means of cooling [either external jacketing, internal cooling coils,
reflux condensors (if the monomer is sufficiently volatile), or a combination
of these]. In practice, initiator feed rate and/or monomer feed rate to the
kettle are controlled in an effort to control the reactor temperature. (Remember that all polymerization reactions are exothermic and that the reaction rates for propagation, initiation, and termination are exponentially
dependent on temperature. Thus, increasing the temperature about 10°F
will normally increase the rate of reaction twofold. And, if the heat removal
is insufficient, the exothermic heat of reaction will result in an increase in
the temperature, again increasing the rate of reaction. Thus, without sufficient heat removal surface area, a polymerization reaction can "run away,"
resulting in degradation of the polymer and possibly destruction of the
reactor.)

    The difference-differential equations of Chap. 2 that describe addition
or free-radical polymerization, for example, are directly applicable to the
design of batch reactors where the kinetics of reaction control. (This means

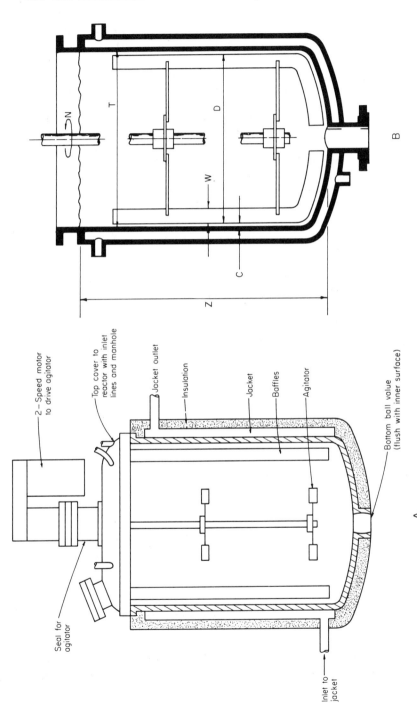

Figure 3.3-1   A: Typical schematic for a suspension batch reactor for the production of PVC. Typically this tank is pressure-vessel-rated at 300 psi at 120°F, with the cooling jacket pressure-rated at 150 psi at 150°F [1].   B: Another version of the jacketed kettle reactor, showing anchor agitator [3].

that sufficient agitation is available throughout the course of the reaction to
minimize diffusion-controlled steps.) Thus, we can use directly the analyses of Liu and Amundson [4] to determine the concentration profiles of various species of growing and dead polymer. In their simplest model, they
assume (1) no volume change owing to the decrease in number of molecules
during polymerization and (2) initial charge of both initiator and monomer.
Furthermore, they assume that all the heat generated by polymerization is
removed by convection to the cooling medium (i.e., an overdesigned heat
removal system). As shown in Figs. 3.3-2 and 3.3-3, the concentrations
of growing polymer approach steady state quite slowly, and for monomer
termination, the concentrations of dead polymer continue to increase with
time. In Fig. 3.3-4, they show the effect of increasing rate of initiation on
the weight fraction at 2 hr into the polymerization reaction. It is apparent
that the rate of initiation must be low if a high-weight-average-molecular-
weight polymer is needed. Increasing the rate of initiation causes rapid
sharpening and left-skewing of the molecular-weight distribution curve. In
Fig. 3.3-5, we see the effect of increasing rate of propagation on the weight
fraction, again at 2 hr into the polymerization reaction. Here increasing
the rate of propagation (all other rates constant) causes a broadening and
right-skewing of the molecular-weight distribution curve. In Fig. 3.3-6,
the effect of increasing the rate of termination is shown, again at 2 hr into
the polymerization reaction and again with all other rates constant. Note

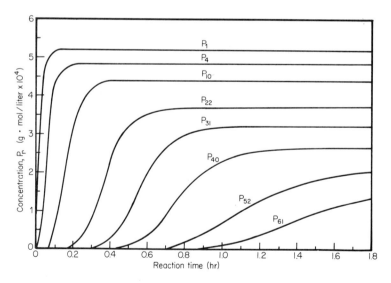

Figure 3.3-2   Batchwise polymerization, addition with monomer termination, active polymer behavior; $(M)_0 = 1.0$, $k_i = 0.03$, $k_p = 60$, $k_t = 1.0$ [4].

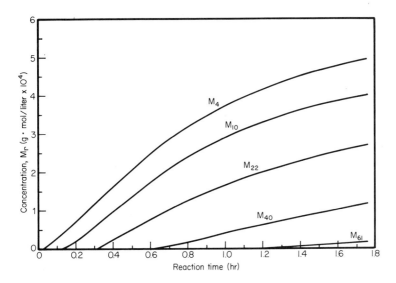

Figure 3.3-3 Batchwise polymerization, addition with monomer termination, dead polymer behavior; $(M)_0 = 1.0$, $k_i = 0.03$, $k_p = 60$, $k_t = 1.0$ [4].

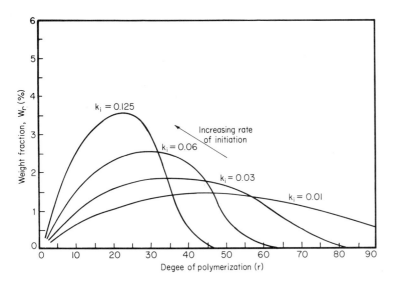

Figure 3.3-4 Batchwise polymerization, monomer termination, addition reaction; effect of initiation rate on molecular-weight distribution after 2 hr; $(M)_0 = 1.0$, $k_p = 60$, $k_t = 1.0$ [4].

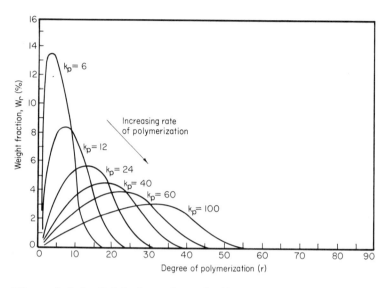

Figure 3.3-5   Batchwise polymerization, monomer termination, addition reaction; effect of polymerization rate on molecular weight distribution after 2 hr; $(M)_0 = 1.0$, $k_i = 0.125$, $k_t = 1.0$ [4].

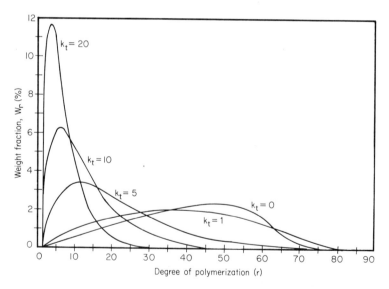

Figure 3.3-6   Batchwise polymerization, monomer termination, addition reaction; effect of termination rate on molecular-weight distribution after 2 hr; $(M)_0 = 1.0$, $k_i = 0.03$, $k_p = 60$ [4].

that with no monomer termination the curve is distributed around a $\overline{DP}_n$ of about 50. As termination rate increases, to $k_t = 5$, for example, the distribution is sharpened and shifted to the left, with a distribution around $\overline{DP}_n$ of about 12 or so. Continuing increase in the monomer termination rate causes dramatic decreases in the percent of high-molecular-weight species. Increasing monomer concentration does not appreciably affect monomer concentration in the reactor after about 2 hr into the reaction, as shown in Fig. 3.3-7. But, in Fig. 3.3-8, note that the dramatic increase in high molecular weight ends with increasing monomer concentration after 2 hr. The effect of reaction time on the molecular-weight distribution curve is shown in Fig. 3.3-9, where after about 2 hr, the distribution remains nearly constant. Liu and Amundson were also concerned about the effect of the termination reaction on the molecular-weight distribution within the reactor. In Fig. 3.3-10, we see that there is little difference in the molecular-weight distribution 2 hr or more into the reaction if termination proceeds by disproportion termination, combination termination, or no termination at all. Furthermore, there is little difference in the distribution if termination occurs spontaneously or by monomer collision. We see that polymer chain interaction to cause termination results in a higher number-average molecular weight and a lower weight-average to number-average

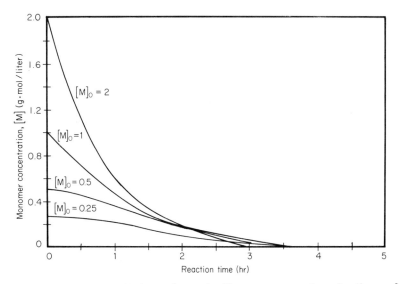

Figure 3.3-7 Batchwise polymerization, monomer termination, addition reaction; effect of initial monomer concentration on monomer concentration as a function of reaction time; $k_i = 0.03$, $k_p = 60$, $k_t = 1.0$ [4].

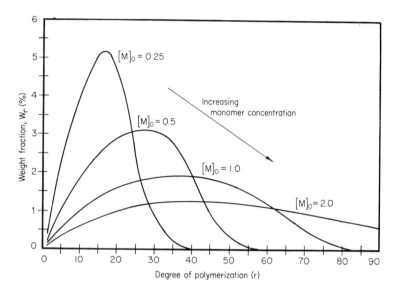

Figure 3.3-8   Batchwise polymerization, monomer termination, addition reaction; effect of initial monomer concentration on molecular-weight distribution after 2 hr; $k_i = 0.03$, $k_p = 60$, $k_t = 1.0$ [4].

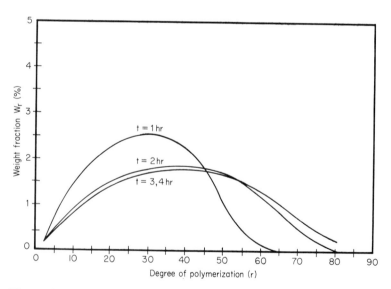

Figure 3.3-9   Batchwise polymerization, monomer termination, addition reaction; molecular-weight distribution as a function of reactor time; $(M)_0 = 1.0$, $k_i = 0.03$, $k_p = 60$, $k_t = 1.0$ [4].

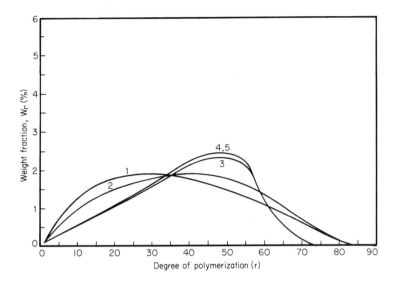

Figure 3.3-10   Batchwise polymerization, addition reaction; effect of termination step on molecular-weight distribution; $(M)_0 = 1.0$, $k_i = 0.03$, $k_p = 60$. Curve 1: spontaneous termination; 2: termination by monomer collision; 3: disproportionation termination; 4: no termination; 5: combination termination [4].

ratio. We can also see now why chemists can distinguish between a monomer-terminated reaction and a combination-terminated reaction but have great difficulty determining whether a two-polymer termination is either disproportionation, combination, or no termination at all. The refinements in chemical analysis are not sufficient to separate these effects via molecular-weight distribution alone.

In the subsequent work discussed in Chap. 2, Zeman and Amundson [5], using the continuous model (rather than choosing to integrate the difference-differential equations), show that for monomer termination reactions there is a dramatic shift in molecular-weight distribution with increasing hold time in the reactor. The importance of the time required to produce a polymer of a finite length is graphically illustrated in Fig. 3.3-11, where high-molecular-weight materials are completely absent from the reactor at low hold times. Zeman and Amundson define a "terminal time" as the time when 99% of the monomer is consumed and thus, according to the definition of degree of polymerization, the reaction is 99% complete. The effects of termination rate and initiation rate on the polymer chain length are shown in Figs. 3.3-12 and 3.3-13. It is immediately apparent that low termination rates and low initiation rates yield high-molecular-weight polymers, thus

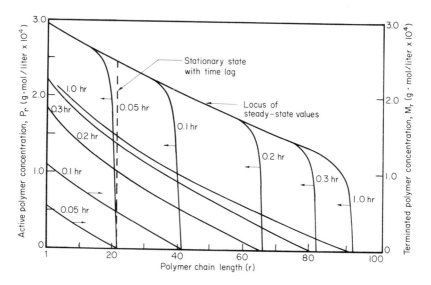

Figure 3.3-11 Batchwise polymerization, addition reaction; monomer termination; active and terminated polymer concentrations as calculated using different analytical models; $k_p = 500$, $k_t = 50$, $k_i = 0.15$, $(M)_0 = 1$ [5].

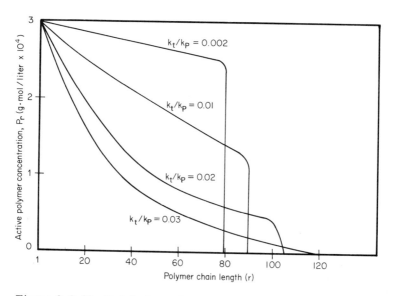

Figure 3.3-12 Batchwise polymerization, addition reaction, monomer termination; effect of termination rate on concentration of active polymers; $k_p = 500$, $k_i = 0.15$, $(M)_0 = 1$ [5].

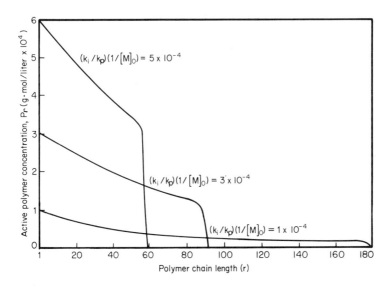

Figure 3.3-13   Batchwise polymerization, addition reaction, monomer termination; effect of initiation rate on active polymers; $k_o = 500$, $k_t = 5$, $(M)_0 = 1.0$ [5].

confirming the original digital computer analyses. They also show that if the polymerization is initiated by a second-order reaction of monomers,

$$M_1 + M_1 \xrightarrow{\ k_i\ } P_2$$

the active polymer concentrations do not reach steady-state values (as presumed from the simple steady-state analyses) but achieve maxima and then decline to asymptotic values. This is shown in Fig. 3.3-14. Also, we see by comparing Figs. 3.3-11 and 3.3-15 that the second-order initiation reaction results in a much shorter hold time to achieve measurable amounts of a polymer with a given chain length. This also exhibits itself as a maximum in the molecular-weight distribution curve, as shown in Fig. 3.3-15. The other parameters, such as termination rate and initiation rate constants, exhibit influences that are similar to those of the first-order initiation reaction.

Amundson and his coworkers have extended their analyses to include many types of batch reactor cases other than the one discussed above. Liu and Amundson consider copolymerization with all types of termination reactions; Zeman and Amundson consider the effect of reversible propagation on homopolymerization, copolymerization with first- and second-order initiation reactions and reversible propagation reactions, and catalyst-initiated

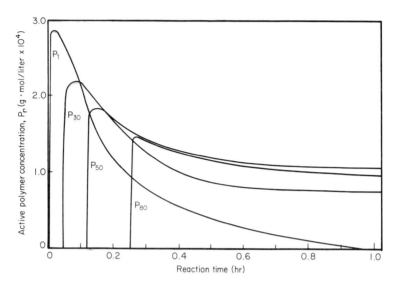

Figure 3.3-14 Batchwise polymerization, addition reaction, second-order initiation, monomer termination; active polymer concentrations as functions of reaction time; $k_p = 500$, $k_t = 5$, $k_i = 0.15$, $(M)_0 = 1.0$ [5].

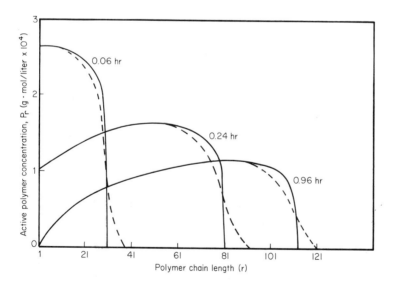

Figure 3.3-15 Batchwise polymerization, addition reaction, second-order termination; comparison of active polymer concentrations for first and second derivative models; $k_p = 500$, $k_t = 5$; $k_i = 0.15$, $(M)_0 = 1$. Note: solid line, first derivative model; dashed line, second derivative model [5].

homogeneous reaction with monomer concentration variation, temperature variation, catalyst concentration variation, and various termination reactions. We shall briefly summarize some of their results and refer the reader to their work for more details.

For copolymerization, they find that the steady-state assumption for active polymer concentration is invalid, particularly in the early stages of reaction and for the high-molecular-weight materials. The appearance of dead polymers is again quite dependent on reaction time, and the growth rate of the dead polymers of any species seems to be nearly linear with respect to time for a monomer termination reaction. An increase in the initiation rate of one of the two monomers causes a sharpening of the molecular-weight distribution with a little shifting in the number-average molecular weight of the copolymer to the lower-molecular-weight species. Increasing the rates of propagation (holding the other reaction rates constant) causes a sharpening and a shifting to lower-molecular-weight species. An increase in the ratio of monomer concentrations (B_1^0/A_1^0) causes a broadening of the molecular-weight distribution curve with little skewing. With no termination, the molecular-weight distribution curve is skewed toward higher-molecular-weight species. With increasing termination rates, the curve skews toward the lower-molecular-weight species. Increasing the holding time causes a broadening of the molecular-weight distribution curve and a slight decrease in the peak number-average molecular weight. And combination termination, disproportionation termination, and no termination reactions produce similar results on molecular-weight distribution curves, causing a high number-average molecular weight and a distribution that is skewed toward higher chain lengths. Monomer and spontaneous termination reactions produce similar curves, characterized by low-number-average-molecular-weight distributions and curves skewed toward lower chain lengths.

We can certainly continue their analyses for the other cases, but perhaps these examples are sufficient to illustrate the behavior of polymerization reactions within a batch reactor.

3.4 The CSTR with Homogeneous Addition Polymerization

Again we turn to the work of Amundson and coworkers to analyze continuous stirred tank reactors (CSTRs) with free-radical addition polymerization reactions having several types of termination reactions. CSTRs are frequently used in series as shown in Fig. 3.4-1 [6]. The reactants are fed continuously into the sump of the first reactor, and the overflow from this reactor is fed, in turn, to the sump of the second reactor, and so on. We assume that the concentrations of all species are totally uniform everywhere within the CSTR reactor, and as a result, the kinetic analyses are couched in the form of difference equations for each reactor (rather than difference-differential equations for the batch reactor). The key to the material analyses is the

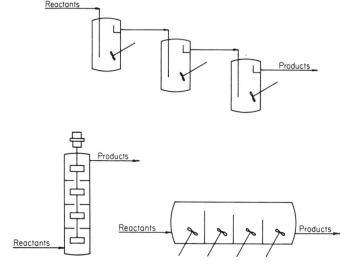

Figure 3.4-1   Typical examples of staged or chained reactors. Top, a typical reactor chain (CSTRs); bottom left, a vertical mixer-settler reactor column; bottom right, a compartmented overflow reactor [6].

simple steady-state material balance, applied to each specie of polymer, active or dead:

$$\text{input rate} - \text{output rate} = \text{accumulation rate} \qquad (3.4-1)$$

(which is zero because of steady-state assumptions). Consider Fig. 3.4-2, where a series of N reactors is shown. If F_{m-1} is the volumetric rate of flow of material into the mth reactor (and it is equal to the volumetric rate of flow of material out of the reactor because of the assumption of steady state), V_m is the volume of the mth reactor, and C_{m-1} and C_m represent the concentrations of a specie entering and leaving the mth reactor, then

$$F_{m-1} C_{m-1} - (F_m C_m + r_m V_m) = 0 \qquad \text{(for a single specie)} \qquad (3.4-2)$$

where r_m is the rate of production of that specie during its time in the mth reactor. Now we know that we can express the rate of reaction in terms of the concentration of the specie (and previous species) as

$$r_m = f(C_m ; C'_m) \qquad (3.4-3)$$

where C'_m represents the concentration of all other species present in the reaction. If we define a residence time as the ratio of the tank holdup to the volumetric flow rate,

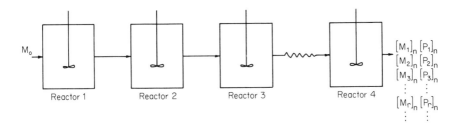

Figure 3.4-2   Schematic for a typical reactor chain of n reactors [7].

$$\theta_m = V_m/F_{m-1} \tag{3.4-4}$$

and if we realize that at steady state $F_m = F_{m-1}$, we can rewrite Eq. (3.4-2) to obtain

$$r_m = -(F_{m-1}/V_n)C_m + (F_{m-1}/V_m)C_{m-1}$$

$$= (C_m - C_{m-1})/\Delta t = f(C_m; C'_m) \tag{3.4-5}$$

or

$$r_m = (C_{m-1} - C_m)/\theta_m = f(C_m; C'_m) \tag{3.4-6}$$

That this equation is a simple algebraic equation for any tank in the series can be seen if we choose the very simple equation

$$A \xrightarrow{\;\;k\;\;} \text{products}$$

For this equation, $r_m = kC_m$, where C represents the concentration of A. Now Eq. (3.4-6) becomes

$$kC_m \theta_m = C_{m-1} - C_m \tag{3.4-7}$$

or, if C_{m-1} is known,

$$C_m = C_{m-1}/(1 + k\theta_m) \tag{3.4-8}$$

If the initial concentration into the first tank (m = 1) is C_0, the solution for the concentration in the mth tank is

$$C_m = C_0 / \sum_{j=1}^{m} (1 + k\theta_j) \tag{3.4-9}$$

and so on. Walas [7] points out that for reactions other than first order, the arithmetic becomes quite complex. To illustrate the CSTR terminology for polymerization reactions, we again turn to the work of Liu and Amundson. For the case of monomer termination (chosen so that we can compare these results with those of the batchwise process), we find that the mass balance in the mth reactor for the active polymer species P_n is given as

$$(P_{n,m-1} - P_{n,m})/\theta_m$$

$$+ M_{1,m}[k_p P_{n-1,m} - (k_p + k_t)P_{n,m}] = 0 \quad n \geq 2 \qquad (3.4\text{-}10)$$

$$(P_{1,m-1} - P_{1,m})\theta_m$$

$$+ M_{1,m}[k_i - (k_p + k_t)P_{1,m}] = 0 \qquad (3.4\text{-}11)$$

Here the first term in each of the equations represents the bulk flow of fluid into and out of the reactor. The second term represents the net rate of production of active species P_n in the mth reactor, as dictated by the kinetics of Chap. 2.

For the dead polymer species M_n in the mth reactor we find

$$(M_{n,m-1} - M_{n,m})/\theta_m + k_t M_{1,m} P_{n-1,m} = 0 \quad n \geq 2 \qquad (3.4\text{-}12)$$

$$(M_{1,m-1} - M_{1,m})/\theta_m - M_{1,m}(k_i + \sum_{n=1}^{\infty} (k_p + k_t)P_{n,m}) = 0 \qquad (3.4\text{-}13)$$

Note here that we have assumed that the reaction rate constants are independent not only of the chain length but also of the reactor stage. This is clearly not the case if one reactor is running at a higher temperature than the rest, but for this example, we shall assume all reactors are isothermal, steady state, and at the same temperature. There are elaborate computer schemes for solving these series difference equations for large values of m. However, more practical is the solution for one- and two-CSTR units. For the concentrations of M_1 and P_1 for the first reactor we can show that

$$\sigma_1 = M_{1,1}k_p/[1/\theta_1 + M_{1,1}(k_p + k_t)] \qquad (3.4\text{-}14)$$

$$P_{n,1} = (k_i/k_p)\sigma_1^n \qquad n \geq 1 \qquad (3.4\text{-}15)$$

$$M_{n,1} = \theta_1 k_t k_i M_{1,1}\sigma_1^{n-1}/k_p \qquad n \geq 2 \qquad (3.4\text{-}16)$$

$$M_{1,1} = -b_1/2a_1 + (b_1^2 + 4a_1 c_1)^{1/2}/2a_1 \qquad (3.4\text{-}17)$$

where

$$a_1 = k_t + \theta_1 k_i k_p + 2\theta_1 k_i k_t \qquad (3.4\text{-}18)$$

$$b_1 = 1/\theta_1 + k_i - k_t M_{1,0} \tag{3.4-19}$$

$$c_1 = M_{1,0}/\theta_1 \tag{3.4-20}$$

The total concentration of free radicals exiting the first reactor in given by

$$\sum_{n=1}^{\infty} P_{n,1} = k_i M_{1,1}/(1/\theta_1 + M_{1,1}k_t) \tag{3.4-21}$$

The importance of σ_1 can be found when recalling the definitions for the weight fraction, number-average, and weight-average chain lengths defined earlier. We can show that the weight fraction $W_{n,1}$ is given as

$$W_{n,1} = n(P_{n,1} + M_{n,1})/(M_{1,0} - M_{1,1}) \tag{3.4-22}$$

where $M_{1,0}$ is the initial monomer concentration to the first reactor. Through proper substitution we find that

$$W_{n,1} = k_i(n\sigma_1^n + \theta_1 k_t M_{1,1} n\sigma_1^{n-1})/(M_{1,0} - M_{1,1}) \tag{3.4-23}$$

If we differentiate $W_{n,1}$ with respect to n and set this derivative to zero to find the maximum weight fraction of polymer, we obtain

$$n_{max,1} = -1/\ln \sigma_1 \tag{3.4-24}$$

Thus σ_1 is revealed to be related to $n_{max,1}$, the most probable degree of polymerization in the first reactor. Furthermore, we can find the number-average chain length, defined as

$$\bar{X}_n = \left(\sum_{n=1}^{\infty} nP_{n,1} + \sum_{n=2}^{\infty} nM_{n,1}\right)/\left(\sum_{n=1}^{\infty} P_{n,1} + \sum_{n=2}^{\infty} M_{n,1}\right) \tag{3.4-25}$$

$$= [1 + \theta_1 k_t M_{1,1}(2 - \sigma_1)]/[(1 - \sigma_1) + \theta_1 k_t M_{1,1}(1 - \sigma_1)\sigma_1] \tag{3.4-26}$$

As shown in Fig. 3.4-3, the concentration profiles of the various species at the exit at the first reactor show some species of high chain length polymers appearing with a reactor hold time of 1.2 hr.

The determination of the concentrations of all the species $P_{n,2}$ and $M_{n,2}$ from the second reactor is somewhat more difficult, and as a result we give here only the final results from Liu and Amundson:

$$\sigma_2 = (M_{1,2}k_p)/[1/\theta_2 + (k_p + k_t)M_{1,2}] \tag{3.4-27}$$

$$P_{n,2} = (k_i/k_p)\sigma_2^n + [k_i/(k_p^2\theta_2 M_{1,2})](\sigma_1^{n+1}\sigma_2 - \sigma_1\sigma_2^{n+1})/(\sigma_1 - \sigma_2) \tag{3.4-28}$$

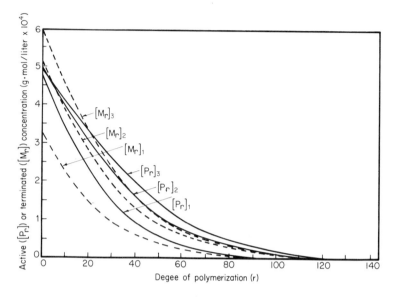

Figure 3.4-3 Continuous addition polymerization, monomer termination; $(M)_0 = 1$, $k_i = 0.03$, $k_p = 60$, $k_t = 1$. Three CSTRs in series. Each residence time 1.2 hr. Solid line, active polymer; P_r; dashed line, terminated polymer M_r [4].

$$M_{n,2} = (\theta_1 k_t k_i M_{1,1} \sigma_1^{n-1}/k_p) + (\theta_2 k_t m_{1,2})\Big\{(k_i \sigma_2^{n-1}/k_p) + [k_i/(k_p^2 \theta_2 M_{1,2})]$$

$$\cdot \ (\sigma_1^n \sigma_2 - \sigma_1 \sigma_2^n)/(\sigma_1 - \sigma_2)\Big\} \qquad (3.4\text{-}29)$$

$$M_{1,2} = -b_2/2a_2 + (b_2^2 + 4a_2 c_2)^{1/2}/2a_2 \qquad (3.4\text{-}30)$$

where

$$a_2 = k_t + \theta_2 k_i k_p + 2\theta_2 k_i k_t \qquad (3.4\text{-}31)$$

$$b_2 = (k_p + k_t)a_1 + 1/\theta_2 + k_i - k_t M_{1,1} \qquad (3.4\text{-}32)$$

$$c_2 = M_{1,1}/\theta_2 \qquad (3.4\text{-}33)$$

And the total concentration of free radicals exiting the second reactor is given by

$$\sum_{n=1}^{\infty} P_{n,2} = \Big(\sum_{n=1}^{\infty} P_{n,1} + \theta_2 k_i M_{1,2}\Big)/(1 + \theta_2 k_t M_{1,2}) \qquad (3.4\text{-}34)$$

Owing to the complexity of the equations, it is difficult to gain insight into the changes in concentration with change in the various parameters (such as hold time θ, initial monomer concentration, initiator, propagation, or termination reaction rates) for multiple reactors in series. Looking at Eq. (3.4-17) for $M_{1,1}$, however, we can see that if we increase hold time, θ_1, we shall decrease $M_{1,1}$ (as expected). If θ_1 is made large enough, θ_1 approaches the ratio of $k_p/(k_p + k_t)$. As long as k_t is not zero, this value is always less than 1. Thus, the concentration of $P_{n,1}$ is always less than $P_{n-1,1}$ (as is also expected). Likewise the total concentration of free radicals leaving the first reactor approaches the ratio k_i/k_t. For most addition reactions this ratio is much less than unity, as is expected. The concentration of dead polymer will increase with increasing hold time, but the relationship will not be linear since $M_{1,1}$ is decreasing with increasing hold time. And as seen before for batch reactions, as the propagation rate increases in relation to the termination and initiation rates, the number-average chain length increases as well (since σ_1 approaches 1, $1 - \sigma_1$ approaches zero in the denominator of $\overline{X}_n$). A similar analysis can be done for the second reactor. We should note there that although Liu and Amundson consider several other forms of termination, including combination termination, spontaneous termination, and disproportionation termination, and derive all the equations necessary to analyze multiple reactor cases, they solve equations for only the first reactor. The primary reason for this is that to determine the monomer concentrations in the effluent of each reactor, it is necessary to solve cubic and even fifth-order algebraic equations (because of the monomer participation in every reaction). It would therefore appear that the appropriate solutions for multiple CSTRs with polymerization reactions and several alternative termination steps are best obtained by digital computation. This method also offers the alternative to change the hold times of various reactors and the reactor temperatures (resulting, of course, in dramatic changes in the reaction rates of the various steps). Also, Eqs. (3.4-10) and (3.4-13) will permit the addition of monomer at various stages in the process, thus enabling some control of the reaction temperature (and the final molecular-weight distribution).

Zeman and Amundson also consider the application of their continuous method of solution for CSTRs and addition reactions. Consider, for the moment, the first reactor. The active polymer concentration equation can be written as

$$P_{n,0} - P_{n,1} + \theta_1 M_{1,1} k_p P_{n-1,1} - \theta_1 M_{1,1}(k_p + k_t)P_{n,1} = 0 \qquad n \geq 2$$

$$(3.4\text{-}35)$$

Now in the initial feed to reactor 1, there is no activated species; thus, $P_{n,0} = 0$. Dropping the ",1" subscripts and rearranging, we get

$$P_{n-1} = (1 + \theta k_t M_1 + \theta k_p M_1)P_n/(\theta k_p M_1) \qquad (3.4\text{-}36)$$

We can show that if we use a backward Taylor series expansion of a continuous form for the activated polymer concentration P(n) we can obtain

$$P(n - 1) - P(n) = -(dP/dn) + (d^2P/dn^2)/2!$$

$$+ \text{ higher-order terms} \qquad (3.4-37)$$

or, in terms of our equation,

$$(d^2P/dn^2) - 2(dP/dn) - 2aP$$

$$+ \text{ higher-order terms} = 0, \qquad n \geq 2 \qquad (3.4-38)$$

where

$$a = (1 + \theta k_t M_1)/\theta k_p M_1 \qquad (3.4-39)$$

As a first approximation, assume that $(d^2P/dn^2) \ll (dP/dn)$ or $P(n)$. Then only two boundary conditions are needed: $P(1) = P_1$ and $P(\infty) = 0$. The latter condition implies that very high-molecular-weight materials are nonexistent in the effluent from the first reactor. The solution to the equation is

$$P(n) = P_1 \exp[-a(n - 1)] \qquad (3.4-40)$$

Note that this is the continuous solution to the discrete set of difference equations given as Eq. (3.4-10). Zeman and Amundson point out that if we had chosen a forward Taylor series expansion the same equation [Eq. (3.4-40)] would hold, except that the term a would be replaced with a', where

$$a' = (1 + \theta k_t M_1)/(1 + \theta k_p M_1 + \theta k_t M_1) = 1 + (k_t/k_p)\sigma_1 \qquad (3.4-41)$$

where σ_1 is discussed in detail above. There is little difference in the shapes of the curves, as Zeman and Amundson point out. Nevertheless, they suggest that the retention of the second derivative in the Taylor series expansion makes even these accurate equations more exact. To obtain a comparison between this continuous solution and the discrete one given above, one need only express the exponential in series form and pick out the nth value.

Again, Zeman and Amundson consider several other cases, including various types of termination reactions, reversible reactions, copolymerization reactions, catalyst-initiated reactions, and block reactions. It is interesting to note that they do not extend their solutions to multiple CSTRs in series. However, there seems to be little difficulty in doing so. Note that the complete solution P(n) is known for the effluent of reactor 1 and that this becomes the concentration profile for $P_{n,1}$ of the second reactor. Thus, the differential equation for the second reactor would appear (using the forward Taylor series expansion) as

$$(d^2P/dn^2) - 2(dP/dn) - 2aP$$

$$+ \text{ higher-order terms} = P_1 \exp[-a(n-1)], \quad n \geq 2 \quad (3.4\text{-}42)$$

This is the nonhomogeneous analog to the previous equation, and the solution is known provided we know P_1. But P_1 is given by the difference model as

$$P_1 = (k_i/k_p)(M_1 k_p)/[1/\theta + (k_p + k_t)M_1] \qquad (3.4\text{-}43)$$

for reactor 1. Again M_1 concentration depends on the solution of a difference equation (3.4-13) (in the case of monomer termination), written here with the infinite summation replaced with the equivalent integral:

$$M_{1,0} - M_{1,1} - \theta_1 M_{1,1} k_i + (k_p + k_t) \int_{n=1}^{\infty} P(n) \, dn = 0 \qquad (3.4\text{-}44)$$

Or, substituting for $P(n)$ and integrating, we get

$$M_{1,0} = M_{1,1}[1 + \theta_1 k_i + (k_p + k_t)P_1/a'] \qquad (3.4\text{-}45)$$

Substituting for P_1 and a', we get an implicit relationship for M_1 in terms of the initial concentration of $M_{1,0}$:

$$M_{1,0} = M_1[1 + \theta k_i + (k_i M_1 \theta)(k_p + k_t)/(1 + \theta k_p M_1)] \qquad (3.4\text{-}46)$$

Of course, we can solve for M_1 by trial and error, and having that we can determine P_1, a constant that is required for the solution of the concentration profiles of all other active species in the effluent of reactor 1 and for the solution of the equation to determine the effluent concentration of active species in reactor 2. And so on. To obtain the concentration profiles of the dead polymer species, we use the backward Taylor series to obtain a differential equation in $M(n)$ beginning with Eq. (3.4-12). Since we know $P(n)$ exactly, the solution is quite easy, and, again, the concentration of monomer leaving the first reactor is known by the implicit solution of Eq. (3.4-46) for M_1.

Compare the two methods of production of addition polymers. Denbigh [8] showed that w, the "unit output" of the CSTR, defined as the quantity of reaction product formed per unit time per unit volume of reaction space,

$$w = (M_{1,0} - M_{1,1})/\theta \qquad (3.4\text{-}47)$$

is exactly comparable to the differential coefficient $d(M_1)/dt$ for the batch reactor. (Denbigh also points out that by measuring the variation of w with M_1 in a continuous reactor it is possible to determine the order of the reaction, although he does not state whether this is applicable for polymerization reactions.) From a comparison of the number-average chain lengths and weight-average chain lengths of both batch and CSTR reactors of the same unit output, Denbigh finds that when the rate of propagation (reflected in the rapid growth rate of the chains) is high when compared with the residence time in the reactor (measured by the unit output), batch reactor products exhibit much broader molecular-weight distributions than CSTR reactor products. Conversely, when the rate of propagation is low when compared with the residence time in the reactor, the batch reactor products exhibit narrower molecular-weight distributions than CSTR reactor products. For most homogeneous-phase addition reactions, according to Platzer, the mean lifetimes of active polymer species in most commercial reactors are much shorter than the residence times, and thus the ideal concept of continuous polymerization should yield narrower molecular-weight distributions (and also, apparently, lower number-average molecular weights) than batch reactions. However, producing the "ideal" CSTR-type process is considerably more difficult than producing the batch reactor, as Platzer and others [2,9,10] point out. Part of the problem lies in "bypassing," where a molecule entering the tank in one moment can exit the tank in the next without participating in the chemical reactions taking place within the tank. As a result, we should consider the concept of "hold time" to represent, in actuality, a mean residence time, comprised of the summation of times of all particles residing within the tank. The qualitative effect of this nonideality is to broaden the molecular-weight distribution somewhat. It should be pointed out, though, that the assumption of perfect mixing (allowing no concentration profile throughout the tank) is as weak for the batch reactor as it is for the CSTR reactor. In both cases, high-speed agitation is necessary to ensure some approximation to perfect mixing, and this can become quite critical as the batch viscosity increases.

3.5 The Tubular Reactor—Homogeneous Reactions

It is perhaps important that we include the third form of reactor, the tubular reactor. Typical types of tubular reactors are shown in Figs. 3.5-1 and 3.5-2. If we were dealing with a fluid that could be pumped at very high rates through a pipe, we would ensure ourselves of high-shear mixing and turbulent flow, and thus the reactor would have a nearly flat velocity profile. The idealized model for the tubular reactor is the plug flow or piston flow reactor. Here the mass flow rate and fluid properties at any point in the reactor are independent of the pipe radius. Further, there is negligible diffusion of material down the pipe. Thus, we can imagine the reactor to be an infinite series of infinitesimal CSTRs, all having the same residence time in the pipe. Denbigh [8] has discussed the implications of this type of reactor and the arithmetic required to obtain the concentrations of various species at the exit of the reactor. Unfortunately, the plug flow assumption

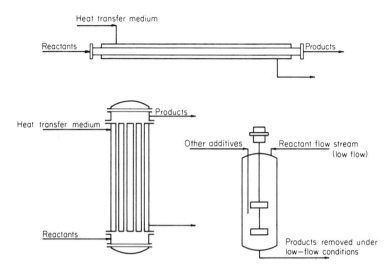

Figure 3.5-1 Reactors that might be classed as tubular or plug flow reactors. Top, typical jacketed pipe reactor; bottom left, shell-and-tube reactor that resembles a shell-and-tube heat exchanger; bottom right, a typical batch or semibatch reactor, where agitation is so low as to allow striation [7].

is not valid for many simple reactions and for all polymerization reactions. Lynn and Huff [11] show that for polymerization in a tubular reactor, the bulk Reynolds number is usually below that for turbulent flow. Thus, they characterize flow as laminar with homogeneous addition reaction. With laminar flow, the velocity profile is characteristically parabolic (assuming Newtonian fluids), and thus the material at the center of the pipe is in residence in the pipe up to twice as long as the average, and the material in contact with the wall theoretically resides there forever. Since the viscosity of the reactive material increases dramatically with extent of polymerization, the fluid in the center of the pipe is considerably more fluid than the average, and that at the wall is extremely viscous. Furthermore, Lynn and Huff point out that polymers are notorious for their low thermal conductivities. Since the only mechanism of heat transfer in laminar flow is conduction, the tubular reactor is a very inefficient heat exchanger (contrary to the implications regarding the stability of reactors given below). They therefore suggest that the tubular reactor be considered an adiabatic reactor (that is, all the energy generated through exothermic polymerization be used to heat the mass, thus increasing the rate of reaction, and so on). They also question the stability of flow in such a reactor. They obtain a tentative relationship between the viscosity of a butadiene syrup and the

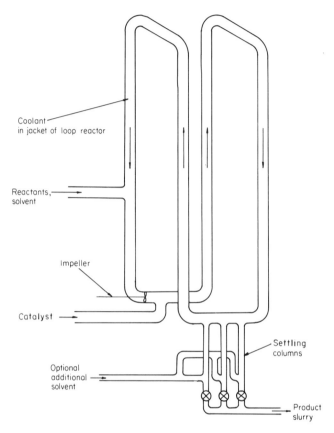

Figure 3.5-2   Typical schematic of tubular reactor system as used by
Phillips Petroleum in their LDPE process [1].

extent of reaction (this information being needed to determine the effect of
reaction on the velocity profile and the shear rate) and then solve the laminar-
flow momentum equations, energy equation, and material balance equations
for several cases, including externally cooled and adiabatic reactors. A
comparison of polymer concentrations, temperatures, and velocity profiles
for two cases (with and without external cooling) are shown in Figs. 3.5-3
and 3.5-4. Their computations indicate that the pressure loss owing to variation in viscosity (and hence distortion in velocity profile) is not significantly
different from that for conventional viscous Newtonian flow in a pipe. This
is probably not surprising when one considers that the effect of velocity is
integrated across the pipe dimension and down the length of the pipe. Integration is notorious as a curve-smoothing device. They also show, again

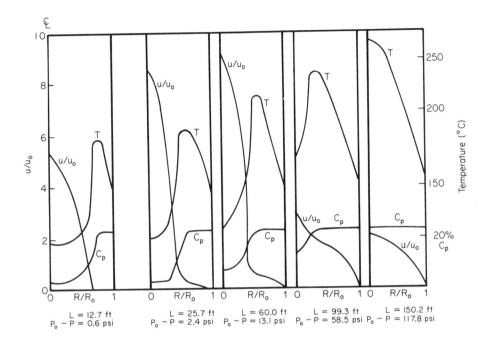

Figure 3.5-3   Radial profiles in a tubular polymerizer (case II). Tube diameter = 2.9 in., tube length = 200 ft, solution flow rate = 3200 lb/hr, inlet monomer conc. = 25%, inlet temperature = 80° C, coolant temperature = 130° C; u/u₀ = relative velocity, T = temperature, C_p = wt% polymer [11].

from computation, that the number-average molecular weight does not vary dramatically with radius across the tube at the reactor outlet. Their data are shown in Table 3.5-1. The variation should be caused by molecular diffusion owing to the concentration gradient across the pipe radius, and hence, they conclude, molecular diffusion is not significant. They conclude that for reactor scale-up, when the diameter of the reactor is constant, the change of concentration or temperature with distance at any point along the reactor is solely a function of the velocity. Reducing the velocity by a factor of 10 increases the longitudinal concentration and temperature gradients by a factor of 10. What is rather surprising about the calculations is that no mention was made of the much longer residence time (and thus supposedly higher molecular weights) near the wall. In fact, the opposite is true; the molecular-weight distribution is nearly constant across the pipe cross-section.

Biesenberger and coworkers [12] also consider addition polymerization in tubular reactors, with slightly different conditions: power-law velocity

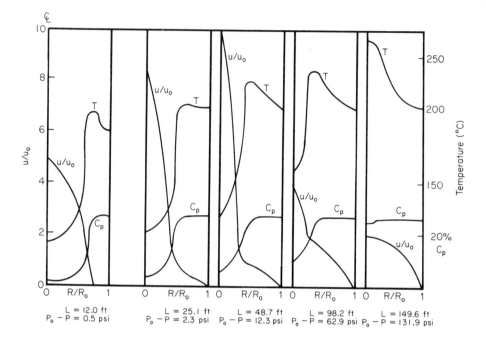

Figure 3.5-4   Radial profiles in a tubular polymerizer (case III). Tube diameter 2.9 in., tube length = 200 ft, solution flow rate = 3200 lb/hr, inlet monomer concentration = 25%, inlet temperature = 80°C, coolant temperature = insulated wall; u/u_0 = relative velocity, T = temperature, C_p = wt% polymer [11].

profiles that are unaffected by temperature; constant wall temperature in contrast to the work of Lynn and Huff, who assumed a finite rate of heat transfer (that is, a heat flux proportional to the thermal driving force); and negligible heat generation owing to viscous dissipation. Their representative curves are very similar to those of Lynn and Huff and show that increasing pipe diameter shifts the radial maximum temperature toward the tube wall and that decreasing values of the flow index (trends toward more "non-Newtonianness") result in less axial convection at the maximum temperature and more axial convection at the tube center. This is shown in Fig. 3.5-5. [Note that "VFD" refers to the volumetric flow rate distribution, or fraction of fluid leaving the reactor that is contributed by the streamlines (laminar flow) located at r/R.] They also consider the effect of "runaway" reactions, owing to the thermal insulation provided by the layer of high-temperature fluid close to the wall. Designers of tubular reactors should be cognizant of the potential dangers of this insulating layer of high-molecular-weight fluid and design accordingly.

Table 3.5-1

Number-Average Molecular-Weight Variation Across Tube at Reactor Outlet [11]

	Radial position						$\dfrac{MW}{MN}$
	0	0.2	0.4	0.6	0.8	1.0	
Case I	52,600	70,800	121,800	140,100	135,100	147,100	1.08
Case III	113,200	117,200	129,800	129,500	128,400	130,000	1.001

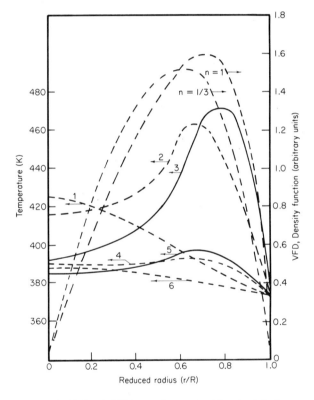

Figure 3.5-5 Addition polymerization in tubular reactors, power-law velocity profile, isothermal walls; computer results. Note: VFD = volumetric flow rate distribution. Curve 1: T_b = 399°K, 17.6% conversion, 308-cm length; 2: T_b = 434°K, 16.6% conversion, 160-cm length; 3: T_b = 428°K, 16.1% conversion, 152-cm length; 4: T_b = 393°K, 6.1% conversion, 120-cm length; 5: T_b = 393°K, 6.2% conversion, 120-cm length; 6: T_b = 383°K, 6.0% conversion, 160-cm length. Note: solid lines, n = 1 (Newtonian); dashed lines, n = 1/3 (power-law) [12].

Albright [1] discusses several commercial processes that use tubular reactors to carry out primary homopolymerization steps. For example,
the Phillips process for HDPE uses the loop reactor shown in Fig. 3.5-2.
A typical reactor uses a solution of hydrocarbons as a vehicle to carry the
polymerizing HDPE. The reactors operate at approximately 100°C and 500
psi, are 20 in. in ID by 300 ft long, and allow for approximately 600 ft^3 of
solution. The velocities are normally 10 to 30 ft/sec, with recirculation of
the solvent occurring every 10 to 20 sec. Note, however, that this type of
reactor is not for bulk polymerization.

Nevertheless, the tubular reactor, in any form, has not been used
extensively for bulk polymerization, first because of the success in designing CSTRs and second because of the occasionally long residence times that
require batch reactors.

3.6 Homogeneous Condensation Reactions in Various Reactor Types

Rather than consider the various reactor types singly for condensation reactions, we shall contrast and compare the reactors for a classic condensation polymer, characterized by the following propagation of bifunctional
species:

$$(AB)_n + (AB)_m \xrightarrow{\ k_p\ } (AB)_{m+n}$$

Flory [13] considered the kinetics of this type of reaction in a batch reactor
in detail. According to Biesenberger [14], if ϕ represents the total population of (AB) species, the rate of depletion of the monomer $(AB)_1$ is given as

$$d(AB)_1/dt = k_p(AB)_1 \sum_{i=1}^{\infty} (AB)_i = -2k_p(AB)_1\phi \qquad (3.6-1)$$

and the growth rate of reactive bifunctional species is given as

$$d(AB)_i/dt = -2k_p(AB)_i\phi + k_p \sum_{j=1}^{i-1} (AB)_j(AB)_{j-i} \qquad (3.6-2)$$

Furthermore, the increase in the total population of (AB) species is obtained
by taking the infinite sum of the reactive species:

$$d\phi/dt = -k_p\phi^2 \qquad (3.6-3)$$

Initially, the concentrations of $(AB)_1$ and $(AB)_i$ for $i \geq 2$ are unity and zero.
The time required to achieve a given degree of polymerization is obtained by
integrating Eq. (3.6-3) to obtain

$$t = (1 - \phi)/k_p \phi \qquad (3.6\text{-}4)$$

We can now replace t, time, in Eqs. (3.6-1) and (3.6-2) with ϕ, yielding

$$d(AB)_i/d\phi - 2(AB)_i/\phi = (1 - i)\phi^2(1 - \phi)^{i-2} \qquad \text{for all i} \qquad (3.6\text{-}5)$$

which, when integrated, yields Flory's most probable distribution for poly-condensation:

$$(AB)_i = \phi^2(1 - \phi)^{i-1} \qquad (3.6\text{-}6)$$

The weight fraction of $(AB)_i$ reactive bifunctionals is given as

$$W_i = i(AB)_i = i\phi^2(1 - \phi)^{i-1} \qquad (3.6\text{-}7)$$

The weight distribution curve as a function of the degree of polymerization is shown in Fig. 3.6-1.

For a single CSTR reactor, where θ is the hold time in the reactor, the equation for the monomer $(AB)_1$ becomes

$$(AB)_1 - 1 = 2\theta k_p (AB)_1 \phi \qquad (3.6\text{-}8)$$

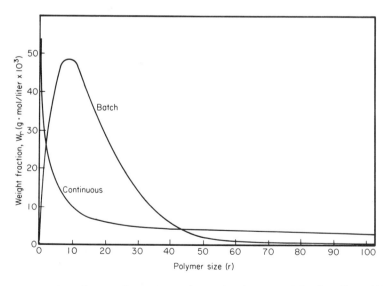

Figure 3.6-1 Condensation polymerization. Weight fraction $(AB)_r$ for degree of polymerization, $P = 0.90$ [14].

and for the higher bifunctionals,

$$(AB)_i = -2\theta k_p (AB)_i \phi + k_p \theta \sum_{j=1}^{i-1} (AB)_j (AB)_{i-j} \qquad (3.6-9)$$

The total number of reactive species exiting the reactor in the overflow is again obtained by taking the infinite series of $(AB)_i$ for all i:

$$\phi - 1 = -k_p \theta \phi^2 \qquad (3.6-10)$$

By rearrangement in the last equation we can solve for the hold time in the reactor as a function of the total concentration of reactive species:

$$\theta = (1 - \phi)/k_p \phi^2 \qquad (3.6-11)$$

Again solving the equations for $(AB)_i$, we obtain

$$(AB)_i = c_i \phi (1 - \phi)^{i-1}/(2 - \phi)^{2i-1} \quad \text{(for all i, including the monomer)} \qquad (3.6-12)$$

where

$$c_i = \sum_{j=1}^{i-1} c_j c_{i-j} \qquad (3.6-13)$$

which can be calculated beginning with $c_1 = 1$. Biesenberger gives a table of c_i but also shows that this series can be approximated by

$$c_i = 2^{2i-1}/(2i - 1)(\pi i)^{1/2} \qquad (3.6-14)$$

for i greater than 4 or 5. The equivalent weight fraction of $(AB)_i$ in the continuous reactor is given as

$$W_i = ic_i \phi (1 - \phi)^{i-1}/(2 - \phi)^{2i-1} \qquad (3.6-15)$$

and this distribution is contrasted with the batch distribution in Fig. 3.6-1 for $\phi = 0.10$.

Smith and Sather [15] consider in detail not only the steady-state polycondensation reaction but the unsteady reactions involved in start-up of a CSTR unit. They find that a dimensionless group, considered to be a measure of the effect of chemical reaction on the concentration profiles, can be defined in such a way that both batch and continuous reactions can be described. Their dimensionless group, called the kinetic effect number, is

$$N = k_p \theta (AB)_1^0 \qquad (3.6-16)$$

where $(AB)_1^0$ is the initial monomer concentration. When N is large (i.e., long hold times or high reaction rates), the reaction terms predominate; when N is small (short hold times, low monomer concentration), the flow terms predominate. In the limit, as $N \to \infty$, we approach a batch process $(\theta \to \infty)$; as $N \to 0$, we have a purging or nonreactive operation. Since their system of equations allows for the unsteady operation of a reactor (either batch or continuous), they find it useful to have a reduced time that is the ratio of actual time and hold time in the reactor. Thus, their curves of weight fraction of reactive species i are plotted against this reduced time, with the value of i given as a parameter, as shown in Fig. 3.6-2, where N = 100. At this value of N the system is approaching a batch reactor. In Figs. 3.6-3 and 3.6-4 we see that the number-average and weight-average degrees of polymerization (or chain length) with N as the parameter. Again, the longer the hold time in the reactor or the higher the reactivity of the bifunctional species, the more rapidly the molecular weight increases. This is in agreement, then, with Biesenberger's analysis that the molecular-weight distribution for polycondensation reactions is much sharper for a batch reactor and that there are more high-molecular-weight ends in a CSTR unit. Smith and Sather give an example at steady state where the weight-average to number-average chain length is given by

$$\overline{X}_w / \overline{X}_n = (1 + 1/N)[(1 + 2N)^{1/2} - 1] \tag{3.6-17}$$

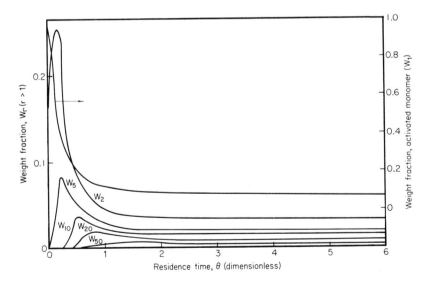

Figure 3.6-2 Concentration profiles as functions of hold time in start-up of single CSTR. Kinetic effect number N = 100, condensation polymerization [15].

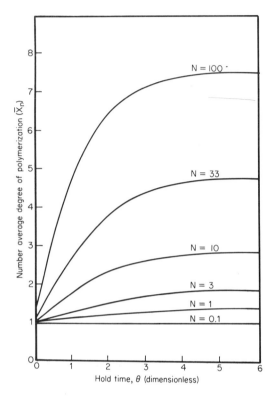

Figure 3.6-3   Number-average degree of polymerization for various values of kinetic effect number N. CSTR, condensation polymerization [15].

For N = 100, the calculated ratio is 13.2 for CSTRs and 2.0 for a batch reaction at complete conversion (Flory's most probable distribution). For N = 10, the ratio is 3.74, but this tight distribution is offset by a very low mean molecular weight. Thus, there seems to be no way of improving molecular-weight distribution by altering hold times or reaction rates, in the way we can with addition reactions.

Mellichamp [16] has carried out calculations for reversible polycondensation reactions in batch reactors. The reversible reactions are probably more characteristic of polycondensation reactions than the propagation equations, but the arithmetic needed to obtain solutions to the equations is quite complex and is best left to the serious reactor design engineers.

As mentioned earlier in the section on addition reactions, the tubular reactor has been considered as an alternative to the CSTR, or continuous flow reactor. Biesenberger and his coworkers [12] have considered polycondensation reactions in tubular reactors with the following constraints:

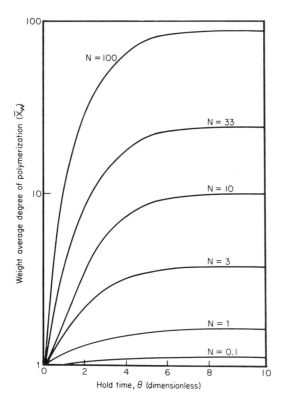

Figure 3.6-4 Weight-average degree of polymerization for various values of kinetic effect number N. CSTR, condensation polymerization [15].

1. The flow is power-law non-Newtonian (in contrast to the work by Lynn and Huff where Newtonian fluids were considered).

2. The "holdback" H is a measure of the importance of residence time distribution within the pipe and can be considered as the fraction of reactor contents that has been in the reactor (held back) for a time θ, the mean residence time, or more. For a power-law fluid, $H(\theta)$ is given as

$$H(\theta) = [1 - (n + 1)/(3n + 1)]^{(3n+1)/(n+1)} \qquad (3.6\text{-}18)$$

where n is the power-law index (defined in the chapter on viscosity). For a Newtonian fluid, n = 1, and $H(\theta) = 0.250$; for a plug flow reactor, we can show that there is no shear of fluid at the wall, and thus n = 0 (an inviscid fluid). Hence for plug flow, H = 0. For most plastic materials, $0 < n < 1$, as we shall see.

3. The flow is isothermal (again in contrast to the work of Lynn and Huff).
4. Both axial and radial diffusion could be neglected (it was included in the work of Lynn and Huff but was shown that for most cases it had no significant effect on the molecular-weight distributions).

They examine the weight-average to number-average chain length ratio (his number-dispersion index $\overline{D}_n$) as a function of the holdback H for both stepwise addition reactions and polycondensation reactions. It is seen in Fig. 3.6-5 that for $0 < H < 0.25$ there is relatively little effect of "non-Newtonianness" on the fractional conversion. The weight-average to number-average chain length ratio for polycondensation polymers increases with increasing non-Newtonianness (n decreasing).

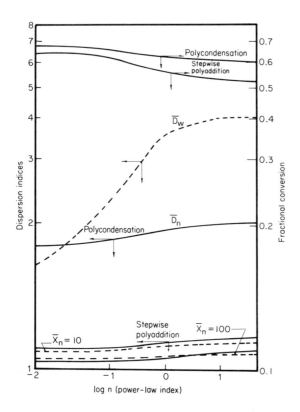

Figure 3.6-5   Effect of polymer non-Newtonianness on weight-average to number-average chain length ratio. Note: solid lines, $\overline{D}_n$; dashed lines, $\overline{D}_w$ [12].

3.7 Mixing Methods in Reactors

Before we consider in detail the control and stability of reactors, or the various problems in suspension, emulsion, and catalyzed reactions, we should point out various problems in mixing and heat removal in reactors. The design of reactors such as the CSTR train depends on the assumption that concentration of all species and temperature are uniform within the reactor. In practice, this is quite difficult to achieve if the contents of the reactor are very viscous or exhibit unusual non-Newtonian flow behavior. Bates et al. [3] review carefully the various types of impellers that can be used in reactors and demonstrate their power correlations and flow patterns. Metzner and his coworkers [17] have concentrated their efforts on understanding the non-Newtonian fluid power requirements, and since their arguments are valid for Newtonian fluids when the power index is unity, we shall discuss their work here.

For studies on highly viscous polymer solutions, they found that the average shear rate was a linear function of the impeller speed:

$$(du/dr)_{av} = KN \qquad (3.7\text{-}1)$$

where $(du/dr)_{av}$ is the average shear rate in the vicinity of the impeller; K is a proportionality constant, apparently a function of the impeller design; and N is the speed of the impeller. As shown in Fig. 3.7-1, the dimensionless power number $N_p = Pg_c/D^5N^3\rho$ [where P is the power delivered to the impeller in ft·lb$_f$/sec (or HP, with proper conversion), g_c is the conversion factor = 32.2 ft·lb$_m$/lb$_f$·sec^2, D is the impeller diameter in ft, N is the rotational speed in rev/sec, and ρ is the fluid density in lb$_m$/ft^3] is linear with the rotational speed of the impeller. Furthermore, when an "apparent viscosity" is defined as the ratio of the shear stress to shear rate at a given shear rate, the impeller power number as a function of the Reynolds number based on the apparent viscosity is seen to follow the Newtonian curve predicted by Rushton [18]. Note that the Reynolds number here is defined as $Re = D^2N\rho/\mu_a$. Figure 3.7-2 is that of Calderbank and Moo-Young [19]. The laminar-turbulent transition apparently occurs at a Reynolds number value of 20 or so, although the transition is not clearly defined. Bates et al. review the literature to determine the proper value for K and find that for the ratio of tank diameter to impeller diameter (i.e., D_T/D_I) and for power-law indices less than or equal to unity, the following values of K are accurate to ±20%:

Impeller	K
Anchor	19
Six-blade turbine	11.6
Two-blade paddle	10
Three-blade propeller	10
Fan turbine	13

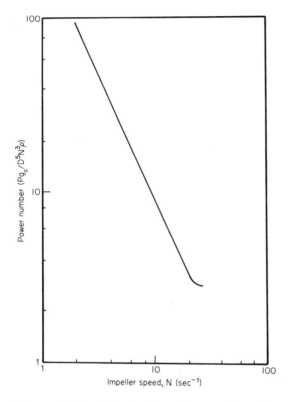

Figure 3.7-1   Effect of impeller speed on dimensionless power number
for 2.5% CMC in water, a non-Newtonian solution. Experimental data not
shown [17].

To obtain the necessary data to use the curve of Calderbank and Moo-Young,
it is necessary to obtain the "apparent viscosity" or more importantly the
stress-shear rate curve using a viscometer.

   Note the implications here, which will be made stronger in Chap. 5
on viscosity when we talk about increasing non-Newtonianness of fluids. For
a power-law fluid, the apparent viscosity decreases with increasing shear
rate. This means that increasing the speed of rotation will decrease the
viscosity, and both effects cause an increase in the Reynolds number. If the
process is already turbulent, increasing the Reynolds number will have
little effect on the power number. Thus, the total power will increase in
proportion to the cube of the impeller speed. If the fluid is in deep laminar
motion (with Re much less than 20, say), then doubling the Reynolds number
will result in reduction of the power number by a factor of 2. Rushton and
Oldshue [20] have shown that for laminar mixing (for either Newtonian or
non-Newtonian fluids)

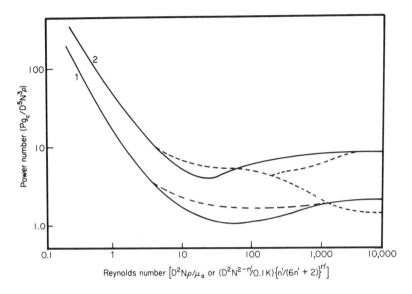

Figure 3.7-2   Effect of Reynolds number on power number for various fluids. Curve 1: for Newtonian fluids; 2: for power-law type fluids. Dotted lines represent theories. Experimental data not shown, but represented by solid lines [19].

$$N_p = 71/Re \quad \text{and thus} \quad P = 71\mu_a D^3 N^2/g_c \tag{3.7-2}$$

Thus, increasing the shear rate in laminar flow will cause a decrease in the viscosity but an increase in the speed of rotation. The net effect on the total power will be an increase with increasing impeller speed to a power somewhat less than 2. Increasing non-Newtonianness will cause a decreasing effect on the total effect of increasing impeller speed on the total power required. Thus, to improve mixing in a highly non-Newtonian fluid, more power is required than that required for the equivalent Newtonian fluid.

Note that scale-up is also related strongly to the fluid characteristics. If we double the diameter of the reactor (thus increasing its capacity eightfold, assuming that the height-to-diameter ratio is held constant, a common practice) and double the diameter of the impeller at the same impeller speed, the shear rate remains constant. As a result the shear stress is also constant. Thus, doubling the diameter causes a fourfold increase in the Reynolds number. In turbulent flow, since the power number is nearly independent of increasing Reynolds number, the increased diameter of the reactor causes the power required to increase by a factor of $2^5 = 32$ times. Thus, the net result is that the power requirement per unit volume of reactor increases by a factor of 4. If we desire to hold the Reynolds number constant, then the ratio of apparent viscosity to impeller speed must increase by a factor of 4. If the fluid is Newtonian, the impeller speed must decrease by a factor of 4; however, if the fluid is power-law non-Newtonian, decreasing the impeller speed decreases the shear rate, which in turn increases the viscosity of the fluid. Therefore, the impeller speed will be decreased by a factor somewhat less than 4, the actual value depending on the power-law index. If we decrease the impeller speed by a factor of 4, say, for illustrative purposes, we see that the power required is now proportional to $2^5/(2^2)^3 = 1.2$. We actually get a net decrease in power required with increasing capacity. In general, however, our extent of mixing will become inadequate to provide uniform concentration and temperature profiles at the lower speeds, and a compromise solution must be worked out, normally through the sacrifice of some uniformity of mixing in exchange for a slight increase in power requirements.

The above analysis is apparently valid for both batch reactors and CSTRs. We should keep in mind, however, that if our reactor is power limited because we have sized the mixing unit for an average viscosity (or a Newtonian viscosity), we shall be sacrificing uniformity in mixing in a batch reactor as we approach the end of the process. This nonuniformity in mixing can cause local hot spots and degradation of product. It is therefore advisable to size the mixing unit on the reactor for the most extreme cases to be encountered. As we have seen, the only mixing that can occur in a tubular reactor is that of molecular diffusion, and according to calculations of Lynn and Huff, this form of mixing is negligible in most cases. One method of promoting mixing in a tubular reactor would be the use of a static mixer such as that manufactured by Kenics Corporation [21]. These devices fit the center of the tubular reactor and periodically divide and mix the flow streams. The obvious improvement in mixing in the tubular reactor may be offset by the added pressure drop needed owing to the presence of additional surface area. Nevertheless, the designer should be cognizant of efforts being made to improve mixing in tubular reactors.

3.8 Heat Transfer from Reactors

There are four main methods of removing heat from a reactor vessel: (1) through external jacketing of the reactor; (2) through coils immersed in the polymerization reaction; (3) through overhead condensers, which condense a volatile material such as the monomer or a solvent and return the condensed fluid to the reaction vessel; and (4) through external loops through double-pipe heat exchangers. In most reactor designs, the coolant is water, and frequently more than one means of cooling is used. For some very stable emulsions, the reaction mass can be cooled by continuously pumping the reacting fluid through an external shell-and-tube exchanger and then returning the cooled fluid to the reactor. This external method of cooling requires a very stable emulsion, but if one is being used, a very high degree of temperature control of the reaction can be obtained.

Most of the work done on heat transfer from reactor contents has been done on Newtonian fluids, and modified Colburn heat transfer equations seem suitable for prediction here [22]: For jacketed reactors,

$$Nu = hD_T/k = 0.36(D^2 N\rho/\mu)^{2/3}(c_p \mu/k)^{1/3}(\mu/\mu_w)^{1/7} \qquad (3.8-1)$$

for internal coils,

$$Nu = hD_c/k = 0.87(D^2 N\rho/\mu)^{2/3}(c_p \mu/k)^{1/3}(\mu/\mu_w)^{1/7} \qquad (3.8-2)$$

Here h is the heat transfer coefficient from the reacting fluid to the tube or reactor wall, D_T is the diameter of the reactor, D_c is the diameter of the internal coil, D is the diameter of the impeller and thus $D^2 N\rho/\mu$ is the mixing Reynolds number, $c_p\mu/k$ is the Prandtl number for the reacting fluid, and $(\mu/\mu_w)^{1/7}$ is the viscosity correction factor, since the viscosity of the fluid at the wall is much higher than that in the bulk (in the case where the reaction mass is being cooled). Some appropriate corrections can be made to this equation if we assume that the viscosity in the mixing Reynolds number and the Prandtl number is that of the "apparent viscosity" defined earlier. Nevertheless, even for Newtonian fluids, these equations are accurate only to ±20% at best. If we are dealing with a CSTR design, we can calculate the overall heat transfer coefficient by calculating the heat transfer coefficient, assuming turbulent flow for the coolant, from a standard equation such as the Seider-Tate equation:

$$h_c D/k = 0.027(\overline{DU}\rho/\mu)^{4/5}(c_p \mu/k)^{1/3}(\mu/\mu_w)^{1/7} \qquad (3.8-3)$$

where D is the characteristic dimension of the coolant chamber (e.g., the channel dimension for the jacketed reactor or the inside pipe diameter for the coil), $\overline{U}$ is the fluid velocity through the coolant chamber, and μ and μ_w represent the bulk viscosity and viscosity at the wall for the coolant. Normally the $(\mu/\mu_w)^{1/7}$ is unimportant for most coolants owing to their low viscosities. Another way of obtaining h is by use of a standard convection chart, such as Fig. 3.8-1. Calculating h for the external surface from the appropriate equation and h_c for the coolant, we can now calculate the overall heat transfer coefficient:

$$1/U = 1/h_c + 1/h(A/A_c) + (D/2k_m) \ln(D/d) \qquad (3.8-4)$$

Here A/A_c is the correction for thick walls in internal coils (it can be assumed to be unity for reactor walls); D is the outside diameter of the internal coil and, conversely, the inside diameter of the reactor; and d is the inside diameter of the internal coil, or the inside diameter + two thicknesses of reactor wall for the reactor. For most reactor systems, A/A_c and D/d are nearly unity, and thus these terms can conveniently be ignored when compared with h_c and h, the two heat transfer coefficients. Once we have

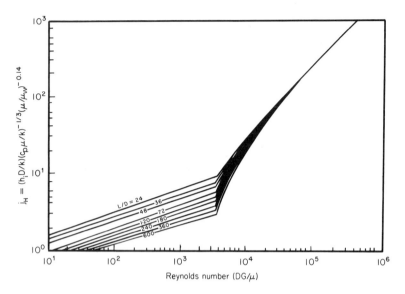

Figure 3.8-1   Convection heat transfer coefficients for various Reynolds numbers; Newtonian fluid flow [22].

the overall heat transfer coefficient U, we can calculate the rate of heat removal from the reactor per unit area:

$$Q/A = U(T_b - T_c) \tag{3.8-5}$$

where T_b is the bulk reactor temperature and T_c is the coolant temperature. Before we consider transient temperatures in batch reactors, we should note that the reactive fluid heat transfer coefficient is proportional to the diameter of the impeller to the 1/3 power and to the impeller speed to the 2/3 power (assuming a Newtonian fluid, for the moment). Doubling the diameter of the reactor (and holding the speed of rotation constant) increases the heat transfer coefficient by 26% (and, remember, increases the power by a factor of 32). Doubling the impeller speed at a constant reactor diameter increases the heat transfer coefficient by nearly 59% but increases the power by a factor of 8. Thus, significant changes in the rate of heat transfer are difficult to make within the constraints of the reaction system.

For batch reactors, we must include the fact that the reaction temperature is changing with time. This is complex, because we must take into account the energy released as polymerization proceeds. We can show that initially this total amount of energy is small, as the reaction is being initiated. And we can show that as the reaction proceeds toward completion the amount of energy generated is also small. Therefore, there is a time period where a maximum amount of energy is generated. It is in this time period where cooling is most important, as we shall see when we discuss instability in batch reactors.

To illustrate how to use the heat transfer coefficients calculated from the above equations, we choose a very simple model where no chemical reaction is taking place. The heat removed from the tank per unit time is just

$$dQ/dt = -Mc_p \, dt_h/dt \tag{3.8-6}$$

where M is the mass of fluid in the tank, in pounds say; c_p is the specific heat of the fluid in the tank; and T_h is the reactive mass bulk temperature. This heat is removed, either through coils or jacket, according to

$$dQ/dt = Wc_p'(T_{2c} - T_{1c}) = UA(LMTD) \tag{3.8-7}$$

W is the mass flow rate of coolant; T_{1c} is the inlet coolant temperature, here assumed to be a constant; and c_p' is the specific heat of the coolant. T_{2c} is the variable exit coolant temperature. U is the overall heat transfer coefficient; A is the heat transfer surface area; and LMTD is the logarithmic mean temperature difference, defined as

$$LMTD = (T_{2c} - T_{1c})/\ln[(T_h - T_{2c})/(T_h - T_{1c})] \tag{3.8-8}$$

Now if $K = \exp(UA/Wc_p')$, we can show that upon integration we get

$$T_{2c} = T_h + (T_h - T_{1c})/K \tag{3.8-9}$$

$$\ln[(T_{1h} - T_{1c})/(T_{2h} - T_{1c})] = (Wc_p'/Mc_p)(1 - 1/K)t \tag{3.8-10}$$

The overall heat transfer coefficient is calculated, knowing h_c and h, as before. If we rearrange this equation, we can show that initially, when $T_{1h} \gg T_{1c}$, we get the greatest effect in heat transfer. As T_h approaches T_{1c}, the effectiveness decays exponentially. We have not included a heat generation term in this equation; as a result, we are seeing only the passive response to heat transfer, and a slow exponential decay to the coolant temperature is not unexpected. For a very high heat transfer coefficient or surface area, K is very large, and thus 1/K is much smaller than unity. Hence, we see that the rate of approach to steady state is controlled by the ratio of heat capacities of the coolant and the reactive fluid, as is expected. This same effect will be seen when we consider the stability of an active reactor with an exothermic reaction driving the system. Kern [22] gives several other reactor cases, none of which involve chemical reactions. He considers external counterflow heat exchangers with an isothermal cooling medium or a nonisothermal cooling medium, external counterflow heat exchanger with an isothermal cooling medium or a nonisothermal cooling medium with liquid continuously added to the tank, external multipass heat exchanger with various cooling modes and liquid added to the tank, and so on.

The design engineer should acquaint himself (or herself) with these alternatives, especially if he (she) has a system that is a stable emulsion with a highly exothermic reaction. The dangers and, surprisingly, the advantages of having a nonagitated batch reactor are also pointed out for several types of heat transfer systems.

For the design of reflux condensers, we shall consider only condensation of a pure organic vapor with some suitable coolant presumably in turbulent flow on the shell side of a heat exchanger. The coolant heat transfer coefficient can be obtained from the Seider-Tate equation (3.8-3). For condensation in the laminar region [i.e., where the "loading" Reynolds number, defined as $Re = 4w/\pi D\mu$ (where w is the condensate rate in lb/hr, D is the tube diameter, and μ is the viscosity of the condensate), is less than 2000], the average heat transfer coefficient for a vertical condenser is given as [22]:

$$h = 0.943[k^3\rho^2(\Delta H_c)g/\mu L(T - T_w)]^{1/4} \tag{3.8-11}$$

where the condensate thermal conductivity k, density ρ, and viscosity μ are determined at the temperature T of the film. ΔH_c is the latent heat of condensation in Btu/lb, and T_w is the wall temperature in °F. g is the gravitational constant, 32 ft/sec². For loading Reynolds numbers in excess of 2000, we have turbulent condensation, and as a result, the average heat transfer coefficient in a vertical condenser is given as [22]

$$h = 1.47(k^3\rho^2g/\mu^2)^{1/3}(Re)^{-1/3} \tag{3.8-12}$$

In general, for condensing organic vapors, the heat transfer coefficients are on the order of 100 to 500 Btu/ft²·hr·°F, with 300 being a good first approximation. Equivalent relationships for shell-side condensation on horizontal tubes filled with coolant are available in Kern [22]. The overall size of the condenser is obtained by calculating the total amount of heat released by the condensing vapor and equating this to the overall heat transfer coefficient, area, and LMTD:

$$Q \text{ (Btu/hr)} = M(\Delta H_c) = UA(LMTD) \tag{3.8-13}$$

Condensers in which a noncondensible is present or in which mixed vapors are condensing are considerably more difficult to design, and the designer should refer to Kern or recent sources [23] for guidance.

Heat transfer from tubular reactors is very similar to standard heat transfer in a double-pipe heat exchanger, and thus the standard heat transfer references can be used to determine the necessary heat transfer surface area required to cool or maintain the temperature in this type of reactor. If the flow is laminar, we should consider the fact that for short reactors (with mixing zones between the reactors) the heat transfer coefficient will depend on the length of the heat exchanger. The classical Graetz heat

Table 3.8-1

Nusselt Number Variation with Graetz Number, X' [23]

Constant wall temperature fully developed velocity profile

X'	Nu_{av}
0	∞
0.001	23.0
0.004	12.6
0.01	9.00
0.04	5.86
0.08	4.89
0.1	4.66
0.2	4.16
∞	3.66

Constant wall heat flux fully developed velocity profile

X'	Nu_{av}
0	∞
0.002	12.01
0.004	9.93
0.01	7.48
0.02	6.17
0.04	5.20
0.1	4.51
∞	4.36

Constant surface temperature developing velocity profile

X'	Nu_{av} (PR = 1)	Nu_{av} (Pr ≥ 10)
0	∞	∞
0.002	22.1	17.1
0.004	17.4	14.4
0.01	12.8	10.0
0.02	9.96	7.8
0.04	7.6	6.4
0.1	5.4	4.95
∞	3.66	3.66

Constant wall heat flux developing velocity profile

X'	Nu_{av} (Pr = 1)	Nu_{av} (Pr ≥ 10)
0	∞	∞
0.0002	51.9	39.1
0.002	17.8	14.3
0.01	9.12	7.87
0.02	7.14	6.32
0.1	4.72	4.51
0.2	4.41	4.38
∞	4.36	4.36

transfer problem where the velocity profile (for Newtonian fluids) and temperature profile are fully developed yields asymptotic values for the Nusselt number for the two cases of constant wall temperature and constant wall heat flux [24]:

constant wall temperature: $Nu_\infty = hD/k = 3.66$ (3.8-14)

constant wall heat flux: $Nu_\infty = 4.36$ (3.8-15)

However, if the temperature profile is developing (and the velocity profile is either developed or is also developing), we find that actual Nusselt number is greater than that given above and is a declining function of the dimensionless group (known as the Graetz number), $X' = (x/R)/RePr$. X' represents a dimensionless tube length, R is the pipe radius, Re is the Reynolds number (less than 2000), and Pr is the conventional Prandtl number. For various cases, the average Nusselt number as a function of X' is listed in Table 3.8-1 [23]. As an example of this asymptotic approach to a constant value of heat transfer coefficient, we see that if the Reynolds number is 1000 and the Prandtl number is 100, for a pipe 0.1 ft in radius the temperature profile will be fully developed (assuming no chemical reaction) at a distance of $X' = 0.2$, or x = 2000 ft. Thus, we can expect that for laminar flow in most tubular reactors the velocity and temperature profiles are not fully developed and thus the heat transfer coefficient is higher than that calculated using the asymptotic values. Under any conditions, however, we must remember that heat transfer in laminar flow is by conduction, and polymer materials have notoriously poor thermal conductivities. The accompanying tables, of course, do not hold for non-Newtonian fluids, reactive fluids, or fluids in which there is a significant amount of heat generated by viscous dissipation. We shall cover heat transfer from these types of fluids in Chap. 6.

3.9 Homogeneous Reactor Stability

We now turn our attention to reactor stability, a critical phase of reactor design. In our discussions on heat transfer, we have seen that the heat removed from a reactor is proportional to the amount of surface area provided and the thermal driving force. The amount of heat generated, on the other hand, depends on the rate of reaction, which, according to standard Arrhenius theory, is exponentially dependent on temperature:

$$k_p = A_p \exp(-E_p/RT)$$ (3.9-1)

where A_p is the preexponential factor (also somewhat temperature dependent) and E_p is the activation energy for propagation of the species. A slight simplification of the Arrhenius equation, as suggested by Foss [25], helps to show the effect of temperature on the reaction rate:

$$k_p = A_p \exp(-E_p/RT) \doteq a \exp(bT) \qquad (3.9-2)$$

As we have seen when dealing with the kinetics of both the batch and CSTR reactors, the CSTR reactor is the easiest to use to illustrate the effect of heat generation and heat removal. However, to be consistent and complete the batch reactor should also be considered. We therefore choose the batch reactor with a very simple kinetic reaction to illustrate the interaction between the rate of heat generation and the rate of heat removal. Consider the reaction

$$B \xrightarrow{\ k_p\ } products$$

where the heat capacities of the products are identical to the heat capacity of the reactant. Consider first the adiabatic reactor (e.g., no heat transfer from the reactor). If the initial concentration of B is B_0, then the first-order kinetic reaction equations for B are

$$-dB/dt = k_p B \qquad\qquad B(0) = B_0 \qquad (3.9-3)$$

where k_p is a function of temperature, which in turn is a function of time. Now, according to Walas [6], the net enthalpy change of reaction equals the enthalpy of the initial reactants minus the enthalpy of unconverted reactants and the enthalpy of the products formed (in this case, consider the product concentration to be C):

$$-(B - B_0)(\Delta H) + (C - B)c_p(T - T_0) = 0 \qquad (3.9-4)$$

If x is the extent of reaction, then $x = C = B - B_0$, according to a simple mass balance. Now the reaction rate equations become

$$dx/dt = k_p(B_0 - x) \qquad\qquad x(0) = 0$$

and the adiabatic energy equation becomes

$$-x(\Delta H) + B_0 c_p(T - T_0) = 0 \qquad (3.9-6)$$

The solution of x depends on knowing the temperature–time relationship. Now $x = B_0 c_p(T - T_0)/(\Delta H)$. Thus, $dx/dt = [B_0 c_p/(\Delta H)] dT/dt$, and the kinetic equation becomes

$$dT/dt = \left\{[(\Delta H)/B_0 c_p][1 - c_p(T - T_0)/(\Delta H)]\right\} a \exp(bT) \qquad (3.9-7)$$

which yields the implicit time–temperature relationship:

$$t = \int_{T_0} e^{-bT}\, dT/(a'' - b''T) \qquad (3.9-8)$$

where

$$a'' = a[(\Delta H)/B_0 c_p][1 + c_p T_0/(\Delta H)]$$

and

$$b'' = a$$

The shape of this curve is shown in Fig. 3.9-1. We see that the temperature climbs slowly with time initially, then accelerates very rapidly as B is
consumed, and eventually reaches a constant value. This is for an exothermic reaction and is characteristic of batch reactor responses.

In a nonadiabatic reactor, the rate of heat removal may take several
forms, including constant rate of heat removal, a rate of heat removal that
is linear with reactor temperature, or a rate of heat removal that is programmed to keep the reactor temperature constant. In each of these cases,
we find an additional heat removal term on the right-hand side of the for-
merly adiabatic energy equation (3.9-6):

$$x(\Delta H) + B_0 c_p (T - T_0) = Q_r \tag{3.9-9}$$

Consider first a linear rate of heat removal. Now, $Q_r = UA(T - T_c)$ and

$$x = [UA/(\Delta H)](T - T_c) + [B_0 c_p/(\Delta H)](T - T_0) \tag{3.9-10}$$

Differentiate to obtain dx/dt:

$$dx/dt = [(UA + B_0 c_p)/(\Delta H)] \, dT/dt \tag{3.9-11}$$

Appropriate substitution into the rate equation and integration yield an implicit time-temperature relationship

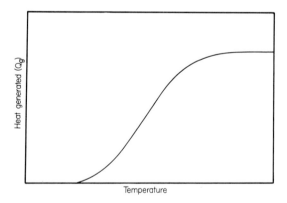

Figure 3.9-1   Batch reactor heat generation curve, owing to single first-
order reaction [6].

$$t = \int_{T_0} \exp(-bT)\ dT/(a''' - b'''T) \tag{3.9-12}$$

where now $a''' = [a(\Delta H)/(UA + B_0 c_p)][B_0 + UAT_c/(\Delta H) + B_0 c_p T_0/(\Delta H)]$ and $b''' = a$, as before. Integrating, we find that the temperature profile will flatten with increasing UAT_c until we have suppressed the reaction. The objective, of course, is to allow the reaction to proceed to completion, and this implies that initially we must carefully specify a proper UAT_c to permit reaction to proceed. Curves for various values of UAT_c are shown in Fig. 3.9-2. Remember here that the coolant flow rate or temperature is not controlled. To control the rate of heat removal so that the reaction mixture remains constant, we note that if the reaction temperature is constant, the reaction rate must be constant. Therefore, we can easily integrate the reaction rate equation to yield the isothermal expression:

$$x = B_0(1 - e^{-k_p t}) \tag{3.9-13}$$

where $k_p = a\ \exp(bT_r)$ and T_r is the constant reactor temperature. Thus, the coolant temperature is given by

$$B_0[1 - \exp(-k_p t)](\Delta H) + B_0 c_p(T_r - T_0) = UA(T_r - T_c) \tag{3.9-14}$$

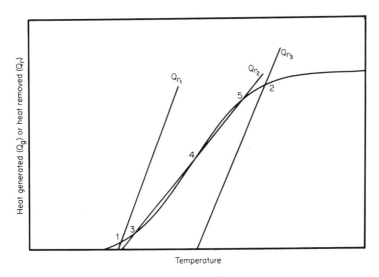

Figure 3.9-2 Batch reactor with heat removal curves shown over heat generation curve; single first-order reaction [6].

For an exothermic reaction, the coolant temperature must be slowly raised
with time as the reaction proceeds to completion so as to maintain a constant
reactor temperature. For appropriate values of the parameters the time-
temperature profile for the coolant is shown in Fig. 3.9-3. Note that it is
also possible to change the flow rate of coolant. This affects the overall
heat transfer rate, but as seen earlier, the heat transfer coefficients are
not very sensitive to flow rate changes. Actually a control using a combination of decreasing flow rate and increasing exit temperature on the cooling
fluid is the common practice.

   The stability of a steady-state CSTR reactor is somewhat easier to
comprehend. If F is the volumetric flow rate of material into and out of the
reactor, the rate at which heat is generated is given as

$$Q_g = -k_p V B(\Delta H) \qquad\qquad (3.9\text{-}15)$$

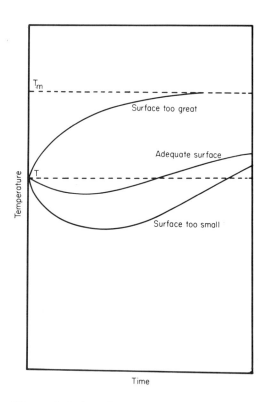

Figure 3.9-3 Schematic of effect of amount of heat transfer surface on
batch reactor temperature. T_m is the temperature of the heat transfer
medium [6].

where B is the exit concentration of reactant, k_p is the reaction rate, V is the tank volume, and (ΔH) is the heat of reaction. We have seen in Eq. (3.4-8) that $B = B_0/(1 + k_p\theta)$, where θ is the hold time of the reactor, defined as

$$\theta = V/F \quad \text{(for a single CSTR unit)} \tag{3.9-16}$$

This means that the rate of heat generation is given as

$$Q_g = -Vk_p B_0 (\Delta H)/(1 + k_p \theta) \tag{3.9-17}$$

where k_p is defined as the exponential function of reactor temperature:

$$Q_g = -VB_0 (\Delta H)/(ae^{-bT} + \theta) \tag{3.9-18}$$

The rate of heat removal includes that removed with the exiting fluid:

$$Q_r = UA(T - T_c) - F\rho c_p (T_0 - T) \tag{3.9-19}$$

where T_0 is the incoming temperature of the feed to the reactor. Thus, the time-dependent increase in reactor temperature is given as

$$V\rho c_p dT/dt + Fc_p \rho(T_0 - T) - UA(T - T_c)$$

$$+ (\Delta H)VB_0/(a^{-1}e^{-bT} + \theta) \tag{3.9-20}$$

At steady state, dT/dt must be zero in order to maintain a constant reactor temperature (e.g., $Q_r = Q_g$). But, again, we see that Q_r is linear with respect to reactor temperature:

$$Q_r = (UA + F\rho c_p)T - (UAT_c + F\rho c_p T_0) \tag{3.9-21}$$

The slope of the coolant line is dictated by the feed rates and heat transfer coefficients. The intercept is dictated by the feed and coolant temperatures. The Q_r curve is shown in Fig. 3.9-4. On the other hand, the amount of heat generated is exponentially dependent on temperature. If the hold time is significantly greater than $1/k_p$, the amount of heat generated is constant. This implies high temperatures (so that k_p is very large, and $1/k_p$ is $\ll \theta$). On the other hand, at low temperatures the amount of heat generated is small and increases slightly with increasing temperatures. We can see therefore that the heat generation curve is sigmoidal in shape, as shown in Fig. 3.9-4. The intercept between the heat generation curve and the heat removal curve represents a stationary state (SS). Note that there can be either one or three stationary states (e.g., where $Q_r = Q_g$) for this relatively simple first-order reaction. Bilous and Amundson [26] and Foss [25] show that the stationary state at low temperatures and the one at high temperatures are stable states but that the center stationary state is unstable.

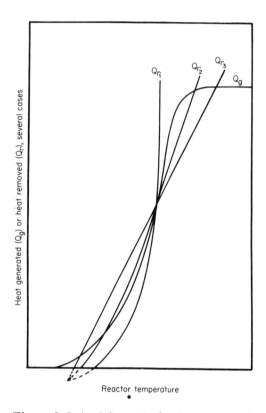

Figure 3.9-4 Schematic for heat generation and heat removal in a stirred tank reactor; single-step reaction [25].

Thus, if we were to achieve this stationary state, slight excursions in temperature above the stationary state will cause the reactor to heat until it reaches the highest stable condition. Conversely, slight excursions below the stationary-state temperature will cause the reactor to cool until it reaches the lowest stable condition. Remember, now, that we are considering open-loop reactor stability and not the control of the reactor at or near this stationary state. Bilous and Amundson [26] consider the stability of a CSTR operating with two exothermic reactions, as

$$B \xrightarrow{\text{k}_{p1}} C \xrightarrow{\text{k}_{p2}} D$$

and find that the heat generation curve has two inflections, thus affording one, three, or even five stationary states. This is seen in Fig. 3.9-5. Numbering the stationary states from the lowest temperature, we find that the odd stationary states are stable and the even stationary states are unstable.

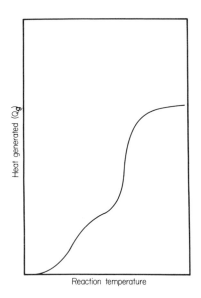

Figure 3.9-5 Typical heat generation curve for addition polymerization reaction with typical termination reaction (e.g., monomer) [26].

Extending this work, Goldstein and Amundson [27-30] published an intensive study on the stability and control of CSTRs with homogeneous addition reactions, heterogeneous addition reactions, catalyst-initiated reactions, and chain transfer models. The forms for the heat removal term remain linear with reactor temperature. For the simple monomer addition polymerization with disproportionation termination, the shape of the heat generation curve is remarkably like that of the simple B $\rightarrow$ C $\rightarrow$ D reaction shown in Fig. 3.9-5. This is shown in Fig. 3.9-6, where the energy for propagation E_p is relatively low and that for termination E_t is relatively high.. In Fig. 3.9-7, we see that the amount of heat generated owing to propagation exhibits a maximum with increasing reactor temperature (curve 5), the amount of heat generated owing to the initiation reaction (curve 7) steadily increases with increasing reactor temperature (but in general is not significant when compared with the amount generated owing. to propagation or termination), and the amount of heat generated owing to the disproportionation termination (curve 3) shows a very rapid increase with increasing temperature. For combination termination, at the same energies of propagation and termination, we see that the curve (curve 2) is sigmoidal and thus similar to the simple case A $\rightarrow$ products, discussed earlier. (Note that curves 4, 6, and 8 represent the heats generated owing to termination, propagation, and initiation, respectively.)

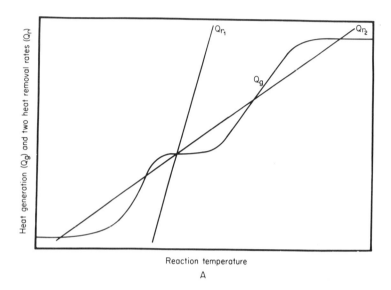

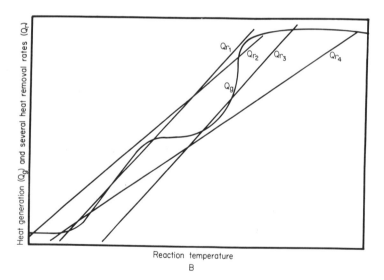

Figure 3.9-6   A: Schematic of addition polymerization reaction heat generation, and heat removal; one SS and five SSs. B: Schematic of addition polymerization reaction, heat generation, and heat removal; three SSs and several heat removal rates [27].

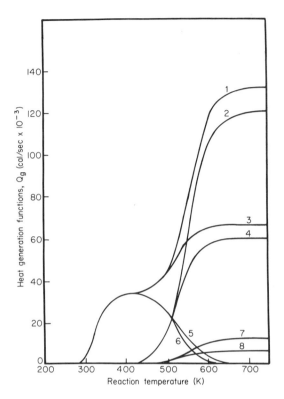

Figure 3.9-7 Heat generation functions for several addition polymerization reactions and two types of addition polymerizations. Curves 1, 3, 5,
and 7 represent slow propagation steps. Curves 2, 4, 6, and 8 represent
slow initiation steps. Curves 1 and 2: total heat generation; 3 and 4: termination reaction heat generation; 5 and 6: propagation reaction heat generation; 7 and 8: initiation reaction heat generation [27].

Note the extremely high rates of heat being generated at these conditions. For the disproportionation termination reaction, for example, at
600°K, the simple polymerization reaction is producing nearly 130,000
cal/sec. If we have a high energy of propagation and a relatively low energy
of termination, the shapes of both curves (for combination and disproportionation termination reactions) take the classical sigmoidal shape and thus
indicate only two stable temperatures rather than the three possible in Fig.
3.9-6 for the reaction that is stopped by disproportionation termination. In
a parametric study Goldstein and Amundson show that eventually for all
types of termination reactions if the energy of activation for the termination
reaction is sufficiently high and/or the energy of activation of the propagation

reaction is sufficiently low, the heat generation curves will characteristically show five stationary states. If the opposite is true, the curves will show characteristically the classical sigmoidal curves with three stationary states.

Phase trajectory is a traditional method of analyzing the approach to the open-loop stationary states. The transient form for the energy equation is given as

$$V \rho c_p \, dT/dt = Q_g - Q_r = Y\left(M_1, \sum_{n=1}^{\infty} P_n, T\right) \tag{3.9-22}$$

where Q_g and Q_r depend on the reaction kinetics and the modes of heat transfer. Furthermore, the concentration profiles of monomer, total number of reactive species, and dead polymer specie of chain length j can all be classified as time-dependent equations, according to the rate expressions in Chap. 2. For example, consider the transient monomer concentration in a CSTR to be given as

$$dM_1/dt = W\left(M_1, \sum_{n=1}^{\infty} P_n, T\right) \tag{3.9-23}$$

and the total concentration of free radicals to be

$$d\left(\sum_{n=1}^{\infty} P_n\right)/dt = X\left(M_1, \sum_{n=1}^{\infty} P_n, T\right) \tag{3.9-24}$$

When solving these equations numerically (and this appears to be the only way, since they are highly nonlinear), it is quite possible to plot, for example, the concentration of $M_1(t)$ against the reactor temperature $T(t)$. The result, as shown in Fig. 3.9-8 for a simple sigmodal heat generation curve, is called a phase-plane plot. This shows, for example, that if we were to begin, in set of numerical calculations, at a temperature of about 325°K and a relatively low concentration of monomer in the reactor, the monomer concentration in the reactor would rapidly build (thus showing an accumulation of monomer owing to more monomer being fed to the reactor than that being consumed to produce product). Gradually, however, we see that the temperature begins to increase to approximately 337°K and the monomer concentration drops gradually until it reaches a stable state at something less than 0.001 mol/liter. If, on the other hand, we were to begin our reaction at 340°K with a high concentration of monomer in the tank (say, 0.003 mol/liter), we would see a rapid drop in monomer concentration and a very rapid increase in temperature to 360°K or so before we would again see an approach to the stable condition of 337°K and a concentration less than 0.001 mol/liter. With five stationary states for a reaction with combination

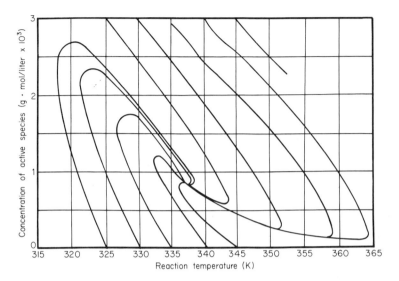

Figure 3.9-8   Phase plane or reaction path for a single stable state (SS) for addition polymerization reaction, open loop [27].

termination the phase-plane trajectories are considerably more complex, as seen in Goldstein and Amundson's curve (Fig. 3.9-9). Here the monomer concentration and the total number of free radicals are plotted against the temperature. Note for the monomer-temperature phase-plane plot that three stable points are indicated: one at a temperature of about 270°K and a monomer concentration of slightly more than 3 mol/liter, one at a temperature of about 430°K and a monomer concentration of about 0.03 mol/liter, and one at a temperature of about 580°K and a monomer concentration of about 2×10^{-4} mol/liter. Note in particular the very critical sensitivity to initial monomer concentration and temperature. For example, if the initial monomer concentration is about 0.01 mol/liter and a reactor temperature less than about 330°K, the reactor temperature will tend to drop and the monomer concentration rise and the system will approach a stable state at about 270°K and 3 mol/liter. If the reactor temperature is somewhat greater than 330°K (and the monomer concentration initially is the same), the system will at first experience an increase in monomer concentration and a slight dip in temperature, and then the temperature will rapidly increase and the monomer concentration will decrease until we reach the second stable state at 430°K and 0.03 mol/liter. We must be quite careful in predicting which stationary state the system will seek. If the trajectories pass by an unstable stationary point, they will approach very slowly, experiencing temperature changes on the order of 1°/10 to 20 sec. On the other hand, if the trajectories are moving away from an unstable stationary

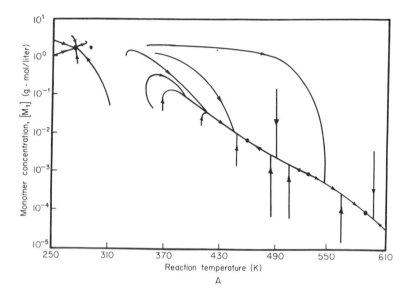

A

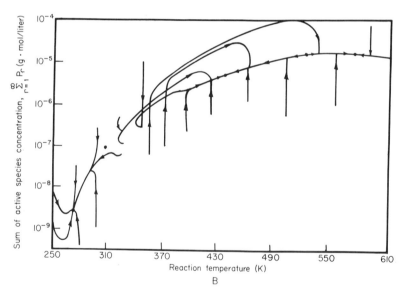

B

Figure 3.9-9   A: Phase-plane trajectories for five steady states, addition polymerization reaction, open-loop stability; monomer concentration M_1 vs. reactor temperature, °K.   B: Phase-plane trajectories for five steady states, addition polymerization reaction, open-loop stability; total active species concentration $\sum_{r=1}^{\infty} P_r$ vs. reactor temperature [27].

point, they will move very rapidly. For the example chosen by Goldstein and Amundson, if the initial temperature is about 340 to 350°K and the initial monomer concentration is increased to about 1.0 mol/liter, a very rapid temperature increase to about 450°K (at the rate of 30 to 50°/sec) will occur. A dramatic decrease in the monomer concentration to less than 0.01 mol/liter will take place, and the system will approach the second stable state at 430°K and 0.03 mol/liter. But increasing the initial monomer concentration to 2 mol/liter (at the same initial temperature) will cause the reaction temperature to increase to about 480°K and the concentration to fall to less than 0.001 mol/liter. Thus, this trajectory is approaching the fourth stationary point from the high (temperaturewise) side and as it passes this unstable point will again slow but will again gain speed as it approaches the third stable point at 580°K and 2×10^{-4} mol/liter. If you have a reactor containing volatile solvents or monomers, a system started at these conditions will result in detonation and/or combustion. Remember, again, that we are considering the stability of an "open-loop" reactor, with no controls on either the temperature or the feed rates. Similar curves are possible with polymerizations with disproportionation termination steps. The design engineer is referred to the extensive work by Goldstein and Amundson [27] for enlightenment.

We shall discuss the stability of reactors with heterogeneous or two-phase reactions later. Understanding the implications of the stationary state and the stable points is central to the understanding of reactor stability and methods of control of such reactors.

As mentioned earlier, in a tubular reactor the mode of heat transfer from the fluid to the wall is conduction. For a naturally insulating material such as a polymer, conduction is a very poor mode of heat transfer. Furthermore, excessive accumulation of the reactants at the wall of the reactor (owing to the much longer residence time there than that in the center of the reactor) will lead to a buildup of undesirable products, referred to in the heat transfer literature as "fouling" materials. Thus, regardless of the efficiency of cooling on the outside of the tubular reactor, the rate of heat transfer is controlled by conduction of heat through the material adhering to the inside wall of the reactor. If there is excessive heat generation owing to viscous dissipation (for very high-viscosity fluids), the temperature in the flowing fluid quite close to the wall will be much higher than the bulk fluid temperature, and thus heat will be transferred from this layer to the bulk fluid. Thus, it is quite possible to have extensive cooling capacity on the outside wall of the reactor and have the temperature of the bulk of the fluid increase owing simply to viscous dissipation. If we couple this phenomenon with exothermic reaction, we see that we can achieve very high temperatures in the reactor, even though we have sufficient cooling at the wall of the reactor. Bilous and Amundson [31] have considered this effect in detail, and their results are shown for a single exothermic reaction A → products in Figs. 3.9-10 and 3.9-11. As is shown, if the walls of the

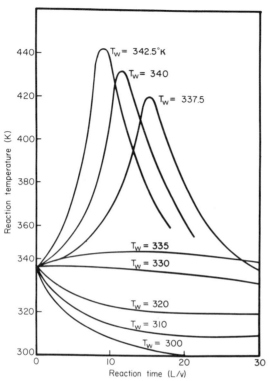

Figure 3.9-10 Effect of wall temperature for reaction time in a tubular reactor. Simple reaction: A → B. E = 22,500, p = 3.94 × 10^{12}, x_i = 0.20, Q = 7300, K = 0.20, T_i = 340°K [31].

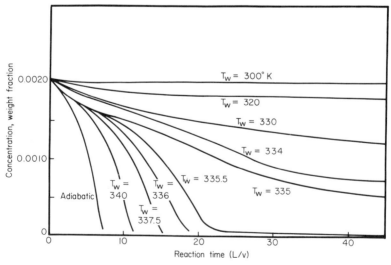

Figure 3.9-11 Concentration profiles in a tubular reactor, showing the effect of constant wall temperature. Properties as given in Fig. 3.9-10. Simple reaction: A → B [31].

reactor are maintained at a constant temperature significantly below the
temperature of the incoming fluid, the reacting fluid will cool and the reaction will slow dramatically. As the reactor wall temperature exceeds a
certain value (which may be many degrees below the incoming reactant temperature), the reaction temperature will rise dramatically until the initial
amount of reactant is completely consumed and the reaction stops. This
analysis is applicable for polyaddition reactions that are monomer concentration dependent as well. Bilous and Amundson show in one example that
an increase in wall temperature of 1°C would result in a decrease in the
time to complete consumption of monomer by more than a factor of 2. In
Denbigh's words, "...there can occur conditions in which it is difficult to
achieve simultaneously both an adequate reaction rate to useful product and
also a safe limitation of temperature" [8]. One recommended method of
control for tubular reactors is the "staging" of the reactors in series such
that each reactor acts adiabatically (without appreciable removal of heat)
and the temperature rise of the reactor is limited only by the amount of
monomer added to that stage. At the end of that stage, the entire mass is
cooled (now possible because the exothermic reaction has been effectively
stopped by the consumption of monomer), and then the cooled mass is fed,
with more monomer, to the next stage. Bulk polymerization of polymethyl
methacrylate, polystyrene, and even phenol-formaldehyde have been successfully carried out in staged tubular reactors [1]. Nevertheless, the
dangers inherent in controlling an overheated polymerization reaction are
amplified when dealing with tubular reactors.

3.10 Control of Single-Phase Reactors

It is really beyond the scope of this text to introduce an exhaustive analysis
of process control. Others have done this in a very complete fashion [27-30].
Nevertheless, to convince the designer that the inherently unstable CSTR
reactor described in the preceding section can be operated safely, we shall
abstract the work of Warden and Amundson [32] on the various control modes
for "closing the loop" on the reactor. They deal with the continuous homogeneous polymerization model described earlier. They consider the linear-
ization of the three differential equations (3.9-22) to (3.9-24) for temperature, monomer concentration, and total free-radical concentration about a
stationary point (either stable or unstable). The objectives in controlling a
reactor are to ensure stability by widening the zone of stability, to make a
naturally unstable reactor stable, or to improve the transient response of
the reactor about a stable point. Of the many variables in the system, the
inlet monomer temperature, the inlet monomer concentration, the coolant
temperature, and the coolant flow rate are probably the easiest to control.
The total number of free radicals does not appear to be directly controllable
for this type of reactor. To the linearized equations, we can add a term

that represents the controllable portion of the equation. In Warden and Amundson notation, for temperature control, we add

$$G_T\, d = [\partial(\partial T/\partial t)/\partial d]_{SS}\, d \qquad\qquad (3.10\text{-}1)$$

and for monomer concentrations,

$$G_M\, d = [\partial(\partial M/\partial t)/\partial d]_{SS}\, d \qquad\qquad (3.10\text{-}2)$$

where Gs are defined by the terms in the brackets, and d represents the deviation of the controlled reactor condition from its value at steady state [33]. In process control language, the Gs represent the controllable feedback portion of the signal, as shown in the block diagram in Fig. 3.10-1. To control the process, we introduce a controller, with notation K, as in Fig. 3.10-2. Now if we wish to control monomer concentration, $G_T = 0$. Likewise if we wish to control temperature, $G_M = 0$. Again in standard process control language, if e represents the deviation of the reactor variable from its steady-state or set-point value, then for each of the three characteristic types of control (proportional, derivative, and integral), we can define the relationship between the deviation of the controlled condition and the error from set-point value:

proportional control: $d = K'_p\, e$ $\qquad\qquad (3.10\text{-}3)$

derivative control: $d = K'_d (de/dt)$ $\qquad\qquad (3.10\text{-}4)$

integral control: $d = K'_i \int e\, dt$ $\qquad\qquad (3.10\text{-}5)$

We can thus collapse the feedback circuit by writing

$$K = -(G_T + G_M)K' \qquad\qquad (3.10\text{-}6)$$

for each type of controller (K_p, K_d, and/or K_i).

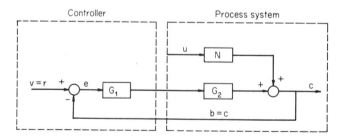

Figure 3.10-1   Typical process control diagram for "closing the loop" on a process reactor [33].

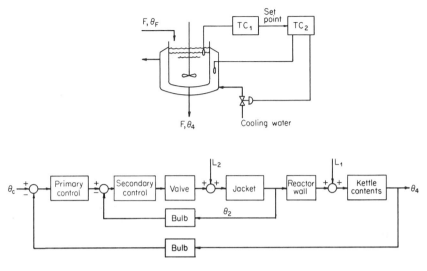

Figure 3.10-2   Cascade process control system as impressed on a reactor kettle [32].

Warden and Amundson then use matrix analysis to determine the stability of the new "closed-loop" reactor system for various combinations of the controllers. They conclude that for control on reactor temperature only ($G_M = 0$), a suitable value for K_p, proportional controller setting, can be found that will stabilize the reactor if the control is placed on coolant flow rate. Derivative control alone will probably result in an unstable reactor if the measuring system exhibits significant time lags, and a similar result is possible for integral control alone. However, proportional control, proportional-derivative (PD) control, or proportional-integral-derivative (PID) control will result in a stable reactor. The size of the stable neighborhood depends on the relative values for the various parameters.

For monomer concentration control ($G_T = 0$), proportional control will lead to instabilities, but there is a range of proportional and derivative control conditions that can stabilize the reactor. In other cases, to control the total number of free radicals, the authors recommend either proportional control or PD control on the coolant flow rate. To control the weight-average molecular weight of the polymer, PD control of the coolant flow rate may control the reactor for very limited values of K_p and K_d, but the authors admit that this is difficult. The same is true if the controls are placed on the monomer concentration. They show that control of addition polymerization reactions having monomer termination, combination termination, spontaneous termination, and disproportionation termination is basically the same. Their most important conclusion, then, is that

proportional control of the coolant flow rate, coolant temperature, and/or
inlet monomer temperature through monitoring reactor temperature will
always make the reactor locally stable, provided that the proportionality
constant is sufficiently large.

Incidentally, the same analysis applies to the control of batch reactors except that the monomer concentration cannot be controlled. Here the
rate of addition of monomer concentration (i.e., the flow rate of monomer
to the reactor) can be controlled using PD controllers provided the proper
settings are chosen for these controllers. Again, the controller settings
will depend on the specific values for the parameters in the given process.

Luyben [34] considers "autorefrigerated" reactors, where refluxing
of a volatile solvent or monomer is the method of heat removal rather than
cooling coils. With this method of heat removal the Q_r line is no longer
linear, as shown in Fig. 3.10-3. W is the boil-up rate of volatiles with a
latent heat of vaporization that is temperature dependent. As can be seen,
the reactor is unstable in open loop. With proportional control on the condensate return to the reactor, using the reactor temperature as the set-point
variable, as long as the system is sufficiently far from the critical temperature, it can be controlled provided that the proportionality constant is sufficiently large. This is shown in Fig. 3.10-4. In Fig. 3.10-5 are given the
phase-plane trajectories of concentration (of monomer) and reactor temperature for $k_p = 100$. Luyben also considers indirect control on the reactor
temperature through control of the condenser flow rate.

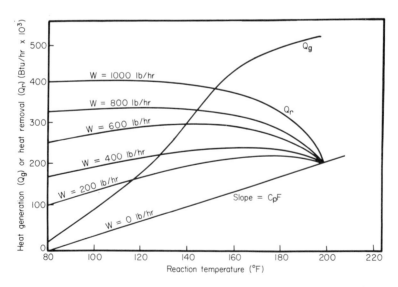

Figure 3.10-3 Open-loop stability of autorefrigerated reactors; heat generation and heat removal curves [34].

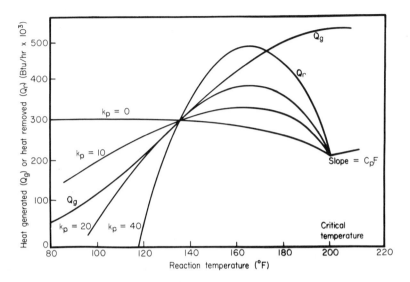

Figure 3.10-4   Open-loop stability of autorefrigerated reactors; heat generation and heat removal curves for variable heat of vaporization [34].

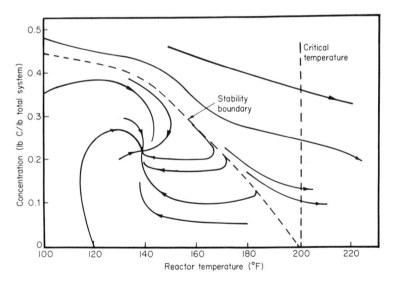

Figure 3.10-5   Phase-plane stability plot of autorefrigerated reactor showing open-loop stability; $k_p = 100$ [34].

3.11 Heterogeneous Reactions in Various Types of Reactors

Until now, we have considered a homogeneous polymerization in either
batch, CSTR, or tubular reactors. The homogeneous polymerization is
suitable for bulk polymerization where the polymer is soluble in the monomer or for solvent-based polymerization, where (1) the diffusion rate of the
monomer in the solvent is not limiting and (2) both the monomer and polymer are completely soluble in the solvent. We shall take up the case of
solvent polymerization where the resulting polymer is not completely soluble
in the solvent in Chap. 14.

    There are three general alternatives to homogeneous polymerization:
(1) emulsion polymerization where the "oil" phase, containing the monomer
or polymer, is held in droplets (the discontinuous phase) in a continuous
"water" phase through the addition of surface-active agents (surfactants)
such as laurate soaps; (2) suspension polymerization, where the "oil" phase
is preserved in droplet form through high-shear mixing of the aqueous
phase; and (3) solid-phase catalyst-initiated reactions. The obvious advantage in using either emulsion or suspension polymerization is that the reactors are microscopic in size and surrounded by an excellent heat sink (the
aqueous phase) with no polymer-reactor interface to offer additional resistance to heat transfer. As a result, the thermal stability of these reactions
is much better than their equivalent bulk reactions. Another advantage of
using these types of reactions is to circumvent various patented bulk and
solution polymerization processes, as Platzer [2] and Bishop [35] illustrate.
Actually, emulsion polymerization was one of the initial ways of making
polymethyl methacrylate and polystyrene [1] and as a result has a long history of analysis. As evidence of the rapidly growing literature in emulsion
polymerization, the serious student should make a careful analysis of the
recent 500-page monograph by Blackley [36]. In addition to detailing the
Smith-Ewart theory, discussed below, Blackley analyzes the roles of dis-
sociative and redox initiators; categorizes micelle-generating substances
and modifiers; considers the roles of other additives, such as dispersing
agents and shortstopping agents; discusses the various types of monomers
and their behavior in various aqueous media; briefly discusses nonaqueous
emulsion polymerization; and looks ahead to newer emulsion polymerization
techniques, such as stereospecific reactions. Blackley's book is extensively
illustrated with case studies and must be considered a major effort in the
developing technology of emulsion polymerization.

    In this section, we shall consider the early and now classical work
of Smith and Ewart [37] on batch reactions of emulsion addition reactions,
discuss briefly the extensions of this work to continuous polymerizations,
and finally discuss the conditions of reactor stability and control when more
than one phase exists (the work of Amundson and coworkers). We shall also
discuss briefly the comparison of emulsion and suspension polymerization
in various types of reactors. And we shall end the chapter by considering
the role of the solid-phase catalyst on initiating polymerization reactions.

3.11.1 The Smith-Ewart Model

The earliest quantitative explanation of emulsion polymerization is attributed to Smith and Ewart [37]. In an extensive series of articles, Gardon has given the limitations and extensions of the Smith-Ewart model in terms of its applicability to many polymer systems in batch operation [38]. Nevertheless, the Smith-Ewart model illustrates one method of approach to the understanding of "micellular" growth of addition polymerization in an emulsion. The qualitative argument is that when soap (such as sodium or potassium laurate) is added to water (or other aqueous materials), micelles on the order of 50 to 100 Å in diameter are formed. As an oil-based monomer is added, the micelles increase in size. Normally, however, there is much more monomer added to the aqueous medium than can be stabilized by the soap, and thus the majority of monomer is present as droplets and much of the soap (with some monomer) is in micelles. According to the early work of Harkins [39], free-radical polymerization occurs in the micelles by having the monomer initially present react with free radicals in the aqueous medium. Polymerization continues by diffusion of monomer through the aqueous medium to the active sites within the micelles until termination occurs by combination with either an active monomer radical or a growing polymer chain from another micelle. Apparently the number of micelles does not increase as polymerization proceeds, thus indicating that polymerization occurs nearly entirely within the micelle. Hawkins supported his work by pointing out that increasing the soap concentration increased the reaction rate dramatically, because more micelles were available as reaction sites, and thus the consumption of monomer became more rapid (as did the increase in polymer chain length). This work was supported brilliantly by Smith and Ewart when they analyzed the rate at which free radicals were generated in the aqueous medium and the rate at which they entered the growing polymer molecule in the micelle. They showed that for most polymerization reactions the rate of termination of the free radical in the aqueous medium was on the order of 10^{-5} times less than its rate of generation. Thus, most of the free radicals entered the growing polymer molecules rather than terminating with another free radical or combining with a monomer molecule to generate an active polymer chain in the aqueous medium. Furthermore, Smith and Ewart showed that any micelle can contain, at best, either one or no radical at any time. Thus, the rate of polymerization within the micelle is best approximated by

$$R_p = k_p (N/2) M_1 \qquad (3.11-1)$$

where R_p is the rate of polymerization (per cc), k_p is the reaction rate for propagation, M_1 is the monomer concentration in the micelle, and N represents the number of micelles. In other words, the rate of polymerization depends on the number of micelles (which is a direct function of the concentration of surfactant added to the water) and is apparently independent of the rate of generation of free radicals by the aqueous-phase initiator. The 1/2 occurs since it is equally likely from a probability viewpoint to have either one or no reactive site in a given micelle. As long as there are sufficient micelles, we get polymerization growth within the micelle. Eventually, of course, the growing micelle will require more and more surfactant to cover its surface, and any micelles that have not grown polymers will be depleted of their surfactant and disappear. Until that time, however, the number of micelles and hence the number of polymer sites remain constant. If R_i is the rate of free-radical generation, obtained from the kinetics of the initiator and monomer reactions and assumed to be known and parametric, Smith and Ewart show that if $\dot{v}$ is the rate of increase of volume of a given polymer micelle and A_t is the combined area of all micellular particles at time t, then

$$A_t = \int_0^t a_{r,t} R_i \, d\tau \qquad\qquad (3.11\text{-}2)$$

where $a_{r,t}$ represents the surface area of a given growing micelle. The notation here is that of Flory [13]. At time t the volume of a micelle with a particle initiated at time τ will be $\dot{v}(t - \tau)$. Thus, the equivalent area $a_{r,t}$ will be

$$a_{r,t} = [(4\pi)^{1/2} \cdot 3\dot{v}(t - \tau)]^{2/3} \qquad\qquad (3.11\text{-}3)$$

The rate of increase in volume of a particle depends directly on the rate of polymerization $k_p M_1/2$ and the volume of the monomer unit (divided by volume fraction of polymer). As a result, substituting $a_{r,t}$ into Eq. (3.11-2) and integrating, we obtain the total area of all particles:

$$A_t = (3/5)[(4\pi)^{1/2} \cdot 3\dot{v}]^{2/3} R_i t^{5/3} \qquad\qquad (3.11\text{-}4)$$

The point where all the soap molecules form a monolayer on the growing polymer micelles is given by knowing that 1 g of soap will yield a surface area a_s one molecule thick. If the concentration of soap is c_s, we then equate A_t with $c_s a_s$ to obtain the time of exhaustion of micellular surface material:

$$t_{ex} = 5^{3/5}/[3(4\pi)^{1/5}](c_s a_s/R_i)^{3/5} \dot{v}^{-2/5} \qquad\qquad (3.11\text{-}5)$$

The number of micelles containing growing polymers is simply the rate of initiation of free radicals times t_{ex}:

$$N = 0.53(c_s a_s)^{3/5}(R_i/\dot{v})^{2/5} = C(c_s a_s)^{3/5}(R_i/\dot{v})^{2/5} \qquad\qquad (3.11\text{-}6)$$

We can therefore show that the rate of propagation of addition polymerization in emulsion reactions is given by

$$R_p = Ck_p(c_s a_s)^{3/5}(R_i/\dot{v})^{2/5}M_1/2 \tag{3.11-7}$$

Experiments have shown that C is somewhat less than theory predicts, and for styrene and butadiene, at least, $C = 0.37$ fits the data better than $C = 0.53$. The number-average chain length can also be shown to be related to the growth rate of a polymer chain, according to

$$\overline{X}_n = k_p NM_1/R_i = Ck_p(c_s a_s)^{3/5}M_1/[2R_i^{3/5}\dot{v}^{2/5}] \tag{3.11-8}$$

We can write the expression for $\dot{v}$, the volumetric rate of growth of the micelle, as

$$\dot{v} = (3/4\pi)(k_p/N_a)(\rho_m/\rho_p)\phi_M/(1 - \phi_M) \tag{3.11-9}$$

where N_a is Avogadro's number and ρ_m and ρ_p are the densities of monomer and polymer, respectively. ϕ_M is referred to by Gardon as the thermodynamic equilibrium volume fraction of monomer in solution at saturation and is assumed to be constant. Gardon quotes values of ϕ_M of 0.6 for styrene and 0.73 for methyl methacrylate. The monomer concentration M_1 is given as $M_1 = V_M\phi_M$, where V_M is the molar volume of monomer.

We can then calculate the number-average molecular chain length as

$$\overline{X}_n = 0.185k_p^{3/5}(c_s a_s)^{3/5}V_M\phi_M^{3/5}(1 - \phi_M)^{2/5}N_a^{2/5}(4\pi/3)^{2/5}/$$
$$[R_i^{3/5}(\rho_m/\rho_p)^{2/5}] \tag{3.11-10}$$

For most practical emulsion polymerization recipes, R_i is on the order of $10^{14}/\text{sec}\cdot\text{cc}$ and $c_s a_s$ is on the order of 10^5 cm^{-1}. According to Gardon's calculations, $\dot{v}$ is about 2.65×10^{21} cc/sec for methyl methacrylate at 55°C and about 7.5×10^{20} cc/sec for styrene. k_p at 30°C is about 35 liter/mol·sec and at 35°C is about 50 liter/mol·sec for polystyrene. A typical monomer concentration is on the order of 5×10^{-3} mol/cc.

3.11.2 Smith-Ewart Model for CSTR Reactors

For CSTR reactors, the work of DeGraff and Poehlin [40] and Funderburk and Stevens [41] utilizes the Smith-Ewart kinetics to determine the steady-state behavior of CSTRs. Funderburk and Stevens, for example, define a population density of growing polymer micelles such that

$$d(n\dot{v})/dV + n/\theta = 0 \tag{3.11-11}$$

where $\dot{v}$ is the volumetric growth rate of the micelles, given by Eq. (3.11-9). V is the volume of a given micelle, n is the number of micelles of a given size range, and θ is the hold time in the CSTR. Note that here the appropriate definition of n is best given as

$$\int_{V_1}^{V_2} n \, dV = \text{total number of micelles between the sizes of } V_1 \text{ and } V_2 \qquad (3.11\text{-}12)$$

We can also define a population density m that is a function of the micelle radius rather than the volume:

$$\int_{r_1}^{r_2} m \, dr = \text{total number of micelles between radii of } r_1 \text{ and } r_2 \qquad (3.11\text{-}13)$$

The equivalent steady-state balance on this population density is given as

$$d(mR_r)/dr + m/\theta = 0 \qquad (3.11\text{-}14)$$

where R_r is the radial micellular growth, which according to the Smith-Ewart theory is given as

$$R_r = r^{-2}\left\{[3/(4\pi)^2](k_p/N_a)(\rho_m/\rho_p)\phi_M/(1 - \phi_M)\right\} \qquad (3.11\text{-}15)$$

Now we can show that for the Smith-Ewart model $d\dot{v}/dV = 0$ and $dR_r/dr = -2R_r/r$. Therefore, we can integrate the population density curves for both n and m, yielding

$$n = n^0 \exp[-(V - V^0)/\theta\dot{v}] \qquad (3.11\text{-}16)$$

and

$$m = m^0 \exp[\ln(r^0/r)^2 - (8\pi/3k_p\theta)(r^3 - r^0)^3] \qquad (3.11\text{-}17)$$

Note that since the rate of nucleation equals the rate of particle removal from the solution, the nucleation rate for the Smith-Ewart model is $dN^0/dt = n^0\dot{v}^0$. We define a dimensionless radius $\sigma = r/R_r^0\theta$, a dimensionless volumetric population density $\eta = n/n^0$, a dimensionless volume, $v = V/\dot{v}^0\theta$, and a dimensionless radial population density $\mu = m/m^0$. Now as the radius increases (or hold time is decreased), the radial population density for the Smith-Ewart model passes through a maximum, and the volumetric population density decreases with increasing volume of the micelles (or as the hold time is decreased). This is shown in Figs. 3.11-1 and 3.11-2. DeGraff and Poehlin, using the Stockmayer set of equations, also shown in Figs. 3.11-1 and 3.11-2, found that the fraction of free radicals generated that actually diffuse into the micelles is a very strong function of the hold time (at reasonable values of θ). They plot this fraction, $F_m = N_p/R_i\theta$ (where

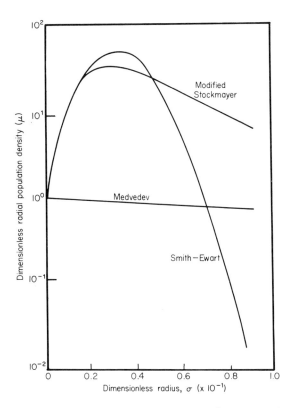

Figure 3.11-1   Heterogeneous emulsion reaction, micellular growth, various radial models [40].

N_p is the number of polymer particles contained in micelles and R_i is the rate of generation of free radicals per unit volume), against θ, the hold time for various rates of initiation, and C_E, the concentration of soap in the solution, as in Fig. 3.11-3. We see that for relatively short hold times (θ less than about 1 hr) nearly all the activated monomer diffuses into the micelle to form new polymer. This confirms the Smith-Ewart model, which assumes this. At high hold times, the fraction of activated monomer used in polymerization decreases linearly with hold time. Thus, to maximize the useful free radicals, the hold time should be relatively short, the soap concentration rather high, and the rate of initiation of monomer rather low. Also, longer residence time causes a broadening in the molecular-weight curve and an increased number of polymers with high chain lengths. Increasing the rate of initiation will decrease the number-average chain length, and as we have seen before, increasing the concentration of soap in the solution increases the number of polymer micelles. This, coupled with short

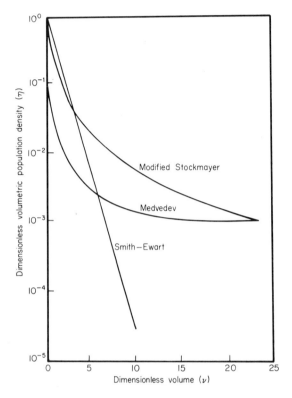

Figure 3.11-2   Heterogeneous emulsion reaction, micellular growth, various volumetric models [40].

residence times, will cause an increase in the number-average chain length. This increase can be counteracted somewhat by increasing initiator reaction rate (without changing polymerization rate).

    Wessling [42] has extended these works on emulsion polymerization to rate-dependent reactions. He points out that the common practice is to add monomer continuously to the reactor (rather than to charge the reactor in a batchwise fashion), and if the monomer concentration remains constant, the Smith-Ewart mechanism seems to predict the experimental data obtained by Gerrens [43] on styrene in a CSTR.

    We can conclude that the utility of the simplistic Smith-Ewart model seems valid in both batch- and CSTR-type reactors. The additional effort required in using other more complex models does not seem warranted for a first design of an emulsion polymerization reactor of either type. No analysis has been attempted on the tubular reactor, but it would seem that coalescence would be a problem in this type of reactor. In addition, the

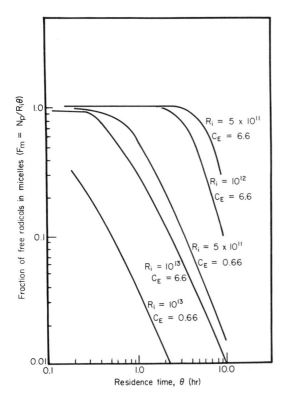

Figure 3.11-3   Effect of hold time on reaction of free radicals generated that diffuse into micelle. C_E is concentration of soap in solution; R_i is the rate of generation of free radicals per unit time; $F_m = N_p/R_i\theta$, where N_p is the number of polymer particles in micelles [40].

stability of the most stable emulsion begins to break down when the total surface area of the polymer growing within the micelles begins to exceed the amount of surfactant available to form a monolayer. This will mean coalescence of the growing polymer micelles and eventual phase separation. This type of phase separation is well known in batch reactors as the reaction proceeds toward completion. Inside a tubular reactor, with no means of agitation, this phase separation can result in fouling of the heat transfer surface, among other things.

3.11.3 Emulsion Processing Aids

We pause here to discuss the various "additives" that have been added to form a stable emulsion. We must begin with an emulsifier, normally a laurate; the monomer, normally in a liquid state; and the aqueous medium, frequently ultrapure water. The catalyst should be a polar-type liquid, soluble in the aqueous medium. This is frequently a peroxide (e.g., H_2O_2). As mentioned, the rate of monomer addition is used for thermal control of the reaction. The operating temperature of the reactor affects not only the degree of polymerization and the various rates of reaction (particularly the rate of initiation) but also the life of the emulsified structure. High temperatures cause the micelle to begin to break down. Thus, a buffering agent or modifier such as a hydrophilic protective colloid (methyl cellulose or polyvinyl alcohol) is used to prevent "aging" of the emulsion and premature coalescence of the polymer. Agitation is necessary to maintain uniformity of the reaction mixture, to provide sufficient fluid movement for effective heat transfer, and to minimize the effects of diffusion of the monomer to the reactive site. The pH affects the degree of hydrolysis of the monomers, and the presence of salts in the aqueous medium (inorganic or organic) decreases the effectiveness of the emulsifying agent. Thus, if the water of aqueous medium contains high concentrations of mineral salts, the emulsions will not be so stable, and as a result larger drop sizes and poorer product quality are found. As a result, there are many recipes for the production of emulsion-based polymerizations. Bishop [35], for example, gives two formulas each for polystyrene with the following characteristics: (1) easy-flow, low-molecular-weight; (2) easy-flow, high-molecular weight; (3) high-heat-distortion; (4) general-purpose, easier-flow; and (5) general-purpose, high-molecular-weight. He also gives four typical suspension formulations, all this for one polymer, polystyrene. Rohm and Haas [44] list about five recipes for polymethyl methacrylate and similar numbers for other homologs of the acrylics. We have not counted the number of PVC formulations or the copolymerizations of polystyrene-butadiene, polystyrene-acrylonitrile, ABS, vinyl-acetate, vinyl-acrylic, and so on, all of which can be made with proper recipes using emulsion polymerization. Again, the reader is directed to Blackley's monograph for the newer processing aids [36].

3.11.4 The Role of Agitation

The effect of agitation on polymerization rate is much more important for suspension polymerization than for emulsion polymerization. However, when the surfactant is completely used and polymerization continues, agitation is necessary to continue to provide sufficient monomer to the reactive polymer sites (thus minimizing diffusion-controlled reactions) and to prevent coalescence of the micelles and separation of the system into two continuous phases. Sweeting et al. [45-47] have shown for emulsion polymerization of vinylidene chloride that early in the reaction the Smith-Ewart kinetics hold (up to a conversion of monomer of about 15%) and that as the soap is used the kinetics become diffusion controlled (their so-called stages II and III). We would expect increases in agitation, intensity, or duration to increase the rate of conversion as a function of time in the batch reactor. This is the case, as shown in Fig. 3.11-4. Note, however, that at conversions

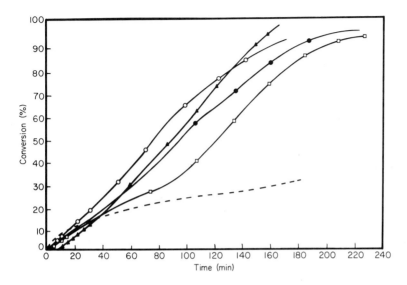

Figure 3.11-4   Emulsion polymerization of vinylidene chloride, showing the effect of stirrer speed on polymerization rate. Dashed line, no stirring; squares, 106 rpm; solid circles, 131 rpm; open circles, 388 rpm; triangles, 756 rpm [45].

below 15% (where the micelles are not coalescing) the increase in agitation causes a <u>decrease</u> in the percentage of conversion. This is popularly re-ferred to as the "Trommsdorf effect" and is described in detail in other sources [13]. In the case of suspension polymerization, no emulsifier is added to the reacting mixture, and thus the Smith-Ewart kinetics are not valid. As a result, we should see an obvious and marked increase in the rate of conversion with increase in agitation for all concentrations of monomer.

3.11.5 Suspension Polymerization

There are relatively few reaction studies in which suspension poly-merization is carried out. PVC is apparently the most important suspension polymer. Bankoff and Shreve [48] were apparently the first to consider the kinetics of this type of reaction, and their work has subsequently been veri-fied experimentally by Talamini [49] and Farber and Koral [50], all for PVC. Bankoff and Shreve argued that the degree of conversion of monomer should be directly proportional to catalyst concentration. As shown in Fig. 3.11-5 (from Farber and Koral), this is verified experimentally for diisopropyl peroxydicarbonate (IPP) (and also for benzoyl peroxide). Farber and Koral argue that if the polymerization <u>rate</u> (% conversion/unit time) is plotted

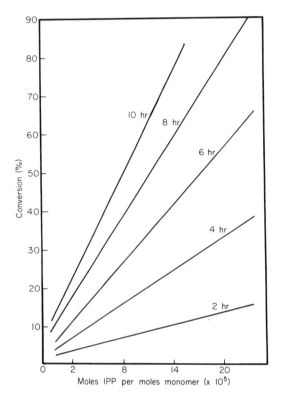

Figure 3.11-5   Polymerization kinetics for PVC using diisopropyl peroxy-dicarbonate at 42.5°C, showing the effect of monomer concentration on % conversion [50].

against the concentration of initiator, the curve is nonlinear. Talamini also argues, theoretically, that if R_d is the polymerization rate in the dilute or monomer-rich phase and R_c is the rate in concentrated or polymer-rich phase, then Q is their ratio:

$$Q = R_c/R_d \tag{3.11-18}$$

R_c and R_d are also the degrees of conversion of monomer present in each phase per unit of time. (Q is sometimes referred to as the Nernst distribution. Normally, however, the Nernst distribution related thermodynamic equilibrium of a single component between two phases.) Now if C is the degree of conversion, the fraction of reaction occurring in the concentrated phase will be AC and that in the dilute phase will be 1 - AC. Thus the overall polymerization rate is

$$dC/dt = R_d(1 - AC) + R_c QAC = R_d + (QA - A)R_d C \qquad (3.11-19)$$

Now, if $q = (QA - A)$, the equation can be rewritten as

$$dC/dt = R_d(1 + qC) \qquad (3.11-20)$$

where $C(0) = 0$. The resulting integration is

$$C = q^{-1}[exp(qR_d t) - 1] \qquad (3.11-21)$$

By the kinetics of steady state in Chap. 2, the polymerization rate can be approximated by

$$R_d \doteq k(I_2)^{1/2} \qquad (3.11-22)$$

and thus

$$C = q^{-1}\left\{exp[qkt(I_2)^{1/2} - 1]\right\} \qquad (3.11-23)$$

Talamini expands this exponential to obtain, for short times,

$$C = q^{-1}[qkt(I_2)^{1/2}] = kt(I_2)^{1/2} \qquad (3.11-24)$$

Thus, at least for short times, the conversion rate should be proportional to the initiator concentration to the square root. Farber and Koral show this to be the case for conversion percentages to 40% (although no data points are shown). See Fig. 3.11-6. This experimental observation is also in agreement with their data obtained using benzoyl peroxide (BPO). Thus, the kinetics of polymerization for suspension polymerization differ significantly from those of either bulk or emulsion polymerization. There seems to be little indication of deviations from these kinetics except at relatively high reaction temperatures, where the kinetics of degradation are important. Farber and Koral show that at long reactor times the overall rate of conversion decreases rather dramatically with slight increases in reactor temperature. For example, for PVC initiated with IPP at a concentration of 3.8×10^{-5} mol/mol of monomer, at 1 hr into the polymerization the percent conversion is about 11% at 60°C and 21% at 65°C. At 6 hr, the percent conversion is 70% at 60°C and 72% at 65°C. It would thus appear that continuing the reaction at 65°C beyond 6 hr would not appreciably increase the percent conversion. As mentioned earlier, the effect of agitation on polymerization has been explored experimentally for emulsion polymerization. We would expect that increased agitation for suspension polymerization, either in batch or CSTR, would yield increased rate of conversion.

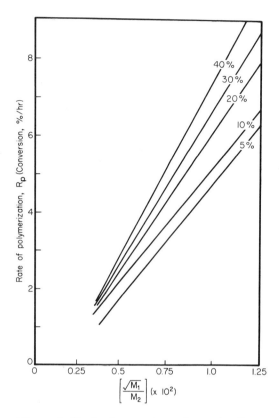

Figure 3.11-6   Suspension polymerization; polymerization rate of PVC for various concentrations of diisopropyl peroxydicarbonate catalyst [50].

3.12   Stability of Multiple-Phase Reactors

When we deal with multiple phases (either in suspension, solution, or late-stage emulsion polymerization), we must consider the fact that as these phases pass through a reactor (either batch or CSTR), they exchange mass and heat in order to support the polymerizations. Goldstein and Amundson [27-30] considered the various aspects of free-radical polymerization in two phases, including complete physical and thermodynamic equilibrium between phases, thermal equilibrium with interfacial mass transfer resistance, and interfacial mass and heat transfer resistance. In their computations, they assumed uniform and complete mixing. Their kinetic models were the standard free-radical propagation with either combination termination (model 1) or disproportionation termination (model 2). Furthermore, they assumed that all polymer species have the same relative solubility in the two phases. This is somewhat restrictive for solvent-precipitated polymerizations, where the solvent is thermodynamically good to the monomer and becomes progressively poorer as the polymer increases in molecular weight. We shall consider the thermodynamic solvents (called theta solvents) in a later chapter. If physical and thermodynamic equilibrium exists between both phases, $T_I = T_{II}$ (the temperatures of the phases are equal), and $c_I = \lambda c_{II}$, where λ (Q in an earlier example) is referred to as the Nernst distribution and c_I and c_{II} are concentrations of a given specie in phases I and II, respectively. λ is a strong function of temperature of the reaction. For this case, the performance of a CSTR (stability and control) is identical to that for a homogeneous or single-phase polymerization, because of the free association and therefore instantaneous response of chemicals across the interface. Therefore we shall not repeat the analysis, except to state that this model is probably most applicable to the early-stage Smith-Ewart emulsion polymerization model or for highly agitated, very low-viscosity reactor fluids.

Probably the most stringent of the assumptions of physical and thermodynamic equilibrium is that of no interfacial mass transfer resistance (e.g., the mass transfer coefficient KS is infinite). The Goldstein-Amundson approach is nonkinetic and nonthermodynamic. They assume that the rate of transfer of a polymeric specie is proportional to the interfacial area S and to the difference in concentration between the two phases. The proportionality constant K is referred to as the mass transfer coefficient:

$$KS(M_n^{II} - \lambda M_n^I) \qquad \text{for } n \geq 1 \tag{3.12-1}$$

$$KS(P_n^{II} - \lambda P_n^I) \qquad \text{for } n \geq 1 \tag{3.12-2}$$

This, in essence, assumes that the Whitman two-film theory of chemical reaction holds. More appropriate methods, such as those detailed by Astarita [51], have not been incorporated, nor has the fact that the rates of mass transfer should depend on the chain length of the molecule. These points are considered briefly later on. Goldstein and Amundson assume that K, the proportionality constant, is a controllable parameter through adjustment of the agitation rate of the CSTR and the temperature of the reacting phases ($T^I = T^{II}$).

Because of the extensive number of cases that can be generated (variation in the energies of activation, either through consideration of various types of reactants or through variation of reactor temperature), we shall consider only trends in their analysis. For example, in Fig. 3.12-1 we see that for no interfacial resistance (KS = ∞), we have the familiar double-S-shaped curve for heat generation Q_g and, therefore, if the heat removal remains linear five stationary states. The first, third, and fifth are stable. If KS is finite, we have as many as nine stationary states, for which the first, third, fifth, seventh, and ninth are stable. Note the

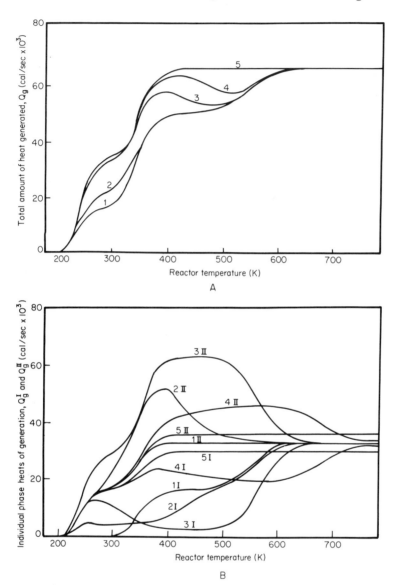

Figure 3.12-1   A: Total amount of heat generated for two-phase system
with mass transfer between two phases, KS as parameter. Activation energy for phase 1, $E_i^I = 34,000$; for phase II, $E_i^{II} = 22,000$. Curves 1-5: KS =
0, 0.1, 10, 100, and ∞, respectively. B: Individual phase heat generation
rate for two-phase system with mass transfer between two phases, KS as
parameter. Curves 1-5 as shown in A; energies of activation as shown in A;
I = phase I; II = phase II [27].

unusual shape of the Q_g curve, showing decreases in heat generated with increase in temperature. In these curves, incidentally, KS was assumed to be constant and independent of temperature. Probably a better analysis would have had KS = f(T), as well as $\lambda = \lambda(T)$. Nevertheless, this illustrates the effect of finite interfacial resistance on the amount of heat.

Turning to phase trajectories, Goldstein and Amundson find that mass transfer is important primarily at lower reactor temperatures, where the reaction rates are lowest. Their computer data indicate that if one could conduct a polymerization in two phases with mass transfer between two phases, the number-average chain length would be significantly increased over that same reaction carried out homogeneously. For one numerical example where KS = 1000, this ratio at the lowest stable "open-loop" condition is about 6.3. The equivalent total weight-average chain length ratio is about 44. Thus, we get an increase in the degree of polymerization, indicating a broadening of the molecular-weight distribution. Incidentally, the phase-plane trajectories give us little new information about the reactor stability that we could not have extrapolated from the earlier work on homogeneous-phase reactions.

To add thermal resistance to the mass transfer resistance at the interfaces, we must define the rate of heat transfer between the two phases as

$$hS(T^I - T^{II}) \tag{3.12-3}$$

where h is the heat transfer coefficient and S is the (usually unknown) interfacial surface area for heat transfer. Owing to the possible differences in temperature between the two phases, the transfer of mass through the interface must include the effects of the transfer of sensible heat. If there is no resistance to mass transfer and if hS is finite, we can show that there are 5 possible stationary states in the heat generation-heat removal vs. reactor temperature curve. Goldstein and Amundson further argue that as KS and hS increase in magnitude, the number of stationary states must decrease. The proof of this is obvious, for if there is neither thermal nor mass resistance at the interface, only 5 stationary states are possible. On the other hand, if hS and KS are both zero (assuming the impossible case of no exchange of mass or heat between the two phases), then there are 5 stationary states for each phase, and a total of 25 stationary states is possible, 9 of which are stable. Goldstein and Amundson demonstrate that small amounts of heat transferred with the transferring mass can usually be neglected and that the effect of heat transfer on the reaction rates is most apparent at lower temperatures where the reaction rates are smallest. It is difficult to demonstrate the Q_g vs. Q_r curves for this two-phase system because there are two reference temperatures. Of interest is the phase-plane trajectories of the two temperatures and of the monomer concentrations in each phase as functions of their respective temperatures. These are shown in Figs. 3.12-2 and 3.12-3 for the condition where KS = 10 and hS = 1000. Note in the T^I vs. T^{II} curve that the temperatures tend toward the 45° line (where $T^I = T^{II}$). As high temperatures in the M_1^I vs. T^I curve, large excursions in temperature and monomer concentration are occurring. The effect of KS = 10, hS = 1000 on the number-average chain length, weight-average chain length, and degree of polymerization is also important. If we increase the total resistance to transfer (either by decreasing KS or by adding a term for hS, or both), we find a decrease in both the number-average chain length and the weight-average chain length. For example, for KS = 10, hS = 1000, the number-average chain length is only 20% of that for KS = 1000, hS = ∞ and only 25% greater than that for the homogeneous-phase reaction. For the weight-average chain length, the value is 33% of the lower-resistance model and only 14.7 times that of the homogeneous-phase reaction. This yields a much broader molecular-weight distribution

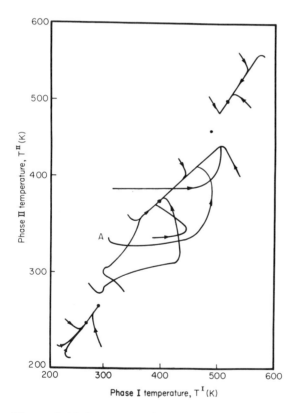

Figure 3.12-2 Phase-plane trajectories for five steady states, open-loop stability, two phases. Note that individual phase temperatures are plotted [27].

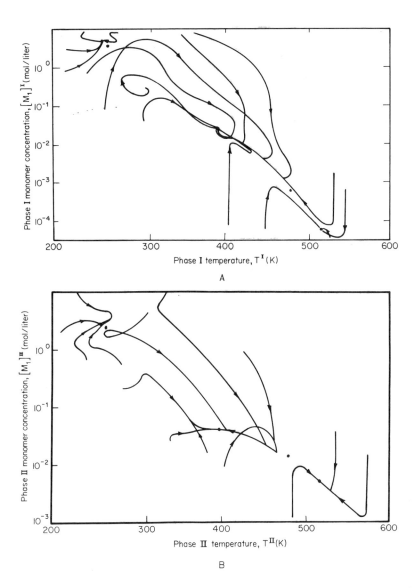

Figure 3.12-3   A: Phase-plane trajectories for monomer concentration in phase I of two phases, open-loop stability, five steady states, KS = 10, hS = 1000.   B: Phase-plane trajectories for monomer concentration in phase II of two phases, open-loop stability, five steady states [27].

than either the homogeneous-phase reactor product or the product with a
much lower interfacial resistance. We can conclude that for multiple-phase
reactions the interfacial resistances must be kept as low as possible.

Note that additives may accumulate at the interface, thus offering
additional mass transfer resistance. Improper mixing of the reactants can
lead to low mass transfer and heat transfer coefficients. In general we can
neglect the sensible heat transfer connected to the mass transfer between
phases, but we cannot neglect the overall heat removal from the reactor.
Allowing extreme differences in temperature between phases can lead to
depletion of monomer and increased polymerization (accompanied by increased energy generation) from one phase. Thus, heat is transferred to
the other phase, cooling the first phase and increasing the rate of reaction
in the second. Note the line marked A in Fig. 3.12-2 where this instability
is demonstrated. Beginning with a reaction mixture where phase I is at T^I
of 300°K and phase II is at T^{II} of about 350°K, we see first a slight drop in
T^{II} to about 340°K, accompanied by a very rapid rise in temperature T^I to
420°K. Then both phases increase in temperature until T^I is about 400°K
and T^{II} is 460°K. Then the temperatures equilibrate at about 440°K and the
reacting mass cools to the stable temperature of 390°K. (Even at this
stable bulk temperature, the phases have slightly different temperatures,
as Goldstein and Amundson point out: $T^I = 393°K$, $T^{II} = 383°K$.)

The authors offer no advice as to the control of these multiphase
reactors, but it is assumed that the process control concepts developed
earlier can probably be used to control these reactors. One of the obvious
problems that must be faced is the selection of the phase that must be monitored so that proper set points can be established. No one has addressed
himself (herself) to the question of proper monitoring and control of these
multiphase reactors, either because the mathematics are so complex as to
preclude analysis or one or the other phases is somewhat volatile and thus
temperature can be controlled through autorefrigeration, as suggested by
Luyben [34].

3.13 A Diffusional Model

As mentioned earlier in this section, the Goldstein-Amundson model considers a mass transfer coefficient based on the Whitman two-film theory.
Astarita [51] shows that the film theory is valid when the overall driving
force is due to molecular transport. Under these conditions, the mass
transfer coefficient is assumed to be directly proportional to the molecular
diffusion coefficient (and inversely proportional to the thickness of the stagnant film). In contrast, the penetration theory assumes that the phase into
which the specie is diffusing is infinitely deep when compared to the lifetime
of the specie (e.g., the time for a monomer to add to the active growing
polymer chain). It yields the following diffusion equation:

$$\partial P_n/\partial t = D_n \partial^2 P_n/\partial x^2 \tag{3.13-1}$$

where P_n is the concentration of the diffusing specie and D_n is the diffusion coefficient for the diffusion of specie n through the medium containing all species, P_n and M_n. Assume that $P_n(0,x) = P_n^0$ and $P_n(t,0) = P^0$. At $x = \infty$, P_n is finite; in other words, the fluid is infinitely deep to the incoming reactive specie. As an illustration, assume no reaction. The solution to this equation is the familiar penetration solution [52], where the instantaneous penetration rate through the interface is given as

$$V_n^0 = -D_n (\partial P_n/\partial x)_{x=0} = (P_n^{0\prime} - P_n^0)(D_n/\pi t)^{1/2} \tag{3.13-2}$$

If we define the average life of the reactive species as t^*, then the average penetration rate of specie n through the interface is given by averaging the above equation for all times to t^*:

$$V_{n\ av}^0 = 2(P_n^{0\prime} - P_n^0)(D_n/\pi t^*)^{1/2} = k_n(P_n^{0\prime} - P_n^0) \tag{3.13-3}$$

And thus the mass transfer coefficient is proportional to the square root of the diffusivity of that specie. Note, however, that this equation can be written as

$$V_{n\ av}^0 = (\text{mol of specie n/unit time} \times \text{unit surface area})$$
$$= k_n(P_n^{0\prime} - P_n^0) \tag{3.13-4}$$

Since the interfacial area for suspension polymerization is not known, the total mass transfer must be written as

$$\sum_{n=1}^{\infty} SV_n^0 = \sum_{n=1}^{\infty} Sk_n(P_n^{0\prime} - P_n^0) = \sum_{n=1}^{\infty} Sk_n(P_n^{II} - \lambda P_n^I) \tag{3.13-5}$$

in accordance with Goldstein and Amundson. Nevertheless, k_n is a function of the diffusivity D_n, which, as shown in Chap. 14, is a function of the chain length n. Including depletion by chemical reaction and using the Zeman-Amundson continuous model, we can write a partial differential equation for concentration profiles of $P(n,x,t)$. The diffusion equation is quite similar to Eq. (3.13-1) except that we have generation and dissipation terms and, if desired, diffusion coefficients that are molecular chain length dependent. The form of the equation is given as

$$\partial P(n,x,t)/\partial t = D\partial^2 P/\partial x^2 + P(n,x,t) + a(\partial P/\partial n) - a(\partial^2 P/\partial n^2)/2!$$
$$+ \text{higher-order terms} \tag{3.13-6}$$

(Here D is assumed to be independent of n, although this restriction can easily be relaxed.) Astarita addresses the majority of his work to the solution of diffusion and reaction of A → B. His work can logically be extended, through the use of the continuous model, to include polymerization.

Again, probably the most important factor in the argument against the film theory model chosen by Goldstein and Amundson is that the mass transfer coefficient is not directly proportional to the diffusivity. This assumption in their model is a significant limitation if the lifetimes of the various species are dramatically changed by reactor conditions (agitation, temperature, etc.) and/or the diffusivities are found to be strongly dependent on the chain length of the molecules. The latter is suspect from chemical kinetic data on high polymer reactive ends. Incidentally, Astarita points out that the actual depth of penetration is at most equal to $(Dt^*\pi)^{1/2}$. For this analysis to hold for bubbles and droplets, the radius of the drop must be greater than the depth of penetration. For monomer diffusion (such as vinyl chloride) and normal reaction times (to form PVC), the minimum radius must be greater than about 10^{-5} cm, a criterion that is met for most suspension polymerization systems. We shall return to this problem of diffusion in polymers in Chap. 14, where removal of monomer from polymer and the rate of diffusion of solvent into polymer are important processes.

3.14 Catalytic Polymerization

Many polymerizations are being carried out using solid Ziegler-Natta or Freidel-Crafts catalysts: butyl lithium, butyl sodium, $AlCl_3$, $TiCl_4$, and so on. Here the catalyst initiates reaction in much the same way as the liquid catalyst. However, the polymer chain(s) grow from the surface of the catalyst, expanding the catalyst effectiveness (in volume) by 10^5 to 10^6 times during the polymerization. These catalysts are not recoverable in the true sense of catalyst-initiated reactions. In many cases, we can consider a rather simple diffusion model for catalysis of this nature. The monomer diffuses through the polymeric material surrounding the catalyst particle until it reaches a reaction site (either the catalyst particle itself or a reactive polymer site), at which point a conventional monomer addition reaction takes place. Schmeal and Street [53] have examined the diffusion characteristics for an isothermal expanding catalyst particle, and Tinkler and Pigford [54] have considered the effect of exothermic reaction (simple A → B reactions) on the efficiency of the catalysis. Schmeal and Street consider two limiting cases of catalysis: the nonexpanding polymeric particle with randomly distributed reactive sites and the solid core model. In the latter case, the polymer accumulates around the catalyst in a spherical fashion, and catalytic initiation of the monomer diffusing through the polymer layer must occur at the polymer surface. The first is the polymeric core model; the second is the flow model. For an isothermal system in the general model, the concentration of monomer around the catalyst sphere is given as

$$\eta^{-2} \, \partial(\eta^2 \, \partial c/\partial \eta)/\partial \eta - a^2 c = \partial c/\partial \tau \qquad (3.14\text{-}1)$$

where η is the dimensionless radius, r/R; c is the monomer concentration, c_s is the monomer concentration at the surface of the catalytic sphere, $a^2 = R^2(kL/D)$, the square of the Thiele modulus, where K is the polymerization reaction rate constant, L is the concentration of reactive sites on the catalyst surface (sites/volume), and D is the diffusivity of monomer in polymer (assuming that the diffusivity of polymer is essentially zero); and τ is a dimensionless time, tD/R^2. The Thiele modulus is a measure of the efficiency of utilization of the reactive layer. In other terms it is an effectiveness factor for catalytic reaction. It is rather apparent that if a is very large, the polymerization rate is high, and the diffusion coefficient is low. Thus, the rate of diffusion will control the reaction. On the other hand, if a is very small, the diffusion coefficient is high, and either the rate of polymerization or the number of reactive sites is low. Thus, the rate of mass transfer is reaction controlled. The Nernst distribution coefficient λ is included in the definition for c_s, the concentration of monomer at the surface of the catalyst.

Schmeal and Street consider several cases. In the steady-state model, $\partial c/\partial \tau = 0$. For this case, the concentration at the surface is

$$c/c_s = \sinh a\eta/\eta \sinh a \qquad (3.14\text{-}2)$$

and the specific reaction rate at the surface of the catalyst is given as

$$f(a) = c_s^{-1}(\partial c/\partial \eta)_{\eta=1} = a \coth a - 1 \qquad (3.14\text{-}3)$$

For large a, $f(a)$ approaches a; for small a, $f(a)$ approaches $a^2/3$. They show that if k^* is defined as a specific polymerization reaction rate,

$$k^* = (3V_p \lambda/R)(DkL)^{1/2} \qquad \text{for large values of the}$$
$$\text{Thiele modulus} \qquad (3.14\text{-}4)$$

and

$$k^* = V_p \lambda kL \qquad \text{for small values of the}$$
$$\text{Thiele modulus} \qquad (3.14\text{-}5)$$

Here V_p is the volume of particles per unit volume of solution. Note that for diffusion-controlled reactions the reaction rate is proportional to the square root of the diffusion coefficient and to the square root of the number of reactive sites and inversely proportional to the catalyst particle radius. For reaction-controlled catalysis, the reaction rate is proportional directly to the number of reactive sites.

For the solid core model, where reaction occurs only on the surface, the accumulation term and the time-dependent term have been considered negligible. The Thiele modulus appears in the boundary condition at the surface of the catalyst:

$$\partial c/\partial \eta = a^2 c/3, \qquad \eta = 1 \tag{3.14-6}$$

The monomer mass flux at the surface of the catalyst particle is given as

$$(\partial c/\partial \eta)_{\eta=1} = a^2 c_s /[3 + a^2(1 - 1/\eta_s)] \tag{3.14-7}$$

where η_s is the radius of the entire sphere, which is increasing in proportion to the amount of generation of polymer. $\eta_s^3 = 1 + v$, where v is the volumetric ratio of polymer to solid catalyst. Now for large values of Thiele modulus a

$$k^* = (3DV_p \eta_s)/[R^2(\eta_s - 1)] \tag{3.14-8}$$

and as $\eta_s \gg 1$, this value tends toward

$$k^* = 3DV_p \lambda/R^2 \tag{3.14-9}$$

Note that the specific rate of reaction is independent of the local polymerization reaction rate, directly proportional to the diffusion coefficient of monomer in polymer, and inversely proportional to the square of the catalyst particle radius. For small values of the Thiele modulus a we find

$$k^* = DV_p \lambda a^2/R^2 \tag{3.14-10}$$

or substituting for a,

$$k^* = V_p \lambda k \tag{3.14-11}$$

Thus, for the hard core model, we find that the specific reaction rate is independent of the radius of the catalyst particle, for a reaction-controlling process.

In the most general case considered, Schmeal and Street consider the flow model. Here reaction sites are free to move or flow outward with the polymer. Thus, we have mass balance equations for spherical flow not only for the monomer and polymer but also for the catalyst sites. If the time-dependent terms are assumed small (e.g., the steady-state model), the equations become

$$\eta^{-2} \partial(\eta^2 \partial c/\partial \eta)/\partial \eta - a^2 \theta c = 0 \tag{3.14-12}$$

$$\eta^{-2} \partial(\eta^2 v)/\partial \eta = \gamma a^2 \theta c = 0 \tag{3.14-13}$$

$$-\eta^{-2} \partial(\eta^2 \theta v)/\partial \eta = \partial \theta/\partial \tau \tag{3.14-14}$$

Here, as before, η is the dimensionless radius, a is the Thiele modulus, and θ now is a dimensionless site concentration, L/L_A where L_A is a reference site concentration (at $\tau = 0$). v is a dimensionless velocity, uR/D, where u is the velocity at which reactive polymer sites are convected around the sphere. γ is the specific polymer volume, referenced to the monomer concentration in the polymer. The boundary conditions are

$$\partial c/\partial \eta = \partial \theta/\partial \eta = v = 0 \qquad \text{at } \eta = 0 \tag{3.14-15}$$

(e.g., symmetry of concentration of reactive sites and monomer concentration and no velocity at the center of the reacting sphere of catalyst)

$$c = c_s \qquad \text{at } \eta = \eta_s \tag{3.14-16}$$

and

$$\theta = c = c_s = 1 \qquad \text{at } \eta = 1 \tag{3.14-17}$$

(e.g., site and monomer concentrations referenced to the surface of the catalyst). For batch reactors,

$$-dc_s/d\tau = 3V_p \lambda (\eta^2 \partial c/\partial \eta)_{\eta=\eta_s} \tag{3.14-18}$$

and for CSTR conditions,

$$dc_s/d\tau = 0 \tag{3.14-19}$$

The above-listed equations can be simplified and rearranged into two first-order partial differential equations that can be solved using the method of characteristics. Their results are illustrated using a diffusion-controlled model, where the Thiele modulus a = 48. In Fig. 3.14-1, the dimensionless polymerization rate as a function of v, the volume of polymer per unit volume of catalyst, initially shows a characteristic decrease with increasing volume (as is the case with the polymer core model described earlier). At high volumes a slight increase in polymerization rate occurs. At high volumes of polymer to catalyst, the flow model predicts more than 10 times the reaction rate that the polymer core model does. The authors point out that for the reaction-controlled cases (small Thiele modulus values) all the sites are equally accessible to monomer, and thus the size of the sphere should not affect the total polymerization rate. Extensive calculations for both the batch reactor and CSTR boundary conditions indicate that the weight-average to number-average molecular weight (dispersity index) for the polymer core model was given as

$$\overline{DI} = \overline{M}_w/\overline{M}_n = (\beta p)^{-1} + a/6 \tag{3.14-20}$$

for large values of the Thiele modulus. Here is the ratio of monomer concentration to site concentration at references, c_s/L_A, and p is the dimensionless moles of polymer per unit volume of catalyst. For small values of the Thiele modulus, the dispersity index is given as

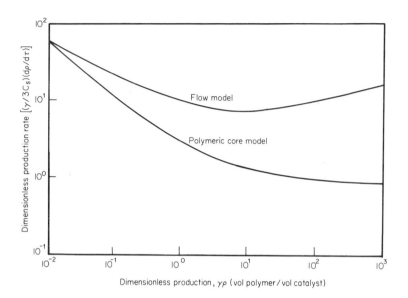

Figure 3.14-1 Comparison of limiting case expansion models for the case of diffusion control. Thiele modulus, $a = R(kL/D)^{1/2} = 48$ [53].

$$\overline{DI} = \overline{M}_w / \overline{M}_n = (\beta p)^{-1} + 1 \qquad\qquad (3.14\text{-}21)$$

Thus, we see that as the polymerization increases, $\overline{DI}$ approaches the Thiele modulus over 6 for large values of a (e.g., diffusion-controlled reactions) and approaches unity for small values of a (e.g., reaction-controlled processes). For the flow model, on the other hand, the authors found that for large values of the Thiele modulus $\overline{DI}$ approached a maximum as the values of p approached 100 and then decreased slightly thereafter. Increasing the Thiele modulus increased the dispersity index, as is expected ($\overline{DI}$ = a). For small values of the Thiele modulus, the results were the same as those predicted by the polymer core model. The results for both the batch reactor and the CSTR were essentially the same in every case investigated. This implies that the molecular-weight distribution depends only on the amount of polymer accumulated around a catalyst particle and the nature of the catalyst particle itself (number of active sites, reactivity, size, etc.) and not on the history of the particle. The authors compare the flow model dispersity indices for the flow model and the polymer core model as functions of the Thiele modulus; see Fig. 3.14-2. Attempts to compare these calculations with experimental observations are obviously very difficult owing to the lack of suitable catalytic reaction data. Nevertheless, the authors point out that Natta, for example, found no effect of catalyst size on polymerization rates of polyolefins for spherical catalyst particles. Others

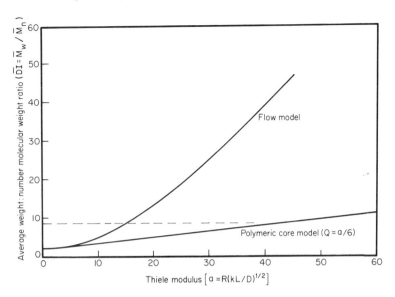

Figure 3.14-2   Weight-average to number-average molecular-weight ratio $\overline{DI}$ as a function of the Thiele modulus, $a = R(kL/D)^{1/2}$, for two expansion models. Note: Experimentally, for polypropylene, $\overline{DI}$ values in the range of 8 to 9 were found for various reaction conditions [53].

have seen that spherical and cylindrical particles give polymers of different molecular weights. For "living polymers," e.g., those initiated with an metallorganic catalyst such as butyl sodium (a solid) or butyl lithium (a viscous liquid), the $\overline{DI}$ values for batch reaction are on the order of 8 to 9, indicating a Thiele modulus of 10 to 50 and a diffusion-controlled reaction.

The importance of this type of study becomes obvious when reactor design engineers attempt to determine molecular-weight distributions and residence times for the catalytic polymerization of new resins. The authors have given sufficient information to the designer to allow the engineer to establish certain upper and lower bounds on his (her) polymerization model and thus design the system around the reactor accordingly.

Petersen [55] defines an effectiveness factor for a single catalytic reaction as

η = reaction flux at the catalyst surface/reaction rate at the surface   (3.14-22)

Petersen shows that the ideal, non-diffusion-controlled reaction rate for a simple first-order reaction A → B is given as

$$R_{ideal} = (4\pi/3)R^3 LkA^0$$ (3.14-23)

where L is the number of reactive sites per unit volume, k is the reaction rate, and A^0 is the concentration of A at some distance from the catalytic surface. If diffusion is included, we can define a reaction time, similar to that of Astarita, for penetration into the catalyst. This time t is given as

$$t = (D/9kL)^{1/2} \qquad (3.14\text{-}24)$$

where D is the diffusion coefficient of A. t represents the penetration time. Using this, the real reaction rate becomes

$$R_{real} = 4\pi R^2 LkA^0 t \qquad (3.14\text{-}25)$$

or upon substitution,

$$R_{real} = 4\pi R^2 LkA^0 (D/9kL)^{1/2} \qquad (3.14\text{-}26)$$

Thus, we can redefine the effectiveness factor as the ratio of real to ideal reaction rates at the surface of the catalyst:

$$\eta = R_{real}/R_{ideal} = (D/kLR^2)^{1/2} = 1/a \qquad (3.14\text{-}27)$$

In other words, the effectiveness of a catalyst is inversely proportional to the Thiele modulus. This gives a better indication of the importance of the Thiele modulus in defining the performance of catalyst-initiated reactions. For simple reactions such as A → B, Tinkler and Pigford [54] have shown that the catalyst effectiveness is strongly dependent on the energy generated during chemical reaction. In fact, owing to accumulation of heat generation within the catalyst (in the case of a porous catalyst), it is entirely possible to have catalyst effectiveness values in excess of unity. If this is in fact the case for polymerization reactions as well, the work of Schmeal and Street must be extended to the nonisothermal reactions.

According to Tinkler and Pigford and to Carberry [56], the maxima in catalyst efficiencies for simple exothermic reactions seem to occur for a Thiele modulus range of $0.4 < a < 1.0$ or so. For a polymerization weighted effectiveness factor, say, $\bar{\eta}_n = \sum_{n=1}^{\infty} n\eta_n / \sum_{n=1}^{\infty} \eta_n$ might be plotted against a weighted Thiele modulus $\bar{a}_n = \sum_{n=1}^{\infty} na_n / \sum_{n=1}^{\infty} a_n$ with a weighted heat generation term $\bar{\delta}_n = \sum_{n=1}^{\infty} n\delta_n / \sum_{n=1}^{\infty} \delta_n$ as a parameter. δ is given as $\delta = (\Delta E/RT_0) \cdot (\Delta H/kT_0) \times$ diffusion rate, where ΔE is the energy of activation of the activated monomer, T_0 is the isothermal temperature surrounding the catalyst particle, and ΔH is the heat of reaction during polymerization. If this type of analysis were carried out, we might find a definite effect on the molecular-weight distribution owing to the nonisothermal behavior of the catalyst. We would anticipate a broadening in the molecular-weight distribution and thus an increase in $\overline{DI}$ at a constant value of the Thiele modulus. And this factor, rather than the flow model concept, might be the reason for the higher $\overline{DI}$ observed by Schmeal and Street during the production of isotactic polypropylene.

The designer must remember that heat buildup can increase the rate of polymerization, and thus the choice between the flow model and the polymer core model of Schmeal and Street is no longer quite so clear-cut. Increasing temperatures can cause degradation in reaction-controlled reaction processes. Perhaps, therefore, it is preferable to have Thiele modulus values that are not zero, but reasonably small. That this is in fact the experimental case is seen in the data discussed by Schmeal and Street. It is probably too early to determine the actual mechanism of polymerization around the Ziegler-Natta- or Friedel-Crafts-type catalysts, but certain hints as to the parameters that might be critical are emerging for the designer.

Amundson has explored catalytic-initiated reactions from a stability viewpoint. Kuo and Amundson [57] show that the rate of heat removal from a polymer core model catalyst is linear with catalyst temperature, whereas the heat generation curve shows the characteristic double-S-shaped curve with temperature for a simple addition reaction with no termination step. There are five stationary states, with the first, third, and fifth being stable. Of course, it seems very difficult to control this type of reaction since it is molecular in diffusion and reaction at the catalyst surface. Certainly, the controls can be placed on the catalyst characteristics and on the medium through which the monomer must diffuse, thereby a priori controlling the reaction rate to some degree. But once the reaction is proceeding and suitable macroscopic agitation and heat removal means have been established, external control seems difficult and most likely ineffectual. Increasing the degree of mixing would certainly decrease the effect of diffusion on the transport of monomer to the catalyst site. The Thiele modulus would increase and so would the molecular-weight distribution. However, this should also increase the rate of heat removal from the area surrounding the growing catalyst particle, and this may lower the reaction rate by shifting the system from a stable temperature to a lower one. Along with broadening molecular-weight distribution, we would get a slowing of the reaction. Going to the other extreme is dangerous, since allowing the temperature to build in the growing catalyst particle will encourage degradation reactions and the formation of insoluble polymer materials. Again, we can only stress that insufficient experimental and theoretical research has been done on the question of nonisothermal catalysis. We can only indicate the trends in this area.

3.15 Some Experimental Observations

It would be impossible to sift through the hundreds of papers that have been published on the polymerizations of all types of resins and select those which illustrate best the applications of the various mechanisms discussed in this chapter. Again, the reader should examine the many examples given by Blackley [36] for emulsion polymerization. As mentioned earlier, Bishop

gives dozens of recipes for the production of various types of polystyrene.
There have been some definitive studies done on specific reactions in various types of reactors, and a brief glance at some of these is appropriate.

Hamielec and his coworkers [58-61] have examined the free-radical
solution polymerization of polystyrene in various types of reactors, both
in steady state and transient, and have compared their results with those
generated by computer. Two methods of examination of the polymer product
are viscosity and gel permeation chromatography. They were concerned
about viscosity effects on the catalyst efficiency and the termination rate
constant, since it has been shown that both factors decrease with increasing
reactor viscosity. They found that increasing the reactor viscosity from
about 1 to 400 cP reduced the effectiveness of the catalyst by about 40% and
the rate of termination by a factor of about 5. Their analysis of molecular-
weight distribution was made with a gel permeation chromatograph. Briefly,
a column is filled with swollen organic or porous ceramic beads. A
solution of resin in solvent is introduced to the column, and the smaller
molecules of resin are absorbed or diffuse into the pores or interstices of
the beads, while the higher-molecular-weight resin passes through with
little resistance. This effects a separation according to molecular weight.
By monitoring the effluent volume as a function of time, a molecular-weight
distribution curve can be obtained, such as that shown in Fig. 3.15-1.

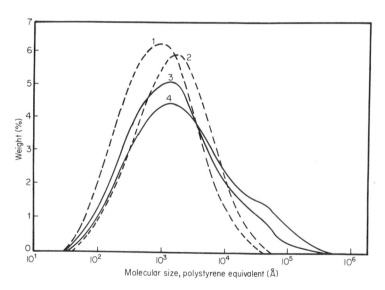

Figure 3.15-1   GPC monitoring of LDPE polymerization. 1: Injection
molding resin, MI = 30; 2: high-clarity resin, MI = 2.5; 3: Extrusion
coating resin, MI = 4; 4: low-clarity, linear resin, MI = 2 [59].

Hamielec and his coworkers used both methods to analyze their experimental data. Examples of their experimental data are shown in Fig. 3.15-2 where the molecular-weight distribution as a function of chain length is given for three CSTRs in a train. Note the excellent agreement between the computed and experimental number-average and weight-average molecular-weight distributions, the dispersity indices, and the extent of conversion. The reactor viscosities were all less than 10 cP. Hamielec also discussed reactors with recycle, with dead volume and bypass flow, and with rather high viscosities. (Incidentally, the viscosity effect on catalyst reaction rate and termination rate referred to earlier is the Trommsdorf effect.) At asymptotically high viscosity (greater than about 1000 cP) with recycle, the theoretical number-average weight of polymer from each of the three reactors was 18,330, 14,800, and 14,390, and the conversion percentages were 12.1, 18.8, and 22.8. Compare these values with the experimental values: number-average molecular-weight distributions (via GPC) were 19,050, 15,900, and 14,970, and the conversion percentages were 12.4, 17.8, and 22.8. The calculated and measured dispersity indices were within 5% for all three reactors in the train. The remarkable agreement here attributable to the accuracy in modeling the system and to the intrinsic averaging process (which represents a "smoothing process") that shows only gross deviations from reality. This work is representative of the careful experimental research required to establish the validity of the exhaustively discussed theoretical models. Until this work is extended to cover, for example, suspension polymerizations, stability of single- and multiple-phase reactors during start-up, and catalyst-initiated reactions, the design engineer will need to rely on the extrapolations of the information produced by computation of suitable mathematical models.

3.16 Polymerization Reactor Design Criteria—A Brief Summary

If you have followed the many byways to this point, a review is in order. Realize that there are an enormous number of variables involved in reactor design, in general. We are concerned not only with the kinetics of the reactions but with mass transfer, heat transfer, mixing, fluid flow into and out of the reactor, thermodynamics and phase stability, and overall reactor stability and control. And all of these variables are interacting in nonlinear fashions. From this extensive variety of variables, we must pick and choose those that most influence the final design of the reactor of our choosing. It is our job, in simple terms, to determine design and operating guidelines for the reaction section of the polymerization step in the polymer process. To do this, we must rely on our industrial and/or academic experience and the tentative guidelines established by others. We have discussed in detail the kinetics of homogeneous polymerization in the previous chapter and choose not to review this at this point, except to point out

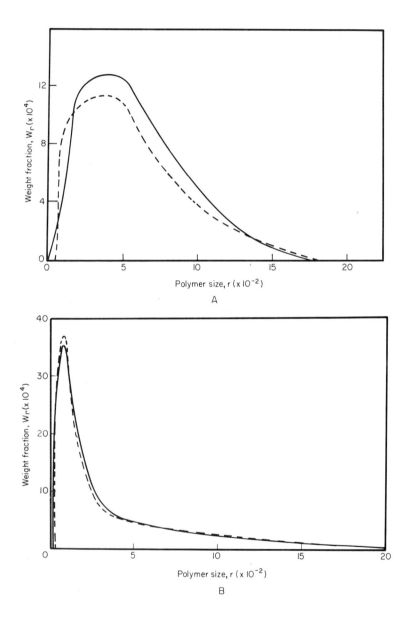

A

B

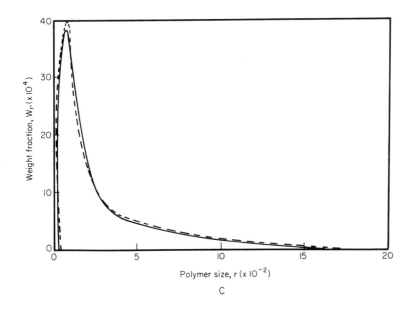

C

Figure 3.15-2   A: CSTR 1, experimental and theoretical molecular-weight distributions. Solid line, theoretical, $\overline{M}_n$ = 45,190, $\overline{M}_w$ = 68,220, $\overline{M}_w/\overline{M}_n$ = 1.51, conversion = 15.1%; dashed line, experimental, $\overline{M}_n$ = 42,770, $\overline{M}_w$ = 67,780, $\overline{M}_w/\overline{M}_n$ = 1.58, conversion = 14.7%. B: CSTR 2, experimental and theoretical molecular-weight distributions. Solid line, theoretical, $\overline{M}_n$ = 14,290, $\overline{M}_w$ = 39,250, $\overline{M}_w/\overline{M}_n$ = 2.75, conversion = 31.5%; dashed line, experimental, $\overline{M}_n$ = 16,400, $\overline{M}_w$ = 44,600, $\overline{M}_w/\overline{M}_n$ = 2.72, conversion = 30.5%. C: CSTR 3, experimental and theoretical molecular-weight distributions. Solid line, theoretical, $\overline{M}_n$ = 13,000, $\overline{M}_w$ = 33,800, $\overline{M}_w/\overline{M}_n$ = 2.60, conversion = 40.8%; dashed line, experimental, $\overline{M}_n$ = 15,070, $\overline{M}_w$ = 37,850, $\overline{M}_w/\overline{M}_n$ = 2.51, conversion = 38.6% [59].

that, for homogeneous systems, we are not primarily concerned with mass transport of material across an interface. We tacitly assume that all the material exists all the time everywhere within the volume of reactor under consideration. For heterogeneous reactions, on the other hand, we must include communication between phases. By this, we mean that some external fluid or solid interface enters into contact with either the monomer or the reactive polymer (or both) in such a way as to increase some useful property of the mass (such as the number-average molecular weight) without instability or destruction of that interface in the process. We have considered several types of heterogeneous polymerizations:

1. Interaction between immiscible or partially immiscible phases where reactive species are present in both phases. Since this type of polymerization involves the interfacial diffusion of one or many reactive species, it is considered heterogeneous in character.

2. Interaction between immiscible phases where one phase is active and the other simply is a carrier for the monomer, for example. This is characteristic of emulsion polymerization in the Smith-Ewart stages.

3. Interaction between a solid catalyst surface and the monomer, initiating a homogeneous reaction. We can use the simple A → B type of kinetics to describe this form of reaction. Here once the reaction is initiated on the catalyst surface, the active monomer diffuses into the polymer phase, where it reacts homogeneously. Thus, the latter portion of the reaction is homogeneous. However, the kinetics of initiation will control this type of reaction.

4. Interaction between a solid or very viscous (and stable) catalyst surface and the growing polymer chains. This is the true heterogeneous catalysis-type reaction, but note that the catalyst is not recovered in most cases, since the polymer chains grow outward from the catalyst particles.

5. Interaction between solid precipitated high-molecular-weight active polymer and the liquid phase containing active monomer. This is characteristic of solution polymerization where the solvent is a good solvent for the monomer and a poor solvent for the polymer once it has reached a certain molecular weight. This type of heterogeneous reaction can probably be treated by the same arithmetic as the Schmael-Street flow model for catalyst that swells with time. The Cosden Corporation bead-precipitated high-impact styrene is produced this way and is unique in that selective precipitation by proper selection of solvents (and very accurate temperature control) can produce extremely uniform high-molecular-weight polymers.

We have seen that mass transfer and heat transfer affect the rate of heterogeneous polymerization early in the reaction and that the Thiele modulus is a good indicator of the effect of diffusion on catalyst-type reactions. We should note that in mass transfer we are concerned with the molecular diffusion of monomer to reactive sites and in the bulk diffusion of materials across a liquid-liquid interface. The active sites of bulky growing polymer chains are significantly less mobile than the active monomer. Nearly all the reactions that occur in reactors are exothermic, and this heat must be removed regardless of the size of the reactor. It is apparent that in conjunction with heat removal we must be able to maintain a suitable reaction temperature. For most reactions, a 5°C change in reactor temperature may make the reactor unstable and/or change polymerization, termination, and degradation rates by factors of 2 or more. We have seen that agitation of reactor products is necessary for uniform heat distribution. Even more critical is the reduction of resistance to molecular diffusion of monomer on a microscopic level. These factors are synergistic. If we have poor heat transfer (with either two phases or catalyst-active systems), we can get increasing temperatures. Increased temperatures lead to degradation of the polymer, the formation of insoluble gel product such as the "fish eyes" that are characteristic in polymethyl methacrylate, or branching owing to chain transfer or some other inherent form of polymerization.

We have also considered several types of polymerization reactions: bulk polymerization, emulsion polymerization, and suspension polymerization. In bulk polymerization we considered the homogeneous type of reaction of the previous chapter. We must remember, however, in reactor stability that heat is being generated volumetrically and removed geometrically. Therefore, the larger the bulk reactor becomes, the more difficult it is to maintain heat stability (and, more important, temperature uniformity). In emulsion polymerization, we have seen that the simple steady-state kinetics of Smith and Ewart are useful in characterizing the various types of reactors. Here the soap forms a micelle into which an active monomer molecule diffuses. As additional diffusion of monomer molecules through the micelle interface occurs, the growing polymer continues to stretch the soap film until it is a monolayer. Beyond that point, the polymer micelles do not have surfactant protection and can coalesce. Agitation inhibits polymerization during the early stages of polymerization and enhances polymerization in the later stages. It appears that the later stages are very similar to suspension polymerization, where the monomer is kept in small droplets by high-shear mixing. It has often been said that suspension polymerization may be considered as comprised of a number of individual bulk polymerizations where the aqueous medium serves as the heat transfer sink.

With regard to reactors, we discussed briefly three types: (1) the batch reactor; (2) the CSTR (continuous stirred tank reactor), sometimes in a train with more than one unit in series; and (3) the tubular flow or pipe reactor. We must realize that polymerization reactions are characterized by times in hours, rather than seconds or minutes, because it takes a finite period of time to form a polymer chain of j units once a polymer chain of j - 1 units is formed, and so on. In batch reactors, we can initially add the monomer and initiator to the tank, begin agitation and cooling, and wait for the reaction to reach an acceptable degree of completion determined by viscosity or molecular-weight distribution or some established standard. More likely, we shall add monomer in a programmed fashion, to prevent overheating of the reacting mixture. Or we shall provide a solvent (or the monomer itself) which will control the reaction temperature by autorefrigeration. The batch reactor is very useful, for example, if we are blocking a copolymer or carrying out the formation of an ABA alloy with a Ziegler-Natta catalyst. We can control the reaction extent and add the proper monomers at the proper time. On the other hand, the CSTR reactor offers continuous flow of product and therefore is desirable for high production runs of "staple" polymers, such as GP, easy-flow polystyrene. The system offers the greatest choice of variables for control and stability in that we can control monomer concentration, flow rate, boil-up rate (in the case of a reflux reactor), coolant temperature, and flow rate. Ideally, it yields a narrower molecular-weight distribution for addition polymerization than batch reactors, but the opposite is apparently true for polycondensation polymerization reactors. The major problem with CSTRs, in practice, is that they do not perform ideally. There is a finite residence time distribution (not a single value as indicated in the ideal model), and there is a fair amount of bypassing. Good agitation is required to minimize these nonidealities.

We have also discussed briefly the tubular flow reactor, which for polymerization reactions is normally in laminar flow. The major difficulty is the enormously long reactors that are required (owing to the long polymerization times). And most types of reactions cannot be carried out in this type of reactor, since the only mode of mixing is molecular diffusion and the only mode of heat transfer is conduction. There are, of course, many other types of reactors that we haven't covered in this section, including the scraped-wall reactors, such as the Luwa unit or the Votator unit [2]. In these units, the surface-to-volume ratio is nearly unity, and thus an accurate control of the reaction temperature should be possible. One major problem is the relatively short residence times of these reactors. Thus, they have been employed in final stages of polymerization, where a polymer syrup is used and the molecular weight of the syrup is to be increased. Scraping or wiping the walls with rotating blades offers periodic surface renewal. Thus the conduction heat transfer is aided by fresh contact with hot fluid, and, of course, mixing flattens the concentration profiles in the reactor.

The batch reactor has been and will continue to be the backbone of the polymer industry. It is useful for the small supplier, for the supplier of specialty resins, for the supplier of various types of co- and terpolymers, and for the research and development team exploring the advisability of producing a new resin. It is simple in construction (and thus can be cleaned with a chipping tool when the reaction goes wrong) and can be equipped with special spargers, in-tank coil heaters, various types of agitators, refluxing systems, external cooling systems, and so on. Furthermore, it is less expensive to dump one bad batch of, say, 1000 gal than to find out after many hours of continuous operation with, say, 5000 to 10,000 gal of material in CSTRs that the material that was fed into the first reactor was impure or in the wrong concentration. Blending off-spec material is also apparently somewhat easier from batch operation than from continuous processing.

Nevertheless, we must consider the consequences of CSTRs and addition polymerization. Amundson and his coworkers found several important conclusions regarding the performances of these units:

1. For CSTRs and homogeneous addition polymerization, if only monomer termination is considered, increasing holding time causes an increase in the number-average molecular weight to a maximum.

2. If combination termination is the mode of termination, a maximum in the number-average molecular weight can occur with increasing holding time in the CSTR. This also means that increasing the hold time beyond a certain time can lead to a decrease in the number-average molecular weight.

3. The initiation rate is the controlling rate. This is seen also from the steady-state models. Small variations in this can produce very large variations in the number-average molecular weight.

4. Increase in the termination reaction rate (for either spontaneous or monomer termination) produces a very rapid skewing of the chain length distribution curve, resulting in rapid decrease in the number-average molecular weight.

5. Multiple paths of propagation (such as chain transfer) apparently produce little changes in the number-average molecular weight (or the stability of the reactor).

They devise a similar set of criteria for CSTRs and heterogeneous polymerizations as well. Notwithstanding all this work, however, the understanding of infinite systems of kinetic reactions and the various constraints on them is just beginning. And yet the reactor design engineer must know what effect a change in catalyst concentration or a change in catalyst type will have qualitatively and quantitatively on the overall operation of his (her) system and the various material and processing parameters of the polymers that are desired by the customer.

The stability of a polymer reactor may be rather unimportant when considered in light of the above considerations. Nevertheless, it is the design engineer's responsibility to outline "open-loop" instabilities in reactor systems and proposed methods of stabilizing the system. It may seem an afterthought to the design engineer, but to the reactor operator it is a fact of life that exothermic reactions can be potentially hazardous and that a seemingly benign reactor may, in fact, be operating near a point of instability. We have seen, therefore, that heat removal from most reactors (except the autorefrigerated reactors, where we have a volatile material that controls internal temperatures) is linear. The slope and intercept of the heat removal curve vs. reactor temperature depend on the overall heat transfer coefficient, the surface area available for heat removal, the feed stream temperature, the coolant temperature, the flow rates to the tank, and the heat capacity and density of the reactants and products. On the other hand, the heat generation curve can either be a simple S-shaped curve (sigmoidal) or have multiple humps. We can show for open-loop reactors that the first, third, fifth, seventh, and so on stationary points (where the heat generated equals the heat removed) are stable and that the even stationary points are unstable. The shape of the curve and the position of the humps depend on the hold time, the initial concentration of reactant, the zero-state reaction rate constants and their ratios, the flow rates of the reactants, the activation energies of the reactions, and the form of termination reaction (spontaneous, monomer, combination, disproportionation). Most reactors operate in a stable mode near the first stationary state. In fact, it should be pointed out that in many actual industrial reactions upper stable steady points cannot be reached owing to the excessive reactor temperature that might cause the polymerizing mass to degrade or, in the case of tetrafluoroethylene (TFE), "unzip" explosively.

If we have multiple phases, we can have stationary points where either interfacial mass transfer or interfacial heat transfer (or both) may be controlling. In the case where these resistances to transfer are very great, two phases may act independently, and thus as many as 25 distinct stationary states are possible.

And, finally, the consideration of control of the simple one-reactor system is necessary. The purposes of control are (1) to make a naturally unstable stationary state stable, (2) to improve the transient behavior of the reactor around a stable stationary state, and/or (3) to enlarge a stable region so that small upsets in temperature and/or concentration (particularly during start-up) will not cause the reactor to go unstable. Conditions such as reactor temperature, monomer concentration, coolant flow rate, and temperature have been considered as means of controlling a single CSTR. From a cursory analysis of more than 60 combinations of conditions and three modes of control (proportional, integral, and derivative), it has been found that proportional control or proportional control with some integral control on many of the controllable conditions will satisfy the control of an exothermic CSTR. Further, for nearly all cases considered, it was found that derivative control on most conditions by itself led to reactor instability, and if the reactor is stabilized by proportional or proportional-integral control on one of the conditions, it is stabilized by proportional-integral-derivative control. The opposite is not true, however.

The reader is warned not to expect to be an expert polymer reactor design engineer after having read this material. We have explored some of the major areas of importance to the design and operations engineers. It should be obvious that the subject is vast and the chapter too brief to gain little more than an appreciation for the complexity of the problem. To intelligently observe the complexity, however, the reader must be aware of the possibility of being involved in extensive calculation, art, guesswork, and empiricism at times to the point of Edisonianism. Nevertheless, we must deal with this problem as a fact of polymer processing and must face the "dirty" industrial problem as a reality. Efforts to establish sound engineering rules will (must?) come, however slowly. Awareness of the various facets of the design problem should be uppermost in the minds of both the design engineer and his counterpart, the operating engineer.

Problems

P3.A For a CSTR with an addition polymerization, using the work outlined in Sec. 3.4, calculate σ_1 if $M_{1,0} = 1$, $\theta = 1$, $k_p = 50$, $k_t = 1$, and $k_i = 0.02$. Plot $M_{1,1}$, σ_1, $\overline{X}_n$, and the total number of free radicals exiting the first reactor, $\Sigma_{n=1}^{\infty} P_{n,1}$ for θ_1.

P3.B For a CSTR train of two reactors, using the values of P3.A, calculate σ_2. Prepare a plot of $M_{1,2}$, σ_2, $\overline{X}_n$, and the total number of free radicals exiting the second reactor, $\Sigma_{n=1}^{\infty} P_{n,2}$, as functions of θ_2 (assuming $\theta_1 = 1$).

P3.C For the data given in P3.A, calculate $P(10)$ for the Zeman–Amundson continuous model, using both a and a' from Eqs. (3.4-39) and (3.4-41). Compare these values with that calculated from Eq. (3.4-15) for $P_{10,1}$.

P3.D For the Zeman–Amundson continuous model, we found that we can write the difference-differential equations in serial form. Following the argument after Eq. (3.4-42), write the differential equation for the dead polymer $M(n)$ and obtain a solution. Assume a simple backward Taylor series. Consider d^2M/dn^2 to be much less than dM/dn or $M(n)$. The boundary conditions are $M(1) = M_1$ and $M(\infty) = 0$. $P(n-1)$ is given as

$$P(n-1) = P_1 \exp[-a(n-2)] \qquad (3.D\text{-}1)$$

P3.E Following Biesenberger's batch reactor calculations beginning with Eq. (3.6-1), integrate Eq. (3.6-5) to obtain Flory's most probable distribution. Make a plot of $(AB)_{10}$ as a function of ϕ. Then plot the ratio of the weight fraction of $(AB)_{10}$ for a batch reactor to that for a CSTR against the total population, ϕ. If $k_p = 50$, find the time required to achieve $\phi = 0.5$ and $\phi = 0.1$ in a batch reactor and in a CSTR. Discuss the relative values in light of the concentration of $(AB)_{10}$.

P3.F Calculate the power required to agitate a reactor 6 ft in diameter by 9 ft high with a six-bladed turbine impeller 3 ft in diameter at a speed of 300 rpm. The apparent viscosity of the contents is 100 P at these shearing conditions with a power-law index of n = 0.8. The fluid density is 60 lb/ft^3.

P3.G Following the argument about mixing and scale-up, work out an equivalent argument for mixing occurring in the deep laminar region. Show that for a constant impeller speed, the power consumption increases by a factor of 8 for a Newtonian fluid if the impeller speed is decreased by a factor of 4. Under these conditions, show that the power consumption is decreased by a factor of only 2.

P3.H Repeat P3.G for a power-law fluid with the power-law index of either n = 0.25 or n = 0.5.

P3.I   Calculate the heat transfer coefficient for a jacketed batch reactor. Assume that the coolant is water with a jacket-side heat transfer coefficient of 500 Btu/ft^2·hr·°F, the surface area for heat transfer is 150 ft^2, the coolant flow rate W is 10,000 lb/hr, the tank is full of reactive fluid with a heat capacity of 0.5 Btu/lb·°F, and the reactive fluid is to be cooled from 400 to 200°F using incoming coolant at 50°F. Calculate the exit coolant temperature at the beginning and end of the cooling operation. Determine the time required to cool the reactive fluid. If the reactor is a CSTR, how much energy can be generated within the reactor owing to the polymerization reaction if the reactor temperature remains at 300°F and the coolant temperature is constant at 50°F?

P3.J In an autorefrigerated reactor, the refluxing mixture is increasing in viscosity as the reaction takes place. A vertical refluxing condenser is used to control the reactor. The refluxing rate remains constant at 1000 lb/hr. The tube diameter is 1 in. and the viscosity ranges from 1 cP initially to 20 cP finally. The latent heat of condensation of the organic vapor is 200 Btu/lb, and the thermal conductivity and density of the condensate are 0.15 Btu/ft·hr·°F and 50 lb/ft^3, respectively. The thermal driving force through the condensing film is 200°F, and the length of the vertical condenser tubes is 10 ft. Calculate the heat transfer coefficient initially and finally for the vertical condenser. Assuming that the coolant is water entering at 50°F and exiting at 75°F, that the water-side heat transfer coefficient is 500 Btu/ft^2·hr·°F, and that the resistance of the tube wall and fouling factors can be neglected, determine the number of tubes required in the vertical condenser.

P3.K A tubular reactor 0.1 ft in diameter by 10 ft long contains a fluid moving at a Reynolds number of 100. The Prandtl number is 100, and the thermal conductivity of the fluid is 0.15 Btu/ft·hr·°F. The reactor inlet temperature is 500°F. The fluid is being cooled with condensing refrigerant at 50°F. The refrigerant heat transfer coefficient is 300 Btu/ft^2·hr·°F, and the thermal resistance of the reactor wall can be neglected. Calculate the overall heat transfer coefficient and the exit temperature of the reactor fluid. If the latent heat of condensation of the refrigerant is 100 Btu/lb, determine the amount of refrigerant required per hour for this reactor.

P3.L Given: an initial concentration $B_0 = 10$ lb/ft^3, with an exothermic heat of reaction of 350 Btu/lb. Its heat capacity is 0.5 Btu/lb·°F, and the initial temperature is 50°F. $k_p = 50$ sec for the first-order reaction. If the energy of activation of the reaction is 20,000 cal/g·mol, make a plot of the rate of heat generation in an adiabatic system as a function of temperature. Then select various values for UAT_c and draw in the programmed rate of heat removal to achieve one and three stable states.

P3.M In the discussion on the rate of heat generation vs. rate of heat removal, it was pointed out that the variation of flow rate of coolant did not dramatically affect the heat transfer coefficient. Explain the conditions under which this statement can be made, if any.

P3.N In the consideration of addition-type polymerization reactions in batch reactors, it was argued that the A → B → C type of reaction mirrored quite well the initiation-propagation-termination polymerization reaction. This allowed the stability work on simple reactions to be applied to the more complex reactions. Discuss the applicability of the simple parallel reaction A → B to any type of polymerization reaction.
     ↘ C

P3.O In the section on simple reactor stability, an argument was given for the open-loop stability analysis known as the phase trajectory. To illustrate this, consider the simple reaction A → B → C. Work out the equations for an isothermal constant-volume reaction and plot the concentrations of B/A_0 and C/A_0 against A/A_0 for various values of k_1/k_2. Outline the method for the case of an adiabatic constant-volume reaction. Discuss the approximate results of plotting A/A_0, B/A_0, and C/A_0 against temperature.

P3.P Luyben considers the stability of an autorefrigerated reactor in some detail [34]. To understand the nonlinear nature of the heat removal curve, derive it for the simple first-order reaction C → products. Then show that for low temperatures Q_R asymptotically approaches the linear curve of Foss and Denbigh. Show also that at high temperatures Q_R is independent of W. Show that there is a temperature above which Q_R is linear again. What is this temperature?

P3.Q Carry out the integration of Eq. (3.11-2) to obtain the total surface area for all growing micelles of the Smith-Ewart model, Eq. (3.11-4). Using the Smith-Ewart model and Gardon's numbers for methyl methacrylate and styrene, make plots of $\overline{X}_n$ against monomer concentration for both materials.

P3.R Make a plot of total growing polymer micelle area as a function of time using the Gardon data for styrene. Use values of $c_s a_s$ of 10^3, 10^4, and 10^5/cm as parameters. Calculate the number-average chain length at the point where all surfactant is used, at 55°C. Note: Determine k_p at 55°C first. Then calculate $\dot{v}$ from Eq. (3.11-9), given the Gardon data for styrene at 35°C and the density ratio of 0.9. Compare this value with Gardon's value of 7.5×10^{20} cc/sec at 55°C. Discuss reasons for the difference, if any.

P3.S When the Smith-Ewart model is applied to CSTRs, as discussed in Sec. 3.11, $d\dot{v}/dV = 0$ and $dR_r/dr = -2R_r/r$, where R is the radial micellular growth. Show that this is correct. Then integrate the population density curves, Eqs. (3.11-12) and (3.11-13), to obtain the total number of micelles between sizes V_1 and V_2 and between radii r_1 and r_2, as given in Eqs. (3.11-16) and (3.11-17). Then show that the radial population density passes through a maximum and the volumetric population density monotonically decreases with decreasing hold time, as shown in Figs. 3.11-1 and 3.11-2. Discuss the meaning of the maximum in the radial population density curve in terms of the relative sizes of micelles available during polymerization. Is this a real conclusion, or is it due entirely to the arithmetic?

P3.T A suspension polymerization of PVC is considered for commercial application. The polymerization rate in the monomer-rich phase is given as 30 min^{-1} and that in the polymer-rich phase is 15 min^{-1}. Assume that 20% of the reaction is occurring in the polymer-rich phase. Make a plot of C, the degree of conversion against time. Then assume that R_d can be approximated by the square root of the initiation concentration, and compare the linear form of C vs. time with the curve obtained earlier. Discuss the differences in the two curves for both short and long times.

P3.U A polymerization is taking place in a York-Schiebel column equipped with mixer-settler zones. In light of the Goldstein-Amundson analysis regarding multiple phases, discuss the phenomena controlling the polymerization of a low-molecular-weight styrene in suspension in this column. What stability problems would we expect in the stagewise column, and what type of process control should we plan for? How would we go about determining the mass and heat transfer rates for this type of reaction? What additional information do we need to know before we can begin an engineering design of this type of polymerizer?

P3.V Astarita [51] derives a diffusional model for mass transfer across a Whitman two-film model. His chemical model is simple: A → products. Using the approach outlined by Zeman and Anderson in Chap. 2, write the diffusional equations for a polymerization with monomer termination. Show that the simple model can be generalized to obtain the concentration for all growing species and that the solution can be extracted for any growing specie by series expansion of the solution to the diffusional model. Set up a similar analysis where the coefficient of diffusivity D_n is a function of the molecular weight of the growing species. Discuss the expected result.

P3.W In the catalytic reaction analysis, Schmeal and Street [53] show that the specific reaction rate at the surface of a catalyst particle is given as

$$f(a) = a \coth a - 1 \tag{3.W-1}$$

Make a plot of f(a) vs. a to show that for large a, f(a) approaches a and that for small a, f(a) approaches $a^2/3$. Determine the Thiele modulus for an organic catalyst coating on a ceramic base, where the polymerization reaction rate k is 20 min^{-1}; L, the number of sites per unit volume, is estimated to be $10^{20}/cm^2$; R, the radius, is 1 cm; and D, the diffusivity of the monomer in the polymer, is 10^{-6} cm^2/min.

P3.X The Schmeal-Street expanding catalyst or polymer core model yields a dispersity index that is a function of β, the ratio of monomer concentration to site concentration at reference, and P, the dimensionless moles of polymer per unit volume of catalyst:

$$\overline{DI} = (\beta p)^{-1} + g(a) \tag{3.W-2}$$

Discuss the meaning of g(a) in terms of the Thiele modulus. Plot $\overline{DI}$ vs. βp for both the diffusion-controlled process (a $\gg$ 1) and the reaction-controlled process (a $\ll$ 1). Would you expect $\overline{DI}$ to increase with _time_ more rapidly for the diffusion-controlled process or for the reaction-controlled process? Base your argument on the full analysis of the Schmeal-Street analysis.

P3.Y When dealing with nonisothermal catalysis, we have seen that it is possible to obtain a maximum in the catalyst efficiency for a simple reaction when the Thiele modulus is between 0.4 and 1.0. Extend this analysis to polymerization by defining a weighted Thiele modulus, a weighted heat generation term, and a weighted effectiveness factor. Is it necessary to acknowledge the presence of the dead polymer M_n in weighted factors? Should the weighted factors be based on a ratio of number-average to weight-average values? Can the Z transform method be used to define the reaction rates of the weighted species? At what point in the nonisothermal analysis can these corrections be introduced? See Carberry's paper for the analysis as applied to a simple slab geometry [56] for a clue to the analysis.

References

1. L. F. Albright, Processes for Major Addition-Type Plastics and Their Monomers, McGraw-Hill, New York, 1974, p. 111.
2. L. Platzer, Ind. Eng. Chem., 62(1):6 (1970).
3. R. L. Bates, P. L. Fondy, and J. G. Fenic, Impeller Characteristics and Power. In Mixing: Theory and Practice (V. W. Uhl and J. B. Gray, eds.), Vol. 1, Academic Press, 1966, Chap. 3.

4. S.-I. Liu and N. R. Amundson, Rubber Chem. Technol., 34:995-1133 (1961).
5. R. J. Zeman and N. R. Amundson, Chem. Eng. Sci., 20:637-644 (1965).
6. S. M. Walas, Reaction Kinetics for Chemical Engineers, McGraw-Hill, New York, 1959, p. 79.
7. S. M. Walas, Reaction Kinetics for Chemical Engineers, McGraw-Hill, New York, 1959, p. 88.
8. K. G. Denbigh, Trans. Faraday Soc., 43:648-660 (1947).
9. J. H. Duerksen and A. E. Hamielec, J. Polym. Sci., 25C:155-156 (1968).
10. K. F. O'Driscoll and R. Knorr, Macromolecules, 2:507-515 (1969).
11. S. Lynn and J. E. Huff, paper presented at 65th National Meeting, AIChE, Kinetics of Polymerization Processes Symposium, May 4-7, 1969, Paper 14b.
12. R. Citron-Cordero, R. A. Mostello, and J. A. Biesenberger, Can. J. Chem. Eng., 46:434-443 (1968).
13. P. J. Flory, Principles of Polymer Chemistry, Cornell University Press, Ithaca, N.Y., 1953, Chap. III, p. 69.
14. J. A. Biesenberger, AIChE J., 11:369-373 (1965).
15. N. H. Smith and G. A. Sather, Chem. Eng. Sci., 20:15-23 (1965).
16. D. A. Mellichamp, Chem. Eng. Sci., 24:125-139 (1969).
17. A. B. Metzner, R. H. Deehs, H. Lopez Ramos, R. E. Otto, and J. D. Tuthill, A.I.Ch.E. J., 7:3-9 (1961).
18. J. H. Rushton, Ind. Eng. Chem., 47:582-594 (1955).
19. P. H. Calderbank and M. B. Moo-Young, Trans. Inst. Chem. Eng., 37:26-30 (1959).
20. J. H. Rushton and J. Y. Oldshue, Chem. Eng. Prog., 49:161-165 and 267-273 (1953).
21. Kenics Corporation Literature, 1 Southside Road, Danvers, Mass.
22. D. Q. Kern, Process Heat Transfer, McGraw-Hill, New York, 1950, p. 289.
23. W. H. Rohsenow and J. P. Hartnett, eds., Handbook of Heat Transfer, McGraw-Hill, New York, 1973, p. 12-39.
24. J. G. Knudsen and D. L. Katz, Fluid Dynamics and Heat Transfer, McGraw-Hill, 1958, p. 368.
25. A. S. Foss, Chem. Eng. Prog. Symp. Ser., 55:47-60 (1959).
26. O. Bilous and N. R. Amundson, AIChE J., 1:513-521 (1955).
27. R. P. Goldstein and N. R. Amundson, Chem. Eng. Sci., 20:195-236 (1965).
28. Ibid., 449-476.
29. Ibid., 477-499.
30. Ibid., 501-527.
31. O. Bilous and N. R. Amundson, AIChE J., 2:117-126 (1956).
32. R. B. Warden and N. R. Amundson, Chem. Eng. Sci., 17:725-734 (1962).

33. D. P. Eckman, Automatic Process Control, Wiley, New York, 1958.
34. W. L. Luyben, AIChE J., 12:662-668 (1966).
35. R. B. Bishop, Practical Polymerization for Polystyrene, Cahners Books, Boston, 1971, Chap. 1.
36. D. C. Blackley, Emulsion Polymerization: Theory and Practice, Halsted-Wiley, New York, 1975.
37. W. V. Smith and R. H. Ewart, J. Chem. Phys., 16:592-602 (1948).
38. J. L. Gardon, paper presented at 65th National Meeting, AIChE, Kinetics of Polymerization Processes Symposium, May 4-7, 1969, Paper 14c; see the references.
39. W. D. Harkins, J. Am. Chem. Soc., 69:1428-1431 (1947).
40. A. W. DeGraff and G. W. Poehlin, paper presented at 65th National Meeting, AIChE, Kinetics of Polymerization Processes Symposium, May 4-7, 1969, Paper 14d.
41. J. O. Funderburk and J. D. Stevens, paper presented at 62nd Annual Meeting, AIChE, Nov. 16-20, 1969, Paper 57b.
42. R. A. Wessling, J. Appl. Polym. Sci., 12:309-319 (1968).
43. H. Gerrens, paper presented to Polymer Division, Am. Chem. Soc., Fall Meeting, 1966; Polym. Prepr., 7(2):699-702 (1966).
44. Rohm and Haas, Technical Bulletin No. 1042 (1965).
45. P. M. Hay, J. C. Light, L. Marker, R. W. Murray, A. T. Santonicola, O. J. Sweeting, and J. G. Wepsic, J. Appl. Polym. Sci., 5:23-30 (1961).
46. J. C. Light, L. Marker, A. T. Santonicola, and O. J. Sweeting, J. Appl. Polym. Sci., 5:31-38 (1961).
47. C. P. Evans, P. M. Hay, L. Marker, R. W. Murray, and O. J. Sweeting, J. Appl. Polym. Sci., 5:39-47 (1961).
48. S. G. Bankoff and R. N. Shreve, Ind. Eng. Chem., 42:270-278 (1953).
49. G. Talamini, J. Polym. Sci. Part A, 4(2):535-541 (1966).
50. E. Farber and M. Koral, SPE ANTEC, Tech. Pap., 13:398-404 (1967).
51. G. Astarita, Mass Transfer with Chemical Reaction, American Elsevier, New York, 1967, p. 62.
52. H. S. Carslaw and J. C. Jaeger, Conduction of Heat in Solids, 2nd ed., Oxford University Press, New York, 1959.
53. W. R. Schmeal and J. R. Street, AIChE J., 17:1188-1197 (1971).
54. J. L. Tinkler and R. L. Pigford, Chem. Eng. Sci., 15:326-331 (1961).
55. E. E. Petersen, Chemical Reaction Analysis, Prentice-Hall, Englewood Cliffs, N.J., 1967.
56. J. J. Carberry, AIChE J., 7:350-351 (1961).
57. J. C. W. Kuo and N. R. Amundson, Chem. Eng. Sci., 22:49-63 (1967).
58. J. H. Duerksen, A. E. Hamielec, and J. W. Hodgins, AIChE J., 13:1081 (1967).
59. A. E. Hamielec, K. Tebbens, and J. W. Hodgins, AIChE J., 13:1087 (1967).

60. A. W. T. Hui and A. E. Hamielec, <u>Polym. Prepr. Am. Chem. Soc. Div. Polym. Chem.</u>, <u>8</u>:353 (1967).

61. J. H. Duerksen and A. E. Hamielec, <u>J. Polym. Sci. Part C</u>, <u>25</u>:155–166 (1968).

4.

Methods of Physically, Chemically, and Mechanically
Characterizing Polymers

4.1 Introduction

The design of reactor systems in which polymerizations take place has been
explored thoroughly, and the actual mechanisms for various polymerizations
of various homo-, co-, and terpolymer systems have been outlined. Consider, now, methods of characterizing polymers.

In Chap. 1, various types of polymers were categorized in terms of
their generic behavior and/or their processing characteristics. The reader
should be aware of the vast source of physical property data for nearly every
commercial resin manufactured today. The American Society for Testing
Materials (ASTM) has many ongoing groups who continually review old testing
procedures for plastics materials and evaluate new ones. Generally, Parts
25-27 contain most of the physical test procedures for plastics materials.
Modern Plastics Encyclopedia No. 10A [1], published every October by
Modern Plastics magazine, lists many of the common ASTM test procedures

and tabulates processing conditions; mechanical, thermal, electrical, and optical properties; and some chemical, fire, and UV properties for the generic categories of most polymers and plastics. Recently, the encyclopedia has added special sections on colorants, fire retardants, stabilizers, foaming agents, and the like. Other trade journals notwithstanding, the encyclopedia is to be considered the primary source for generic data on plastics.

In addition, it should be mentioned that the encyclopedia offers brief descriptions of most resins and processes and contains an extensive list of suppliers, fabricators, machinery builders, and mold makers.

Ogorkiewicz has recently edited three collections of papers on the engineering properties of thermoplastics. The first book [2] contains an extensive collection of impact and creep property data, with polyolefins dominating the collection. The second book [3] attempts to bridge the gap between the properties of the resin prior to processing and the behavior of the subsequent article fabricated of that resin. In this regard, the authors are not entirely successful. A third edited monograph continues the extension of the work to specific processing effects, with uneven results [4]. Nevertheless, certain insights as to the extension of properties to application can be gained. This is particularly true in the area of long-term durability, where the extensive literature on creep behavior can be effectively used. Along this line, the reader should be aware of the earlier collections of papers edited by Brown [5] and Baer [6]. Throughout this chapter, we shall draw on this vast collection of insights for directions in attempting to characterize polymeric material behavior.

In each discipline, very intense efforts are made to understand a specific segment of that discipline. Crystalline thermoplastic polymers have received a disproportionate amount of attention in the polymer field. There are rational reasons for this. Amorphous materials have seemingly no order at normal use temperatures. GP styrene and polycarbonate are typical examples of materials that remain glassy at room temperature. Thus, the classical methods of microscopic examination of the materials before and after processing—optical, polaroscopic, transmission X-ray, diffraction X-ray, and other techniques such as Raman spectroscopy—cannot be used on amorphous materials. Thus, we simply do not know how amorphous molecules behave during processing. With polycrystalline materials, such as isotactic polypropylene, polyethylene terephthalate, polyacetal, and others, these techniques work and, more importantly, enable us to "see" the effects of deformation (film stretching, filament drawing, shear-induced crystallization) on the molecular structure. Thus, monographs such as those by Samuels [7], devoted almost entirely to isotactic polypropylene, and Geil [8], on single polymer crystals, have become working sources for the behavior of polycrystalline polymers.

And an astute student of polymer process engineering will note that whereas amorphous thermoplastics have received less attention than crystalline ones, the efforts to understand the behavior of thermosetting resins are even less evident. This is surprising in a way when one considers that the "plastics" industry of today owes its allegiance to the thermosets. However, the analysis problem for thermosets is considerably more difficult, owing to the reactivity of the materials and the lower economic incentive to develop the extensive data bank, as mentioned in Chap. 1.

The reader should note that the effort expended in classification and characterization of polymer types is extensive. No single chapter in a text primarily devoted to processing of polymers can do justice to the extensive literature in this field. Enormous sums of money are spent each year by hundreds of companies, universities, and private laboratories in perfecting new analytical techniques. Developments such as the electron probe scanning microscope, optical and laser birefringence, NMR spectroscopy, and various types of chromatographic analytical techniques are required so that we can better understand the molecular structures of these complex materials. The author has made extensive use of the above references and, in particular, of Miller's rational approach to the subject of characterization [9] in distilling material for this chapter.

As noted, there are many excellent texts that have been devoted to the structure of various polymeric materials and even monographs devoted to specific polymers. We shall not detail the physico-organic chemistry of polymers. Instead, it is important to realize that there are testing methods available, first in the chemist's laboratory and second in the plant, by which the progress of a reaction, the molecular-weight distribution, and the extent of branching can be evaluated. Furthermore, realize that certain changes in reactor conditions, catalysts, level of impurities, and so on will have rather dramatic effects on the structure of the final polymer as it issues from the system. But, most importantly, realize that there is an important relationship between the polymer structure <u>as a resin</u> and the mechanical properties of the plastic <u>as a final part.</u> This relationship is <u>never</u> obvious. It will depend on such intangibles as amount and type of internal processing agent, thermal stabilizer concentration, ultraviolet stabilizer concentration, fire-retardant type and concentration, external lubricant characteristics, the effect of polyblending of materials with differing molecular-weight distributions, time-temperature history of the resin in the melt state, extent and intensity of shear on the resin, sensitivity to degradation either thermally, mechanically, or in the presence of oxidative agents, the concentration and type of colorants, the effect of fillers and extenders, the extent of regrind in the processed resin, the extent of nucleating agents in a polycrystalline material, and so on. Obviously, it is not possible today to make simple extrapolations from the physical properties of a virgin polymer to the processing characteristics of the resin or to the final mechanical properties of a part, particularly where the part is undefined.

As an alternative to discussion of the global consideration of the various aspects of polymer characteristics and how they relate to the global application of each of the thousands of resin blends, we shall characterize polymers according to some of the properties that are characteristic to the processing of the resin: molecular-weight distribution, viscosity, polymer structure (characterized as "morphology"), glass-transition temperature. The physical characteristics and mechanical behavior of a plastic during processing and ultimately in a final part will depend a great deal on the molecular-weight distribution and the dispersity index of the resin. Certain properties such as toughness will depend on the extent of crystallization of the polymer and the form that crystallization takes (e.g., large, well-formed spherulites; small, fine-grain spherulites; dendritic crystal growth into predominantly amorphous material; or no spherulitic growth at all). We shall also look briefly here at some of the techniques that have been devised for determining the relationship among the "morphology" or structure of the virgin polymer, its "rheology" or ability to flow, and its mechanical characteristics under load; the techniques will be outlined here in a few illustrative cases. Extensive works such as Miller [9] and Billmeyer [10] should be examined in detail for examples and illustrations.

4.2 The Viscosity of Dilute Solutions

Many early measurements of molecular-weight distribution were made indirectly by measuring the viscosity of a polymer in dilute concentration in a suitable solvent. The viscosity of the polymer solution was then determined as a function of concentration, and information about the molecular weight was determined by extrapolating to zero concentration. This method is still used extensively to determine the concentration of a reacting polymer. Thus, knowledge of the terminology is useful (see Table 4.2-1). Staudinger [11] predicted that the reduced viscosity, η_{sp}/c, of a polymer should be

Table 4.2-1

Current term	Defined by	"Legislated" definition
Relative viscosity	η_r = flow time of solution/ flow time of solvent	Solution-to-solvent viscosity
Specific viscosity	$\eta_{sp} = \eta_r - 1$	
Reduced viscosity	η_{sp}/c; c in g/dl	Viscosity number
Inherent viscosity	$\eta_{inh} = 2.3 \log \eta_r/c$	Logarithmic viscosity number
Intrinsic viscosity	$\{\eta\} = \lim_{c \to 0} \eta_{sp}/c = \lim_{c \to 0} \eta_{inh}$	Limiting viscosity number

proportional to its molecular weight. Einstein [12], on the other hand, showed that for spheres in a solvent the viscosity should be proportional to the square root of the molecular weight. Experimentally,

$$\{\eta\} = kM^a \tag{4.2-1}$$

where k and a values, as measured, apply only to a given polymer-solvent system over a specified molecular-weight range and for a particular temperature. Plotting the log of the intrinsic viscosity against the log of the molecular weight yields k as the intercept (c = 0) and a as the slope of the line. Flory [13] shows many examples of this type of experimental data and shows that $0.5 < a < 1$ for most polymers and that the usual value of a is $0.6 < a < 0.8$. Typical values of k are $0.5 \times 10^{-4} < k < 5 \times 10^{-4}$. Billmeyer [10] warns that these relationships are probably valid only for linear polymers and that increasing the length of long branches does not appreciably increase the intrinsic viscosity, even though the molecular weight increases. The inherent and reduced viscosities for polystyrene in benzene are shown in Fig. 4.2-1 [10]. The effect of molecular weight on intrinsic viscosity for polyisobutylene in two solvents is shown in Fig. 4.2-2 [13].

Incidentally, the molecular weight measured this way is referred to as the "viscosity-average" molecular weight. Recall the general definition for a molecular-weight distribution, Eq. (2.5-5). The viscosity-average molecular weight is given as

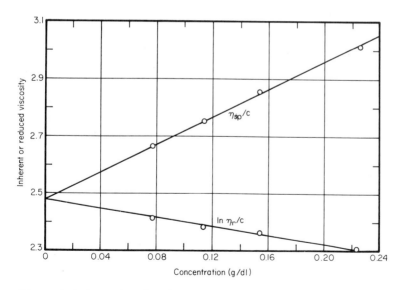

Figure 4.2-1 Reduced and inherent viscosity-concentration curves for polystyrene in benzene at 25°C [10].

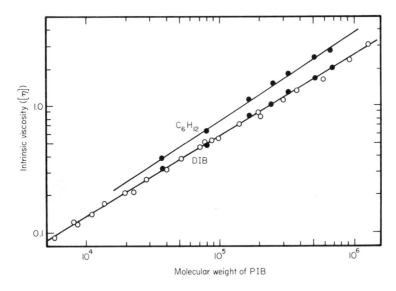

Figure 4.2-2 Intrinsic viscosity as a function of molecular weight for polyisobutylene in diisobutylene (DIB) at 20°C and in cyclohexane at 25°C [13].

$$\overline{M}_v = \left[\sum_{i=1}^{\infty} N_i M_i^{1+a} \Big/ \sum_{i=1}^{\infty} N_i M_i \right]^{1/a} \qquad (4.2\text{-}2)$$

When $a = 1$, the viscosity-average molecular weight is equal to the weight-
average molecular weight. For most polymers, a is less than unity, and
thus the viscosity-average molecular weight is about 15 to 30% below the
weight-average molecular weight but above the number-average molecular
weight. This is one of the limitations with experimental methods. The
data do not yield theoretically verifiable results.

Other methods have been devised to measure molecular-weight distributions. Gel permeation chromatography, with careful calibration, is an
excellent direct measure of the differential weight distribution in the polymer. If the weight fraction of polymer W(x) is given as a function of the
fraction of polymer x, the weight-average chain length and the number-
average chain length can be obtained as

$$\overline{X}_n = \sum_{x=1}^{\infty} W(x)\, dx \Big/ \sum_{x=1}^{\infty} [W(x)/x]\, dx \qquad (4.2\text{-}3)$$

and

$$\overline{X}_w = \sum_{x=1}^{\infty} xW(x) \, dx \Big/ \sum_{x=1}^{\infty} W(x) \, dx \tag{4.2-4}$$

Before the GPC system was perfected and calibrated for various materials and gels, the same data were obtained by vacuum distillation by fractionation of the polymer from a dilute solution. The major problems with this type of fractionation, according to Rodriguez [14], are the tendency for phase separation and the long times required to separate the very high-molecular-weight ends.

Two other techniques are osmometry and ultracentrifugation. In osmometry, a solution of polymer in a good solvent is placed on one side of a sermipermeable membrane that is not susceptible to swelling or deterioration, and a solvent containing no polymer is placed on the other side. Owing to the thermodynamic chemical potential that is established, the solvent will diffuse through the membrane, which has been specifically selected to prevent polymer diffusion until a thermodynamic equilibrium has been established. Owing to diffusion, pressure is built up on the nonpolymer side of the membrane. This pressure can be quite large and is thermodynamically relatable to the number-average molecular weight of the polymer:

$$(\pi/c)_{c=0} = RT/\overline{M}_n \tag{4.2-5}$$

π is the osmotic pressure, c is the concentration of polymer, T is the absolute temperature, R is the gas constant, and $\overline{M}_n$ is the number-average molecular weight. Flory [13] discusses various relationships between π/c and polymer concentration and concludes that a power series truncated after three terms (e.g., a_3c^2) is probably sufficient to describe the concentration profile. Thus, through manipulation of equations,

$$(\pi/c)^{1/2} = (RT/\overline{M}_n)^{1/2}(1 + a_2\overline{M}_n c/2) \tag{4.2-6}$$

where a_2 represents the proportionality constant of the second term of the power series. Thus, plotting $(\pi/c)^{1/2}$ vs. c, the polymer concentration, and extrapolating the curve to zero concentration, we obtain the value $(\pi/c)_{c=0}$ and thus the number-average molecular weight of the polymer. This type of experiment is rather tedious to run since establishment of thermodynamic equilibrium may take hours but is usually quite accurate in determining the number-average molecular weight. Typical plots for polyisobutylene in two solvents are shown in Fig. 4.2-3.

Alternatively, ultracentrifugation can be used. Here a dilute solution of polymer is rotated at a very high speed (100,000 rpm). The larger polymer molecules, because of their higher mass, tend to settle at a faster rate than the smaller molecules, thus segregating the polymer into layers of varying molecular weight. From a thermodynamic viewpoint, either the rate of settling or the equilibrium concentration profile can be used to obtain

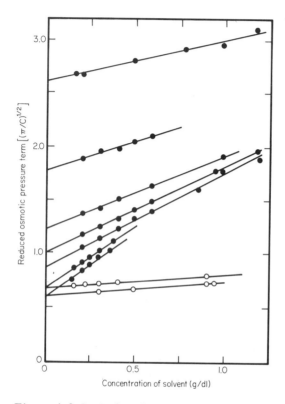

Figure 4.2-3   Reduced osmotic pressure and effect of concentration on value. Data are for polyisobutylene in cyclohexane (solid circles) and benzene (open circles). Curves represent different molecular-weight fractions [13].

molecular-weight information. To observe the concentration gradients, methods such as refractive index can be used. Extrapolation of the data to infinite dilution is required, and the difference in index of refraction between the solvent and the polymer must be large. This restricts the selection of solvent, because most good solvents, from a thermodynamic viewpoint, have refractive indices very close to the indices of the polymer. Flory [13] gives an excellent description of the data reduction and the type of information possible from the ultracentrifuge.

4.3 Crystallinity and Glass Transition

Many plastics have crystalline structures in their solid forms: TFE Teflon, HDPE, Delrin polyacetal, nylons, and so on. As a result, if these solid materials are heated slowly to a liquid state, a distinct melting temperature associated with a given material can be identified within $1°C$ or so. Delrin, for example, melts at $316 \pm 0.1°F$. To observe a melting point, a linear resistance heater on a microscope hot stage should be used. The material to be melted should be either sheet stock (normally formed by compression molding into a strip) or previously melted and solidifed plastic. Take care that the melted material is not cooled too rapidly (normally a rate of $1°/min$ is maximum), or the appearance of crystal structure will be delayed owing to the rate at which crystals grow. Crystal growth is a rate-dependent process, since the long chains of the molecules must diffuse through the other long chains to the growing site. In addition, as the polymer cools, there is a period of time at the crystallization temperature where the molten materials are not growing into crystals but are being ordered in preparation to growth. McKelvey [15] refers to this as the induction time, and the induction time is strongly dependent on temperature:

$$t_i = \alpha(T_m - T_q)^{-e} \tag{4.3-1}$$

where t_i is the induction time; T_m and T_q are the melt temperature and quench temperature, respectively; and e has values ranging from 2.5 to 9, depending on the polymer. The rate of crystallization, once the induction period has passed, has been accurately described (for isothermal crystallization) by Avrami [16] and presented best in Mandelkern's work [17]. If V_∞, V_t, and V_0 represent the specific volumes of polymer at times $t = \infty$, t, and $t = 0$, then the rate of volume change is given as [14]

$$\ln(V_\infty - V_t)/(V_\infty - V_0) = -kt^n/w_c \tag{4.3-2}$$

where w_c is the weight fraction of material crystallized, k is the kinetic rate of crystallization, t is time from the beginning of crystallization, and n is an exponent that depends on the type of nucleation and growth process.

Note that this equation can be rewritten slightly, in accordance with McKelvey, by allowing $\phi = 1 - (V_t - V_0)/(V_\infty - V_0) = 1 - \Delta V/\Delta V_\infty$ and combining w_c into the reaction rate constant,

$$\ln \phi = -Kt^n \tag{4.3-3}$$

or, since ϕ is always less than or equal to unity,

$$\ln(-\ln \phi) = \ln K + n \ln t \tag{4.3-4}$$

Thus, a plot of $-\ln \phi$ vs. t on log-log paper [or equivalently, $\ln(-\ln \phi)$ vs. $\ln t$ on linear paper] should yield the proper isothermal values of K and n. Remember that K is a rate constant and therefore highly temperature sensitive. Thus, slight changes in quench temperature would cause very large changes in the rate constant. Mandelkern showed that for linear polyethylene very little crystallinity was exhibited after 360 min at a quench temperature of $129°C$ (the melting temperature of linear polyethylene being about $136.5°C$), but when the quench temperature was lowered $5°C$ to $124°C$,

crystallization was apparent in 6 min. His data are shown in Fig. 4.3-1.
Thus, probably an Arrhenius-type equation for the reaction rate constant
dependency on temperature is warranted. Mandelkern showed that for
homogeneous nucleation of three-dimensional crystals n = 4, for two-
dimensional crystals n = 3, and for one-dimensional crystal growth n = 2.
For heterogeneous nucleation such as that occurring on solid catalysts, the
corresponding values for n are 3 to 4 for three-dimensional growth, 2 to 3
for two-dimensional growth, and 1 to 2 for one-dimensional growth. Thus,
this information can be used in reverse, by determining the exponent on the
time for crystallization and then using this to determine the type and structure of crystal growth. An example of a shift in crystallization mechanism
is shown in Fig. 4.3-2 for polyethylene terephthalate [18]. This rate-
dependent equation will be used in the discussion of the cooling rates of
injection-molded highly crystalline materials in a later chapter.

Recently, Ostapchenko has found for melt-polymerized polyethylene
terephthalate that highly oriented bundles of crystalline polymer exist unmelted at temperatures 30 to 40°C above the bulk melting point for PET.
These crystalline materials can make up 1% of the weight of the resin and
can nucleate crystal growth, thus dramatically reducing the induction and
half-times for PET crystallization [19].

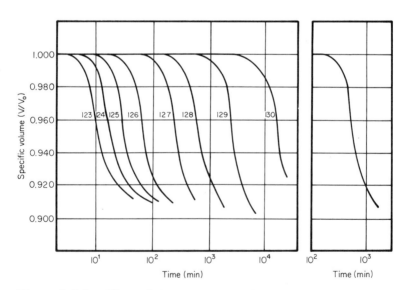

Figure 4.3-1   Effect of time on the specific volume of crystallizing polyethylene. Note that data on right represent superposition of data at all
temperatures to that at 128°C [14].

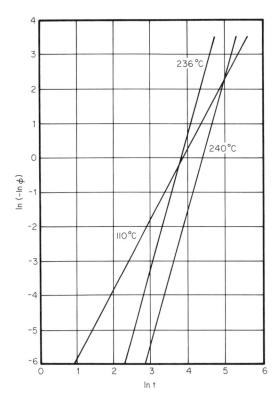

Figure 4.3-2   Effect of quench temperature on the crystallization mechanism of polyethylene terephthalate. Note that time t is in minutes [15].

        Much of the polymer literature in recent years has concentrated on
the forms and structure of the crystal as it grows in the isothermal medium.
This area of study is referred to as "morphology" and is most important in
determining interrelationships between the mechanical properties of a
polymer and the rate and type of crystal that is formed during cooling.
Much effort has been oriented toward understanding the mechanism of crystal growth. While this knowledge helps explain fatigue in highly crystalline
materials that have been carefully prepared in the laboratory, most processing of crystalline plastics takes place in uncontrolled conditions under
high shear rates and very rapid quenching conditions. These conditions are
not conducive to growing the perfect crystal. Much of the crystal growth
is sporadic and dendritic into the amorphous regions, which eventually
crystallize (if the material is held at a high enough temperature for a long
enough period of time) into crystals that may be entirely different in structure from those that were formed earlier. The more extreme the processing

conditions become and the more nonuniform the conditions are (as in the case of heat transfer), the more disorganized the crystal growth becomes. Thus, the interrelationships between crystal structure and mechanical properties of the polymer are no longer reliable. In addition, note that the discussion has centered only on the virgin pure polymer. The addition of noncrystalline materials can provide nucleating sites for crystal growth (as in the case of metallic particles added either for color or mechanical strength) or kinetic rate inhibiting factors [as would be the case if the materials added were completely amorphous (noncrystalline), such as lubricating agents, and must be diffused through and pushed out of the way by the growing crystal chains]. And remember that higher-molecular-weight materials usually have a higher melting temperature than the lower-molecular-weight materials. This means that selective crystal growth according to molecular weight occurs as the temperature is lowered on a broad molecular-weight distribution resin. Thus, dendritic growth can move into amorphous regions, even though cooling rates are sufficiently slow to ensure "isothermal cooling." Thus, there are many reasons for knowing about the degree of crystallinity and the rate of crystal growth, but the processing engineer must face the obvious problems when attempting to process a highly crystalline material in, say, an injection molding machine. Do not rely solely upon the interrelationships between the mechanical properties of a polymer and its crystalline texture. Use the information advisedly to determine a trend for the effect, for example, of slowing the injection rate or increasing the extent of cooling of a material. This may seem strong, but many polymer processing manuals overemphasize the crystallinity of a pure polymer under isothermal conditions and underemphasize (or neglect completely) the effect of the actual processing conditions on the degree of crystallinity.

     Do not forget that many polymers have no significant crystallinity at all. PVC, for example, shows no X-ray diffraction pattern, but certain cooling conditions indicate that crystallinity is present. It has been estimated that PVC has about 20% crystallinity. LDPE crystallization depends on the extent and character of its branching. The lower the branching, the more linear the molecule, and thus the more crystalline it becomes. This is reflected, for example, in the sharpening of a melting point and an increasing melting point. Other materials, such as the atactic polystyrenes, polycarbonates, PMMA, and the styrene-based co- and terpolymers, exhibit no crystallinity at all. Therefore, these materials simply soften as heated, in much the same way as glass softens when heated. A melting temperature cannot be defined for a noncrystalline or amorphous polymer. It simply does not solidify. One characteristic temperature that is important in amorphous polymers and to a lesser extent in crystalline polymers is the glass-transition temperature. The glass-transition temperature is best defined as the temperature below which a polymer changes from a rubbery substance to a glassy substance. This transition temperature is of extreme importance to the processor and the designer, because it can be used to determine the suitability of a polymer for specific purposes. For example, given in Table 4.3-1 are the glass-transition temperatures of some common plastics. Note that some of these materials are crystalline (Nylon 66, polypropylene), while others are amorphous (butadiene, styrene). Note, for example, that polybutadiene is rubber-like at room temperature since it would be above its glass-transition temperature, whereas polystyrene is glass-like because it is below its glass-transition temperature. By forming copolymers of polystyrene and polybutadiene, yielding impact sytrene, the glass-transition temperature can be depressed below room temperature to obtain a tough polymer. For crystalline materials, the glass-transition temperature is also important, for crystallization can occur only between T_g and T_m. It is fairly well established that the ratio of melt temperature to glass-transition temperature for most crystalline materials is between 1.4 and 2.0, where the temperatures are in °K. Thus, the T_g for homopolyacetal is about -60°F, and the melting point is about 316°F (e.g., the ratio is about 1.9).

One characteristic about the polymer above and below the glass-transition temperature is the specific volume change. For any "phase transition," either a melting point or a glass-transition temperature, the specific

Table 4.3-1

Glass-Transition Temperatures

Polymer	T_g (°F)
Linear	-188
TFE Teflon	-143
Polybutadiene	-121
LDPE polyethylene	-90
Polyacetal (formaldehyde)	-40 to -76
Polyvinylidene chloride	+2
Atactic polypropylene	7
Polypropylene	41
Polyvinyl acetate	82-90
Nylon 66	117
PVC	158-176
Acrylonitrile	176-219
Polystyrene	181-201

volume dependence with temperature is measured using, for example, a dilatometer. Normally the specific volume dependence on temperature is nearly linear with temperature away from "phase transitions" but exhibits abrupt slope changes in the vicinity of the change. It is thought by thermodynamicists that melting is a first-order transition and therefore is accompanied by an abrupt step in volume with temperature. For small molecules, such as water, this specific volume change is on the order of 100 times at the boiling point and 10% at the melting point. This discontinuity is due to energy absorption required to change the phase at a given temperature. Since the heat capacity is defined as the rate of change of enthalpy or thermodynamic energy with temperature, at a phase change such as melting, a discontinuity in the specific heat curve (e.g., an infinite specific heat) should occur at the melting temperature. This is shown, for example, for Nylon 66 in Fig. 4.3-3. There is a second-order phase transition called the

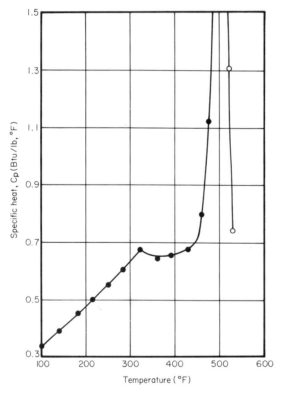

Figure 4.3-3   Characteristic specific heat curve for crystalline Nylon 66, showing second-order "Ehrenfest transition" and melting point.

"Ehrenfest transition," characterized by a change in the slope of the various thermodynamic variables with temperature at a given temperature. This second-order transition is shown in Fig. 4.3-4 for specific heat for a butadiene-styrene copolymer at about 240°K and in Fig. 4.3-5 for a homologous series of methacrylates. In Fig. 4.3-5, the temperatures at which changes in specific volume slopes take place are seen to be depressed with increasing molecular weight of the monomer. Thus, it seems that the glass-transition temperature, defined as the point of specific volume slope change with temperature, is a second-order or Ehrenfest transition. The exact value of the glass-transition temperature depends on the accuracy of measurement of, say, specific volume or specific heat. With rapid tests, the error can be on the order of 2 to 5°C. Normally, the accuracy of determining the glass-transition temperature is about the same as that for determining the melting temperature. The glass-transition temperature T_{gc} of a copolymer (such as styrene-butadiene) can be determined from either of the following two equations:

$$1/T_{gc} = W_1/T_{g1} + W_2/T_{g2} \qquad (4.3-5)$$

or

$$T_{gc} = W_1 T_{g1} + W_2 T_{g2} \qquad (4.3-6)$$

The accuracy in using these very empirical equations is quite poor. For example, for styrene in methyl acrylate, the linear equation seems to work well. T_{g1} (for methyl acrylate) = 278°K; T_{g2} (for styrene) = 363°K. For

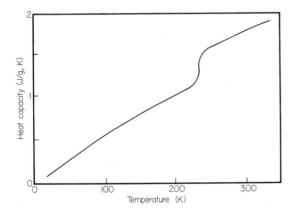

Figure 4.3-4   Effect of temperature on heat capacity of styrene-butadiene copolymer [9].

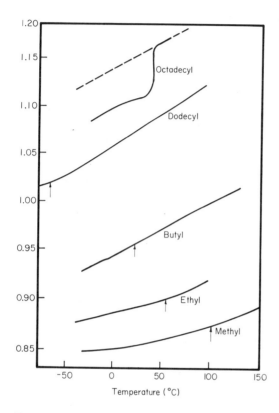

Figure 4.3-5   Effect of temperature on specific volume of homologous series of alkyl methacrylates; arrow denotes T_g [9].

$W_2 = 0.40$ (40 wt% polystyrene in methyl acrylate), the measured $T_{gc} = 315°K$. From Eq. (4.3-5), $T_{gc} = 307°K$, and from Eq. (4.3-6), $T_{gc} = 312°K$. For acenaphthylene in styrene, on the other hand, the linear calculation is in error. T_{g2} for acenaphthylene is $570°K$. For $W_2 = 0.35$ (35 wt% acenaphthylene), the experimental $T_{gc} = 415°K$. The corresponding calculated values of T_{gc} are $415°K$ and $435°K$, respectively. These curves are shown in Figs. 4.3-6 and 4.3-7. Note also that the glass-transition temperature can be affected by low-molecular-weight materials. The glass-transition temperature increases dramatically with increasing molecular weight up to values on the order of $\overline{M}_n$'s of 10^5 or so, before asymptotically approaching the tabulated values. This is shown in Fig. 4.3-8 for poly-methyl methacrylate and polystyrene (both amorphous materials). Therefore, if a very low-molecular-weight material is blended into a polymer for the purposes of improving its flow characteristics in the molten state, it

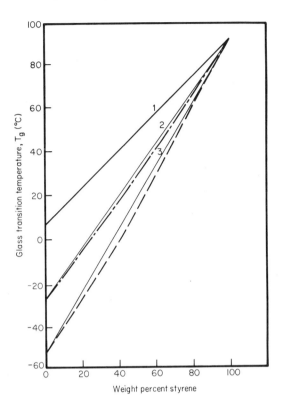

Figure 4.3-6   Effect of amount of styrene on glass-transition temperature
for three copolymers. 1: Methyl methacrylate; 2: ethyl acrylate; 3: n-butyl
acrylate. Dashed lines represent experimental data; solid lines represent
linear law [9].

may affect its glass-transition temperature. Also, its ability to form a
crystalline material may be dramatically altered, as mentioned earlier.
A dramatic example of the latter effect is the random copolymerization of
50 wt% linear polyethylene-50 wt% isotactic polypropylene; the resulting
polymer is a rubbery, amorphous polymer with a glass-transition temperature of -76° F.

Newer techniques used to identify various polymers, polyblends, co-
and terpolymers, additives, and the like include infrared spectroscopy,
nuclear magnetic resonance, differential thermal analysis, differential
gravimetric analysis, scanning electron microscopy, and others. Consider
only two of these techniques: IR spectroscopy [20] and differential thermal
analysis [21]. For transparent materials, such as the acrylates, the crystal styrenes or the polycarbonates, transmission spectroscopy is used.

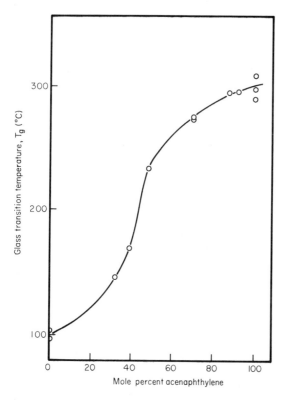

Figure 4.3-7 Effect of acenaphthylene concentration on glass-transition temperature of styrene-acenaphthylene copolymers; T_g, glass-transition temperature, °C [9].

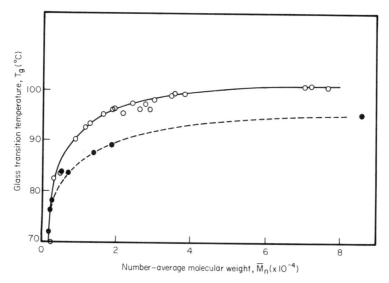

Figure 4.3-8 Effect of molecular weight on glass-transition temperature of two homopolymers; open circles, polymethyl methacrylate; closed circles, polystyrene [9].

For opaque materials, such as the impact styrenes, polyethylenes, and others, a technique known as attenuated total reflectance, or ATR, is used [9]. Here the infrared radiation is incident on the boundary between the sample and the optical material of the instrument. Part of the radiation penetrates beneath the reflecting surface of the sample. Internal reflection occurs, and at appropriate ratios of the refractive indices of both materials, the reflected radiation returns to the reflecting surface. This returned radiation is modified by absorption in the sample of the radiation at certain wavelengths, depending on the chemical structure of the resin. Thus, the thickness of the sample is of no significance, and the spectral intensity is independent of sample thickness. For transmission spectroscopic methods, the attenuation of radiation is a function of the thickness of the material (according to Lambert's law) and also the molecular structure of the plastic, and thus the thickness of the material must be known rather precisely [22]. Infrared spectra of materials are frequently used as evidence of various molecular modifications to the material. In Fig. 4.3-9, the effect of changing the chemical composition of styrene-based materials is shown. In the top figure, butadiene-styrene shows characteristic absorption peaks at 6.8, 13.2, and 14.4 μm. In the butadiene-acrylonitrile specimen, the

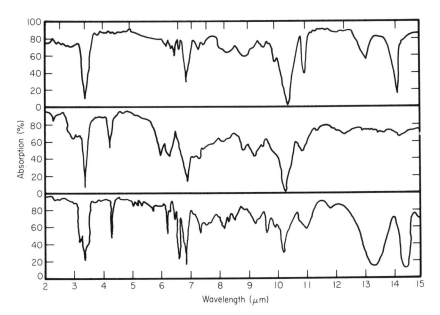

Figure 4.3-9 Infrared spectra for (top to bottom) butadiene-styrene copolymer, butadiene-acrylonitrile copolymer, and butadiene-styrene-acrylonitrile terpolymer [20].

characteristic peaks are at 4.4 and 7 μm. In acrylonitrile-butadiene-styrene (ABS), the presence of all five characteristic peaks is seen: 4.4, 6.8, 7.0, 13.2, and 14.4 μm.

Not only is this technique useful for determining the chemical structure of the polymer (for from the spectra, the concentrations of each of the polymers in the co- and terpolymer or polyblend can be determined), but it can be used to fingerprint various changes in chemical composition of plastics in quality control of incoming resins. The technique can be used to establish a differential spectrometer of sorts. First the infrared fingerprint of a material that works particularly well in the processing equipment and gives useful final parts is obtained. When a competitive resin or a new shipment of resin is to be tested, a fingerprint of this resin is compared against the standard in a differential fashion. Thus, any differences in chemical composition (or even molecular configuration) can be detected, as shown in Fig. 4.3-10 for types of isoprenes vs. natural rubber.

Miller [9] discusses several other uses for IR spectroscopy, including the determination of crystallinity of a given polymer at an isothermal temperature. These data are valuable in determining the effects of quench

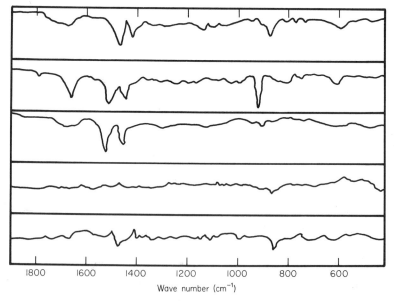

Figure 4.3-10   Comparative infrared absorption spectra for several rubbers. Top to bottom: cis-1,4-polyisoprene, 3,4-polyisoprene, cyclopolyisoprene, differential spectra of natural rubber-synthetic cis-1,4-polyisoprene, and differential spectra of natural rubber-cyclopolyisoprene [20].

temperature or amorphous additives on the extent of crystallinity of a given material. IR spectroscopy is also rather easy to run, and thus a small unit could be considered an important adjunct to a processing laboratory. And IR spectrographs yield information about the thermoform processing characteristics of transparent plastics, as will be shown.

As mentioned above, the transition conditions of polymers can be obtained by measuring either volume changes or changes in heat capacity as functions of temperature. Differential thermal analysis (DTA) and its more accurate cousin differential gravimetric analysis (DGA) are extensively used to determine these various transition temperatures. Basically, the systems depend on heating of a sample and a known reference material at the same temperature rate. Any absorption of energy in the polymer sample owing to changes in its structure (e.g., glass transition) causes the temperature of the polymer sample to lag behind that of the reference material. This lag is displayed using a differential thermocouple, as shown in Fig. 4.3-11. When reactions take place, as is the case with cross-linking resins or resins that thermally degrade, energy is given off by the polymer sample, and the temperature of the sample will be in excess of that of the reference material. As shown in Fig. 4.3-12, changes in enthalpy, heat capacity, and temperature for first- and second-order transitions are easily detected, even at heating rates of 2 to 10°C/min. The thermograms for the three polyethylenes (Fig. 4.3-13), show that there are sharp heat absorption peaks at 138°C for Phillips Marlex 50 and at 132°C for Koppers Super Dylan, both high-density, linear polyethylenes with high degrees of

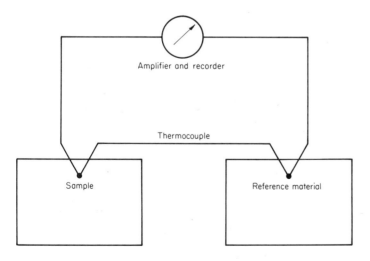

Figure 4.3-11   Basic differential thermal analysis circuit [21].

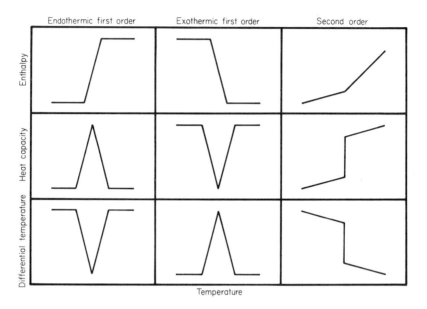

Figure 4.3-12   Schematic changes in enthalpy, heat capacity, and temperature for characteristic endothermic first-order, exothermic first-order, and second-order transitions [21].

crystallinity, but a rather poorly defined melting point for Union Carbide DYNH low-density polyethylene with a peak at about 112°C. The data also indicate that the Phillips material had a higher molecular weight (and therefore greater crystallinity) than the Koppers material, since it exhibited a higher transition temperature. That this is true is shown in Fig. 4.3-14, where the degree of crystallinity for all three materials is displayed as a function of temperature. These data are obtained using IR spectroscopy. The DTAs of nylons show rather well-defined melting points (about 227°C for Nylon 6 and 262°C for Nylon 66) (Fig. 4.3-15), and the crystalline form of polystyrene (isotactic polystyrene) shows a melting point at about 233°C. But the amorphous atactic polystyrene (crystal styrene) shows no dramatic absorption peaks, although a slight change in slope is noted at about 90°C, indicating perhaps that a second-order transition is occurring. Of course, this second-order transition is the glass transition for crystal styrene (as well as isotactic polystyrene) (Fig. 4.3-16).

The utility of the DTA method extends beyond seeking the transition conditions, as shown in the copolymerization of Nylon 610 with Nylon 66 (Fig. 4.3-17). The homopolymers exhibit very strong melting point peaks at 262°C for Nylon 66 and at 230°C for Nylon 610. As more and more Nylon 66 is copolymerized into the Nylon 610, the peak first occurs at lower

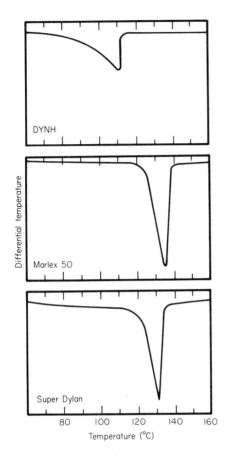

Figure 4.3-13 Thermograms of several types of polyethylenes [21].

temperatures and then begins to broaden and flatten until at 40% Nylon 66–60% Nylon 610 there seems to be no first-order transition. What remains is a second-order transition at about 50°C (the glass-transition temperature for nearly all nylons) and a weak second-order transition at about 250°C. Ultimately, as 100% Nylon 66 is achieved, this peak sharpens and increases in temperature until a true melting temperature occurs. But at the 40:60 point, noncrystalline nylon is predicted. It can also be shown, by other means, that this nylon should be optically transparent. And the very dramatic solid-phase transition of polybutylene (polybutene-1) is also observable using DTA. This is a very inexpensive polymer with some excellent properties that has not seen significant commercial application until recently owing to a long-term solid-phase transition. In Fig. 4.3-18, immediately after quench, the material exhibits a melting point and thus

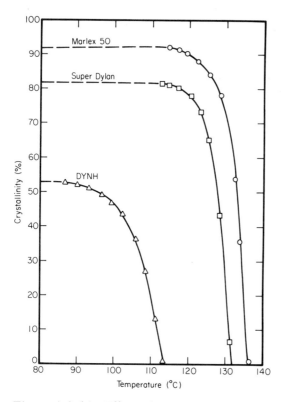

Figure 4.3-14   Effect of temperature on percent crystallinity for three polyethylenes, shown in Fig. 4.3-13 [21].

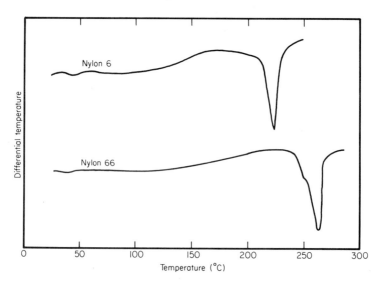

Figure 4.3-15   Typical thermograms of Nylon 6 and Nylon 66 [21].

230

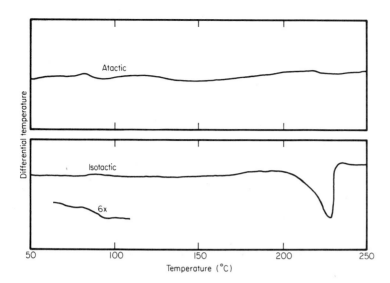

Figure 4.3-16   Thermograms of atactic (amorphous) and isotactic (crystalline) polystyrene. Note the weak second-order effect near 90°C for crystalline polystyrene [21].

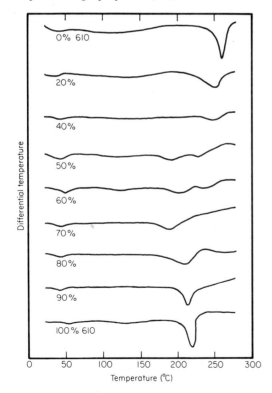

Figure 4.3-17   Thermograms of copolymers of hexamethylenediamine-adipic acid (Nylon 66) and hexamethylenediamine-sebacic acid (Nylon 610). Note the disappearance of the characteristic melting point at 50:50 [21].

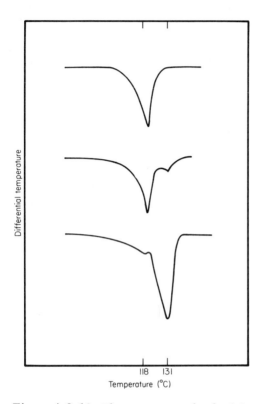

Figure 4.3-18 Thermograms of poly-1-butene showing time-temperature crystalline transition. Top to bottom: quenched from melt, annealed at room temperature for 16 hr, and annealed at room temperature for 64 hr [21].

a crystalline structure at 118°C. After 16 hr, the beginning of a second crystalline structure, having a melt temperature of 131°C, is seen. After nearly 3 days of evaluation, nearly 100% conversion of the polymer from one crystalline structure to the second has occurred. Unfortunately, this solid-phase transition is accompanied by significant volumetric changes, and thus the parts molded to specifications on Monday, say, are completely off-spec by Thursday. Significant effort has been made to force the polymer directly into the second phase without much success until now. Witco has recently found suitable modifiers, processing conditions, and molecular-weight distributions that will allow extrusion of polybutylene hot water pipe in which the material retains its final crystalline state indefinitely [23].

As mentioned earlier, the DTA can be used to determine various types of thermal degradation and cross-linking. In Fig. 4.3-19, the effect

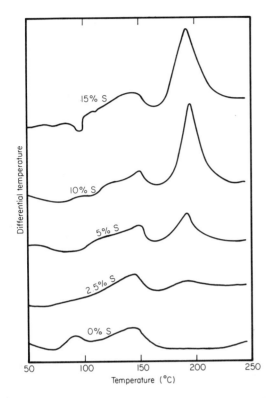

Figure 4.3-19   Thermograms of vulcanized polybutadiene showing the effect of various levels of sulfur on structure near 190°C [21].

of adding sulfur to polybutadiene for the purpose of cross-linking the polymer into a three-dimensional rubber-like structure is shown. The increasing sulfur concentration means more cross-linking is taking place, and thus an increase in the amount of energy produced at the reaction temperature is expected. Thus, at temperatures of about 190°C, vulcanization of polybutadiene rubber takes place in situ. On the other hand, heating polyethylene in various atmospheres at elevated temperatures can lead to thermal or oxidative degradation. This is shown in Fig. 4.3-20 for high-density polyethylene in the presence of an inert (N_2) and an oxidative (O_2) atmosphere. The peaks at around 130°C represent the melting points of the material. Oxidative degradation is somewhat endothermic and seems to occur in the region of about 330°C in the presence of O_2. There is no peak there for the N_2 atmosphere. At temperatures in excess of 500°C, however, thermal degradation is taking place, and the type of atmosphere makes little difference in the extent of degradation above this temperature. Note that thermal

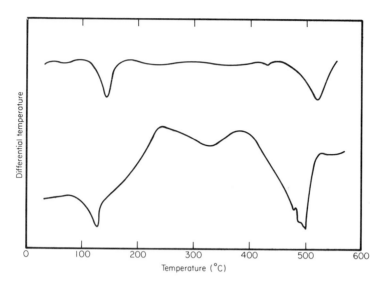

Figure 4.3-20   High-temperature thermograms of polyethylene showing the effect of oxidative degradation in air above 450°C. Top curve, N$_2$; bottom curve, air [21].

degradation is endothermic; that is, it requires energy. Furthermore, although the presence of antioxidants can minimize the amount of oxidative degradation that takes place in an O$_2$ environment, no amount of antioxidant can prevent thermal degradation. Under certain processing conditions, both types of degradation can take place (rotational molding or thermoforming, for example). The mechanical consequence of thermal degradation is discoloration and loss of mechanical strength, and thus regrind, for example, might have lower physical properties than the virgin resin.

Thermomechanical analysis (TMA) is another fingerprint method that is becoming popular [24]. Here the specimen is subjected to a constant stress while the temperature is increased slowly. The strain on the specimen is measured, and the resulting strain-temperature curve gives an indication as to how the material will perform under different loadings at different temperatures. It is not directly applicable as a creep-measuring device, however; these devices will be considered later. The strain on the specimen is normally measured by its rate of elongation. TMA curves for various types of polymers are shown in Fig. 4.3-21. Amorphous polymers exhibit significant increases in rate of elongation near the glass-transition temperature with this rate of elongation continuing to increase with temperature until break occurs. Crystalline polymers, on the other hand, show an increase in rate of elongation at a glass-transition temperature and then no further appreciable increase in rate of elongation until the melt

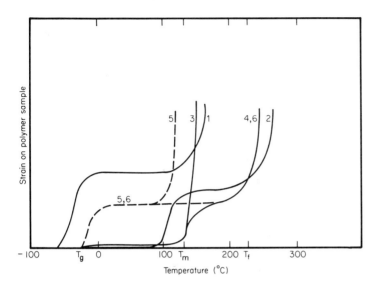

Figure 4.3-21   Schematic representation of thermomechanical responses of various types of polymers. Curves 1 and 2: amorphous polymers; 3 and 4: crystalline polymers; 5 and 6: supercooled crystalline polymers [24].

temperature is reached. Note that for some materials, exceeding the melt temperature does not produce immediate rate of elongation to infinity, as would be the case if the fluid had rather low viscosity. As shown in Fig. 4.3-22 for Nylon 66 that has been supercooled by quenching, as the melt temperature is exceeded, a finite amount of elongation rate is noted, but infinite rate of elongation (or fluidity) does not occur until the temperature reaches the flow temperature, more than 50°C above the melt temperature. This can be due to the extremely high viscosity of the fluid or possibly some structure that remains in the material above the melt temperature. Recall the earlier discussion of PET. It must be remembered that melting, like crystallization, is a kinetic process and that rapid heating of the sample may not completely randomize the crystalline structure until the temperature is many degrees above the melt temperature. Some of the other important uses for TMA analysis are shown in Figs. 4.3-23 through 4.3-27. In Fig. 4.3-23, the TMA shows the effect of cross-linking on the extent of elongation for a typical polyether urethane. Cross-linking here is nearly complete at about 400°C, and thus the rate of elongation drops to zero. At temperatures in excess of 600°C, thermal degradation of the material again makes it fluid, and elongation rates increase very rapidly. It was mentioned earlier that DTA could be used to determine the effect of antioxidants on a polymer. In Fig. 4.3-24 a thermal stabilizer (or antioxidant) is seen to increase the usefulness of the amorphous polymer polystyrene. For

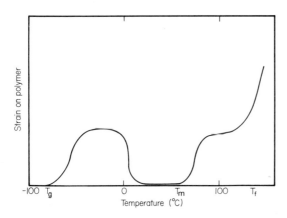

Figure 4.3-22 Typical thermomechanical strain behavior of a supercooled polymer that rapidly crystallizes during heating [24].

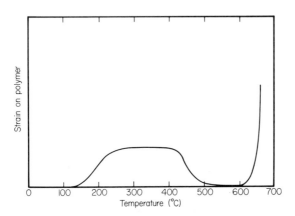

Figure 4.3-23 Typical thermomechanical strain behavior of a cross-linking polymer (above 450°C) during heating [24].

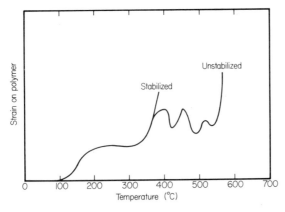

Figure 4.3-24 Typical thermomechanical strain behavior of stabilized and unstabilized polymer during heating; typical TMA behavior of poly-propylene [24].

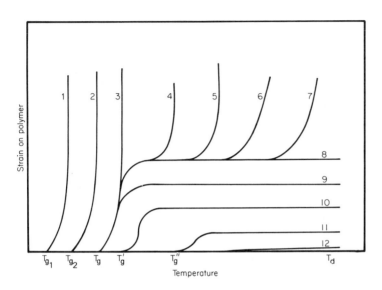

Figure 4.3-25   Typical thermomechanical strain behavior of a homolog polymer. Curves 1-7: increasing molecular weight; 8-12; increasing degree of cross-linking [24].

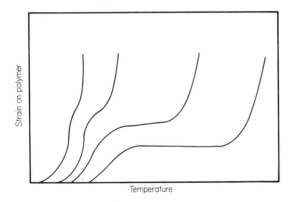

Figure 4.3-26   Typical thermomechanical strain behavior of a polymer with increasing levels of plasticizer, with curves moving to left [24].

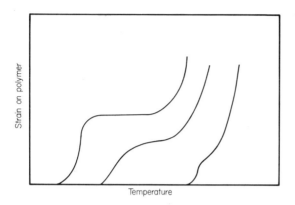

Figure 4.3-27 Typical thermomechanical strain behavior of polymer in which plasticizer flexibilizes at high temperature; increase in plasticizer levels, with curves moving to left [24].

materials such as acrylics, the rates of elongation can be dramatically changed by increasing the molecular weight, as shown in Fig. 4.3-25 for curves 1-7. If cross-link density is increased, the material is toughened at essentially the same molecular weight, and thus the rate of elongation decreases with increasing cross-link density. Additives and their effects on the rates of elongation of a polymer can also be identified using TMA, as shown in Fig. 4.3-26 for a soft plasticizer in PVC and in Fig. 4.3-27 for a rigid plasticizer. The key to the use of the TMA is the application of a constant load, a highly accurately controlled thermal environment, and a rather slow increase in temperature with time. For thick samples, the heat must be transferred by conduction completely through the sample. Thus, if the specimen is heated too rapidly, uneven thermal gradients will yield incorrect strain measurements and poor reproduction in the data.

4.4 Extensions of Molecular Considerations to Processing Properties

Moving away from the details concerning the molecular structure and toward the physical properties needed to determine the processing characteristics of the plastic, we find fewer theoretical bases upon which to draw solid conclusions. For example, extensive work has been done on the relationship of morphology of a pure polymer and the microscopic stress-strain relationships that the polymer sees when subjected to various controlled deformations or temperatures. For example, Collier [25] and others have shown that for a highly crystalline material the size and orientation of the spherulites are strongly dependent on the method of processing of the material, and that, in turn, affects dramatically the stress-strain behavior of the final material. As shown in Fig. 4.4-1, highly crystalline isotactic polypropylene was heated into its melt region at 180°C for 5 sec and quenched and then for 60 sec and quenched. For the short annealing time, very small spherulites were formed, and thus the stress-strain curve shows a ductile yield and extensive elongation to break. For the long anneal time, the spherulites were large, and thus the material was strong and brittle, exhibiting little yield before breaking brittlely. The operating temperature of the specimen can be quite influential in determining whether the specimen will fail in a ductile manner or a brittle manner. Billmeyer [10] shows that several generalizations can be made about the effect of spherulite size on mechanical properties. These are presented in Table 4.4-1 (showing the effect of decreased spherulite size). As shown in Figs. 4.4-2 and 4.4-3, increasing the temperature for both an amorphous polymer and a crystalline polymer causes a shift in the performance of the polymer from a strong brittle material with a relatively high breaking stress and low elongation to a ductile material with a low breaking stress, a low yield, and a long elongation to break.

Now we shall try to relate the molecular performance of the polymer with, for example, its ability to flow, or its rheological characteristics. This subject has been considered by Middleman [26], Combs and coworkers [27], White and Tokita [28], and many others. The primary means of measuring viscosity in a commercial laboratory is the melt indexer, ASTM D-1238. We shall have more to say about this means of viscosity measurement in the next chapter. At this point we consider it only as a measure of the ability of a fluid to flow at a given temperature and at a given shear rate (although the latter is not quite correct). The melt index is defined in

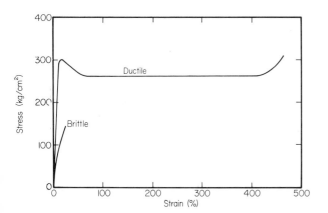

Figure 4.4-1 Typical stress–strain curves for ductile (top curve) and brittle (bottom curve) polymers [24].

Table 4.4-1

Effect of Decreased Spherulite Size [10]

Mechanical property	Polypropylene	Nylon 66	Polyoxymethylene
Impact strength	Increased	Increased[a]	Increased
Yield stress	Decreased	Increased	Decreased
Ultimate elongation	Decreased	Decreased	Increased

[a]Only for thin films.

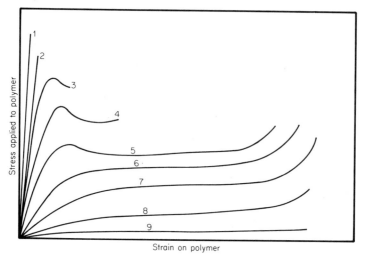

Figure 4.4-2   Typical stress-strain diagrams for an amorphous polymer with temperature as a parameter; temperature increases with increasing curve number [24].

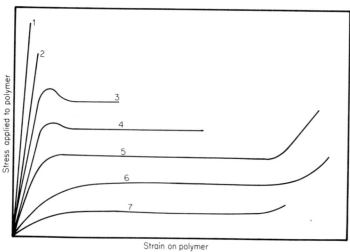

Figure 4.4-3   Typical stress-strain behavior of a crystalline polymer showing the effect of temperature; temperature increases with increasing curve number [24].

terms of the amount of polymer that flows through a die in a given period of time. The higher the flow rate and thus the lower the viscosity, the higher the melt index. Therefore, for an "easy-flow" plastic, the material should have a high melt index, low crystallinity, nonlinearity (in the case of polyethylene), and a low molecular weight. In the last case, it has been found that for linear polymers with very sharp molecular-weight distributions there is a good relationship between the shear stress and the weight-average molecular weight:

$$\tau = K\bar{M}_w^{a} \tag{4.4-1}$$

where $a = 3.8 \pm 0.4$ and K is a constant representative of the polymer. As the molecular-weight distribution increases, the accuracy of the above equation decreases such that for dispersity indices greater than 10 or so, a can be as low as 2. Furthermore, long-chain branching can lead to significant deviations from this rule. Kontos [29] states that the following molecular or structure properties can influence the viscosity of rubbery polymers in particular: absolute molecular weight, molecular-weight distribution, branching, stereoisomerism, monomer distribution and ratio, type of unsaturation if any, chain stiffness and polarity, and glass-transition temperature T_g. The absolute effects of each of these features are not specified. In Combs et al., a similar analysis is made for polyolefins [27]. They state that for proper processibility it is necessary to specify the desired properties as functions of operation. For example, a high melt index (low viscosity) is desired for injection molding, but a low melt index (high viscosity) is desired for blow molding. For four major processes, the desired physical properties are given in Table 4.4-2. The effect of

Table 4.4-2

Properties Desired for Certain Polymer Processes [27]

Operation	Desired property
Film extrusion	Low melt index, high melt strength, high critical shear rate
Extrusion coating	Low energy of activation for viscous flow, high melt index, high melt strength, high critical shear rate
Injection molding	High melt index, high degree of non-Newtonian-ness, high energy of activation for viscous flow
Blow molding	Low melt index, low energy of activation for viscous flow, high melt strength

of molecular-weight distribution on the processing characteristics of polypropylene is shown in Table 4.4-3, where increasing the dispersity index
results in a decrease in the melt index and thus an increase in the viscosity
even though the solution viscosity is relatively constant. Note also the
dramatic decrease in the critical shear rate. The critical shear rate is
defined as that shear rate where the extrudate issuing from a cylindrical
die becomes rough and irregular. This critical shear rate is important to
the extruders of film, filament, and pipe, because exceeding this shear rate
in the extruder die will result in undesirable surface properties. It is not
considered to be critical in the injection mold process. For polyethylenes,
an increase in the average molecular weight (found by solution viscosity
methods) results in a significant decrease in the melt index. This is shown
in Fig. 4.4-4. Branching, another molecular structural feature, results in
dramatic changes in the melt index. Increasing the length of the branches
causes a dramatic increase in the melt index (and thus a significant decrease
in the process viscosity of the melt). Therefore, LDPE is expected to have
a much lower processing viscosity than HDPE at the same molecular weight.
This is the case. A tabulation of the effect of other molecular features on
the flow properties of polyethylenes is given in Table 4.4-4.

Many others have studied other aspects of the interaction between
polymer structure and flow. As an additional example, consider the work
of Blyler [30] on various types of polyethylenes. The properties are shown
in Table 4.4-5. (Note: Molecular-weight determinations were carried out
with GPC.) As shown in Fig. 4.4-5, Blyler found that for linear PEs, increasing molecular weight shifts the flow curves to lower shear rates. However, the curves are not parallel, nor do they follow power-law behavior
(see Chap. 5). The effect of branching is seen when linear polyethylene
(designated "2") is compared with branched polyethylene. The flow behavior
of branched polyethylene is shown in Fig. 4.4-6 to be continuous with

Table 4.4-3

Effect of Molecular Weight Distribution on Processing Characteristics
of Polypropylene [27]

$\overline{M}_w/\overline{M}_n$	$\{\eta\}$	Melt index, 230°C	Melt strength (in.)	Critical shear rate (sec^{-1})	E_a, Energy of activation for viscous flow
4.6	1.48	9.58	2.9	76	9.9
12.1	1.68	5.86	4.4	43	11.3
15.8	1.59	7.60	3.7	58	14.7
25.6	1.75	5.35	4.7	27	13.0

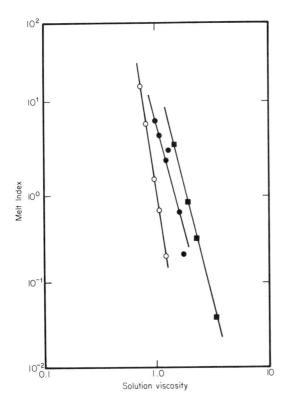

Figure 4.4-4   Relationship between solution viscosity and melt index for three olefinic classes. Open circles, LDPEs; solid circles, HDPEs; solid squares, PPs [27].

increasing shear stress, even though the extrudate is very rough. For LPE-1, however, there is a definite discontinuity in the stress-strain curve that is very shear stress dependent. Furthermore, the branched material energy of activation for viscous flow E_a is much more shear stress dependent than that for the LPE-1. To demonstrate the effect of short- and long-chain branching on the flow characteristics, Blyler randomly copolymerized acrylic acid along the chain. The pendant carboxyl groups on the acid groups hydrogen-bond to one another, and since hydrogen bonding is a weak form of cross-linking (sometimes referred to as quasi-cross-linking), this dramatically affects the flow characteristics. This is shown in Fig. 4.4-7. Blyler then esterified 85% of the carboxyl groups, thus reducing the extent of hydrogen bonding, and the result is a dramatic decrease in the viscosity of the resin. He thus concludes that hydrogen bonding is similar in mechanism to entanglements in long-chain branching that occur in polyethylene.

   For condensation polymers, Combs and coworkers [31] found that for amorphous materials, at a given inherent viscosity and processing

Table 4.4-4

Effect of Other Molecular Parameters on Flow Characteristics of Polypropylene [27]

Rheological parameters	MW Inc.	$\overline{M}_w/\overline{M}_n$ Inc.	Short-chain branch		Long-chain branch	Temp. Inc.	Shear stress Inc.
			Number	Length			
Melt flow	-	-Low shear +High shear	-	+	-	+	+
Melt recovery	0	+	0	0	+	+High shear	+
Critical shear stress	-	-	0	-	0	+	x
Power-law index	+	+	-	+	-	-	x
E_a, energy of activation	+(?)	0	+	-	0	0	-

Table 4.4-5

Properties of Polymers Investigated [30]

Polymer	Type	Supplier	Melt index[a]	$\overline{M}_w$[b]	$\overline{M}_n$[b]	CH_3[c]/100°C
LPE-1	Linear polyethylene	Phillips Petroleum Co.	0.1	—	—	—
LPE-2	Linear polyethylene	Phillips Petroleum Co.	0.2	150,000	4,820	—
LPE-3	Linear polyethylene	Phillips Petroleum Co.	0.9	86,200	3,020	—
LPE-4	Linear polyethylene	Phillips Petroleum Co.	5.0	57,500	1,500	—
LPE-5	Linear polyethylene	Phillips Petroleum Co.	—	40,000	20,000[e]	—
BPE	Branched polyethylene	duPont Co.	3.0	—	—	1.7
EAA	Ethylene–acrylic acid copolymer, 5.3 mol % acrylic acid[d]	Union Carbide Corp.	4.8	—	—	1.5–2.0

[a] ASTM D-1238–65T.
[b] Determined by gel permeation chromatography.
[c] Determined by infrared measurements.
[d] Determined by titration.
[e] Estimated.

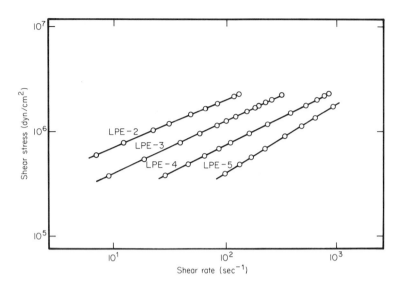

Figure 4.4-5   Shear stress-shear rate curves for four linear polyethylenes (of different molecular weights) at 160°C; data on polymers given in Table 4.4-5 [30].

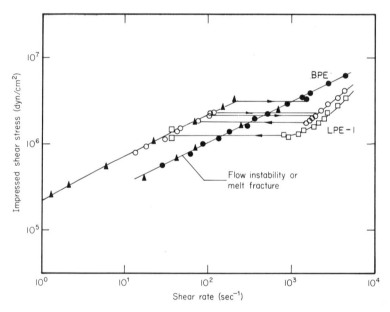

Figure 4.4-6   Shear stress-shear rate curves for one linear polyethylene (LPE-1) and the branched polyethylene (BPE) at 160°C in a capillary rheometer; data on polymers given in Table 4.4-5. Closed, open circles, capillary radius = 0.2307 cm; closed triangles, capillary radius = 0.04821 cm; squares, capillary radius = 0.01470 cm [30].

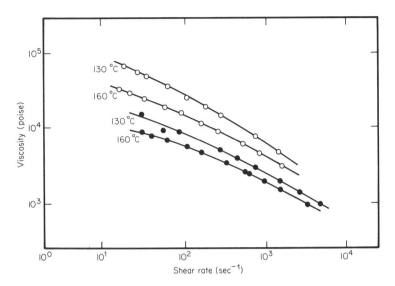

Figure 4.4-7   Effect of hydrogen bonds on linear polyethylene. Open circles, randomly polymerized acrylic acid on linear polyethylene; closed circles, esterification of 85% of the acrylic acid (thus removing the hydrogen bonding effect) [30].

temperature, increasing the glass-transition temperature decreased the melt flow and thus the ease of processing. They also found that this holds for crystalline materials, and thus good processing conditions and high heat distortion temperatures are obtainable from crystalline polymers with low glass-transition temperatures. Note, however, that low glass-transition temperatures usually indicate low melting points.

4.5 Extensions of Molecular Features to Long-Term Properties

Very frequently, we are faced with mechanical or physical failure of plastics products. In an effort to trace the source of the problem, information about the molecular features of the virgin plastic is needed. Before considering such relationships between the molecular features and the long-term mechanical properties, recall that many chemical and physical changes can occur to the resin on its way from the reactor to the final product. For example, to achieve specific processing conditions, additive, fillers, lubricants, stabilizers, and polyblends may have been added. Furthermore, to reduce scrap losses, regrind may have been added. Colorants, fire retardants, and so on may affect the resin's molecular features as well as its processability. And, most importantly, the actual processing of the resin, including excessive and uneven temperature history, high-shear history, and frozen strains, may contribute more to the failure of a plastic part than changes in the chemical composition of the plastic or its resin base. Also be aware that long-term properties differ significantly from short-term properties. For example, to obtain a tough melamine-formaldehyde product, the reaction is usually stopped before it reaches completion. At high cross-link densities, melamine is quite brittle. Therefore, while the product, initially, may be tough, subjection of that product to repeated high temperatures will, in essence, continue the reaction until the product is brittle. This is a well-documented phenomenon with melamine dishware, for example, that has endured extensive high-temperature dishwasher service.

Therefore, even though much work has been directed toward attempting to relate mechanical failure with morphological characteristics of the polymer, only a few insights are possible. For example, a weak relationship among the molecular-weight distribution, the extent of branching, and the environmental stress-cracking time is shown in Tables 4.5-1 and 4.5-2 for polyethylene. Therefore, small changes in the flow characteristics of polyethylene can yield very dramatic changes in one specific property, the environmental stress-cracking resistance. Note in Fig. 4.5-1 the effect of stress as a function of hours to failure for low-density polyethylene (0.918) in Igepal at 60°C. Again, melt index is an inverse measure of the molecular weight. This is cross-plotted to show the extent of failure at various stresses, with the melt index again a parameter, as shown in Fig. 4.5-2. These are rather short time failure tests (less than 500 hr). Longtime failures under constant load are primarily categorized as creep rupture failures. For examples, in Fig. 4.5-3, longtime failure curves are given for polyethylene in water at 60°C. Note the abnormal brittle failures of highly crystalline polyethylene in the 1000- to 10,000-hr usage category. Also note that beyond 100 hr the tensile stress to failure for a 0.93-density PE with a melt index of 10 is lower than that for a melt index of 1.2. Before that time, the opposite is true. Thus, selection of PE for its tensile strength based on a short time test would be incorrect if longtime exposure were important. Note also that changing the environment will also change the material's ability to withstand longtime loads. For example, in Fig. 4.5-4, ABS has slightly better flexural stress-cracking resistance than high-impact styrene in air. (Characteristic S-N curves for plastics are shown in Fig. 4.5-5.) However, the dramatic loss of strength of high-impact polystyrene in vegetable oil negates its use in this type of environment. Therefore, using this material in an oil environment after having determined its S-N curve in, say, air or nitrogen environment is courting product failure. Table 4.5-3 gives another example of environmental effects, ultraviolet absorption by PVC over a 5-year period of time (10^6 min). Note that although the tensile strength at break and the modulus of elasticity are relatively unchanged by weathering, a significant amount of

Table 4.5-1

Change in Stress-Crack Resistance with Melt Index [32]

Melt index[a] (g/10 min)	Environmental stress-cracking time (hr)[b]			
	Series A[c]	Series B[c]	Series C[d]	Series D[d]
0.08	>500	>500		
0.15				>500
0.19				>500
0.20				>500
0.21			>500	300
0.22	>500			>500
0.23		>500		>500
0.24				425
0.25			>500	
0.26				140
0.27			>500	
0.28			6	
0.29			25	
0.33			3	
0.35	>500			
0.40				70
0.46			1.5	
0.60	>500	5		
0.83		3		
1.3	>500			
1.4		2		
3.2		0.5		
7.8	<0.5			
15		<0.5		
16	<0.5			

[a]ASTM D-1238, Condition E.
[b]ASTM D-1693.
[c]20% failure point (F_{20}) for samples as molded.
[d]50% failure point (F_{50}) for samples conditioned 7 days at 70°C before testing.

Table 4.5-2

Change in Stress-Crack Resistance with Molecular Weight [32]

Melt index[a] (g/10 min)	Intrinsic viscosity[b]	Environmental stress-cracking time (F_{20}) (hr)[c]
0.08	1.35	>500
0.22	1.17	>500
0.35	1.04	>500
0.60	1.10	>500
1.30	0.980	>500
7.8	0.940	<0.5
16	0.895	<0.5

[a] ASTM D-1238, Condition E.
[b] In xylene at 85°C.
[c] ASTM D-1693.

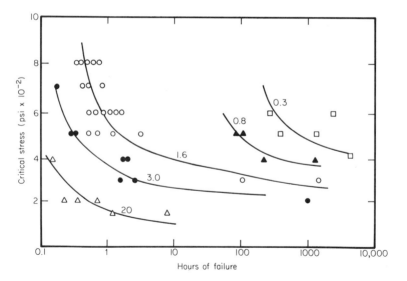

Figure 4.5-1   Effect of environmental stress-cracking agent (Igepal, 60°C) on low-density polyethylenes (0.918 sp. gr.); failure stress as a function of time to failure, with melt index as parameter [32].

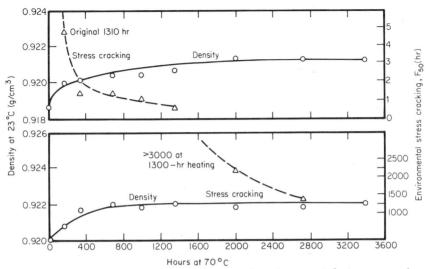

Figure 4.5-2   Changes in LDPE density and environmental stress-crack-ing hours to failure for resins of two different melt indices. Top curves, MI = 0.67; bottom curves, MI = 0.27 [32].

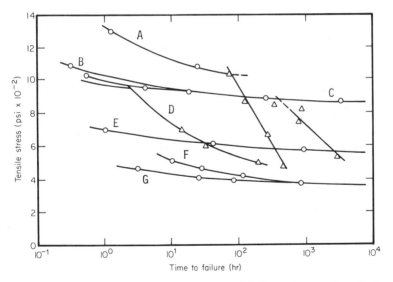

Figure 4.5-3   Effect of tensile stress of failure time and mode for various types of polyethylenes in water at 60°C. Curve A: 0.96 density, 0.6 MI; B: 0.95 density, 0.1 MI; C: 0.95 density, 0.03 MI; D: 0.93 density, 10 MI; E: 0.93 density, 1.2 MI; F: 0.918 density, 0.3 MI; G: 0.918 density, 2.0 MI. Open circles, ductile failure; open triangles, brittle failure [33].

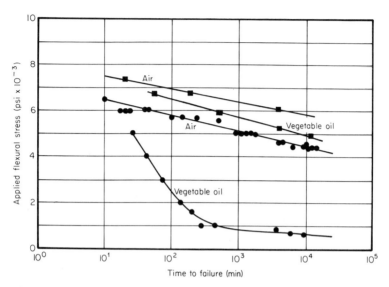

Figure 4.5-4 Flexural creep rupture owing to environmental stress cracking for two resins (HIPS and ABS) and two environments (air and vegetable oil). Solid squares, ABS; solid circles, HIPS [33].

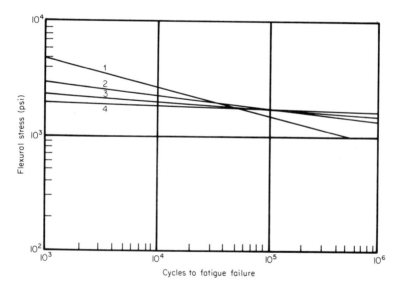

Figure 4.5-5 Cyclical fatigue failure curves for typical polymers at room temperature. Curve 1: polystyrene; 2: ABS; 3: rigid PVC; 4: PP [33].

Table 4.5-3

Effect of Outdoor Weathering at Ottawa on Mechanical Properties of PVC Sheeting[a] [33]

Original color	Tensile strength at break (psi × 10⁻³)		Elongation at break (%)		Modulus of elasticity (psi × 10⁻⁵)	
	Control	Weathered	Control	Weathered	Control	Weathered
White	5.2	5.0	230	50	1.5	1.5
Green	6.0	5.3	260	60	1.6	1.7
Light green	5.4	5.1	230	36	1.5	1.6
Orange	6.3	5.2	250	52	1.5	1.6
Coral	5.4	5.0	250	42	1.5	1.6
Yellow	5.0	5.0	190	57	1.5	1.6

[a]The white sample was opaque; all others were initially translucent. Samples were weathered for 5 years; epoxidized soybean oil was used as a plasticizer.

embrittlement in the PVC is noted by the dramatic decrease in the elongation at break (from about 230% to an average of 45%).

Another physical property that can be misused during material evaluation is tensile strength. Recognize that tensile strength, flexural strength,
stiffness, and modulus are all temperature sensitive. As shown in Fig.
4.5-6, the stiffness for amorphous atactic polystyrene is the same as that
for isotactic or crystalline polystyrene at room temperature. However,
above the glass-transition temperature of about 90°C, amorphous polystyrene has no strength, whereas crystalline polystyrene retains much of its
strength until a temperature a few degrees below the melting point. Even
more important are the changes in relative strengths of the materials with
temperature, as shown in Fig. 4.5-7. Note that at 100°F filled phenolic
and steel have nearly the same tensile strength that they had at room temperature (73°F) but that high-impact styrene has lost about 20% of its
strength. At 180°F, styrene encounters the glass-transition temperature,
and the strength drops precipitously. The measurement of the tensile yield
strength can also influence the actual value, as shown in Fig. 4.5-8. Note
that at strain rates on the order of 0.1 cm/sec and slightly higher there is
little change in the tensile yield strength of polyethylene terephthalate. At
rates on the order of 100 cm/sec a very rapid rise in the apparent strength
of the material to 40% more than that at the lower speeds is seen. Likewise,
if the material is subjected to very low strain rates, on the order of 10^{-4}
cm/sec, the material has only about 80% of the strength it exhibits at 0.1
cm/sec. Most plastics exhibit similar curves. Thus, care must be taken
to prevent selection of a plastic physical property, such as tensile strength,

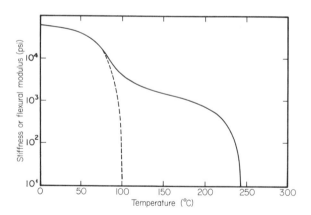

Figure 4.5-6   Effect of temperature on the stiffness or flexural modulus
of an amorphous polystyrene (dashed line) and a crystalline polystyrene
(solid line) [33].

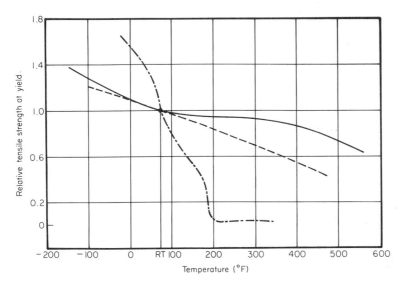

Figure 4.5-7   Effect of temperature on the relative tensile yield strength of several materials; all curves normalized to room temperature. Solid line, mild steel (78,000 psi at RT); dashed line, woodflour-filled phenolic (6500 psi at RT); dash-dot line, HIPS (3100 psi at RT) [33].

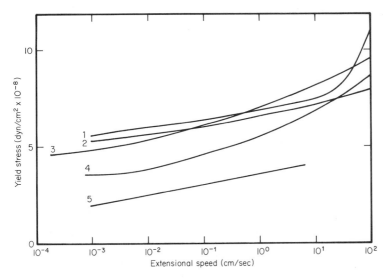

Figure 4.5-8   Effect of extensional speed on tensile yield stress at 20°C for several polymers. Curve 1: amorphous polyethylene terephthalate; 2: polyacetal copolymer; 3: PVC; 4: Nylon 66 (ambient humidity); 5: poly-4-methyl-1-pentene [34].

based on a value at one set of conditions with the assumption that it can be extrapolated to a second set of conditions. This, of course, is one of the reasons that so much effort is spent in trying to relate these end properties to molecular phenomena occurring in the material.

4.6 Miscellaneous Mechanical Testing Procedures

Each year the American Society of Testing Materials reviews and evaluates many pragmatic tests on materials. Some of these tests are concerned with the flow properties as methods of characterizing a resin, and thus the "melt indexer" utilizes the ASTM D-1238 testing procedure. Others are concerned with the flammability characteristics of the material in a building application, and thus the Underwriter's Laboratory Tunnel Test carried as ASTM number, E-84. It was mentioned earlier that information about the tensile strength of the specimen could give information about its toughness or brittleness at a given operating temperature.

4.6.1 Tensile Tests

The characteristic methods for tensile tests are described either by ASTM D-638 for large tensile test bars or ASTM D-1708 for "microtensile test bars" or bars that are 1 1/2 in. in length. The standard tensile testing under short-term loading can distinguish five principal modes of polymer behavior: uniform extension, cold drawing, necking rupture, brittle fracture, and necking rupture of a second kind [34]. At extension speeds of about 0.5 in./min, the recommended maximum speed for elongation according to the ASTM standards, the uniform extension mode is best exemplified by a vulcanized rubber (either synthetic or natural). The characteristic curve shows that the stress is a rapidly increasing function of the draw ratio (the length divided by the original length). According to Treloar [35], it can be shown from thermodynamics that increasing orientation of the macromolecules in the direction of draw requires increasing extension energy. The maximum extension ratio (at which the extensional energy becomes infinite) is proportional to the square root of the cross-link density of the rubber. It should be noted that highly oriented rubber shows a very high degree of crystallinity in the direction of draw, and thus this form of uniform hardening is referred to as orientation hardening, strain hardening, or reinforcement.

Cold drawing is characteristic of a crystalline material such as isotactic polypropylene, high-density polyethylene, Nylon 66, or Teflon TFE. This type of load-extension curve is characterized by a knee occurring at a rather low extension and by extension at low loading continuing to very high levels before fracture occurs. This is called cold drawing of a crystalline material [34]. The elastic phase of the material exists until the "yield point" is reached, and then the material behaves plastically, much like the load-elongation curves of many metals such as aluminum. Physically, cold drawing is characterized by a very rapid necking of the polymer to a rather constant cross-sectional area. This necking region is extended as the plastic is elongated until the entire draw region has, in essence, been completely necked. Then failure occurs. What apparently happens is that molecular orientation continues in a region of maximum local stress until a certain level of orientation is achieved. At this point, additional orientation requires considerably more energy than that required to orient the unoriented material around it, and thus the diameter of the neck stabilizes, and thinning in the adjacent parts proceeds. Once the entire specimen is thinned, the neck fails. The knee in the curve is apparently that minimum amount of energy required to orient the first molecules. See Fig. 4.6-1.

PVC is a material that exhibits necking rupture [34]. Here it is probably due more to the development of microvoids around the superagglomerates that make up the bulk of PVC than to the orientation of molecules. Characteristically, the development of voids in materials causes the material to appear white in that area, and this is the case with PVC. In brittle fracture, which occurs with general-purpose polystyrene below its glass-transition temperature, there is a linear, Hookean relationship between the stress and rate of elongation until fracture. The fracture patterns are quite similar to the fracture patterns of any glassy substance, including inorganic glasses.

A second type of necking fracture is characterized by the slow drawing of polymethyl methacrylate. Here the neck continues to draw down but at a much slower rate than the material surrounding the neck. The tensile test bar is usually wasp-waisted at fracture. Rapid extension of PMMA will lead to a brittle fracture, however.

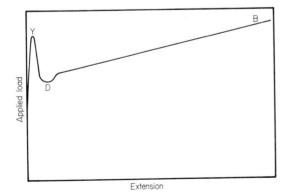

Figure 4.6-1   Cold drawing of isotactic polypropylene at 20°C°. Point Y, yield stress; D, drawing stress; B, fracture [34].

4.6.2 Impact Tests

Very frequently, the impact strength of a given part is needed. Izod and Charpy are two types of tests that have been used extensively for measuring this strength. Consider the variants in these tests as they pertain to energy absorption around the fracture area and as this pertains to the physical characteristics of the plastic material. Brighton [36] points out that if the strength of plastics were determined according to thermodynamic theory, breaking strengths would be hundreds of times greater than those measured by conventional testing procedures. The reasons for the lower values are twofold: (1) The tests tend to concentrate the energy at a local point, and (2) imperfections and microvoids in the material significantly weaken the material in an impact mode. Griffith [37] has shown that impact energies are concentrated at the microvoids and cracks, and thus

$$T = (2E\sigma/\pi C)^{1/2}$$

(4.6-1)

where E is the elastic modulus, σ is the specific surface energy of the plastic, T is the breaking strength, and C is the length of the defect. This is a linear Hookean equation. Unfortunately, it is difficult to measure such values as the length of the defect and the specific surface energy of the plastic. Therefore, a very simple theory has been developed by Buchdahl [38] which seems to work for several plastics, including impact styrene, polyethylenes, nylons, and others:

$$T = E/30$$

(4.6-2)

where T is the tensile strength and E is the elastic modulus. This also works for some metals, such as low-carbon steel and some aluminums.

In the Izod test (ASTM D-256), a notched bar is held in a vertical position in a cantilever position beneath a potential-energy pendulum. This test is identical to the standard metallurgical tests, but the interpretation of the results is considerably more difficult. The chief reason for the uncertainty in this test is the fabrication of the notch. It has been determined that to break a notched Izod bar, the breaking energy is used to initiate fracture, to propagate the fracture through the specimen, to deform the specimen plastically, to propel the fractured section away from the break, to vibrate the specimen and the apparatus elastically, and to overcome friction in the mechanism. The machining of the notch in the specimen must be done properly and at the right machining temperature. This is particularly true with highly notch-sensitive materials such as polycarbonate, polyacetals, and most glass-filled plastics. For example, Riddell [39] shows that materials such as PMMA and melamine exhibit some increase in impact strength (measured in ft·lb_m) with increasing notch radius (Fig. 4.6-2). Melamine, for example, increases in strength from about 0.5 ft·lb_m at a notch radius of 5×10^{-3} in. to about 1.0 ft·lb_m at a notch radius of 100×10^{-3} in. Nylon, on the other hand, increases from 0.5 to

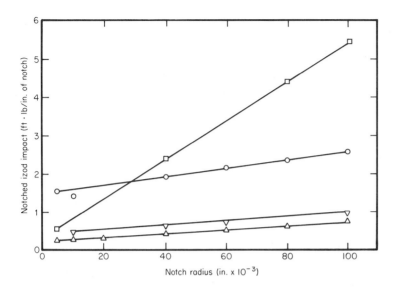

Figure 4.6-2   Effect of notch radius on Izod impact values for several polymers. Open squares, dry Nylon 66; open circles, 30% glass-reinforced Nylon 66 (dry); diamonds, melamine; triangles, PMMA [39].

5.5 ft·lb$_m$ over the same notch radius range. Polyacetals show an even more dramatic increase in impact strength over this radius range. The effect of notching on crystalline materials is seen when the notch is molded in, rather than machined in. The impact strengths of polyacetals increase by a factor of 10 when the notch is molded in during injection molding rather than machined in, because the crystalline structure at the surface of the injection-molded piece is much stronger (and of entirely different morphology, being transcrystalline in nature) than that in the interior of the specimen. When the outer material is machined away, the spherulitic crystalline material in the center has less strength and more microvoids (for polyacetals, not for all crystalline materials) and therefore fails at a much lower impact level. To avoid this problem of notching specimens, many resin suppliers will report "unnotched" Izod data. This type of data is valid only when comparing materials of homologous species and cannot be used to determine performances of the materials under notched Izod testing procedures. Furthermore, it must be realized that the material must be sufficiently rigid to break under impact and not simply bend. Flexible PVC, for example, has no notched Izod value, simply because it bends under impact.

　　The Charpy test is also a metallurgical test that has been adapted to testing of plastics. Here the notched specimen is held in a horizontal position against two supports, and impact occurs on the side of the specimen opposite the notch. In nearly all cases, the Charpy data are as suspect to notch sensitivity, platform vibration, and so on as the Izod test (see Fig. 4.6-3). Even more important, the thickness of the sample strongly influences the test results. The thickness of the sample also influences other types of impact tests such as the falling–ball and falling–dart tests, both of which are ASTM tests. Here a sheet of material is held loosely in a form, and a steel ball (normally 1/2 or 1 lb, depending on the test) or a steel dart is dropped from increasing heights until the sheet fails. Rather sophisticated routines are established to determine the statistical impact strength. Even with these internal checks on data, errors as large as 100 to 200% can be expected. For example, for a PVC [39], the impact strength using a Charpy test is on the order of 0.20 ft·lb$_m$. For a sample 0.10 in. thick,

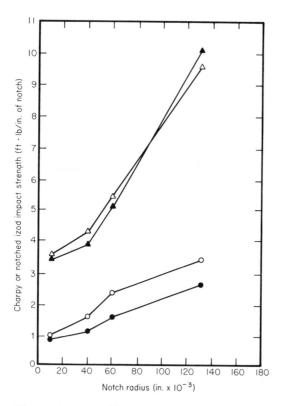

Figure 4.6-3   Effect of notch radius on Charpy and notched Izod impact values. Solid points, Izod; open points, Charpy; triangles, ABS; circles, HIPS [40].

the falling-ball impact strength is on the same order of magnitude. For a sample 0.040 in. thick it is about 8 ft·lb$_m$, for a sample 0.060 in. thick it is about 13 ft·lb$_m$, and for a sample 0.100 in. thick it is about 28 ft·lb$_m$ (Fig. 4.6-4). Furthermore, impact strength is strongly dependent on temperature, as shown in Fig. 4.6-5. For the same specimen 0.060 in. thick, the falling-ball impact strength at temperatures in excess of 20°C is nearly constant at about 13 ft·lb$_m$. There is a dramatic decrease in strength, however, as the temperature decreases from 20 to 0°C. At 0°C, the impact strength is on the order of 0.8 ft·lb$_m$. Below 0°C, the material impact strength declines very slowly with temperature. The characteristic difference in the type of fracture is evident also. Most materials with very low falling-ball impact strengths exhibit brittle failure, while those with high falling-ball impact strengths fail in a ductile mode. The advocates of the falling-ball method of testing state that the test closely approximates more actual situations than does Charpy or Izod and that it is also much more sensitive to failures caused by anisotropy in the material. Nevertheless, the impact testing remains an art, since the size of the weight (its impact dimension), the actual weight of the test ball, and the fixturing of the specimen are all variables in the test procedure. For example, if a specimen is to withstand 5-ft·lb$_m$ impact, it would be possible to find many materials that could pass a test where a 5-lb$_m$ weight is dropped from 1 ft. Many,

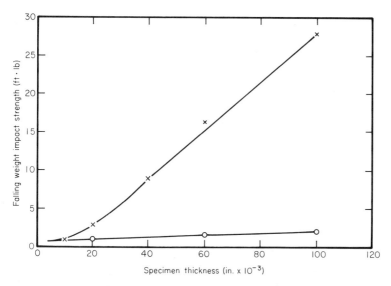

Figure 4.6-4   Effect of specimen thickness of falling-weight impact strength for two polymers. Xs, rigid PVC; open circles, impact polystyrene. Tests at 23°C [40].

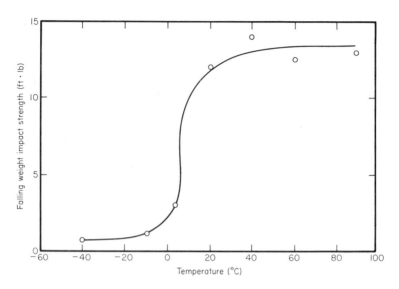

Figure 4.6-5   Effect of temperature on the falling-weight impact strength
of rigid PVC (0.060 in. thick). Note the brittle–ductile transition around
15°C [40].

however, would fail when a 1-lb$_m$ weight is dropped from 5 ft or a 0.5-lb$_m$
weight is dropped from 10 ft.

    Recently, Rheometrics, Inc. has developed a driven impact machine
that has strain gages mounted in the driving head. In this way, the time-
of-flight amount of energy required to penetrate a plastic sample can be
obtained directly. According to Rheometrics [41], the device is much less
dependent on the method of fixturing the plastic specimen. Furthermore,
the device can be used to test impact on molded plastic parts, thereby
minimizing the amount of extrapolation presently required from test results
to final product performance. However, the requisite correlations between
present test data on resins in plaque or bar form as measured with standard
ASTM methods discussed above and those obtained on the Rheometrics device
have not been completed.

    Brighton [36] compares the impact performance (at 23°C) of various
materials with various tests. This comparison is given in Table 4.6-1.
It can therefore be generally argued that increasing the Charpy value has a
corresponding effect on the falling-ball impact strength but that comparing
materials performance using different tests is illogical.

    In many instances not only impact strength but also penetration
strength is important. Thus, a falling dart is used, for example, to test
film and sheet for penetration or puncture strength. Again, this is a relative test in which relationships back to the physical characteristics of the

Table 4.6-1

Different Impact Test Results [36]

Material	Charpy impact strength (ft·lb/in. of notch)	Falling-ball impact strength (ft·lb$_m$)
PVC	0.20	14.7
High-impact BVC	0.54	18.7
Styrene-acrylonitrile (SAN)	1.38	12.6
Cellulose nitrate	1.61	8.0
Cellulose acetate	0.36	6.3
Polystyrene (GP)	0.07	0.21
Impact polystyrene	0.25	1.76
Polymethyl methacrylate	0.11	0.32

plastic are quite tenuous owing to the complex mechanisms that occur in the testing procedure. As shown in Fig. 4.6-6 [39], the rate of impact (measured in velocity of the projectile at the point of impact in inches per minute) is not easily related to the type of polymer. The acetal and the polypropylene are both crystalline materials, whereas the polystyrene is amorphous. The molecular explanation of this type of experimental data is quite lacking.

4.7 Mechanical Testing for Product Liability

Note that many of these tests are one-time testing procedures. Many applications require multiple impact testing. For example, one would expect that a single falling-dart test on, say, PVC tile would not test the durability of a PVC floor tile subject to thousands of impacts from shoe heels. And one would not expect the impact testing data on polystyrene, regardless of the type of test of the extent of notching, to be applicable to drop tests on products ranging from portable radio cases to toilet seats. In fact, Heater and Lacy [42] have found that there is no apparent relationship between the material's resistance to a single impact and to its resistance to repeated impact blows. While certainly this area needs extensive work, it does appear that the rate of loading is more important to the resistance to repeated impacts than the amount of loading. Thus, if a product is being designed to be exposed to repeated impacts but the rate of loading is very

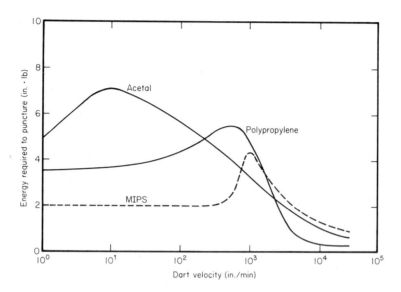

Figure 4.6-6   Effect of impact velocity on the puncture strength of several polymers (sample thickness at 0.010 in.) [39].

low, it may withstand many more impacts before failure than an equivalent product with a very high rate of loading. As a result, care must be taken to ensure that in the laboratory testing the rate of loading, in a very pragmatic sense, is very nearly the same as that experienced in actual use. This is perhaps why material that passes a 5-ft·lb_m test using a 5-lb ball dropped from 1 ft will fail when a 1-lb ball is dropped from 5 ft.

Product testing is so varied and "pragmatic" that only certain cases can be discussed. These can be neither generalized nor reduced to formulas for calculating the product performance from known materials properties. There are many pragmatic tests that must be carried out to determine if a product can withstand the mechanical environment for its expected life. The reader is warned, however, that simply testing the product in an "actual-use" situation is not enough. For example, for household appliances, the product may be bumped or dropped repeatedly at room temperature during testing and may pass with great margins only to have significant field failures, because the testing laboratory did not reckon with storage in unheated warehouses and shipment in containers that subjected the product to gravitational forces tens or hundreds of times in excess of that experienced in "actual use." This is not to suggest that overdesign of the part or overtesting should be of major emphasis in the production of a device or product. Simply stated, neglecting the full range of environments experienced by the product may lead to product failure that at first is automatically attributed to off-spec resin or other equally difficult-to-determine physical properties of this material. Hagan and Isaacs [43] and Throne [33] have discussed the problem of attempting to relate product performance with resin properties, and the mutual conclusions can be summarized in the following ways:

1. Intuition and hunches about the product performance do not work.
2. Extrapolation from experience with metals and/or ceramics is invalid. There are too many plastics, too many additives, and certainly too many process variables to allow this type of engineering to work.
3. As seen in Chap. 1, plastics are not cheap substitutes for more expensive materials. In fact, plastics are some of the most expensive raw materials available in quantity. Even the "staple" plastics cost 2 to 3 times the price per pound of steel and more than 10 times that of ceramics.
4. Determine the performance requirements of the product. Here make the requirements reasonable. Do not demand that the material meet the conditions established by testing a product made of steel or ceramic or other conventional material.
5. Determine the expected environmental conditions. At the same time, determine how long the product is to remain in these conditions. If a part is exposed to extreme conditions for less than 5% of its life, for example, determine whether to design to meet that limitation. Determine also what fraction of the product will be used in what types of environmental conditions. For example, a product designed to protect a boat against salt spray deterioration may be overdesigned if its primary market is in freshwater boats.
6. Determine the expected manufacturing processes. These processes will be discussed in detail below. The reason for determining this at this point is that frequently product designers will ask to hold tolerances that are impossible to achieve for a given process. Thermoforming, for example, yields a much wider variation in part wall thickness than does injection molding. This is commented on in detail when the comparative economics of the various processes are considered in Chap. 16.
7. Do a preliminary design with plastics in mind. Do not call out dimensions that are specifically for die casting or machining. Plastics shrink and have tendencies to distort and show sink marks. In addition, try to determine the stresses that might occur at various points in the part, and try to determine how the part will perform under these stresses. Sharp corners and variable wall thicknesses are notorious for initiating points of mechanical failure.
8. Determine the proper choice of materials, keeping in mind the environmental conditions and the fact that the materials must be

processed into the product shape. If the conditions chosen are too stringent, the materials required will be more expensive than those actually needed. Thus, the product may not be economically attractive. At the extreme, the conditions may be so strict that no commercially available plastic material can meet them. At this point, the project may be abandoned and your opportunity lost. Your naivete will become painfully evident when you find that your competition's material choices may work well in the marketplace. On the other hand, however, do not scrimp on the materials used. If the environmental conditions and the use of the part dictate an expensive material, do not relax the constraints at this point and choose a cheaper material. The attractiveness of cheaper materials (and thus greater profits, ROIs, and the like) will be offset by consumer litigation costs. Be your worst critic; if you think the product is cheap and will probably fail, change the materials and/or the processes now. Don't wait for the consumer to legally require you to do it.

9. Early in the program establish proper quality control on incoming materials (all of them) and proper AQL levels on finished parts. Furthermore, make certain that the test procedures are sufficiently sensitive to variations in materials properties by deliberately molding off-spec products and testing them in blind tests with run-of-the-mill parts. If off-spec materials cannot be distinguished from the quality products, perhaps the material requirements are too stringent or, more importantly, perhaps the testing procedures for catastrophic materials failures are not severe enough.

10. Design and fabricate prototype parts. Be aware that all prototype parts are the best that can be done. Undoubtedly you will reject all bad prototype parts, and mechanical and design testing procedures will be concentrated on only the good ones. Keep in mind, therefore, that production quality is usually poorer than prototype quality. If the material is particularly sensitive to a specific condition (for example, highly notch sensitive or highly sensitive to water absorption), production quality will demonstrate this sensitivity far more than will prototype parts. Make sure that all prototype parts are subjected to all critical environmental extremes and tested against every performance requirement, regardless of how insignificant that performance requirement may seem at that time. Do not forget field testing. This does not mean "mother-in-law" surveys or "president's wife" surveys but full-blown blind testing. Do not be afraid of criticism. Improvements in products can, in many cases, result in improvements in design and reduction in costs. And it is probably not worth mentioning, but do not forget the federal, state, and local

governmental agencies. To get government approval, in many cases, certain codes must be met, and therefore months or years of testing must be done prior to approval. This of course is the case with the Federal Food and Drug Administration, where requirements on materials that contact foodstuffs require certain limitations on plastics additives. Furthermore, keep in mind that not only must these products meet standards on a regular consistent basis, but as the level of requirements is raised, they must meet the increasing standards as well. There are some logical ways to avoid the "Chinese army" syndrome. Do not try this bag and that bag of resin without a planned sampling and evaluation program. Remember that most simulated life testing programs, as discussed in detail above, are based on the assumption that there is a correlation between the simulation and the actual in-field performance. In the majority of cases, this is just not true. Field performance should be integrated into testing as soon as practical to help in writing specifications.

11. Once prototype design and performance are satisfactory, preparations for production should begin. At this point, the design and materials should be locked in. Make no major design or materials changes from this point until extensive in-use product performance dictates some changes. If major changes are made while in production, product failures and/or callbacks could lead to disaster. Make certain that not only are the prototype testing records in order but that testing and certification of production materials and products are kept up to date and witnessed. Thus, any product liability cases can be traced to suspect resin and/or process conditions. Careful screening of materials and an accurate and impartial in-plant evaluation of the important physical properties of the plastics will help minimize resin problems. Careful record keeping of lot numbers from given machines by given operators on given days can help pinpoint operator error. The "Monday morning-Friday afternoon" automobile that is received by the customer in a semifinished state is a prime example. This can occur in plastics production as well. In addition, watch for production pieces that are made during process start-up, because the process conditions have dramatic effects on materials properties. Accurate records of this nature are particularly useful if personal injury or death can be traced to degradation of material or adulteration in material added to the machine. For example, seepage of hydraulic oil into the screw/barrel region of an extruder owing to the partial failure of a sealing ring can lead to mechanical weakness of thermoplastic elastomers in much the same way as vegetable oil can affect the mechanical properties of high-impact polystyrene.

Thus, a critical part may fail catastrophically, and its ensuing litigation may require tracing back to the given machine before the source of the problem can be isolated.

12. Never be afraid to admit that plastics can fail. Regardless of all the wonderful tests that can be invented, someone is certain to try to use the product in some manner other than its intended use. Common sense prevails. If in doubt, list precautions on the label. It looks good in court. If, in fact, your material did cause an injury or death owing to mechanical failure of a product, take immediate action to determine if it was a once-in-a-lifetime incident or if others are also failing in this way. If the latter is the case, recall the bad parts. The plastics industry has been given a black eye by those people who underdesign, underspecify, and underfabricate a product that fails catastrophically in a very short statistical time. The industry does not want and does not need additional emphasis on the "cheapness" of the material either through shoddy workmanship or poor public relations on the part of the manufacturer.

The importance of these interrelationships in plastics product design has now been recognized by the recently established semimonthly trade journal Plastics Design Forum [44]. The journal offers a broad-based vehicle for dialogues between the various groups that make up the plastics process industry: resin suppliers, additive suppliers, mold makers, processors, product designers, quality assurance and control people, and end users. This journal should be required reading for anyone considering developing a new plastic product.

4.8 The Obvious Conclusions

It should be obvious that there are many levels of relationship between physical and mechanical properties of plastics and the structural features of the polymer that constitute the majority of the resin. Only a few of these levels have been covered here, and only a few of the physical properties that are required in the final product have been detailed. More of this will be considered in a later section on the design of plastic parts. Note, for example, that the effect of fillers or additives on the abrasion resistance of materials was not considered—nor was the fire resistance of materials, nor a criterion proposed for hardness of the material. Exigencies require that the processing behavior be considered also since it, too, is important to the interrelationship between the mechanical properties of the final plastic parts and the structural features of the polymer. For additional information on these and other tests, the reader is referred to articles in Polymer Engineering and Science or Plastics Engineering or one of the many excellent trade journals for the latest specific information and to one of the many fine books on the structure of plastics for more comprehensive, digested information.

Problems

P4.A The Huggins equation for reduced viscosity as a function of concentration and intrinsic viscosity is given as [45]

$$\eta_{sp}/c = \{\eta\} + k'\{\eta\}^2 c \qquad (4.A-1)$$

Obtain values for k' and $\{\eta\}$ from the data for polystyrene in benzene in Fig. 4.2-1.

P4.B Rework the data of P4.A to show that a Huggins equation in terms of the relative viscosity is given as

$$(\ln \eta_r)/c = \{\eta\} - k''\{\eta\}^2 c \qquad (4.B-1)$$

Show that $k' + k'' = 0.5$. Obtain values for k'' and $\{\eta\}$ from the data in Fig. 4.2-1.

P4.C Make a plot of intrinsic viscosity as a function of molecular weight for polystyrene in benzene at 25°C, knowing that the coefficients in the intrinsic viscosity equation (4.2-1) are $k = 1.03 \times 10^{-4}$ and $a = 0.74$. These values are obtained from osmometry. Is the molecular weight determined in this way closer to number average or weight average? Using the values given in P4.A, cross-plot molecular weight as a function of concentration. How can a plot of this be used in quality control?

P4.D The induction time for crystallization of one form of polypropylene is 1.8 min at a quench temperature of 123°C and 90 min at a quench temperature of 142°C. If the induction time is inversely proportional to the temperature difference below the melting point to the fourth power ($e = 4$), determine the induction times at quench temperatures of 135 and 100°C.

P4.E From dilatometric measurements, the volume change in crystallizing polypropylene at 135°C is given as

$(V_\infty - V_t)/(V_\infty - V_0)$	Time (min)
0	11
0.05	30
0.1	37
0.2	48
0.3	56
0.4	64

$(V_\infty - V_t)/(V_\infty - V_0)$	Time (min)
0.5	71
0.6	80
0.7	90
0.8	100
0.9	125
0.95	165

Determine the Avrami constants. What type of crystalline growth is taking place? If the energy of activation for crystallinity of polypropylene is E_c = 22 kcal/mol and if the mechanism of crystallinity remains the same, determine the time to crystallize 90% polypropylene at 100°C, including the induction time. Determine the time to crystallize 50% polypropylene at 142°C, including induction time, and compare your results with the experimentally measured time of 550 min. Discuss how the Avrami equation can be helpful in determining quench conditions for transparent polypropylene blown film production.

P4.F (McKelvey [15]) Using the data given in Fig. 4.3-2, determine the time required for polyethylene terephthalate to become 50% crystalline and 95% crystalline at 110°C. The time in Fig. 4.3-2 is in minutes.

P4.G A melt blown process uses polyethylene terephthalate to form a nonwoven fabric. It strips the 0.010-in.-diameter filaments with 25°C air at a speed at 1000 ft/sec. The filaments fall 2 ft to a continuous belt. What is the extent of crystallinity of the filaments at the point of contact with the belt? List all process assumptions before answering. Is the answer realistic?

P4.H Plot T_{gc}'s from the empirical equations (4.3-5) and (4.3-6) for the copolymer acenaphthylene-styrene, and compare your results with Fig. 4.3-7. Propose a new form for T_{gc} based on an exponential law of mixtures.

P4.I (Rosen [46]) The following data were obtained for a batch of polymethyl methacrylate in acetone at 30°C:

η_{rel}	c (g/100 ml)
1.170	0.275
1.215	0.344
1.629	0.896
1.892	1.199

For polymethyl methacrylate in acetone at 30°C, $\{\eta\} = 5.83 \times 10^{-5}(\overline{M}_v)^{0.72}$. Determine $\{\eta\}$ and $\overline{M}_v$ for the sample and k', the constant in Eq. (4.A-1).

P4. J Brighton [36] states that there is no correlation between Charpy impact strength and falling-ball impact strength. Examine carefully the data of Table 4.6-1 to substantiate or repudiate his argument.

References

1. Modern Plastics Encyclopedia No. 10A, issued each October by Modern Plastics, McGraw-Hill, New York.
2. R. M. Ogorkiewicz, ed., Engineering Properties of Thermoplastics, Wiley-Interscience, New York, 1970.
3. R. M. Ogorkiewicz, ed., Thermoplastics: Properties and Design, Wiley-Interscience, New York, 1974.
4. R. M. Ogorkiewicz, ed., Thermoplastics: Effects of Processing, Iliffe (Plastics Institute), London, 1969.
5. W. E. Brown, ed., Testing of Polymers, Vol. 4, Wiley-Interscience, 1969.
6. E. Baer, ed., Engineering Design for Plastics, Van Nostrand Reinhold, New York, 1964.
7. R. J. Samuels, Structured Polymer Properties, Wiley-Interscience, New York, 1974.
8. P. Geil, Polymer Single Crystals, Wiley-Interscience, New York, 1963.
9. M. L. Miller, The Structure of Polymers, Van Nostrand Reinhold, New York, 1966, Chaps. 6, 9, and 10.
10. F. W. Billmeyer, Jr., Textbook of Polymer Science, Wiley-Interscience, New York, 1962, Chap. 3.
11. H. Staudinger and W. Heuer, Ber., 63:222 (1930).
12. A. Einstein, Ann. Physik., 33:1275 (1910).
13. P. J. Flory, Principles of Polymer Chemistry, Cornell University Press, Ithaca, N.Y., 1953, Chap. 7.4.
14. F. Rodriguez, Principles of Polymer Systems, McGraw-Hill, New York, 1970, p. 127.
15. J. M. McKelvey, Polymer Processing, Wiley, New York, 1962, p. 149.
16. M. Avrami, J. Chem. Phys., 6:1103 (1939).
17. L. Mandelkern, Growth and Perfection of Crystals, Wiley, New York, 1958, p. 467.
18. L. B. Morgan, Philos. Trans. Roy. Soc. London Ser. A, 247:21 (1954).
19. G. Ostapchenko, duPont, personal communication, Oct. 1976.
20. I. Kössler, "Infrared-Absorption Spectroscopy." In: Characterization of Polymers (N. M. Bikales, ed.), Wiley-Interscience, New York, 1971, p. 125.
21. B. Ke, "Differential Thermal Analysis." In: Characterization of Polymers (N. M. Bikales, ed.), Wiley-Interscience, New York, 1971, p. 191.

22. R. C. Progelhof, J. Franey, and T. W. Haas, J. Appl. Polym. Sci., 15:1803 (1971).

23. Witco Chemical Company, Witron Division, personal communication, 1976.

24. V. A. Kargin and G. L. Slonimsky, "Mechanical Properties." In: Mechanical Properties of Polymers (N. M. Bikales, ed.), Wiley-Interscience, New York, 1971, p. 1.

25. J. R. Collier, Ph.D. thesis, Case Institute of Technology, Cleveland, 1966.

26. S. M. Middleman, The Flow of High Polymers, Wiley-Interscience, New York, 1968, p. 142.

27. R. L. Combs, D. F. Slonaker, and H. W. Coover, Jr., SPE ANTEC, Tech. Pap., 13:104 (1967).

28. J. L. White and N. Tokita, J. Appl. Polym. Sci., 11:321 (1967).

29. E. G. Kontos, Rubber Chem. Technol., 44:1082 (1971).

30. L. L. Blyler, Jr., Rubber Chem. Technol., 42:823 (1969).

31. T. F. Gray, Jr., R. L. Combs, D. F. Slonaker, and W. C. Wooten, Jr., SPE ANTEC, Tech. Pap., 13:370 (1967).

32. J. B. Howard, "Fracture-Long Term Testing." In: Mechanical Properties of Polymers (N. M. Bikales, ed.), Wiley-Interscience, New York, 1971, p. 73.

33. J. L. Throne, paper presented at Product Testing and Failure Analysis, University of Wisconsin-Extension, Milwaukee, May 1973.

34. P. I. Vincent, "Fracture-Long Term Testing." In: Mechanical Properties of Polymers (N. M. Bikales, ed.), Wiley-Interscience, New York, 1971, p. 105

35. L. R. Treloar, The Physics of Rubber Elasticity, Oxford University Press, New York, 1958, Chap. 2.

36. C. A. Brighton, "Impact Resistance-Theory." In: Mechanical Properties of Polymers (N. M. Bikales, ed.), Wiley-Interscience, New York, 1971, p. 175.

37. A. A. Griffith, Philos. Trans. Roy. Soc. London Ser. A, 221:163 (1920-1921).

38. R. Buchdahl, J. Polym. Sci., 28:239 (1958).

39. M. M. Riddell, Applications and Design Properties of Plastics, SPE Educational Seminar Paper, presented May 9, 1973, SPE ANTEC.

40. H. Burns, "Impact Resistance-Test Methods." In: Mechanical Properties of Polymers (N. M. Bikales, ed.), Wiley-Interscience, New York, 1971, p. 185.

41. Joseph Starita, Rheometrics, Inc., Union, N.J., personal communication.

42. J. R. Heater and E. M. Lacey, Mod. Plast., 41:124 (1964).

43. R. S. Hagan and J. L. Isaacs, paper presented at SPE ANTEC, EPSDIV Business Meeting, May 1971.

44. Plastics Design Forum, semimonthly publication of Industry Media, Inc., 1129 E. 17th Ave., Denver, Mel Friedman, ed.

45. M. L. Huggins, J. Am. Chem. Soc., 64:2716 (1942).

46. S. L. Rosen, Fundamental Principles of Polymeric Materials for Practicing Engineers, Barnes & Noble, New York, 1971, p. 57.

5.

Viscosity and an Introduction to Rheology

5.1 Introduction

More than 300 years ago, Isaac Newton defined solids and fluids according to their behavior toward imposed shear stresses. Fluids, basically, deform under shear and remain deformed when the shear is removed. Furthermore, the rate at which they deform is related to the intensity of shearing. Solids, on the other hand, deform under shear and return to their original shape when the shear is removed. And the extent of deformation of a solid is related to the intensity of shearing.

Unfortunately, this simple viewpoint does not describe the majority of materials behavior toward shearing forces. For example, we consider copper to be a solid. Yet if we bend a copper wire into a circle, it will remain in that circle when we remove the stresses used to bend it. A kitchen screen-door spring doesn't keep the door closed as tightly in September as it did in May. On the other hand, we consider mayonnaise to be a fluid, since we can spread it onto bread with a knife. But we can't pour it from a container. Glass is considered by most people to be a solid, yet in very old houses we find that the thickness of window glass is greater at the bottom than at the top of the pane, indicating long-term flow. Or consider another common material, chewing gum. It stretches and snaps like a rubber (and therefore has the elastic character of a solid), but it also flows under very low shear (and since it does not retain its shape indefinitely, it must be a fluid). And even fluids that appear to be fluids exhibit peculiar characteristics, as any housewife knows. Egg white (albumin) will appear to climb up mixer shafts. Characteristically, fluids should form vortices around mixer shafts.

What does this have to do with plastics production? Because of the enormously long molecules with branching, cross-linking, and other effects, these fluids do not behave as Newton wished they would. Instead they exhibit material behavior that at times is solid-like and at other times is fluid-like. Even more important, their abnormally high viscosities in the molten state (10^3 to 10^{10} times that of water) make analyses of their behavior extremely complex. For example, TFE Teflon has a viscosity at its melt temperature of approximately 10^7 poise (P) (the viscosity of water is 0.01 P). When a solid block of this material is heated to the melting temperature and beyond, it does not flow. It is therefore processed as one would process a solid, even though it is a liquid. The behavior of a given plastic will depend on such external effects as temperature, time, and most important, the severity and extent of shear. The study of the flow behavior of materials is "rheology," and dozens of texts and thousands of papers have been devoted to this field. The Society of Rheology [1], a division of the American Institute of Physics, is devoted entirely to the definition and understanding of materials flow behavior. We shall not attempt here to summarize all the work done in the field but shall give a glimpse into the more important aspects of rheology as they pertain to processing of plastics materials.

5.2 The Simple Models for Plastics [2]

Because of the complexity of the materials and the enormous amount of additives and fillers required to process a given resin in a given application, it is certain that the perversity of inanimate objects will prevent the theorist from ever adequately explaining in equation form all the peculiar behaviors of all plastics materials. Nevertheless, many theorists will continue to work toward this goal. Basically, we can divide materials into several general categories (with very broad overlaps, of course). A truly rigid solid is one where no deformation takes place regardless of the extent of shear (or, the elastic modulus is infinite). A Hookean solid is defined as one in which there is a linear relationship between the extent of deformation and the severity of shear. The ratio of stress to extent of deformation (or strain) is called the elastic modulus or shearing modulus (or in the case where tensile strength is used to pull the sample, the tensile modulus):

$$\tau = G\gamma \tag{5.2-1}$$

where τ is the shearing stress, defined as the shearing force per unit area, N/M^2; G is the shearing modulus of elasticity, or simply modulus; and γ is the extent of deformation, in radians. A Hookean solid exhibits no creep, plasticity, yield point, or hysteresis in the stress-strain curve.

At the other end of the spectrum, we have the inviscid fluid, or a fluid with no viscosity. An infinitesimal amount of shear will cause the fluid to move, and it will continue to move indefinitely. Since the fluid has no viscosity, it cannot be stopped once it is moving. While this idealization seems to yield a useless fluid mechanical model, nevertheless inviscid fluid models are commonly used in aerospace applications, including rocket nozzle design and shock tube analysis. The apparent reason for this is that at very high speeds only the fluid in a microscopic layer near a solid wall exchanges energy and momentum with the wall; the bulk of the fluid acts as if it has no viscosity. Incidentally, we alluded to this concept in our discussion on tubular reactor design in Chap. 3. If the fluid is moving through the tube in plug flow, the presence of the tube wall offers no resistance to the fluid (e.g., no shearing effect), and thus the fluid must be inviscid. We shall make no use of the inviscid fluid model from this point on.

The most common fluid behavior is that described by Newton and named for him, the Newtonian fluid. There is a direct relationship between the severity of shearing stress on the fluid and the rate of deformation of the fluid (or the rate at which the fluid is stretched):

$$\tau = \mu\dot{\gamma} = \mu(d\gamma/dt) \tag{5.2-2}$$

where τ is the shearing stress and $\dot{\gamma} = d\gamma/dt$ is the time rate of deformation of the fluid (in radians per second, or simply sec^{-1}). μ is the proportionality constant known as the Newtonian viscosity. If τ is in N/m^2 or pascals, the units on μ are $N\cdot sec/m^2$ or pascal·sec. If τ is in $lb_f/in.^2$, the units on μ are $lb_f/in.^2\cdot sec$. Unfortunately, as with all unit systems, these logical units are not acceptable. Instead viscosity units can be in centipoise, $lb_m/ft\cdot hr$, "centistokes," "Saybolt seconds," and so on. It is not facetious to state that there are probably more obscure ways of measuring and reporting viscosity than alibis for infidelity. And even more aggravating, there are probably two theoretical methods for interpreting the data for each method of measuring them. And remember here that we are only talking about Newtonian fluids. A proper definition for Newtonian fluids is given in Eq. (5.2-2), where the defined viscosity is constant and independent of severity of shearing stress or shearing history, and thus no hysteresis is seen. Furthermore, when the shearing stress is removed, the fluid remains deformed and exhibits no behavior to return to an earlier shearing state (e.g., no recoil behavior). The Hookean solid and Newtonian fluid responses to imposed shearing stress are shown in Figs. 5.2-1 and 5.2-2.

We shall consider Newtonian fluid behavior to be the asymptote for plastics flow behavior. We would expect, for example, at very low rates

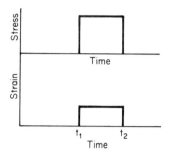
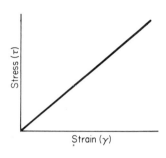

Figure 5.2-1 Stress-strain diagrams for ideal Hookean solid; stress applied at t_1, removed at t_2 [2].

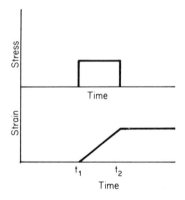
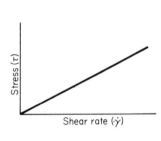

Figure 5.2-2 Stress-strain diagrams for ideal Newtonian fluid; stress applied at t_1, removed at t_2 [2].

of shear (creep, gravity flow, and the like) that the plastics materials deform Newtonianly. We might also consider for certain materials that at very high rates of deformation the materials might behave as if they were solids (not necessarily Hookean, but possibly deforming plastically). In between the Hookean solid and the Newtonian fluid, however, is the vast realm of fluids studied by rheologists [3]. Here are fluids which have abnormal characteristics, as mentioned earlier. Silicone materials flow as fluids as long as the shearing stresses are relatively low but fracture as brittle solids at very high shearing stresses. Here, too, are the Bingham-type plastics. At low shear, these materials deform elastically. However, once a critical shear stress is exceeded, these materials flow as Newtonian fluids. A colloidal latex (rubber or plastic) is an example of a Bingham plastic. We concoct Bingham plastic materials for special purposes. Coating applications, for example, require that a material be very fluid at high shear application conditions but not drape or run when placed on a vertical surface, where the only shear is that caused by gravity. There are also time-dependent materials. Quicksand, for example, becomes more fluid with time at constant shear rate, as do certain types of latexes. Certain printing inks become more viscous with time at constant shear rate.

Some materials have viscosities that are shear sensitive. Some slurries (including certain PVC resins) have viscosities [as defined by Eq. (5.2-2)] that increase with increasing shear stress. These are referred to as dilatent. Dilatency leads to mixing and blending problems, where the extent of dispersion of a pigment, say, is proportional to the intensity of shear. Most plastic melts exhibit a decrease in viscosity with increasing shear stress, and these materials are referred to as pseudoplastic. We can best illustrate this shear dependency through redefinition of the viscosity:

$$\tau/(d\gamma/dt) = \eta[(d\gamma/dt)] \tag{5.2-3}$$

In this form of Newton's equation, the viscosity is not constant. Such materials with shear-dependent viscosities are called non-Newtonian. Incidentally, this equation does not allow inclusion of the elastic character of the material into the viscosity. Thus, this is a "viscous-only" equation. Furthermore, in conventional flow problems, in three physical dimensions, there are nine possible stresses that can be applied to the fluid element (six shearing stresses and three normal stresses). This equation allows only one. We shall discuss the difficulties that arise when these factors are included in the model for stress and strain on a fluid element in a later section.

The earliest attempt (and most widely used to date) to include the shear dependency into the stress rate of strain equation is that of Ostwald and de Waale, and the equation is commonly known as the power-law equation [3]:

$$\tau = K(d\gamma/dt)^n = [K(d\gamma/dt)^{n-1}](d\gamma/dt) \tag{5.2-4}$$

where K is the viscosity coefficient and n is the viscosity index, both empirical and dependent on the materials being tested. To determine the viscosity of a Newtonian fluid, we need only one measurement of shearing stress and corresponding rate of deformation. Here we need two: K and n. This piece of information is extremely important in the evaluation of various methods of measuring viscosities of plastics, as we shall see later. An example of a power-law fluid is shown in Fig. 5.2-3.

There are several limitations on the power-law model. These should be mentioned since much of the work on fluid flow and heat transfer is based on power-law fluid. Recall in Chap. 4 that Blyler's work showed that for high shear rates of polyethylenes, the power-law model did not adequately describe the shear rate dependence of the viscosity [4]. K and n are not constants since they too depend on shear stress. Furthermore, many materials such as LDPE and polystyrene exhibit non-power law

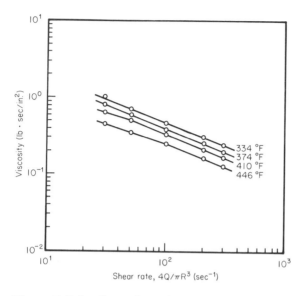

Figure 5.2-3   Shear-dependent viscosity for 1.7-MI LDPE at various melt temperatures; 0.019-in.-ID tube, L/D = 9, flat entry.

behavior at very low shear rates. Chee and Rudin [5] have shown that polystyrene exhibits a shear-independent viscosity at shear rates on the order of 10^{-4} sec^{-1}. Since a shear-independent viscosity is a Newtonian viscosity, this means that polystyrene is Newtonian at low or zero shear rates. As seen from Eq. (5.2-4), if n is less than unity, as it is for pseudoplastics, the viscosity at zero shear must approach infinity. Polystyrene, for one plastic, is thus not power law over all of its shear range. A typical viscosity curve for nylon is shown in Fig. 5.2-4.

The engineer is cautioned that the models using power-law viscosities may in fact be wrong, but obviously not so wrong as the use of Newtonian values. We also comment on "apparent viscosity." There is nothing apparent about viscosity; it simply represents the ratio of applied stress to rate of strain, as given by Eq. (5.2-3). Just because the viscosity varies with shear rate, temperature, and other factors does not make it "apparent." To paraphrase, "a viscosity by any other name would still represent the resistance of a fluid to shearing stresses and thus still be a viscosity."

To take into account the dependency of K and n on shearing conditions for certain fluids, Rabinowitsch and Mooney [6], independently, found that if certain manipulations were performed, the stress rate of strain curve (and thus viscosity) could be defined for a given value of stress independent of the type of equation used to describe viscosity. Thus, we can obtain a viscosity at a given shear stress without resorting to an equation. The

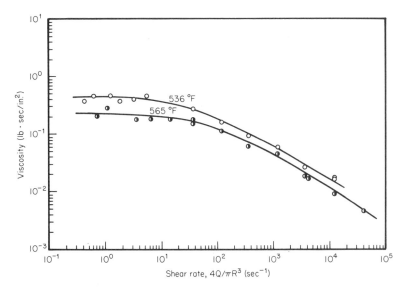

Figure 5.2-4   Shear-dependent viscosity for Nylon 66 at two melt temperatures; 0.005- to 0.047-in.-ID tubes, L/D = 16, flat entry.

stress rate of strain equation is referred to as the "constitutive equation of state" for a fluid. To see how this is done, consider steady flow of a fluid with unknown viscous properties through a horizontal tube of circular cross-section. The volumetric flow rate is Q:

$$Q = \pi R^2 \int_0^R r u \, dr / \int_0^R r \, dr \tag{5.2-5}$$

where R is the radius of the pipe and u is the radius-dependent fluid velocity. Recalling the rule of "integration by parts," given as

$$\int_1^2 a \, db = ab \Big|_1^2 - \int_1^2 b \, da \tag{5.2-6}$$

we can rewrite Eq. (5.2-5) as

$$Q = 2(r^2 u/2) \Big|_0^R - 2\pi \int_0^R r^2 \, du/2 \tag{5.2-7}$$

Consider the first term on the right. When r = R, the velocity at the tube wall is zero. When r = 0, the term is also zero. Thus, the first term on the right is always zero. For the second term on the right, recognize that du = (du/dr) dr by the chain rule. Now by definition, du/dr is the rate of deformation, $d\gamma/dt = \dot\gamma$. Furthermore, the shear stress τ anywhere in the fluid can be related linearly to the shear stress at the wall by

$$\tau = \tau_w(r/R) \tag{5.2-8}$$

Solve this for $r = R\tau/\tau_w$, differentiate for $r[dr = (R/\tau_w)\,d\tau]$, and substitute this into the second term on the right side of Eq. (5.2-7):

$$Q = (-\pi R^3/\tau_w^3) \int_0^{\tau_w} \dot\gamma\tau^2\,d\tau \tag{5.2-9}$$

where $\dot\gamma = d\gamma/dt$. Now, using Leibnitz' rule, differentiate this equation with respect to τ_w. This yields the following equation:

$$(1/\pi R^3)[\tau_w^3\,(dQ/d\tau_w) + 3\tau_w^2 Q] = -\dot\gamma_w\tau_w^2 \tag{5.2-10}$$

But $\tau_w = R\,\Delta P/2L$, by definition, where ΔP is the total pressure drop of the fluid over the length L. Thus, Eq. (5.2-10) becomes

$$-\dot\gamma_w = (1/\pi R^3)[3Q + \Delta P(dQ/d\,\Delta P)] \tag{5.2-11}$$

This equation is classically called the Rabinowitsch-Mooney equation. Note that it gives the shear rate as a function of pressure drop and volumetric flow rate (and of course the rate of change of flow rate with pressure drop). Since we know τ_w directly from pressure drop measurements, we can plot τ_w against $\dot\gamma_w$. The slope of the curve is, by definition, the viscosity. Thus, we can obtain viscosity without resorting to any equation of state. In some sources, a viscosity is defined as

$$\tau_w/\dot\gamma_w = K'(\dot\gamma_w)^{n'-1} \tag{5.2-12}$$

where K' and n' are called the local viscosity coefficient and local viscosity index, respectively. This is not correct. It is better to realize that the slope of any stress rate of deformation curve is the viscosity at that value of stress and not try to relate this Rabinowitsch-Mooney model to a fictitious power-law model. Incidentally, it is comforting to note that for a power-law fluid the nonlinearity factor $dQ/d\,\Delta P$ is zero and that the Rabinowitsch-Mooney equation reduces to the power-law model (as well as the Newtonian model when n = 1). This does not mean that this model yields an equation of state; it is much more than that. It yields the entire viscosity curve irrespective of the characteristics of the fluid. It works, for example, on solid-liquid slurries and Bingham plastics as well as on Newtonian fluids. Note, however, that this method requires many more experiments to establish the viscosity curve than, say, the power-law model, because of the need to establish the $dQ/d\,\Delta P$ curve in addition to the standard flow rate-pressure drop data.

5.3 More Complex Models—An Evaluation

As mentioned above, the stress applied to a fluid element is more complex than that represented simply above as τ. In fact, this really represents only one type of stress on a fluid. Referring to sources such as Pearson [7], Bird et al. [6], and others [3,8-10], we find that the stress applied to any fluid element can be broken into nine components:

$$T \longrightarrow \begin{pmatrix} \tau_{xx} & \tau_{xy} & \tau_{xz} \\ \tau_{yx} & \tau_{yy} & \tau_{yz} \\ \tau_{zx} & \tau_{zy} & \tau_{zz} \end{pmatrix} \qquad (5.3\text{-}1)$$

Thus, the stress is referred to as a tensor of order 2. The diagonal terms are normal to the direction of stress and are therefore referred to as "normal stress tensor components." The off-diagonal terms are the shearing stresses. For example, τ_{xy} represents the x-direction effect on the fluid element owing to stress applied in the y-direction. Thus, these off-diagonal terms represent the shearing stresses discussed above. The rate of deformation, which we earlier called $\dot{\gamma}$, is also a tensor of order 2 and is given as

$$e_{ij} = (1/2)(\partial v_i / \partial x_j + \partial v_j / \partial x_i) \qquad (5.3\text{-}2)$$

where i and j can take values of 1, 2, or 3. For example, for the rate of deformation term that corresponds to the shearing stress τ_{xy}, we write $e_{xy} = (1/2)(\partial v_x / \partial y + \partial v_y / \partial x)$. The relationship then between the tensor of stress and the tensor of the rate of deformation is referred to as the constitutive equation of state. In its simplest form, it defines Newtonian viscosity. In more complex forms, it defines other types of viscosities, such as elongational viscosity or cross-viscosity. Shortly we shall show how various constitutive equations can define these properties. First, we wish to consider general problems in dealing with complex fluids.

It has been shown over and over again that if we write all the equations describing fluid motion and count up all the variables that are involved, we find that we are always one equation short. That equation connects stress, deformation, and rate of deformation. In its simplest case it is the Newtonian equation. However, as mentioned before, complex fluids are rarely described using simple equations to model their behavior. This, of course, is the genesis of the field of rheology. Consider for a moment the goals of the rheologists and their direction of endeavor. Keep in mind our objective, prediction of processing conditions of a given plastic from measurements that can be made in a laboratory.

5.3.1 Eulerian-Lagrangian Considerations

Consider the experiences of a blob of fluid during processing. Choose extrusion through a die as an example. All the experiences are transient insofar as the fluid is concerned: the compression of the solid particles, the shearing that leads to melting, the elongation as the fluid is stretched along the flights of the extruder, the intermittent contact with the walls or flights of the extruder, the mixing of the blob with other blobs, the buildup of pressure prior to the screen pack, the flow through the screen or sieve pack, the orientation and heating of the blob as it passes through the die geometry, the uncontrolled relaxation of the fluid as it issues from the die, and the drawing and cooling of the blob in the ancillary equipment. Now if we were dealing with a simple fluid, regardless of its viscosity, we could directly translate the experiences of the fluid (known as Eulerian or material) to our observations of the entire process (known as Lagrangian or laboratory). In other words, changes to the simple fluid are felt immediately and its experiences a fraction of a second ago are forgotten. As a result, for simple fluids we need not concern ourselves with the motion of the fluid blob in its material coordinates as translated to our laboratory coordinates.

We have already mentioned some effects (egg white climbing the beater, for example) that lead us to believe that many fluids cannot be considered as simple fluids. There are many other effects, such as die swell at the exit of an extruder die, recoil, melt fracture when our extrusion rates exceed a critical value, formation of fibers (viscous fluids such as corn syrup cannot be drawn into fibers), orientation of the melts to achieve anisotropy, and others. All of these effects depend to a great degree on the history-dependency of the fluids. Macroscopic measurements of, say, pressure drop between two points may depend not only on flow rate and fluid properties such as viscosity but also on the history of the fluid prior to the first measuring station.

5.3.2 Invariants in Purely Viscous Fluids

Eringen [11] has listed many principles that all constitutive equations must satisfy before they can truly describe the fluid. Among these are several "invariant" conditions. These are conditions that must remain unchanged regardless of the orientation of the fluid blob either in a material sense or a laboratory sense. If we let T be the stress tensor and E be the rate of deformation tensor, we can write that for simple fluids

$$T = F(E) \tag{5.3-3}$$

F is a functional that is independent of orientation of the fluid blob. It may contain information about the history of the fluid; e.g., F(E) may have time dependency from time at $-\infty$ to the present. In this notation, a simple Newtonian (or more appropriately a Stokesian) fluid is represented as

$$T = -p\underline{1} + a_1(II_e, III_e)E + a_2(II_e, III_e)E^2 \tag{5.3-4}$$

The notation used here is as follows: p is the hydrostatic pressure, and $\underline{1}$ is referred to as the unit tensor [that is, it has a value of unity when the diagonal terms (the xx, yy, zz terms) are considered and zero when the off-diagonal terms are considered]. There are three terms I_e, II_e, and III_e, that are called the "rate-of-deformation strain invariants" and that have similar forms in structural mechanics. I_e is simply the trace of the rate-of-deformation tensor and in this case is zero. II_e is the sum of the cofactors of the matrix formed by the rate-of-deformation tensor, and III_e is the determinant of the rate-of-deformation tensor. To find methods for calculating these values, see Spiegel [12].

This constitutive equation of state can be illustrated with a very simple example. Consider shear between two parallel planes to illustrate the kind of information that can be derived from the model. The shearing model is shown in Fig. 5.3-1. The stress tensor for this equation can be written as [13]

$$T = \begin{Bmatrix} \tau_{11} & \tau_{12} & 0 \\ \tau_{21} & \tau_{22} & 0 \\ 0 & 0 & 0 \end{Bmatrix} \tag{5.3-5}$$

The importance of this equation lies not in the explicit values for τ_{11}, τ_{12}, τ_{21}, or τ_{22} but in the fact that to produce a shearing flow between two parallel infinite plates it is necessary to have normal stresses τ_{11} and τ_{22} in addition to the shearing force $\tau_{21} = \tau_{12}$. In many simple fluids, the models yield the information that $\tau_{11} - \tau_{22} = 0$. Experimentally, however, this is not always true. Incidentally, the finite values for τ_{11} and τ_{22} are referred to as the Poynting effect [8]. Without this effect, the plates will draw together under shear.

If a_1 and a_2 are independent of the invariants, we have the Reiner-Rivlin equation that exhibits no memory and no elasticity and is homogeneous

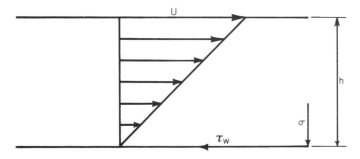

Figure 5.3-1   Planar shear at constant velocity U showing h, gap width; τ_w, wall shear stress; and σ, normal shear stress.

and isotropic. If $a_1 = 2$ and $a_2 = 0$, the fluid is Newtonian. Pearson gives many other forms for these constitutive equations, and we shall point out several more. For example,

$$T = -p\underline{1} + C_1 E + C_2 E^2 \qquad (5.3-6)$$

where $C_1 = C_1(II_e)^a$. This is the power-law model in its tensoral form.

5.3.3 Models with Memory

Consider also a very simple model that contains storage characteristics. Many authors have considered the "spring and dashpot" models of Maxwell and Voigt, and we shall not go into the details regarding them. Entire theses have been developed on the working of these very simple models for very complex materials [14]. Basically, in the Maxwell model, an elastic element (e.g., a spring) is connected in series with a viscous element (e.g., a dashpot or "shock absorber"). Thus, the total strain or extent of deformation on the Maxwell element is the sum of the strain on the spring and the rate of strain of the dashpot:

$$\gamma_m = \gamma_s + \dot{\gamma} = \tau/G + \tau/\eta \qquad (5.3-7)$$

The Maxwell element is generally classified as a viscoelastic solid. On the other hand, Voigt proposed placing the spring and dashpot in parallel (that is, the elastic and viscous components in parallel). For this model, the total stress on the element is given as the sum of the stress on the viscous element and the stress on the elastic element:

$$\tau = \eta\dot{\gamma} + G\gamma_s \qquad (5.3-8)$$

The Voigt element is classified as a viscoelastic fluid and has been the subject of many theses and texts [15] on the linear characteristics of such energy-absorbing materials as polymers. The representative stress-rate of strain curves for the Maxwell and Voigt models are shown in Figs. 5.3-2 and 5.3-3. The major problem in the application of these very simple models is that when they are applied engineers inevitably find that the terms such as viscosity and spring constant are not single variables. They can only be characterized by a statistical distribution. Thus, we must have many Maxwell or Voigt elements in series or parallel or combinations thereof, and this of course makes evaluation of physical constants very tedious (if at all possible). The example of the tensor form for the Maxwell model is given by Pearson [7] as

$$T = -p\underline{1} + c_1 E + c_2 (D_c E/D_c t) + c_3 E^2 \qquad (5.3-9)$$

The second term on the right-hand side represents the application of the principle of invariance, since the Maxwell model describes the stress-strain

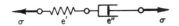

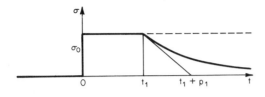

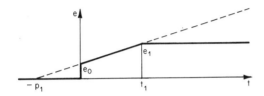

Figure 5.3-2   Maxwell series spring and dashpot linear viscoelastic model; diagrams showing response to instantaneous strain.

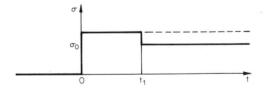

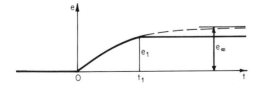

Figure 5.3-3   Voigt parallel spring and dashpot linear viscoelastic model. The schematic on the right is a modified constrained model; diagrams showing response to instantaneous stress and strain.

behavior for the element in material coordinates and we must interpret it
in spatial coordinates. $D_c E/D_c t$ therefore represents the convected derivative, representing the fact that the material is moving and we are not.

5.3.4 Simple Fluid Models

Classical mechanists have pointed out that even these tensoral models are too limited. In particular, they restrict the amount of memory that
can be included in the model, and experimentally, many plastics show extremely long memories. Truesdell [16] and Coleman et al. [17] have pointed
out that for very restrictive flow situations (known as viscometric flows)
certain functional relationships exist between T and E. For the very simple
linear shearing problem that we considered above for the illustration of the
normal stresses, they show that the equivalent shearing and normal stresses
are given as

$$\tau_{12} = \mathop{T}_{s=0}^{\infty} [g(s)] \tag{5.3-10}$$

$$\tau_{11} - \tau_{33} = \mathop{S_1}_{s=0}^{\infty} [g(s)] \tag{5.3-11}$$

$$\tau_{22} - \tau_{33} = \mathop{S_2}_{s=0}^{\infty} [g(s)] \tag{5.3-12}$$

where $g(s)$ is a function that depends on the rate of change of the velocity as
a function of time. T, S_1, and S_2 are scalar functionals that are independent
of the orientation of the coordinate system and depend only on the material
used. They are "material functionals," and determination of them completely characterizes the mechanical behavior of "Noll simple fluids" in
viscometric flow. If the fluid is in steady flow, these relationships reduce
to the familiar shear and normal stress relations obtained earlier.

We should point out that at this point in the development of the field
of rheology, there is a schism between the theoretical model builders and
the people who need information about the flow behavior of plastic materials. Very frequently, for example, the plastics engineer needs to know
(within engineering accuracy) the value of viscosity at one set of stress-
strain conditions from information obtained at a second set of conditions.
The accuracy of the extrapolation is secondary to the completion of a design
or establishment of initial operating conditions. Many mathematical models are therefore of little use to him. For example, the simple power-law
model, while not mathematically accurate, requires determination of only
two constants, whereas the Rabinowitsch-Mooney model requires sufficient
data to plot a family of curves, a relatively sophisticated constitutive equation of state such as the Bird-Carreau model needs 6 constants to be useful,
and the Giesekus equation requires evaluation of 9 constants (or 10, depending

on the interpretation) [3]. It is apparent that the engineer must weigh the value of extensive rheological determinations of constants for a given material against the variation in material content and the accuracy of the information needed. In the majority of the cases, it seems that the study of rheology is best left to the physicist and theoretician and the measurement of viscosity to the engineer. The engineer must determine for himself (herself) what types of measurements are necessary in order to determine whether a material will process satisfactorily in a given piece of equipment. More important than the argument of the suitability of a given model is the incontrovertible fact that the properties we wish to measure, such as stress-rate-of-deformation curves, viscosity, normal stresses, and the like, are <u>material</u> properties and that these properties should be independent of the type of measurement. The various methods of viscosity measurement are discussed in the next section.

5.4 Measuring Viscosity and Other Fluid Properties

Some ground rules for the determination of fluid material properties must first be established. As mentioned above, the properties must be independent of the method of measurement. Furthermore, they must be sufficiently extensive so that the stress-rate-of-deformation curve can be established, since both the power-law and Rabinowitsch-Mooney models require the shape and slope of this curve. And finally we must select a form for the constitutive equation of state and construct a measuring system that enables unique determination of the necessary parameters. There is an old rule that says that the simpler the experiment geometry, the easier the analysis. We would not expect, therefore, to obtain immediately useful experimental (and theoretical) results from, say, stress-strain measurements in the sigma-mixer attachment of a Brabender Plasticorder. This means, in essence, that most processing equipment is not suitable for obtaining flow behavior data, and therefore we must construct "special" viscometers (or rheogoniometers). We can probably categorize mechanical processing according to material experiences as

1. Short-time processing: typical of flow through gates and egressing of jets from spinnerettes or extruder dies
2. Longtime processing: typical of flow through pipes

It is desirable to understand both the long- and short-time characteristics of the fluid. Most viscometers operate on the longtime basis and therefore are steady-state viscometers. We can separate these steady-state viscometers into three categories, according to the solids-fluid interface:

1. External, where the fluid is confined in an outer surface:
   a. Rotational types, such as concentric cylinders
   b. Translational types, such as pipe flow

2. Internal, where the fluid surrounds the solid surface:
    a. Falling objects, such as balls, cylinders, pins, darts, and so on
    b. Flat plate or surface of revolution moving in an infinite reservoir of fluid
3. Discontinuous, where the surface-fluid interface stops abruptly:
    a. Melt indexer, fluid jet, capillary tube
    b. Finite or infinite plate being withdrawn from a bath of fluid
    c. Spinnerette or extensional flow

We shall consider the nonconstant viscometer later. First consider the simplest viscometer used, the melt indexer. There are many models and many manufacturers of this device; all are designed to meet the ASTM 1238 test "Method of Measuring Flow Rates of Thermoplastics by Extrusion Plastometer." Basically, the plastic material to be tested is placed in a heated chamber, held at an isothermal temperature (determined by the melt or softening temperature of the plastic), and when thoroughly heated to the proper temperature, forced through a capillary die or orifice with a diameter of 0.0825 in. and a length of 0.315 in. The amount of material (in grams) that can be forced through this orifice at a constant pressure (determined by applying 21.6 kg on a piston 0.3730 in. in diameter) in a 10-min period is given as the "melt index." Obviously the more material forced through the orifice, the higher the melt index and the lower the viscosity of the material. Note that the shear stress remains nearly constant (since the pressure on the fluid is fixed by the weight on the fluid and the slight changes in hydrostatic head as the fluid flows from the chamber) but the rate of deformation changes with changing viscosity. Thus, the device gives us a measure of the "flowability" of the plastic at only one shear rate. We know nothing about the shape of the stress-rate-of-deformation curve and thus nothing about the variation of viscosity with shear rate. We have seen that polydispersity, branching, and other factors can dramatically affect the shape of the viscosity-shear rate curve. Thus, it is quite conceivable to have two polyethylenes, as an illustration, that have identical melt indices at a given temperature but completely different flow characteristics. Furthermore, it is possible to have a plastic with a high melt index (at testing shear rates) exhibit much higher viscosity at actual processing conditions than another with a low melt index, simply because the viscosity curves cross at some intermediate shear rate. This is shown in schematic form in Fig. 5.4-1. Thus, melt index data are worthless in the evaluation of competitive resins.

A modification of the melt indexer can be used, however, to yield some necessary information. It is apparent that this device can be used to make gross measurements on volumetric flow rates and pressure drops. The capillary rheometer has evolved from the melt indexer. Basically, the melt index tube is replaced with a capillary tube of diameter 0.010 to 0.030 in. and a length-to-diameter ratio of 10:1 to 50:1. The plastic is

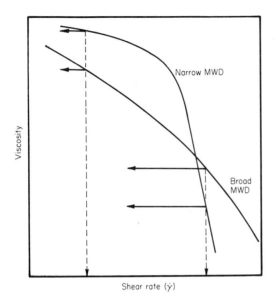

Figure 5.4-1   Characteristic crossover of viscosity-shear rate curves with changes in MWD for HDPEs of similar MIs.

melted in a heated cylinder having a diameter 30 or more times that of the capillary tube. The melt is then forced through the capillary tube either using a constant pressure gas source, a plunger operating at a constant force, or, more recently, a plunger that operates at a constant rate (thus fixing Q, the flow rate) while the resulting pressure in the cylinder is measured. Knowing the pressure drop and the flow rate, the data needed for the Rabinowitsch-Mooney form of the equation of state can be obtained. If there is no slip of the plastic melt at the cylinder wall, no time dependency in the fluid, no secondary flow effects around the entrance to the capillary tube, a constant flow pattern all along the tube, isothermal flow, no viscous dissipation, and incompressible melt, no corrections need be made to the raw data. However, this idyllic situation is rarely achieved, as Bagley and others have confirmed [18]. If we know nothing about the fluid, the Rabinowitsch-Mooney equation can be used to characterize it. If we assume (a priori) that the fluid is a power-law fluid, the Poiseuille equation for steady flow in a pipe can be used to describe the flow rate-pressure drop relationship.

Nearly every book on polymer fluid flow [7,8,19] gives a derivation of laminar steady flow of a power-law fluid in a pipe. We choose not to be different. Consider a force balance made on a fluid element in the center of the circular pipe, as shown in Fig. 5.4-2 [20]. The hydrostatic forces are countered by shearing forces. Thus,

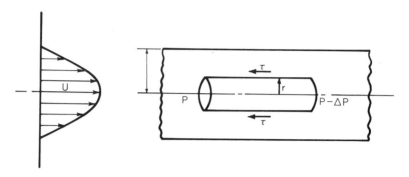

Figure 5.4-2   Schematic for steady, fully developed flow in a pipe; capillary viscometer.

$$[\pi r^2 P + \pi r^2 (dP/dz)\,dz] - \pi r^2 P = 2\pi r \tau\,dz \tag{5.4-1}$$

where the first two terms on the left-hand side represent the pressure times the cross-sectional area at point $z + dz$. The third term is the same except at point z, and the term on the right-hand side is the shearing force times the surface area ($2\pi r\,dz$) over which it is applicable. Thus, we can show that

$$\tau = (r/2)(dP/dz) \tag{5.4-2}$$

or, integrated across the radius of the pipe,

$$\tau_w = (R/2)(dP/dz) = (R\,\Delta P/2L) \tag{5.4-3}$$

A relationship for shear stress-shear rate is

$$\tau = \eta\dot{\gamma} \tag{5.4-4}$$

where $\eta = \eta(\dot{\gamma})$. But in the case of laminar flow in a pipe, $\dot{\gamma} = du/dr$. Therefore, Eq. (5.4-4) is

$$\tau = K(du/dr)^n \tag{5.4-5}$$

Substituting this expression into Eq. (5.4-2),

$$K(du/dr)^n = (r/2)(dP/dz) \tag{5.4-6}$$

Rearranging and integrating,

$$u = u_0[1 - (r/R)^{(n+1)/n}] \tag{5.4-7}$$

where u_0 is given by using the boundary condition that $u(r = 0) = 0$:

$$u_0 = -[nR/(n + 1)][(R/2K)(dP/dz)]^{1/n} \tag{5.4-8}$$

The negative sign appears because our pressure drop (dP/dz) was defined such that it increased with z, thus yielding the negative sign [8]. Flow is in the left to right direction in Fig. 5.4-2. The total volumetric flow rate is obtained by integrating the velocity of Eq. (5.4-8) across the cross-section:

$$Q = 2\pi \int_0^2 ru\ dr = [(n + 1)/(3n + 1)]\pi R^2 u_0$$

$$= [n\pi R^3/(3n + 1)](R\ \Delta P/2KL)^{1/n} \qquad (5.4-9)$$

To determine the shear rate $\dot{\gamma}_w$ at the wall, differentiate the velocity profile with respect to r and evaluate it at $r = R$. This yields

$$\dot{\gamma}_w = -[(3n + 1)/n](Q/\pi R^3) \qquad (5.4-10)$$

Note from Eq. (5.4-9) that Q is proportional to ΔP to the 1/n power. Thus $dQ/d\ \Delta P = Q/n\ \Delta P$, a constant, or in the Rabinowitsch-Mooney equation,

$$-\dot{\gamma}_w = (1/\pi R^3)[3Q + Q/n] = [(3n + 1)/n](Q/\pi R^3) \qquad (5.4-11)$$

In the standard capillary tube flow, then, the shear stress is determined directly from pressure drop without recourse to the power-law model. The shear rate, on the other hand, must be calculated from the flow rate, and the determination of the viscosity index is determined by a log-log plot of Q vs. ΔP. The slope of the line is the reciprocal of the viscosity index. Having n, the viscosity coefficient can be obtained by calculation from Eq. (5.4-9) for Q vs. ΔP, since K is the only unknown value. In many cases, if we are determining the viscosity of the fluid in the region of processing, we can simply draw a straight line through the Q vs. ΔP data (even though it might curve slightly) and get approximate values for n and K that are suitable for engineering calculations. The technique avoids the Rabinowitsch-Mooney form of calculation, which, while not very complex, still cannot be easily used in established fluid flow and heat transfer equations where the power-law information is needed.

There are many other types of viscometers on the market. Although the book written by Van Wazer and his coworkers in 1962 [21] is quite outdated, the basic viscometric principles that they describe are the bases for nearly all viscosity-measuring devices today. Coleman et al. [17] consider the following five types of measuring devices to be laminar shearing devices, and thus they should yield identical information:

1. Steady laminar flow in a capillary tube (just illustrated)
2. Simple shear flow between parallel plates (considered briefly in the section on more complex rheometers)
3. Flow between coaxial cylinders with one cylinder turning

4. Shearing flow between cone and plate
5. Torsional deformation between parallel disks

Since several books have been devoted entirely to this subject, we shall not
explore in depth the various equations that describe each of these devices.
We shall consider, however, some popular viscosity-measuring devices in
order to fit them into one of the above categories. The reason for this is
obvious; if we are measuring a physical property on a device, we want that
physical property to be relatable to values measured by others on other
devices.

An example of the type of device that is in common use in commercial laboratories is the "cup-and-bob" viscometer. The most popular is the
Brookfield viscometer. Here a "bob" of some geometry (usually a spindle
having a disk with one of several available diameters attached) is fastened
to a constant-speed spindle that is equipped with a torsion spring. The "bob"
is then immersed into a "cup" containing the fluid to be tested and the
constant-speed motor is turned on. The drag of the fluid (or better, its
resistance to the shearing action of the bob) "winds" the spring to a fixed
tension, and the resultant torque is registered on a dial. Normally the dial
is calibrated directly in "centipoise." For Newtonian fluids, such as silicone oil, petroleum products, glycerine, waxes, molten metals, and glass,
the geometry of the bob is of little importance, and as a result the readings
are easily interpreted. For polymers, however, the results require careful analysis before it can be stated that the readings found on the dial in
centipoise are in fact relatable to the viscosity of the fluid at that shear
condition. In general, the device comes with several bobs, some of which
resemble flat disks of various diameters with thin spindles through the centers. Others are simply cylinders or "can lids." The major problem with
most bobs is that the geometries cannot be described mathematically for
materials that are non-Newtonian (or even power law). Recognizing this
problem, some manufacturers are now providing bobs of various diameters
at additional cost. Spherical bobs are available, and direct torque readout
is now an optional extra on some machines. Many liquid resin suppliers
(epoxies, polyesters, urethanes, etc.) can provide the user with viscosity
data in the following general form: the fluid centipoise value at 60 rpm (the
spindle speed) with a number 6 spindle (a small-diameter disk mounted on
a thin spindle) and the fluid centipoise value at 6 rpm with a number 6
spindle. The ratio of these two values is referred to as the "thixotropic
index," and the larger it is, in general, the more non-Newtonian the fluid
is. But beware. The spindle geometry is so complex that to attempt to
determine if a 10-fold increase in rotational speed is equivalent to a 10-fold
increase in shear stress is, in many cases, an impossible task. Therefore, use this information only as a relative measure of the non-Newtonianness of the fluid, and place no value on the "centipoise" value given by the
dial of the viscometer.

For cylinders rotating in cylindrical vessels (concentric cylindrical geometry), the rate of deformation of the fluid is

$$r \, d\Omega/dr = -\dot{\gamma} = f(\tau) \qquad\qquad (5.4\text{-}12)$$

where Ω is the angular velocity of the cylinder in radians per second, $\dot{\gamma}$ is the shear stress, and $f(\tau)$ is a defined function (= τ/μ for a Newtonian fluid). The geometry is shown in schematic form in Fig. 5.4-3. The inner cylinder is driven with a constant angular velocity Ω by the application of a torque T. The torque must balance the torque exerted by the fluid on the face of the inner cylinder. Hence, $T = 2\pi R^2 L \tau_R$, where R is the radius of the inner cylinder, L is its length, and τ_R is the shear stress exerted on the inner cylinder. For this shear field, $r^2\tau$ = constant for all $r \leq R$. By differentiation,

$$r^2 d\tau + 2r\tau \, dr = 0 \qquad\qquad (5.4\text{-}13)$$

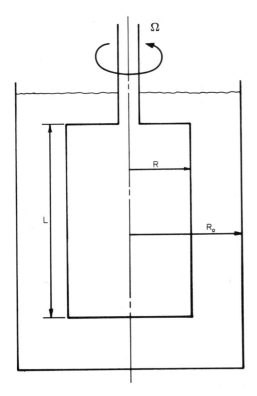

Figure 5.4-3   Schematic for cup-and-bob viscometer.

Thus, Eq. (5.4-12) can be written as

$$f(\tau) = 2\tau \, d\Omega/d\tau \tag{5.4-14}$$

Now $\Omega = 0$ at $r = R_0$ (the radius of the outer cylinder), and $\tau = \tau_0$ at that point as well. Integrating Eq. (5.4-14) from $r = R$ to $r = R_0$ yields

$$\Omega = (1/2) \int_{\tau_0}^{\tau_R} f(\tau)\tau^{-1} \, d\tau \tag{5.4-15}$$

Applying Leibnitz' rule for the differentiation of an integral, we obtain

$$d\Omega/d\tau_R = (1/2)\{f(\tau_R)/\tau_R - [f(\tau_0)/\tau_0](d\tau_0/d\tau_R)\} \tag{5.4-16}$$

Now $r^2\tau$ is a constant. Then, $d\tau_0/d\tau_R = (R/R_0)^2 = s^2$. The above equation becomes

$$2\tau_R \, d\Omega/d\tau_R = f(\tau_R) - f(s^2\tau_R) \tag{5.4-17}$$

We know how torque varies with shear stress ($T = 2\pi R^2 L \tau_R$). Thus, all the terms in Eq. (5.4-17) can be obtained from measurement of torque vs. rpm data. For very tight-fitting bob-in-cup arrangements, s is very close to unity, and $f(\tau_R)$ must be determined by trial and error from the $d\Omega/d\tau_R$ data, since $f(\tau_R)$ is implicit in Eq. (5.4-17). Note that if s is zero, as would be the case if the cylinder were revolving in an infinite medium, Eq. (5.4-17) allows exact solution for $f(\tau_R)$:

$$f(\tau_R) = \dot{\gamma}_R = 2\tau_R \, d\Omega/d\tau_R = 2\Omega(d \ln \Omega/d \ln T) \tag{5.4-18}$$

Thus, the shear rate from this equation can be obtained by measuring the change in torque with rpm. The shear stress can be calculated directly from knowing the torque. Of course once we have this we can plot shear rate against stress. The slope of this curve is the viscosity. Note that the torque must be measured directly and that therefore dials calibrated in viscosity units are worthless. Some manufacturers have seen this and are converting their machines over to direct torque readout, as mentioned.

There are certain cases in which the bob and cup are nearly the same diameter. As we can see from Eq. (5.4-18), we can get a much higher shear stress under this condition for the same spindle rpm. Letting $m = d \ln \Omega/d \ln \tau_R$, Eq. (5.4-18) can be approximated as [19]

$$f(\tau_R) = \dot{\gamma}_R = (\Omega/-\ln s)[1 - m \ln s + 1/3(m \ln s)^2] \tag{5.4-19}$$

This approximation is suitable as long as $-m \ln s < 0.5$. Thus, we have an explicit, albeit approximate, expression for the shear rate, and the torque yields the shear stress. With the appropriate plot we have the viscosity.

As Middleman points out [19], the results should always be checked with the known mathematical models. For a power-law fluid, for example, m = 1/n, and Eq. (5.4-18) becomes

$$f(\tau_R) = \dot{\gamma}_R = 2\Omega/m(1 - s^{2/n}) \qquad (5.4\text{-}20)$$

And for a Newtonian fluid, m = n = 1,

$$f(\tau_R) = 2\Omega/(1 - s^2) \qquad (5.4\text{-}21)$$

The Brookfield engineering staff has found that they can use this information to establish a correction factor for their "Newtonian" viscosity readings. As long as the fluid is truly power law, the dial readings (in Newtonian centipoise) can be corrected to power-law viscosities (which are shear dependent) by dividing the dial reading by the following correction factor:

$$g(m,s) = m(1 - s^2)/(1 - s^{2m}) = (1 - s^2)/n(1 - s^{2/n}) \qquad (5.4\text{-}22)$$

As an example, if the radii ratio is 0.95 = s and the fluid has a power-law viscosity index of 0.5, the correction factor g(m,s) has a value of 1.05. Still, the most accurate way of using the concentric cylinder information is to use direct torque and rpm readout. There are many literature references to correction factors that should be applied to the concentric cylinder viscometer. For example, there is an upper limit on the stability of the fluid being sheared between the cylinders. At very high shear rates, secondary flow effects known as Rayleigh waves are formed, and any correlation of data in this range is obviously invalid. Furthermore, the bottom and the top of the cylinder are not in concentric shear. And as a result, very frequently, cylinders of increasing length are used to eliminate the "end effects." And we have not included the effects of viscous dissipation and the heating of the fluid owing to shear. Since viscosity is exponentially dependent on temperature, a few degrees increase in the fluid temperature can lead to a dramatic decrease in the measured viscosity.

A cup-and-bob viscometer can be purchased, fully equipped with x-y plotter for stress-strain curves, for less than about $3000. If the stress-strain curve is to be extended over several decades of shear (10^{-4} to 10^1 sec^{-1}, say), a more sophisticated (and thus more expensive) viscometer is needed. The common viscometer with a wide range is a cone-and-plate viscometer. For this device, the subject fluid is placed between a flat plate and a cone having an angle of not more than 4°. In most models, the cone is rotated at a constant rpm, and torque is measured by the tendency of the plate to torque. Many people have studied the arithmetic of the cone-and-plate viscometer, and their results are highlighted here, following Middleman [19]. If θ is the angle measured from the vertical and ϕ is the cone angle [then $\sin \theta = \sin(\pi/2 - \phi)$], a force balance on the rotating fluid between the cone and plate is obtained:

$$r^{-1} \, d\tau/d\theta - (2 \cot \theta/r)\tau = 0 \tag{5.4-23}$$

where τ is the shear stress and r is the radius outward from the center of the cone apex. Integrating this,

$$\tau = C/\sin^2 \theta \tag{5.4-24}$$

where C is the constant of integration. Now, as before, there is a direct relationship between the torque T and the shear stress at the surface of the cone:

$$T = \int_0^R \tau \, 2\pi (r \sin \theta)^2 \, dr \quad \text{(evaluated at } \phi) \tag{5.4-25}$$

After integration [by substitution of τ from Eq. (5.4-24)] the value for C is obtained:

$$C = 3T/2\pi R^3 \tag{5.4-26}$$

and

$$\tau = 3T/2\pi R^3 \sin^2 \theta \tag{5.4-27}$$

For very small values of ϕ, $\sin \theta$ is approximately unity. Therefore, the shear stress across the distance between the cone and the plate is nearly constant. If this is the case, then the shear rate must also be nearly constant. The shear rate is given as

$$-\dot{\gamma} = r^{-1} \, dv/d\theta \tag{5.4-28}$$

If $\dot{\gamma}$ is the constant, we can integrate to obtain the velocity profile:

$$v = -r\dot{\gamma}(\theta - \pi/2) = r\dot{\gamma}\theta \tag{5.4-29}$$

But at the surface of the cone, $v = r\Omega$. Substitution obtains

$$\dot{\gamma} = \Omega/\phi \tag{5.4-30}$$

Thus, you see, if we measure the torque, we can obtain the shear stress. If we measure the speed of rotation, we can obtain the shear rate. And again, having these two values, we can determine the slope of the curve, which of course is the viscosity (regardless of what values of shear rate we are working at). Unfortunately, the upper limit of this device is determined by the ability of the fluid to remain in the gap at high shear rates. When fluid is lost, the data are wrong. The upper limit is considered to be 1 to 10 $\sec^{-1}$.

   For many years it was thought that the capillary flow data (normally giving shear rates in the range of $10^3 \sec^{-1}$ and up) and the cone-and-plate viscometry data (at the lower shear rates) could not be reconciled because the data in the middle range of shear were missing. Recently, however, a device known as an "orthogonal rheometer" has been developed that has a useful range from about 1 to 10^5 sec^{-1}. The instrument has two parallel plates, one of which is rotated at a fixed rpm. The other, which can either be fixed or allowed to rotate freely, contains strain-gage circuitry to measure torque transmitted through the fluid. At very low shear rates the plates can be operated as conventional parallel plates. At higher shear rates, however, the bottom plate is moved off-center with respect to the top plate. The eccentric wobbling that is induced subjects the plastic to alternating extension and relaxation of the imposed stresses. Although this device does not quite meet the conditions of a viscometric measuring machine, the results are easily interpreted and fit both the cone-and-plate data and the capillary flow data at both ends of the spectrum [22]. Furthermore, the use of the strain gages enables determination of normal stresses in the fluid. As detailed earlier, normal stresses are those over and above the standard hydrostatic pressure. These stresses are apparently the cause of the die swell of an extruding elastic thermoplastic or molten plastic climbing the rotating shaft of a stirring rod. Normal stresses are also important in determining the amount of material that is needed is a blow-molded object. Under any conditions, however, normal stresses until recently have been quite difficult to measure. The older viscometers had no provision for measuring the "pushing" force. The newer viscometers, however, are equipped with strain-gage bridge circuitry sensitive enough to measure the force exerted by the fluid on the restraining plates while the plates are moving in rotational shear motion. Thus, examination of normal stresses of polymer melts is underway on many fronts.

5.5 Other Types of Viscosities

For a long time, it was thought that thermosetting materials such as phenol-formaldehyde and melamine-formaldehyde had no true viscosity, because as the materials were heated to the "soft" molding condition they would continue their cross-linking reaction. Thus, they would become more and more viscous until, of course, the reaction was nearly complete and the material became a rigid solid. Other liquid, catalyzable thermosetting materials, such as epoxies, polyesters, and polyurethanes, were known to have viscosities that increased abruptly at one point during the cross-linking reaction. Thus, while the viscosities of the uncatalyzed materials were easy to determine, the viscosities of the reacting liquids remained for the most part unmeasured. Recently, however, newer resins have been developed for the injection molding of thermosets. Because long times are required in the processing of these materials, the resins such as phenol-formaldehyde must have long periods where their "softness" remains nearly constant and, more important, known. Plastic Engineering Co. (Plenco), among other companies, has initiated extensive rheological studies of their

thermosetting resins in an attempt to characterize the materials during
this fluid state. Early results indicate that the materials behave as shear-
dependent non-Newtonian fluids that can best be described by power-law
models. Typical viscosity time curves for phenolics and epoxies are
shown in Figs. 5.5-1 and 5.5-2. Many batch operations processing steps
have been developed around thermosetting resins (compression molding,
casting, spray up, and so on). Therefore, data on the viscosity of the
catalyzable resins are also important. Recently Kamal and coworkers [24]
have been examining the viscosity-time curves of these thermosets as they
approach the gel points, and Craig [25] has studied similar curves for the
curing of an injection-moldable cross-linking rubber (EPDM). While
cross-correlation between the extent of reaction and the viscosity of the
reacting mixture is still absent for most materials, it appears that here,
too, the fluids can be described as power-law fluids (where the viscosity
coefficient is time dependent). These data are discussed in Chap. 7.

In addition to squeezing or shearing reactive materials, we may
wish to stretch softened plastic. In thermoforming, for example, a sheet
of plastic is softened until pliable and then stretched into a mold. It is
held there until cold. The stretching of soft plastic sheets involves another
set of viscosities known as elongational viscosities. Any time we have a
deformable plastic, elongating it requires energy to overcome its "melt
strength." Drawing a highly viscous Newtonian fluid, such as molasses or
silicone oil, results in breaking the fluid strand at very low levels of drawing strength. Materials such as nylon, on the other hand, can be drawn
as liquid to very small diameters by simply applying sufficient tension on
the fluid. Trouton [26] first characterized the strength of a fluid as an
"elongational viscosity," that is a resistance of the fluid as it is being
stretched. Trouton found in 1906 that this elongational viscosity for simple
stretching in one dimension was directly proportional to the shear viscosity
for a Newtonian fluid, where the proportionality constant was 3:

$$\eta_e = 3\eta_n \tag{5.5-1}$$

η_e is the elongational Newtonian viscosity, and η_n is the shear viscosity.
Recently, Denson and Gallo [27] have identified three forms for extensional
viscosity: a uniaxial extensional viscosity (e.g., Trouton's viscosity); a
uniform biaxial extensional viscosity, where $\eta_b = 6\eta_n$; and a pure shear
extensional viscosity, where $\eta_s = 4\eta_n$. The shear viscosity here is apparently the power-law viscosity in the cases where the fluid being elongated
is non-Newtonian and can be treated as power law. There is some disagree-
ment here, though, as shown in Fig. 5.5-3, where for PMMA the measured
elongational viscosity is independent of tensile stress, that for LDPE shows
a nearly two-decade increase in value over a 10-fold increase in tensile
stress, and that for polypropylene also shows a nearly two-decade decrease
in value over a 10-fold increase in tensile stress. And recently, Hoover
and Tock [29] have carefully examined the stress-deformation behavior of

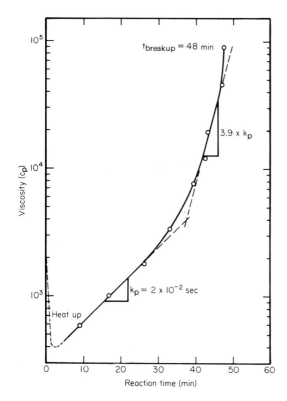

Figure 5.5-1   Cone-and-plate viscosity for phenol-formaldehyde reactive
system; T = 100°C [23].

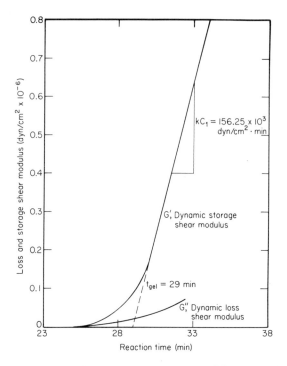

Figure 5.5-2   Dynamic storage and loss moduli for Epon 828/HHPa, 1:1, epoxy system; catalyst: TEA, 1%; 10 sec^{-1} frequency at 20% amplitude and 115°C. Note the gel time of 29 min, obtained by extrapolation to zero G', elastic shear storage modulus [23].

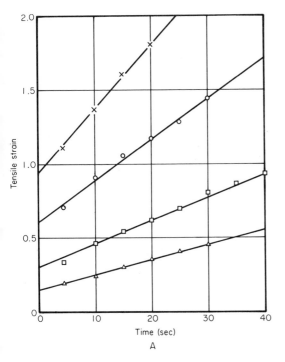

Figure 5.5-3A: Elongational strain; LDPE, MI = 0.3, 150°C. X, 4.2 x
10^5 dyn/cm; circle, 2.3 x10^5; square, 1.1 x 10^5; triangle, 0.51 x 10^5
dyn/cm.

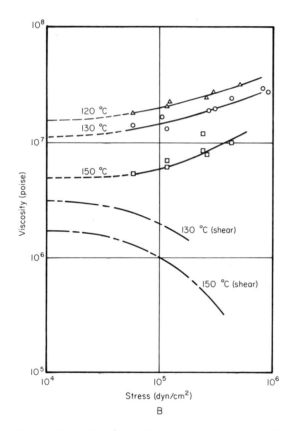

Figure 5.5-3B: Elongational and shear viscosities for two temperatures; LDPE, MI = 0.3.

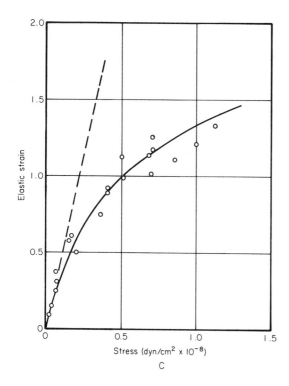

Figure 5.5-3C: Elastic strain for PMMA at 120°C [28].

of polypropylene and found no rate-of-deformation behavior below the melt temperature. Funt [30] considers that polypropylene below its melt temperature can be better described by characteristic large deformation stress-strain behavior of solids than by viscoelastic behavior of fluids. White [31] warns that the potential for stress-induced crystallization with PEs and PP must not be overlooked. This effect would cause an artificial increase in viscosity.

Nevertheless, we can get some idea of the effect of viscosity on the drawing ability of a filament by making a simple force balance:

$$\pi r^2 \tau_{xx} = F_t \qquad\qquad (5.5\text{-}2)$$

r is the variable radius of the filament, τ_{xx} is the normal stress on the fluid in the direction of draw, and F_t is the tensile force on the filament, which, in this case, is assumed to be constant. The value of the radius r can be related to the velocity profile in the draw direction by using the conservation of mass equation:

$$v(dr/dx) = -(r/2)(dv/dx) \qquad\qquad (5.5\text{-}3)$$

Let dv/dx be the shear stress caused by tension in the x-direction. It can be written as $\dot{\gamma}_{xx}$. Introduce the uniaxial elongational viscosity as

$$\tau_{xx} = \eta_e \dot{\gamma}_{xx} = K(\dot{\gamma}_{xx})^{n-1} \dot{\gamma}_{xx} \qquad\qquad (5.5\text{-}4)$$

Furthermore, $v(dr/dx) = dr/dt$, and therefore

$$K\pi r^2 (-2/r)^n (dr/dt)^n = F_t \qquad\qquad (5.5\text{-}5)$$

or, rearranging,

$$r^{(2-n)/n} \, dr/dt = -(1/2)(F_t/\pi K)^{1/n} \qquad\qquad (5.5\text{-}6)$$

Upon integrating (with the condition that when $t = 0$, $r = R$),

$$[1 - (r/R)^{2/n}] = n^{-1}(F_t/\tau KR^2)^{1/n} \qquad\qquad (5.5\text{-}7)$$

Or, put in terms of the tensile force,

$$F_t = \pi KR^2 n \{[1 - (r/R)^{2/n}]\}^n \qquad\qquad (5.5\text{-}8)$$

Remember, now, that this is for power-law fluids. It appears that perhaps power law is not the best model for describing the viscosity of elongation. Note if $a = \pi r^2$, the cross-sectional area at any point (and $A = \pi R^2$, the initial cross-sectional area) for a Trouton fluid (where the elongational viscosity is constant), the tensile force is directly proportional to the draw-down area ratio, as has been assumed in most designs for windup machines. Note that if the viscosity index n is less than unity, the early draw forces would be greater than those for a Troutonian fluid. At maximum draw (a/A = 0), for n = 0.5, however, we would need only about 70% of the force required to draw a Troutonian fluid. On the other hand, if n is greater than unity, the early draw forces are also greater than those for Troutonian fluids, but they remain that way even at maximum draw, where for n = 2 the forces are four times that for the Troutonian fluids. Probably the power-law model is not applicable here; more appropriate would be a viscosity model that exhibited two Newtonian regions and a power-law region between. Thus, this problem should probably be reworked using a four-constant Ellis model.

The above equation is of course restricted in its applicability in practical situations. For example, in most spinning operations, the filaments are cooled while being drawn. The coefficient K in the above equations should reflect the temperature dependence of viscosity. The resulting equation, where K has been replaced with an Arrhenius-type expression, and a proper energy equation used to calculate the filament temperature as a function of distance down the filament have not been fully explored, except to note that the results are nonlinear and not experimentally verified. The major effort in fiber spinning is being conducted in Europe by Ziabicki and his associates [32-34]. Their work includes not only the melt draw-down zone but the cool draw-down area, where the material orients in the direction of draw and if crystallizable, solidifies. They are also working extensively on the problem of drawing of cold filaments. Some recent work in the United States has been done on the cooling and orientation of film sheets, but this work seems highly empirical and not extrapolatable to filaments.

The other two types of elongational viscosities mentioned by Denson and Gallo [27] should have great interest to plastics processors. The biaxial elongational viscosity is that viscosity that relates stress and rate of deformation of fluids being stretched in two dimensions. This, then is apparently the viscosity that we should use to determine fluid flow characteristics of blown films during processing. The data on this viscosity and its shear dependency are very limited. However, Denson believes that the shear dependency of any elongational viscosity can be predicted as a first approximation as power law. The shear extensional viscosity for polyisobutylene is shown in Fig. 5.5-4, along with Denson's method of obtaining this property. He simply inflates (in an isothermal medium) a sleeve of softened plastic and measures the rate of deformation of a square scribed on its surface. Vinogradov et al. [35] have also measured axial expansion on polyisobutylene, and their data are shown in Fig. 5.5-5. The shear stretching of a polymer melt occurs in injection molding of a thermoplastic into a symmetric cavity and in transfer molding of a liquid (or softened) thermoset. Essentially, we are stretching the fluid as it fills the cavity. This approach to the interpretation of viscosity as applied to various types of plastics processing is quite timely and appears very fruitful.

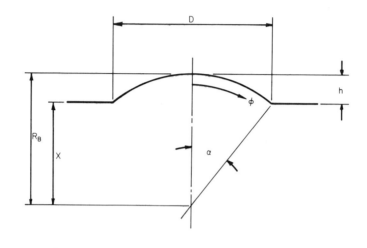

A

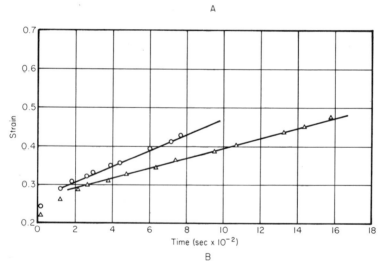

B

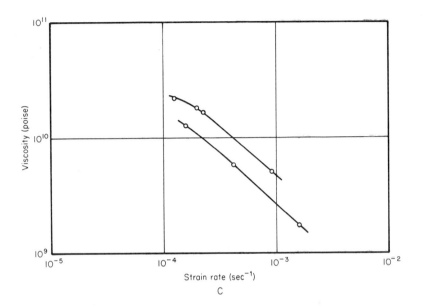

Figure 5.5-4   A: Coordinates for biaxial sheet inflation, elongation.
B: Strain curves for polyisobutylene. Triangles, strain rate, 1.32×10^{-4}
sec^{-1}; stress, 2.9×10^{6} dyn/cm^2; circles, strain rate, 2.10×10^{-4} sec^{-1};
stress, 3.5×10^{6} dyn/cm^2.   C: Biaxial extensional viscosity for two types
of polyisobutylene [27].

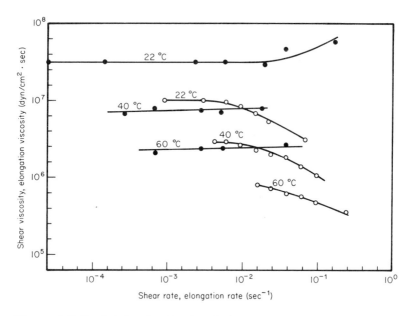

Figure 5.5-5   Axial extensional and shear viscosity for polyisobutylene.
Solid circles, axial extensional viscosity; open circles, shear viscosity [35].

5.6 Comments About Viscosity

There is no question that viscosity is one material property that is required
to determine, among other things, pressure drop-flow rate relationships.
The problem is that this property must be measured. The simple methods,
such as the cup-and-bob viscometer or the melt indexer, are simply not
adequate to determine the shear dependency of viscosity. More sophisticated machines have been developed according to certain guidelines established from the understanding of the interrelationship of stress and rate of
deformation in the fluid element. With these machines, the viscosity profile measured on one machine can be related to that measured on another.
More sophisticated machines have the capability of determining normal
stresses, which are unique to polymeric materials and which play important
roles in phenomena such as melt fracture, die swell, and elongational behavior of the fluids.   The entire field of viscometry has not been explored
here.   Others have done this in much more elaborate ways.   We have illustrated some of the viscometers that might be encountered in processing
laboratories. And finally other types of viscosities that may be of importance
in determining processing conditions have been reviewed.

Problems

P5.A A Bingham plastic is described according to the following equation:

$$\tau = \tau_c + \eta(du/dr) \qquad\qquad\qquad (5.A-1)$$

Follow the arguments leading to the Rabinowitsch-Mooney equation (5.2-11) for a Bingham plastic. Discuss the types of experiments needed to obtain the τ_w vs. $\dot{\gamma}_w$ data.

P5.B Green and Weltman [36] outline the equations for a concentric cylinder rotational viscometer with a Bingham plastic as the measured fluid. Repeat the analysis leading to the bob-in-cup equation (5.4-17). Then determine $\dot{\gamma}_w$ and τ_w knowing the following information about quartz in Nujol at 30°C:

rpm	Torque
0	0.66×10^5 dyn $\times$ cm
80	1.4
160	2.15
240	2.85
320	3.45
400	4.2

Can a correction factor be established for correcting these values to "Newtonian" viscosity readings? Why or why not?

P5.C Frizelle and Norfleet [37] have shown that thermoset phenolics exhibit power-law viscosities when processed through a special-purpose capillary rheometer. For Plenco 509, for example, at 235°F, the viscosity at 10 sec^{-1} is 36 lb·sec/in.^{2} and at 1000 sec^{-1} is 0.48 lb·sec/in.^{2}. Obtain the values of K and n at this condition. Consider that the viscosity is exponentially dependent on temperature according to

$$\eta = \eta_0 \exp(-bT/T_0) \qquad (T \text{ in } °R) \qquad\qquad (5.C-1)$$

and is exponentially dependent on the extent of reaction. Consider the extent of reaction to be reflected in an exponential increase in η_0 with time, according to

$$\eta_0 = \eta_\infty \exp(at) \qquad\qquad\qquad (5.C-2)$$

For a typical phenolic, b = 0.68 and a = 0.0462. Make a series of plots of η vs. shear rate for a Plenco-type resin over the temperature range of 230 to 270°F and the shear rate range of 10 to 1000 sec^{-1}.

P5.D The nylon shown in Fig. 5.2-4 is extruded at 565°F through a laboratory pipe 2 cm in diameter by 100 cm long under a pressure drop of 150 psi. What assumptions are necessary to scale this laboratory pipe to one of 25 cm in diameter by 250 m long operating under the same pressure drop? Hint: Determine the proper form for the Rabinowitsch-Mooney equation first.

P5.E (Following McKelvey [8]) The Arrhenius equation describing the relationship between viscosity and temperature depends on interpretation of the activation energy of the given material:

$$\eta = A \exp(E/RT) \tag{5.E-1}$$

It has been found that the activation energy E can be defined either at constant shear stress (E_τ) or at constant shear rate $(E_{\dot\gamma})$. Thus, Eq. (5.E-1) should be written as

$$\eta = A \exp(E_\tau/RT) \quad \text{or} \quad \eta = A \exp(E_{\dot\gamma}/RT) \tag{5.E-2}$$

For a power-law fluid, $E_\tau = (1/n)E_{\dot\gamma}$, where n is the power-law index. Show by maniuplation of Eqs. (5.E-2) that

$$\ln(\eta_2/\eta_1) = -(E_{\dot\gamma}\, \Delta T/RT_1 T_2)$$
$$\ln(\eta_2/\eta_1) = -(E_\tau\, \Delta T/RT_1 T_2) \tag{5.E-3}$$

Find the relationship between the logarithmic values on the left-hand sides of the equations for a power-law fluid. In a constant-stress experiment, the viscosity of a non-Newtonian fluid decreases from 100 to 50 P as the temperature increased from 150 to 170°C. The fluid has a power-law index of 0.6. Compute E_τ for the material, and estimate its viscosity at 190°C for a constant-shear-rate experiment.

P5.F In Eqs. (5.3-7) and (5.3-8) are given the extent of deformation for discrete Maxwell and Voigt spring-dashpot elements. First combine the Voigt model with a spring in series and obtain the total stress on the model. Then combine the Voigt and Maxwell models in series and obtain the total stress on this composite model. Draw the stress-rate-of-deformation curves for each composite model.

P5.G As mentioned in the text, the simple models of Maxwell and Voigt do not reliably mirror the response of a simple linear viscoelastic fluid. Derive the equation describing the extent of deformation of a series of Maxwell elements with n moduli values G_n and n viscosities η_n. Can this series model be extended to an infinite set of Maxwell elements? Describe how this might be done.

P5.H Since the Maxwell spring and dashpot describes a very simple viscoelastic system, it can be rewritten in terms on energy dissipation and energy storage terms. Rewrite Eq. (5.3-7) in terms of τ as

$$\tau = \eta\dot\gamma - (\eta/G)\dot\tau \tag{5.H-1}$$

Then let $\lambda = \eta/G$, and show that for constant strain ($\dot{\tau} = 0$) the solution of Eq. (5.H-1) is

$$\tau/S = \exp(-t/\lambda) \tag{5.H-2}$$

where S is the applied stress at time t = 0. What is the physical meaning of λ? Introduce a periodic strain $\gamma = \gamma_0 \sin(wt)$, where $w = 2\pi f$ and f is the frequency of strain (sec^{-1}). Solve Eq. (5.H-1) for τ as a function of time, allowing the system to be periodic for a long period of time. This allows the transient terms to go to zero. Show that the solution is given as

$$\tau = [G\gamma_0 w\lambda/(1 + w^2\lambda^2)][w\lambda \sin(wt) + \cos(wt)] \tag{5.H-3}$$

Show that if the strain has only real components that that part of Eq. (5.H-3) that is in phase with the strain is

$$\tau'(t) = G\gamma_0 w^2\lambda^2 \sin(wt)/(1 + w^2\lambda^2) \tag{5.H-4}$$

Show also that that portion that is out of phase with the strain is represented as

$$\tau''(t) = G\gamma_0 w\lambda \cos(wt)/(1 + w^2\lambda^2) \tag{5.H-5}$$

Then, if G' is defined as the ratio of the real components of stress to strain and is called the storage modulus and if G" is defined as the ratio of the complex components of stress to strain and is called the loss modulus, show that they are represented as

$$G''/G = w\lambda/(1 + w^2\lambda^2) \qquad G'/G = w^2\lambda^2/(1 + w^2\lambda^2) \tag{5.H-6}$$

Make plots of G"/G and G'/G against $x\lambda$, and show that G"/G has a maximum when $w\lambda = 1$. Discuss the shapes of the curves in terms of the mechanical behavior of a viscoelastic material. Outline an experiment to measure G"/G and G'/G over a wide range of shear rates.

P5.I Macosko and Mussatti [23] devised a method for measuring the loss and storage moduli for thermosetting resins in terms of their extent of cross-linking. They point out that the complex dynamic shear modulus $G^* = G' + iG''$ can be directly related to the cross-link density of the resin as

$$G^* = 2RTX \tag{5.I-1}$$

Consider the reaction kinetics of a thermosetting resin to be first order:

$$dX/dt = kX \tag{5.I-2}$$

where k is the reaction rate constant, R is the gas constant, and T is the absolute temperature. Now show that G^* can be related to the time and G_0^*, the initial dynamic shear modulus, as

$$G^* = G_0^* \exp(kt) \qquad\qquad\qquad (5.I-3)$$

Obtain the individual values of G' and G'' from Eq. (5.I-3). Macoski and Mussatti measured G' and G'' for epoxy at a fixed frequency of $10 \ sec^{-1}$ and a fixed shear strain amplitude of 20%. They found an energy of activation of 9.3 kcal/mol and a reaction rate constant of $16.2 \times 10^{-7} \ mol/cm^3 \cdot min$ at $100°C$. Calculate G^* assuming that G'' is much less than G', and compare the results with their measured values of $G' = 0.16 \times 10^6 \ dyn/cm^2$ at 60 min and $G' = 1.78 \times 10^6$ at 75 min. Discuss the assumption that $G'' \ll G'$ for an epoxy system.

P5.J Using the appropriate terms for the loss and storage moduli, obtain the maximum amount of energy stored by a viscoelastic material at any time in a given cycle. Obtain the total amount of energy dissipated by the material during a given cycle.

References

1. Transactions of the Society of Rheology is a quarterly journal published by the society, 605 Third Avenue, New York.
2. J. A. Brydson, Flow Properties of Polymer Melts, Van Nostrand Reinhold, New York, 1970, Chap. 1.
3. A. S. Lodge, Elastic Liquids, 2nd ed., Academic Press, New York, 1972, Chaps. 2, 3.
4. L. L. Blyler, Polym. Eng. Sci., 14:806 (1974).
5. K. K. Chee and A. Rudin, Can. J. Chem. Eng., 48:362 (1970).
6. R. B. Bird, W. E. Stewart, and E. N. Lightfoot, Transport Phenomena, Wiley, New York, 1960.
7. J. R. A. Pearson, Mechanical Principles of Polymer Melt Processing, Pergamon Press, Elmsford, N.Y., 1966, p. 9.
8. J. M. McKelvey, Polymer Processing, Wiley, New York, 1962, p. 15.
9. C. D. Han, Rheology in Polymer Processing, Academic Press, New York, 1976.
10. K. Walters, Rheometry, Chapman & Hall, London, and Wiley, New York, 1975.
11. A. C. Eringen, Nonlinear Theory of Continuous Media, McGraw-Hill, New York, 1962, p. 160.
12. M. R. Spiegel, Theory and Problems of Vector Analysis, Schaum Publishing Co., New York, 1959, p. 171.
13. M. Reiner, Deformation, Strain and Flow, Wiley-Interscience, New York, 1960.
14. J. D. Ferry, Viscoelastic Properties of Polymers, Wiley, New York, 1961, Chap. 2.
15. W. Flugge, Viscoelasticity, Ginn/Blaisdell, Waltham, Mass., 1967, Chap. 1.

16. C. Truesdell, J. Rational Mech. Anal., 1:125-300 (1952).

17. B. D. Coleman, H. Markovitz, and W. Noll, Viscometric Flows of Non-Newtonian Fluids: Theory and Experiment, Springer-Verlag, New York, 1966.

18. E. B. Bagley, J. Appl. Phys., 28:624 (1957); see McKelvey [8], pp. 85-97, for additional details.

19. S. Middleman, The Flow of High Polymers: Continuum and Molecular Rheology, Wiley-Interscience, New York, 1968, p. 8.

20. J. M. Kay, An Introduction to Fluid Mechanics and Heat Transfer, Cambridge University Press, London, 1963.

21. J. R. Van Wazer, J. W. Lyons, K. Y. Kim, and R. E. Colwell, Viscosity and Flow Measurement, Interscience, New York, 1963.

22. B. Maxwell and R. P. Chartoff, Trans. Soc. Rheol., 9:41 (1965).

23. C. W. Macosko and F. G. Mussatti, SPE ANTEC Tech. Pap., 18:73 (1972).

24. M. R. Kamal, S. Sourour, and M. Ryan, SPE ANTEC Tech. Pap., 19:187 (1973).

25. D. F. Craig, General Electric Technical Information Series 71-MAL-2 (1971), Schenectady, New York, 1971.

26. F. T. Trouton, Proc. Roy. Soc. London Ser. A, 77:426 (1906).

27. C. D. Denson and R. J. Gallo, Polym. Eng. Sci., 11:174 (1971).

28. F. N. Cogswell, Trans. Plast. Inst., 36:110 (1968).

29. K. C. Hoover and R. W. Tock, Polym. Eng. Sci., 16:82 (1976).

30. J. M. Funt, Polym. Eng. Sci., 15:817 (1975).

31. J. L. White, personal communication, Nov. 1974.

32. A. Ziabicki and K. Kedzierska, Kolloid-Z., 171:51 (1960).

33. A. Ziabicki and K. Kedzierska, Kolloid-Z., 171:111 (1960).

34. A. Ziabicki, Kolloid-Z., 175:14 (1962).

35. G. V. Vinogradov, B. V. Radushkevich, and V. D. Fikhman, J. Polym. Sci. Part A, 2(8):1-17 (1970).

36. R. N. Weltman, Rheology of Pastes and Paints, In: Rheology, Vol. 3 (F. R. Erich, ed.), Academic Press, New York, 1960, p. 2010.

37. W. G. Frizelle and J. S. Norfleet, SPE ANTEC Tech. Pap., 19:538 (1973).

6.

Heat Transfer to Solid and Molten Polymers

6.1 Introduction

In nearly all plastics processing operations, a plastic must be heated in order to move it easily from one area of the process to another. Sufficient
energy must be provided to make it sufficiently pliable to attain the shape of
the desired part. And it must be cooled so that it retains that final part
shape. If the material is a thermosetting resin, we should be concerned
with the rate of cross-linking reaction that occurs when we elevate the material temperature. The rate at which the reaction approaches completion
must be known to determine the optimum time for removal from the mold,
for example. We should also be aware of the excellent thermal insulating
properties of plastics in both the solid and fluid states. Thus, not only are
the fluids of high viscosity (meaning that they flow laminarly), but the mode
of heat transfer during laminar flow is conduction. Since plastics have
such poor thermal conductivities, transfer of heat to or from molten plastics in flow and solid plastics in static cases is very difficult. In addition,
the high viscosities of those plastics that are pumpable require the use of
large amounts of power in transfer from point to point in the process. Thus,
excess power is transformed into heating the material through viscous dissipation or the conversion of mechanical energy to thermal energy through
shearing of the fluids. Many processes, such as extrusion, rely heavily on
shear heating as a primary mode of heat transfer.

In this chapter, we shall concentrate on the three forms of heat transfer—conduction, convection, and radiation—as they relate to plastics processing. Convection implies fluid motion, of which there are two major types: laminar and turbulent. Because of the extremely high viscosities of the molten plastics, we shall consider only laminar motion. The criterion for laminar-turbulent transition for steady flow in a horizontal circular pipe is the value of the Reynolds number. When the Reynolds number exceeds a value of about 2000, the flow is turbulent. As an example of the order of magnitude of the Reynolds number for polymers, consider the flow of LDPE in a standard circular runner of 1/4-in. diameter under standard injection molding velocity of about 1 ft/sec. The viscosity of LDPE at an extrusion temperature of 400°F is approximately 0.1 $lb_f \cdot sec/in.^2$, and its density is 57.6 lb_m/ft^3. The Reynolds number (calculated without regard to the Newtonianness of the fluid) is

$$Re = DU\rho/\mu = \frac{(1/48) \text{ ft} \times 57.6 \text{ lb}_m/\text{ft}^3 \times 1 \text{ ft/sec}}{0.1 \text{ lb}_f \cdot \text{sec/in.}^2 \times 144 \text{ in.}^2/\text{ft}^2 \times 32.2 \text{ ft} \cdot \text{lb}_m/\text{lb}_f \cdot \text{sec}^2}$$

$$= 0.0026$$

It is apparent that the flow for most plastics is laminar and that the principal mode of heat transfer is conduction. And since the thermal conductivity of plastics is in general very low, heat transfer in convection flow will be controlled by conduction through the plastic.

There are two types of conduction of importance in plastics processing. The first is a simple steady-state conduction and is very easy to analyze. The more difficult form of conduction, transient or time-dependent conduction, is unfortunately the type that occurs most of the time. This type of heat transfer problem is analyzed in some detail also.

And finally, we shall consider radiation, which is the most difficult form of heat transfer to understand and to analyze in terms of plastics processing. In particular, radiative heating is being used extensively to heat thermoplastic sheet to forming temperatures, to condition surfaces to accept certain types of adhesives, to "rapid-dry" printing on surface of plastics, and so on. Thus, it is necessary to understand the rudimentary concepts of radiation heat transfer in order to design this type of equipment.

6.2 Conduction Heat Transfer

Probably the most important area where steady-state conduction heat transfer is used in plastics processing is in the cooling of plastic parts by conduction of heat through a mold. Mold cooling designs will be considered in some detail in the chapter on injection molding, but here some of the groundwork for that material is outlined. Conceptually, if a heated surface is brought into contact with a block of material of finite thickness that is kept cool at a constant temperature on the opposite side, conduction heat transfer will quickly heat the block nearest the hot surface first. This wave of heat will progress through the block until steady state is reached. For example, if the hot surface has temperature T_h, and the cold surface has temperature T_c, and the block is X units thick, the temperature profile through the block at steady state is linear:

$$(T_h - T)/(T_h - T_c) = x/X \qquad (6.2-1)$$

where x is the distance from the hot surface and T is its temperature. The term on the left is an "unaccomplished temperature change." It is used extensively in transient heat conduction. This equation can be written in terms of the rate of heat flow into the block at any surface:

$$Q/A = -k \, dT/dx\Big|_{x=0} = -k \, dT/dx\Big|_{x=X} \qquad (6.2-2)$$

e.g., what is conducted in must be conducted out to maintain steady state. Q/A is the heat flux (Btu/ft$^2 \cdot$hr), and Q is the total rate of heat flow (Btu/hr). From Eq. (6.2-2),

$$Q/A = k(T_h - T_c)/X = (T_h - T_c)/(X/k) = \Delta T/R \qquad (6.2-3)$$

Q/A is thought of as a current of heat, ΔT as the thermal driving force, and $R = X/k$ as the resistance to heat flow. This has direct analogies to electrical circuit theory; $I = E/R$. Other instances of this analogy will be seen later.

For composite structures (such as insulation layers covering electrical band heaters, which in turn supply energy to metal extruder barrels), recall that whatever heat flows into a structure must flow out to maintain the steady state [1]. Therefore, a heat balance across several layers of insulative material, shown in Fig. 6.2-1, is

$$Q/A = (T_1 - T_2)/R_1 = (T_2 - T_3)/R_2$$

$$= (T_3 - T_4)/R_3 = \cdots = (T_n - T_{n+1})/R_n \qquad (6.2-4)$$

And by eliminating the internal temperatures (T_2, T_3, etc.) the overall heat transfer is

$$Q/A = (T_1 - T_{n+1})/\sum_{i=1}^{n} R_i \qquad (6.2-5)$$

If, as in the case of an insulated heater band, a heat source is located within layers of insulative material, then a heat balance must be made across that heater band layer. If Q' is the heat source strength per unit volume per unit time (Btu/ft$^3 \cdot$hr), determined by knowing the amount of I^2R power

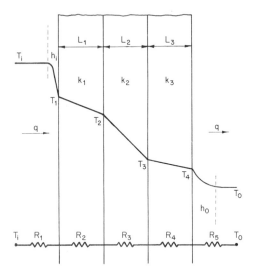

Figure 6.2-1   Steady-state temperature profile through composite wall;
equivalent electrical circuit shown below [1, p. 40].

into the heat and the dimensions of the heater, then the heat transfer equation that describes temperature profile in the heater 2L units thick is given
as

$$-k \, d^2T/dx^2 = Q'$$ (6.2-6)

where $T(x = 0) = T_0$ and $T(x = 2L) = T_0$. The last term can be relaxed
easily. The solution to the equation yields the temperature profile:

$$T - T_0 = (Q'L^2/2k)[2x/L - (x/L)^2]$$ (6.2-7)

The heat flux at the surface (Q/A at $x = 0$) is given by differentiation to yield

$$Q/A = Q'L$$ (6.2-8)

The way to handle the insulative layers is to realize that "heat runs down-
hill." Thus, if we know the maximum temperature within the heater band,
we can make a Q/A heat balance from that point. The maximum temperature occurs in this case at the center ($x = L$), and the temperature difference
between the surface and the center is given as

$$T_c - T_0 = Q'L^2/2k$$ (6.2-9)

or, rearranging into the form for the passive insulative layers in series,

$$Q/A = Q'L = (T_c - T_0)/R_c \qquad\qquad (6.2\text{-}10)$$

where $R_c = L/2k$. This equation and Eq. (6.2-5) can be used to determine
the energy requirements to a barrel heater. We determine how much "leaks"
out through the insulation and how much is actually conducted through to the
plastic.

Very frequently the steady-state conduction heat transfer problem is
two-dimensional. Methods of estimating and improving heat transfer in the
injection molds are detailed in the section on mold cooling in Chap. 10.
However, consider here potential-field theory. Recall that the simple one-
dimensional heat conduction equation had a direct analogy to electric circuitry. The same is true for steady two-dimensional heat transfer. Take
a very simple geometry, such as that shown in Fig. 6.2-2. We can draw
lines of constant temperature parallel to the surfaces. These are isotherms.
Perpendicular to these lines are lines of constant heat flow (for we see that
$Q/A = -k \, dT/dx$). These lines are called heat flow lines. If the heat flow
lines and the isotherms are spaced the same distance apart, so that the
space enclosed is a small square, the change in heat flux (between two

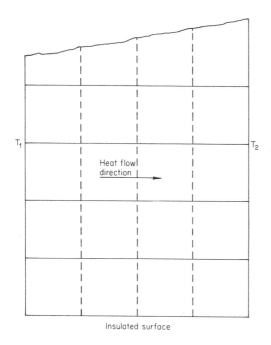

Figure 6.2-2 Isotherms and constant heat flux lines in one-dimensional
wall. Dashed lines, isotherms; solid lines, heat flux lines [1, p. 98].

isotherms) is related to the change in temperature (between two heat flux lines):

$$\Delta Q = -k \, \Delta T \qquad\qquad\qquad (6.2-11)$$

This is independent of the sizes of the squares. Therefore, if we construct a field of isotherms, the heat flow lines must cross each isotherm at right angles. Let M be the number of squares of isotherms between any two walls and N be the number of heat flow tubes at right angles to these isotherms. Then a measure of the conduction heat transfer for any shaped object is obtained simply by counting:

$$Q = (N/M)k(T_1 - T_2) = Sk(T_1 - T_2) \qquad\qquad (6.2-12)$$

S, the ratio of the number of heat flow tubes to the number of isotherm squares, is the shape factor. The calculation of the shape factor is simple and can give a good indication of the efficiency of the conduction process. Obviously, the higher the value of S, the greater is the amount of heat transferred for a given thermal driving force. Krieth [1] gives a step-by-step procedure for plotting these isotherms and flux lines freehand. Other techniques will be discussed later; the Handbook of Heat Transfer [2] gives extensive tables for shape factors of various geometries. There are also numerical methods for obtaining the temperature net for two-dimensional steady conduction problems. The reader is referred to either Kreith's book or the Handbook of Heat Transfer for the principles of techniques such as relaxation and finite differences.

Consider the more difficult but more common case of transient heat conduction. A typical plastics processing problem is the immersion of a sheet of thermoforming plastic, initially at room temperature, into a convective isothermal oven at, say, T_∞. Given the material properties and the thickness of the sheet, the time for the centerline of the sheet to reach a new temperature T_V, this temperature being the Vicat softening temperature, is required. (For melting a crystalline polymer, the temperature desired is T_m.) The problem can be reversed to consider a molten plastic at temperature T_∞ that is suddenly brought in contact with a constant temperature surface at T_0. Here we desire the time it takes for the centerline to reach a temperature that is cool enough to allow us to remove it from the mold without distortion, warping, or sink marks. In general we shall be heating or cooling a plastic that is passive. No chemical reactions or massive absorptions of heat take place during temperature change. This is not always the case, however, since in compression molding of thermosets, the cross-linking reaction does exotherm as it proceeds, thus distorting the temperature profile. Crystallization is also an exothermic process. Foaming is another example of transient heat conduction with unusual physical processes taking place to distort the simple models. Here, the cross-linking reaction that causes the thermoset foam to become rubbery is also exothermic, and the increase in temperature causes the blowing agent to vaporize.

This produces a gas. This gas expands the foam, forming very small cells and coincidentally decreasing the thermal conductivity of the mass in nearly direct proportion to the decrease in foam density. Thus, internal energy generation and foam formation occur, and heat removal can take place only through the surface of the foam that now has a time-dependent thermal conductivity. While the analysis of cooling of foams of variable densities across their cross-section is relatively straightforward, the added problem of understanding foam structure is beyond the subject matter in this section.

Thus, there seems to be sufficient motivation for studying the basic equation of heat conduction in a slab, a cylinder, and, on occasion, a sphere. The basic equations for time-dependent (transient) heat conduction in a slab and in a cylinder are [1, p. 152]

$$\rho c_p (\partial T/\partial t) = \partial (k \, \partial T/\partial x)/\partial x \qquad (6.2\text{-}13)$$

$$\rho c_p (\partial T/\partial t) = r^{-1} \partial (kr \, \partial T/\partial r)/\partial r \qquad (6.2\text{-}14)$$

where T is temperature (°F), ρ is density (lb_m/ft^3), c_p is specific heat ($Btu/lb_m \cdot °F$), x is the distance from the centerline for a slab, and r is the radius outward for a cylinder. There are hundreds of analytical solutions for these equations. Many can be found in Carslaw and Jaeger [3] or Crank [4]. Analytical solutions will not be considered except as illustrations of the form for the transient portion of the equations.

Consider first putting the slab equation into a dimensionless form as follows. Define a thermal diffusivity as $\alpha = k/\rho c_p$ and a dimensionless temperature, as before:

$$\theta = (T_f - T)/(T_f - T_i) \qquad (6.2\text{-}15)$$

T_i is the uniform initial temperature of the slab and T_f is its final uniform temperature, and a dimensionless thickness $X = x/L$, where L is the half-thickness of the slab (assuming heat treatment on both sides of the slab simultaneously). Define a dimensionless time $Fo = \alpha t/L^2$. Fo is the symbol for Fourier number. Note that Fo contains the thermal diffusivity and the half-thickness of the slab squared. Now the dimensioned equation (6.2-13) becomes

$$\partial \theta/\partial Fo = \partial^2 \theta/\partial X^2 \qquad (6.2\text{-}16)$$

The solution of any equation depends on the selection of boundary conditions. There are hundreds of practical boundary conditions that could be chosen. For example, one surface of the slab might be exposed to a constant temperature and the other to a sinusoidal temperature with convection heat transfer occurring simultaneously, or we might have an irregular initial temperature profile. For very unique boundary conditions, numerical solutions are normally required. For many problems, however, a compromise

between the practical conditions and known solutions to the equation with
conditions that are not quite the same is necessary. The reason for this is
obvious; for most engineering problems, accuracy is needed to within ±10%
at most. Therefore, a tabulated or graphical solution yields a much quicker
answer than does a computer analysis. There are hundreds of graphical
solutions to the above equation given by Schneider in [2, Chap. 3], and
therefore these should be considered first.

There are two major boundary conditions that should be considered:
the isothermal condition (taken care of in the definition of the dimensionless
temperature) and the convective boundary condition. With convection heat
transfer, energy is transferred through a slow-moving fluid layer near the
solid surface. The proportionality between the heat flux Q/A and the thermal
driving force between the bulk of the fluid and the solid surface is a heat
transfer coefficient. This was discussed earlier. Thus, if the plastic is
being cooled by forcing coolant over the surface, as is the case with a quench
water stream or chilled air (or equivalently if it is being heated in a similar
manner), then the resistance to heat transfer at the surface can also affect
the temperature profile within the material. The condition at the surface
relates the rate of heat removal owing to conduction from the interior of the
plastic to the film coefficient at the interface between the plastic and the
coolant. In this way, another dimensionless term, the Biot number, is used.
It is defined as $Bi = hL/k$, where h is the heat transfer coefficient of the
coolant $(Btu/ft^2 \cdot hr \cdot °F)$, L is the half-thickness of the slab, and k is the
thermal conductivity of the plastic.

Most texts on heat transfer give graphical displays of the solution
of Eq. (6.2-17) subject to step changes in surface temperature or in heat
flux at the surface. θ is plotted as a function of Fo for various values of X.
Most graphs have the Biot number as a parameter, as shown in Figs. 6.2-3
and 6.2-4 for slabs and cylinders, respectively. Consider the injection
molding of a slab of LDPE 1/2 in. thick initially at 470°F into a mold cavity
at a constant 70°F. Determine the time required to reduce the centerline
temperature of the slab to 170°F. Thus, $\theta = (70 - 170)/(70 - 470) = 0.25$.
The thermal diffusivity for LDPE is approximately 5×10^{-3} ft²/hr, and
thus Fo = 5×10^{-3} ft²·hr·t (hr)/$(1/48)^2$ ft² = 11.5t (hr). At the centerline,
X = 0. Furthermore, since the fluid is in the direct contact with the mold,
h = ∞, and the Biot number is also infinite. Thus, from Fig. 6.2-3, the
value for Fo is 0.65 = 11.5t, or t = 0.0565 hr = 204 sec. Note that we
could have quenched the LDPE sample with cold water (70°F) at essentially
the same rate, since the heat transfer coefficient for water is on the order
of 500 Btu/ft²·hr·°F, and the thermal conductivity of plastic is on the order
of 0.14 Btu/ft·hr·°F. Thus, the Biot number would be Bi = 500 × (1/48) ft/
0.14 = 74.4. The reciprocal is 0.0134. As seen from Fig. 6.2-3, the same
time would show only a slight increase (less than 5%). In fact, the Biot
number for the cooling of most slab stock is usually extremely large. Cooling of film, on the other hand, will lower its value, since it is directly

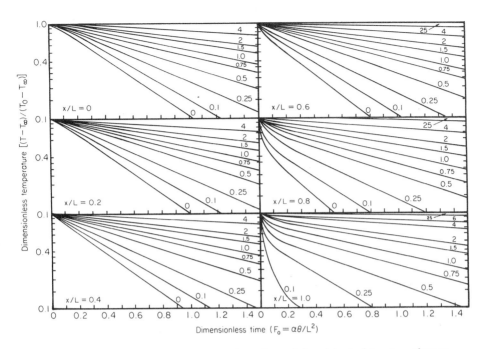

Figure 6.2-3 Dimensionless temperature in a slab subjected to step change in surface temperature. X/L = 0 is insulated centerline; x/L = 1 is surface; parameter is surface conductance, $Bi^{-1} = k/hL$ [1, p. 152].

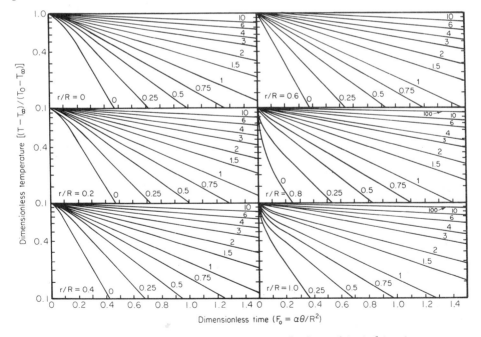

Figure 6.2-4 Dimensionless temperature in cylinder subjected to step change in surface temperature. r/R = 0 is centerline; r/R = 1 is surface; parameter is surface conductance, $Bi^{-1} = k/hR$ [1].

proportional to the thickness of the film. Even more important is the cooling of film by extrusion into quiescent air. The heat transfer coefficient for this case would have an upper limit of about 10 $Btu/ft^2 \cdot hr \cdot °F$. For the same conditions ($\theta = 0.25$), in the cooling of a 5-mil-thick film, the Biot number becomes

$$Bi = hL/k = 10 \ Btu/ft^2 \cdot hr \cdot °F \times (0.0025 \ in./12 \ in./ft)/0.14 \ Btu/ft \cdot hr \cdot °F$$

$$= 0.0149$$

The reciprocal is 67.1. From an extended Fig. 6.2-5, the Fourier number required to achieve this degree of cooling is approximately 50. Thus,

$$t = FoL^2/\alpha = 50 \times (2.08)^2 \times 10^{-8} \ ft^2/5 \times 10^{-3} \ ft^2/hr = 4.33 \times 10^{-4} \ hr$$

$$= 1.56 \ sec$$

The times required to cool centerlines of plastics having other shapes are given in Fig. 6.2-6. The surface resistance is zero ($Bi = \infty$), and thus information from this figure could be used to determine centerline temperatures of injection-molded disk-shaped plastic parts or spherical parts and so on.

Probably the most important factor in understanding the use of the heat conduction charts is that the Fourier number is proportional to the reciprocal of the square of the slab (or cylinder or sphere) thickness. Thus, if one section of a plastic part is twice as thick as another, the time required to cool it to the same temperature will be four times longer than that for the thinner section. Therefore, the thickest section of a plastic part will control the cooling portion of, say, an injection molding cycle. This is why production engineers try to encourage designers to design parts with uniform wall thickness. This fact will also be very important when we consider "gate freezing" of molded parts.

As an example of the effect of exothermic reactions on the cooling rates of plastic parts, consider the recent work of Collier and his associates. They have studied the cooling rates of a crystallizing plastic such as Celcon or, in their example, Penton (the chlorinated polyether, 3,3-bis-chloromethyl-oxycyclobutane). They solved numerically the heat conduction equation with a heat generation term. Such an equation would be

$$\rho c_p \ \partial T/\partial t = k \ \partial^2 T/\partial x^2 + r(\Delta H) \tag{6.2-17}$$

where r is the rate of reaction (or rate of crystallization) and ΔH is the heat of reaction (or heat of crystallization). In the case of crystallization, they used the Avrami equation to calculate the rate of crystallization. This equation has been discussed in detail in Chap. 4. The material has a melting temperature and thus a crystallizing temperature of 250°F. An example of their calculations is shown in Fig. 6.2-7, where the initial temperature is 575°F, the isothermal quench temperature is 200°F, the sample half-thickness

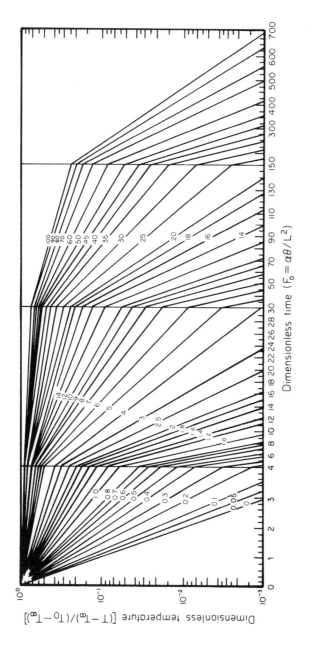

Figure 6.2-5 Centerline temperature in slab subjected to step change in surface temperature; extended scale for parameter, surface conductance, $Bi^{-1} = k/hL$ [1].

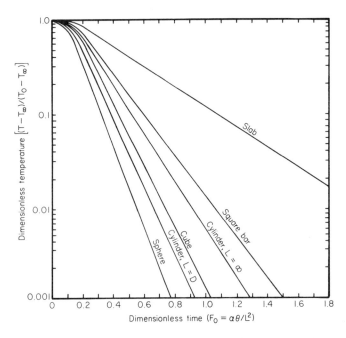

Figure 6.2-6   Centerline temperatures for various shapes subjected to step change in surface temperature; no surface conductance, $Bi^{-1} = 0$ [5].

is 0.016 in., and the heat transfer coefficient is estimated to be 120 Btu/ $ft^2 \cdot hr \cdot °F$. Note that the surface and the internal temperatures fall rapidly to values of 201 to 204°F. This is expected with this very thin film and the very high heat transfer coefficient. At these temperatures, the material is sufficiently subcooled to initiate very rapid crystallization, causing the surface and centerline temperatures to rise very rapidly until crystallization is complete (about 27% crystallinity at completion). From the data given, the thermal conductivity is k = 0.140 Btu/ft·hr·°F (they compute their curves on k = 0.076), and the thermal diffusivity is $\alpha = 4.6 \times 10^{-3}$ ft^2/hr. To get experimental verification of the calculated values, Collier and his coworkers heated a 0.483-in.-thick slab to 450°F and then quenched the top surface in water at 73°F. The bottom surface was insulated. They estimated the heat transfer coefficient to be about 500 Btu/ $ft^2 \cdot hr \cdot °F$. Under these conditions, the Biot number would be Bi = 500(0.483/12)/0.14 = 144, or essentially infinite. They were not able to maintain a uniform temperature for their slab (the insulated bottom surface was about 40°F lower than the top surface at the beginning of immersion). Nevertheless, they were able to show that the rate of cooling was delayed by the crystallization. In this case, with a thicker sample, they did not experience an increase in temperature but

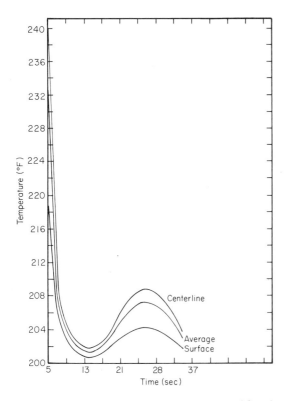

Figure 6.2-7 Cooling temperature profiles for Penton, chlorinated poly-ether. Initial temperature of melt, 575°F; mold temperature, 200°F; surface heat transfer coefficient, $h = 120$ Btu/ft$^2 \cdot$hr$\cdot$°F; thickness = 0.032 in. [6].

simply a slower cooling rate. For example, at an X value of 0.24, for an unaccomplished temperature change of $\theta = (73 - 250)/(73 - 450) = 0.469$, with no crystallization effects, the Fourier number should be Fo = $\alpha t/L^2 = 0.39$. Thus,

$$t = 0.39 \times (0.483/12)^2/4.6 \times 10^{-3} = 0.134 \text{ hr} = 483 \text{ sec}$$

Experimentally, a value of 650 sec was observed, which shows the effect of crystallization. Keep in mind that the initial temperature was not constant across the sample, and thus the simple application of the standard heat conduction charts is not reliable. For a value of X = 0.73, the time to cool to a $\theta = 0.469$ without crystallization effects is 135 sec, whereas the experimental value is about 240 sec. Collier and his coworkers force the computed and experimental values to agree by adjusting the thermal conductivity of the plastic to 0.140 Btu/ft$^2 \cdot$hr$\cdot^\circ$F (a more reasonable value, and the value that we have used) and by lowering the surface heat transfer coefficient from 500 to 60 (there seems little motivation for this since a decrease of a factor of 8 in the Biot number at these values does little to increase the Fourier number).

We would expect the lag on cooling to have a counterpart in a lead in heating if heat is generated by way of chemical reaction or equivalent. The effect of heat of crystallization can be seen in another way when the effect of thickness on the cooling rate is plotted. The sample thicknesses in Fig. 6.2-8 are varied from 0.032 to 0.128 in. Doubling the thickness increases the time to cool to a given temperature by a factor of 4, assuming that the fluid has no heat sinks or sources. As is seen, doubling the thickness for the crystallizing material increases the cooling time by a factor of about 3. This deviation is directly attributable to the heat sink in the material. One can surmise that if the material exhibited a heat source rather than a heat sink, doubling the thickness would increase the cooling time by factors in excess of 4. We have already discussed the effect of slow cooling or rapid quench on the crystalline structure of the material. Thus, only the rate of heat transfer is of concern here.

In the next chapter reactive materials during heating are considered, with emphasis on compression molding and casting in batch operation.

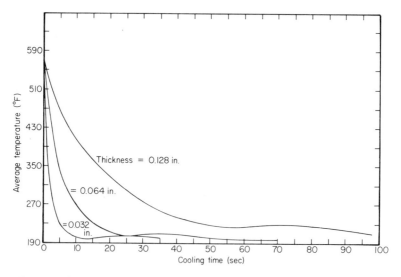

Figure 6.2-8   Average cooling temperature profiles for various thicknesses of Penton, chlorinated polyether. Initial melt temperature, 575°F; mold temperature, 200°F; surface heat transfer coefficient, h = 120 Btu/ft$^2 \cdot$hr$\cdot^\circ$F [7].

Frequently we need to know the temperature response of a metal mold to changes in the environmental temperature. The easiest analysis of this is a "lumped-parameter" transient heat conduction model. The thermal conductivity of the material is considered to be large, and the temperature through the material at any time is nearly uniform. Thus, a very simple heat balance is

$$
\begin{array}{c}
\text{The change in internal energy of} \\
\text{the mass during time dt}
\end{array}
=
\begin{array}{c}
\text{The net heat flow into the mass} \\
\text{from the environment} \\
\text{during time dt}
\end{array}
$$

$$\rho c_p L \, dT = h(T_a - T) \, dt \qquad (6.2\text{-}18)$$

ρ is the density of the mass (lb_m/ft^3), c_p is the specific heat (Btu/lb_m), L is the characteristic dimension of the material (V/A, where V is the volume of the mass in ft^3 and A is its surface area for heat transfer in ft^2), h is the heat transfer coefficient, T_a is the ambient temperature (°F), and T is the instantaneous mass temperature. If $T = T_0$ when $t = 0$, the solution to the differential equation is

$$(T_a - T)/(T_a - T_0) = \exp(-FoBi) = \exp(-hat/Lk) \qquad (6.2\text{-}19)$$

where $Fo = \alpha t/L^2$, the Fourier number, and $Bi = hL/k$, the Biot number. While this equation is used primarily for calculating metal responses to transient conditions, it can be used if the film coefficient is very small (as in the case of natural convection) and if the film thickness is also very small (as in the case of blown film of 0.5-mil thickness or less). Some representative values for α/k for metals are given in Table 6.2-1.

Table 6.2-1

Thermal Diffusivity to Conductivity Ratios for Several Metals at 500°F

Material	α/k (at 500°F) ($ft^3 \cdot °F/Btu$)
Aluminum	0.0025
Copper	0.00208
Nickel	0.00188
Steel	0.00196

6.3 Radiation Heat Transfer

Radiation is perhaps the least understood and most misapplied mode of heat
transfer. In plastics processing, radiant heating is the primary mode of
heat transfer in thermoforming. Here, plastics sheets or films are placed
beneath or between banks of infrared heaters. For all opaque and many
translucent plastics, the radiant energy emitted by the heaters is absorbed
on the surface of the plastic. The absorbed energy is then transmitted
through the body of the plastic by conduction. In some cases, convection
heat transfer from the surface of the sheet accompanies the radiant heat
transfer. Note that in these cases only the condition at the surface is affected. Thus, the equation to be solved is that given by Eq. (6.2-16) for a
slab with new boundary conditions that include radiation. The Handbook of
Heat Transfer [2] has many cases for combinations of convection and radiation. One of these, for the centerline of a slab where the ratio of initial
temperature to source temperature (in °R) is 0.25, is shown in Fig. 6.3-1.
The dimensionless parameter here is a radiative Biot number, $Bi_r =
\sigma F T_s^3 L/k$. σ is the Stefan–Boltzmann constant (= 0.172×10^{-8} Btu/ft^2·hr·°R^4),
T_s is the source temperature (in °R), F is an emissivity factor (discussed
below), L is the slab half-thickness, and k is the thermal conductivity. In
conventional radiation heat transfer, there is an interchange of energy between the source (the heated surface) and the sink (the cooler surface). The
interchange can be written as

$$Q/A = F\sigma(T_s^4 - T^4)$$
(6.3-1)

where T is the temperature of the sink, say the plastic in the case of radiant heating. F, as mentioned, is the emissivity factor, which is a function
of the geometries of the two objects and their emissivities. The emissivity
of a black body is 1.0. If the amount of energy emitted by an object is less
than the amount it would emit if it were a black body, the emissivity would
be less than unity. For most heater materials, the emissivities are 0.9 to
1.0. Likewise we find for most plastic materials that emissivities are also
in the range of 0.9 to 1.0. There are tables of emissivities—and methods
of calculating F, the emissivity, "view," or "shape" factor—in most books
on heat transfer, including McAdams [5], Brown and Marco [8], and the
Handbook of Heat Transfer [2]. For the simple case of parallel infinite
planes (typical of heating of thermoplastic sheet for thermoforming), the
emissivity factor F is given as

$$F = (1/e_1 + 1/e_2 - 1)^{-1}$$
(6.3-2)

If, for example, $e_1 = e_2 = 1$, F = 1. If $e_1 = e_2 = 0.9$, F = 0.82. Thus, a
slight reduction in the emissivity of either the plastic materials, owing to
changes in pigmentation, say, or the heater materials, owing to oxidation

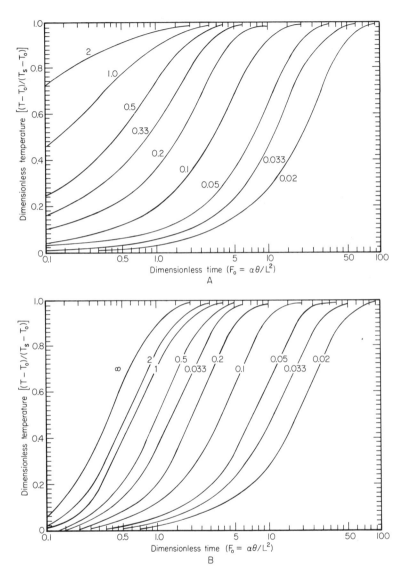

Figure 6.3-1   A: Temperature response to step change in surface temperature-radiative energy input. $x/L = 0$, surface heating. Parameter is radiative Biot number, $Bi_r = \sigma F T_s^3 L/k$; $T_0/T_s = 1/4$ (absolute). B: Temperature
response to step change in surface temperature-radiative energy input.
$x/L = 1$, centerline temperature. Parameter is radiative Biot number,
$Bi_r = \sigma F T_s^3 L/k$; $T_0/T_s = 1/4$ (absolute) [2, Chap. 3].

or deterioration of the heater surface, say, would cause a noticeable decrease in the total amount of energy interchanged by radiation. If one of the
surfaces (with area A_1) is completely enclosed by a second surface of area
A_2, the emissivity factor F for the heat flux (Q/A_1) is given as

$$F = [1/e_1 + (A_1/A_2)(1/e_2 - 1)]^{-1} \qquad (6.3-3)$$

As A_1/A_2 approaches zero, $F = e_1$. This type of geometry is sometimes
encountered when passing a rod, wire, or pipe through a radiant heater pri-
or to coating or embossing, for example.

The radiation Biot number can thus be easily calculated if the geometry
of the system is known. Then knowing the dimensionless temperature required, the Fourier number can be calculated, thus giving the time required
to heat the centerline to the desired temperature. Consider heating a sheet
2 in. thick initially at 550°R (90°F) with a radiant source at 2200°R (1740°F)
(thus, the ratio is 2200/550 = 0.25). Assume an emissivity factor $F = 0.9$.
The sheet thermal conductivity is 0.14 Btu/ft·hr·°F.

$$Bi_r = 0.172 \times 10^{-8} \times 0.9 \times (2/12) \times (2.2)^3 \times 10^9/0.14 = 19.6$$

To heat the centerline to 850°R (390°F), θ = (390 - 90)/(1740 - 90) = 0.182,
and from Fig. 6.3-1 a Fourier number of 0.171 is obtained. Thus, t =
$FoL^2/\alpha = 0.171 \times (2/12)^2/5 \times 10^{-3} = 0.948$ hr. If the sheet were 0.2 in.
thick, the radiative Biot number would be 1.96, and the Fourier number
would be about 0.25. The heating time would then become t = 0.0138 hr =
50 sec. Incidentally, from Fig. 6.3-1, for the surface temperature for this
0.2-in.-thick sheet, when the centerline temperature reaches 390°F, the
value for θ at the surface is 0.84. This means that the surface temperature
is theoretically 1475°F (which is probably far above its ignition temperature).
Thus, a common way of preventing this excessive surface temperature is to
cool it with convected room-temperature air. The ratio of maximum to
minimum temperature in the sheet has been referred to as an "evenness
index" EI or a "uniformity index." McKelvey's uniformity index [9] is maximum for short times and thick slabs and approaches unity for long times
and thin films. The evenness index of Progelhof and his coworkers [10] is
initially unity and approaches zero for long times. Probably the best way to
handle the problem of the complex radiation-convection boundary condition
is to introduce a fictitious heat transfer coefficient for radiation such that
the Biot number can be defined as Bi = h'L/k, where h' is the sum of the
convective heat transfer coefficient and the fictitious radiation heat transfer
coefficient. This aspect of heat transfer is considered in more detail in
Chap. 12.

For a slightly more complex but equally practical problem, consider
those materials such as Lucite (PMMA), Lexan (polycarbonate), or transparent ABS that are somewhat transparent to a large portion of incident

radiation. If these materials are heated using radiant energy, the rate of
heating is found to be much slower than that calculated using Fig. 6.3-1. The
reason is apparent; some of the incident radiation passes through the material without being absorbed, and thus the efficiency of heating is lower. If the
materials were completely transparent to incident radiation, they would not
heat at all. Most materials exhibit "windows" in their absorption curves, as
shown for polymethyl methacrylate in Fig. 6.3-2. Thus, if the predominant
portion of incident radiation has a wavelength band in this window region,
heating will be quite slow. Most plastics, fortunately, are semitransparent
in that the radiant energy is absorbed volumetrically as well as on the surface
of the material. Thus, if we know the infrared spectrum of the material (obtained in experiments outlined in Chap. 4), we can shape the spectrum of
the heating source to maximize the radiation absorbed by the material. And
with the alternatives of surface convective cooling and "on-off" control of
the heater banks, very uniform temperature profiles can be achieved in all
semitransparent plastics. This aspect of heating is also discussed in Chap. 12.

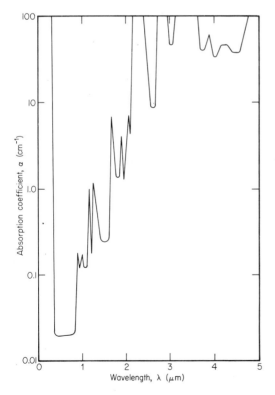

Figure 6.3-2   Radiation absorption spectrum for PMMA [11].

6.4 Convection Heat Transfer in Circular Tubes

It is perhaps a little strange that in this chapter we consider laminar flow heat transfer in circular tubes. Until very recently this was the only geometry that experimenters could use to verify some of the predicted temperature profiles generated by the vast array of theorists [12].

As mentioned earlier, when considering other forms of laminar flow, we are faced with a situation of developing velocity profile and developing temperature profile. In addition, the effects of viscous dissipation and compressibility effects of the fluid must be included.

Toor [13], as outlined by McKelvey [9, p. 81], shows that compressibility of plastics is not to be neglected. Skelleland [14, p. 363] considers the following equation to fully describe the temperature profile in a steady-state flow in a circular tube:

$$\rho c_p u \, \partial T/\partial x = (k/r) \, \partial[r(\partial T/\partial r)]/\partial r \tag{6.4-1}$$

For developing velocity profile, $u = u(r,x,T)$; ρ, c_p, and k are assumed to be temperature dependent but only to a limited extent. In reality, as has been shown by Spencer and Gilmore, plastics are compressible fluids, and thus decreasing the pressure on the fluid will cause it to expand and cool in much the same way as CO_2 exiting a nozzle of a pressure tank will cool and form "dry ice." This point is amplified in Chap. 14. As a result, compressibility must be included in the above equation:

$$d\rho = -\kappa \rho \, dT + \beta \rho \, dP \tag{6.4-2}$$

κ is the coefficient of thermal expansion, $= -\rho^{-1}(\partial \rho/\partial T)_P$, β is the compressibility, $= \rho^{-1}(\partial \rho/\partial P)_T$. When the effect is included, the momentum equation, which heretofore could be solved without reference to the energy equation, is now in fact coupled to the energy equation. The momentum equation, including the pressure- and temperature-dependent density, is

$$\rho u^2 [\kappa(\partial T/\partial x) - \beta(\partial P/\partial x)] = -(\partial P/\partial x) + r^{-1}(r \, \partial \tau/\partial r) \tag{6.4-3}$$

The second term on the right-hand side is the rate of change of shear stress. The energy equation has added terms, as well, to account for the changes in temperature owing to compressibility:

$$\rho c_p u(\partial T/\partial x) = (k/r) \, \partial[r(\partial T/\partial r)]/\partial r - T(\partial P/\partial T)_\rho (\partial u/\partial x) + \phi \tag{6.4-4}$$

where ϕ represents a viscous dissipation term, discussed shortly. The term $(\partial P/\partial T)_\rho = \kappa/\beta$ is a function of temperature and pressure, in general. The viscous dissipation term ϕ in the above equation is, in essence, a very complex set of derivatives of velocity with position [15]. If the fluid is Newtonian, a good approximation for ϕ would be

$$\phi = \mu(\partial u/\partial r)^2 \tag{6.4-5}$$

For power-law fluids, the viscous dissipation can be written as

$$\phi = K(\partial u/\partial r)^{n+1} \tag{6.4-6}$$

Porter [16] points out that most studies either neglect this factor or include it as a Newtonian dissipation factor. Metzner [17] cautions that the dissipation term should probably contain higher orders of shear stresses (and also different types of shear stresses, including normal shear stresses). The theoretical data indicate that ϕ as defined above predicts viscous dissipation greater than that measured. Incidentally, Porter's paper tabulates all the theoretical models and experimental data taken on laminar flow in capillary tubes.

We shall explore this concept of viscous dissipation or "viscous heating" a little more to understand the apparent discrepancies between the experimental evidence and the theories. In Fig. 6.4-1 for a very simple geometry, where shear is occurring on one surface of a slab of fluid (the other surface being held at a velocity of zero), the velocity (assuming no compressibility or pressure effects, e.g., $\kappa = \beta = 0$) is proportional to the distance from the moving shear surface: $u = (x/b)V$. The viscosity is assumed to be temperature independent. Now the volume heat source resulting from viscous dissipation is

$$S_V = -\tau(du/dx) \tag{6.4-7}$$

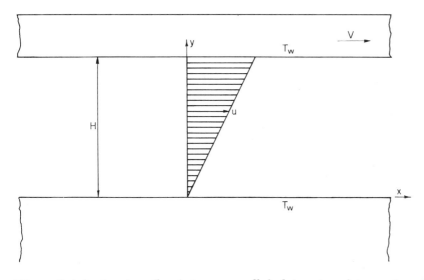

Figure 6.4-1   Laminar flow between parallel plates, top plate moving at constant velocity V [9].

where τ is the shearing stress. If the temperatures at the two surfaces are $T(x = 0) = T_0$ and $T(x = b) = T_b$, respectively, a very simple energy balance (assuming steady state) says that

$$qA\big|_x - qA\big|_{x+\Delta x} + A \, \Delta x S_v = 0 \qquad\qquad (6.4\text{-}8)$$

where A is the surface area in the z-direction. Substituting,

$$dq/dx = \tau \, (du/dx) \qquad\qquad (6.4\text{-}9)$$

Assuming a power-law equation of state between shear stress and rate of deformation,

$$\tau = K(du/dx)^n \qquad\qquad (6.4\text{-}10)$$

or, substituting this into Eq. (6.4-9),

$$dq/dx = K(du/dx)^{n+1} \qquad\qquad (6.4\text{-}11)$$

Remember, now, that dq/dx is the energy generated solely by viscous dissipation. q, the heat flux, is defined as before: $q = -k \, dT/dx$. Thus, this equation can be integrated twice, using the two temperature conditions, to obtain

$$\theta = (T - T_0)/(T_b - T_0) = (x/b) + (Br/2)(x/b)[1 - (x/b)] \qquad (6.4\text{-}12)$$

Br is the Brinkman number, defined as $Br = [KV^{n+1}/kb^{n-1}(T_b - T_0)]$. Note that for a Newtonian fluid, $K = \mu$, $n = 1$, and $Br = (V^2\mu)/(T_b - T_0)k$, e.g., the ratio of inertial energy to conduction energy. Thus, in addition to the Prandtl number $c_p K/k$ and the Reynolds number (based on some definition for the viscosity which we shall consider shortly), the Brinkman number should appear as a measure of the extent of viscous dissipation. We shall return to this point shortly.

As Porter and Skelleland point out, perhaps the first important contribution to the understanding of the temperature profile in a circular pipe was by Christiansen and Craig [18]. They used a power-law shear model with a viscosity that was exponentially dependent on temperature in an Arrhenius fashion. They placed rather severe restrictions on the form of their energy equation, allowing neither viscous dissipation (Br = 0) nor thermal expansion nor compressibility effects ($\kappa = \beta = 0$). Thus, the momentum equation (6.4-3) can be solved independently of the energy equation. A solution can be obtained:

$$u = R(\tau_w/K)^{1/n} I_1 \qquad\qquad (6.4\text{-}13)$$

where I_1 is a dimensionless integral, defined as

$$I_1 = \int_{r/R}^{1} (r/R)^{1/n} \exp(-\Delta H/\overline{R}T) \, d(r/R) \qquad (6.4\text{-}14)$$

ΔH is the energy of activation for viscous flow, and $\overline{R}$ is the gas constant. (Units on $\Delta H/\overline{R}T$ must be consistent since the exponential must be unitless.) Now if Q is the total mass flow rate in the pipe, given by

$$Q = 2\pi\rho \int_0^R ru\ dr = 2\pi\rho R^2 (\tau_w/K)^{1/n} I_2 \qquad (6.4\text{-}15)$$

where I_2 is the integral of I_1,

$$I_2 = \int_0^1 I_1(r/R)\ d(r/R) \qquad (6.4\text{-}15)$$

the velocity at any point in the pipe (under the above restrictions) is given as

$$u = (Q/2\pi R^2 \rho)(I_1/I_2) \qquad (6.4\text{-}17)$$

Define the Graetz number, a measure of reciprocal length, in the following way:

$$Gz = Qc_p/kx \qquad (6.4\text{-}18)$$

where x is the distance down the pipe. The energy equation can be rewritten as

$$c_p(Q/2\pi R^2 \rho)(I_1/I_2)(k/Qc_p)[\partial T/\partial(1/Gz)] = (k/R^2)y^{-1}\ \partial[(y\ \partial T/\partial y)]/\partial y \qquad (6.4\text{-}19)$$

where $y = r/R$. After appropriate cancelling of terms, we obtain the final form for the Craig-Christiansen equation:

$$\partial T/\partial(1/Gz) = 2\pi(I_2/I_1)y^{-1}\ \partial[y(\partial T/\partial y)]/\partial y \qquad (6.4\text{-}20)$$

The obvious method of solution of this equation is by computer, since I_1 and I_2 must be obtained simultaneously with the temperature profile. This has been done many times since the early work of Craig and Christiansen with variations and permutations on the boundary conditions [16]. Incidentally, note that the velocity profile is assumed to be fully developed prior to heat transfer. Further, isothermal temperature has been assumed along the walls. Subsequent investigators have considered the other easy alternative of constant heat flux along the walls [16]. There seems to be some question as to the applicability of either of these ideal cases. Nevertheless, as shown in Fig. 6.4-2 for the Nusselt number as a function of the Graetz number (e.g., the inverse of the length down the pipe—thus the larger Gz is, the shorter the pipe is), for values where $\Delta H/\overline{R}T$ is zero (viscosity independent of temperature), there is little significant difference between the power-law

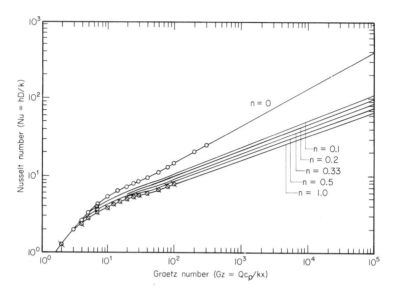

Figure 6.4-2   Isothermal heat transfer (Nusselt number) as a function of
reciprocal distance (Graetz number) for laminar flow in circular tubes.
Open circles, Graetz' theoretical solution, n = 0; slashed circles, Graetz'
theoretical solution, n = 1 (Newtonian). Viscosity is independent of temperature [18].

fluids and the Newtonian fluids insofar as heat transfer is concerned.   The
Nusselt number is defined here as $Nu = hD/k$.   Including the influence of
temperature on viscosity for $n = 1/3$, as shown in Fig. 6.4-3, $\psi(H)$ [= ($\Delta H/$
$\overline{R}T_0 - \Delta H/\overline{R}T_W$), where T_0 and T_W represent the initial bulk temperature
and the wall temperature, respectively] increases the rate of heat transfer
as it increases.   Of course these values assume that the fluid is being heated
by heat transfer from the wall (the values are positive).   Similar computer
calculations with negative values of $\psi(H)$ should show the opposing trend.
       Note that full solutions of the energy equation across the entire pipe
are not always necessary.   It is only when the velocity profile is fully developed initially that the above approach to the solution of the energy equation is valid.   The computations of Lyche and Bird shown in Fig. 6.4-2 are
fully developed power-law fluids.   For a very short pipe (the length of the
pipe will be discussed shortly), a technique known as the Leveque solution
can be used.   Leveque assumed that the velocity profile was linear with
distance from the wall of the pipe [18] and thus recommended the use of his
method for conditions where the familiar parabolic profile had not fully
developed.   More important, at high values of Gz, nearly all heat transfer
occurs in the thin layer near the wall.   Thus, even the parabolic profile for

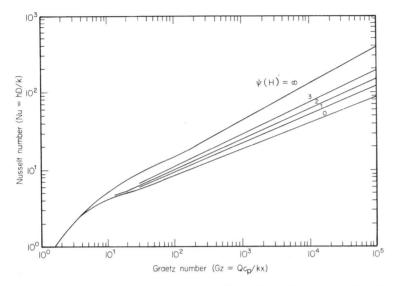

Figure 6.4-3   Effect of temperature-dependent viscosity on isothermal heat transfer coefficient for power-law fluid, n = 1/3 [18].

velocity can be assumed to be linear in this region. Incidentally, it is fairly well established [16] that the entrance length for the development of a steady velocity profile (assuming no major effects owing to compressibility or dramatic temperature effects on viscosity) is $L/D = 0.03Re$, where for Newtonian flow $Re = DV\rho/\mu$ and for power-law flow $Re = D^n V^{2-n} \rho \cdot 8[n/(6n + 2)]^n/K$. In general, $Re = 8\rho V^2/\tau_w$. Extension of this work to the fully developed thermal profile has shown that

$$L/D = 0.05\ RePr \tag{6.4-21}$$

where again Re is the Reynolds number and Pr is the Prandtl number, defined as $Pr = c_p\mu/k$ for a Newtonian fluid and equivalently for a power-law fluid. Note that for rather viscous fluids, such as polymers, Re's can be on the order of 1 or less. Thus, the development of the velocity profile is rather rapid, occurring in less than 0.03 pipe diameter (at Pr = 1). On the other hand, the Prandtl number (containing the viscosity in the denominator) might be on the order of 10^3 or more. This means that to obtain the fully developed thermal profile the pipe might need to be more than 50 pipe diameters long. The ratio of the thermal to hydrodynamic entry length is given approximately by

$$L_t/L_n = 1.75Pr \tag{6.4-22}$$

Thus, it is quite possible to have a fully developed velocity profile in a very short pipe and a developing thermal profile in a very long pipe. Pigford [19] extended Leveque's work to power-law fluids as follows:

$$Nu = 1.75[(3n + 1)/4n]^{1/3} Gz^{1/3} \qquad (6.4\text{-}23)$$

If we examine this equation in conjunction with Fig. 6.4-2 with the complete profile, we see that until we reach a value of $Gz = 10$ the slope of the curve (regardless of the viscosity index) is $1/3$. Beyond that point (Gz less than 10) the slope approaches that computed by Graetz for a parabolic velocity profile. As an example of the order of magnitude of the Graetz number, consider the experimental data of Griskey [20]. For a 0.957-in.-diameter pipe, at 167.5 g/min, with $k = 0.14$ and $c_p = 0.61$ $Btu/lb_m \cdot °F$ (LDPE), the Graetz number at 9.525 ft from the inlet to the pipe was 10.7. Thus, at this long distance ($L/D = 120$) the flow was just beginning to develop a parabolic profile, and the Leveque solution was becoming invalid.

Gee and Lyon [21] included viscous dissipation and a volumetric expansion term κT_m that was assumed to be constant. They also solved the problem for Newtonian fluids, and thus their viscous dissipation term was $\phi = \mu(du/dr)^2$. Their computer results were rather surprising in that at an $L/D = 48$, the reduced temperature, $\theta = (T_w - T)/(T_w - T_c)$, where T_c is the centerline temperature at $L/D = 48$, was 4.25 at a distance $y = r/R = 0.8$. Actually, the maximum temperature of the perspex was 267°C, the wall was maintained at 250°C, and the centerline temperature was 254°C. The inlet temperature of the plastic was about 252°C, and thus the plastic was heating dramatically rather than cooling as it would with no viscous dissipation. The Brinkman number was not calculated, but from the temperature profile calculated earlier for simple shear, the maximum temperature can be obtained and back-calculated to find the value. We get a Newtonian $Br = 3.3$ (in order to have a maximum in the temperature profile within the fluid cavity, $Br \geq 2$).

Toor [13] replaced the viscous dissipation term with a power-law term, although he kept the viscosity temperature independent (as did Gee and Lyon). He found essentially the same types of curves as Gee and Lyon, and a set of his curves for various values of n and κT, the compressibility factor, are shown in Fig. 6.4-4 and 6.4-5. Note that as the pipe gets longer, for $\kappa T = 0$, $n = 1$, the effects of viscous dissipation cause the temperature to increase at a reduced radius $r/R = 0.8$. This confirms the work of Gee and Lyon. With increasing non-Newtonianness and increasing compressibility cooling, a reverse situation occurs. The centerline temperatures fall rapidly owing to high compressibility cooling and little viscous dissipation (in the center of the pipe the shear stress is zero), whereas the viscous dissipation in the region near the wall is sufficient to overcome any compressibility cooling. Note that in Fig. 6.4-4, $Y = \alpha x/U_0 R^2$, where U_0 is the bulk velocity. Thus, as x increases down the pipe, Y gets larger (contrary to the definition of the Graetz number). Additional mathematical modeling, to include

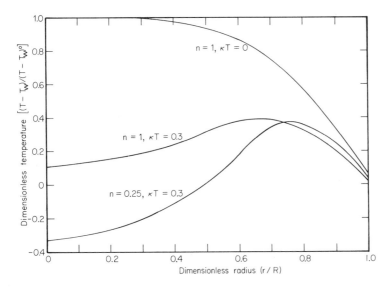

Figure 6.4-4   Effect of compressibility (κT) on temperature profile in Newtonian (n = 1) and power-law fluids (n = 1/4) flowing in fully developed laminar flow in a pipe [13].

temperature-dependent physical properties, has been carried out by Forsyth and Murphy [22].

Experimental examination of these effects has been carried out by Griskey and his coworkers [20]. Their objective was to determine how significant viscous dissipation and compressibility were to heat transfer of molten polyethylene and polypropylene. They used pipes ranging in diameter from 3/8 to 1 in. (nominal) and in length to 12 1/2 ft. To eliminate extruder surge, they fed the polymer through an auxiliary gear pump at the exit of the extruder. They heated the pipes with oil. The oil was hotter than the bulk temperature of the polymer at all points in the heat exchanger and was considered to be held constant. The inlet temperature of the plastic was held to 365°F ± 1.0°F. It should be noted that Griskey uses a different definition for the Brinkman number: $Br_G = 0.0234 \tau_W n/[kR(3n + 1)(T_1 - T_W)]$, where $\tau_W = R \Delta P/2L$. If the fluid is power law, $\tau_W = K(4V/R)^n[(3 + n)/4n]^n$. Thus, $Br_G = 0.0234 K n^{1-n} V^n/kR^{1+n}(3n + 1)^{1-n}(T_1 - T_W)$, where T_1 is the initial temperature of the fluid. The difference between Griskey's Brinkman number and that defined earlier ($Br = KV^{n+1}/kR^{n-1} \Delta T$) is a volumetric flow rate term:

$$Br_G = [0.0234 n^{1-n}/VR^2(3n + 1)^{1-n}] \times (V^{n+1}K/R^{n-1} \Delta T)$$

$$= [0.0234 n^{1-n}/VR^2(3n + 1)^{1-n}]Br \qquad (6.4-24)$$

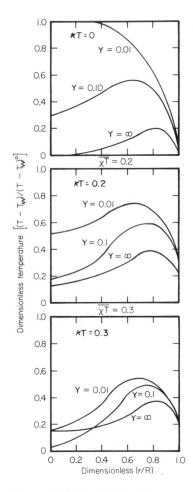

Figure 6.4-5   Effect of compressibility (κT) on temperature profile of a power-law fluid in a circular tube (n = 1/2). Note that $Y = \alpha x/U_0 R^2$, a dimensionless distance down the pipe. $(T - T_w)^\circ$ is incompressible flow [13].

As a result, we would expect Griskey's Brinkman number values to be much smaller than those used by others, and in fact his values are on the order of 0 to -0.5 (where the negative sign appears because the wall temperature is greater than the initial temperature, and thus ΔT is negative). As shown in Fig. 6.4-6, with a flow rate of 410 g/min in a 1-in.-diameter pipe, there is very little evidence of viscous dissipation of the order indicated by Gee and Lyons. And the curve shows much more dissipation within the volume

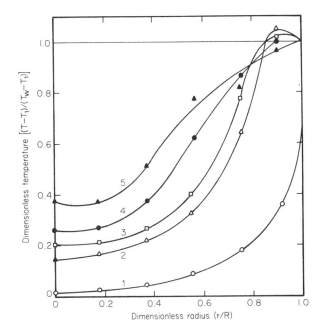

Figure 6.4-6 Experimental temperature profiles along pipe length. Graetz number (Gz) is dimensionless reciprocal distance down pipe. Curve 1: Gz = 474; 2 = 70; 3 = 37.8; 4 = 26.1; 5 = 19.8 [20].

of the fluid than predicted by Gee and Lyons. In addition, the effect of compressibility cooling, as espoused by Toor, is not so strong as predicted. In Fig. 6.4-7, Griskey compares the experimental data for an experimental $Br_G = -0.048$ with calculated profiles for $Br_G = 0$ and -0.5. Note that this is for a Graetz number of 8.3, or where the agreement between theory and experiment for incompressible Newtonian fluids with temperature-independent physical properties and parabolic velocity profiles has been excellent. And in Fig. 6.4-8, Griskey gives a plot of the Nusselt number-Graetz number relationship for various calculated values of the Brinkman number. It is apparent that the agreement between Toor's theory and Griskey's experiment is poor. At low flows there seems to be little viscous dissipation (Br = 0), and with increasing flow rates, the effect of viscous dissipation exceeds that predicted by Toor and Forsyth and Murphy [22]. The tentative conclusion is that it is not sufficient to include only one term in the viscous dissipation. Higher-order terms, coupled with the compressibility cooling factor that may induce added normal stresses, appear necessary to bring theory and experiment into line. This is a difficult task, at best, but necessary if we are to understand the relative roles of viscous dissipation and compressibility cooling in the processing of plastics.

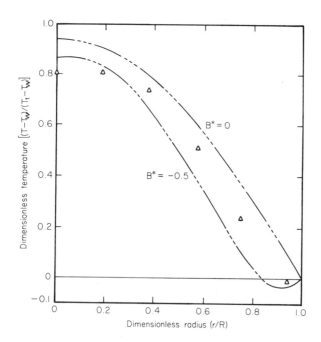

Figure 6.4-7 Comparison of experimental and theoretical temperature profiles for compressible polyethylene flowing in a heated pipe. Values at Gz = 28.6; experimental value for Brinkman number, $B^* = -0.048$ [20].

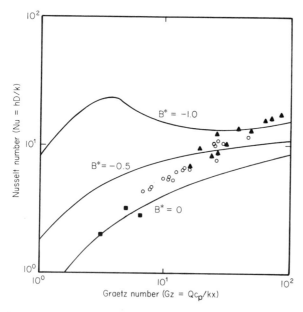

Figure 6.4-8 Comparison of experimental dimensionless heat transfer coefficient (Nusselt number) with theory. Circles, $B^* = -0.043$; triangles, $B^* = -0.141$; squares, $B^* \simeq 0$ [20].

Of course this inhibits execution of a heat transfer analysis with any degree of confidence. Note, however, that Griskey's results seem to coincide with the earlier computer work of Christiansen and Craig, as extended by Pigford to include power-law effects. It would appear that the Christiansen-Craig curve labeled $\psi(H) = \infty$ coincides with Griskey's experimental data fairly well. This of course has little meaning except perhaps implying that the temperature dependency of viscosity is more important than either the effects of viscous dissipation or compressibility cooling, or possibly that the last two effects, having opposite effects, tend to cancel each other.

Why is this information critical to the designer? Consider for the sake of argument that the Toor analysis in Fig. 6.4-4 is correct and that $Y = \infty$. If the melt temperature is 470°F and the wall temperature is 70°F, a temperature drop of 300°F occurs between the centerline and the wall if the fluid is incompressible and Newtonian. Assume now that the fluid is Newtonian but that the compressibility factor is $\kappa T = 0.03$. Now the centerline temperature drops to $0.12 \times 300°F = 36°F$ above the wall temperature (or a centerline temperature of 106°F). Yet at an r/R of 0.7 the temperature remains high at $0.39 \times 300°F = 117°F$ above the wall temperature (or 187°F). This is more than 80°F above the centerline temperature. Assume further that the fluid is highly non-Newtonian, where $n = 0.25$, and that it is compressible. Now the centerline temperature drops below the wall temperature ($-0.38 \times 300°F = -108°F$ above the wall temperature) or $-38°F$ actually. At this time the region at or near the r/R value of 0.8 is at a temperature of $0.36 \times 300°F = 108°F$ above the wall temperature or 178°F. Thus, the centerline temperature can be more than 200°F below the highest temperature in the flow stream. Viscous dissipation degradation caused by overheating of the fluid can occur. Materials can scorch in the volume (note that this thermal degradation does not occur at the wall of the container and as a result the thermally degraded product can move along with the good material into the final part). Furthermore, if the material is crystalline, low centerline temperatures can induce crystallization (particularly when shear is also taking place), and thus some of the material entering a part can be partially crystallized. Some amorphous materials that exhibit these processing problems are acrylonitrile, styrene-acrylonitrile, butadiene-based materials, ABS, polymethyl methacrylate, and polycarbonate. Acetal and terephthalates are crystalline materials that are typical candidates. In the case of SAN and PMMA, these thermally degraded materials appear as yellowed streaks in clear materials. If, in fact, the temperature dependency of viscosity shapes the velocity and temperature profiles, any cooling of the fluid in the center would result in a plug of much higher viscosity than that near the walls (which would become considerably less viscous owing to the increased temperature of viscous dissipation). Thus, a simplistic consideration of the Leveque problem where plug flow ($n = 0$) occurs is not a bad alternative to the design of long pipes. According to Christiansen and Craig [18], the form for the plug flow Leveque problem ($Gz \geq 10$) is

$$Nu = 2.54 + 1.27 Gz^{1/2} \qquad\qquad (6.4\text{-}25)$$

A curve with this slope fits Griskey's data quite well at Gz's from 10 to 100, and again this would probably correspond with the $\psi(H) = \infty$ curve of Christiansen and Craig.

Note that the analysis of convective heat transfer has been restricted to circular pipes. Some mathematical models have been developed for other geometries, such as annuli, square pipes, elliptical pipes, and so on. In general, however, few (if any) have had the extensive experimental effort applied to them that has been applied to flow in a circular pipe. Furthermore, most lack rigor in dealing with fluids that might exhibit elastic effects.

Natural convection has also been neglected herein, although it is quite important in many plastics processing cases. We consider natural convection in various processing techniques. However, there is very little experimental investigation of natural convection effects in polymeric materials. It is normally added in as a correction factor when a priori theory and experimental data do not agree. Skelleland amplifies these comments in his section on combined forced and natural convection [14].

6.5 Melting of Plastics

While we shall go into "plastication" or the shear melting of plastics in more depth in Chap. 8, we shall consider here the interrelationship between conduction and convection that occurs not only in this type of process but in other melting processes, where, for example, solid plastic slabs are pressed against rod heaters for melting. A schematic of the melting process is shown in Fig. 6.5-1, where the plastic is pressed against the heated surface, and the molten plastic is sheared away from the interface. Thus, if a crystalline material is heated as it is sheared, the molten plastic will carry away sensible heat with it. Furthermore, the heat needed to melt the solid must be conducted through the layer of convecting fluid. An early study by Ross [24] on Newtonian fluids gives some indication about the movement of the melt front as a function of time. Recently, Griffin [23] and Griffin and Skelleland [14] examined the melting process of a crystalline material with either zero-shear Newtonian viscosity such as nylon, with data that should agree with that of Ross, or power-law viscosity for materials such as HDPE. Griffin lists the three equations that describe heat transfer:

1. The conduction equation of heat from the melting layer (assumed to be at the melt temperature T_m) into the solid
2. The convection equation of heat from the hot metal wall (assumed to be at temperature T_w) into the fluid and the sensible heat flow leaving with the melting fluid
3. The heat transfer equation at the interface between the fluid and solid that determines the rate of melting of the plastic

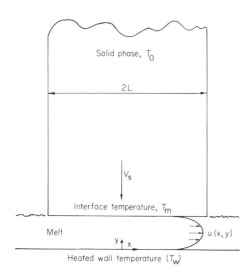

Solid phase, T_0

2L

V_s

Interface temperature, T_m

Melt

$u(x,y)$

$y \uparrow x$

Heated wall temperature (T_w)

Figure 6.5-1   Contact melting schematic [23].

He then uses an integral energy method to describe both the velocity and temperature profiles in the fluid and determines a form for the heat transfer rate in terms of the Nusselt number $Nu = hL/k_f$. k_f is the thermal conductivity of the fluid, and L is the length down the heated surface, the Prandtl number of the fluid, defined as $Pr = c_p\mu/k$, where μ is the local viscosity. The Reynolds number is defined in terms of the local viscosity as $Re = \rho Ub/\mu$, where ρ is the fluid density; U is the average fluid velocity; b is the gap width at $x = L$, the edge of the melting frame; and μ is the local viscosity. For power-law fluids, $\mu = K(\dot{m}/\rho L^2)^{n-1}$, where $\dot{m}$ $(x = L)$ is the mass flow rate of fluid per unit width of melting surface. It can also be shown that $Re = \dot{m}(x = L)/\mu$ and therefore represents a loading parameter.

From Griffin's analysis, an overall heat transfer rate (in terms of the Nusselt number, etc.) can be written as

$$Nu/RePr = \dot{m}(x)[A^{-1} + \theta_B(x)]_{x=L} + B/A \qquad (6.5-1)$$

$\theta B = (T_w - T)/(T_w - T_m)$; $\dot{m}$, as before, is the melting rate; $A = c_{pf}(T_w - T_m)/\Delta H$, the ratio of sensible heat of the fluid to the latent heat of melting of the solid; and $B = c_{ps}(T_m - T_0)/\Delta H$, the ratio of sensible heat of the solid to the latent heat of melting of the solid. T_0 is the initial temperature of the plastic (at a distance from the melting surface); ΔH is the latent heat of melting.

Note that B/A represents the ratio of sensible heat of the solid to sensible heat of the fluid. If the fluid has no latent heat ($\Delta H = 0$ for an amorphous plastic, for example), then $B/A = 1$ and $Nu/RePr = 1$ in accordance with heat transfer to a convective fluid.

As mentioned earlier, Ross proposed a simple conduction heat transfer model [24]. Griffin compares his model with Ross's model (Fig. 6.5-2)
and points out that the convection effects occur when A becomes quite large.
This is essentially a "shielding effect," to use Griffin's words, because
much of the energy is convected away. As shown in Fig. 6.5-3, the agreement between the calculated and experimental heat transfer coefficients for
a Newtonian fluid is fair, with the calculated values about 10% below the
experimental values. Griffin points out that this heat transfer coefficient
was determined by

$$h = [c_{pf}\dot{m}(x = L)/L](Nu/Re/Pr) \qquad\qquad (6.5\text{-}2)$$

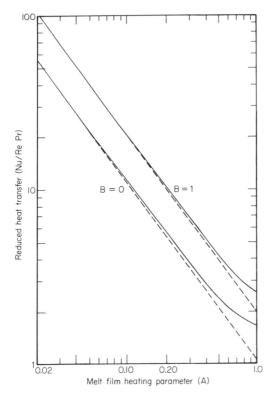

Figure 6.5-2 Dimensionless heat transfer as a function of A, the ratio of
sensible heat of fluid to latent heat of melting. Solid line, integral heat
balance; dashed line, conduction solution. B is ratio of sensible heat of
solid to latent heat of melting [23].

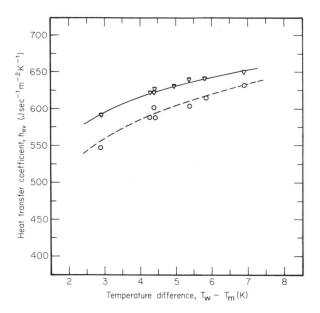

Figure 6.5-3 Experimental and theoretical heat transfer coefficients for conduction-convection melting. Solid line, experimentally determined; dashed line, theoretical, based on conduction without convection [23].

Griffin points out that the values of A (and B also) range between 0.5 and 3.0 for many plastics, including high-density polyethylene and polyoxymethylene.

As we shall see when we get to extrusion, the melting of pelletized resins can be directly related to Griffin's "convected melter." Thus, his work can be compared with that of Tadmor, Marshall, and Klein.

6.6 Conclusion

Our purpose in this chapter was to define certain heat transfer terms, such as might be found in typical conduction, convection, and radiation heat transfer processes. This does not mean that mastering the material in this chapter will enable the reader to solve all the heat transfer problems in plastic processing. We have not, for example, considered cooling with volume change, or natural or forced convection cooling of films or extrudates, or plastication melting in extruders, or sinter melting in rotational molding, and so on. We certainly have not exhausted all the types of heat transfer either. Certainly, we cannot do justice to the analyses presented in the

Handbook of Heat Transfer, Kreith, McAdams, McKelvey, Jacob, Knudsen and Katz, Sparrow and Cess, and many other texts on heat transfer. Certain tie points, however, between the conventional methods of heat transfer and heat transfer to plastics materials are needed. As we consider the individual processes we shall consider other types of heat transfer in application.

Problems

P6.A The following Reynolds number has been proposed for power-law non-Newtonian fluids:

$$Re = D^n (V)^{2-n} \rho / KB \qquad (6.A-1)$$

where

$$B = 2^{n-3} (3 + n^{-1})^n \qquad (6.A-2)$$

If $K = 3$ lb·sec^n/in.^{2} and $n = 0.32$, calculate the non-Newtonian Reynolds number. Compare your value with that given in Eq. (6.1-1). Discuss any differences. Ans.: 0.0043.

P6.B A 1-in. insulation blanket is placed over the heater bands on an extruder. The extruder heater band temperature is isothermal at 250°C, and room temperature is 25°C. The thermal conductivity of the insulation is 0.1 Btu/ft^2·hr·°F/in. Determine the heat loss from an extruder 12 in. in diameter by 30 ft long. Determine the improvement in thermal efficiency if the insulation blanket is increased to 2 in. The thermal conductivity of the second layer of insulation is 0.5 Btu/ft^2·hr·°F/in. Does it make a difference which layer is placed against the heater bands?

P6.C Phenolic is being compression-molded between isothermal oil-heated platens. The plastic is initially at 70°F, and the platens are 350°F. Phenolic must be held at or above 250°F for 60 sec in order to cross-link into a three-dimensional part. The part thickness is 0.5 in. If the thermal diffusivity of phenolic is 5×10^{-3} ft^2/hr, determine the total time required to cure the part at the centerline. Make a plot of the temperature of the part 0.2 in. in from the heated surface as a function of time until cure.

P6.D The phenolic compression molding described in P6.C is found to take too much time to be economical. Therefore, it is proposed that the phenolic powder be heated to 200°F in a dielectric heater prior to introduction to the molding machine. Now how long does it take to cure the part to the centerline?

P6.E 0.060-in. sheet stock initially at room temperature (RT = 70°F) is being heated in a hot air isothermal oven at 450°F. For the sheet to be thermoformable, the average temperature must be greater than 250°F and less than 350°F. To prevent internal cracking during forming, the center-line temperature must be greater than 200°F. Can the sheet be properly heated if it has a thermal diffusivity of 4×10^{-3} ft^2/hr and a thermal conductivity of 0.10 Btu/ft$^2 \cdot$hr$\cdot$°F/ft and if the heat transfer coefficient is 1 Btu/ft$^2 \cdot$hr$\cdot$°F? If so, how long does it take? What is the surface temperature at the time of forming? To satisfy these criteria and minimize the heating time, what is the maximum rate of heat transfer to the sheet in terms of the heat transfer coefficient?

P6.F A sheet metal mold 1/16 in. thick is to be used for rotational molding. The isothermal oven temperature is 1000°F, and the mold is to be inserted at 70°F. Determine the time required to heat the mold to 550°F if the effective heat transfer coefficient within the oven is 5 Btu/ft$^2 \cdot$hr$\cdot$°F. If a radiant heater bank is added to the oven, thus effectively doubling the heat transfer coefficient, how much time is saved?

P6.G Consider the addition of radiant energy in P6.F. Using the concept of an effective Biot number described in Sec. 6.3, determine the approximate temperature of the radiant heater bank if the mold is completely enclosed by the heaters and if it has an effective emissivity of 0.9. Note all assumptions. Note that only one side of the mold is being heated.

P6.H Experiments have shown that the following constitutive equation of state is suitable for Marbon ABS over a short temperature range:

$$\rho = 6 \times 10^{-4}(T - 260) + [0.9945 + 5.875 \times 10^{-6}(P - 2000)] \qquad (6.\text{H-1})$$

Calculate the coefficient of thermal expansion and compressibility at 240°C and 1000 psi. Calculate these for 200°C and 10,000 psi. Ans.: 6×10^{-4}; 5.87×10^{-6}.

P6.I In Chap. 14, extensive use will be made of an equation of state for polymer melts. If the following equation describes many plastics, find

$$(1/\rho + b)(P + w) = AT \qquad (6.\text{I-1})$$

appropriate forms for the coefficient of thermal expansion and compressibility.

P6.J LDPE is being extruded at 100 lb/hr through an isothermal sizing die 12 in. in length by 6 in. in width by 0.050 in. in thickness. The material has a density of 57.2 lb/ft^3, a heat capacity of 0.4 Btu/lb$\cdot$°F, and a thermal conductivity of 0.1 Btu/ft$\cdot$hr$\cdot$°F. It has a power-law index of 0.3. Plot the heat transfer coefficient along the sizing die using the Leveque method. Obtain an average heat transfer coefficient for the die. If the sheet stock average temperature throughout the length is about 300°F, how much heat is being removed by the water?

P6.K It has been argued that for melting of amorphous plastics Nu/RePr = 1. Follow Griffin's arguments in Sec. 6.5 to arrive at this argument. How much error is introduced if the material being melted has a very broad melting range? Will Griffin's method work if the latent heat of melting is replaced with an effective specific heat over the region melting? Try to extend Griffin's work to that of Collier et al. in the melting and solidifcation of Penton (Sec. 6.2).

References

1. Frank Kreith, Principles of Heat Transfer, 2nd ed., International Textbook Co., Scranton, Pa., 1965.
2. W. M. Rohsenow and J. P. Hartnett, eds., Handbook of Heat Transfer, McGraw-Hill, New York, 1973.
3. H. S. Carslaw and J. C. Jaeger, Conduction of Heat in Solids, Oxford University Press, New York, 1947.
4. John Crank, Mathematics of Diffusion, Oxford, University Press, New York, 1956.
5. W. H. McAdams, Heat Transmission, 3rd ed., McGraw-Hill, New York, 1954.
6. W. Sifleet, N. Dinos, and J. R. Collier, Polym. Eng. Sci., 13:10 (1973).
7. Y. T. Tam, N. Dinos, and J. R. Collier, SPE ANTEC Tech. Pap., 19:601 (1973).
8. A. I. Brown and S. M. Marco, Introduction to Heat Transfer, 2nd ed., McGraw-Hill, New York, 1951.
9. J. M. McKelvey, Polymer Processing, Wiley, New York, 1962.
10. R. C. Progelhof, James Quintiere, and J. L. Throne, J. Appl. Polym. Sci., 17:1227 (1973).
11. R. C. Progelhof, J. Franey, and T. W. Haas, J. Appl. Polym. Sci., 15:1803 (1971).
12. M. Necati Ozisik, Boundary Value Problems of Heat Conduction, International Textbook Co., Scranton, Pa., 1968.
13. H. L. Toor, AIChEJ., 4:319 (1958).
14. A. H. P. Skelleland, Non-Newtonian Flow and Heat Transfer, Wiley, New York 1967.
15. R. B. Bird, W. E. Stewart, and E. N. Lightfoot, Transport Phenomena, Wiley, New York, 1960.
16. J. E. Porter, Trans. Inst. Chem. Eng., 49:1 (1971).
17. A. B. Metzner, Adv. Heat Transfer, 2:1 (1965).
18. E. B. Christiansen and S. E. Craig, Jr., AIChEJ., 8:154 (1962).
19. R. L. Pigford, Chem. Eng. Prog. Symp. Ser., 51:79 (1955).
20. R. G. Griskey, SPE ANTEC Tech. Pap., 18:98 (1971).
21. R. E. Gee and J. B. Lyon, Ind. Eng. Chem. Ind. (Int.) Ed., 49:959 (1957).
22. T. H. Forsyth and N. F. Murphy, Polym. Eng. Sci., 9:22 (1969).
23. O. M. Griffin, Polym. Eng. Sci., 12:140 (1972).
24. T. K. Ross, Chem. Eng. Sci., 1:212 (1952).

7.

Batch Processing

7.1 Introduction

Many processes that begin as batch or hand operations become economically unattractive as volume increases. Thus, most processes are eventually automated in some fashion. In this chapter batch processing is identified and explored. Cyclical processing is considered in detail in Chaps. 10 and 11. In the later chapters the cycle time is assumed to be controlled by mechanical means of cavity filling or heat transfer, and thus the machine is occupied for long periods of time in process. Here, however, we are concerned with those processes that are characterized by high labor input and machines that are, in the main, not being used efficiently. In other words, if the cycle time for the machine is much longer than the time needed for manual or hand labor, these processes are considered to be cyclical. If it's the other way around, the process is characterized as a batch process. For example, thermoforming of very large objects such as camper tops or outdoor turnpike gasoline signs should probably be considered as a batch operation. And compression molding (or its more advanced process, transfer molding) of, say, electrical outlet components or face plates requires very little manual labor. Thus, this type of compression molding should probably be considered cyclical.

However, compression molding is considered here to be a batch process, together with casting of epoxies or polyesters, spray up or hand lay-up of fiberglass-impregnated resins, and foaming of styrene and urethanes.

Traditionally, very large pieces of molded thermosets are compression-molded, and much hand labor is needed to fill the mold, remove the final part, fixture it, and clean the mold prior to refilling. Motor housings, washing machine agitators, and lavatories are still molded in this fashion.

In this chapter, then, we shall concentrate on the molding, in batch quantities, of reactive materials such as the epoxies, urethanes, polyesters, and formaldehyde-based resins. Thus, we shall look at the character of dealing with viscous, reactive fluids and their various problems in processing.

7.2 Viscosity Chracteristics of Reactive Liquids

Recall in the discussion on the batch production of thermosetting resins that the reaction is stopped during production at a predetermined time, normally by dropping the reaction temperature. Therefore, heating these reactive materials will cause the reaction to continue to completion. Note that even at room temperature reaction is continuing, but the rate of reaction is normally exponentially dependent on temperature, and thus at room temperature any changes in physical properties of the resin would take years to detect. The viscosity of the resins is characterized first. As mentioned earlier, these resins are non-Newtonian. The degree of non-Newtonianness, analogous to the viscosity index n, is referred to as the "thixotropic index" by the resin suppliers of polyester resin. Let us consider, briefly, then, why this "thix index" is important to the resin user. Consider the operation of fiberglass-reinforced polyester (FRP) spray up. Here the resin (about 50% polyester and 50% styrene) is mixed with catalyst [methyl ethyl ketone peroxide (MEKP) at 1 to 4%] and sprayed through an atomizing nozzle. The droplets, in reality small chemical reactors, are mixed with chopped fiberglass roving (expelled from another portion of the same hand gun), and this admixture impacts a prepared mold surface. It is apparent that to get maximum velocity and flow rate through the spray unit the resin must have a very low viscosity at these very high conditions of shear. On the other hand, once this mass has reached the surface, the plastic should stay in the mixture of fiberglass and plastic even if the surface is vertical. This means that the viscosity at this very low shear condition must be very high. Thus, the material must have a low value of viscosity index. Fortunately, there are many ways of changing the shear thinning characteristics of these materials through the addition of low-viscosity solvents (such as increasing the concentration of styrene in the resin) or addition of "thixing" agents such as fumed silica or calcium carbonate. The latter materials give the material a zero-shear viscosity or, in some cases, cause the material to behave as a solid at low shears (e.g., a Bingham plastic) and yet allow for easy flow at the high shear rates experienced in a spray-up nozzle.

As mentioned earlier, one of the more popular ways of measuring the non-Newtonianness of a resin is by using a variable-speed viscometer such as a Brookfield viscometer. The viscosity of the resin in taken first at a low rpm reading (usually 5 or 6 rpm) and then at a high rpm reading (50 or 60 rpm) using the same Brookfield spindle. The ratio of the two viscosities is usually greater than 1 and can be as high as 5 or 6. The higher the ratio, the more non-Newtonian the fluid. This means that for a decade change in the shear rate the viscosity of the resin decreases by a factor of 5 or 6. When the changes in shearing stress during this increase are taken into account, the power-law viscosity index is found to be less than 0.5.

This should also yield some information about the shear conditions required to disperse pigments and fillers in non-Newtonian liquid resins. A high-shear mixer such as a Banbury or Hobart mixer should be used to get maximum dispersion in minimum time. Hand mixing is usually ineffici- ent and yields poor dispersion. Incidentally, this is apparently true for the dispersion of low-viscosity solvents and blowing agents as well. The details of dispersive mixing here are given in texts on mixing [1, 2, Chapter 12]. An example of the effect of various parameters, such as temperature time of mixing and speed of mixing (and thus shear), is shown in Table 7.2-1 for the dispersion of ZnO in GRS rubber [2, p. 336]. This is typical of data for the dispersion of pigments and the compounding of lubricants, fillers, and other additives in various plastic materials.

Examples of the viscosity characteristics of catalyzed resins are shown in Figs. 7.2-1 and 7.2-2 for an epoxy system and a polyester system [3]. Note that each set of curves shows little change in viscosity with time until the onset of gelation, at which point the viscosity of the resin increases extremely rapidly. Early evidence indicates that there is relatively little buildup in normal stresses in the fluid until the gelation point is reached. Both materials can be characterized as power-law fluids, except at very low shear rates where for polyester resins, at least, there is a zero-shear New- tonian viscosity region. This information is, of course, of great use when methods of processing the resins are considered. Surprisingly, partially reacted thermosets such as urea-formaldehyde, melamine-formaldehyde, and phenolics can also be characterized according to their viscosities. With the newer methods of processing thermosets that are solid at room tempera- ture (and thus in a powder or preform state when introduced to the mold), additives have been developed that, in effect, retard the reaction for a con- trolled period of time. Thus, at a given processing temperature, it is pos- sible to measure the viscosity of the resin. One popular device for doing this is the sigma-mixer attachment to the Brabender Plasticorder. Unfor- tunately the torque-rpm measurements here are not entirely relatable to a viscosity because the materials exhibit intense shear heating in the mixing section [4]. The spiral mold flow path has similar objections in that it is difficult to relate a time-dependent phenomenon to a viscosity, a property

Table 7.2-1

Dispersion of Zinc Oxide in GRS Rubber [2, p. 336]

Jacket temperature (°C)	Final temperature (°C)	Time of mixing (min)	Rotational speed (rpm)	Energy consumed (Whr)	Rotor tip speed (ft/min)	Relative rating
50	92.5	3	69	398	89	1
65	100.0	3	69	380	89	2
80	107.5	3	69	379	89	3
50	80.0	4	35	286	89	3
50	92.5	3	69	398	89	1
50	105.0	2	137	532	174	2

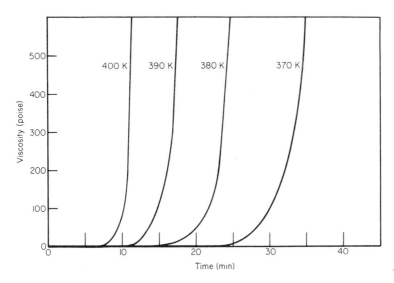

Figure 7.2-1   Effect of reaction time on epoxy viscosity at various isothermal temperatures; data points not shown [3].

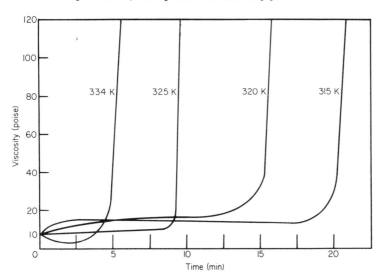

Figure 7.2-2   Effect of reaction time on polyester viscosity at various isothermal temperatures; data points not shown [3].

that is characterized as the ratio of two steady-state variables. Do not
discount the worth of these tests, however. With the torque-rpm plots from
the Brabender Plasticorder, it is possible to characterize the time when
sufficient cross-linking has taken place to cause the material to effectively
stop flowing. As shown in Fig. 7.2-3 for a general-purpose phenolic [5],
the torque-time curve shows, characteristically, a long region where the
torque gradually rises in a rather smooth fashion. Then a rather abrupt
increase in both the torque and the torque oscillation with time occurs.
Again, characteristically, this slope change is associated with sufficient
cross-linking in various regions of the fluid that have a "slip-stick" effect
on the remaining fluid in the mixer. Goodrich and Porter present a rather
useful method of converting torque-rheometer data to rheological data for
thermoplastics. This is shown in Figs. 7.2-4 and 7.2-5 for polypropylene
and general-purpose polystyrene [6]. They emphasize, however, that for
thermoplastics the capillary rheometer data are preferred and considered
more accurate. Similar analyses have not been completed for thermosets.

Insofar as capillary rheometry on formaldehyde based thermosets
is concerned, recent work by Frizelle and Norfleet indicate [7] that these
materials, too, are best described by power-law equations of state.

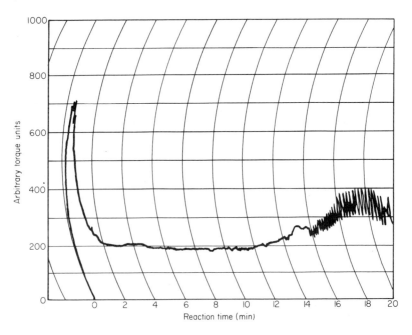

Figure 7.2-3   Torque rheometry of type 308 injection molding grade Plenco Two-Stage Phenolic showing processibility window; T = 300°F [5].

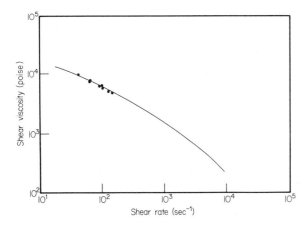

Figure 7.2-4   Comparison of viscosity-shear rate data for Instron capillary (solid line) and Brabender torque rheometer (data points); material: Chevron PP 9094 [6].

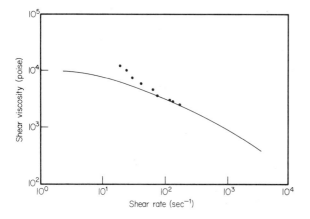

Figure 7.2-5   Comparison of viscosity-shear rate data for Instron capillary (solid line) and Brabender torque rheometer (data points); material: Dow Styron 666 [6].

See Figs. 7.2-6 and 7.2-7. Keep this information in mind when considering compression molding. Also keep in mind the data from the torque-rheometer show the effects of viscous dissipation buildup over long periods of time.

Owing to the complex interaction between the extent of cross-linking and the extent of shearing on the fluid, no rheological models have been proposed for thermosets. Construction of a heuristic model is relatively

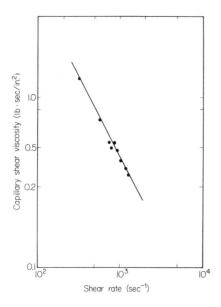

Figure 7.2-6   Capillary rheometer data for formaldehyde-based thermoset resin. Resin: Plenco 482; resin temperature: 235°F; capillary L/D = 4 [7].

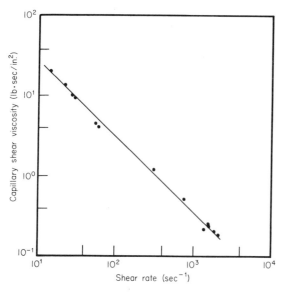

Figure 7.2-7   Capillary rheometer data for formaldehyde-based thermoset resin. Resin: Plenco 509; resin temperature: 235°F; capillary L/D = 4. Compare with data of Fig. 7.2-6 [7].

easy, however. Consider two distinct models, that of the non-Newtonian power-law fluid and that of an elastic solid. The fluid is characterized as being Arrhenius temperature dependent in the traditional sense, as has been shown in Figs. 7.2-1 and 7.2-2. The point at which the fluid becomes a solid occurs when a certain critical value of cross-linking has been achieved. Since the extent of cross-linking is determined by chemical reaction, and thus is time and temperature dependent, the critical value is also time and temperature dependent. Thus,

$$\tau = K(T)(\dot{\gamma})^{n-1}\dot{\gamma} = K\dot{\gamma}^{n} \qquad (f \le f_c) \qquad\qquad (7.2\text{-}1)$$

where f is the fraction of cross-linking sites converted and f_c is the critical value of cross-linking sites at which point the material becomes solid-like. The above equation is thus suitable for flow problems in compression and transfer molds as well as in injection molds, provided proper data are available about the rate of reaction of cross-linking sites and the critical value of cross-linking sites. As we shall see shortly, with some effort this information can be generated.

Before considering the reaction kinetics of thermosets in, say, compression molding, consider one other effect of viscosity on thermosets. Whenever liquid epoxies or polyesters are cast into molds, either for encapsulation of specimens or for fabrication of such items as decorator "synthetic marble" slabs, bubbles that have been entrained during the dispersive mixing phase are retained. One obvious way of minimizing bubble encapsulation in the final cast part is to allow the mixed, catalyzed resin to stand either at room temperature or elevated temperature to allow the bubbles to rise to a free surface. Another way is to apply vacuum to the casting. This is done frequently when making the silicone casting resin for the fabrication of a rubber mold of a model maker's pattern. If the resin has a zero-shear Newtonian viscosity, the rate of rise of bubbles in the fluid can be easily calculated. Simply equate the drag force around the bubble with the buoyant force due to the difference in density between the gas in the bubble and the surrounding liquid:

$$u_t = (\rho - \rho_b)gD^2/18\eta_n \qquad\qquad (7.2\text{-}2)$$

u_t is the terminal velocity of the bubble, ρ is the density of the liquid surrounding the bubble, ρ_b is the density of the gas in the bubble, and η_n is the zero-shear Newtonian viscosity of the liquid. For polyesters, for example, where there is a definite zero-shear viscosity region (at very low shear rates), application of this equation is correct. If the fluid is a non-Newtonian fluid, however, additional terms are frequently required to correct for the shear-dependent viscosity in the above equation. Interested readers should read the work by Hirose and Moo-Young [8]. In examination of the above equation, note that the rate of rise of bubbles is proportional to the square of the bubble density. Thus, even if encapsulation occurs before all

the bubbles have reached the free surface, those remaining will be very
small. Apparently, the presence of these small bubbles in materials such
as "synthetic marble" do not seem to dramatically affect or weaken the
structure. Obviously, if and when the thermoset liquid gels, the viscosity
will rapidly rise to infinity, and bubbles will remain encapsulated. It is
therefore the case that for most stable gels bubbles remain in the final
product. Composites such FRP admixtures must also be considered with
regard to bubble elimination. Here, the material behaves as if it is Bingham plastic in that a certain critical shear stress must be applied to the
fluid before it will move. It can be shown that the buoyant force (or static
force) on a sphere is given as

$$F_s = (4/3)\pi R^3 (\rho - \rho_b) \tag{7.2-3}$$

Further, the shearing force (or "drag force") on the sphere is given as the
integral of the shearing stress across the surface of the sphere. The surface area of a sphere is $4\pi R^2$. Thus, the force is given as

$$F_d = 4\pi R^2 \tau_c \tag{7.2-4}$$

τ_c is the critical shear stress, defined previously. Thus, by equating the
buoyant force, or that force that exists on the bubble regardless of whether
it is moving or not, and the force required to initiate movement owing to
shear, we have

$$\tau_c = R(\rho - \rho_b)g/3 \tag{7.2-5}$$

Hence, the minimum radius of a bubble moving in a gel-like medium where
a critical shear stress is present is directly proportional to the critical
shear stress and inversely proportional to the difference in the densities of
the fluid and the gas. A form drag term has been omitted. It is not important
unless the fluid is actually moving.

7.3 Reaction Characteristics of Thermosets

Until recently, little was known about the rates of reactions of thermosetting
resins. Heuristically, it is expected that the reaction rates be represented
by the fraction of reacted sites remaining and that this fraction should be
sigmoidal in shape. Craig [9] has found that this is the case of EPDM,
thermoplastic elastomer. Kamal and his coworkers [3] have found similar
relationships for polyesters and epoxies. Craig's form, which also fits the
data for epoxies of Kamal, is

$$df/dt = K(T)f(1 - f) \tag{7.3-1}$$

where f is the fraction of unreacted sites. This is shown in Fig. 7.3-1. Kamal rewrites this equation in terms of the relative degree of cure. This is defined as the ratio of heat generated by the reaction, from its beginning until time t, to the total amount of heat upon completion of the reaction. Kamal cautions, however, that the latter quantity is not the ultimate or maximum amount of heat that could be generated in an entirely adiabatic system. K(T) is a reaction rate constant that is exponentially dependent on temperature. Prime [10] used DTA to determine reaction rates for two epoxy-amine thermosets and found that the following kinetic model was desirable:

$$da/dt = K(T)(1 - a)^n \qquad (7.3-2)$$

a is defined as the ratio of moles reacted at time t to the number of moles originally present, and n is on the order of 1.1 to 1.3. Prime found, however, that the best fit to the data required that K, the reaction rate constant, be time dependent as well as Arrhenius temperature dependent.

Craig assumes that for relatively low temperature ranges K(T) can be described by a simple polynomial in temperature:

$$K(T) = k_1 + k_2 T + k_3 T^2 + \cdots \qquad (7.3-3)$$

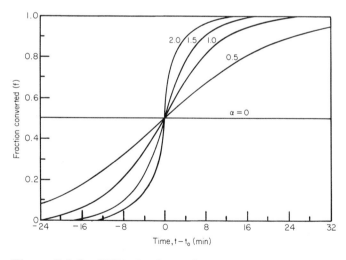

Figure 7.3-1 EPDM isothermal rubber cure rate; no shear, time references to 50% cure. Parameter is first-order reaction rate constant; composite of experimental data [9].

Actually the Arrhenius form for the reaction rate constant is correct:

$$K(T) = K_0 \exp(-E_a/RT) \tag{7.3-4}$$

where E_a is the energy of activation and K_0 is a preexponential factor. For example, for Kamal's polyester, $K_0 = 2.72 \times 10^9$ and $E_a = 17.4$ kcal/mol. The units on $K(T)$ are min^{-1}. There is a somewhat simplified form for Eq. (7.3-4) that is easier to program on an analog computer:

$$K(T) = K_0' \exp(bT) \tag{7.3-5}$$

where K_0' and b are empirical constants determined by plotting the curves given by Eq. (7.3-4).

As seen in the heat transfer section, the proper form for transient heat conduction without chemical reaction is given by Eq. (6.2-13). If the reactive materials are being heated or cooled, however, an internal heat generation term must be added:

$$q''' = (df/dt) \Delta H \qquad (Btu/ft^3 \cdot hr) \tag{7.3-6}$$

where ΔH is the heat of reaction. The heat of reaction has been included in the definition of the extent of cure with Kamal's work. Note, however, that the internal heat generation term is a strong function of temperature and time. Thus, if this term is added to the transient heat conduction term, a solution cannot be displayed using the graphical form of Chap. 6. The proper form for the equation is

$$\rho C_p \frac{\partial T}{\partial t} = k \frac{\partial^2 T}{\partial x^2} + q''' \tag{7.3-7}$$

subject to appropriate boundary conditions. These conditions depend on the molding process. If heat is removed from the surface of the reacting slab by conduction with a mold surface at a constant temperature, $T(x = 0) = T_w$. If it is placed in contact with another material that conducts the heat away at a finite rate, then the heat flux at the surface is given as

$$q_r = -kA(dT/dx)_{x=0} \tag{7.3-8}$$

And if the casting is quenched by placing it in water or blowing air across the surface, a convective heat flux occurs at the surface:

$$q_r = hA(T_a - T) \tag{7.3-9}$$

where T_a is the ambient temperature of the quenching medium. Note that the amount of heat generated is a function of the volume of material, whereas the rate of heat removed is a function of the surface area of the material. As shown in Fig. 7.3-2 for no heat removal (e.g., the adiabatic reaction), the temperature excursion becomes quite large. Note that for any regular

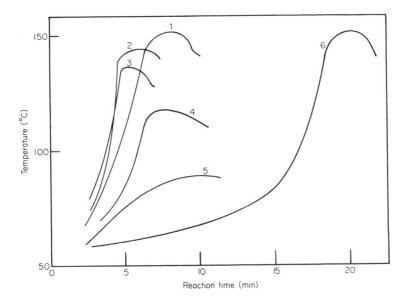

Figure 7.3-2   Effect of monomer type of exotherm profiles for polyester.
1: Styrene; 2: vinyl acetate; 3: acrylonitrile; 4: methyl methacrylate;
5: diallyl phthalate; 6: triallyl cyanurate.

geometry the ratio of surface area to volume is inversely proportional to
the characteristic length. Thus, for a slab of thickness t, the ratio of heat
generated volumetrically to that removed geometrically is inversely proportional to t. Thus, the thicker the slab becomes, the higher the internal
temperature becomes. This is an important fact if we are casting very
thick slabs of epoxy or polyester. For polyester, a free-radical cross-
linking reaction takes place, and thus the catalyst concentration is used to
control the rate of reaction. To keep the reaction going much more catalyst
must be used for a very thin casting than for a thick casting. These extremes are quite critical. In very thin castings it is frequently necessary
to keep the reaction going by placing the material in an oven or under heat
lamps, because the rate of heat removal due simply to convection with ambient conditions is higher than the rate of heat generation. On the other
hand, adding too much catalyst to a very thick casting can cause excessive
temperature excursions, thus cracking the casting or in extreme cases
degrading the polymer and setting the casting afire.

As an example, Progelhof and Throne [11] recently applied the standard transient heat transfer equation with heat generation [Eq. (7.3-7)] to
the Kamal et al. [3] reactive equations for epoxies and polyesters to demonstrate the thickness sensitivity to applied mold surface temperature. As

shown in Fig. 7.3-3, the centerline temperature for slabs of half-thicknesses of 0.1 and 0.2 in. for a mold surface temperature of 150°F shows
little effect of exothermicity. At 200°F mold surface temperature, the 0.1-
in. slab centerline temperature follows the characteristic nonreactive transient heating. However, note that the 0.2-in. slab shows a rapid temperature excursion to more than 500°F at a Fourier number of about 2.5. For
a mold surface temperature of 250°F, temperature excursions for both slab
thicknesses are seen, with the thicker slab showing a peak temperature of
570°F at a Fourier number of about 1.2. For the properties of polyester,
the peak temperature occurs at 3 min for the 0.2-in. slab and at 1.3 min
for the 0.1-in. slab. The corresponding degree of cure curves are shown
in Fig. 7.3-4, where we see that above a mold surface temperature of 200°F
the center can cure more rapidly than the surface. More importantly,
Progelhof and Throne found that Kamal's isothermal heat generation rates
underestimate the nonisothermal values by more than a decade for the cases
studied.

One of the direct applications for new technologies in reaction molding is in the formation of large parts. The developments in liquid reaction

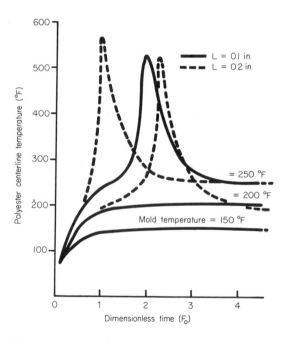

Figure 7.3-3   Effect of a step change in surface temperature on the centerline temperature of slabs of half-thicknesses of 0.1 and 0.2 in. Material
is reactive polyester resin [11].

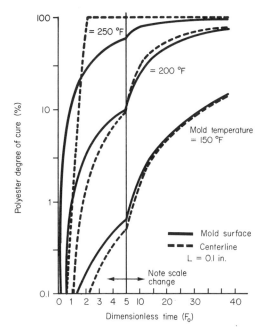

Figure 7.3-4   Degree of cure of reactive polyester responding to a step change in surface temperature. See Fig. 7.3-3. Note that centerline can cure more rapidly than surface [11].

molding or reaction injection-molding are paralleling early technological developments in thermoplastic injection molding. Much of the technology is concerned with mixing and dispensing moderately viscous liquids at very high rates in order to ensure deposition of the material in the mold cavity before reactivity increases the viscosity to the unprocessible state. The primary resin being studied today is isocyanate-polyol urethane. The technology can be directly adapted to the bisphenol A-epichlorohydrin two-component epoxies, although the higher resin cost and longer curing times of the epoxies make them considerably less economically attractive. Thermosetting phthalic-acid-based polyester resins (polyester-styrene solutions) cannot be processed in this manner today, owing to the low ratio of catalyst to resin (on the order of 1:100) and to the normally higher exotherm temperatures. Liquid injection casting techniques, in which the resin is mixed at the nozzle with catalyst and then pumped into a closed mold, bear great similarity to spray-up techniques, and frequently the conventional spray gun is simply fitted with a mixing nozzle and manually held against the mold sprue during filling. To get the reinforcement so desired with polyester resins, the mold must be manually fitted with glass matt or cloth.

The polyester resin injection systems are thus quite rudimentary
today.

In competition with these systems is reactive rotational molding.
Recent research indicates that with control on and intelligent programming
of the time and temperature of the reactive liquid successful hollow parts
can be made of most reactive systems. Progelhof and Throne [12] have
examined the coupling between the time-temperature response of the metal
mold used in rotational molding and the time-temperature response of
liquid in contact with that mold. As shown above, high reactivity of the
resin enables the resin temperature to quickly exceed that of the mold,
heating it to temperatures in excess of the isothermal environment. Thus,
although the purpose of heating the reactive mixture is to shorten the cycle
time of the process, in effect, the initially higher environment temperature
can be exceeded by the reactive resin.

The two primary problems facing the reactive rotational molders
are air entrainment and wall uniformity. The latter problem is also faced
by the conventional rotational molders, as is seen in Chap. 11. Categoriza-
tion of the flow regimes for flow of liquids in a horizontally rotating cylinder
have recently been studied by Deiber and Cerro [13]. They relate the ability
of the fluid to maintain a uniform coating across the mold surface to two
dimensionless groups: the Reynolds number, defined as

$$Re = wb^2/\nu \qquad\qquad\qquad (7.3\text{-}10)$$

and the Froude number, defined as

$$Fr = w^2R/g \qquad\qquad\qquad (7.3\text{-}11)$$

where w is the angular velocity (sec^{-1}), b is the film thickness (cm), ν is
the kinematic viscosity (= η/ρ, where the units on ν are cm^2/sec), R is the
inside radius of the cylinder (cm), and g is the gravitational constant (cm/
sec^2). As shown in Fig. 7.3-5, Deiber and Cerro identify several regimes
in which the fluid remains continuous on the cylinder surface. For example,
for solid body rotation, the speed of rotation is sufficient to prevent the
resin from forming a pool near the bottom of the cylinder. For very low
values of Re and Fr, as would be the case for simple rotational molding of
highly viscous fluids, the region formed is called the falling film region.
A process that begins with a pool in the bottom of the mold requires that
the resin be drawn from that pool and that the material coating the mold
wall not fall away from the mold as it moves through the horizontal plane
and back toward the pool. Deiber and Cerro show that the maximum amount
of lifting of material from the pool by the moving wall is given by

$$Re = Fr \qquad\qquad\qquad (7.3\text{-}12)$$

This also defines the maximum flow rate of material from the pool. In essence, then, we can use this equation to determine a relationship among the

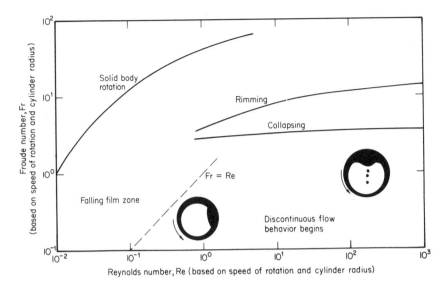

Figure 7.3-5 General flow regimes for rotational flow in a horizontal rotating cylinder. Solid body rotation implies a no-flow condition. Rimming
implies that some liquid is beginning to break away from the main body of
fluid. Collapsing indicates that the fluid is breaking away from the mold
surface at or near the top of the rotation. The falling film zone implies that
liquid is being drawn from a stable pool at or near the bottom of the cylinder. Two types of discontinuous solutions are also shown [13].

fluid viscosity, the speed of rotation, and the maximum permissible amount
of material remaining on the wall:

$$b = (wR\nu/g)^{1/2} \qquad (7.3\text{-}13)$$

Although this work has been developed for a time- and temperature-independent fluid, the information can be used for approximaton of these parameters for even very hot systems such as polyesters. The Deiber-Cerro
model has been tentatively verified for high-viscosity polyester and silicone
resins using ultrasonic thickness measurements [14]. Throne [15] has also
shown that Bingham plastic fillers, such as fumed silica, can dramatically
increase the stable value for b.

    Of course, the instabilities in flow phenomena can lead to stable pool
formation near the bottom of the rotating mold. This pool slowly rotates
owing to the shear of the fluid-containing rotating mold. At the fluid-covered
mold-pool interface, the air boundary layer is dramatically interrupted,
causing air to be intermittently drawn into the pool. As was seen earlier,
bubbles in highly viscous systems move very slowly, even in a uniform

gravity field. In a rotating field, the bubbles are quite stable in location within the resin. As a result, the bubbles, when drawn with resin to the mold wall, simply remain with the resin throughout the gelling and curing stage. Therefore, most reactive rotational molding parts exhibit densities that are effectively less than that of the pure, air-free cast resin.

Thus far we have considered only those reactions that exhibit excessive exotherms. Many thermosets, such phenolics or melamines, are very sluggish in their reaction rates, and even though these materials have exothermic cross-linking reactions, their heats of reaction are very low. To process these materials, they must be placed in a heated mold to initiate and keep the reaction going. Under these conditions, the heat transfer problem is essentially uncoupled from the reaction problem. Thus, the transient heat conduction graphs of Chap. 6 are adequate to determine temperature profiles within, say, a compression molding of phenolic. The type of kinetics as espoused by Craig and Kamal and others can be used simply to determine the degree of cure at any point in the casting. This simplification is considered in the section on compression molding below.

Little effort has been given to the understanding of the flow of reactive fluids of this nature. At present, it seems that this is not a major problem in either casting, spray up, compression molding, or even transfer molding. However, the understanding of the flow of reactive liquids is critical to rotational molding of, say, urethane and possibly quite important to the injection molding of certain thermosets. Thus, the effort to solve a fluid flow problem with heat transfer and chemical reaction should not be overlooked; there are practical applications.

7.4 Compression Molding

Before considering the details about the engineering phenomena that occur within the compression molding machine, we shall discuss compression molding in general. It is usually the case that the materials to be molded are thermosetting resins, although this process has been used to mold large thermoplastic pieces. With the new advances in injection molding of thermosetting resins, compression molding is becoming restricted to the processing of filled resins, such as iron-filled phenolics, woodflour-filled urea-formaldehyde, or alpha-cellulose-filled melamines. The process dates back to the first molding of plastic materials in the mid-nineteenth century. Normally a hydraulic vertical ram applies pressure to two platens. Between the platens are the mold halves; the upper half is the force, and the lower half is the cavity. The molds are heated electrically, with steam, or with hot oil. Three basic types of molds are in use: the flash mold, where the parting line on the part is flush with the edge of the cavity; the positive mold, where the force is recessed into the cavity when the mold is closed; and the semipositive mold, where excess material is allowed to flash around

the force in a controlled manner. These three types of mold designs are
shown in Fig. 7.4-1. The semipositive mold is used predominantly in the
molding industry. However, all three types of mold design are used to
advantage.

Simons [16] states that there are several advantages to choosing
compression molding over other forms of molding of thermosets: (1) material flow is shorter, reducing residual stresses in the molded parts;
(2) since there are no gates in the mold, excessive shearing is avoided,
thus minimizing loss in mechanical properties in the final part; (3) part wall
uniformity can be controlled to a much closer degree than with any other
form of molding; (4) gases, generated during the cross-linking reaction,
can be released through proper design in the force-to-cavity clearance
(this minimizes venting or "breathing" molds, as is frequently done with
transfer molds); (5) even though very corrosive or erosive materials are
molded, mold wear is minimized owing to the low shear applied to the materials; (6) since gating is not necessary, mold costs are lower than with
the equivalent transfer mold; and (7) mold design is not dependent on sprue
and runner design.

Molding temperatures for most conventional thermosetting resins
are 300 to 400°F. To ensure that the material is sufficiently heated during
molding, preheating is a common practice. With most materials dielectric
heating is the most uniform way of heating. Dielectric heating can only be
used with polar resins. It is a general rule of thumb that for relatively flat
parts pressures on the order of 2000 to 4000 psi (based on the projected
surface area of the part) are required to produce even flow across the mold.
Thus, a 10-in.-diameter dinner plate would require a press with a clamping
or hydraulic pressure capacity of more than 100 tons to achieve 2500-psi
pressure on the surface.

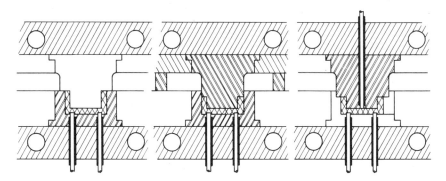

Figure 7.4-1   Three characteristic compression molds. Left, typical
flash-type mold; center, typical positive seal mold; right, semipositive or
partial vent mold, in common use today [16].

To picture the compression molding cycle, consider Fig. 7.4-2. Assume that the molding material is at a uniform temperature throughout, that it can be handled as if it were a solid, and that when sheared it flows as a fluid. Uniform temperature indicates either that the material is loaded into the mold at room temperature or that it has been uniformly preheated using a dielectric preheater. Assume that because of the nature of dielectric heating the material reaches a uniform "molding" temperature very rapidly with little chemical reaction. Furthermore, consider that the closing of the platens of the mold is sufficiently rapid to prevent significant additional reaction from taking place. This effectively assumes that the flow problem, filling the mold with plastic, can be separated from the heat transfer problem, curing the plastic for a given period of time. That time is determined by the optimum mechanical properties, such as flexural or tensile strength, desired in the final part.

To consider the fluid flow, assume that the plastic viscosity is independent of the rate of flow, e.g., is Newtonian [17]. Assume that for this simple case a constant force F can be applied on the material. And assume that the platens are so long in the y-direction that flow along this direction is disregarded when compared to flow along the x-axis (thus giving a one-dimensional model). Thus, the standard relationship between the shear on the fluid and the amount of pressure required to flow it is

$$d\tau/dz = (dP/dx) \qquad (7.4-1)$$

z is the distance between the platens and x is the distance outward from the centerline. Now for a Newtonian fluid, $\tau = \eta\, du/dz$, where u is the velocity in the x-direction and η is the viscosity. At any time t and z = 0, there is no fluid flow (u = 0) because fluid contacts the wall. Further, at z = h/2,

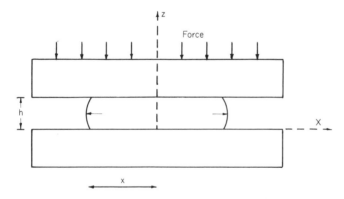

Figure 7.4-2   Schematic for one-dimensional compression molding model [17].

the velocity profile is symmetric, and thus $du/dz = 0$. Under these conditions, Eq. (7.4-1) can be written as

$$\eta d^2 u/dz^2 = (dP/dx) \qquad (7.4-2)$$

or, integrating,

$$u = -(1/2\eta)(dP/dx)(hz - z^2) \qquad (7.4-3)$$

The volumetric flow rate is thus given as

$$\int_0^h u\, dz = (h^3/12\eta)(dP/dx) \qquad (7.4-4)$$

Now, however, h is the variable distance between the platens. Thus, the volumetric rate of flow outward through any plane perpendicular to the x-axis (per unit length of y and unit time) is given as $-x(dh/dt)$, where t is time. Thus,

$$x(dh/dt) = (h^3/12\eta)(dP/dx) \qquad (7.4-5)$$

Rearranging,

$$(12\eta/h^3)(dh/dt) = (dP/dx)/x = -2K \qquad (7.4-6)$$

since the first term is a function of time only and the second of x only. Integrate the second term in the following fashion. Assume the material in the mold is of constant volume, V. Thus, per unit length of material, L, in the y-direction, at any time, $V = 2hxL$. $P = 0$ at $x = X$, e.g., at the edge of the material as it flows outward. Therefore, upon integration,

$$P = K(X^2 - x^2) \qquad (7.4-7)$$

Integrating this pressure over the entire surface of the advancing material, the total force on the platens is obtained:

$$-F = -2XLP = 2L \int_0^X P\, dx = 2KL \int_0^X (X^2 - x^2)dx = 4KX^3L/3 \qquad (7.4-8)$$

Solving for K,

$$K = -6Fh^3L^2/V^3 \qquad (7.4-9)$$

Now substitute this into Eq. (7.4-6) for K and integrate for h, assuming that h_0 is the initial distance between the plates and h is the distance at any time t:

$$(\eta/h^3)(dh/dt) = Fh^3L^2/V^3 \qquad (7.4-10)$$

$$(1/h_0)^5 - (1/h)^5 = -5FL^2t/\eta V^3 \qquad (7.4-11)$$

Now assume that $h_0 \gg h$ for long times. Then $(1/h_0)^5$ is very small when compared with $(1/h)^5$, and it can be neglected for long times. Thus,

$$h \doteq (\eta V^3/5FL^2t)^{1/5} \tag{7.4-12}$$

We shall consider several variations on this approach shortly. Note, however, that the force applied on the material has been assumed to be uniform and constant (e.g., F is a constant load, in dynes or lb_f). Thus, h decreases with increasing time. Note further, however, that the pressure applied to the load is a function of the thickness: $F = 2PLX = PV/h_f$, where h_f is the final thickness. If the pressure is constant, and thus the force is dependent on the surface area, the above equation can be rewritten to read

$$h_f \doteq (\eta V^2/5PL^2t)^{1/4} \tag{7.4-13}$$

Now consider some modifications on the above simple compression molding model. For example, it is frequently the case that the gap width is closed initially at a finite rate. Thus, rather than having F constant, dh/dt should be constant. To show this, define $h = h_0(1 - at)$, where h_0 is the initital gap width, as before, and a is a constant, defined as proportional to the rate of closing: $dh/dt = h_0a$. It has the units of sec^{-1}. Thus, from Eq. (7.4-10), the force can be written as

$$F = (V^3a/5L^2h_0)(1 - at)^{-5} \tag{7.4-14}$$

Or, define F_0 as occurring when $h = h_0$. Then

$$F/F_0 = (1 - at)^{-5} \tag{7.4-15}$$

This function is plotted as Fig. 7.4-3. Note that as at approaches unity, F/F_0 approaches infinity. Normally, the pressure on the platens builds very rapidly as h gets very small, and this is what is seen. Placing an upper limit on the force determines h. This is seen in the example below. At this upper limit of force, the model reverts back to that given earlier for a constant force, and the time begins at zero in Eq. (7.4-11).

As an example of the applicability of this model, consider a phenolic resin with a Newtonian viscosity of 10^3 poise (P), with a volume of 100 cm^3 as a cake L = 10 cm, x = 10 cm, h_0 = 1 cm, and X = 100 cm. a = 1 sec^{-1} (e.g., a closing speed of 1 cm/sec). $F_0 = 4 \times 10^7$ dyn. The final area LX = 100 cm^2. For a final pressure of 2400 psi, a force of 1.6×10^{10} dyn is needed. Or F/F_0 = 400. From Fig. 7.4-3, this occurs at at = 0.7. Since $h = h_0(1 - at)$, we have a thickness of 0.3 cm. Closing the press with F = 1.6×10^{10} dyn force will require approximately 250 sec to achieve a final thickness h_f = 0.100 cm. Note that this analysis is not restricted to Newtonian fluids, for in Eq. (7.4-1) $\tau = m (du/dz)^{n+1}$ can be chosen, in accordance with power-law theory. This makes the final form of the equation more difficult arithmetically, but the result is basically the same.

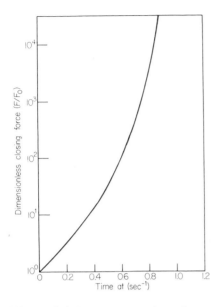

Figure 7.4-3   Dimensionless force of compression mold closing as a function of time. a is constant and proportional to the mold closing rate.

This basic model can also be used to advantage for variable viscosity where the variation is owing to temperature or reaction. For example, we can write $\eta = \eta_0 \exp(bt)$, where b depends on the rate of reaction and/or the temperature of the material. A more proper way would be to write η as dependent on temperature and then include an energy equation that is time dependent. Perhaps this model will suffice to illustrate the point, however. Introducing the time-dependent viscosity, for constant applied force in Eq. (7.4-10),

$$dh/h^6 = FL^2 \exp(-bt) \, dt/\eta_0 V^3 \tag{7.4-16}$$

Integrating,

$$(1/h)^5 - (1/h_0)^5 = [5FL^2(1 - e^{-bt})/b\eta_0 V^3] \tag{7.4-17}$$

Again assume $h_0 \gg h$, and replace F by pressure:

$$h \doteq [b\eta_0 V^2/5PL^2(1 - e^{-bt})]^{1/4} \tag{7.4-18}$$

Note that as with Eq. (7.4-12) h is infinite when t = 0 (assuming constant pressure). But note that h does not approach zero as t approaches infinity as it does in Eq. (7.4-12). Instead it approaches a constant value: $h_f =$

$(4b\eta_0 V^2/5PL^2)^{1/4}$, and this value may not be the final value of $h_f = V/XL$ desired. Note that the larger the value of b, the faster h approaches its final value and the larger that final value is. Note also that b could be negative, in which case h approaches zero at a much faster rate than that predicted by the isothermal Newtonian model. Thus, with a highly reactive material, we may not be able to achieve a final thickness desired before the material has fully reacted. This might be the case if the material is preheated too much prior to molding or if insufficient pressure is supplied during the final stages of molding.

Incidentally, Bikerman [17] has considered this problem in the pressing of adhesives and has included pressure-dependent viscosities in his equations. He has also considered radial flow, as might be encountered if the cake is placed at the center of a symmetric mold. Roller [18,19] has considered a similar analysis for the resin flow of a curing laminate. In his treatment, ⲙⲩ ⲓⲓⲓⲓⲓⲉ ⲓ.ⲓ, ⲣⲟⲇ ⲛ ⲣ-ⲥⲧⲁⲅⲉⲇ epoxy-impregnated glass cloth that is being heated and compressed in a simulated sheet molding ⲟⲓⲡ ⲓⲓⲓⲓⲓⲓⲓⲓ Ꭰⲟⲣⲓⲛⲛⲓⲛⲅ ⲱⲓⲧⲏ a force balance similar to that of Eq. (7.4-5) but extending it to elliptical geometry and maintaining a time-dependent pressure function, he found that the relationship among pressure, viscosity, and the gap distance was

$$\frac{1}{h^2} - \frac{1}{h_0^2} = \frac{2(a^2 + b^2)}{3a^2b^2} \int_0^t \frac{P(t)}{(t,T)} \, dt \qquad (7.4\text{-}18a)$$

where for circular laminates R = a = b and the geometric factor before the integral becomes $4/3R^2$. His early experiments show that care must be taken in properly scaling the pressure term during processing. Large experimental errors are to be expected if the standard SPI Test procedure is followed, since recommended pressures do not take into account the effect of viscosity on the squeezing behavior of the resin. Other processing insights are considered in the problems at the end of this chapter.

It is certainly the case that the outward flow of material in contact with a heated mold should include the energy equation. Nevertheless, this rather difficult problem has not been considered in the literature. A companion problem, that of radial flow of a thermoplastic in contact with a cold wall with stationary platens, has had considerable attention, as is seen in the section on injection molding. Therefore, these two phenomena are separated here. Consider only the heat transfer to a reacting material that is held between stationary platens. Solving the Craig equation for a constant temperature

$$(f + 1) = 2 \exp(KT)/[1 + \exp(Kt)] \qquad (7.4\text{-}19)$$

When t = 0, f = 0, since f is the fraction of reacted ends. There are very few studies on reactive materials. There are observations on the effects that take place just after preheating the reactive mixture to temperature T_0

(at which point $K = K_0$) and to the mold surface temperature T_m, at which point $K = K_m = K_0 \exp[c(T_m/T_0 - 1)]$. For various values of Kt, f can be obtained, as given in Table 7.4-1. If, for example, $K = 0.5$ min^{-1} for a preheat temperature $T_0 = 200°F = 365°K$ and if the constant $c = 8.0$, the reaction rate constant at the mold surface with a temperature of $350°F \doteq 450°K$ is 3.2 min^{-1}. Thus, if the material at the mold surface is instantaneously raised in temperature to $350°F$, while the center material remains at $200°F$, and if we desire a material that has more than 98% of its reactive sites cross-linked, the surface will reach this condition in $Kt = 5.0$, or $t = 5.0/3.2 = 1.56$ min, whereas the centerline material will reach it in 10 min. More importantly, assuming that the centerline temperature remains equal to the preheat temperature, at 1.56 min into the reaction only 37.2% of the active sites at the centerline have been reached. It is apparent, further, that because of the dependency of the temperature profile on the Fourier number, which in turn is dependent on the thickness of the slab and its thermal conductivity, very thick slabs of reactive materials must either remain in the mold for abnormally long periods of time to achieve a sufficient degree of reactivity or we must be satisfied with parts that have nearly fully reacted surfaces and centers that exhibit considerably less reaction.

This is a very sketchy viewpoint of compression molding reactions in that the time-dependent nature of the temperature at any place in the slab has not been considered, nor has the effect of the heat of reaction on the temperature profile. In the former case, the temperature-time relationship can be included as in Chap. 6, since θ can be shown to be exponentially dependent on the Fourier number. In turn, then, the reaction rate constant should be exponentially dependent on temperature, and finally the extent of

Table 7.4-1

Relationship Between Dimensionless
Time and Extent of Reaction

Kt	f
0	0
0.1	0.050
0.2	0.0997
0.5	0.245
1.0	0.462
2.0	0.762
5.0	0.986
∞	1.000

reaction should be exponentially dependent on temperature through the reaction rate constant. The latter assumption is more important. Certainly, inclusion of the heat of reaction would make the model more realistic. This means simultaneous solution of the energy equation and the reaction equation. This has not been done for compression molding, but it has been alluded to in the section on the stability of reactors in Chap. 3. Many of the reactive materials such as phenolics or melamines are solid at room temperature. They require heat to keep the reaction rate at an optimum point. This implies that these types of reactions are low-energy reactions and possibly even zero-order reactions. In the latter case, the rate of reaction is simply a constant, independent of concentration of reactive species. Therefore, the reaction rate constant (coupled with the heat of reaction) is represented in the heat conduction equation simply as an energy-generating source, and thus the arithmetic is very similar to that considered earlier in this chapter.

Consider two additional facets of the molding process. First, it is common practice to remove a part from the mold and fixture it while it is cooling to ambient conditions. Fixturing a molded part is frequently necessary to prevent part distortion as it cools. This is partly due to the continuing chemical reaction that takes place inside the part while it is cooling. If the part is allowed to cool in quiescent conditions, the heat transfer from the surface of the part is controlled by the natural convection heat transfer coefficient. As seen from Chap. 6, the controlling dimensionless group is the Biot number $Bi = hL/k$. Thus, for a heat transfer coefficient of $h = 3$ Btu/ft$^2 \cdot$hr$\cdot$°F and a material thermal conductivity of 0.15 Btu/ft$\cdot$hr$\cdot$°F, for a slab 0.240 in. thick (assuming cooling on both sides) the Biot number is only $Bi = 0.2$. Hence, if the surface-to-centerline temperature profile at the time of mold opening is assumed to be linear, the surface temperature is gradually cooling to ambient conditions, and the point of maximum temperature is progressing through the part to the centerline. This means that if additional reaction is needed to "fully react" the material in the center of the specimen, this reaction will take place while the part is out of the mold and is cooling to ambient conditions. This can cause material problems, since if the surface material is soft, the reaction in the interior of the slab will be taking place at essentially zero pressure. Since most thermosetting materials release low-molecular-weight by-products such as water or formaldehyde, the part will blister. If the pressures are sufficiently great and the surface is sufficiently cured so that it is rigid, the blister will become a crack or chip. Thus, it is generally the case that the optimum cycle time (and fixture time) is governed on the low side by the appearance of surface defects, distortion, blistering, or cracking and on the high side by economics and possible thermal degradation of the resin and/or discoloration of the surface of the part.

The second facet of the molding of thermosetting materials is in the retention of the low-molecular-weight by-products, such as water. For most materials, there exists sufficient hydrogen bonding energy to keep nearly all the water in a "bound" state throughout the life of the part. Certain resins, such as the cellulose-filled resins, have a sink for the reaction products in the filler. Thus, some water is always available or in a "free" state for diffusion through the solid slab to the surface. There are no firm data on the diffusion coefficient for water in, say, 20% alpha-cellulose-filled melamine, but from data on diffusion coefficients of other low-molecular-weight materials in solids (He in Pyrex, for example) the diffusivity is estimated to be on the order of 10^{-11} to 10^{-12} cm^2/sec [20]. If c_a is the concentration of water anywhere in the solid in g·mol/cm^3, the rate of mass transfer of water from the solid is given as

$$N_a = -D \, dc_a/dx \quad \text{(at the surface)} \tag{7.4-20}$$

For the simplest model, steady-state diffusion through a solid, the concentration profile is linear with respect to distance into the solid. Assume that the concentration at a distance 1 cm into the solid is 1 g of water per 100 g of plastic, where the specific gravity of the plastic is 1.8 g/cm^3. This is equivalent to a concentration of 10^{-3} g·mol/cm^3. Using a diffusion coefficient of 10^{-11} cm^2/sec, the rate of mass transfer from the surface (assuming that the surface concentration is zero) is given as $N_a = 10^{-14}$ g·mol/cm^2·sec. For a surface area of 100 cm , the part would lose roughly 6×10^{-4} g of water per year. The size of this value is not so critical as the implications it invokes. For many materials the diffusion coefficient is dependent on temperature to the 3/2 power. Thus, increasing the temperature increases the rate of water removal from the part. If the part shrinks linearly with decreasing moisture content, a 0.1% change in dimension per year is indicated. Note also that diffusion can occur the other way. If this material is used in a very-high-moisture application, the surface moisture concentration would be significantly greater than that in the interior, and thus water would diffuse into the sample. This would cause it to increase in dimension slightly. This expansion and contraction of the plastic is again due to the diffusion of free water within the interstices of the plastic. Probably then the most severe test of these materials would be to expose the material on one surface to boiling water and that on the other surface to high-temperature air at zero humidity. Undoubtedly, this should result in significant part distortion as well as variation in mechanical properties owing to the moisture concentration gradient. Irrespective of the actual concentration gradients and rates of water removal, the above analysis indicates that we are dealing with a material whose moisture content, overall dimension, and therefore mechanical properties are dependent on prevailing environmental conditions. We must allow for this in the design and proper application.

7.5 Conclusions

In this chapter we have only outlined a portion of the various types of batch processing that the plastics engineer will encounter. The broad topic of atomization has not been covered. It is a subject that is needed in order to determine the sizes of the droplets of polyester that are sprayed during FRP spray-up operations. Foam casting has not been discussed. Foaming in general is a relatively new study for engineers and is very complicated, owing to the simultaneous heat, mass, and momentum transfers that take place while a bubble is being formed. Some inroads are being made, however, and the interested reader is referred to the literature for information [21].

And the view of the process of compression molding has been sketchy. Hopefully, however, some insight to the process by which a material is squeezed between platens and then held there until fully reached has been gained. It is certain that a more complete understanding of the processing of thermosetting or reactive resins is necessary. In many areas, including injection and rotational molding, the lack of original data or even good hypotheses as to why process variables change the way they do is truly hindering the development and exploitation of these materials. Certainly more fundamental work is needed to gain an engineering understanding of the processing of these materials.

Problems

P7.A Examine the current literature to determine if data are available on calcium-carbonate-filled thermosetting polyester viscosity as a function of shear rate and time. If no substantial data are found, formulate a research program for obtaining such data when the polyester is catalyzed and reactive.

P7.B From Frizelle and Norfleet's data on phenolics, Figs. 7.2-6 and 7.2-7, obtain values for K and n, the viscosity index and the power-law index, respectively. Use the following equation:

$$\tau = K(\dot{\gamma}/\dot{\gamma}^{\circ})^{n} \qquad\qquad (7.\text{B}-1)$$

Let $\dot{\gamma}^{\circ} = 1 \ \text{sec}^{-1}$. Then follow through the arguments regarding compression molding-spreading of a power-law fluid up to Eq. (7.4-12). Put the final equation into dimensionless form, and identify the meaning of the dimensionless variables. Then plot the dimensionless mold height against the dimensionless time for the power-law fluid.

P7.C There have been several attempts to relate the data from the Brabender Plasticorder to the stress rate of deformation data. Blyler and Deane [22] outline a procedure for obtaining the power-law index n as follows. A plot of torque vs. roller rpm yields a straight line on log-log paper. Thus,

$$M_{T_0} = C_0 S^a \qquad (7.C\text{-}1)$$

where M_{T_0} is the torque at temperature T_0, C_0 is a proportionality constant, S is the roller speed, and a is a constant. If a distribution function for shear rates is defined as $f(\dot{\gamma})$, where $f(\dot{\gamma}) \, d\dot{\gamma}$ is the fraction of roller surface exposed to shear rates between $\dot{\gamma}$ and $\dot{\gamma} + d\dot{\gamma}$, then the amount of shear stress on the roller owing to the incremental range of shear stress is

$$\delta_\tau = K\dot{\gamma}^n f(\dot{\gamma}) \, d\dot{\gamma} \qquad (7.C\text{-}2)$$

The total torque for one roller is found by summing all torques due to this distribution:

$$M_i = \int_{\dot{\gamma}_{min}}^{\dot{\gamma}_{max}} 2\pi R_e^2 L K \dot{\gamma}^n f(\dot{\gamma}) \, d\dot{\gamma} \qquad (7.C\text{-}3)$$

Here $\dot{\gamma}_{max}$ and $\dot{\gamma}_{min}$ are the maximum and minimum roller shear rates at the roller surface. R_e is an equivalent roller radius, and L is the roller length. Blyler and Daane then consider the roller chamber to be analogous to concentric cylinders and relate $\dot{\gamma}_{max}$ and $\dot{\gamma}_{min}$ to the roller speed:

$$\dot{\gamma}_{min} = k_2 S \quad \text{and} \quad \dot{\gamma}_{max} = k_1 S \qquad (7.C\text{-}4)$$

They point out that k_1 and k_2 are weakly dependent on n but for a first approximation can be considered constant. Furthermore, the distribution function probably is a power-law function of the roller rpm as

$$f(\dot{\gamma}) = A\dot{\gamma}^b \qquad (7.C\text{-}5)$$

where $A = A(S)$, a function of the roller rpm. Since the distribution function summed over all shear rates must be unity, $A(S)$ can be determined as

$$A(S) = \frac{b+1}{(k_1^{b+1} - k_2^{b+1})S^{b+1}} \qquad (7.C\text{-}6)$$

With proper substitution and recognition that the total torque M is a sum of the torques on the individual rollers and that one roller turns at 3/2 the speed of the other, the total torque is

$$M = 2R_e^2 L[3/(2^n + 1)][(b+1)/(n+b+1)] \frac{k_1^{n+b+1} - k_2^{n+b+1}}{k_1^{b+1} - k_2^{b+1}} KS^n \qquad (7.C\text{-}7)$$

or

$$M = C(n)KS^n \qquad (7.C\text{-}8)$$

where K and n are the power-law coefficient and index, respectively. Obtain the form for Eq. (7.C-7) from the above analysis. Show that the log-log slope for the M-S curve is identical to that for the τ-$\dot{\gamma}$ curve at constant temperature obtained from a capillary rheometer. Discuss the significance of flow activation energies obtained from the Brabender torque rheometer data. Given a plastic such as linear polyethylene with a known shear stress-shear rate curve, obtain the corresponding C(n):

$$\text{linear PE (T = 190°C):} \quad \begin{array}{l} \text{at } \tau = 10^5 \text{ dyn/cm}^2, \ \dot{\gamma} = 10 \text{ sec}^{-1} \\[2mm] \text{at } \tau = 10^6 \text{ dyn/cm}^2, \ \dot{\gamma} = 1000 \text{ sec}^{01} \end{array}$$

Is this value of C(n) universal, applicable to PE only, or applicable to this resin only?

P7.D In most standard fluid mechanics texts [20], force balances are used to derive the terminal velocity of a bubble. Assume that the fluid is Newtonian and arrive at Eq. (7.2-2). Then calculate the rate of rise of a nitrogen bubble 10 μm in diameter in a polyester of 10^3-P viscosity and 1.1-g/cm^3 specific gravity. Outline the method of solution for a power-law fluid, and discuss the pros and cons of the practicality of using the model for most polymer systems.

P7.E Gels are difficult to categorize. For example, Get-Set, a popular hair setting gel, is distributed with all sizes of stationary bubbles within its volume. Two sources give Get-Set critical shear stress values of 1000 and 10,000 dyn/cm^2, respectively. Calculate the minimum bubble diameter for each of these critical shear stresses and then compare your calculated results with the real material on the drugstore shelf.

P7.F Kamal and his coworkers [3] show that the proper form for the extent of reaction for polyesters is

$$\frac{d\alpha}{dt} = 2.72 \times 10^9 \exp(-17.4/RT)\alpha^{1/3}(1-\alpha)^{5/3} \tag{7.F-1}$$

where α is the fraction of unreacted sites. Make a plot of α as a function of time for T = 150°C, 200°C, and 250°C. Compare qualitatively the shapes of these curves with that of Craig for EPDM in Fig. 7.3-1. Then determine the form for the transient one-dimensional heat conduction equation with internal heat generation. As mentioned in the text, this form for the equation cannot be solved analytically. Investigate methods of solving this equation, with isothermal mold surface temperatures.

P7.G Consider a simplification of the transient one-dimensional heat conduction equation (7.3-7). Assume that the internal heat generation q''' is proportional to the temperature of the plastic as

$$q''' = a(T - T_w) \tag{7.G-1}$$

where a is a proportionality constant and T_w is the wall temperature. Then show that the following substitution reduces Eq. (7.3-7) to the sourceless heat conduction equation examined carefully in Chap. 6:

$$T - T_w = \theta \ \exp(at) \qquad\qquad (7.G-2)$$

For the condition where $a = 0.1$ Btu/ft$^3 \cdot$hr, obtain a graph for the unaccomplished temperature change as a function of the Fourier number. Discuss the implications of the role of internal heat generation. For example, what order of chemical reaction is indicated by this heat generation term?

P7.H Verify Eq. (7.4-13). Determine a proper set of units for each of the terms of this equation. Then make the equation dimensionless, and plot the dimensionless thickness against the dimensionless time. Compare your results with the dimensionless form of Eq. (7.4-12).

P7.I In Craig's work, he relates the degree of reaction to the viscosity of the thermoset in the following way:

$$\eta = \eta_0 + (\eta_\infty - \eta_0)g \qquad\qquad (7.I-1)$$

where η is the instantaneous viscosity, η_0 is the viscosity of the unreacted material, and η_∞ is the viscosity at complete reaction. Assuming isothermal conditions, integrate Eq. (7.3-1) to obtain f(t), and then obtain the time-dependent form for the viscosity. Then introduce this expression into Eq. (7.4-10), and obtain a form for the mold height as a function of time for constant applied force. Compare your results with Eq. (7.4-18).

P7.J The rate of diffusion of moisture from a slab corresponds to the rate of heat removal in that the forms for the equations are the same. In Chap. 6, curves for unaccomplished temperature change as a function of dimensionless time were presented for several geometries. Construct an unaccomplished concentration change where the moisture content initially in a phenolic thermoset is 10^{-3} g$\cdot$mol/cm^3 and the moisture content at equilibrium everywhere around the thermoset is 10^{-5} g$\cdot$mol/cm^3. Determine the time required for the centerline to reach 10^{-4} g$\cdot$mol/cm^3 if the slab half-thickness is 1 cm and the diffusional coefficient through the material is 10^{-11} cm^2/sec. Conversely, determine the amount of moisture at the centerline after 1000 hr if the environmental moisture content is 10^{-1} g$\cdot$mol/cm^3.

References

1. V. W. Uhl and J. B. Gray, eds., Mixing: Theory and Practice, 2 vols., Academic Press, New York, 1966, 1967.
2. J. M. McKelvey, Polymer Processing, Wiley, New York, 1962.
3. M. R. Kamal, S. Sourour, and M. Ryan, SPE ANTEC Tech. Pap., 19:187 (1973).

4. B. Schreiber, Kunststoffe, 59:362 (1969).
5. Plastics Engineering Co., Sheboygan, Wisc., Type 308 Injection Molding Grade Phenolic.
6. J. E. Goodrich and R. S. Porter, Polym. Eng. Sci., 7:45 (1967).
7. W. S. Frizelle and J. S. Norfleet, SPE ANTEC Tech. Pap., 19:538 (1973).
8. T. Hirose and M. Moo-Young, Can. J. Chem. Eng., 47:265 (1969).
9. F. E. Craig, SPE ANTEC Tech. Pap., 18:533 (1972).
10. R. B. Prime, SPE ANTEC Tech. Pap., 19:205 (1973).
11. R. C. Progelhof and J. L. Throne, Polym. Eng. Sci., 15:690 (1975).
12. R. C. Progelhof and J. L. Throne, Polym. Eng. Sci., 16:680 (1976).
13. J. A. Deiber and R. L. Cerro, Ind. Eng. Chem. Fund., 15:102 (1976).
14. J. L. Throne and J. Gianchandani, Polym. Eng. Sci., in press.
15. J. L. Throne, SPE ANTEC Tech. Pap., 20:367 (1974).
16. N. D. Simons, Mod. Plast. Encyclopedia, 49(10A):546 (1972).
17. J. J. Bikerman, Science of Adhesive Joints, 2nd ed., Academic Press, New York, 1968.
18. M. B. Roller, Polym. Eng. Sci., 16:687 (1976).
19. M. B. Roller, Polym. Eng. Sci., 15:406 (1975).
20. R. B. Bird, W. E. Stewart, and E. N. Lightfoot, Transport Phenomena, Wiley, New York, 1960, Chap. 16.
21. C. J. Benning, Plastic Foams, 2 vols., Wiley, New York, 1969.
22. L. L. Blyler, Jr. and J. H. Daane, Polym. Eng. Sci., 7:178 (1967).

8.

Extrusion—A Continuous Process

8.1 Introduction

The extruder is the heart of most extrusion and injection and blow molding
operations today. It is also being used to process thermosetting resins and
for the hybrid process of injection blow molding. The primary purpose of
the extruder is to take room-temperature plastic resin in the form of pellets,
beads, or powders and convert the resin to a molten polymer at sufficiently
high pressure to allow the highly viscous melt to be forced through a nozzle
into a mold, in the case of injection-type processes, or through a die, in the
case of blow molding or continuous extrusion. Therefore, within the extruder, the solid plastic pellets, beads, or powders must be conveyed and
compressed. They must be heated until soft or fluid through transfer of
heat from the barrel walls, from the friction of pellets shearing against one
another, and/or solid-liquid contact of the resin with its melt. The molten
polymer must then be pumped through channels to build the pressure prior
to flow through the nozzle or die. Before concentrating on the engineering
concepts of extrusion, we shall examine a typical extruder. Figure 8.1-1,
is a schematic of a typical single-screw extruder, with components labeled
[1]. As shown, the unit is comprised of a motor drive normally with a siz-
able torque capacity, a gear train, and a screw keyed into the gear reducing
train. The screw is normally allowed to float free at the die end, and the
fluid layers between the screw flights and the barrel wall keep the screw

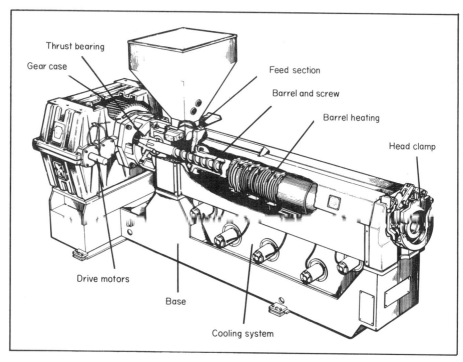

Figure 8.1-1 Breakaway diagram of a typical single-screw extruder with characteristic electric heater bands on the barrel. (Used by permission of Modern Plastics Magazine [1].)

centered. The drive train normally has a continuously variable speed of 0 to 150 or 200 rpm. The barrel is usually electrically heated in all but the oldest units. "Dowtherm" heat transfer oil was used extensively in the late 1940s and early 1950s and was either piped to the barrel from a reservoir or contained in condenser-reboiler units attached directly to the barrel. The electric heaters of today are simple band heaters with either "on-off" control or proportional control. Thus, the barrel is "zoned" according to the number of controllers on the heater bands. In smaller extruders, the barrel may have only three zones; on large vented or two-stage machines with L/D ratios of 30:1 or more, the barrel may have as many as six zones. The screw may also be cored to allow for either cooling or heating in certain areas. In some cases, electric cartridges are placed at the die end or tip of the screw to aid in the pumping phase of extrusion. At the end of the barrel, there is frequently a screen pack (or sand pack in an old machine). This pack helps even the flow and prevents unmelted resin particles from entering the die section. Sometimes a breaker plate follows the screen pack; this plate supports the screen pack and also aids in evening the flow. Surging that occurs with fluid flow from the screw flight tip will be discussed later. Beyond the breaker plate may be a backflow ball valve. This valve is used to control the flow rate through the die that follows the valve. When a screen pack is very new and clean, it is frequently necessary to limit the flow by partially closing the valve. As the screen pack accumulates resin insolubles and foreign debris, the valve is gradually opened to allow the flow to remain constant. When the valve is completely open, it is time to change the screen pack or suffer the loss in flow through the die.

Attached to the front of the extruder is the die. It may be a simple annulus producing a pipe profile. It may be quite complex in shape to allow for extrusion of, say, hollow intricately shaped profiles for window structures, for example. It may contain a drawing mechanism for wire coating. It may be a very-large-diameter ring with a very small cross-section to allow for air injection into the center of the ring, thus producing blown film. Or it may contain many tiny holes called spinnerettes through which fine filaments known as threadlines are produced. Normal instrumentation includes a pressure gage on the barrel after the breaker plate-screen pack and before the ball valve and a thermocouple in the hot melt in the same region. Normally for noncritical operation, pressure, temperature, and rpm are monitored, and flow rate is measured by weighing a sample taken over a measured period of time. As the demand for more uniform profiles, filaments, and the like has increased, more controls and more accurate controls are being touted. It is now common to have several pressure transducers and thermocouples along the barrel controlling barrel heaters and/or auxiliary heaters such as the above-mentioned cartridge heater in the tip of the screws. Furthermore, controls on die geometry are becoming common; for example, on large fishtail sheeting dies, traveling beta gages measure the film thickness and feed signals back to servos on the die lips, thus controlling film thickness to $\pm 2\%$ or better. On highly instrumented research machines, thermocouples have been included on the screw as well as in the barrel wall. Thus, temperatures in the conveying flights can be measured quite accurately. It should be pointed out, however, that these signals must be conveyed from the screw through slip rings. These can be quite expensive and are excellent sources for measurement error.

Nevertheless, the process of extrusion is undoubtedly the most thoroughly studied and understood of all plastics processing. Computer designs of screws are commonplace today, thanks largely to the intensive research efforts of McKelvey [2], Tadmor et al. [3,4], and Squires [5], to name a few. We shall examine their work in order to understand the process.

In this chapter we shall be primarily concerned with screw extrusion and plastication. This does not mean that plunger or ram-type extrusion should not be studied. The application, in straight extrusion of profiles, is quite limited, however. Ultrahigh-molecular-weight polyethylene (UHMWPE) and TFE Teflon are two materials that are best processed into

profiles by ram extrusion, because at the melting temperature of these
materials, their viscosities are so high [e.g., TFE Teflon viscosity at its
melt temperature is 10^{10} to 10^{11} poise (P)] that the materials appear solid-
like and cannot be pumped in the traditional sense of the word. We shall
therefore concentrate on the most important aspects of extrusion, screw
extrusion, at the neglect of this type of processing.

Consider several "rules of thumb" [1] with regard to pumping capacity. Many extruders will plasticate and pump a maximum of 10 to 15 lb/hr/
hp. However, if mixing, working, high throughput with little barrel heat
input (e.g., adiabatic extrusion), or high temperatures are needed, a given
machine may yield as little as 3 to 5 lb/hr/hp. Thus, on an easy material
such as nylon (with a relatively low-melt viscosity), a 7-hp machine could
process 70 lb/hr. If the material is a thermoplastic elastomer, it might
process only 25 lb/hr. Note further that only single-screw extruders will
be considered. If you have a major mixing problem, a thermally sensitive
material or are dealing with a highly plasticized material, you would probably go to a double-screw system (e.g., two intermeshed screws). Rigid
PVC is best extruded in twin-screw machines. A constant pitch metering
screw is recommended unless the material requires extensive mixing. A
material that requires intensive mixing to ensure melt uniformity (or color
dispersion) might require a two-stage screw, with a let-down section in the
center of the screw, and one of several intensive mixing devices might be
included at or near the tip of the screw. If the material requires venting, a
two-stage screw with venting at the letdown section is needed, as shown in
Fig. 8.1-2. Under any conditions, if mixing and/or venting is needed, a
machine should have a large length-to-diameter ratio (L/D) on the order of
30:1 or 36:1. If the material melts easily and/or flows easily, a shorter
extruder (L/D of 20:1 or even 16:1) can be used. Normally the barrels of
extruders are designed to withstand extrusion pressures of 10,000 psi. This
is sufficiently high for most thermoplastics; at times, however, pressures
(particularly if the screw is also used to inject the plastic) can exceed this
limit. It is not uncommon to register extrusion pressures of 30,000 psi for
FEP Teflon, and obviously under these conditions special barrels are needed.
Many barrels are of S-7 tool steel with Xaloy linings and/or nitriding at
wear points. Screws are frequently 4140 steel with high-density nickel-
chrome plating and Stellite #6 used at wear points. Extrusion and/or plastication of filled materials (regardless of the base resin) will normally wear
the screw flights at a much faster rate than unfilled resins, and thus extra
precautions must be taken at the outset to minimize wear and thus minimize
backflow and attendant pressure loss at the tip of the screw. Extensive research into high-performance flight materials is required owing to the increasing interest in filled, reinforced, and high-performance resins.

Note that the hopper is part of the extruder, and very little is known
about efficient hopper design. This is discussed in the solids conveying
section later. Note here, however, that the hopper throat is frequently

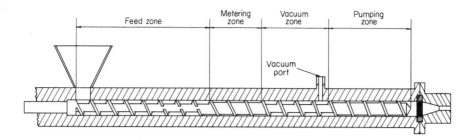

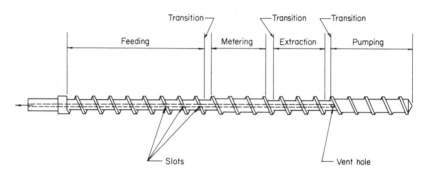

Figure 8.1-2   Typical cross-section of vented barrel extruder and screw design [6].

actively cooled (with tap water or in extreme cases with ethylene glycol refrigerant). This cooling section on the barrel abuts the heater bank, and thus the efficiencies of both the cooled section and the heated section are significantly lower than the efficiencies in other sections of the extruder. Many of the resins being extruded today require arid conditions for processing. Polycarbonates, nylons, terephthalates, and modified acrylonitriles are some of the plastics that require moisture contents to be less than 0.05% for void-free extrusions. Thus, hoppers can be heated with conditioned air, nitrogen-blanketed, and/or evacuated. As we shall see, the solids conveying aspects are thought by many extrusion experts to be the keys to successful surge-free extrusions.

As mentioned, screw design has been computerized to a great degree, and the actual screw configuration will depend, within reason of course, on the characteristics of the material being extruded. Nevertheless, some generalities can be made about screw extruder design. As shown in Fig. 8.1-3, the section of the screw immediately ahead of the gear train is the solids metering or feed zone [7]. This is characterized by a rather deep

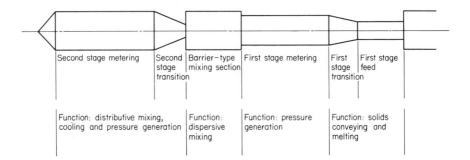

Figure 8.1-3   Typical screw root profile for two-stage screw [7].

channel between the root of the screw and the barrel wall. The transition
or plasticating zone of the screw is immediately ahead of the solids metering zone. Here the channel narrows very quickly. It is in this zone that
the most intense friction and melting of the plastic take place. The section
near the tip of the screw is the melt pumping or melt metering zone. Here
the pressure is built, and the melt is homogenized and raised in temperature
for extrusion or injection molding. As will be seen, the screw can be considered to be a nearly positive displacement pump. The ratio of the solids
metering channel depth (sometimes called the flight depth) to the melt pump
channel depth is the compression ratio. It can be as small as 2:1 for LDPE
or as large as 6:1 for some nylons. The screw flights enable the screw to
move the plastic down the length of the barrel, and they are set on a pitch
angle that can vary from 12 to 20°. In most (but not all) screws, the pitch
angle is held constant; a pitch angle of 17.61° (the condition where the flight
equals the diameter) yields a single-flighted screw. This screw design is
frequently called a "general-purpose" screw design and is the most popular
and easiest to fabricate. It should be pointed out here that there is a growing minority of special-purpose screws, with overlapping flights, double
converging flights, variable pitch flights, interrupted flights, and so on.
Most of these are developed using computer programs and/or common sense
and are considerably more expensive to fabricate. Therefore, the use of
a special-purpose screw must be economically justified on the basis of significantly improved physical properties of the extrudate, on much improved
throughput per horsepower-hour, and/or on high-volume processing of low-
cost resin.
    The distance between the flights is the lead, and it is approximately
equal to the ID of the barrel for a single-flighted screw (assuming that the
flights have negligible thickness, which they do not have). Owing to the
thermal expansion and contraction of the metal parts, the screw cannot fit
tightly into the barrel. The radial clearance between the flight tip and the
barrel (at room temperature) is on the order of 0.001 in./in. of barrel ID.

This clearance cannot be too great, or molten polymer will flow back along the barrel, thus reducing the effectiveness of the melt pumping zone.

The analysis of multiple-flighted screws is treated in the same manner as single-flighted screws. The channels are considered to be in parallel. Furthermore, our analysis assumes that the solids conveying zone/ plasticating zone/melt pumping zone/die flow model is a series of stages. Others believe that the entire extruder must be modeled without considering which "zone" will control the process. There are valid arguments for the latter case. Chung states that "a screw cannot pump out more than it can convey and melt, or vice versa, and these three functions (excluding die flow) depend on each other. The overall performance of a screw extruder can be predicted only by combining the solid-conveying, melting, and metering theories into a complete plasticating screw extrusion theory" [8]. Also, the flow through the die must be matched to that exiting the extruder before a complete design can be developed. Nevertheless, for illustrative purposes, it is advisable at this time to consider the zones as discrete. However, would-be extruder screw designers should consider even the most sophisticated models to be only guidelines and use common sense and practical experience in addition to the computer program. This agrees entirely with the philosophy of Kruder and Kim [9].

Having now set forth the limitations of mathematical modeling, we wish to consider the various effects that take place in the extruder barrel and screw flights. Remember that the material that follows is to be used for illustrative purposes and demonstrations of the methods of prediction. To understand the progress being made in screw extruder design and the variations on these illustrations, the reader should begin with the text by Tadmor and Klein [3].

8.2 Solids Conveying and Hopper Design

It is important to note that, literally speaking, the design of the screw extruder begins and ends with the solids conveying zone. The purposes of this zone are to bring the plastic resin from the hopper into the barrel; to compress the resin pellets, beads, or powders sufficiently well to express entrapped air back through the hopper; and to break up lumps and agglomerates of particles so that efficient melting can be accomplished. Much more is known about the processes of melting and melt pumping than about solids conveying, since the science of solids mechanics is very weak. Recently much attention has been given to this zone of the extruder, and certain guidelines for design are emerging. It is now believed that the solid moves down the flights in nearly plug-like flow, at least in the early few turns of the zone. To get a better picture of the flow of materials in the flights, we need to "unwrap" the solid from the screw, as Donovan has done [10]. (See Fig. 8.2-1.) The nomenclature generally associated with the extruder

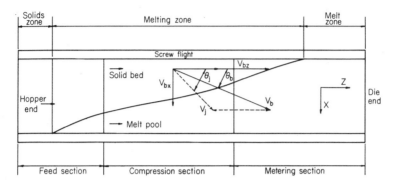

Figure 8.2-1   Schematic of extruder screw as unwrapped. Note relationship between velocity in flights and bulk velocity through extruder [10].

screw is shown in Fig. 8.2-2 [5]. Note that one can either slide the screw
and hold the barrel stationary or slide the barrel and hold the barrel stationary. Note, however, that the relative motion is not parallel to the
flights but at an angle to them. This is shown in Fig. 8.2-3 [5]. The angle
is the helix angle of the flight ϕ. A vector diagram shows that when the
screw axis is horizontal, the forward velocity of the plug is given as the
velocity down the flights divided by the sine of the helix angle. If the forward velocity of the plug is U_z, p is the number of flights in parallel, e is
the width of the flight, and h is the depth of the channel, the flow rate of
the solids is given as

$$Q_s = U_z \rho \int_{R_s}^{R_b} (2\pi R - pe/\sin \phi) \, dR \qquad (8.2\text{-}1)$$

where R_s and R_b are the radii of the root of the screw and the inside of the
barrel, respectively. Integrating,

$$Q_s = \rho U_z [\pi (D_b^2 - D_s^2)/4 - peh/\sin \phi] \qquad (8.2\text{-}2)$$

A value of U_z is needed. Note that the velocity of the plug in the down-
channel direction at the barrel surface is $U_{zp} = U_z/\sin \theta$, where θ is the
angle of advance of the plug relative to the screw axis. The friction forces
acting between the plug and the barrel depend on the angle θ at which the
plug moves relative to the barrel (or screw) axis. This angle is a function
of the force and torque on the plug. However, it is perhaps worth looking
at some simple extremes before considering the more difficult problem.
Note that if the frictional forces between the polymer and screw are sufficiently large so that all the polymer adheres to the screw, the plug does
not move relative to the screw. Thus, $U_z = 0$ and $Q_s = 0$. If the frictional

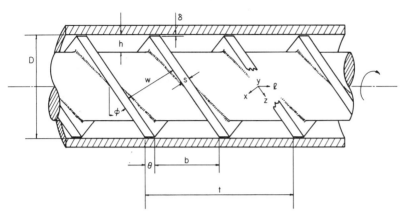

Figure 8.2-2   Characteristic dimensions and terms for a single-screw extruder (double-flighted) [5].

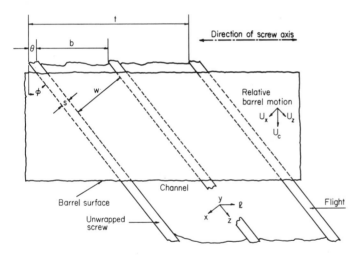

Figure 8.2-3   Additional nomenclature for a double-flighted single-screw extruder.   Note the differences in notation between this figure and Figure 8.2-1 [5].

force between the polymer and the screw is zero, the plug acquires a down-channel velocity equal to that at the barrel surface. Thus, $\phi = 90° - \theta$. This is the maximum plug velocity. Thus, as a rule of thumb, the maximum solids conveying flow rate can be achieved by polishing or coating the screw and roughening the barrel, thereby minimizing friction between the polymer and the screw and maximizing it between the polymer and the barrel.

Normally ϕ is much closer to the upper limit than to the lower limit. Tadmor and Klein [3] and others [10] carry through the detailed force and torque balances on the plug to determine ϕ as a function of the forces and pressures on the plug and show that the above simplistic analysis is in fact correct in the limit.

U_{zp} is a function of U_b, the barrel velocity, in the following form [11]:

$$U_{zp} = U_b \tan \phi \tan \theta/(\tan \phi + \tan \theta) \tag{8.2-3}$$

But $U_b = ND_b$, where N is the rpm of the screw. Defining,

$$W = (\pi/p)(D_b - h) \sin \phi - e \tag{8.2-4}$$

where W is the average channel width, Eq. (8.2-2) can be rewritten as

$$Q_s = \rho \pi^2 NhD_L (D_L - h)[W/(W + e)](\tan \phi \tan \theta)/(\tan \phi + \tan \theta) \tag{8.2-5}$$

Normally $W \gg e$, and thus the term $[W/(W + e)] \doteq 1$. Note that the density ρ used here is the compacted powder density and not the density of the molded resin. According to Griffith [11], ϕ as given earlier is $17.61°$ and θ is on the order of 10 to $25°$. For the idealistic case of no friction between the screw and the solid plug, the pressure drop along the solids transport portion of the screw is zero, and $\tan \phi = \cotan \theta$. For this extreme case,

$$Q_s' = \rho \pi^2 NhD_b (D_b - h) \sin \phi \cos \phi \tag{8.2-6}$$

For this case, with $\phi = 17.61°$, the maximum possible solids flow rate is obtained:

$$Q_s' = \rho \pi^2 NhD_b (D_b - h)(0.2884) \tag{8.2-7}$$

Any real solids flow rate can be referred to this by ratio:

$$Q^* = Q_s/Q_s' = 0.2884(\tan \phi \tan \theta)/(\tan \phi + \tan \theta) \tag{8.2-8}$$

For example, if $\phi = 17.61°$ and θ is $17.61°$, the ratio is $0.2884 \tan \phi/2 = 0.0458$. For the range of θ from 10 to $25°$, the range on the ratio is 0.0327 to 0.0545. Thus, for practical ranges of θ, the actual throughput is normally less than 6% of the maximum. Griffith's data, shown in Fig. 8.2-4, show this ratio Q^* as a function of the force applied (in this case using a spring-loaded plate on the exit of a model glass-walled extruder). Note the remarkable drop in Q^* with only a mild increase in the force against the material. In addition to this, the push S that a self-conveying solid plug can exert on an obstruction can be calculated. If S_0 is the friction between the plug and the screw channels, F is the frictional force per unit area between the barrel and the plastic solids, and L_s is the length of channel filled with solids,

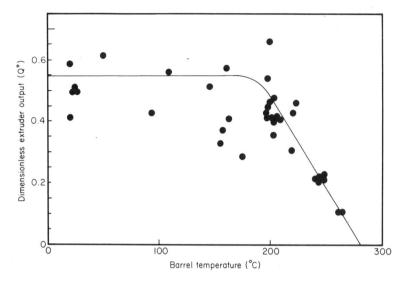

Figure 8.2-4 Comparison of dimensionless polymer output (compared with theoretical screw capacity) with barrel temperature. Note rapid drop in extruder output as barrel temperature exceeds 200°C [11].

$$S = F \pi D_b L_s \cos (\phi + \theta) - S_0 \qquad (8.2-9)$$

If the plug is non-self-conveying, it offers resistance S_n, defined as

$$S_n = 2 \pi D_b R \qquad (8.2-10)$$

where R is the non-self-conveying friction per unit surface. Thus, as Griffith argues, for effective solids conveying, S must exceed S_n, or, in other words,

$$[FL_s \cos(\phi + \theta)/2R - S_0/2\pi D_b R] \gg 1 \qquad (8.2-11)$$

From Griffith's work, shown in Fig. 8.2-5, it can be seen that more significant improvement in solids conveying throughput can be achieved with pellets than with powders. This has been attributed to the lower coefficient of friction of powder particles with the barrel, although there seems to be equally sufficient data to indicate that the coefficient of friction of the powder with the screw is lower as well [12]. This argument can be directly applied to Eq. (8.2-11), where, at a constant frictional coefficient between the material and the screw, increasing F (e.g., increasing the coefficient of friction between the material and the barrel) leads to more effective conveying. Also, increasing the feed section length and the barrel diameter increase solids conveying capacity.

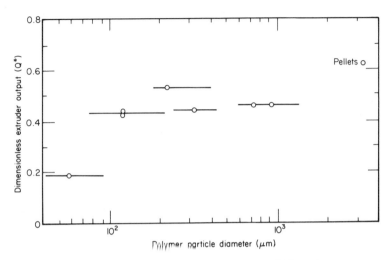

Figure 8.2-5   Relationship between dimensionless extruder output Q^* and particle size of polymer; screw speed constant at 80 rpm [11].

Schneider [12] also shows that in many cases the coefficient of friction increases rather dramatically with temperature and with applied load, but the data scatter so that the obvious conclusion that heating the roughened barrel and chilling the highly polished screw will increase the solids conveying capacity cannot be drawn with sufficient accuracy today. Certainly the effect of applied load may be more important than the surface condition of either the screw or the barrel and thus obviate extensive engineering efforts on the treatment of the barrel and screw surfaces.

In addition to the solids conveying flow rate considerations we should also gain an appreciation for the heat generated through mechanical processing. Griffith gives the following equation for mechanically generated heating:

$$H_s = FN\pi^2 D_b^2 L_s \tag{8.2-12}$$

Note that the amount of mechanically generated heat added to a pound of material conveyed is given as

$$H_s/Q_s = FD_b L_s (\tan \phi + \tan \theta)/\rho h(D_b - h)[W/(W + e)](\tan \phi \tan \theta) \tag{8.2-13}$$

Thus, to obtain some mechanically generated heat due to friction, a large frictional force between the barrel and the solids, a relatively long feed section, and a relatively shallow channel (e.g., small values of h) are needed. Thus, a significantly shallower channel is needed to mechanically heat powders than that needed to mechanically heat pellets to the same en-thalpy increase. (See Fig. 8.2-5.)

Before considering the concepts of hopper design, note that much research effort is being concentrated in the area of solids conveying. Again, Kruder and Kim and Chung et al. [13] caution that the extruder should not be considered as sections in series but rather as a device in which solids conveying, plasticating, and melt pumping occur throughout the flight path and that surging and other flow instabilities can be traced in many cases to the flow and compaction of powder, beads, or pellets in the solids conveying section. This, again, is why emphasis is being placed on this particular area at this time. Kruder and Kim further point out that the addition of regrind of arbitrary shape and fillers, including chopped glass fibers, talc, and mineral wool, compounds the problem of dealing with virgin stock of known shape and surface conditions. All of these "added materials" increase the difficulty in predicting the flow rates of the solids conveying section and in particular in predicting the fluctuations (or perturbations) in flow rates that might be caused by changes in back pressure, agglomeration, bridging, low hopper feed, plating out of surface lubricants on barrel surfaces, and so on. There is far more known about the melting and pumping of a plastic than about the solids conveying aspects, and yet most of the designs in those sections depend on the input from the solids conveying section.

From a rather careful force and momentum balance analysis, Tad-mor and Klein [3] show that the pressure in the solids conveying zone rises exponentially with distance down the flight. In their notation,

$$P = P_1 \exp[(B_1 - A_1 K)z_s/(B_2 + A_2 K)] \tag{8.2-14}$$

where P_1 is the initial pressure at the base of the hopper, z_s is the distance down the channel, and B_1, A_1, B_2, A_2, and K are functions of the coefficients of friction between the solid and the barrel, the solid and the screws, the helix angle, the angle of movement of the solid, the geometry of the flight, and so on. What is most significant about this equation is that the pressure at any point along the flight channel depends on the initial pressure at the base of the hopper. It is apparent that if the material in the hopper was a liquid, the pressure would be the hydrostatic pressure, given as $P_1 = \rho g H$, where H is the height of the column of liquid and ρ is its density. For powders or pellets, however, this linear relationship is invalid. Janssen [14] developed a semitheoretical pressure profile for solids which indicates that the pressure at the bottom of a hopper is dependent on the height of material above it. However, the relationship is highly nonlinear. Consider a tall rectangular cylinder of diameter D [15]. There is a vertical pressure and a horizontal pressure that can apply force in the vertical direction. The force in the vertical direction due to pressure differential in the vertical direction is

$$(\pi D^2/4) \Delta p_v \tag{8.2-15}$$

The friction force on the side walls of the cylinder is given as

$$\pi D \, \Delta x \, \mu_w p_h \tag{8.2-16}$$

acting upwards, where Δx is the thickness of the slice of material in the x-direction, μ_w is the friction coefficient between the material and the wall, and p_v is the horizontal pressure of the powder on the wall. Of course, in any force balance the weight of the slice acting downwards must be included:

$$(\pi D^2/4)\rho_b \, \Delta x \tag{8.2-17}$$

where ρ_b is the bulk density of the powder (which may be a function of x if the powder compacts significantly). According to Janssen, p_h, the horizontal pressure, is proportional but somewhat smaller than the vertical pressure. Therefore, $p_h = Kp_v$, where K is a proportionality constant that is less than unity and represents the ratio of horizontal to vertical pressure. Assuming the slice to be in equilibrium, the force balance becomes

$$dp_v/dx + (4\mu_w K/D)p_v = \rho_b \tag{8.2-18}$$

Assume that $p(x = 0) = p_{v0}$, e.g., that the pressure at the top of the bin is known. Equation (8.2-18) can then be integrated to obtain the Janssen equation:

$$p_v = (\rho_b D/4\mu_w K)[1 - \exp(-4\mu_w Kx/D)] + p_{v0} \exp(-4\mu_w Kx/D) \tag{8.2-19}$$

If, as is the case of open-top plastics extruder hoppers, $p_{v0} = 0$, the second term on the right-hand side of Eq. (8.2-19) is zero. Note that as x becomes large, the pressure approaches a constant value, independent of bed depth:

$$p_{v\infty} = D\rho_b/4\mu_w K \tag{8.2-20}$$

Note that if $\mu_w K$ has the value of about 0.2 that for bin with an L/D = 5, the term $\exp(-4\mu_w Kx/D)$ becomes 0.0183, or the vertical pressure at the bottom of the bin is within 98% of the maximum pressure. Experimental evidence of the Janssen equation is well documented elsewhere [15]. Rudd [16], for example, shows that for 1/8-in. polystyrene cubes in a bin 10 in. in diameter, with a measured $\mu_w K = 0.272$ and a bulk density of 39 lb/ft^3, the pressure at the base approaches 30 lb/ft^2 when the bed height exceeds about 35 to 40 in. (e.g., an L/D of 3.5 to 4). His data are shown in Fig. 8.2-6. Beyer and Spencer [17] consider the effect of $P_{v\infty}$ on the pressure at the base of the hopper and show dramatically that the second term on the right-hand side of Eq. (8.2-19) is also correct. In Fig. 8.2-7 is plotted the ratio of pressure on the top of the bed to that at the bottom as a function of the depth of the bed. They also note that addition of surface lubricant to the resin particles could dramatically reduce the value of $\mu_w K$, as would be expected. In Table 8.2-1 is shown the effect of external lubricant on $\mu_w K$.

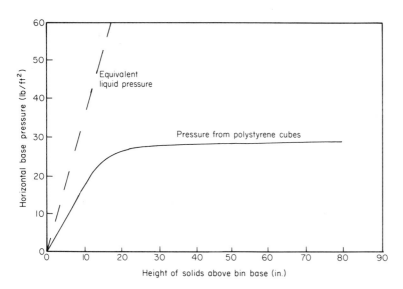

Figure 8.2-6 Effect of solids height on base pressure in cylindrical bin. Material, 1/8-in. polystyrene cubes; bulk density, 39 lb/ft^3; bin radius, 5 in. [16].

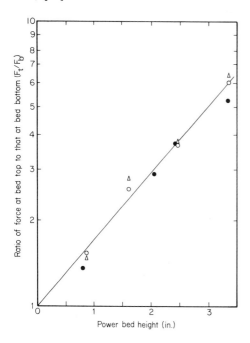

Figure 8.2-7 Ratio of the pressure at the top of a static column of plastic powder to that at the bottom, as a function of the bed height, for three polystyrenes [17].

Table 8.2-1

Effect of External Lubricant on Solids Flow Parameter

ppm Lubricant	$\mu_w K$
0	0.262
150	0.200
300	0.174
500	0.142
750	0.106

Note that additives that reduce $\mu_w K$, allowing materials to flow easily from hoppers, also reduce the friction between the material and the screw and the friction between the material and the barrel. Excesses of lubricants can cause blemishes on extruded sections or injection-molded parts (e.g., streaking, plating out, etc.). Furthermore, the Janssen equation is specifically designed for static loading in a rectangular cylinder hopper. Most hoppers are conical for the obvious reason that the most material is to be put in the hopper in the shortest possible head space and still must be able to feed a small cross-section barrel inlet. Several investigators have considered pressure profiles in conical hoppers. Lenczner [18] concluded that the pressure is independent of wall friction and is linear with height above the bottom of the hopper:

$$p_v = \rho_b x \qquad\qquad (8.2-21)$$

As shown in Fig. 8.2-8, this seems valid for relatively short hoppers. Thus, the pressure profile is very similar to the hydrostatic head pressure equation for liquids.

Two additional points should be made about hopper design. The first is that $\mu_w K$, considered constant in the Janssen derivation, is probably a function of the height of material in the bin. For example, Lenczner shows that for cement powder $\mu_w K$ increases linearly with increasing depth of fill until bin diameters exceed 3 or 4. This can, of course, be included in the mathematical model presented earlier (where a constant value was tacitly assumed). This will make some changes in the form of Eq. (8.2-19). Furthermore, flowing powders exhibit a $\mu_w K$ somewhat different from static powders. Again, Lenczner points out that this is probably due to a significant increase in the frictional coefficient between the material and the walls of the container. This increase overcomes the measured lower loads carried by the walls (e.g., lower values of K), and thus $\mu_w K$ probably increases with increasing emptying rate.

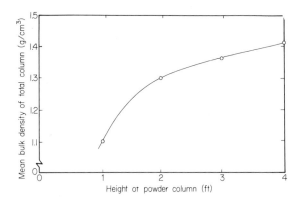

Figure 8.2-8   Effect of column height on bulk density of powder in column. Note: Curve represents measured data for cement powder; column diameter, 10 in. [18].

In actuality, hoppers attached to extruders do not fit the categories of static hopper-bins or freely emptying hopper-bins. The flow rates in extruder hoppers are controlled by the feed rate of the extruder. It is important only that the feed rate from the hopper be equal to that of the extruder output, that it match the extruder output at any rate, that no pressure fluctuations occur in the hopper (owing to variation in hopper level), and that the materials in the hopper be freely flowing. Also, remember that any air entrained with the particulates flowing into the extruder must come back through the hopper or be dissolved in the melt. Brown and Richards [15] discuss in detail the flow from hoppers and include a very brief discussion on "percolation" or backflow of air through the discharging powder or pellets. Apparently this problem becomes more serious as the particulate size decreases. Another problem with the handling of particulates is the adsorption of moisture on the surface of the material. The surface area of any particle is defined as

$$S = 6bD_p^2 \tag{8.2-22}$$

where D_p is the geometric dimension of a particle and b is a geometric constant. The volume of any particle is

$$V = aD_p^3 \tag{8.2-23}$$

Thus, the surface-to-volume ratio is given as

$$S/V = D_p/6(b/a) \tag{8.2-24}$$

where b/a is frequently referred to as a shape factor. For regular materials (cubes, spheres, and cylinders with $L = D_p$), b/a is unity. For sand, for example, $b/a \doteq 1.1$, and for mica (or metal) flakes, $b/a \doteq 3.6$ or more. The total surface area is related to the shape factor, the mean diameter of the particles, and the density of the particle (not the bulk density) by

$$A = (6b/a \rho_p) \int_0^1 d\phi/D_p \qquad\qquad (8.2\text{-}25)$$

ϕ is the fraction of particles having geometric dimensions D_p or larger. A is the surface area per unit mass of sample. Thus, increasing the mesh size decreases the geometric dimension D_p and increases the surface area proportionally. Note that Eq. (8.2-25) can be applied to mixtures with distributions of particle sizes. More importantly, the adsorption of moisture on the surface increases with the decreasing mesh size. To remove the adsorbed moisture, therefore, hoppers can be heated, blanketed with inert gas, or evacuated. Heating the outside of a hopper does little to remove moisture, since the thermal conductivity of plastics powders is very low [19] and the material in the center of the hopper moves through at a much faster rate than that at the wall [15]. An alternative heating method uses warm desert-dry air or nitrogen moving up through the bed from a porous collar at the base of the hopper. As seen from the heat transfer section, the effectiveness of heat transfer (and similarly mass transfer) is usually directly related to the velocity of the fluid moving past the solid surfaces. For moisture removal, for example, in laminar flow, the dimensionless group is the Sherwood number, defined as the mass transfer coefficient times a characteristic dimension divided by the diffusion coefficient: Sh = kD/D_m. For low flow rates, the Sherwood number is approximately proportional to the square root of the velocity through the "porous medium." Accordingly, therefore, we can determine the pressure drop-flow rate relationship for flow through a porous bed [20]. The porosity e of the packed bed of resin particles is defined as the volume of particles per unit volume of the container. Thus, the pressure drop-flow rate equation for laminar flow is written in terms of the porosity of the bed, the viscosity of the fluid μ, the height of the bed L, and the surface area of the particles per unit volume of the vessel S_V as

$$\Delta P/L = K S_V^2 \mu u/e^3 \qquad\qquad (8.2\text{-}26)$$

K experimentally has the value of about 5 for many materials, including plastic powders. This equation, called the Carman-Kozeny equation, can be rewritten in terms of the shape factor (b/a) and the volume surface mean diameter, defined as $D_{vmp} = 6b/a\rho_p A$:

$$\Delta P/L = 180 (b/a)^2 (1 - e)^2 \mu u/e^3 D_{vmp}^2 \qquad\qquad (8.2\text{-}27)$$

The major concern then is the rate of moisture removal. Assume that k, the mass transfer coefficient, represents this rate of removal. If the difference in concentration of moisture on the surface of the particle and that in the moving air is constant, the rate of removal is proportional to the square root of the pressure drop. Putting it another way, to double the rate of moisture removal, the velocity in the bed must be increased by a factor of 4, and this immediately increases the pressure drop by the same factor. Care must be taken, however, in making such a step, because increasing the pressure at the bottom of the hopper by counterflow of air can fluidize the bed. When the bed becomes buoyant, very little powder flow from the bottom of the bed occurs, and thus all flow into the extruder is reduced. Evacuating the hopper, therefore, is a better alternative to moisture removal, since there is relatively little fluid motion, and the concentration gradient is nearly the same as it would be with 0%-humidity air. There are obvious problems with evacuated hoppers in sealing at the hopper-barrel interface and in feeding material to the hopper during processing. But these are not unsurmountable, and thus evacuated hoppers are used rather extensively. Countering the evacuated hoppers, however, is the heated air hopper, which although limited in its moisture removal (when compared with evacuated hoppers) does raise the temperature of the resin. This, of course, aids in solids conveying and will affect plastication as well.

8.3 Plasticating or Melting

The most critical aspect of changing the compacted solid resin particles into a homogeneous melt is the relatively poor thermal conductivity of powder or melt. As a result, intensive efforts are being made in design of screws to accomplish this conversion process in a shorter period of time (e.g., fewer feet of screw flight length). In addition to heat conduction from the barrel to the resin, intense mixing of the melt with the unmelted resin and of shear heating between the resin and the barrel surface is necessary. The larger the amount of viscous dissipation within the melt/solid admixture, the lower the amount of energy that must be supplied to the barrel for transfer to the admixture. In fact, materials such as ABS can be heated from a resin state to a melt without addition of heat along the barrel. This mode of heating is known as adiabatic heating and is one asymptotic condition of the arithmetical analysis applied to screw design. Another is isothermal heating, where the barrel, screws, and resin are held at the same temperature, and the shearing energy simply converts the resin from solid to melt. Remember now that the relative motion of the barrel to the screw flights is at an angle to the screw flights. Therefore, the material is not transported parallel to the flights but at an angle. This means that the material must be pushed against the back of the lead flight in a channel as it moves down the channel. The best picture of the melting of granular materials was given first by Street [21] and in greater detail recently by Tadmor [22] and Chung [23]. Shown in simplistic fashion in Fig. 8.3-1 is the melting phenomenon. Because of the "cross-flow" effect, melt is generated and flows across the screw channel at the barrel surface toward the rear of the channel where it accumulates in a circulating melt pool. Obviously, as the channel gets progressively thinner, the remaining solids are forced against the barrel surface, and shear heating and melting are intensified until the entire solid bed has reached a melt condition. This rather simplistic picture does not allow for infiltration of the melt into the bed, for breakup of the bed into many small beds, or for flow of the melt beneath the bed. These characteristics are evident in certain cases in the excellent experimental work done by Tadmor et al. [24]. Present sophisticated computer models can be changed to include these effects, and most certainly second- or third-generation computer programs will include them.

Consider, however, the rather simplistic concepts as outlined by Street, Tadmor, and others. To illustrate the several sections of the model, assume that the fluid is Newtonian. If y is the direction from the top of the solid bed to the barrel surface and U_z is the velocity of melt flow in the down-channel direction (e.g., parallel to the screw flights), the velocity profile in this melt layer of thickness δ is given as

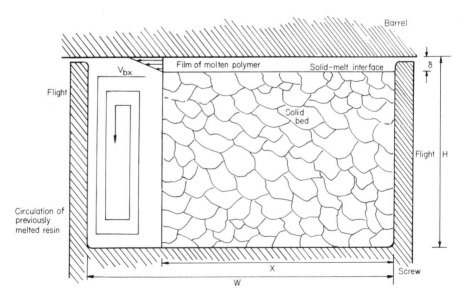

Figure 8.3-1 Idealized interpretation of the cross-section of a screw channel in the plasticating region of a single-screw extruder [8].

$$d^2U_z/dy^2 = 0 \tag{8.3-1}$$

If $U_z(0) = 0$ at the surface of the "stationary plug" and $U_z(\delta) = U_z'$, the velocity component of the barrel in the down-channel direction, Eq. (8.3-1) can be integrated to obtain

$$U_z = (y/\delta)U_z' \tag{8.3-2}$$

A very simple conduction equation, including viscous dissipation, is given as

$$k_m \, d^2T/dy^2 + \eta(dU_z/dy)^2 = 0 \tag{8.3-3}$$

where k_m is the thermal conductivity of the melt and η is its viscosity. Recall that the primary mode of heat transfer in laminar flow, as mentioned in Chap. 6, is conduction. We know that $T(y = 0) = T_m$, the melt or softening temperature, and that $T(y = \delta) = T_b$, the temperature of the barrel. Using these two conditions and Eq. (8.3-2) for the velocity, the temperature profile through the melt is

$$(T - T_m)/(T_b - T_m) = [\eta U_z'^2/2k_m(T_b - T_m)](y/\delta)(1 - y/\delta) + y/\delta \tag{8.3-4}$$

The term in the first parentheses on the right-hand side is dimensionless and is called the Brinkman number. As mentioned in Chap. 6, it is a measure of the importance of viscous heat generation to heat conduction resulting from the imposed temperature difference, $T_b - T_m$. $Br = \eta U_z'^2/2k_m(T_b - T_m)$. The heat flow from the melt into the solid bed below it can be found by differentiating Eq. (8.3-4) and evaluating it at $y = 0$, giving

$$-q(y = 0) = k_m(T_b - T_m)\delta + \eta U_z'^2/2\delta \tag{8.3-5}$$

Note that the object is to transfer as much heat through the melt to the solid as possible. From Eq. (8.3-5) it is apparent that this can be accomplished by keeping $T_b - T_m$, η, and U_z as large as possible and/or δ as small as possible. This can be accomplished by large temperature difference, high viscosity, and/or high extrusion speed.

In the solid bed, a simple conduction problem exists. The surface temperature is given as $T(0) = T_m$. The temperature of the bed at some large distance into the bed is T_s, a reference temperature that could be the screw base temperature. Note here that the bed is assumed to be moving with no velocity profile. This assumption is being challenged and corrected in more sophisticated computer models of screws. Nevertheless, for this model argument it is satisfactory. The heat conduction equation into the solid bed is given as

$$\rho_s c_s V_{sy} \, dT/dy = k_s \, d^2T/dy^2 \tag{8.3-6}$$

where V_{sy} is the velocity of the bed in the y-direction. Integrating,

$$(T - T_s)/(T_m - T_s) = \exp(V_{sy}y/\alpha_s) \tag{8.3-7}$$

where ρ_s, k_s, c_s, and α_s are the density, thermal conductivity, specific heat, and thermal diffusivity of the bed, respectively. Note here that y is decreasing from zero toward $-\infty$. Thus, the temperature profile in the solid is exponentially decreasing from its melt temperature at the surface to some value T_s in the interior of the bed.

A way of coupling the heat transfer to the solid bed to that transmitted through the liquid, e.g., a relationship between δ and V_{sy}, is still needed. The rate of melt entering the region above the solid bed is approximately equal to the rate of mass flow leaving the melt pool. If X is the average solid bed width, as shown in Fig. 8.3-1,

$$\Omega = V_{sy}\rho_s X = V_{bx}\rho_m \delta/2 \tag{8.3-8}$$

Ω is the rate of melting per unit down-channel distance, ρ_s is the density of the solid, V_{bx} is the velocity of the barrel in the screw axis direction relative to the screw, and ρ_m is the density of the melt. We can further relate V_{sy} and δ through a heat balance at the melt-solid surface. The heat flux into the solid is obtained by differentiating the heat conduction equation (8.3-7) at y = 0:

$$-q(y = 0) = \rho_s c_s V_{sy}(T_m - T_s) \tag{8.3-9}$$

The overall heat balance at the interface is obtained by noting that the rate of heat conducted into the interface minus the rate of heat conducted out of the interface must equal the rate of melting times the heat of fusion at the interface. If H' is the heat of fusion of the solid, this is obtained in equation form as

$$[k_m(T_b - T_m)/\delta + \eta U_z^{'2}/2\delta] - \rho_s c_s V_{sy}(T_m - T_s) = V_{sy}\rho_s H' \tag{8.3-10}$$

Combining this equation with Eq. (8.3-8), we obtain the thickness of the melt and the speed of rotation (and thus the velocity of the solid bed):

$$\delta = \left\{ \frac{[2k_m(T_b - T_m) + \eta U_z^{'2}]X}{V_{bx}\rho_m c_s(T_m - T_s) + H'} \right\}^{1/2} \tag{8.3-11}$$

$$\Omega = \left\{ \frac{V_{bx}\rho_m k_m(T_b - T_m) + \eta U_z^{'2}/2}{2[c_s(T_m - T_s) + H']} \right\}^{1/2} X^{1/2} = \phi X^{1/2} \tag{8.3-12}$$

Consider the second equation for a moment. ϕ represents the ratio of the rate of heat being supplied for melting to the rate of heat required to heat the solid through conduction to the melting temperature and convert it to liquid. We would like ϕ to be as large as possible to get the maximum rate of melting, again indicating that the thermal driving force is as large as possible, that screw speeds are as large as possible, and that high viscosity is needed to aid in viscous dissipation. Note that this equation holds if the resin is amorphous, where H' is zero. For amorphous materials, such as polystyrene, all that is needed is the enthalpy of the plastic between T_s and T_m.

As Tadmor and Klein point out, we need to know now how X varies with the length of the channel. The rate of melting is proportional to the square root of the width of the solid bed. Further, the volumetric rate of solid flow is proportional to the rate of melting:

$$-\rho_s V_{sz}\, d(hX)/dz = \Omega \qquad (8.3\text{-}13)$$

where V_{sz} is the velocity of the solid bed in the down-channel direction and h is the height of the solid bed. Assume that δ, the thickness of the melt film, remains rather small and nearly constant during the melting phase and thus that h_s, the height of the bed, is nearly proportional to and nearly equal to the height of the channel. Therefore, although Eq. (8.3-13) is not quite correct, it will illustrate the point rather well. Consider a tapered channel, described as

$$h = h_0 - az \qquad (8.3\text{-}14)$$

where h_0 is the height of the channel at the beginning of melting and a is the slope of the channel for a linearly decreasing channel depth. Substituting Eqs. (8.3-12) and (8.3-14) into Eq. (8.3-13), a simple differential equation is obtained:

$$(h_0 - az)\, dX/dz = aX + (\Omega\phi/\rho_s V_{sz})X^{1/2} \qquad (8.3\text{-}15)$$

At the beginning of melting, $X(z = 0) = W$, the channel width. Furthermore, the mass flow rate of material is defined as $G = V_{sz}h_0 W\rho_s$. Therefore, the equation can be normalized by defining a new dimensionless group ψ to be

$$\psi = \phi W^{1/2}/(G/h_0) \qquad (8.3\text{-}16)$$

Thus, the solution to Eq. (8.3-15) becomes

$$X/W = \left\{ \psi/a - (\psi/a - 1)[h_0/(h_0 - az)]^{1/2} \right\}^2 \qquad (8.3\text{-}17)$$

The total melting length Z is given when X = 0, e.g., all the solid is melted. From Eq. (8.3-17), Z is given as

$$Z = (h_0/\psi)(2 - a/\psi) \qquad (8.3\text{-}18)$$

As shown in Fig. 8.3-2, X/W can be plotted against z/Z for various values of a/ψ. Further, from Eq. (8.3-18), increasing the channel width (possible by using a larger extruder barrel diameter) reduces the length of melting. In addition, increasing the barrel temperature, the viscosity of the melt, and screw speed all decrease the length of melting. Recall that increasing the temperature of the solids entering the screw was possible by drying the solids using low-humidity warm air. Increasing the temperature of the solids increases the enthalpy of the solids and thus increases the value of φ, which in turn decreases the melting length. For the limiting case where a/ψ = 1, the length of melting is half that for a screw with a constant channel depth (at the same value of ψ). Tadmor and Klein also integrate Ω over the length of the channel between z = 0 and z = Z to show that the average rate of melting is

$$\Omega_{av} = \phi W^{1/2}(2 - a/\psi)$$ (8.3-19)

It is apparent that many simplifying assumptions have been made to obtain these conclusions. Nevertheless, this should give a feeling for the performance of a screw in the melting zone. Tadmor and Klein have extended this simple analysis to include non-Newtonian fluids and nonconstant

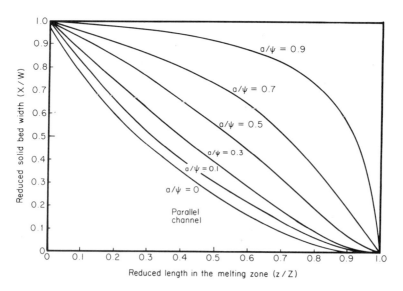

Figure 8.3-2   Effect of screw flight tapering on solid bed profile X/W. Note: Parameter is a/ψ, where a is the rate of change of channel depth and ψ is defined by Eq. (8.3-16) [3].

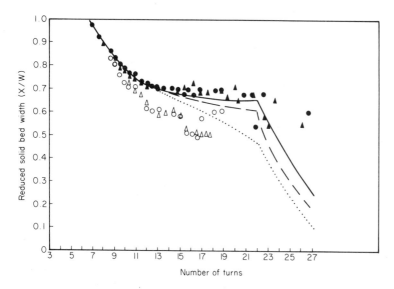

Figure 8.3-3   Calculated and measured solid reduced bed width profiles for LDPE; $T_b = 300°F$; 60-rpm screw speed; $P = 3000$ psi; output, 136.1 lb/hr. Dotted curve, Newtonian model; dashed curve, non-Newtonian model assuming zero flight clearance and no curvature; solid line, non-Newtonian model, assuming finite flight clearance, curvature [3].

properties and in particular variable channel depth (other than linear taper). As an example of the ability to predict X/W for various lengths, consider Fig. 8.3-3, in which three models are used for prediction: Newtonian as just presented, non-Newtonian, and non-Newtonian including the effect of flight clearance and channel curvature. As can be seen, the accuracy in prediction is quite good for even the Newtonian model presented above.

8.4 Melt Pumping

Once the plastic has been completely melted it is necessary to pump it through an additional length of channel in order to build the pressure sufficient for injection or extrusion through dies into profiles or parisons and to thermally homogenize the melt. Since the early paper by Squires [5], much effort has been extended toward the understanding of the melt pumping phenomenon [3,23,25,26]. Kruder and Kim [9] have examined the mathematical models for melt pumping and have found that the very early one proposed by Squires seems satisfactory for most cases (the major exception being screws with very high back pressures). Perhaps the most careful presentation of the Squires or refined drag flow model is that by Carley [27].

McKelvey [2] discusses a simplified version of the drag flow model also. As an example, consider the simple relationship between the pressure drop in the down flow channel and the velocity profile:

$$dp/dz = \eta \; d^2 v_z/dy^2 \tag{8.4-1}$$

where η is a Newtonian viscosity (assumed for illustration), v_z the velocity component in the down channel or z-direction, and dp/dz the pressure drop in that direction, assumed to be a constant. y is the distance above the root of the screw. Again, $v_z(y = 0) = 0$, since the velocity of the melt at the surface of the screw is zero relative to the movement of the screw, and $v_z(y = h) = V_z$, since the velocity of the melt at the barrel surface is the z-component of the barrel speed relative to the screw. Integrating Eq. (8.4-1),

$$v_z = yV_z/h - y(h - y)(dp/dz)/2\eta \tag{8.4-2}$$

The volumetric flow rate is obtained by integrating the velocity profile across the channel height as

$$Q = W \int_0^h v_z \, dy \tag{8.4-3}$$

yielding

$$Q = V_z Wh/2 - Wh^3(dp/dz)/12\eta \tag{8.4-4}$$

Squires points out that the first term on the right-hand side represents drag flow induced by the moving barrel surface and that the second term represents the pressure flow induced by the axial pressure gradient in the extruder. He proposes a more general form for Eq. (8.4-4) in which the shape of the rectangular channel can be included in the equation. Booy extends Squires' analysis to include channel curvature [28]. Under these conditions, a more general form for Eq. (8.4-4) is

$$Q = F_d[(\pi^2/2)(\sin \phi \cos \phi)D^3 hN] - F_p[(\pi/12) \sin^3 \phi \; h^3 (D/L)(P/\eta)] \tag{8.4-5}$$

ϕ is the helix angle, defined earlier; D is the barrel diameter; N is the rpm of the screw; and L is the length of the melt zone of the screw in the axial direction. F_d and F_p are complex functions of W/h, and n is the number of parallel flights. Booy plots F_d and F_p as functions of h/D, the relative channel depth for various helix angles. They are shown in Figs. 8.4-1 and 8.4-2, respectively. F_d and F_p are the drag flow and pressure flow shape factors, respectively. Booy shows that for zero helix angle these curves reduce to those of Squires et al. [6]. Note the difference in notation, since Booy's curves are based on a relative channel depth and Squires'

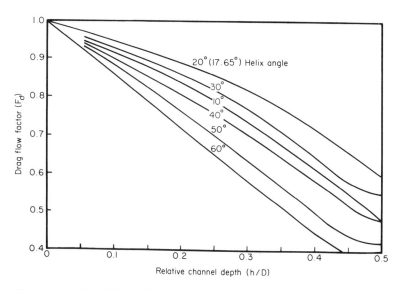

Figure 8.4-1   Effect of channel depth and helix angle on the drag flow factor F_d [28].

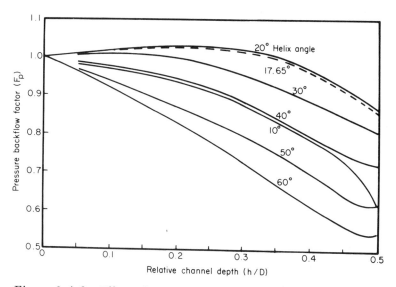

Figure 8.4-2   Effect of channel depth and helix angle on the pressure backflow factor F_p [28].

curves are based on a height-to-width ratio. Further, $F_p = F_d = 1$ for melt
flow in very narrow channels, and thus with the exception of a helix angle
factor, the arithmetic reduces to Eq. (8.4-4). This is usually an excellent
approximation for large-diameter screws (6 in. and over). The experimental data of Squires, with F_p and F_d plotted as functions of h/W, are
shown in Fig. 8.4-3, thus confirming the application of the simple model
for the case where backflow pressure is not a major factor. Kroesser and
Middleman [29] extend this work to include the power-law model for viscosity. They choose $G = (h/V_z)(hP/KL)^{1/n}$ as a parameter, where K and n
are the power-law coefficients, and repeat the analysis that leads to Eq.
(8.4-4). Their results, a plot of $Q/V_z Wh$ vs. G^n with n, the power-law
index as a parameter, are shown in Fig. 8.4-4, for n = 0.5, with several
superpositions of pressure flow included. No effect of helix angle is included.
Additional work shows the shape factors F_p and F_d as functions of n, the
power-law index. They are shown in Figs. 8.4-5 and 8.4-6 [30]. Again,
note the accuracy of the data in predicting the pressure flow shape factor.

     Consider, again, the general equation (8.4-5). Assume that the
non-Newtonianness can be taken care of in the manipulations of F_p and F_d.
In the solids conveying section, $Q = V_z Wh$. Thus, it is apparent that if Q is
to remain constant throughout the solids metering, melting, and melt pumping zones, h must change from a deep channel (h/W large) to a narrow

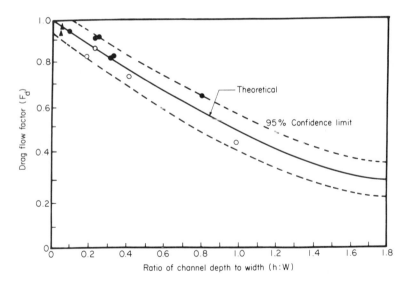

Figure 8.4-3   Experimental and theoretical verification of drag flow factor
F_d for two samples of polyisobutylene and one of LDPE.   Open, closed circles, PIB; triangles, LDPE [6].

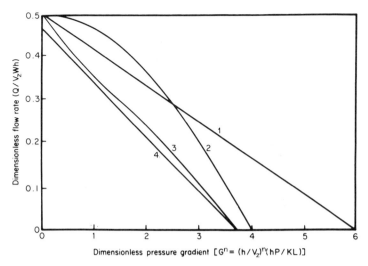

Figure 8.4-4   Effect of dimensionless pressure gradient on flow rate as calculated for several mathematical models. 1: Superposition of pressure and drag flow with Newtonian equations; 2: Superposition of pressure and drag flow with non-Newtonian equations; 3, combined flow, excluding cross-channel flow; 4: combined flow, including cross-channel flow; power law where applicable; n = 0.5 [29].

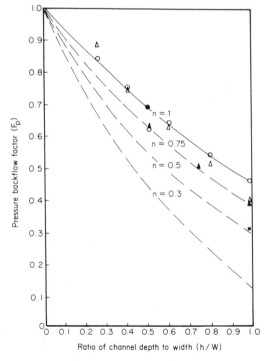

Figure 8.4-5   Back-pressure factor F_p as experimental function of channel depth-to-width ratio. Note: Power-law values shown as dashed lines, Newtonian theory as solid line. Open, closed circles, n = 1; triangles, n = 0.75; solid squares, n = 0.5 [30].

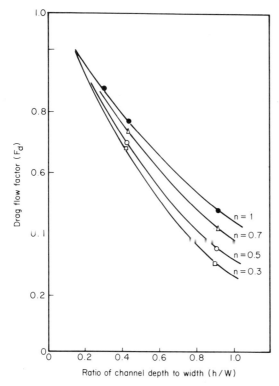

Figure 8.4-6   Shape factor for pure drag flow F_d as computed for non-Newtonian flow in square ducts [30].

channel (h/W small). Most of the pressure buildup in an extruder occurs in the melt pumping zone. Thus, Eq. (8.4-5) represents a pressure drop-flow rate relationship within the extruder as a whole. Note that $V_z = \pi DN$ $\cos\phi$. For the Newtonian case, then, pressure drop is a linear function of the flow rate as well as the geometry. For a non-Newtonian fluid, we would expect a power-law model similar to that analyzed for simple flow in a pipe in Chap. 5.

Do not consider the above analysis as a detailed presentation of the design of the melt pumping section. Rather, let it serve as a guideline to the type of analysis that you will find extended to isothermal non-Newtonian fluids, nonisothermal Newtonian fluids, and even nonisothermal non-Newtonian fluids. Nevertheless, the simplistic model serves as a point of reference, for regardless of the complexity of the mathematical model, it must still be capable of recovering the isothermal Newtonian model when asked. This point has been amplified by Carley [27].

8.5 Matching the Die to the Extruder

As early as 1954, Carley [31] was concerned with flow through various types
of dies that could be mounted on the melt end of the extruder barrel. In
particular, Carley broached the problem of flow down an end-fed film or
"cross-head die." Independent of the specific application, however, he
pointed out the importance of matching the pressure drop-flow rate relationship in the die to that in the melt pumping section of the screw. This assumes, of course, that the melt pumping flow rate controls the design of
the screw. Only by matching the performance of the screw with that of the
die can flow rates through the extruder be determined. Note that any resistance to flow beyond the tip of the screw flights—whether screen pack;
flow valve; nozzle, in the case of injection molding; mandreled die, as in
the cases of blow, extrusion, and blown film molding; or multiported die,
as in the case of a spinnerette head for extrusion of filaments—will cause a
loss in pressure. The more restrictions placed in the flow path beyond the
flight tips, the higher the pressure at the flight tips must be in order to
maintain a constant or given flow rate.

In this section, we shall consider two types of dies that are of major
economic importance today—the coat hanger sheeting die and the pipe die—
in order to ascertain the pressure drop-flow rate relationships for them.
The equations for these dies should then be matched with the similar equations for flow in the melt pumping section of the screw, again assuming that
the generation of pressure is largely controlled by the melt pumping section
of the screw, in order to determine the appropriate flow rate equations for
the matched extruder-die.

The coat hanger sheeting die has largely replaced the cross-head die
for the production of sheet. This can partially be attributed to the inherent
unevenness of flow in the end-fed die. Carley showed that for the end-fed
die one could define an evenness factor or uniformity index E. It essentially
is the ratio of the flow rate through the die land at the end of the pipe (farthest from the extruder) to that through the die land at the entrance of the
pipe (nearest the extruder). For isothermal power-law fluids, Carley
found that

$$E \doteq 1 - 0.525(aL)^2 \tag{8.5-1}$$

where

$$a = [(1/2\pi)(3n + 1)/(2n + 1)(H^{(2n+1)/n}/Y^{1/n}R^{(3n+1)/n})]^{n/(n+1)} \tag{8.5-2}$$

where n is the power-law index, H is the die lip spacing, Y is the die land
length, and R is the radius of the delivery pipe. He further shows that if
the fluid is nonisothermal but isobaric (e.g., that no pressure drop occurs
along the delivery tube),

$$E \doteq 1 - b_{\dot\gamma}(T_{max} - T_{min})/n \tag{8.5-3}$$

where b, is a temperature-dependent coefficient for a given plastic [2].
When $T_2^\gamma - T_1 = 1/b$, $\eta_2 = \eta_1/e$, where e is the exponential. While it is
obvious that the effects of temperature and pressure are not necessarily
additive, for the case where the evenness index is nearly unity, the additive
effect can be used. Thus, for the case where nonisothermal, non-Newtonian
power-law fluids are issuing from an end-fed or cross-head die, the evenness of flow can be determined by multiplying the right-hand sides of Eq.
(8.5-1) and (8.5-3). It is rather apparent, however, from examination of
Eq. (8.5-1) that to minimize uneven flow for a given length of die we should
have "land control." That is, R should be as large as possible, and H/Y
should be as small as possible.

     As mentioned, the end-fed die is being replaced by the coat hanger
die. Here the flow is to the center of the die and is spread in a fan shape
so that the cross-section of the flow channel changes from circular to thin
rectangular. Klein and Klein [32] have developed a rather extensive computer program for the design of this type of die for nonisothermal, non-
Newtonian fluids. The equations used in their analysis are not readily available, however. Carley's illustration of fan-shaped (or fishtail) dies can be
used to illustrate the flow rate-pressure drop relationship. The fan-shaped
die is shown in schematic form in Fig. 8.5-1. Normally, in practice the
flow is controlled across the fan shape by a "straining bar," which is shown
in Fig. 8.5-2. The objective is to determine the position of the straining
bar as a function of distance across the land from the center of the distributing
pipe. In this case, however, Carley chooses to determine the lead up thickness (ahead of the land) and the land length L_2 in order to produce a sheet
W units wide by h_2 units thick. Note that h_2 is the desired film thickness
multiplied by the draw ratio on the film. The pressure in the pipe ahead of
the die is P, and the pressure drop down the center is p_1. The pressure
drop across the die land is p_2. The unevenness of flow will normally cause
p_1 to be a very weak function of θ, the half-angle of the fan. Klein and Klein
include this fact in their computer program. Carley just notes that it is a
small correction, at best. If E is the desired uniformity index, then the
pressure ratio p_1/p_2 can be obtained as a function of it as

$$g = p_1/p_2 = (1 - E^n)/(\sec \theta - 1) \qquad (8.5\text{-}4)$$

where n is the power-law index. Now $g = W_e/W$, where W_e is an effective
average slit width and W is the actual slit width. Furthermore,

$$p_1 = gP/(1 + g) \qquad (8.5\text{-}5)$$

$$h_1 \doteq [15KQW^{(1-n)/n}(\cot \theta)^{1/n}/p_1^{1/n}]^{n/(1+2n)} \qquad (8.5\text{-}6)$$

$$p_2 = P - p_1 \qquad (8.5\text{-}7)$$

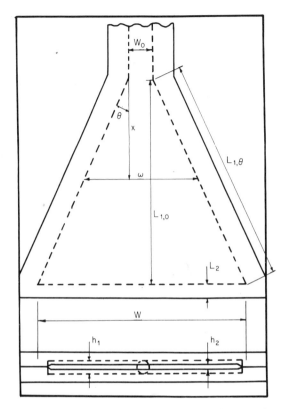

Figure 8.5-1   Schematic of a typical fishtail sheeting die [31].

and

$$L_2 = (Wh_2^{(1+2n)/n} p_2^{1/n}/3KQ \cdot 2^{(1+n)/n})^n \qquad (8.5\text{-}8)$$

As Carley and Klein and Klein point out, holding h_2 constant increases the
length of the fan. It would be much better to contour the die so that this
additional expense in materials and machining can be reduced. According
to Klein and Klein, the use of a straining bar between the manifold entry
pipe and the land is important if the die is to be used with more than one
resin or is to be operated at temperatures other than that specifically used
to design the land length in Eq. (8.5-8). As an example of the changes in
the straining bar nip, Fig. 8.5-3 shows the straining bar setting as a function of distance from the center of the die with the manifold angle as a
parameter. The die shown is operating at 450°F with a design condition for
350°F with a 2° manifold angle on an ABS.

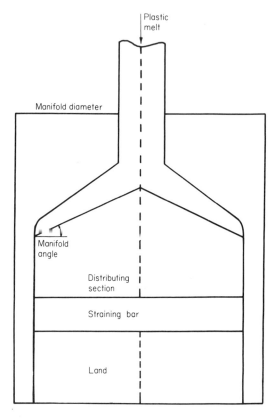

Figure 8.5-2 Schematic diagram of a coat hanger die with straining bar [32].

As a second type of flow at the extruder end of the die, consider pipe die flow. Here, simply stated, the objective is to make a hollow pipe of known, constant dimensions. To accomplish this, it is necessary to place a mandrel or torpedo in the center of the cylindrical die; to hold the mandrel in place, it is necessary to anchor it to the cylindrical die walls with legs or "spiders." The geometry of flow through the die, assuming that the spider legs are not factors in flow design, is fairly well understood, described by Frederickson and Bird [33] and discussed shortly. However, to place their work in proper perspective, it is necessary to consider the effect of the presence of the spiders holding the mandrel in place. The mandrel can be and frequently is heated, as is the cylindrical die. The heating mode of the mandrel is cartridge heaters, whereas for the cylindrical die band heaters are used. In both cases, the controllers are on-off thermostatically controlled. The spider acts as a flow divider. Separation

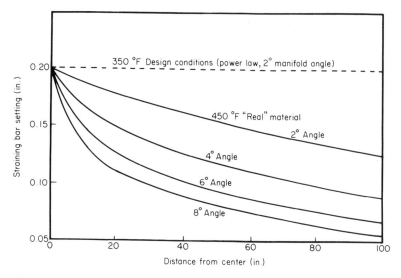

Figure 8.5-3   Effect of manifold angle on the straining bar setting for a coat-hanger-type die as a function of the distance from the center [32].

of plastic melt streams is frequently the cause of defects in the final plastic part which have been called weld lines. So it is with flow around the spider, for the spider causes the flow at the spider wall to slow, and thus there is little mixing between the separated streams, and a definite downstream velocity gradient is established. This can be readily seen in extruded pipe. To overcome this, several "tricks" are in use. Frequently a "smear" ring is placed upstream of the mandrel. This ring applies back pressure on the fluid in the region of the mandrel and helps remix the separated streams. Another method is to set the spider legs on an angle to the fluid flow direction (most spiders are parallel to the fluid flow direction). This enables the spider to "twist" the fluid and imparts a secondary flow to it, thus improving its mixing. Unfortunately, this helical flow cannot completely eliminate the markings of the spider, and thus the final extrusion shows helical markings in addition to the drag marks due to the uneven velocity gradient. Other designs that have met with marginal success include "airfoiling" the spider leg and heating it, thus decreasing the viscosity of the fluid in the immediate vicinity of the leg, which allows this fluid to flow faster than the bulk fluid, thus flattening the downstream velocity gradient. Nevertheless, the implications in poor tubing cross-section uniformity and resulting mechanical weakness require additional effort in the area of mandrel location and mounting.

Assume now that the mandrel problem, discussed above, is secondary to the understanding of fluid flow and pressure drop and can in fact be

solved somehow. Then we want to know how to predict extruder flow rates
at given extruder pressures, so that the die can be "married" to the extruder for optimum performance. Frederickson and Bird considered flow
of several ideal types of plastic materials in flow in annuli (or between concentric cylinders). Among their more restrictive assumptions is that the
flow path be sufficiently long so that end effects may be neglected. This is
not necessarily valid in light of the above discussion about spider marks.
Nevertheless, their analysis is most useful in determining the performance
of various materials between concentric cylinders. In cylindrical coordinates, they write the pressure equation as

$$(1/r) \, d(r \tau_{rz})/dr = (p_0 - P_L)/L + \rho g_z \tag{8.5-9}$$

where p_0 is the static pressure at $z = 0$, P_L is the static pressure at $z = L$,
g_z is the gravitational constant, ρ is the fluid density, τ_{rz} is the shear stress
in the annulus and the limitations on r are $kR \leq r \leq R$, where k is less than
unity. Integrating Eq. (8.5-9) yields

$$\tau_{rz} = (P/2)[r - (aR)^2/r] \tag{8.5-10}$$

where a is the constant of integration. When $\tau_{rz} = 0$, $r = aR$. There are
many expressions for the relationship between shear stress τ_{rz} and the
velocity profile dv/dr. Frederickson and Bird discuss several of these in
dimensionless form, including Bingham plastic; Newtonian; pseudoplastic,
where the viscosity index is less than unity; and dilatent, where it is greater
than unity. To obtain the flow rate, first integrate Eq. (8.5-10) using the
proper constitutive equation of state (power law, say) to obtain the velocity
and then integrate the velocity across the cross-section of the annulus:

$$Q = 2\pi R^2 \int_k^1 v(r/R) \, d(r/R) \tag{8.5-11}$$

Consider the power-law case here. (An appropriate model for extrusion of
foam, for example, would probably be the Bingham plastic model.) The
velocity profile is

$$v = R(PR/2K)^{1/n} \int_a^{r'} (a^2/r' - r')^{1/n} \, dr' \qquad (k \leq r' \leq a)$$

$$= R(PR/2K)^{1/n} \int_{r'}^1 (r' - a^2/r')^{1/n} \, dr' \qquad (a \leq r' \leq 1) \tag{8.5-12}$$

where $r' = r/R$. To obtain a, it is necessary to solve two simultaneous integral equations. Thus, one obtains a as a function of k and 1/n, as shown
in Fig. 8.5-4. Having this, the flow rate-pressure drop relationship can be
obtained:

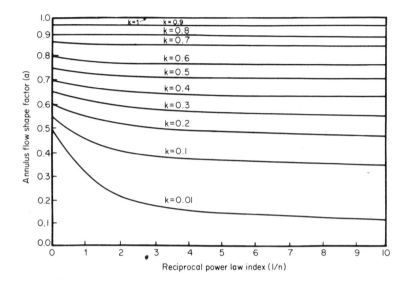

Figure 8.5-4   Analytical power-law result for fully developed flow in an annulus. Analysis yields results for Newtonian flow in an annulus and power-law flow in a cylindrical die as well [33].

$$Q = \pi R^3 (PR/2K)^{1/n} \int_k^1 \left| a^2 - r'^2 \right|^{1/n+1} r^{-1/n} \, dr' \qquad (8.5\text{-}13)$$

Recognize the limiting cases of the power-law model and this equation. First, it should fit for Newtonian flow in a cylindrical die ($n = 1$, $k = 0$). In this case, $K = \eta$, the Newtonian viscosity, and the equation becomes simply

$$Q = \pi R^4 P / 8\eta \qquad (8.5\text{-}14)$$

as is expected. It also reduces to power-law flow in a cylindrical die (without a mandrel):

$$Q = [\pi R^3 n/(3n + 1)](PR/2K)^{1/n} \qquad (8.5\text{-}15)$$

as indicated earlier. More important, however, is the power-law flow in a slit with limiting cross-section. Here k is approaching unity, as would be the case for large-diameter pipes of relatively thin cross-section (medical tubing, swimming pool tubing, and the like). The designing equation reduces to

$$Q^* = [\pi R^3 n/(2n + 1)](PR/2K)^{1/n}(1 - k)^{2+1/n} \qquad (8.5\text{-}16)$$

In the more general case, of course, it is necessary to obtain a value for the integral on the right-hand side of Eq. (8.5-13). Frederickson and Bird do this analytically, but graphical results are shown in Fig. 8.5-5. Here $Y(1/n, k)$ is defined as

$$Q = Q^* Y(1/n, k) \tag{8.5-17}$$

where Q^* is defined as in Eq. (8.5-16) for the limiting case where k is nearly unity.

Probably the most important fact that can be gathered here is that analysis of flow rate-pressure drop relationships in pipe dies results in equations that show in essence that Q is proportional to the pressure drop to the $1/n$ power. This is characteristic of flow in capillary dies as well. The fishtail die characteristics [Eqs. (8.5-5) to (8.5-8)] show a similar relationship. Within limitations, an appropriate general relationship between flow rates and pressure drops in dies (regardless of their geometry or use) can be written as

$$Q_d = C'P^{1/n} \tag{8.5-10}$$

where P represents the pressure drop and C' represents all the correction factors and geometric parameters necessary to design the die. Further, in Eq. (8.4-4), developed for Newtonian fluids, and the subsequent discussion

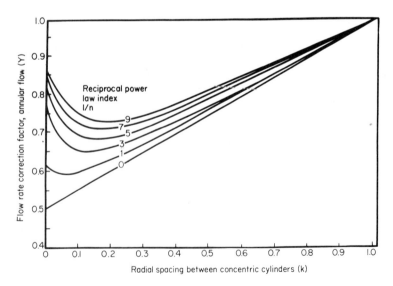

Figure 8.5-5   Graphical result of the effect of annulus spacing and power-law index on reduced flow in an annulus [33].

regarding correction factors for power-law fluids, at least for the melt pumping section of the extruder, the flow rate-pressure drop relationship can be heuristically written as

$$Q_e = AN - B'P^{1/n} \tag{8.5-19}$$

where N is the speed of rotation of the screw, A and B' are geometric factors, and P again represents the pressure drop in the melt pumping section. Now the only way the extruder die can function properly is if the flow rate from the extruder equals the flow rate through the die, as McKelvey points out. Therefore, $Q_d = Q_e$, and

$$P = [AN/(C' + B')]^n . \tag{8.5-20}$$

or

$$Q = C'AN/(C' + B') \tag{8.5-21}$$

Do not forget that A, B', and C' contain factors other than the geometric factors and that this bit of rearranging assumes that the melt pumping section of the extruder is controlling. Keep in mind that the entire screw must be functioning at the same flow rate of material and therefore that the implications about the appropriate values for A, B', and C' are more profound than they appear in this simplistic analysis. Nevertheless, note the importance of screw speed on flow rates (a direct proportionality). Note also that the ratio of B' to C' represents a measure of the relative resistances to flow of the melt pumping section to the die section. The greater the value of C' (in relation to B'), the greater the capacity of the machine at a given speed and the lower the extruder pressure. From Eq. (8.5-16) for annular flow, the increase in C' can be accomplished by narrowing the gap width or decreasing the temperature. While this appears to be opposite to the type of effect desired, remember that the relative changes in C' to B' are being considered. Certainly increasing C' will increase the pressure required to flow the material at a given rate. Thus, it is not correct to assume that all the control on flow rate occurs in the flow through the die. The melt pumping rate can dramatically affect the efficiency of the die performance. McKelvey discusses this point quite well [2].

There are, of course, other implications given by the matching equations (8.5-20) and (8.5-21). For example, increasing melt temperature will lead to a decrease in the pressure buildup at a given flow rate in the melt pumping section, whereas it will lead to an increase in the flow rate at a given pressure in the die section. Thus, increasing the temperature of the die at a constant melt pumping temperature will result in a decrease in pressure drop and an increase in flow rate, whereas increasing the melt temperature in the melt pumping section of the extruder at constant die temperature will result in a decrease in both the pressure drop and the flow rate. Thus, the delicate balancing of temperatures in the die and in the melt pumping section can lead to significantly increased flow rates at given pressure drops or reduced pressure drops at the same flow rates or can lead to anomalous drop behavior that only a Solomon could decipher. Remember that the speed of extrusion has been assumed constant and that the effects of solids feeding and conveying and plastication or melting do not significantly contribute to this portion of the argument.

8.6 Conclusion

The various aspects of continuous extrusion design have been briefly outlined in this chapter. Spinnerette head design or extrusion into the nip region of a calendering die have not been discussed. Further, venting and two-stage screw design and the newer screw designs, such as the Barr screw, have been bypassed, as has twin- and two-screw plastication, although these are becoming very important in the formulation of foamable extrusion materials and various grades of vinyls. In this chapter we have given an overall view of the interaction of the various zones of a screw extruder and of the importance of designing the screw extruder die to match the capacity of the machine. Do not consider this material to be the last word in extrusion. Excellent texts by McKelvey, Tadmor and Klein, Klein and Marshall, and others should be consulted for the in-depth information.

Problems

P8.A The linear coefficient of thermal expansion of high-alloy steel is 12×10^{-6} $(°C)^{-1}$. If an ABS-SBR extruder is to be operated such that the barrel remains at 100°F and the screw attains a steady-state adiabatic temperature of 450°F, determine the minimum flight clearance required when both are at room temperature (77°F).

P8.B Chang and Daane [34] recently showed that over all normal operating temperatures PVC powders have a lower coefficient of friction than PVC pellets. For pellets, the coefficient of friction at 55°F is 0.278 and at 123°F is 0.175. For powders, at 70°F it is 0.23 and at 150°F it is 0.13. Using Griffith's work leading to Eq. (8.2-10), determine the conditions satisfying the inequality. Set the temperature conditions for the screw and barrel based on this information. Chang and Daane also show that the isothermal coefficient of friction drops rapidly with pressure on the powder. At 104°F and 20 psi, the coefficient of friction is 0.25; at 175 psi, it is 0.19. Discuss how a hopper might be redesigned to increase pressure on the powder.

P8.C Make a plot of the Janssen equation (8.2-19) as a function of height x for a hopper design with an effective diameter of 12 in., $\mu_w K = 0.200$, and a bulk density of 30 lb/ft^3. At which point is the vertical pressure at the bottom of the bin within 98% of the maximum pressure? Compare your results with those calculated from Lenczner's equation (8.2-21). Discuss the applicability of the latter equation. From Rudd's data of Fig. 8.2-6, determine $\mu_w K$. Design an experiment to measure vertical bin pressure in both static and dynamic conditions.

P8.D A major corporation mistakenly purchased a solids blender that meters the various powders and pellets volumetrically. The corporation expected the blender to meter materials by weight. According to the information given in Sec. 8.2, is there any way in which the hoppers of the powder dispensers can be modified to achieve a pseudovolumetric measurement? How accurate would your modification be?

P8.E How much gas pressure is needed to force air from compressing powder in the screw out through the hopper when the feed material is 0.060-in. cubes with a bulk density of 40 lb/ft^3, a material density of 70 lb/ft^3, a 40-in.-high hopper with a bed porosity of 0.3 = e, and a mean air velocity of 1 ft/sec.

P8.F (Following McCabe and Smith [35]) The Janssen equation can be derived from force balances made on a circular bin with vertical walls. If the inside radius of the bin is r ft and the total height of the solids is Z_T ft, show that the vertical pressure p_v at a level Z in the bin is

$$p_v = F_v / \pi r^2 \qquad (8.F-1)$$

where F_v is the concentrated vertical force. Show then that the differential downward force at any layer is given as the difference between the force of gravity dF_g and the frictional force dF_f:

$$dF_v = dF_g - dF_f \qquad (8.F-2)$$

If the frictional force is the product of the coefficient of friction and the lateral force F_L and if the lateral force is a product of the lateral pressure p_L and the area, show that

$$dF_v = \pi r^2 \, dp_v = \pi r^2 \rho_b (g/g_c) \, dZ - \mu'(2\pi r p_L \, dZ) \qquad (8.F-3)$$

If $p_L/p_v = K' = (1 - \sin \alpha_m)/(1 + \sin \alpha_m)$, where α_m is the angle of internal friction, approximately equal to the angle of repose, show that when Eq. (8.F-3) is integrated from 0 < Z < Z_T the result is the first term on the right-hand side of Janssen equation (8.2-19). Discuss the origin of the second term on the right-hand side. What terms must be added to the force balance to achieve Eq. (8.2-19)?

P8.G A silo contains 100,000 lb of 1/16-in.-diameter LDPE pellets. The silo is 7 ft in diameter by 40 ft high. The bulk density of the pellets is 40

lb/ft^3. What are the lateral and vertical pressures at the base of the silo
if the coefficient of friction is 0.20 and if the angle of repose is 45° ?

P8.H Experiments show that Cosden bead-precipitated HIPS flows through
a square hopper opening approximately in proportion to the cube of the opening characteristic dimension. For what kind of non-Newtonian fluid would
the mass flow rate in steady laminar flow vary as the cube of the characteristic dimension? In other words, what is the value of the viscosity index n
for an equivalent power-law fluid?

P8.I Using Eq. (8.3-4) for heat conduction through the melt bed, make a
plot of the temperature profile as a function of the dimensionless distance
y/δ for various values of the Brinkman number Br. Discuss the significance
of Br = 0 and Br = ∞. Can Br be negative? Why or why not?

P8.J Make an equivalent plot for the temperature profile into the solid bed
using Eq. (8.3-7), together with Eqs. (8.3-11) and (8.3-12) in proper dimensionless form, and discuss how the dimensionless groups V_{sy}/ω_g and the
Brinkman number influence the thickness of melting δ and the speed of rotation of the screw Ω.

P8.K Analyze Eq. (8.4-5) in terms of the development of the simple equation (8.4-4) which represents the relationship between fluid flow and pressure drop down the melt portion of the screw flights. Can dp/dz be either
positive or negative? What is the meaning of pressure decreasing throughout the melt zone? How can this be controlled? Is it necessary?

P8.L One-inch-diameter garden hose is to be produced from a standard
profile die. The material is LDPE with n = 0.33, K = 4.7, and sp. gr. =
0.921. The wall thickness is to be 0.075 in., and the extrusion rate is to
be 500 lb/hr. Describe the preliminary design parameters for the screw
and the die. List all assumptions involved.

P8.M In a sheeting die, the manifold pipe diameter is 2 in. The cross-
head die is 60 in. in length and fed from one end. The material is Nylon 66
with a power-law coefficient of 0.5 and a viscosity at 100 sec^{-1} of 0.1 lb·sec/
in.^{2} at 565°F. The die lip spacing is 0.010 in., and the land length is 0.765
in. Calculate the Carley uniformity index E for this extrusion. Discuss
how the uniformity index might be increased. Consider the Klein correction
factor if this die were to be replaced with an equivalent fantail die, Eq.
(8.5-4). How can the Carley uniformity index be used here? What is an
appropriate form for the half-angle of the fan?

References

1. D. C. Craft, Mod. Plast. Encyclopedia, 49(10A):564 (1972-1973).
2. J. M. McKelvey, Polymer Processing, Wiley, New York, 1962, Chap.
   11.

3. Z. Tadmor and I. Klein, Principles of Plasticating Extrusion, Van Nostrand Reinhold, New York, 1969.

4. Z. Tadmor and D. I. Marshall, eds., Computer Programs for Plastics Engineers, Van Nostrand Reinhold, New York, 1968, Chap. 6.

5. P. H. Squires, Soc. Plast. Eng. Trans., 4(1):7 (1964).

6. J. B. Paton, P. H. Squires, W. H. Darnell, F. M. Cash, and J. F. Carley. "Extrusion," in: Processing of Thermoplastic Resins (E. Bernhardt, ed.), Van Nostrand Reinhold, 1959, Chap. 4.

7. R. B. Gregory, Plast. Polymers, 120 (Apr. 1970).

8. C. I. Chung, Polym. Eng. Sci., 11:93 (1971).

9. G. A. Kruder and J. T. Kim, Soc. Plast. Eng. J., 29(3):49 (1973).

10. R. C. Donovan, SPE ANTEC Tech. Pap., 16:561 (1970).

11. R. M. Griffith, SPE ANTEC Tech. Pap., 13:843 (1967).

12. K. Schneider, Kunststoffe, 59:97 (1969).

13. C. I. Chung, R. J. Nichols, and G. A. Kruder, SPE ANTEC Tech. Pap., 19:115 (1973).

14. H. A. Janssen, Z. Ver. Dtsch. Ing., 39(35):1045 (Aug. 1895).

15. R. L. Brown and J. C. Richards, Principles of Powder Mechanics, Pergamon Press, Elmsford, N.Y., 1970.

16. J. K. Rudd, Chem. Eng. News, 32(4):344 (1954).

17. C. E. Beyer and R. S. Spencer, Rheology in Molding. In: Rheology, Vol. 3 (F. R. Erich, ed.), Academic Press, New York, 1960, p. 527.

18. D. Lenczner, Mag. Concr. Res., 15(44):101 (July 1963).

19. M. A. Rao and J. L. Throne, Polym. Eng. Sci., 12:237 (1972).

20. J. M. Kay, An Introduction to Fluid Mechanics and Heat Transfer, 2nd ed., Cambridge University Press, New York, 1963, Chap. 12.

21. L. F. Street, Int. Plast. Eng., 1:289 (1961).

22. Z. Tadmor, Polym. Eng. Sci., 6:185 (1966).

23. C. I. Chung, Mod. Plast., 58(9):178 (1968).

24. Z. Tadmor, I. J. Duvdevani, and I. Klein, SPE ANTEC Tech. Pap., 13:813 (1967).

25. Z. Tadmor, Polym. Eng. Sci., 6:203 (1966).

26. R. M. Griffith, Ind. Eng. Chem. Fund., 1:180 (1962).

27. J. F. Carley, Mod. Plast., 60(2):77 (1971).

28. M. L. Booy, Soc. Plast. Eng. Trans., 3(7):176 (1963).

29. F. W. Kroesser and S. Middleman, Polym. Eng. Sci., 5:231 (1965).

30. S. Middleman, Trans. Soc. Rheol., 9:83 (1965).

31. J. F. Carley, J. Appl. Phys., 25:1118 (1954).

32. I. Klein and R. Klein, Soc. Plast. Eng. J., 29(7):33 (1973).

33. A. G. Frederickson and R. B. Bird, Ind. Eng. Chem., 50:347 (1958).

34. H. Chang and R. A. Daane, SPE ANTEC Tech. Pap., 20:335 (1974).

35. W. L. McCabe and J. C. Smith, Unit Operations of Chemical Engineering, 2nd ed., McGraw-Hill, New York, 1967, p. 804.

9.

Cooling of Plastic Outside the Extruder

9.1 Introduction

Once the plastic has been fully melted and forced under pressure from the appropriate die to form a profile, whether it is a film, a sheet, a wire sheath, a decorative molding, a pipe or tube including corrugated tubing, a filament or fiber, or a solid rod, it is necessary to cool the profile while retaining its shape. There are three popular methods of cooling plastic outside the extruder die: water contact, air contact, and contact of the plastic with a highly conductive surface such as a chill roll. In this chapter these three modes of cooling are discussed only insofar as the heat transfer aspects are concerned.

The stretching and orientation of films and filaments includes profound rheological problems as well as heat transfer problems. Some of these problems are discussed in this chapter, but all the technical questions are not answered to date. Thus, the presentation is somewhat heuristic.

Probably the earliest lesson to be learned in dealing with the cooling of plastic either in an injection mold or in a specially designed cooling chamber attached to the extrusion end of an extruder die is that heat conduction is the primary controlling mode of heat transfer from the plastic. Thus, unless the rod or sheet is very thin (films or filaments), the majority of resistance to heat transfer will not be at the surface of the plastic.

Increasing heat transfer rates in the coolant fluids by increasing air velocities or water agitation around the sheets or rods will have little effect on
the cooling rate of the plastic. The material will behave as if the surface
temperature had been set immediately to the coolant temperature (e.g., as
if $h \simeq \infty$). This is, of course, of great benefit to the engineer in determination of centerline temperatures of thick rods and sheets but of continuing
frustration to the production engineer who is continually trying to increase
his (her) production rates.

9.2 Cooling Thick Sheets and Large-Diameter Rods

Probably the most important single need for cooling of sheets or rods of
very large dimension is in the production of "cell-cast" acrylic materials.
Here, a syrup of polymethyl methacrylate and monomer is poured into a
"cell" with adjustable sides. The cell is heated to initiate (or actually continue) the polymerization. When the material is fully polymerized, the cell
is cooled either through water fog spray on the surface or with cooling coils
through the sides of the cell. Large-diameter rods of polymethyl methacrylate and crystal styrene are extruded "uphill" into ambient air. The direction of extrusion is important since as the plastic solidifies it shrinks, and
voids would be formed in the center of the rod unless filled by the additional
flow of plastic into the center from the extruder. Gravity helps hold the
solidified plastic in place against the force from the extruder.

Recall that in the heat transfer equation we discussed in detail the
transient heat transfer between slabs, cylinders, spheres, and the like and
the cooling or heating medium. This analysis can be applied now to the
cooling of profiles, thick sheets, and large-diameter rods. The describing
equation for a cylinder is

$$\rho c_p (\partial T / \partial t) = (1/r) \, \partial (rk \, \partial T / \partial r) / \partial r \qquad (9.2\text{-}1)$$

where all the terms are well known. Note, however, that time can be directly related to the velocity of the rod as it moves into the cooling medium.
Thus, $t = x/U$, where x is the distance down the axis of the rod and U is the
velocity of the rod. Equation (9.2-1) can then be written as

$$\rho c_p U(\partial T / \partial x) = (1/r) \, \partial (rk \, \partial T / \partial r) / \partial r \qquad (9.2\text{-}2)$$

The plots given earlier in the heat transfer section can be used provided
the Fourier number is calculated correctly. In many cases, with water
quench or direct metal conduction, the surface condition can be considered
to simply be a step change between the known extrusion melt temperature
and the coolant temperature. Thus Fig. 6.2-5 with zero surface resistance can be used to determine the centerline temperature as a function of
distance down the coolant line. From the analysis of the Fourier number,

it is apparent that when cooling to the same centerline temperature, doubling the diameter of the rod increases the cooling length by a factor of 4. This is, of course, true for any profile (planar sheets, square, elipsoidal). If, however, the sheet is cooling in ambient air, it may be necessary to include a surface resistance. As a result the heat transfer coefficient for the coolant must be estimated. As a first estimate, assume an average heat transfer coefficient in the following manner. If, for example, the fluid were stagnant, the Reynolds number for the fluid would be based on the speed of the moving rod. From standard heat transfer books [1], the Nusselt number ($Nu = hD/k$) is calculated from

$$Nu = hD/k = 0.648(Re_D)^{1/2}Pr^{1/3} \qquad (9.2\text{-}3)$$

where $Re_D = D\rho U/\mu$ and $Pr = c_p\mu/k$. Here k, μ, c_p, and ρ are the physical properties of the cooling medium. Now having a value of h, the Biot number $Bi = hD/k_p$ can be obtained, where k_p is the thermal conductivity of the plastic. k_p will always be very small because we are dealing with plastic. Obviously, if h is very large, the surface resistance is nearly zero, and from the equations for zero surface resistance the appropriate information about conduction heat transfer can be obtained. If Bi is not enormously large, then the tables that show the effect of surface cooling on conduction heat transfer are needed. Regardless of the magnitude of Bi, however, the problem is quite soluble.

Weiske [2] has considered the effects of air cooling vs. water quench cooling on the morphological structure of a crystalline polymer. He shows the effects of quenching Nylon 66 in polarized micrographs. Note, however, that although Weiske's heat transfer analysis is wrong, the general implications are correct. Weiske also discusses the effects of crystallinity on density changes during cooling and concludes that there is a maximum rate of extrusion of thick rods beyond which it is not possible to prevent void formation. His argument supports statements made earlier about "uphill" extrusion.

9.3 Cooling Thin Filaments and Films

The assumptions made in the previous section regarding the seemingly negligible surface resistance hold only for rather thick rods and sheets. When extruding filaments or films, it is necessary to reconsider this assumption. Sakiadis [3], Griffith [4], Tsou et al. [5,33], Griffin and Throne [6], and Erickson et al. [7] have considered the problem of heat transfer from a film or filament that is issuing from an orifice or slot in a continuous manner into a stagnant medium. All show that the heat transfer coefficient at the point where the material issues from the slot or orifice is infinite, since the thermal gradient owing to conduction in the material is infinite.

As a result, the use of Eq. (9.2-3) for the calculation of the heat transfer
coefficient is not quite correct when dealing with very thin elements. The
proper approach is through "boundary-layer theory" as applied to a moving
coordinate system (see [3]). Thus, the equations of fluid motion and heat
transfer (and if spinning of the fiber or filament takes place into a solvent
bath, mass transfer) must be solved simultaneously. For example, for a
moving continuous flat surface, as shown in Fig. 9.3-1, the basic boundary-
layer equations are

$$u(\partial u/\partial x) + v(\partial u/\partial y) = \nu(\partial^2 u/\partial y^2) \tag{9.3-1}$$

$$(\partial u/\partial x) + (\partial v/\partial y) = 0 \tag{9.3-2}$$

Appropriate boundary conditions are u = U; v = 0 at y = 0, the surface of the
film; and u → 0 as y → ∞. Sakiadis shows that the drag on the film is
given as

$$C_f = 0.888(Re_x)^{-1/2} \tag{9.3-3}$$

where Re_x is the local Reynolds number, defined as $Re_x = xU\rho/\mu$, and where
x is the distance from the die lips, U is the velocity of the film, and ρ and μ
are the physical properties of the ambient fluid. Further, Erickson et al.,
Tsou et al., Griffith, and Griffin and Throne found experimentally and theoretically that for heat transfer from a nearly isothermal surface,

$$Nu_x = hx/k = CRe_x^{1/2}Pr^n \tag{9.3-4}$$

Griffith argues that through boundary-layer theory n = 1/2. Erickson et
al. argue that C cannot be determined independently of the solution of the

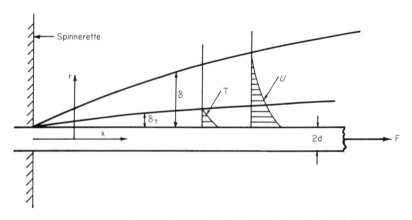

Figure 9.3-1   Coordinates for single-filament spinning [3].

boundary-layer equation for fluid motion. Nevertheless, Erickson provides us with a table for $H = Nu_x/Re_x^{1/2}$ as a function of Prandtl number (Table 9.3-1). Thus, C for air in Eq. (9.3-4) attains the value of 0.44 [as it should, since it can be shown theoretically that the coefficient of the local heat transfer equation should be half that of the drag coefficient equation (9.3-3)]. According to Erickson et al., the value of n is somewhat greater than 0.5, and perhaps n is a function of the momentum boundary-layer conditions as well. As shown in Fig. 9.3-2, the heat transfer experimental values agree with the theory (even though the coefficient on Pr in the original work is incorrect).

As regards the drag coefficient data, the stress on a drawn cylinder in Fig. 9.3-3 is somewhat less than the theoretically calculated stress, particularly at some distance from the die exit. It should be pointed out here that Griffith's work represented a simulation of the actual fiber spinning process in that his drawn element was an "O" ring of fixed cross-section. This geometry is quite different from that experienced by drawing a fiber from a spinnerette head through a solvent bath, and thus his work can only be used as a guide to the effect of drag on a drawn filament. Furthermore, evaporation of solvent and excessive mass transfer can dramatically influence the overall heat transfer from the filament (as is demonstrated by the extensive calculations of Erickson et al.). Some of the mechanics of fiber spinning are considered briefly in the next section, particularly as they relate to heat transfer and drag on the ambient fluid.

Before delving into fiber spinning mechanics, however, consider another work on heat transfer from moving surfaces. Kase and Matsuo [8, 9] have analyzed a simplified version of heat transfer across and along the surface of a fiber. In their work they omit any resistance to heat conduction through the cross-section of the fiber and ignore elongational effects (although including temperature dependency on the tensile viscosity). The equations they propose are

$$v(\partial T/\partial x) + \partial T/\partial t = 2\pi^{1/2}h(T - T^*)/c_p A^{1/2}\rho \qquad (9.3-5)$$

Table 9.3-1

Effect of Prandtl Number on $H = Nu_x/Re_x^{1/2}$

$H = Nu_x/Re_x^{1/2}$	Pr
0.444	1.0
1.68	10
9.08	100

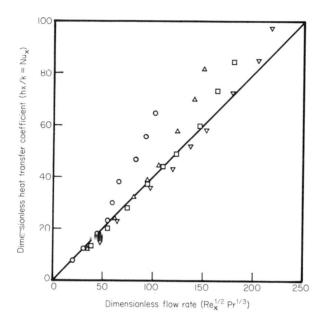

Figure 9.3-2 Experimental and theoretical heat transfer coefficient for boundary-layer growth on a continuous flat surface. Circles, 2.86 ft/sec; triangles, 6.60 ft/sec; squares, 8.92 ft/sec; deltas, 13.9 ft/sec [6].

$$\partial v / \partial x = F/bA \qquad (9.3-6)$$

$$v(\partial A / \partial x) + (\partial A / \partial t) = -A(\partial v / \partial x) \qquad (9.3-7)$$

In the above equations, v is the velocity of the filament as a function of position and time, x is the distance from the spinnerette head, h is the heat transfer coefficient, T is the temperature of the melt (assumed independent of radius but a function of x and t), t is time, T^* is the ambient temperature, and A is the cross-section of the filament (also a function of x and t). F is the unknown force applied to the filament to draw it, and b is the tensile viscosity. Kase and Matsuo point out that these equations can be solved analytically for the steady-state case where the time derivative drops out, provided that the heat transfer coefficient and the tensile viscosity are known as functions of temperature, cross-sectional area, velocity, and distance. We shall return to the discussion of the determination of tensile viscosity properties in a moment. Now focus on the heat transfer coefficient measurements and empirical relationships presented by Kase and Matsuo. In computing the heat transfer coefficient, they use the air velocity relative to the running filament surface. Thus, the heat transfer-flow rate relationship for flow parallel to a cylinder, as given by Simmons [10] and others,

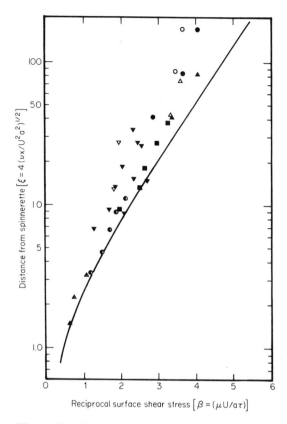

Figure 9.3-3   Experimental and theoretical shear stress on a spun filament [4].

$$Nu_p = 0.42Re^{1/3} \tag{9.3-8}$$

is corrected through the addition of a coefficient $1 + K$ that takes into account the effect of transverse air velocity. Note that the coefficient on the
Reynolds number for this expression for parallel laminar flow is 1/3 rather
than 1/2 as obtained in earlier analyses for continuous sheets. For Re =
DG/μ in the range of 0.1 to 1000, McAdams [11] recommends an empirical
Nusselt number-Reynolds number relationship of

$$Nu = 0.32 + 0.43Re^{0.52} \tag{9.3-9}$$

for laminar cross-flow to a horizontal cylinder. The exactness of the form
for these empirical correlations will affect only the parametric influences
and not the overall analyses of Kase and Matsuo.

As a result, Kase and Matsuo found experimentally that

$$h = 0.21k(\pi/A)^{1/2}[2G/(\pi A)^{1/2}\rho v]^{1/3}(1 + K) \qquad (9.3\text{-}10)$$

where $G = \rho A v$, and ρ, k, and v are fluid properties. The correction is shown in Fig. 9.3-4, where v_y is the transverse air velocity and $v_x = v$, the velocity of the filament. Thus, if the transverse air velocity is nearly equal to the filament velocity, the heat transfer coefficient will be nearly double that for parallel flow. This then substantiates a longtime practice of forcing air across a filament or sheet in addition to flow along the sheet. Enhanced cooling is the result.

Kase and Matsuo further pursue this effect of transverse velocity cooling by curve-fitting $1 + K$ to the velocity curve of Fig. 9.3-4. They find that

$$1 + K \doteq [1 + (8v_y/v)^2]^{1/6} \qquad (9.3\text{-}11)$$

Furthermore, they state that the physical property terms $k(\rho v)^{-1/3}$ are approximately constant at 1.22×10^{-4} (with appropriate units) for all materials in the melt spinning temperature range. This statement is not inconsistent

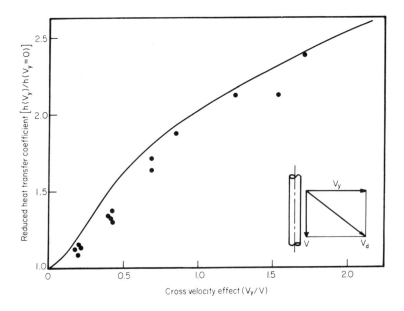

Figure 9.3-4   Enhancement in heat transfer from spun filament owing to cross-flow; single filament [8].

with earlier statements about thermal diffusivity values for plastics. Thus, the heat transfer coefficient expression, Eq. (9.3-10), is reduced to

$$h = 0.473 \times 10^{-4} A^{-1/3} [v^2 + (8v_y)^2]^{1/6} \qquad (9.3\text{-}12)$$

(cgs units on h). This curve fits their data.

Before including the arguments of others in fiber spinning, consider how temperature will vary with distance in a fiber spinning operation, even with the empirical approach of Kase and Matsuo. From simple analyses of tensile viscosity data, b, the tensile viscosity, is assumed independent of shear stress but exponentially dependent on temperature in the form

$$b = b_0 \exp(E/RT) \qquad (9.3\text{-}13)$$

E for polypropylene is on the order of 7 kcal/g·mol. They further assume that the heat of crystallization can be lumped into a fictitious specific heat and that all time derivatives are zero. Thus, their energy equation reduces simply to

$$dT/dx = -(2\pi^{1/2}/GC_p)(T^* - T)A^{1/2}h \qquad (9.3\text{-}14)$$

and

$$dA/dx = -(F\rho/Gb)A \qquad (9.3\text{-}15)$$

where G is the volumetric flow rate, $= \rho Av$, a constant. As an example of the ability of the Kase-Matsuo technique to predict cross-section and filament temperature as a function of distance from the spinnerette, we cite their data, shown as Fig. 9.3-5, for polypropylene.

The numerical method of attack is as follows. If G is a constant, it can be substituted into Eq. (9.3-14) and (9.3-15). The resulting equations can be integrated between A_n, the cross-section at the spinnerette head, and A_f, the cross-section at the freeze line. This gives F, which is a constant. Knowing F, Eq. (9.3-15) can be integrated to obtain A as a function of x, and with this Eq. (9.3-14) yields T as a function of x. Thus, the tension, the rate of change of the area of the filament, and the temperature change are found as functions of position down the filament. Incidentally, although the "Barus effect" or die swell at the exit of the spinnerette has been neglected here, Kase and Matsuo show how to include this in their calculations.

Kase and Matsuo reason that, from the steady-state equations, it can be shown that

$$F \alpha\, bv_y^{1/3}/\rho c_p \qquad (9.3\text{-}16)$$

and

$$dA/dx \alpha\, v_y^{1/3} A/Gc_p \qquad (9.3\text{-}17)$$

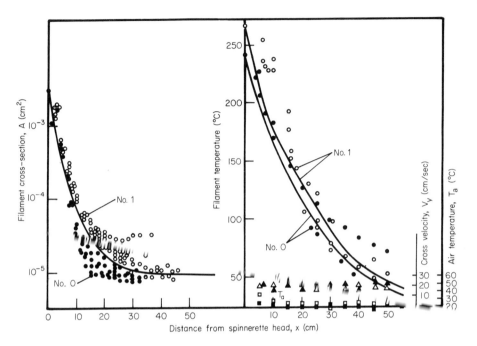

Figure 9.3-5   Effect of two different cross-flow velocities and ambient temperatures on filament cross-section and temperature. Solid line is theoretical [9].

Since $G = \rho v A$, Eq. (9.3-17) can be rewritten as

$$dA/dx \; \alpha \; v_y^{1/3}/\rho c_p v \qquad\qquad (9.3\text{-}18)$$

Thus, as the speed of the filament increases, as would be the case if drawing down increased rapidly owing to dropping temperatures, a very rapid flattening in the dA/dx curve would be experienced (e.g., the area would quickly approach a constant value).

What is probably most surprising in the Kase-Matsuo analysis is the predictability of the theory. They show, for example, that for a 7.5-denier yarn (1 denier = 1 g/9000 m) the temperature differential between the surface of the yarn and the centerline of the yarn is about 10% of that between the filament surface and the ambient air. Thus, the heat transfer rate to the surrounding fluid controls the rate of heat removal from the filament and is thus the antithesis of the thick-sheet problem.

The Kase-Matsuo approach will also be useful in blown film cooling. However, we continue for a moment to discuss filament drawing and cooling. Much interest is at present centered around the tensile or elongational viscosity b, as discussed in an earlier section. Consider arguments as to why determination of the proper constitutive equation for the elongational viscosity is necessary. One of the early criteria for fabrication of plastic filaments and fibers was "spinnability." Ziabicki and his coworkers [12-16] have spent significant effort in attempting to understand not only the elongational effects but in attempting to model the entire spinning problem. Most recently, they have considered that the elongational viscosity (assumed constant by both Ziabicki and his coworkers in earlier works and by Kase and Matsuo) is in fact a function of elongational rate. Lodge [17] states that for a fluid to be spinnable, the elongational viscosity must not only increase with increasing elongation rate but must increase at a rate sufficient to prevent drawing down of the filaments in the region of smallest diameter. Thus, Lodge puts two criteria on the elongational viscosity: First, it must increase with increasing elongational rate, in agreement with some experimental evidence of Meissner and others, and second, it cannot be rate limited, contrary to Cogswell and Lamb's argument. Recently Chang [18] established a criterion for spinnability in terms of an extension of the Lodge argument. From his simplistic model, he shows that the area ratio must achieve a constant value. This is possible only if the materials possess some elastic characteristics in addition to viscous components. Thus, the simple constant value for elongational viscosity or even a power-law model for the elongational viscosity as a function of the elongational rate is insufficient to model spinnability. This is summarized in Fig. 9.3-6, where a power-law model shows a continuously decreasing cross-sectional area in the thinnest section of the fiber as compared with the constant viscoelastic model. It must be remembered here, however, that most rheological models, including those of Lodge, do not include the contributions of temperature and transverse velocity. At this point, it cannot be said what approach is best. The model that seems best able to predict engineering observations and generate engineering extrapolations must be used. The work of Ziabicki, Kase and Matsuo, and White [19] and his coworkers, discussed below, is useful to the spinning technology.

　　To compromise the problems of elongational viscosity and momentum and heat transfer, White [19] and his coworkers have recently considered the spinning of low-density polyethylene and an experimental polystyrene. From their isothermal viscosity measurements, they found that the apparent elongational viscosity for LDPE did, in fact, increase with increasing elongational rate (given as dU/dx), but for polystyrene in the range of 0.03 to 2 sec^{-1}, the elongational viscosity decreased with increasing elongational rate. This, of course, is contrary to the above discussed rheological theory. As a clarifying point, Cogswell [20] has extensively reviewed available elongational flow melt rheology data and concluded that amorphous polymers of very high viscosity show typically rubbery behavior, with Hookean behavior at low stress and strain hardening at high stress. Below the melt point of a resin, the typical stress-strain behavior depends very little on

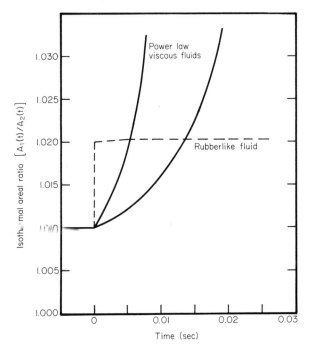

Figure 9.3-6   Isothermal areal ratio $[A_1(t)/A_2(t)]$ response of a power-law viscous fluid and a rubber-like fluid to step changes in applied drawing forces.   Note that power-law fluid areal ratio continues to decrease with time. $\eta = 5 \times 10^5$ P, $\tau_1 = 0.1$ sec [18].

time-dependency terms.   A careful examination of the available data on commercial polystyrenes indicates that all data approach the Troutian viscosity for uniaxial and biaxial elongations at high tensile stresses.   Branched polyolefins show increasing viscosities with increasing tensile stresses, but linear polyolefins show elongational viscosity decreasing with increasing tensile stress.   Cogswell attributes this to molecular flow for the linear polymers and to cluster or cellulite flow for the branched or bulky polymers. From a heat transfer viewpoint, White and his coworkers chose to solve the Sakiadis boundary-layer theory and the integral energy equation rather than use the Kase-Matsuo approach of obtaining an empirical heat transfer coefficient.   Their theoretical results were rather disappointing when compared with their experimental data and that of others, including that of Kase and Matsuo.   This is shown in Fig. 9.3-7, where the Nusselt number is plotted against the reciprocal of the local Reynolds number in the following form: $Re = xU/v$, and $Re' = Re(R^2/x^2)$. $Nu = hR/k$ for White's analysis. White shows empirically that for $1/Re'$ less than 200, $Nu \doteq 0.72$, a constant, and that for $1/Re'$ greater than about 500, Nu is proportional to Re' with the

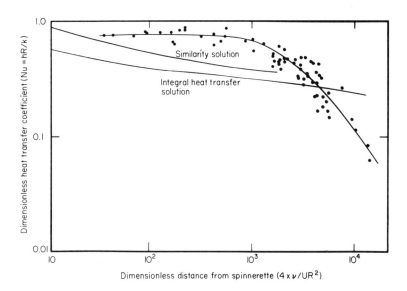

Figure 9.3-7   Experimental heat transfer data from spun LDPE filaments compared with two theories [19].

experimental proportionality constant being 220. Again, this is empirical, and there is little to choose between this empirical model and that of Kase and Matsuo discussed earlier. Most important is the inability of the rheological theory or the boundary-layer theory to support the experimental evidence.

Consider again the typical time–temperature curves of cooling filaments as shown from White's work in Fig. 9.3-8. The open data are for the crystalline LDPE, and the solid data are for polystyrene. The plateau evident for LDPE indicates a region of crystallinity in the typical temperature range of 130 to 135°C. Significant drawing takes place in this region ahead of the "frost line," as evidenced by the experimental work of Dowd [21] and Ast [22] for blown films. Polystyrene shows no such plateau. Recently, Spruiell and White [23] have concentrated on the morphological behavior of the spinning filament of crystalline PP. They find that the onset of plateauing in the temperature curve coincides with stress induction of crystallization, as measured by on-line wide-angle X-ray scattering photography. As much as 60% of the crystallinity occurs prior to the end of the plateau. They find that their half-times for crystallization lie between the very short rotational viscometry shear-induced crystallization half-times of Haas and Maxwell [24] and the very long quiescent, isothermal half-times reported in other literature. This fundamental understanding of morphological behavior of a spinning threadline under various stress levels and thermal environments is required reading for the student interested in film and fiber technology of crystalline polymers.

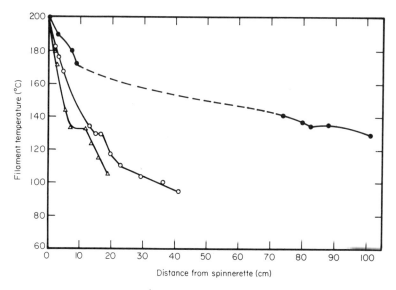

Figure 9.3-8   Experimental temperature profiles from spun LDPE filaments. Areal change not noted. Note crystalline plateau. Triangles,
open circles, mass flow rate = 0.580 g/min; solid circles, mass flow
rate = 5.53 g/min [19].

9.4 Cooling Blown Films

We are again confronted with a combined rheological and heat transfer
problem in the formation of films. Basically, the films are extruded
through a ring die with die gaps on the order of 5 to 10 times that of the
desired final film thickness. To accomplish the thinning of the film, the
melt is elongated in the "machine" direction by applied tensile forces and
at the same time inflated through air pressure applied from within the melt
bag. Thus, instead of elongational stretching of the material, the film
experiences biaxial orientation. Biaxial orientation can also occur in the
stretching of a thermally softened sheet for the thermoforming process,
and additional discussion of this type of "rheological" argument is given
later. A schematic of blown film geometries is shown in Fig. 9.4-1.
Pearson and Petrie [25-28] give a very good summary of the mechanical
processes involved in the biaxial orientation. The objective is to determine
$a(x)$, the film thickness. They select a segment on the film bubble surface
on which to apply a stress rate of strain balance. Because the geometry
of the film bubble does not correspond to the laboratory coordinates shown
in Fig. 9.4-1, they define a new coordinate system z_1, z_2, z_3, in which z_1
is in the direction of draw, z_2 is normal to the bubble (e.g., in the direction

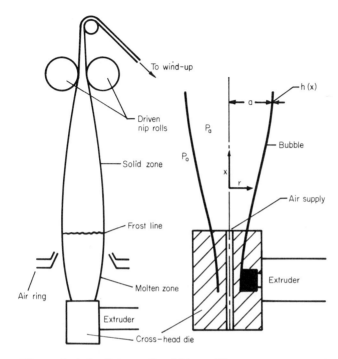

Figure 9.4-1 Schematic of blown film processing with typical coordinate system shown [25].

of the radius of the bubble), and z_3 is the circumferential dimension. Then the three velocities associated with these coordinates are

$$v_1 = v_1(z_1), \qquad v_2 = e_{22}z_2, \qquad v_3 = e_{33}z_3 \qquad (9.4\text{-}1)$$

where e_{22} and e_{33} are the strain rates in the z_2- and z_3- directions, respectively. Furthermore, from continuity,

$$dv_1/dz_1 = -(e_{22} + e_{33}) \qquad (9.4\text{-}2)$$

indicating that e_{22} and e_{33} are functions of z_1-direction only. Further,

$$v_1(dh/dz_1) = he_{22}, \qquad v_1(da/dz_1) = ae_{33} \qquad (9.4\text{-}3)$$

to account for the biaxial orientation of the axes. Now, from geometry,

$$dx/dz_1 = [1 + (da/dx)^2]^{-1/2} \qquad (9.4\text{-}4)$$

And substituting into Eq. (9.4-3),

$$da/dx = (ae_{33}/v_1)[1 + (da/dx)^2]^{1/2} \qquad (9.4-5)$$

and

$$dh/dx = (he_{22}/v_1)[1 + (da/dx)^2]^{1/2} \qquad (9.4-6)$$

Thus, once e_{22} and e_{33} are known functions of x, h(x) and a(x) can be determined from these equations. Furthermore, the stresses can be calculated either from a constitutive equation or from a gross force balance. In the latter case, assume that the hoop tension acts only in the z_3-direction and the in-line or draw tension acts only in the z_1-direction. Thus,

$$F_h = \int_{-h/2}^{h/2} \tau_{33}\, dz_2, \qquad F_t = \int_{h/2}^{h/2} \tau_{11}\, dz_2 \qquad (9.4-7)$$

where τ_{33} is the stress in the hoop direction and F_h is its gross force integrated across the sheet thickness and where τ_{22} is the stress in the draw direction and F_t is its gross force integrated across the sheet thickness. Pearson further shows that the equilibrium radius of the bubble can be determined by balancing these forces against the pressure differential:

$$p_a - p_0 = F_h/R_1 + F_t/R_3 \qquad (9.4-8)$$

where p_a is the pressure within the bubble; p_0 is ambient pressure; R_1 is the radius of curvature of the sheet in the (z_2, z_3) plane, given as

$$R_1(z) = a(x)[1 + (da/dx)^2]^{1/2} \qquad (9.4-9)$$

and R_3 is the radius of curvature of the bubble in the (z_2, z_3) plane, given as

$$R_3(x) = -[1 + (da/dx)^2]^{3/2}/(d^2a/dx^2) \qquad (9.4-10)$$

A force balance (neglecting gravity and inertia) between planes where x = constant is given as

$$(2\pi aF_t)/[1 + (da/dx)^2]^{1/2} - \pi a^2(p_a - p_0) = \text{constant} \qquad (9.4-11)$$

As with the case of a spinning fiber, an energy equation is needed. Here Pearson suggests use of the simple transient heat conduction equation. However, the Kase-Matsuo or Ziabicki approach is probably more correct, since the films are very thin, and most of the resistance to heat transfer will occur in the air layer near the plastic surface. Thus, the temperature profile through the film is assumed to be essentially constant with respect to z_2, and therefore is a function only of z_1. Since we know how x varies with z_1, the heat transfer equations can be converted from the z_1-coordinate system to the x-coordinate system at will.

There are three sets of experimental data that can be used to discuss the Pearson analysis. The first is that of Dowd [21] on HDPE with a very small blowup ratio (1.1:1). His die gap was 35 mils, and the final sheet thicknesses were 2.3, 3.2, and 4.4 mils. He measured surface velocities and temperatures along the draw direction. Note that since there is very little blowup, the z_1-direction is very nearly the x-direction. This means that da/dx is approximately zero. From Eq. (9.4-11), F_t, the tension in the draw direction, can be written as

$$F_t = \text{constant} + a(p_a - P_0)/2 \qquad (9.4\text{-}12)$$

According to Eq. (9.4-7), the hoop force is

$$F_h = be_{33}h \qquad (9.4\text{-}13)$$

where $\tau_{33} = be_{33}$ and b is assumed independent of shear rate (e.g., a fluid with Troutonian behavior). Now the thinning rate can be written as

$$d(\ln h)/dx = e_{22}/v_1 \qquad (9.4\text{-}14)$$

As already seen, e_{22} must be a function of x only according to the continuity equation. Thus, defining $G = hv_1$ and recognizing that v_2, as defined earlier, is constant, we see that $e_{22} = c/h$. Equation (9.4-14) is rewritten as

$$dh/dx = c\rho h/G \qquad (9.4\text{-}15)$$

Most certainly the constant c is a function of the tension on the film, and most certainly it is inversely proportional to the elongational viscosity. Thus, Eq. (9.4-15) can be written another way as

$$dh/dx = -c'F\rho h/bG \qquad (9.4\text{-}16)$$

where the negative sign indicates that h is decreasing with x. c' represents an unknown proportionality between the elongation rate in the z_2-direction and the tension in the x-direction, denoted here as F. b is the elongational viscosity, as before. Integrate this equation for isothermal conditions to obtain h(x). As expected, it is exponentially dependent on x:

$$h = h_0 \exp(-c'F\rho x/bG) \qquad (9.4\text{-}17)$$

More important, perhaps, is the fact that velocity should be increasing exponentially with x owing to the continuity equation:

$$v_1 = -(dh/dx)^{-1} = bG/c'F\rho h = (bG/c'F\rho h_0) \exp(c'F\rho x/bG) \qquad (9.4\text{-}18)$$

A thermal effect can be imposed on this simplistic model in the manner of Kase and Matsuo, but since the model already includes several heuristic

analyses, assume simply that only the elongational viscosity is affected by
the temperature change. Examination of Dowd's data and that of White et
al. for 1/Re' less than 200 reveals that the temperature profile is nearly
linear with respect to x for a great portion of the temperature position
curve. Thus, write the Arrhenius expression for the elongational viscosity
b as

$$b = b_0 \exp(rT) \tag{9.4-19}$$

The "inverse" form for the Arrhenius expression is used here. Now if $T = sx + T_0$, Eq. (9.4-19) can be rewritten as

$$b = b_0' \exp(-Sx) \tag{9.4-20}$$

where s and r have been combined into S and T_0 has been included in the
preexponential coefficient b_0'. Now substitute this, assuming that F and
density are constant, into Eq. (9.4-15) to obtain

$$dh/h = (-c'F\rho/b_0'G) \exp(Sx) \, dx \tag{9.4-21}$$

with the boundary condition that $h = h_0$ when $x = 0$. Integration yields

$$h = h_0 \exp (K/S)[1 - \exp(Sx)] \tag{9.4-22}$$

where $K = c'F\rho/b_0'G$. A plot of h/h_0 for some representative values of K
and S is shown as Fig. 9.4-2. As expected, a dramatic reduction in h with
x is seen. Likewise v_1 can be plotted against x by using the proper form
for Eq. (9.4-18). A similar dramatic rise in velocity at some distance
from the die lips is expected. Examination of Dowd's data, shown as Figs.
9.4-3 and 9.4-4, shows just such a profile for velocity. The expression
for v_1 is

$$Kh_0 v_1 = \exp\{Sx - (K/S)[1 - \exp(Sx)]\} \tag{9.4-23}$$

A comparison of $Kh_0 v_1$ vs. x is given with one set of Dowd's velocity data
as Fig. 9.4-5. Note that this is simply a heuristic comparison to illustrate
the approach and is not meant for prediction.

Examine the meaning of K and S in Eq. (9.4-22). S represents the
combination of a cooling rate and the energy of activation for the elongational
viscosity. The larger the cooling rate (e.g., the higher the air flow around
the film or the faster the extrusion rate into ambient air), the larger the
value of S and the more rapid the reduction of h with x. Likewise, the
higher the energy of activation for elongational viscosity, the more temperature sensitive the viscosity becomes and the more rapid the draw-down
rate. On the other hand, the larger the magnitude for the elongational viscosity, the slower the draw-down rate. And the higher the in-line drawing
force F, the more rapid the draw-down rate, as expected. Thus, this
heuristic argument seems to be valid in predicting the proper direction to

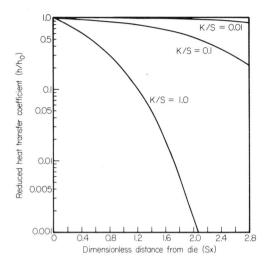

Figure 9.4-2 Decrease in heat transfer coefficient with increasing draw speed and cooling rate. K is a dimensionless draw rate, and S is proportional to the rate of cooling of melt.

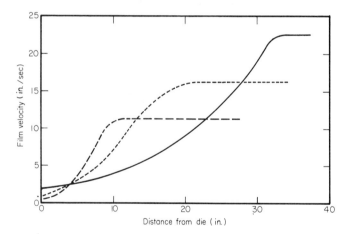

Figure 9.4-3 Increase in film velocity with distance above die for three extrusion rates. Solid line, 140 lb/hr; dotted line, 98 lb/hr; dashed line, 68.4 lb/hr [21].

be taken in improving the draw-down rate. The Dowd data offer much more to the study of blown films than at first meets the eye.

As is apparent, Dowd's data were collected for a very small blowup ratio of 1.1:1. The work of Ast and Han and Park, on the other hand [22,

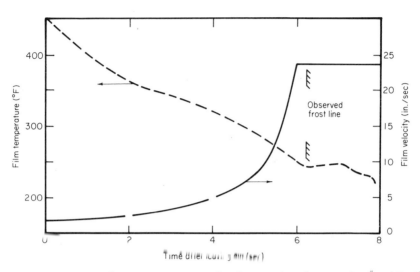

Figure 9.4-4   Film temperature and velocity after leaving die for 140 lb/ hr extrusion rate.   Draw speed is 103 ft/min to produce 2.2-mil film [21].

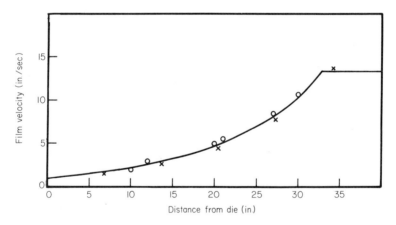

Figure 9.4-5   Comparison of Dowd's [21] experimental data (hollow points) with Eq. (9.4-23) for K/S = 0.01, S = 0.073 in.$^{-1}$ (X's).

29,34,35], was carried out on processes with blowup ratios of 3.5:1. There-fore, variations on the data of Dowd owing to the biaxial stretching effect should be observable. As shown in Fig. 9.4-6, Ast's temperature profile for various film velocities for LDPE with a freeze temperature of 114 to 116°C is very similar to that of Dowd (Fig. 9.4-4). The possible exception is in the first 5 cm (2 in.) above the die where radial velocity becomes significant. There is little cooling in Ast's experiment. It would appear, then, that for

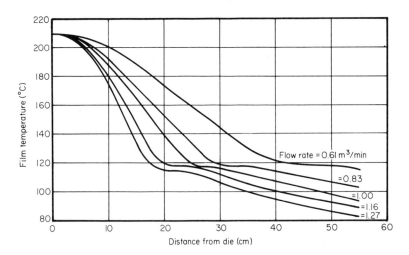

Figure 9.4-6 Experimental film temperatures for 4-mil LDPE film. Drawing speed is fixed. Note crystalline plateau around 120°C. Data points not shown [22].

this "induction period" an isothermal biaxial orientation might be valid. Beyond this period, the film temperature can be assumed to decline linearly with distance until near the freeze line. Ast has measured both radial and axial velocities for the biaxial elongation, and his results are shown for one velocity as Fig. 9.4-7. Note that there is very little contribution to the magnitude of the velocity by the radial velocity. Even more surprising is Fig. 9.4-8, where Ast shows for one velocity that although the radius of the film bubble is increasing rapidly with distance above the die, the film thickness and the velocity of the film assume essentially the same profiles as obtained by Dowd (Fig. 9.4-4). One might therefore argue, using Ast's heat transfer data and his experimental temperature and velocity profiles, that the simplistic view of the blown film process as observed by Dowd for the 1.1:1 blowup ratio and as discussed heuristically in the sections beginning with Eq. (9.4-12) seems to describe the more complex blown film process so adequately measured by Ast. Surprisingly, Han and Park have repeated much of Ast's work using LDPE and HDPE resins and a blowup ratio of 3.5:1. They choose to correlate their work with that of Pearson and Petrie and show excellent agreement with their theory for both the height-dependent radius ratio and reduced film thickness and both films. More importantly, the Han-Park data for LDPE film thickness show striking exponential dependence on the height above the die. Note that this is the result obtained from the simplistic model leading to Eq. (9.4-17). This again confirms the close agreement between simplistic theory and actual blown film practice.

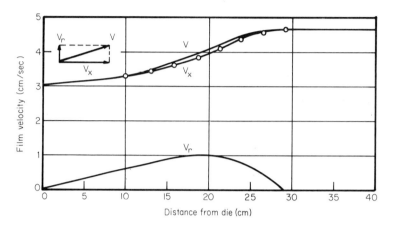

Figure 9.4-7 LDPE radial and axial film velocities along the bubble for a flow rate of 1 m³/min [22].

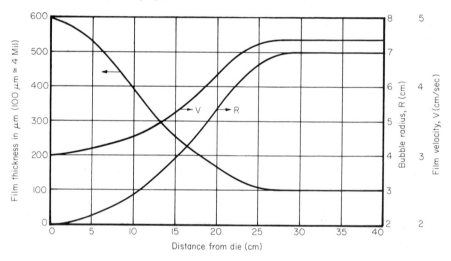

Figure 9.4-8   LDPE film thickness, bubble radius, and film velocity along the bubble for a flow rate of 1 m³/min; 3.5:1 expansion ratio [22].

Recall again that the appropriate approach to the modeling of the blown film process combines the best of Pearson's rheological analysis with the heat transfer calculations and observations of White et al. and/or Kase and Matsuo's empirical approach. Again Dowd's data must be considered most important to the understanding of the process of blown film processing.

9.5 Cooling Films by Direct Contact

In many sheet or film forming operations, the continuously formed sheet is cooled by direct contact with chill rolls. (See Fig. 9.5-1.) Several of the major processes that utilize chill rolls for heat removal include extrusion coating, where the melt such as LDPE is applied continuously to a substrate such as Kraft paper by directly feeding the melt from an extruder to the nip region of calender rolls. From the calender rolls, the sheet then passes to the chill rolls and then on to windup. Impregnation of cloth with PVC plastisol has been a standard way of making "synthetic leather" for a decade or more. More recently, coextrusion of two or three sheets of plastic has been made commercially acceptable. The melts are simultaneously fed to calender rolls for sizing and then while still molten are fed to a set of nip-and-chill rolls. The interfaces are bonded since the materials are still molten, and the outer surfaces are quenched so that further handling of the sheet is possible without delamination. Heat transfer from the plastic to the chill roll surface is relatively straightforward. First, straighten out the surface of the chill roll in order to determine the length of contact between the roll and the plastic. This information is obtained by knowing the degree of wrap around the roll and the diameter of the roll. Then determine the velocity of the sheet and from this the time of contact of the sheet with

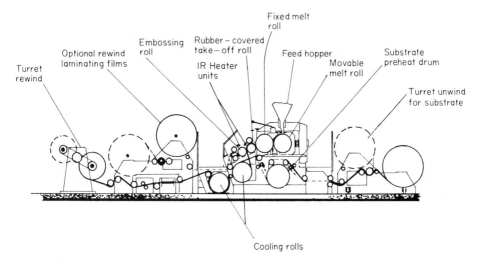

Figure 9.5-1 Characteristic setup for a two-roll melt coater with optional laminating film attachments. (Photo courtesy of <u>Modern Plastics Magazine</u>, Zimmer plastic GmbH [30].)

the roll. Normally it can be assumed that the rear or exposed surface of
the sheet is not losing significant amounts of heat during this period of contact time. Thus, analysis yields a simple transient heat conduction problem
with the rear surface insulated and the front surface set to the temperature
of the chill roll. Assume that there is good surface contact between the
plastic and the roll and that no appreciable slippage occurs in this region.
The appropriate equation is

$$\rho c_p (\partial T/\partial t) = k(\partial^2 T/\partial y^2) \tag{9.5-1}$$

where y is the distance from the chill roll and t is measured from the moment of contact to the moment of release. The solution to this equation has
been discussed several times, with the primary discussion found in Chap.
6. Note, however, that if y is relatively thick and/or t_f, the time of release
of the film from the chill roll, is rather short, an excellent approximate
solution for Eq. (9.4-5) can be obtained by assuming that the sheet in infinitely thick in the y-direction. If $T_{y=0} = T_0$ and $T_{y=\infty} = T_\infty$, the total
amount of heat removed from the sheet in the time period to t_f is

$$Q/A = 2k(T_\infty - T_0)(t_f/\pi\alpha)^{1/2} \tag{9.5-2}$$

where k is the thermal conductivity of the sheet and α is the thermal diffusivity of the sheet (cm^2/sec). Equation (9.5-2) gives the heat flux per
unit width removed from a differential thickness of the sheet as it travels
along the chill roll from t = 0 to t = t_f. Now the total amount of heat contained in this area A is given by

$$Q/A = \rho c_p (T_0 - T_\infty) \tag{9.5-3}$$

and thus the ratio of Q/A of Eq. (9.5-2) to that of Eq. (9.5-3) is fraction of
heat removed by one chill roll. If a second chill roll follows, it can be assumed that there is sufficient time between the chill roll contacts to allow
the temperature, which is exponentially dependent on y at t = t_f, to approach
a constant value. This is determined by subtracting Q/A of Eq. (9.5-2)
from that in Eq. (9.5-3) and recalculating a new plastic temperature T_0'.
If a series of chill rolls are involved, simply calculate the heat removed
from Eq. (9.5-2), subtract that from Eq. (9.5-3), recalculate another T_0'',
and repeat until the temperature of the film coming from the last chill roll
has been determined. As mentioned, this is an approximate solution, since
it assumes that the film is essentially infinite in thickness. The correct
solution for more than one chill roll requires numerical computation of the
heat conduction equation. This is particularly true if the sheet is cooled on
alternate sides in subsequent chill rolls. Nevertheless, we would not expect this simplistic calculation to be significantly different in order of magnitude from the more exact but more complex solution of multiple equations.

Another area that makes extensive use of direct contact heat transfer is fiber yarn heat treatment. Pai and Hyman [31,32] have written an extensive analysis of this problem. Basically, fiber yarn is heated by taking several wraps around internally heated rotating rolls at speeds of 300 to 1000 m/min. The convection heat transfer from the heated drum to ambient air can be calculated from the analysis on continuous surface heat transfer given in Sec. 9.3. Pai and Hyman solve the transient heat conduction equation (9.5-1) in two dimensions. The important change in modeling is in the boundary condition around a single filament in contact with the roll and the air layer adjacent to it. Pai and Hyman assume a heat transfer coefficient derived from that for the continuous surface heat transfer analysis. Their limited experimental data agree well with their model. It is interesting to note that only 10% of the heat load on the roll is transferred directly to the wrapped yarn, the rest being used to heat the surrounding air.

9.6 Conclusions

There are other aspects of postprocess cooling that haven't been covered in this chapter. For example, hot melt stamping or foil stamping has been omitted, as has the processing that occurs in the nip of calender rolls [33]. The heating of plastic outside of the extruder as in the case of thermoforming has been deferred to a later chapter. Considered here has been the cooling of plastic by forced convection, quenching, and direct contact with cold surfaces. The complex and somewhat confusing areas of filament forming and blown film forming have been outlined. Hopefully there is sufficient information here to allow good engineering judgment to prevail.

Problems

P9.A A polyester filament used in forming fabric for paper machines is drawn from a single spinnerette head at a take-up reel speed of 500 ft/sec. The filament is 0.010 in. in diameter at take-up, and the spinnerette diameter is 0.016 in. The spinning temperature for polyester is 550°F, and the freezing temperature is 465°F. The thermal conductivity of polyester is 0.15 Btu/ft·hr·°F, the specific heat is 0.45 Btu/lb·°F, and the density is 70 lb/ft^3. Determine the time required to cool the polyester centerline to the freezing temperature if it is extruded into a water bath at 70°F. Determine the distance between the spinnerette head and the take-up reel for a water bath. Determine the time required to cool the polyester centerline to the freezing temperature if it is extruded into ambient air at 70°F. What is the distance between the spinnerette head and the take-up reel now?

P9.B Consider the above-mentioned spinning of a polyester filament into air. Now consider forced convection cooling, where an air jet at 200 ft/sec is impinged against the filament at an angle of 45° to the direction of draw. Now what is the distance between the spinnerette head and the take-up reel? List all assumptions.

P9.C Kase and Matsuo conclude that for a 7.5-denier yarn the temperature differential between the surface of the yarn and the centerline of the yarn is about 10% that between the filament surface and ambient air. Under what conditions is this true? Is there a minimum filament diameter? Is there a maximum in the take-up speed?

P9.D In deriving Eq. (9.4-22), a linear form for the temperature profile was assumed. Discuss the appropriateness of this assumption based on the discussion of a heat balance around a filament given in Sec. 9.2. Show that for medium values of the Fourier number the unaccomplished temperature change is perhaps best represented as exponentially dependent on the distance:

$$Y = \exp(-aX) \tag{9.D-1}$$

Can this form of the temperature profile be used to arrive at a modified solution of Eq. (9.4-22)? Hint: Try expanding Eq. (9.D-1) in series form first.

P9.E Follow through the argument leading to Eq. (9.4-23), and obtain this result. Then take Dowd's data, shown in Figs. 9.4-3 and 9.4-4, and obtain the results shown in Fig. 9.4-5. Apply the same analysis to Ast's data shown in Fig. 9.4-6. Do these data reduce as clearly as Dowd's? If not, why not?

P9.F Propose a series of experiments to verify the parametric relationships given in Eq. (9.4-23). Carefully specify the ambient conditions and the rate of heat transfer and blowup of the film. What will be your major sources of error?

P9.G A 40-mil film is extruded at 470°F at the rate of 100 ft/min. Its thermal diffusivity is 2×10^{-3} cm^2/sec. The heat is removed by wrapping the sheet around chill rolls 24 in. in diameter. The chill rolls are kept at 70°F. How many rolls are needed to reduce the average temperature of the sheet to 180°F? To 240°F? What happens if the chill rolls are 36 in. in diameter? The wrap angle around each chill roll is 185°.

References

1. W. M. Rohsenow and J. P. Hartnett, eds., Handbook of Heat Transfer, McGraw-Hill, New York, 1973, Sec. 8.
2. C. D. Weiske, SPE ANTEC Tech. Pap., 13:676 (1967).

3. B. C. Sakiadis, AIChE J., 7:26, 221, 467 (1961).
4. R. M. Griffith, Ind. Eng. Chem. Fund., 3:245 (1964).
5. F. K. Tsou, E. M. Sparrow, and R. J. Goldstein, J. Fluid Mech., 26:145 (1966).
6. J. G. Griffin and J. L. Throne, AIChE J., 13:1210 (1967).
7. L. E. Erickson, T. Fan, and V. G. Fox, Ind. Eng. Chem. Fund., 5:19 (1966).
8. S. Kase and T. Matsuo, J. Polym. Sci. Part A, 3:2541 (1965).
9. S. Kase and T. Matsuo, J. Polym. Sci. Part A, 11:251 (1967).
10. A. C. Simmons, Trans. AIChE, 33:613 (1942).
11. W. H. McAdams, Heat Transmission, 3rd ed., McGraw-Hill, New York, 1954, p. 260.
12. A. Ziabicki and R. Takserman-Krozer, Kolloid Z., 198(1):60 (1964).
13. A. Ziabicki and R. Takserman-Krozer, Kolloid Z., 199(1):9 (1964).
14. A. Ziabicki and K. Kedzierska, Kolloid Z., 171(1):151 (1960).
15. A. Ziabicki, Kolloid Z., 175(1):14 (1960).
16. A. Ziabicki, Kolloid Z., 179(2):116 (1961).
17. A. S. Lodge, Elastic Liquids, Academic Press, New York, 1964, Chap. 7.
18. H. Chang, Ph.D. thesis, University of Wisconsin, Madison, 1973.
19. J. L. White, personal communication, Nov. 1974.
20. F. N. Cogswell, Appl. Polym. Symp., 27:1 (1975).
21. L. E. Dowd, Soc. Plast. Eng. J., 28(3):22 (1972).
22. W. Ast, Kunststoffe, 63:427 (1973).
23. J. E. Spruiell and J. L. White, Appl. Polym. Symp., 27:121 (1975).
24. T. W. Haas and B. Maxwell, Polym. Eng. Sci., 8:225 (1969).
25. J. R. A. Pearson, Mechanical Principles of Polymer Melt Processing, Pergamon Press, Elmsford, N.Y., 1966, p. 123.
26. J. R. A. Pearson and C. J. S. Petrie, J. Fluid Mech., 40:1 (1970).
27. J. R. A. Pearson and C. J. S. Petrie, J. Fluid Mech., 42:609 (1970).
28. J. R. A. Pearson and C. J. S. Petrie, Plast. Polym., 38:85 (1970).
29. C. D. Han and J. Y. Park, J. Appl. Polym. Sci., 19:3291 (1975).
30. Mod. Plast. Encyclopedia, 49(10A):546 (1972).
31. V. K. Pai and D. Hyman, Text. Res. J., 42:633 (1972).
32. V. K. Pai and D. Hyman, Text. Res. J., 42:639 (1972).
33. F. K. Tsou, E. M. Sparrow, and R. J. Goldstein, Int. J. Heat Mass Trans., 10:219 (1967).
34. C. D. Han and J. Y. Park, J. Appl. Polym. Sci., 19:3257 (1975).
35. C. D. Han and J. Y. Park, J. Appl. Polym. Sci., 19:3277 (1975).

10.

Injection Molding—An Example of Cyclical Molding

10.1 Introduction

Frequently, once a plastic has been melted or softened, we wish to make discrete things from the melt. The largest process for making discrete objects from plastic melt is injection molding. It represents cyclical molding. As mentioned earlier, compression and/or transfer molding can be considered to be a form of cyclical molding, although they have been categorized here as batch processes. Thermoforming, rotational molding, and blow molding, considered in subsequent chapters, are also forms of cyclical molding.

While there are many manufacturers of injection molding machines, there are only two basic types of processes. Schematics of these are shown in Figs. 10.1-1 and 10.1-2. Cold resin is placed into a chamber where it can be heated until softened or melted; then this melt is pushed into a suitably reinforced mold where it is cooled into the form of the mold cavity. The older-style injection molding machine uses a plunger and a barrel. The plastic pellets are dropped from the hopper into a barrel that has heating elements around it. The plunger (or ram) is held in the "full retract" position. When the plastic has been heated to its processing temperature,

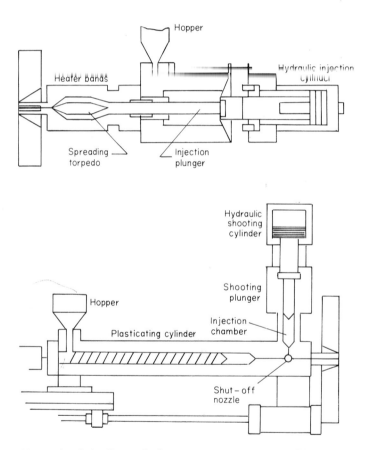

Figure 10.1-1   Typical plunger-type injection molding machine schematics. Top: Single-stage plunger with torpedo constriction between accumulator and sprue. Bottom: Typical screw-ram plunger system showing separate accumulator and rotating shutoff nozzle. This machine led to development of low-pressure foam molding via Union Carbide process [1].

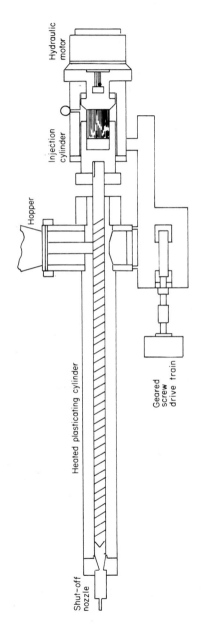

Figure 10.1-2 Typical reciprocating screw injection molding machine schematic as developed in the 1950s. (Courtesy of Modern Plastics Magazine [1].)

the plunger is moved forward either mechanically or hydraulically. Owing to a restriction at the end of the barrel, pressure is built up as the plastic is squeezed between the face of the ram and the front of the barrel. At sufficiently high pressure, the melt is forced through the restriction and into the mold sprue-runner-gate-cavity system. Pressure is applied until the mold is full and sufficient cooling has taken place to prevent formation of vacuum bubbles and/or abnormal differential shrinkage and warpage of the molded part. Then the plunger is retracted, thus opening the hopper and allowing fresh plastic pellets to fall into the cavity. It is normally the case that the ram machine never empties the cavity. In fact many plunger machines have 10 times more plastic in the barrel than that being dispensed in each ejection.

To improve the temperature distribution in ram machines, a "torpedo" is frequently used in the accumulator section of the barrel ahead of the ram. The torpedo is a cigar shaped cylinder fastened to the barrel walls with two to six legs (known as spider legs) and may contain cartridge heaters for supplemental heating. As the melt moves forward during injection, the narrow annulus between the torpedo and the barrel wall requires the machine to exert high shear on the plastic. As mentioned earlier, shear heating occurs because of the very high viscosities of plastic. Thus, not only does the plastic achieve good thermal mixing, but it also increases in temperature owing to shear heating. Many materials (such as rigid PVC) are thermally sensitive and cannot be held at temperatures 50 to 100°C above the melt point for long periods of time without degrading. The use of the torpedo thus increases the melt temperature of the resin without adding to the residence time of the material at these temperatures.

The newer machine is the screw injection molding machine. The plunger has been replaced with a plasticating screw for the purpose of increasing the plasticating capacity of the machine. The description of this process is begun at the point of completion of injection and compaction. The screw (as with the plunger for the ram-type machine) is in the "full forward" position. As the screw is moved back (either hydraulically or mechanically), the screw turns, thus feeding polymer from the hopper, through the solids conveying zone of the screw, through the plasticating section, into the melt pumping section, and finally into the accumulator section ahead of the screw. When the screw reaches the "full retract" position, the accumulator ahead of the screw is filled with plastic melt, and thus there is no time required for melting of the plastic resin prior to injection. The melt is ready immediately to be injected. To inject, the screw is stopped, and hydraulic or mechanical force is applied to it to move it against the melt in the accumulator. Thus, the screw now acts as a ram in the same sense as with the plunger injection molding machine.

Comparison of these types of machines should be made with economics and type of resin to be processed in mind. Suffice it to say here that the screw injection molding machines heat the plastic at a faster rate than the

plunger machines but have a higher initial cost for the same amount of
plastic dispensed per shot and are more temperamental than the plunger
machines. Furthermore, the plunger machine can achieve much higher accumulator pressures than the screw machine owing to the backflow along
the flights of the screw during forward movement.

In the United States, machines are rated according to their shot
capacity (normally in ounces of GPS) and their clamping tonnage (U.S. tons).
Understand that both these ratings are ideal. In reality a processor may
not be able to use all the available clamping tonnage (the mold might be very
small in comparison to the platen sizes), or the processor may be injecting
a highly filled high-density resin (such as 40% glass-filled nylon) and the
specific gravity may be sufficiently high to allow injection of more material
than rated capacity. In most screw machines, it is common practice to
allow 10 to 20% reserve capacity in the accumulator. Thus, if the machine
is rated at 50 oz, the maximum molding shot should be kept around 40 oz.
If a part in GPS (sp. gr. = 1.2) weighs 1200 g (42.3 oz) and has a volume of
1000 cc, the same part in LDPE (sp. gr. = 0.92) would weigh 920 g (32.4
oz). In 40% glass-filled nylon (sp. gr. = 1.56), the part would weigh 1560 g
(55 oz).

The clamping capacity of the machine is usually related to the upper
limit on injection pressure allowed in the machine. For example, if the
projected area of the mold cavity is 5 ft^2, in a 1000-ton machine the injection pressure exerted against the clamping mechanism can be no more than
200 tons/ft^2 (about 2700 psi). In a 1500-ton machine, it can be as high as
300 tons/ft^2 (about 4000 psi). Thus, not only is the shot size of the machine
needed, but the maximum allowable injection pressure must be known. For
this the types of materials that will be processed on the machine should be
known. High clamping pressures are needed for polycarbonates, rigid
PVCs, UHMWPE, and FEP Teflon. Low pressures are needed for nylons,
acetals, LDPE, and GPS. A group of Cornell University, led by J. F. Stevenson, has been carefully considering methods for predicting injection
pressure and clamp force for regularly shaped thin parts [2-4]. Although
their approach mirrors those of others, discussed in detail in Section 10.2,
their results merit consideration in this section. They have reduced the
complex analyses required to a series of graphical design tables and charts.
The majority of the charts represent correction factors to the simplest flow
model, that of isothermal Newtonian radial flow. Recall that the force exerted on a mold is given as the integral of the local pressure over the area
occupied by the plastic. The force exerted against a clamp depends on the
projected area of the plastic. Thus, higher clamping forces are developed
in molding 10-in.-diameter dinner plates than in molding an equivalent number of cups having the same wall thickness and polymer weight. The Cornell group effort is commendable in setting goals for determining guidelines
for calculating developing forces, but it is still too early in the development
of these mathematical models to implicly rely on their results.

There are many ways of holding the mold halves together. Two of the most popular are hydraulic pressure (either as part of the closing system or as a separate "pancake" unit that swings into place after the mold is closed) and mechanical linkage. Schematics of these are shown in Fig. 10.1-3. Normally, the mold halves are opened by pulling one of the halves

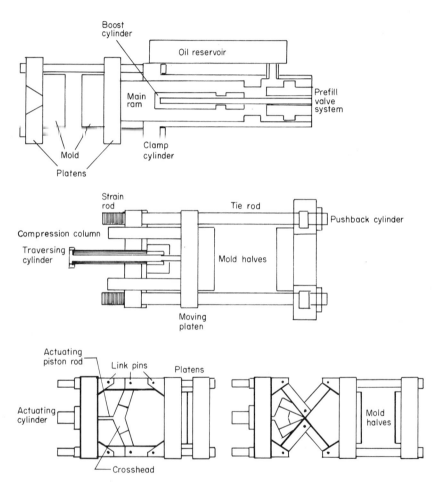

Figure 10.1-3 Typical mold closing methods. Top: Full hydraulic clamp with single hydraulic ram. Center: Typical schematic of hydromechanical clamp with hydraulic traversing cylinder and clamping against compression columns. Bottom: Typical schematic of fully mechanical or toggle clamp; bottom left, toggles in mold-closed position; bottom right, toggles in mold-open position. (Courtesy of Modern Plastics Magazine [1].)

(known in mold drawings as the "B" portion of the mold) away from the stationary (or "A" half) part of the mold that is fastened to the stationary platen. "Pull-to-open-push-to-close" is the popular way of opening a mold, and this can be done either with hydraulic action or mechanical toggle. The hydraulic system is very similar to the piston on the plunger injection portion of the machine. Hydraulic pressure is exerted against the B portion of the mold and platen, forcing the section against the A portion of the mold. The hydraulic pressure is infinitely controllable, and thus the mold can be closed at a very low pressure to avoid "slamming." The hydraulic closure is somewhat slower than the mechanical toggle method of closure. The latter can be pictured as a scissors arm which is pulled from a folded configuration (open mold) to an extended configuration (closed mold) by a mechanical linkage. Frequently the extended configuration is an "overlock," meaning that the knees of the linkage are pushed beyond the equilibrium position, thus ensuring that the mold can come open only by mechanical means.

The mold itself will be considered in more detail later. Note that the primary objective of the machine is to produce money. To produce money, it is necessary to produce parts as cheaply and efficiently as possible that meet customer's specifications as to dimension, stress distribution, surface condition, extent of shrinkage and distortion, distribution of color, minimization of weld lines and streaking, and so on. Regardless of the type of automatic control, the manufacturer of the machine, the type of injection system, the clamping mechanism, and the reputation of the mold maker or pattern designer, the objective is production of as many good parts as possible for the lowest expenditure. In simplistic terms, then, the "optional extras" of the machine must be balanced against the basic considerations of the process. Think of the process in the simplistic terms of an unsteady, nonisothermal hydraulic process using plastic as the hydraulic fluid. We must have a machine that can withstand the repetitive forces exerted by the hydraulic forces required to push the hydraulic fluid into the mold cavity and remain versatile enough to handle odd-sized molds, filled plastics, thermally sensitive plastics, and human errors such as double shots and foreign matter.

10.2 Plastic Flow in the Mold

Consider the mechanism of flow of plastic from the accumulator to the mold cavity. As shown in Fig. 10.2-1, the plastic normally flows from the accumulator through a constriction (which could be a positive shutoff valve) and a sprue bushing. If a single part is being made, the melt would normally flow directly from the sprue into the mold cavity. In a single-cavity mold, there is no apparent gate. The constriction at the nozzle-sprue interface represents the gate. The purpose of a gate is twofold. First, it controls the flow of melt into the mold cavity. Second, and more important,

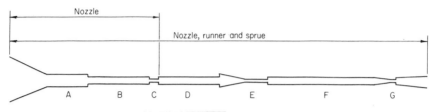

Section	Diameter	Length
A	0.50 in.	2.17 in.
B	0.37	3.11
C	0.25	0.49
D	0.19	3.05
E	0.19	1.18
F	0.37	5.33
G	0.19	0.50

Figure 10.2-1   Injection molding experimental flow path [5].

it prevents appreciable flow of melt back out of the mold cavity after injection pressure has been removed and the melt is cooling. Recently, much attention has been directed toward understanding the mechanism of the filling of a mold cavity through a sprue-runner-gating system. The earliest work that examined the flow in detail was that of Ballman et al. [6-8]. Their rather extensive experimental work was based on earlier incomplete observations by Spencer and Gilmore [9] and Beyer and coworkers [10]. They used a rectangular mold cavity 1 × 12 in. long with three depths: 0.150, 0.050, and 0.075 in. The mold cavity was end-gated using a 1/2-in.-wide fan gate 0.120 in. thick. The mold was gate controlled in that the applied pressure was constant during cavity filling. In some runs, the system resistance was so low that the delivery system could not deliver the material fast enough to maintain a constant pressure. In this case, the pressure depends on the delivery system, and Ballman et al. refer to this as "machine controlled." They excluded all machine-controlled runs from their experiments. They measured short shots into the cavity by varying the ram forward time and found that the velocity of material into the cavity fell exponentially with time as the cavity filled. An example of their data is shown as Fig. 10.2-2. To try to explain this effect, they proposed analysis of the system as a series of resistances to flow. For unidirectional flow, they propose

$$du/dr = -A(T)\tau^m \qquad (10.2-1)$$

where u is the velocity in the x-direction and r is the distance across the cross-section of the channel. $m = 1/n$ in conventional power-law terminology, and $A = (1/K)^n$. $A(T)$ is a viscosity having an Arrhenius form:

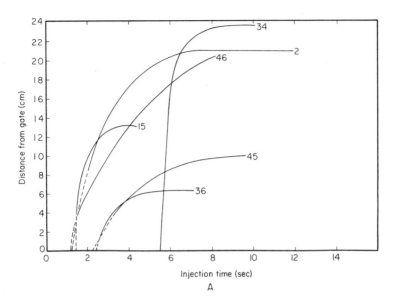

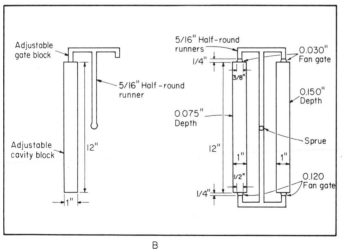

Figure 10.2-2   A: Cavity fill rates for several conditions of injection pressure and mold configuration. Numbers refer to conditions tabulated in reference. B: Experimental configurations for injection mold fill experiments [6].

$$A(T) = A_0 \exp[-b/(T - T_0)]$$ (10.2-2)

The standard force balance for flow between parallel plates is

$$\tau = k'r(-dP/dx)$$ (10.2-3)

where k' is a constant dependent on the geometry of the system. After integration of Eqs. (10.2-1) and (10.2-3),

$$V = K'd^{m+1}I(-dP/dx)^m$$ (10.2-4)

where V is the mean velocity of the wave front, K' is a constant, d is the width of the channel, and I is the nonisothermal term, given as

$$I = (m + 2) \int_0^1 Z^{m+1}A(Z)\, dZ$$ (10.2-5)

and $Z = 2r/d$. I represents the mean value of the rheological constant A across the stream. As a first approximation, assume that the polymer is essentially incompressible. Thus, the mean velocity V must be the same for all x at any time. Equation (10.2-5) can be integrated to obtain the mean velocity in terms of the pressure at the cavity inlet at any time t:

$$V = \left\{ K'I'd^{m+1} \right\} (P_0/X)^m$$ (10.2-6)

where P_0 is the inlet cavity pressure and X is the distance from the cavity inlet to the wave front. I' is a function of x since the viscosity can be nonisothermal, and thus I' is the effective mean value of the parameter A across and along the channel at any time t:

$$I' = \left\{ \int_0^1 dx[I(x)]^{1/m} \right\}^m$$ (10.2-7)

Incidentally, for isothermal flow, this expression is

$$V = K'Ad^{m+1}(P_0/X)^m$$ (10.2-8)

Note that in Eq. (10.2-6) a relationship between force and resistance can be extracted as follows:

$$V = [P_0/XR(t)]^m$$ (10.2-9)

where $R(t) = [K'I'(t)d^{m+1}]^{1/m}$. R(t) is a resistance to flow per unit length of filled channel. This should remain constant in an isothermal channel but increase with time in a nonisothermal channel owing to the decrease in the overall melt temperature. Ballman et al. point out that Eq. (10.2-9) is not correct at time t = 0, because X = 0, and thus at a finite injection pressure the initial velocity is infinite. They propose addition of a resistance in the system delivering the fluid to the cavity. In this way the filling of an initially empty cavity must be considered in terms of its interaction with the delivery system (sprue-runner-gate). The simplest delivery system is a large isothermal reservoir fitted with a ram. The two-channel system consisting of the isothermal reservoir and the nonisothermal channel can be analyzed as follows. The velocity in the reservoir is given as V_r:

$$V_r = [(P_r - P_0)/X_r R_r]^m \tag{10.2-10}$$

where P_r is the ram pressure. If the reservoir is long relative to the length of travel of the ram, X_r is essentially constant, and since the system is isothermal, R_r, the resistance in the reservoir, is also constant. Now according to a simple mass balance,

$$V_r/V = A/A_r = S \tag{10.2-11}$$

Thus,

$$V = S^{-1}[(P_r - P_0)/X_r R_r]^m \tag{10.2-12}$$

If there are a number of filled channels (with varying cross-sectional areas) ahead of and in series with the cavity, then Eq. (10.2-12) is

$$V = [(P_r - P_0)/R_T]^m \tag{10.2-13}$$

where R_T is the total resistance in the system ahead of the cavity. If R_T is known at the instant flow enters the cavity, the cavity inlet pressure can be determined in terms of the velocity V_0 at the instant flow enters the cavity:

$$P_0/P_r = 1 - [R_T(t)/R_T(0)](V/V_0)^{1/m} \tag{10.2-14}$$

[Obviously, if the flow is isothermal to this point, $R_T(t) = R_T(0)$, and the ratio ahead of the velocity ratio is unity.] Having P_0, the cavity pressure, it can now be substituted into Eq. (10.2-9) to obtain

$$V = \left\{ P_r/[XR(t) + R_T(t)] \right\}^m \tag{10.2-15}$$

If the system ahead of the cavity is isothermal, this equation can be written as

$$V/V_0 = [V_0^{1/m} XR(t)/P_r + 1]^m \tag{10.2-16}$$

If the cavity is also isothermal, then $R(t) = (K'Ad^{m+1})^{1/m}$, a constant. In terms of V/V_0, as used by Ballman et al. in their analyses, Eq. (10.2-15) can be rewritten as

$$V/V_0 = R_T(t)/[XR(t) + R_T(t)]^m \tag{10.2-17}$$

The denominator of this equation represents the total resistance of the entire system at any time. As mentioned, the experimental evidence indicates that for a cavity-controlled injection the velocity falls exponentially with time. They express their results as

$$V/V_0 = \exp(-t/B) \tag{10.2-18}$$

where B is a constant dependent on resistance at the gate, on mold temperature, and on melt temperature but apparently independent of injection pressure [as it should be according to Eq. (10.2-17)]. From Eq. (10.2-18), the position of the wave front is given as

$$X = BV_0[1 - \exp(-t/B)] \tag{10.2-19}$$

and at infinite time, $X_\infty - V_0 B$. Thus,

$$V/V_0 = 1 - X/X_\infty = 1 - X/V_0 B \tag{10.2-20}$$

This relationship is in good agreement with the theoretical form of V/V_0 if the form in Eq. (10.2-17) is expanded in a binomial form:

$$(1 + x)^{-m} = 1 - mx + m(m + 1)x^2/2 \mp \cdots \tag{10.2-21}$$

Thus,

$$V/V_0 = [1 + (XR/R_T)]^{-m} \doteq 1 - X/(R_T/Rm) \pm \cdots \tag{10.2-22}$$

As can be seen, $R_T(t)/R(t)m \doteq V_0 B$. It is too much to expect the simplistic theoretical model to give the exact values, but note that in Fig. 10.2-3 for experiment 2 the values for B and V_0 were 1.69 sec and 12.2 cm/sec, respectively, for a 0.150-in.-thick cavity and for experiment 4 about 0.31 sec and about 25.5 cm/sec, respectively, for a 0.050-in.-thick cavity. If the cavity were isothermal (along with the delivery system), R would be proportional to $d^{(m+1)/m}$. R_T should remain nearly constant since only the cavity resistance is changing. For m = 2.7 for the polystyrene used, the cavity resistance would increase about 4.5 times, again assuming isothermal conditions. For nonisothermal conditions, the values would be dependent on balancing freezing times with flow and thus not accurately calculated. Note that the term $V_0 B$ decreases from 20.6 to 7.9 cm, about 2.6 times. Thus, the Ballman et al. model does in fact predict the proper behavior for flow in a simple cavity with a flow-controlling delivery system.

The next obvious step in the analysis of a single-cavity mold with a delivery system is the investigation of the pressure losses through the gate. Paulson and Frizelle [5,11] measured pressure profiles along a delivery system and in a single cavity mold using a now-patented pressure transducer

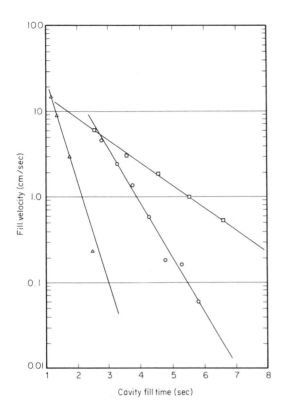

Figure 10.2-3   Fill velocity for various conditions. Squares, experiment 2, 0.150-in.-thick cavity, T_{melt} = 350°F, T_{mold} = 70°F; triangles, experiment 4, 0.050-in.-thick cavity, T_{melt} = 350°F, T_{mold} = 70°F; circles, experiment 36, 0.075-in.-thick cavity, T_{melt} = 400°F, T_{mold} = 80°F [6].

3/8 in. in diameter. The cavity-delivery system is shown in Fig. 10.2-4. In Fig. 10.2-5 is a typical pressure profile, showing a near-constant injection pressure, a rapidly increasing nozzle pressure, completion of filling of the cavity at relatively low pressure (note also that the pressure profile in the cavity at the pressure transducer is increasing linearly with respect to time during filling), and then the rather rapid hydrostatic pressure build up throughout the system during the packing stage and the decline in pressure as the part cools. The effect of gate area on both static and dynamic pressure losses is shown in Fig. 10.2-6 for several gate geometries. Further analysis of the losses in the gate area indicates that the gate can be replaced with an equivalent entrance correction factor that is shear rate dependent. This is shown for one gate geometry and two melt temperatures in Fig. 10.2-7. While Frizelle and Paulson gave no analytical form for

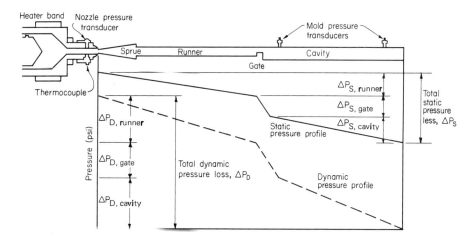

Figure 10.2-4   Schematic of dynamic and static pressure profiles in instrumented injection mold [11].

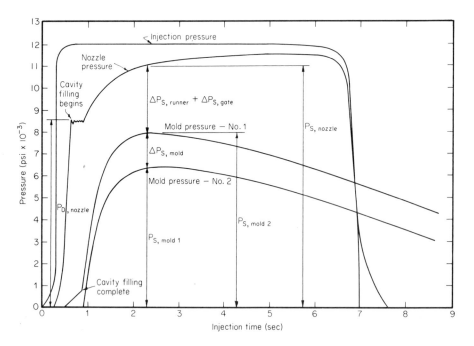

Figure 10.2-5   Typical mold pressure profile showing static and dynamic pressure losses and pressure differential across injection mold after filling is completed [11].

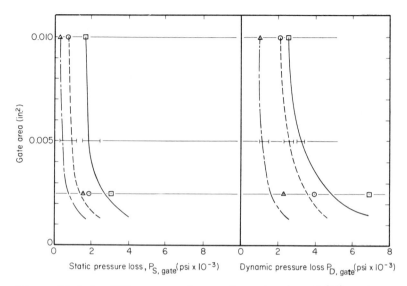

Figure 10.2-6   GPS pressure losses through gates of different cross-sections. Dash-dot line and triangles, 8000-psi injection pressure; dashed line and circles, 12,000-psi injection pressure; solid line and squares, 16,000-psi injection pressure; $T_{melt} = 450°F$; $T_{mold} = 60°F$ [11].

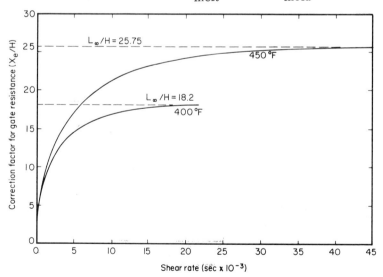

Figure 10.2-7   Length correction for GPS for a flat land gate design [11].

this factor, their data indicate that the following expression should fit rather well:

$$X_e/R = (X_\infty/R)[1 - \exp(-4bQ/\pi R^3)] \tag{10.2-23}$$

where X_e is the equivalent length, R is the hydraulic radius of the runner, X_∞ is an asymptotic length when the shear rate is very large, and b is a gate parameter, = 0 when the gate has the same radius as the runner and becomes large as the gate area decreases. Q is the volumetric flow rate in the runner, and $4Q/\pi R^3$ is the shear rate in the runner. It is also apparent that b is melt temperature dependent and probably mold temperature dependent as well. Thus, the gate resistance can be added to the Ballman et al. resistance equation as part of the total resistance. In this case, however, it is added as an equivalent length of runner. As a further heuristic confirmation of the Ballman et al. theory, Frizelle and Paulson give the dynamic pressure loss per inch of cavity length as a function of cavity depth (Fig. 10.2-8). For an isothermal system, it can be shown that the pressure loss should be proportional to the cavity depth d to the (m + 2)/m power. If m = 2.7, according to Ballman et al., the pressure loss should decrease about 3.4 times with a doubling of the cavity depth. According to Fig. 10.2-8 for melt temperature of 400°F, the pressure decreases 4.2 times by doubling the depth from 0.060 to 0.120 in. and 4.8 times from 0.040 to 0.080 in. This supports the theory adequately, considering the assumption of isothermal temperatures and the fact that a value for m has been assumed. m = 1.7 gives a value of 4.5 for a doubling of cavity depth.

So far, a macroscopic analysis of the mold-filling process has been considered. Several major efforts have been undertaken to determine mold filling using the phenomenological or transport equation approach. Barrie [12-14] bases his rather simple analysis on the spreading-disk theory of Cogswell and Lamb [15] for isothermal flow of a power-law fluid. Barrie's model is primarily applied to radial or divergent flow from a center gate into a disk-shaped mold. He states that the pressure effect in shear is given as

$$P_s = \frac{2K(6Q)^n}{x^{1+2n}(2\pi)^n} \frac{(R^{1-n} - R_0^{1-n})}{1 - n} \tag{10.2-24}$$

where K is the viscosity factor and n is the viscosity index for a power-law fluid, x is the thickness of the disk, and R and R_0 are the radius of the wave front and the sprue-gate radius. Q is the volumetric flow rate.

The pressure effect in tension (e.g., the hoop stress caused by stretching the melt) is given as

$$P_H = \left(\frac{\lambda Q}{4\pi x}\right)\left(\frac{1}{R_0^2} - \frac{1}{R^2}\right) \tag{10.2-25}$$

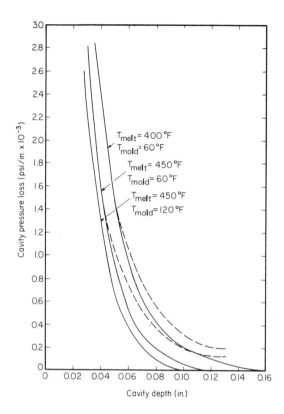

Figure 10.2-8   Effect of cavity depth on cavity pressure loss per unit
length. Solid line, static pressure loss; dashed line, dynamic pressure
loss, GPS [11]

λ is the tensile viscosity, which may or may not be shear dependent, depending on the material and the interpretations of the experimenters' data.
This point was discussed in detail in Chap. 5. To correct for the nonisothermal character of the process, Barrie chooses to include a freezing
zone, defined empirically as

$$dx = C(dt)^m \tag{10.2-26}$$

where dx is the thickness differential of the frozen plastic, dt is the time
since injection, and C and m are arbitrary empirical constants. According
to simple heat conduction (see Chap. 6), m should have a value of 1/2 since
the thermal wave front proceeds into a material proportional to the square
root of time. Barrie finds, however, that m = 1/3 fits his experimental
data more accurately. The reason for this necessary empiricism is seen

by comparing Barrie's data for m = 1/2 and m = 1/3 with the pressure drop-flow rate relations of Figs. 10.2-9 and 10.2-10A. The data used to obtain these results are given in Table 10.2-1 for an ICI propylene-ethylene copolymer with a 3 melt index. The power-law index n = 0.22 is the reciprocal of that used in the earlier work of Ballman et al. The major criticism of Barrie's work is the empirical inclusion of the nonisothermal effects.

Harry and Parrott [16] extended this type of analysis to include the differential equation form for power-law flow. They assumed a linear pressure drop and one-dimensional flow into a rectangular cavity, thus ignoring

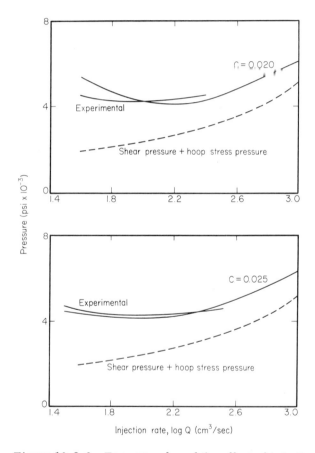

Figure 10.2-9   Two examples of the effect of injection rate on cavity pressure. Experimental data show that true pressure is greater than sum of shear-induced pressure and hoop stress pressure. C is empirical coefficient to account for "freezing" [12].

the nonisothermal effect on pressure drop and the edge effects in the rectangular cavity. Nevertheless, as can be seen from their computed curves in Figs. 10.2-10B and 10.2-10C, their filling rates and velocities for filling are very similar to the curves obtained by Ballman et al. Replotting the velocity data, for example, on semilogarithmic paper yields straight lines. Further, note that the position of the wave front is nearly exponential with time, as predicted by Eq. (10.2-19). Pearson [17] proposes a differential method for solving the radial flow problem that reduces the equations to sets of difference equations but leaves them for the reader to solve, even though the method of solution entails much iteration and simultaneous equation solving.

Kamal and Kenig [18,19] considered the Pearson approach for edge flow into a 180° disk, as shown, along with the other popular geometries, in Fig. 10.2-11. They used a Dow polyethylene for their experimental observations and for the material properties needed in their theoretical model. Briefly, they considered the nonisothermal flow of a power-law fluid and included the continuity, momentum, and energy equations in the following forms:

continuity: $\partial(\rho V r)/\partial r = 0$ (10.2-27)

motion: $A_r(\partial P/\partial r) = Re^{-1}[\partial(\partial V/\partial z)^n]/\partial z$ (10.2-28)

energy: $(\partial T/\partial t) + A_r V(\partial T/\partial r) = (PrRe)^{-1}(\partial^2 T/\partial z^2)$

$$+ (Br/RePr)(\partial V/\partial z)^{1+n}$$

(10.2-29)

All the terms in the above three equations are dimensionless. r is the ratio of the local radius to the overall radius of the cavity (R_0), z is the ratio of the distance from the centerline to the half-thickness of the cavity (h), and t is defined as $t = t'V_e/h$, where t' is real time and V_e is the kinematic velocity, $V_e^2 = P_f/\rho_m$, where P_f is the injection pressure at the entrance to the cavity at the end of filling and ρ_m is a reference density. $V = V'/V_e$, where V' is the velocity in the radial direction, $V' = V_r(z,t)$. P is the ratio of actual pressure to P_f, T is a dimensionless ratio of unaccomplished temperature change (see Chap. 6 for this type of dimensionless ratio), and $A_r = h/R_0$, the ratio of cavity half-thickness to cavity radius. There are three dimensionless groups in Eqs. (10.2-27) to (10.2-29) which need definition:

Re (Reynolds number) $= V_e^{2-n} \rho_m h^n/K$ (10.2-30)

Pr (Prandtl number) $= Kc_p h^{1-n}/kV_e^{1-n}$ (10.2-31)

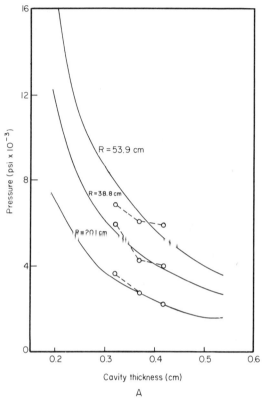

A

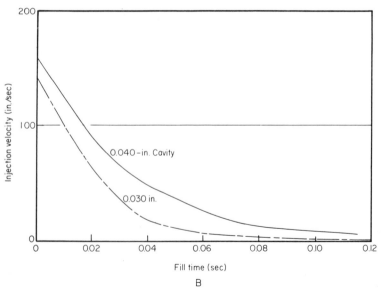

B

476

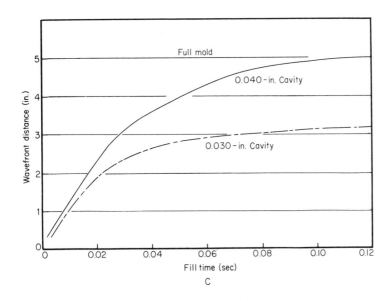

Figure 10.2-10   A: Comparison of calculated and experimental pressure profiles as functions of cavity thickness. Dashed lines and dots, experimental; solid lines, calculated values [13].   B: Computed average fill velocity as a function of fill time for 0.040- and 0.030-in. Cavity thicknesses [16].   C: Fill distance as a function of injection time for two cavity thicknesses. Note that these are computed curves [16].

Table 10.2-1

Physical Data for ICI Propylene-Ethylene Copolymer [14]

Experimental parameters	Material parameters
$R = 22.9$ cm	$K_T = 2.58 \times 10^5$ cgs units
$A = 1650$ cm^2	$n = 0.22$
$x = 0.25$ cm	$\eta_T = 1.7 \times 10^4$ P
$R_0 = 0.24$ cm	$\alpha = 9 \times 10^{-4}$ cm^2/sec
$T = 260°$C	$T_0 = 165°$C
$\theta = 30°$C	

Q is the experimental variable (cm^2/sec).

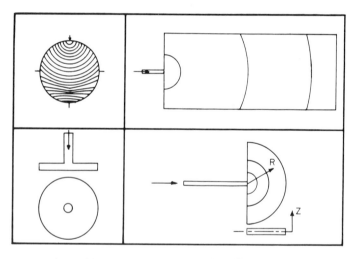

Figure 10.2-11 Characteristic mold-filling patterns used in theoretical analysis of mold filling. Note that the lower two represent mathematically simpler filling patterns than the top two [18].

$$\text{Br (Brinkman number)} = KV_e^{1+n}h^{1-n}/k(T_1 - T_0) \qquad (10.2\text{-}32)$$

T_1 and T_0 are the initial melt temperature and the mold temperature, respectively, and K and n are the power-law constants, respectively. k is the thermal conductivity of the melt, and c_p is its specific heat. Kamal and Kenig examine many of the possible types of boundary conditions that can be placed on these equations and determine that the following are relevant (and yield tractable computer solutions):

1. The velocity of the melt at the mold wall is zero.
2. The velocity profile is symmetrical at $z = 0$, the centerline.
3. At the interface between the polymer and the wall, there is a convective thermal resistance. Thus, there is a finite (and constant) heat transfer coefficient.
4. The melt temperature at the entrance to the cavity is constant.
5. The temperature profile is symmetrical about the centerline, $z = 0$.
6. In the region around the melt front, the radial temperature gradient is zero.
7. If the melt temperature is below the solidification temperature, the velocity is zero.
8. The pressure at the entrance to the cavity is known as a function of time.
9. The pressure at the advancing front is equal to atmospheric pressure.
10. A heat balance including the latent heat of fusion is constructed only if the materials are crystalline.

Kamal and Kenig construct a difference form for these equations and obtain theoretical temperatures, pressures, and wave-front positions for the half-round disk. They then compare their computer results with experimental data obtained using Dow polyethylene, a mold wall temperature (T_0) of 80°F, a melt temperature of 350°F (case 1) or 400°F (case 2), and a final filling pressure at the entrance of 450 psi (case 1) or 250 psi (case 2). The experimental curves of $P_f(t)$ are shown as Fig. 10.2-11. Note that these pressure curves are very similar to those of Frizelle and Paulson discussed earlier. The rate of filling of the cavity (R/R_0 vs. time) is shown for the two cases in Figs. 10.2-12 and 10.2-13. As can be seen, the theory predicts longer filling times as the melt temperature is decreased. The error of prediction of the filling rates in the early stages of filling is probably due to the absence of the inertial term in the equation of motion and the absence of the hoop stress due to extensional stresses. The latter factor was an important argument put forth by Barrie in his analysis.

In Figs. 10.2-14 and 10.2-15 are the experimental and predicted pressure profiles at the end of filling for the two cases. As can be seen, the theory does not predict a linear pressure drop with distance, as has been assumed by Harry and Parrott and implied by the other workers, including Ballman et al. Surprisingly, the Kamal and Kenig model does not match the very nonlinear form for the experimental data either. It would seem, therefore, that the effect of heat removal from the cavity during filling has not been sufficiently accounted for here also, for with a very

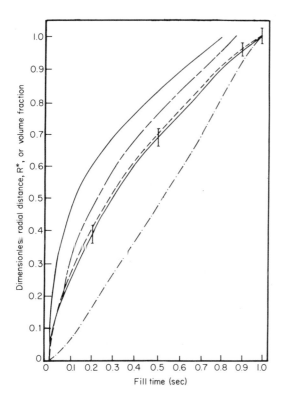

Figure 10.2-12   Progression of melt wave front for radial flow in half-round mold. Solid line, theoretical; dotted line, theoretical with $R^* = 0.2$; dashed line, experimental photography; solid line with vertical bars, experimental with pressure transducer detection; dash-dot line, volume fraction of fill [18].

rapid rate of heat removal, the cavity should be filling more slowly than predicted, and the pressure profile should be considerably more depressed near the gate than predicted.

   Gogos and coworkers [20, 21] have returned to the earlier work of Ballman et al. for guidance in establishing the initial conditions for flow into a disk-shaped mold. Recall that Kamal and Kenig were required to specify $P_0(t)$, the pressure at the entrance to the cavity. Gogos and co-workers now include the isothermal runner in their analysis, as shown in Fig. 10.2-16. In their analysis, they neglect normal stresses in both the radial and angular directions (the latter in the momentum equations only) and axial heat conduction. Thus, their three equations become

continuity: $(\rho/r)\ \partial(rv)/\partial r = 0$ (10.2-33)

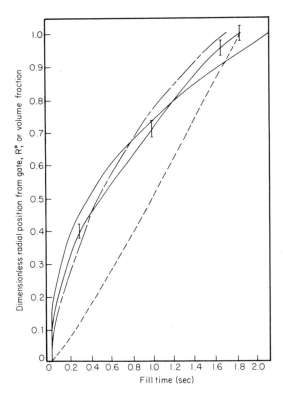

Figure 10.2-13 Progression of melt wave front for radial flow in half-round mold. Solid line, theoretical; dashed line, experimental via photography; solid lines with vertical bars, experimental with pressure transducer detection; dash-dot line, volume fraction of fill [18].

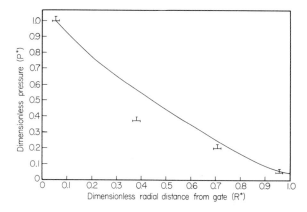

Figure 10.2-14 Computed and experimental pressure profiles at the instant of mold fill for half-round mold. Solid line, theoretical; vertical-horizontal bars, experimental pressures obtained with pressure transducers [19].

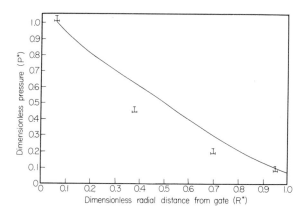

Figure 10.2-15 Computed and experimental pressure profiles at the instant of mold fill for half-round mold. Note: Injection speed approximately half that for Fig. 10.2-14. Solid line, theoretical; vertical-horizontal bars, pressures obtained experimentally with pressure transducers [19].

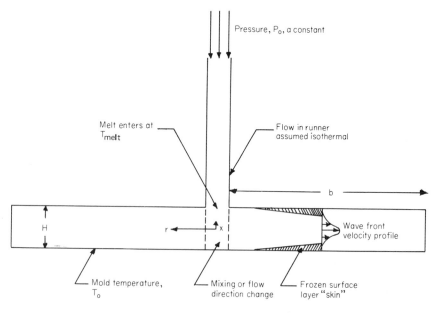

Figure 10.2-16 Characteristic geometry for well-head flow into a radial disk mold. Note the assumption of a developing frozen surface layer and the resulting distortion of the melt wave front [20].

momentum: $\partial \tau_{rx}/\partial x = -\partial p/\partial r + (\rho v^2/r)$ (10.2-34)

energy: $\rho c_p [\partial T/\partial t + v(\partial T/\partial r)]$ (10.2-35)

$$= k(\partial^2 T/\partial x^2) - \tau_{\theta\theta}(v/r) - \tau_{rx}(\partial v/\partial x)$$

where $\tau_{rx} = -K(\partial v/\partial r)^n$ and $\tau_{\theta\theta} = \eta(\partial v/\partial x)$. The term τ_{rx} is the shear
stress, assumed here to be power-law related to the velocity gradient. The
term $\tau_{\theta\theta}$ is the normal stress in the angular direction that gives rise to
hoop stresses, and Gogos assumes it to be directly proportional to the
velocity gradient in the x-direction, e.g., a Newtonian elongational viscosity.
It is perhaps worth a moment to examine the differences in these equations
and those of Kamal and Kenig. Gogos has included the radial momentum
convection term on the right-hand side of the momentum equation to help
compensate for the inertial effects near the gate (for small radii). The
second term on the right-hand side of Gogos' energy equation represents the
effect of hoop stresses. The third term on the right-hand side of Gogos'
energy equation is in essence the same as the second term on the right-hand
side of Kamal and Kenig's energy equation.

More important is the difference in boundary conditions. While most
of Kamal and Kenig's conditions are the same as those used by Gogos, Gogos
assumes no thermal resistance between the melt wave front and the mold
wall and instead assumes heat transfer between the wave front and the air
ahead of it. Kamal and Kenig assume no such heat transfer. Furthermore,
Gogos assumes that P_0 at the entrance to the runner feeding the disk mold
is constant, in the tradition of Ballman et al. In Fig. 10.2-17, it is shown
that Gogos' computer simulation predicts very similar profiles for wave-
front location regardless of the mold wall temperature. These curves have
a shape similar to that predicted by Kamal and Kenig for a time-dependent
injection pressure. Again, since the two systems are quite a bit different
(Kamal and Kenig having melt in contact with the edge of the cavity throughout the flow path and Gogos having no edge contact), it is difficult to compare pressure profiles. However, in Fig. 10.2-18, note that Gogos predicts
a fourfold decrease in filling time with a twofold increase in injection pressure, whereas Kamal and Kenig have found that a 1.8-fold increase in injection pressure decreases the fill time only by a factor of 2. Note also
that Gogos is using PVC physical properties and pressures of 4000 to 15,000
psi and Kamal and Kenig used polyethylene at 250- to 450-psi pressures.

Temperature profiles might give some indication of the freezing
effect that is occurring in the mold. Note that Gogos' melt front advance
seems to be nearly independent of mold temperature. One reason for this
is shown in Fig. 10.2-19 for the temperature profile at the end of filling.
Note that the computed temperature near the entrance to the cavity and
close to the wall of the mold is above the melt temperature entering the
cavity. This is due to the extensive viscous dissipation of the melt as it is

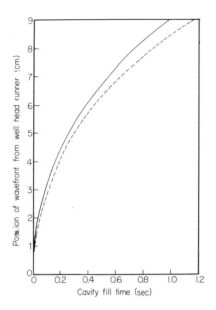

Figure 10.2-17   Position of melt wave front in well-head radial flow as function of time. Theoretical results. Solid line, nonisothermal flow in cavity (isothermal runner); dashed line, isothermal flow in cavity; mold temperature, $T_0 = 30°C$; material, PVC at $T_{melt} = 202°C$; injection pressure, 15,000 psi; cavity thickness, $H = 0.3175$ cm [21].

entering the cavity. This dramatic increase in temperature would tend to remelt any material that would freeze against the mold surface in that region. Thus, viscous dissipation is countering the sensible heat loss to the mold, and the material would behave as if it were isothermal. Now compare these temperature profiles with those of Kamal and Kenig in Fig. 10.2-20. There is no evidence of viscous dissipation, and thus one would expect significant freezing to take place. This effect should be seen in decreased filling time and a very nonlinear pressure profile through the disk at the instant of complete fill, and as mentioned, this is what is being seen in Kamal and Kenig's data. Further support of this interpretation is given by Gogos in Fig. 10.2-21, where the thickness of the frozen layer at 15,000 psi is nearly independent of the mold wall temperature. Two other extensive studies should be noted. As mentioned earlier, the Cornell group is concentrating on the prediction of pressure and force prediction during the filling of simple disk molds. Although they are approaching the problem following the Kamal-Kenig models, their results are depicted in graphical form for the mold designer and machinery builder. Lord and Williams, on the other hand, have concentrated on coupling the cavity-filling arithmetic

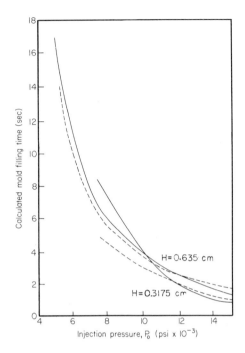

Figure 10.2-18 Filling times for well-head radial flow as function of injection pressure for two values of cavity thickness. Material, PVC, 202°C; cavity radius, 9 cm; solid line, nonisothermal flow, $T_0 = 30°$C mold temperature; dashed line, isothermal flow, $T_0 = T_{melt}$. Note: These are theoretical values [21].

to flow in the sprue and runner system [22,23]. For the cavity model, they consider the solution of the coupled heat and flow equations as they apply to flow in a runner-flow-controlled end-gated plaque mold. Thus, their work should be the advanced analysis of the earlier Ballman et al. As shown in Fig. 10.2-22, they compare their prediction with experimental data for ABS. It would be unfair not to point out that several of their computer analyses, where mold temperature, plaque thickness, type of resin, and melt temperature have been changed, yield much closer agreement with experimental data. Nevertheless, the dotted line shown in Fig. 10.2-22 represents application of the empirical equation (10.2-13) with a power-law index of 0.25. From data points for two plaque thicknesses, Lord and Williams show a pressure differential decrease of about 2.6, with a doubling of the plaque thickness. Applying the Ballman et al. analysis, a pressure differential decrease of 2.4 is predicted. Furthermore, the Lord-Williams model predicts an increase in pressure differential beyond an injection rate of about

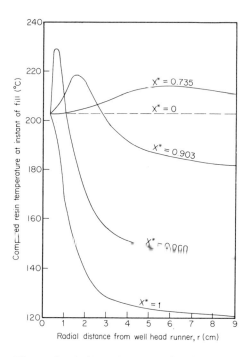

Figure 10.2-19   Theoretical values for the temperature field along the radial direction and across the thickness of a well-head radially filled mold. Material, PVC at 202°C melt temperature; mold temperature, $T_0 = 30°$C; injection pressure, P_0, 15,000 psi; mold radius, 9 cm; injection time, 0.97 sec; cavity thickness, H = 0.3175 cm [21].

2 in.^{3}/sec on their plaque.   Experimental data show either continuing decrease in pressure differential or a plateauing.   Lord and Williams give no argument regarding this disparity.   Incidentally, the most dramatic argument against the increasing pressure differential effect is in structural foam molding, where pressure differentials continue to drop with increasing injection speed to the point where a 0.25 in. thick plaque filled with GP styrene foam at 30.9 in.^{3}/sec shows a 400-psi pressure differential, whereas the same plaque filled at 540 in.^{3}/sec shows a 280-psi pressure differential in 12 in. [24].   Arguments have been forwarded that indicate that the characteristic flow mechanisms are changing from the characteristic "volcano" flow, assumed by all modelers discussed in this section, to plug flow.   If these arguments can be substantiated, it would appear that the application of the above models will be limited to very slow injection speeds.   To the contrary, the injection molding industry seems to be moving toward ever-increasing speeds in order to improve production economics.

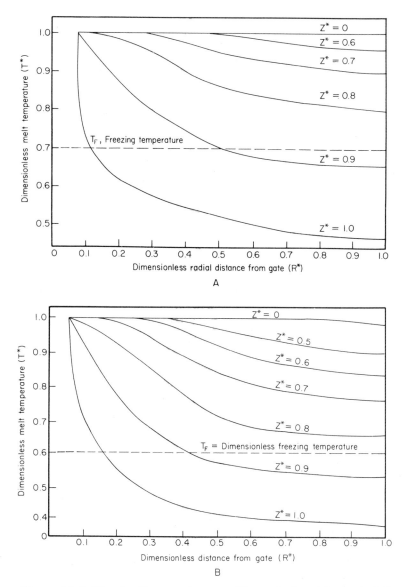

Figure 10.2-20 A: Temperature profiles calculated for radial flow from gate at instant half-round mold is filled. Dashed line represents freezing temperature T_F in dimensionless units. B: Temperature profiles calculated for radial flow from gate at instant half-round mold is filled. Slower injection rate than for A. Dashed line represents freezing temperature T_F in dimensionless units [20].

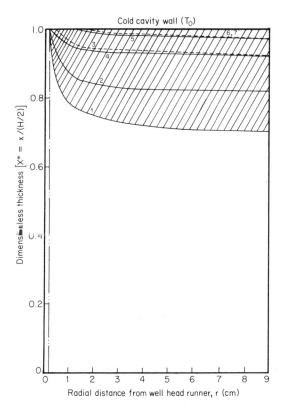

Figure 10.2-21 Computed skin thickness as a function of distance along the radius in a well-head radial mold. Case 1: $P_0 = 5000$ psi, $H = 0.635$ cm, $T_{melt} = 202°C$, $T_0 = 30°C$. Case 2: Same as case 1, but $P_0 = 7500$ psi. Case 3: Same as case 1, but $P_0 = 11,600$ psi. Case 4: Same as case 1, but $T_0 = 100°C$. Case 5: $P_0 = 15,000$ psi, $H = 0.3175$ cm, $T_{melt} = 202°C$, $T_0 = 30°C$. Case 6: Same as case 5, but $T_0 = 100°C$. Case 7: $P_0 = 7500$ psi, $H = 0.635$ cm, $T_0 = 220°C$, $T_0 = 30°C$ [21].

        What, if any, conclusions can be drawn from the work being done in simulation of the filling process at this point? Two basic geometries have been studied in some detail: the rectangular cavity filled from one end and the disk mold and its variants. It is apparent that these geometries are basically one-dimensional: rectangular coordinates in the case of the rectangular cavity and radial in the case of the disk mold. It would seem that, contrary to Kamal and Kenig's viewpoint, the entire system, including accumulator, nozzle, sprue, runner, gate, and cavity, must be considered. If this is not done, the pressure at the gate end of the cavity must be

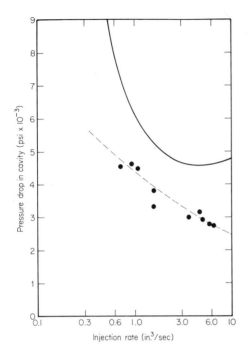

Figure 10.2-22   Comparison of computed and experimental pressure drop in plaque mold cavity for ABS; $T_{melt} = 400°F$, $T_{mold} = 110°F$, $2L = 0.125$ in. (plaque thickness). Solid line, computed results; closed points, experimental data; dashed line, empirical equation (10.2-13) with $n = 0.25$ [23].

accurately measured for each cavity geometry. Furthermore, omission of the inertial terms at the gate can lead to erroneous results, as Kamal and Kenig admit and Gogos has shown. However, Gogos' model has not been verified experimentally. The rectangular cavity offers essentially constant elongational strain at the wave front, whereas the Barrie and Gogos geometries show a rapidly decreasing strain rate (proportional to $1/r^2$) at the wave front. Thus, there is no way of relating the latter geometry to the Ballman et al. and the Harry and Parrott results. The effect of shear heating is graphically illustrated by Gogos in that freezing seems to be inhibited, yet Ballman et al. demonstrate quite graphically that the rate of advance of the wave front decreases exponentially and that the rate is a strong function of the mold temperature.

Keep in mind that the major objective of any simulation is to predict the point of filling of the cavity (any cavity) as a function of melt temperature, mold temperature, gate restriction, injection pressure, volumetric rate of injection, and material properties. The overall cycle time depends on this information, and, more importantly, the morphology of the material is strongly affected by injection speed, mold temperature, melt temperature, and the other parameters that are included in most models. Gogos' thermal profiles, if experimentally supported, indicate that the processing conditions of thermally sensitive materials such as PC, ABS, PVC, and PMMA must be carefully regulated to prevent discoloration from thermal degradation owing to the high temperatures of viscous dissipation. And certainly, not all production parts are characterized by rectangular or circular geometries and constant wall thickness.

10.3 Manifolding in Injection Molds

Before considering the packing or orientation of the melt in the cavity, recognize that the majority of injection molds are multiple-cavity molds. Consider first an even number of cavities having identical shape and volume in the mold block in a single-nozzle machine. The objective in mold design is to place as many cavities in as small a mold base as possible and then have all the cavities fill at essentially the same rate. In this way, the material characteristics should not vary from one cavity to another, and shrinkage should be predictable from every cavity. Certainly, with 2 cavities, the balancing of flow into the cavities is relatively easy. With 4 cavities, the normal procedure is to place all four equidistant from the sprue. With 8 cavities, however, unless the cavities are very small relative to the mold surface, the standard practice is to put 2 cavities at the end of each of 4 runners. In essence, then, there are short runner spurs at the end of the main runners. Sixteen cavities can be handled in a similar manner, except that at the end of each of the main runners we have 2 short spurs and 2 additional spurs extended from each of these runners. The configurations for 32, 64, 128, and so on are very similar, as shown in Fig. 10.3-1 [25]. This type of runner system is known as a "balanced runner system," and sizing of the runners is relatively easy. The objective, of course, is to make the overall pressure drop per unit length of runner in each runner the same regardless of the number of branches. Assume steady flow:

$$[(P_i - P_0)/4K]^{1/n} = (x/R_0)^{1/n}(4Q/\pi R_0^3)/B^{1/n} \qquad (10.3\text{-}1)$$

where P_i is the injection pressure, P_0 is the pressure at the open runner ahead of the flow front, x is the distance down the runner, R_0 is the radius of the runner, Q_0 is the volumetric flow rate into the runner at the injection point, K and n are the power-law indices, and $B^{1/n} = n \cdot 2^{2+1/n}/(3n + 1)$ [27]. The objective is to maintain $(P_i - P_0)/x = dP/dx$. Rearranging Eq. (10.3-1),

$$dP/dx = K'Q_0^n/R_0^{3n+1} \qquad (10.3\text{-}2)$$

where $K' = (4/\pi)^n(4K/B)$, a constant that is independent of the number of splits in the flow stream. Further rearrangement, holding dP/dx constant, yields

$$R_0 = K''Q_0^{n/(3n+1)} \qquad (10.3-3)$$

where $K'' = [K'/(dP/dx)]^{1/(3n+1)}$. If the volumetric flow is split into two side runners, $Q_1 = Q_0/2$, e.g., the flow rate Q_1 in each runner equals half that in the main runner. Therefore, the radius of the side runner must be related to that of the main runner in accordance with

$$(R_0/R_1) = (Q_0/Q_1)^{n/(3n+1)} = 2^{n/(3n+1)} \qquad (10.3-4)$$

For a Newtonian fluid, $n = 1$, and thus the radius in the side runner should be about 80% of that in the main runner. For a highly non-Newtonian fluid, such as LDPE where $n = 1/3$, the side runner is about 90% of the main runner. Note that any change in flow direction consumes additional kinetic energy, as does expansion and contraction. Therefore, if the same radius is maintained in the side runner as in the main runner, the pressure loss will be less, but an expansion loss must be included. These losses are considered shortly. The important point is that if the flow is split in order to fill a 16-cavity mold, there are 4 changes in runner radius. Thus, the radius of the side runner into the cavity must be exactly half that in the main runner for a Newtonian fluid and about 60% of that for a non-Newtonian fluid with $n = 1/3$. If the runners are not round, R represents a hydraulic radius, defined as

$$R_h = \text{(cross-sectional area)/(wetted perimeter)} \qquad (10.3-5)$$

If cooling of the runner system is uniform across the mold, nonisothermal effects are distributed uniformly throughout the system and should not act to "unbalance" the flow.

There are many applications, however, that require unbalanced flow. One example would be a two-cavity mold, where the volumes of the two cavities are considerably different. To fill the two cavities simultaneously, it is necessary to unbalance the runner system. With reference to Ballman et al., one way is to increase the length of the runner on the smaller cavity, thus causing more resistance in that side of the mold. Another way is to change the runner dimension on the smaller cavity if changing the runner length is not feasible. As an example of this, consider the smaller cavity to be exactly half the volume of the larger cavity and the lengths of runners to be the same. Equation (10.3-4) and the results given below the equation can be applied immediately. For a Newtonian fluid, the runner on the larger cavity should be about 20% larger than that on the smaller cavity. For $n = 1/3$, it need be only about 10% larger. See Fig. 10.3-1B.

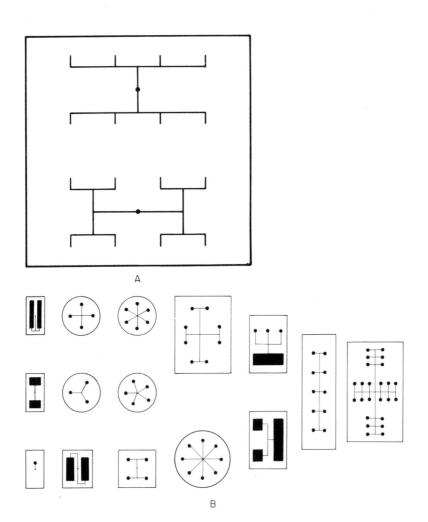

A

B

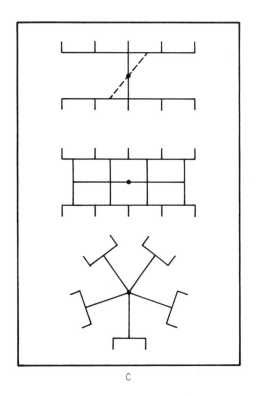

C

Figure 10.3-1 A: Two examples of eight-cavity molds fed from a sprue
and runner system. The top example is a typical unbalanced "H" runner
system with filling of cavities nearest the gate occurring ahead of those
farthest from the gate. The bottom example is a typical balanced runner
system characterized by equal flow distances from sprue to each cavity [26].
B: Typical cavity configurations for balanced and unbalanced runner systems. Note, for example, that eight cavities can be displayed either radially or in cross-pattern to achieve balancing [28]. C: Three examples of
10-cavity molds fed from a single sprue and several variations on runner
designs. The top example is an "H" unbalanced runner system, with the 2
cavities opposite the main runner filling first. A variation, to achieve
more uniform filling, is shown as a dotted main runner line. The second
example attempts to correct unbalanced flow by using a main runner parallel to the "H" runners. This design is wasteful of material in the runners.
The third design utilizes a balanced runner system characterized by equal
flow distances from sprue to each cavity [26].

Consider now one way of effectively increasing runner resistance without decreasing runner radius. It is normally quite difficult to balance a runner system with unbalanced volume cavities simply by changing the radii of the runners. However, the information obtained by Frizelle and Paulson on gate resistances can be used to increase the effective length of the runner. Reworking Eq. (10.3-2) for a constant runner radius and a variable runner length at a constant pressure drop,

$$L = K'''/Q_0^n \qquad\qquad (10.3-6)$$

where $K''' = K'/(\Delta P)(R_0^{3n+1})$. Recalling Eq. (10.2-23), note that the gate adds an effective length to the runner, so that Eq. (10.3-6) can be written as

$$L_t = L_R + L_g = K'''/Q_0^n \qquad\qquad (10.3-7)$$

where

$$L_g = L_\infty [1 - \exp(-4Q_0 b/\pi R^3)] \qquad\qquad (10.3-8)$$

or, in somewhat simplified terms,

$$L_g = L_\infty [1 - \exp(-BQ_0)] \qquad\qquad (10.3-8a)$$

where $B = 4b/\pi R^3$. Unfortunately, this is a transcendental equation in Q_0, and thus the relationship between the gate parameters b and L_∞ and the flow rate cannot be explicitly extracted. Plots of Q_0 vs. B for several values of L_∞/L_R and n are shown as Fig. 10.3-2 to 10.3-5. The effect of gate restriction is dramatic. This supports the hypothesis that gate control of flow rates into unbalanced cavities is quite important in achieving uniform fill and minimum filling times.

Consider now flow into an unbalanced runner system. This is a manifold runner system, and an example is shown as Fig. 10.3-6 [29]. One of the earliest studies on manifolds was that of Keller [30] in which the consideration of pressure rise in manifolds was considered for gas burners. His analysis, extended by Acrivos et al. [31], is based on simple application of the Bernoulli principle to flow in a pipe. If s is the distance from the dead end of the manifold, the pressure rise is given as

$$dP = -\rho d(V^2)/2g + f(ds/D)(\rho V^2) \qquad\qquad (10.3-9)$$

The first term represents the deceleration, and the second term is the frictional contribution. V is the longitudinal velocity in the manifold, ρ is its density, f is the coefficient of friction for the fluid in the manifold, and D is the manifold diameter. Rearranging,

$$dP/ds = -(\rho/2g)(dV^2/ds) + (f/D)(\rho V^2) \qquad\qquad (10.3-10)$$

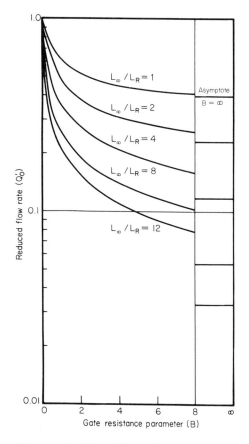

Figure 10.3-2 Effect of gate resistance on flow rate Q_0'. Note: Constant pressure drop per length of runner assumed; $n = 1$, Newtonian fluid.

Assume for the moment that the holes can be considered continuous, such as a slot. Then the volume discharged through a length of slot ds is equal to the decrease in quantity flowing in the manifold:

$$dV = (kwV_1/A)\ ds \qquad (10.3\text{-}11)$$

where k is the coefficient of discharge through the slot (or hole continuum), w is the width of discharge slot in the manifold, V_1 is the velocity of discharge from the manifold, and A is the cross-sectional area of the manifold. If the pressure at the end of each portion of slot in the manifold is constant (or zero), the total pressure drop is given as

$$dP = \rho d(V_1^2)/2g \qquad (10.3\text{-}12)$$

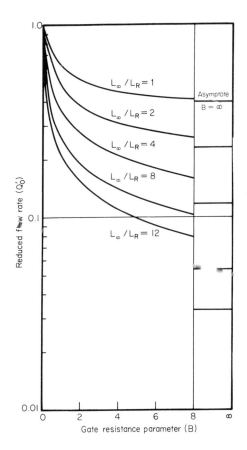

Figure 10.3-3   Effect of gate resistance on flow rate Q_0'. Note: Constant pressure drop per length of runner assumed; n = 0.75, power-law fluid.

Through substitution, we find that

$$(A/kw)^2 (dV/dx)(d^2V/ds^2) = -V(dV/ds) + (fgV^2/D) \qquad (10.3\text{-}13)$$

While this equation cannot be directly integrated, it can be integrated numerically, as shown in Fig. 10.3-7. Furthermore, Keller shows that for no friction, f = 0, the velocity of discharge is given simply by

$$V_1 = [V_0/\sin(kR)]\cos(kRs/L) \qquad (10.3\text{-}14)$$

where V_0 is the inlet velocity, k is again the coefficient of discharge, R is the ratio of the sum of areas of all discharge openings to the cross-sectional area of the manifold, and s/L is the dimensionless distance from the dead

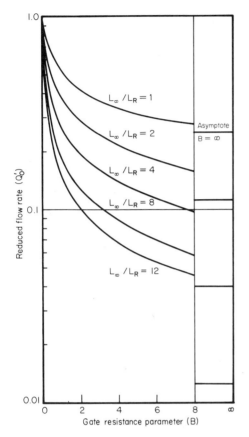

Figure 10.3-4 Effect of gate resistance on flow rate Q_0'. Note: Constant pressure drop per length of runner assumed; $n = 0.5$, power-law fluid.

end. Obviously, the pressure drop for each hole can be obtained by substituting this equation into Eq. (10.3-12) and integrating, thus yielding

$$\Delta P = \rho [V_0/\sin(kR)]^2 \cos^2(kRs/L)/2g \qquad (10.3-15)$$

Keller recommends replacing the diameters with hydraulic diameters in the case of noncircular manifolds. Acrivos et al. elaborate on this model for turbulent flow.

Consider now the effect of additional resistance beyond the manifold hole, as is the case with melt flowing down the manifold and into the various runners. Consider Fig. 10.3-8, which represents the injection point

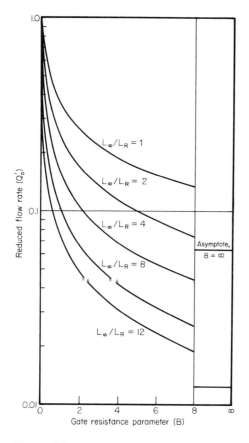

Figure 10.3-5 Effect of gate resistance on flow rate Q_0'. Note: Constant pressure drop per length of runner assumed; $n = 0.25$, power-law fluid.

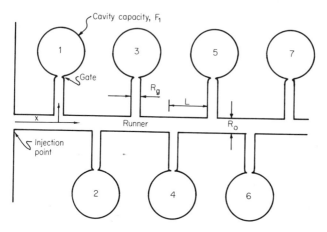

Figure 10.3-6 Schematic for manifold runner system [29].

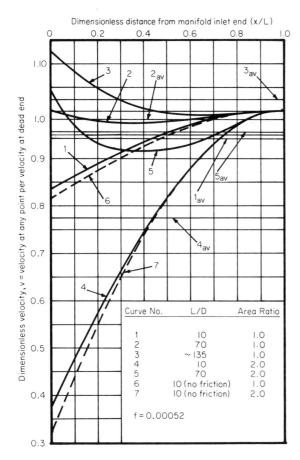

Figure 10.3-7   Velocity of discharge for a Newtonian turbulent fluid flowing into a dead-end manifold. Note the dramatic effect of L/D [30].

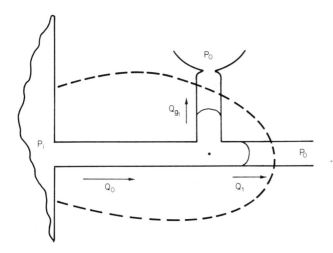

Figure 10.3-8   Schematic for mass and momentum balance on manifold cavities [29].

and the first side runner-main intersection. In the runner prior to the first port, the momentum balance is given as before as

$$[(P_i - P_0)/4K]^{1/n} = (x/R_0)^{1/n}(4Q_0/\pi R_0^3)/B^{1/n} \qquad (10.3\text{-}1)$$

Note that in this equation the kinetic energy loss owing to turning of the fluid into the side runner or "port" has not been included. This means, in essence, that the first term on the right-hand side of Eq. (10.3-9) is neglected. This implies that the change in velocity is small compared with the frictional losses. This interpretation is opposite of that of Keller and Acrivos et al. but seems justified for the very high viscosities of these fluids. The solution of the differential form of the equations has not been attempted to date for power-law fluids. As the fluid flow front passes the first port, the fluid stream splits. The overall mass balance around the port-runner intersection can be written as

$$Q_0 = Q_{g1} + Q_1 \qquad (10.3\text{-}16)$$

At the port-runner intersection, momentum balances can be made for the port and runner:

$$[(P_g - P_0)/4K]^{1/n} = (x_{g1}/R_{g1})^{1/n}(4Q_{g1}/\pi R_{g1}^3)/B^{1/n} \qquad (10.3\text{-}17)$$

and

$$[(P_1 - P_0)/4K]^{1/n} = (x_1/R_1)^{1/n}(4Q_1/\pi R_1^3)/B^{1/n} \qquad (10.3\text{-}18)$$

Here x_{g1} is the effective length of the port [that is, Eq. (10.3-7)], R_{g1} is the radius of the port, x_1 is the additional distance down the runner beyond the first port, and R_1 is the radius of the runner beyond the first port. As a simple illustration, consider $R_0 = R_1 = R_2 \ldots$ and the distances between ports 1 and 2, 2 and 3, and so on as the same, $0 < x_i < L$. At the intersection point, $P_{g1} = P_1$. Thus Eqs. (10.3-18) and (10.3-17) are equal, and we can write

$$[(P_{g1} - P_0)/4K]^{1/n} = (x/R_0)^{1/n}(4Q_1/\pi R_0^3)/B^{1/n}$$

$$= (x_{g1}/R_{g1})^{1/n}(4Q_{g1}/\pi R_{g1}^3)/B^{1/n} \qquad (10.3\text{-}19)$$

Assume isothermal flow and introduce some convenience notation: $P_k' = B(P_k - P_0)/4K$, $X = x/R_0$, $X' = L/R_0$, $G_k = (4Q_k/\pi R_0^3)$, $x_k' = x/x_{gk}$, and $r_k = R_{gk}/R_0$. Here the subscript "k" applies to the port numbers (such as "g1") and runner lengths (such as "n"). P_k' represents a bulk shear stress and G_k the volumetric flow rate. Now Eq. (10.3-19) becomes

$$(P'_{gi})^{1/n} = x^{1/n} G_1 = (X/x'_1)^{1/n} G_{g1} / r_1^{3+1/n} \qquad (10.3-20)$$

Then

$$G_{g1} = G_1 (r_1^{3n+1} x'_1)^{1/n} \qquad (10.3-21)$$

and Eq. (10.2-16) can be rewritten as

$$G_0 = G_{g1} + G_1 = G_1 [1 + (r_1^{3n+1} x'_1)^{1/n}] \qquad (10.3-22)$$

And since G_1 can be found in terms of G_0, the shear stress P'_{g1} can be written in terms of the shear stress at the inlet to the runner system:

$$P'_{g1} = G_0^n \{ X/[1 + (r_1^{3n+1} x'_1)^{1/n}] \}^n \qquad (10.3-23)$$

Now split the overall pressure drop between the first port-runner intersection and the inlet to the runner as

$$B(P_i - P_0)/4K = B(P_i - P_{g1})/4K + B(P_{g1} - P_0)/4K \qquad (10.3-24)$$

or

$$P'_{io} = X'G_0^n + P'_{g1} \qquad (10.3-25)$$

or upon substituting for P'_{g1} from Eq. (10.3-23), the pressure drop between the inlet and first port-runner intersection ($0 < X < X'$) is obtained:

$$P'_{io} = G_0^n \{ X' + X/[1 + (r_1^{3n+1} x'_1)^{1/n}]^n \} \qquad (10.3-26)$$

Moving to the second port, the pressure drop-flow rate relationship between the inlet and the first port-runner intersection is given by Eq. (10.3-20) with X replaced by X'. Furthermore, the pressure loss down the second port is given by an expression similar to Eq. (10.3-20):

$$(P'_{g2})^{1/n} = X^{1/n} G_2 = (X/x'_2)^{1/n} G_{g2} / r_2^{3+1/n} \qquad (10.3-27)$$

But $G_1 = G_{g2} + G_2 = G_2 [1 + (r_2^{3n+1} x'_2)^{1/n}]$, and thus

$$P'_{g2} = G_1^n \{ X/[1 + (r_2^{3n+1} x'_2)^{1/n}]^n \} \qquad (10.3-28)$$

Again,

$$B(P_i - P_0)/4K = B(P_i - P_{g2})/4K + B(P_{g2} - P_{g1})/4K + B(P_{g1} - P_0)/4K \qquad (10.3-29)$$

Through appropriate substitution,

$$P'_{io} = G^n X' \left\{ 1 + 1/[1 + (r_1^{3n+1} x'_1)^{1/n}]^n \right\}$$
$$+ G_0^n X / \left\{ [1 + (r_1^{3n+1} x'_1)^{1/n}][1 + (r_2^{3n+1} x'_2)^{1/n}] \right\}^n \qquad (10.3\text{-}30)$$

Continuing in this manner a general expression for pressure drop in a manifold runner system as the flow front passes the mth port can be found, assuming that all cavities are still filling:

$$P'_{io} = G_0^n X' \left(1 + \sum_{k=1}^{n-1} \left\{ \prod_{j=1}^{k} [1 + (r_j^{3n+1} x'_j)^{1/n}]^{-n} \right\} \right)$$
$$+ G_0^n X \left\{ \prod_{j=1}^{m} [1 + (r_j^{3n+1} x'_j)^{1/n}]^{-n} \right\} \qquad (10.3\text{-}31)$$

Furthermore, the pressure drop down the kth port is given by:

$$P'_{gk} = G_0^n X' \left(1 + \sum_{j=k+1}^{m-1} \left\{ \prod_{i=1}^{j} [1 + (r_i^{3n+1} x'_i)^{1/n}]^{-n} \right\} \right)$$
$$+ G_0^n X \left\{ \prod_{k=1}^{m} [1 + (r_k^{3n+1} x'_k)^{1/n}]^{-n} \right\} \qquad (10.3\text{-}32)$$

Note that nowhere in this derivation has the form for x'_k been found. As seen from Eq. (10.3-8), the equivalent length for a port-gate system is an implicit function of Q, or in this derivation, of G_{gk}, and thus an extensive computer trial and error of the algebraic equations is necessary to extract the solution for P'_{io} as a function of position. If some simplifying assumptions about the fluid and the types of port-gate systems are made, the relative effect of the port as the fluid passes it can be observed. Assume Newtonian flow (n = 1), $r_k = r_1 = r_2 = \cdots = r$ and $x'_1 = x'_2 = x'_3$. Now Eqs. (10.3-31) and (10.3-32) become

$$P'_{io} = G_0 X' \left[1 + \sum_{k=1}^{m-1} (1 + r^4 x')^{-k} \right] + X G_0 (1 + r^4 x')^{-m} \qquad (10.3\text{-}33)$$

and

$$P'_{gk} = G_0 X' \left[1 + \sum_{j=k+1}^{m-1} (1 + r^4 x')^{-j} \right] + G_0 X (1 + r^4 x')^{-m} \qquad (10.3\text{-}34)$$

Consider these in terms of resistances. Thus, the terms $(1 + r^4 x')$ represent resistances to flow. This follows the work of Ballman et al. in their analysis of flow in a single runner with several restrictions. Probably more important is the relationship between the flow rates into adjacent cavities. As is seen from Eq. (10.3-21),

$$G_{g1} = G_1 (r_1^{3n+1} x_1')^{1/n} \qquad (10.3-21)$$

and similarly for G_{g2}:

$$G_{g2} = G_2 (r_2^{3n+1} x_2')^{1/n} \qquad (10.3-35)$$

But G_1 is related to G_2 by

$$G_2 = G_1 / [1 + (r_2^{3n+1} x_2')^{1/n}] \qquad (10.3-36)$$

Hence,

$$G_{g1}/G_{g2} = (r_1/r_2)^{3n+1/n} (x_1'/x_2')^{1/n} [1 + (r_2^{3n+1} x_2')^{1/n}] \qquad (10.3-36a)$$

To see this more clearly, assume that the r's and x's are equal and the fluid is Newtonian $(n = 1)$. Then

$$G_{gi}/G_{gi+1} = (1 + r^4 x') \qquad (10.3-37)$$

As long as the runner has any length [according to Eq. (10.3-37)] and a finite radius $(r \neq 0)$, the flow into the cavity closest to the sprue will always be greater than that into the second closest, and so on. Thus, for the mth cavity,

$$G_{g1}/G_{gm} = (1 + r^4 x')^{m-1} \qquad (10.3-38)$$

for a Newtonian fluid. It is rather apparent then that to even the flow additional resistance to the flow into the cavities closest to the sprue must be provided. It is not easy to differentially change runner radius (e.g., make r a function of distance from the sprue), but it is rather easy to change the effective length of the port by changing the gating dimensions, as has been seen earlier in this section. By changing the gate parameter b such that much more restriction occurs in the gate feeding the cavities closest to the sprue and by opening the gates in the regions near the end of the manifold, the flow can be made more uniform throughout the system. This is the best way to "balance" an otherwise unbalanced flow system.

The cursory attempt above illustrates only one approach to manifold flow. Certainly the kinetic energy losses and, probably more important, a temperature-dependent viscosity at each port should be included. One way of including the temperature-dependent viscosity would be to assume a temperature profile in the material down the length of the manifold. In this way, an Arrhenius form for the viscosity could be used:

$$\phi = (K_T/K_i) = \exp[E\theta/(T_i - T_i\theta)] \tag{10.3-39}$$

Now Eq. (10.3-1) would become

$$[(P_i - P_0)/4\phi K_i]^{1/n} = (x/R_0)^{1/n}(4Q_0/\pi R_0^3)/B^{1/n} \tag{10.3-40}$$

Probably a more propitious approach would be to consider the microscopic forms for the equation (e.g., the transport equations, including momentum, continuity, and energy) for the system.

Lord and Williams [22,23], as mentioned in Sec. 10.2, coupled their mathematical analysis of flow in a plaque mold with flow in the sprue and runner system. As part of their analysis, they included flow in unbalanced manifolds, with three double cavities branched from the main runner. They support their mathematical model with pressure measurements in the runner system. The ABS data and theory agree quite well (unlike their pressure measurements in the cavity), but for PVC, their theory predicts pressure drops of 40 to 60% of that actually measured. They compare their manifold filling theory with short shot measurements on the three double-cavity runner system mentioned above. For the following dimensions, R = 0.500 in., r = 0.375 in., x = 2 in., and X = 4 in., they show a weight ratio at 36% fill of 1.85 between cavities 2 and 3, 1.70 between cavities 1 and 2, and 3.67 between cavities 1 and 3. Again, at 50% fill, the Lord and Williams weight ratios are 1.60, 1.67, and 2.67, respectively. If short shot data can be relied upon to yield useful filling rates, we can see from Eq. (10.3-37) and (10.3-38) that the ratios using the simplistic theory should be 1.63, 1.63, and 2.67, respectively. Lord and Williams then add resistance to the flow channels to cavities 1 and 2 in order to achieve uniform balancing of the filling rates. They find a 2-to-3 cavity runner radius ratio of 0.5 and 1-to-3 runner radius ratio of 0.33 from balancing pressures throughout the runner system. Note, however, that only 1.25 in. of the 2 runner was reduced in radius. The simplistic radius ratio for balancing cavities on uniform filling rates can be obtained from Eq. (10.3-36a). With appropriate choices for equivalent length of the side runner owing to the constriction, we can show that a balanced system occurs when the 2-to-3 cavity runner radius ratio is 0.6 and the 1-to-3 ratio is 0.36. Thus, the Lord and Williams' analysis can be supported with the simpler model.

A note of caution in data interpretation is appropriate here, however. The process of cavity filling is dynamic. Short shots represent the static state of the plastic in each cavity after flow has stopped. Stopping the fluid short of the end of the cavity is an extreme example of varying flow speeds during filling. Changing flow speeds during injection will change the pressure profiles throughout the manifold runner system and thus the individual

cavity filling rates. Stopping the flow entirely will allow rapid imbalance
in flow rates until the kinetic energy drops to zero. We would thus expect
that those regions of the mold experiencing higher pressure to be more in-
fluenced by the rebalancing of flow during short shots than others. The
most reliable way of checking short shot weights in manifold runners is with
positive-acting valve gates.

10.4 Packing in the Mold Cavity

As seen from the pressure profiles of Paulson ([5], Fig. 10.2-5), even
though the injection pressure is maintained constant once the cavity is filled,
there is a steady decrease in the pressure within the cavity. This decrease
is caused by the shrinkage of the plastic as it cools. Thus, it is not sufficient to just fill the cavity. Additional material must be pushed into the cavity
to compensate for the volume decrease during cooling. This pushing is
referred to as "packing." To understand the concept of packing, first realize that all polymer melts are compressible at the normal injection pressures. As seen in Chap. 14, one early equation of state for amorphous
polymers is the Spencer-Gilmore equation [32]:

$$(P + \pi)(V - w) = R'T \tag{10.4-1}$$

where P is the hydrostatic pressure, V is the specific volume, T is the
melt temperature, and π, w, and R' are constants for specific polymers.
As seen later in this chapter, the density of the part affects the finished
part dimensions, the surface quality, the residual stresses in the regions
of weld lines, the distribution of internal stresses in the part as a whole,
the extent of sink marks and warpage, and molecular orientation. In the
last case, this can affect the mechanical properties of the part, such as impact strength, flexural strength, and tensile strength. Following the argument of Paulson, if the melt temperature is constant in the cavity immediately upon injection, then the effective specific volume $V - w$ is inversely
proportional to the effective applied pressure $P + \pi$. Once the gate is frozen,
the effective specific volume $V - w$ is constant, and the effective applied
pressure $P + \pi$ drops in proportion to the temperature. As seen from the
pressure transducer response curves, the pressure increases quite rapidly
once the cavity is filled. This is a typical hydraulic pressure curve for a
slightly compressible fluid. If the fluid were not compressible, the pressure would jump immediately to a constant value. Kamal and Kenig used a
modified form of the Spencer-Gilmore equation of state to determine the
actual pressure profiles during packing, and their data are shown as Figs.
10.4-1 and 10.4-2. During the packing, the bulk melt temperature in the
cavity is dropping owing to transient heat conduction to the cold mold walls.
Thus, their curves exhibit not only the effects of compressibility but also
the cooling of the melt. Their pressure profiles show definite maxima and

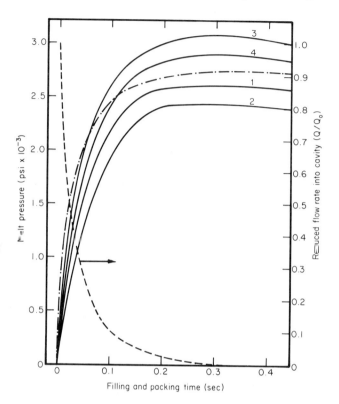

Figure 10.4-1   Development of pressure during packing stage in injection mold filling. Solid lines, experimental via pressure transducers; curves 1 to 4, measured pressures at dimensionless radius values of R* = 0.06, 0.38, 0.70, and 0.95; dash-dot curve: theoretical; dashed line, dimensionless flow rate, Q/Q_0 [19].

are in fair agreement with the pressure profiles obtained using pressure transducers in the mold cavity. Recall that r is the ratio of the local radius to the radius of the cavity (R_0), and thus there are inversions in the pressure profile as we move from near the gate (r = 0.06) to the point near the far wall of the cavity (r = 0.95). Paulson and Frizelle found a steadily dropping pressure from the gate to the end of the cavity. Kamal and Kenig show only one experimental pressure profile and offer no explanation of this pressure inversion effect. One possible explanation is that the temperature profiles in the material as functions of radius at the end of filling are so nonlinear that local pressures away from the gate area can be considerably higher than that at the gate area. As is seen below, abnormal local pressures can lead to frozen-in stresses, and thus uniform pressure is desired at the end of the packing stage.

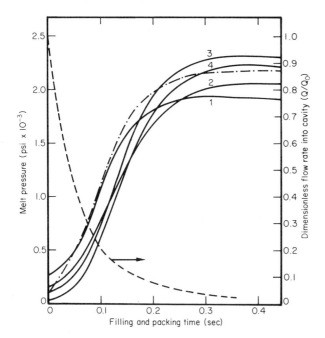

Figure 10.4-2 Development of pressure during packing stage in injection mold filling of half-round cavity. Solid lines, experimental via pressure transducers; curves 1 to 4, measured pressures at dimensionless radius values of R* = 0.06, 0.38, 0.70, and 0.95; dash-dot curve, theoretical; dashed line, dimensionless flow rate Q/Q_0 [19].

10.5 Cooling the Melt

It is very difficult to separate filling rates, flow in runners, flow through gates, and packing effects from the ever-present transient heat transfer problem. Consider the simplistic viewpoint that the material is injected in an isothermal manner (e.g., at very high speed to prevent significant conduction heat transfer). Then the standard transient heat transfer solutions discussed in Chap. 6 can be used. A linear time plot of decrease in mean temperature is given in Fig. 10.5-1 for Kamal and Kenig's HDPE, assuming a thermal diffusivity of 4×10^{-3} ft^2/hr and a half-thickness of 0.032 in. $T_{mean} = 120°F$, assuming a melt temperature of 400°F and a mold temperature of 80°F and that no mold-melt interface resistance, occurs at about 5 sec. This corresponds with the zero-pressure point of Kamal and Kenig in Figs. 10.5-1 and 10.5-2. It is obvious that this is too simplistic for the very complicated process. Kamal and Kenig have shown that for a crystalline material such as HDPE (in their case about 60% crystalline) the latent heat of fusion of the resin (60 to 100 Btu/lb) represents approximately

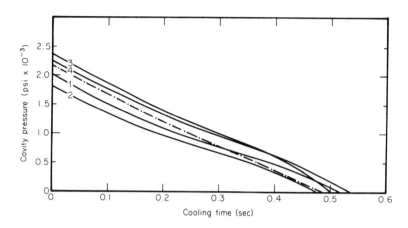

Figure 10.5-1   Experimental and theoretical pressure profiles during cooling portion of injection molding of half-radial mold. Solid lines, experimental values obtained via pressure transducers; curves 1 to 4, pressure values obtained at reduced radial values of R* = 0.06, 0.38, 0.70, and 0.95; dash-dot line, theoretical prediction [19].

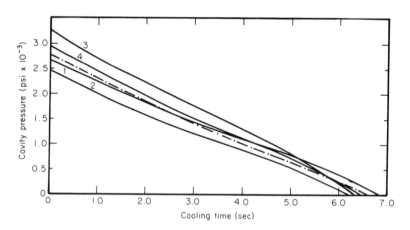

Figure 10.5-2   Experimental and theoretical pressure profiles during cooling portion of injection molding of half-radial mold. Solid lines, experimental values obtained via pressure transducers; curves 1 to 4, pressure values obtained at reduced radial values of R* = 0.06, 0.38, 0.70, and 0.95; dash-dot line, theoretical prediction [19].

40% of the total enthalpy change between 400 and 100°F. Thus, for crystalline materials, this heat of fusion must be taken into account. Kamal and
Kenig [33] have taken into account the effects of crystallinity by modifying
the temperature profiles for heat capacity, thermal conductivity, and density. They then distribute the extent of crystallization over a rather broad
temperature range and in this way need not consider the movement of the
liquidus-solidus interface into the material with time. Their experimental
and theoretical time-temperature profiles for both the mold and the plastic
are shown in Figs. 10.5-3 and 10.5-4 for one set of initial conditions. It
should be pointed out, however, that they have assumed an interfacial resistance between the mold and the polymer melt. This, they claim, follows
tradition. The interfacial resistance is represented as an effective convection heat transfer coefficient having an approximate value of 100 Btu/ft^2·hr·°F.
Carley [26], however, considered convective flow heat transfer in a tube,
and not conductive resistance between the static melt and the mold wall.
Thus, it would appear that the interfacial resistance of Kamal and Kenig is
unique to their work and not attributable to Carley. One might argue that
the interfacial resistance can be caused by poor contact between the polymer
molecules and the rough surface of the mold and/or by gases trapped at the

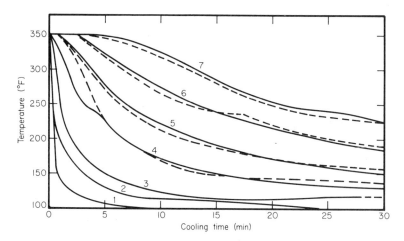

Figure 10.5-3   Temperature profiles across a slab of plastic initially at
350°F, 10,000 psi, in contact with a mold wall initially at 100°F. Solid
lines, theoretical temperature profiles during cooling portion of the injection molding process; dashed lines, experimental via thermocouples; curve
1, coolant-mold interface temperature; curve 2, mold temperature; curve
3, polymer-mold interface temperature; curve 4, 0.9 in. from centerline;
curve 5, 0.76 in. from centerline; curve 6, 0.58 in. from centerline; curve
7, centerline temperature [33].

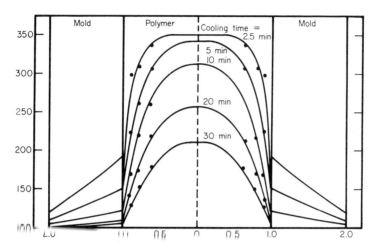

Figure 10.5-4 Temperature profiles across a cooling polymer slab as functions of time; experimental and theoretical. Initial polymer temperature, 350°F; initial polymer pressure, 10,000 psi; initial mold temperature, 100°F; coolant temperature constant at 100°F; solid lines, theoretical calculation; solid circles, experimental data via thermocouple measurements [33].

interface during filling and heating. One would not expect this effect to be significant during conventional injection molding where pressures of 100 atm or more would tend to keep the gas in solution.

Collier, Dinos, and their colleagues [34,35] have examined the effect of cooling of a polymer in terms of the kinetic rate of crystallization. Using the physical data for Penton chlorinated polyether, they included the Avrami rate equation,

$$(1 - \theta) = e^{-Kt^n} \tag{10.5-1}$$

where θ is the relative crystallinity, t is time, and K and n are constants peculiar to the resin, into the standard transient heat conduction equation through the addition of a heat generation term:

$$\partial T/\partial t = \alpha(\partial^2 T/\partial x^2) + g/\rho c_p \tag{10.5-2}$$

where g is the heat generation term that arises because of the heat required for crystallization. For quenching a 0.032-in.-thick sample initially at 575°F in a 200°F constant-temperature bath, assuming a heat convection coefficient of 120 Btu/ft^2·hr·°F, they predicted a time-temperature profile as shown in Fig. 10.5-5. Their experimental results were somewhat

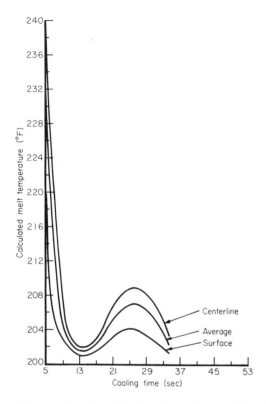

Figure 10.5-5   Calculated cooling profiles for 0.032-in.-thick Penton. $T_{melt} = 575°F$; $T_{mold} = 200°F$; h = 120 Btu/ft^2·hr·°F [35].

disappointing in that the increase in temperature caused by the generation of heat of crystallization was difficult to identify. See Fig. 10.5-6. Nevertheless, the very dramatic temperature drop was rather abruptly halted in the time period where the increase was supposed to occur. Note, however, that Collier, Dinos, and coworkers did not include the effects of temperature on specific heat, thermal conductivity, and density, as did Kamal and Kenig. Thus, one might expect that the experiment and the theory would not agree over the entire time-temperature curve. Collier and Dinos bring their theory into agreement with the experiment through adjustment of the thermal conductivity, but this technique requires that the published values be doubled, and this magnitude of correction does not seem justified. Recent work [36] indicates that the Collier-Dinos method cannot be applied for most crystalline materials because the Avrami rate of crystallinity for most materials is unknown and the induction time preceding crystallinity is not predictable. The preferred solution includes the heat of crystallinity

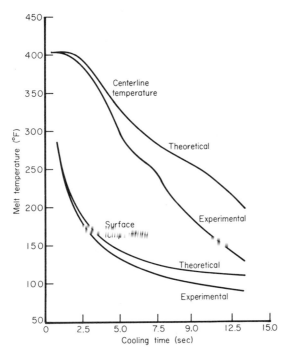

Figure 10.5-6   Calculated and experimental cooling profiles for 0.483-in.-thick Penton, with k = 0.076 Btu/ft·hr·°F, C_p = 0.35 Btu/lb·°F, h = 500 Btu/ft^2·hr·°F, and T_{quench} = 73°F [35].

as an excess heat capacity over a short temperature range on either side of the known melting temperature. Owing to the insensitivity of the rate of cooling curve to small changes in local heat capacity, there seems to be little to choose from in either the spike form or plateau form for the excess heat capacity. There is certainly a major reason for considering the variation in temperature profiles within a part. The morphology or crystal structure formed is highly dependent on temperature and time-temperature history, and thus slight changes in these parameters can lead to remarkable changes in the mechanical properties of the material.

10.6 Heat Removal from the Injection Mold

Before considering the effect of packing and cooling on the mechanical characteristics of the material in the mold, consider the method of cooling the mold itself. As Kamal and Kenig have pointed out, the ratio of thermal diffusivities of mold material such as steel to plastic material is on the order of 100 or more. For copper-beryllium materials (CuBe), the ratio can be as high as 2000. Therefore, for most purposes there is relatively little increase in mold surface temperature as the plastic melt cools. Furthermore, it must be considered that the relative masses of material (mold to melt) are also on the order of 1000 to 10,000:1, and thus the effect of a single shot of plastic into a mold is essentially confined to the mold surface. Even without cooling on a steel mold, runs of 1 to 4 hr can be made before appreciable heat buildup is felt on the external surface of the mold. Until a few years ago, very little attention was paid to the placement of coolant lines in the mold block. An early paper by Temesvary [37] pointed up the importance of placing the water lines as close to the part surface as possible and emphasized complete redesign of deep cores and bubblers in order to "increase the surface of the cooling channel" and "allow the coolant to come closer to the hot plastic in the mold cavity." Recently, Prasad [38] has given a series of guidelines for optimizing cooling in mold cavities. In this section, the two pieces of information needed to determine first the effectiveness of the present cooling system and second the ways of improving it to increase cooling capacity are considered.

Note that all the heat removed from the plastic must be transferred to the coolant, assuming that the mold block remains isothermal. Neglecting Kamal and Kenig's resistance heat transfer coefficient for the moment and assuming slab geometry, a desired centerline temperature from Fig. 10.6-1, the transient heat conduction equation, can be obtained. To determine an equilibrated temperature, Fig. 10.6-2 is used. From this, the thermal diffusivity of the material and the half-thickness of the part, always using the thickest section for this calculation, the minimum time required to remove the necessary heat is determined. The total heat content of the plastic is given as

$$Q_0 = V' \rho c_p (T_i - T_m) \tag{10.6-1}$$

where V' is the volume of plastic, ρ is its density, c_p is its specific heat, and $T_i - T_m$ is the temperature difference between the incoming melt and the mold temperature. From Fig. 10.6-3, that fraction of heat that must be removed from the part in the minimum time previously calculated is obtained. This heat must be removed by the coolant, according to standard heat transfer equations,

$$Q = U'A \, \Delta T \tag{10.6-2}$$

where U' is an overall heat transfer coefficient, A is the surface area of the coolant system, and ΔT is the thermal driving force between the mold surface and the coolant. Now in general the process is transient in that the surface of the mold alternately heats and cools, depending on the presence or absence of material in the mold. Furthermore, there is a temperature increase between entrance and exit of the cooling line, and thus a simple average is normally not correct. Probably, a weighted average, such as a

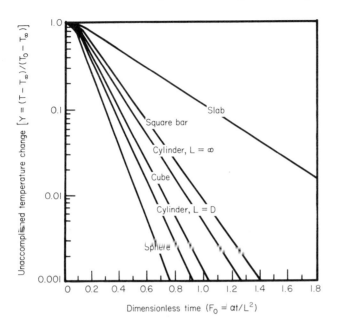

Figure 10.6-1   Centerline temperatures with no surface resistance for various shapes subject to step change in surface temperature.

logarithmic mean temperature difference, is appropriate, but because of the very difficult geometries of conventional coolant lines, it is not always clear how to make this weighted calculation. Therefore, assume an average thermal driving force. The overall heat transfer coefficient per unit length of coolant line is given as

$$1/U' = 1/Sk_m + 1/\pi Dh + ff + 1/h_p \qquad (10.6\text{-}3)$$

where S is the so-called shape factor [30]; k_m is the thermal conductivity of the mold material; D is the diameter of the coolant line; h is an individual convective film heat transfer coefficient; ff is a fouling factor included because coolant lines can collect rust, lime, and mineral deposits and oily film; and h_p is the Kamal-Kenig conduction resistance coefficient, to account for the unexplained resistance between the melt in the cavity and the mold wall. This factor will be neglected henceforth.

To calculate the individual convective film heat transfer coefficient, the Reynolds number and Prandtl number of the coolant and a correlation between the Nusselt number (hD/k) and Re and Pr are needed. The easiest to use for laminar, turbulent, and transition is the Colburn j-factor, as shown in Fig. 10.6-4. Since 1/h is a measure of resistance to fluid flow, it is desirable to have this resistance as low as possible (e.g., h as large

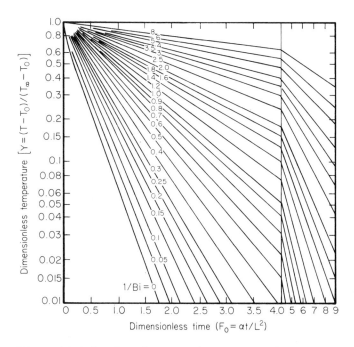

Figure 10.6-2 (See Chap. 6). Average transient temperature for slab response to step change in surface temperature. Parameter is $1/Bi = k/hL$.

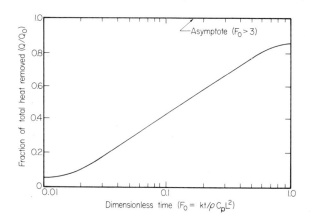

Figure 10.6-3 Fraction of total available heat removed from slab as function of dimensionless time.

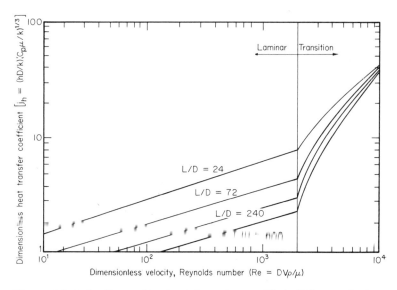

Figure 10.6-4   Convection heat transfer coefficient for coolant as a function of dimensionless velocity.

as possible). As shown from Fig. 10.6-4, this is achieved by turbulent flow in the coolant passage. Normally, there are many coolant passages in a mold and a single water line. This means that the passages are all connected in series. The pressure drop in series flow for turbulent flow is given as

$$\Delta P = 0.158 LV^{7/4} \eta^{1/4} \rho^{3/4} / gD^{5/4} \qquad (10.6\text{-}3a)$$

whereas for laminar flow it is

$$\Delta P = 32 LV \eta / gD^2 \qquad (10.6\text{-}4)$$

Here η is the viscosity of the coolant, and g is the gravitational conversion. If pressure drop is limited, as would be the case if city line pressure were used, the velocity for laminar flow decreases in proportion to the length of coolant flow channel and increases in proportion to the cross-sectional area of the coolant flow channel. For turbulent flow, the velocity decreases in proportion to the 4/7 power of the length of the channel and increases in proportion to the 5/14 power of the cross-sectional area. Furthermore, the heat transfer coefficient for fully turbulent flow is found to be proportional to the velocity to the 0.8 power, whereas for laminar flow the proportionality is to the 1/3 power. Thus, there seems to be great advantages for maintaining turbulent flow in the coolant passages. Contrary to this, however, most molders do not do this because they do not recirculate their water and thus restrict flow rate to minimize cost, they cannot tolerate the low mold temperatures associated with high flow rates through the mold, and/or they have never considered mold cooling to be an important aspect of injection molding.

If turbulent flow occurs in rather large-diameter coolant passages (on the order of 7/16 to 1/2 in. or more), the convective resistance to heat transfer is minimized. For most coolant systems, a fouling factor of 0.002 is considered average.

This leaves the consideration of the conductive resistance L/Sk_m. k_m is, of course, a property of the mold (25 Btu/ft$^2 \cdot$ hr$\cdot°$F/ft for steel, 200 for CuBe, and so on). The shape factor S, on the other hand, is a function of mold design. As Temesvary points out, moving the coolant closer to the hot plastic improves the heat transfer. To see this, examine Fig. 10.6-5 for shape factor as a function of coolant passage diameter, distance between the coolant passage and the melt, and the distance between the coolant passages. Doubling the coolant passage diameter from an initial P/d = 2 and D/d = 2 increases the shape factor value from 1.2 to 3.2. Adding a second row of coolant passages, decreasing P/d from 4 to 2, decreases the shape factor from 2 to 1.2 but doubles the surface area for heat transfer, thus effectively increasing the heat transfer by 10% assuming other resistances are negligible. In Figs. 10.6-6 through 10.6-10 are given other shape factors. Some results are startling. For example, in Fig. 10.6-10 the addition of a second row of coolant passages increases the shape factor value from 2.10 to 3.41 and the surface area for coolant by a factor of 2. Thus, a second row of cooling passages has resulted in an effective increase in heat transfer of 325%. Note that, contrary to intuition, flooding the bottom of the cavity does not effectively increase the shape factor, although the surface area for heat transfer is increased, and thus the overall heat transfer removal effectiveness is increased. In Fig. 10.6-8, increasing the diameter of the coolant line directly under the melt yields a 41% increase in shape factor value and doubles the surface area for cooling, thus yielding an 82% increase in heat transfer. From studies such as these it is apparent that placing coolant lines under flat surfaces such as runners yields significant increases in values for shape factors [39]. Incidentally, simple analog methods for determining two-dimensional shape factors are given in Kreith [40], and some novel three-dimensional analog methods are discussed elsewhere [41]. An example of an electric analog is shown in Fig. 10.6-11. Here a conductive paper is cut to the shape of the heat conduction field and the "source" (e.g., the plastic melt) and the "sink" (e.g., the coolant line) are connected to a dc power supply through a voltage divider and null detector. By dividing the voltage into intervals and probing the conductive surface for the curve that yields this voltage, the entire set of curves of constant temperature can be determined. Then by drawing lines of constant heat flux at right angles to the curves of constant temperature, thus dividing the field

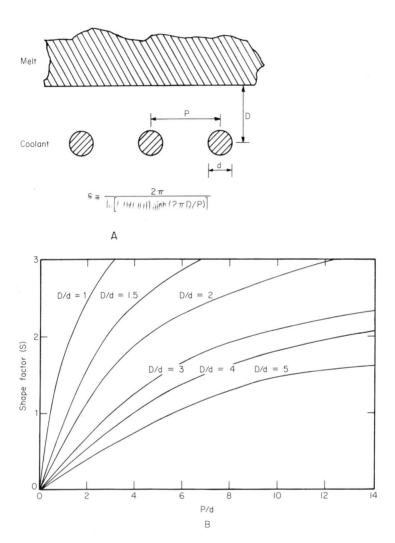

Figure 10.6-5   A: Shape factor geometry for one configuration of coolant lines in a mold block.   B: Effect of coolant line location on shape factor value.

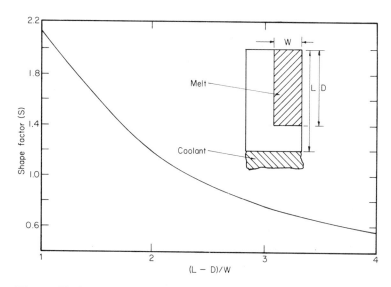

Figure 10.6-6   Dimensional effect of square cavity on shape factor value for flooded coolant channel.

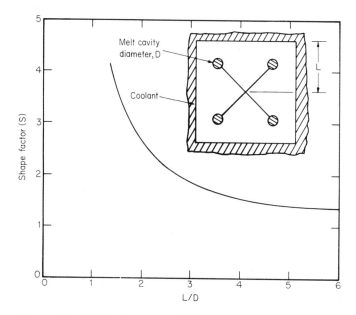

Figure 10.6-7   Effect of cylindrical cavity dimensions on shape factor value for flooded coolant channel.

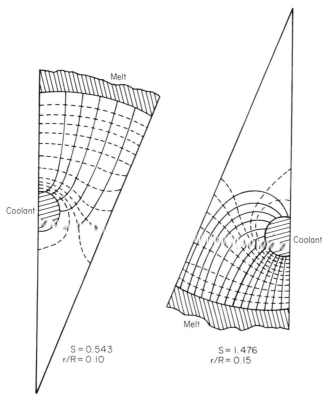

Figure 10.6-8   Effect of coolant line size and location on shape factor value in core pin or cavity plug.   Dashed lines are isotherms.

into curvilinear squares, and simply counting the number of heat flux spaces and dividing by the number of spaces of constant temperatures, the shape factor value can be determined.   This technique is rapid and allows for changes in coolant passage before the mold design is firmed up.

Note that minimizing only one of the resistances in Eq. (10.6-3), or in fact all but one of the resistances, will not reduce the resistance below that of the largest value.   In other words, if a very high heat transfer coefficient occurs in the large coolant passages and the fouling factor and the Kamal-Kenig resistance factor are negligible, the overall resistance will be controlled solely by the shape factor value and the thermal conductivity of the material.   But, in the same sense, optimizing coolant line location and using CuBe inserts to cool critical areas while using laminar flow coolant will cause the overall heat transfer from the mold to be limited primarily by the convective film coefficient.   Others [38,39] have advocated replacement of external hoses and couplings with larger fittings to eliminate

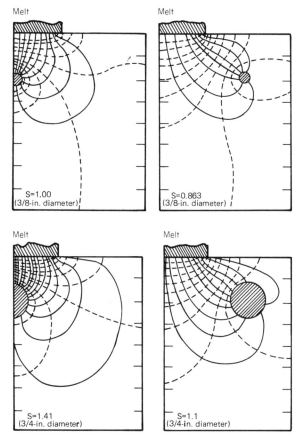

Figure 10.6-9 Effect of coolant line size and location on shape factor value near runner. Dashed lines are isotherms.

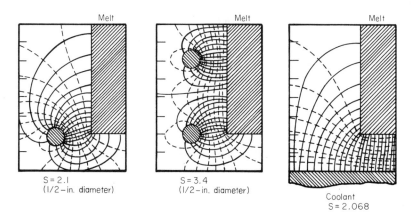

Figure 10.6-10 Comparison of single and double coolant lines and flooded
coolant channels for square cavity. Dashed lines are isotherms.

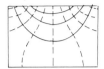

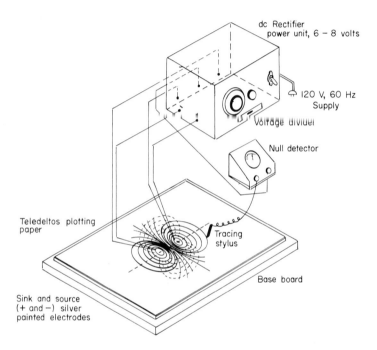

Figure 10.6-11   Experimental setup and typical field plot for the conductive paper experiment described in the text. In the field plot, note that the solid lines, isotherms, radiate from a point source and that the dotted lines, heat flux lines, are at right angles to the isotherms. (The Analog Conductive Paper Plotter is courtesy of Sunshine Scientific Instrument Co.) [40].

resistances in these areas, thus increasing flow rate for a given pump delivery pressure. It should be also apparent that once the mold is designed and the maximum flow rate is selected the objective is still to remove the heat Q given in Eq. (10.6-1). If an attempt is made to decrease the total time, thus decreasing the cooling time, to improve production, it is necessary to remove more heat in a given period of time through Eq. (10.5-2).

If flow rates, materials, and/or shape factors cannot be improved, the temperature differential must go up. This means that if the parts are to be removed with a specific centerline or equilibrated bulk temperature, the coolant temperature must be lowered. Therefore, throughout the industry very short injection cycles (0.8 sec or less, according to Huskey, NPE, Chicago, 1973) require extensive refrigeration of high-pressure coolant (up to 50% ethylene glycol in water at 100 psi) and are achieving coolant temperatures of 0 to 5°F. In many cases, at these low temperatures, it is necessary to insulate the external mold surfaces to prevent condensation or frost on the surfaces.

10.7 Mold Materials

Not all the materials and mold designs that have been used for injection molding will be categorized here. Others are much more qualified to do this well [42,43]. However, to give some continuity to the discussion on material properties, a brief description of the materials that can be and have been used for injection molds is included. A comparison of materials for conventional injection molds and low-pressure structural foam molds is also included.

It should be pointed out immediately that conventional injection mold die design technology is, for the most part, directly applicable to structural foam molding (SFM) die design technology. Most SFM parts are very similar in design to conventional injection-molded (CIM) parts as regards parting line locations, ejector pin locations, techniques for moving cores, stripper plate designs, etc. Some differences occur in mold cooling techniques, gating designs, venting, and material property control, but these are in degree rather than kind. The most important differences between SFM and CIM are in the greater variety of materials that can be used for molds and the potentially corrosive nature of the blowing agent residues. The changes in technology required to design a mold suitable for low-pressure SFM are also discussed.

As molding pressures are decreased from 1000 atm in conventional injection molding (CIM) through 10 to 20 atm for low-pressure structural foam molding (SFM) to 1 to 5 atm for rotational molding and thermoforming, the number of candidate materials for mold forms increases very rapidly. For example, for CIM or high-pressure SFM, tool steel such as S7 or R14 is recommended for long runs. A hard aluminum alloy such as 2024 can be used for short-run or prototype tooling but will show excessive wear rather rapidly.

In this section, the low-pressure SFM processes will be discussed, and the suitability of many candidate materials will be critically analyzed. Throughout this analysis, keep in mind the ultimate life of the tool. In many SFM operations, mold cost is amortized on a per-piece basis. Thus, the

the optimized tool life must be based on the projected piece production and
not necessarily on the ease of tool fabrication or the optimum production
cycle of the tool and machine. There are certain other constraints (some
subjective) that must be placed on material optimization. Some of these are

1. Erosive character of the plastic (e.g., glass reinforcing)
2. Corrosive nature of the plastic or residue of suitable blowing
    agents
3. Effect of localized high-pressure spots in sprue or runner areas
4. Type of finish desired on final part
5. Complexity of part (side cores, unscrewing devices, parting line
    complexity, inserts, multiple-cavity molds, multiple-gated single-
    cavity mold, number of obstructions in flow stream)
6. Skill of the operator in removing stuck parts and in cleaning the
    mold

It is apparent from this list that one or more of these constraints can directly affect material selection, eliminating less expensive mold materials.
Nevertheless, the general suitability of candidate mold materials based on
fabrication of a rather simple part, such as a bookshelf or equipment cabi-
net, will be discussed.

10.7.1 Plastics

Plastics have been used extensively for molds in thermoforming.
Normally glass-reinforced thermosetting polyester and epoxies, either
glass filled or aluminum filled (up to 70% weight), are used. Some CIM
prototype tools have been made of aluminum-filled epoxy (Devon C or Epo-
cast), but the life of the mold rarely exceeds 10 to 100 parts [44]. The
major problems with plastics are that they are thermal insulators and not
very strong when compared with metals. Thus, cooling is critical, but
cast-in water lines cannot be placed very close to the mold surface without
risking compressive failure of the mold. The parts molded in plastic tools
are normally rather small in size because of the potential mold failure due
to excessive forces on the mold. Some additional life can be obtained by
making the mold rather thin and inserting it into a standard steel die block.
The plastic mold insert can then either carry its own cast-in water lines,
or the space between the insert and the die block can be flooded.
    It should be noted, however, that the lower thermal conductivity of
the plastic (as compared with metal) can lead to material properties in SFM
foam that differ from those obtained when foam is injected against high-
thermal-conductivity molds. The melt foam will cool at a slower rate against
plastic than against metal. Thus, the skin of the foam may be thinner and
the core density may be higher for the same bulk density. The resulting
part may exhibit a lower tensile strength and a higher impact strength than
one molded against a metal surface. This can be quite critical to prototype
part evaluation prior to development of a production part and should be kept
in mind when working with plastic tools.

The lower thermal conductivity also implies that heat removal from the mold volume will be slower than that for steel, and thus injection molding cycle times will be significantly longer. Economic evaluations based on these times will obviously be biased, and this must also be recognized in development of production parts.

In working with metal-filled epoxy molds, one normally casts the material around a properly dried plaster (Hy-Cal C) or ceramic pattern. Shrinkage from a pattern is essentially nil (less than 0.1%). In mixing and casting epoxies, it should be noted that many people are sensitive to the gases and resins, and as a result mixing should be done in a well-ventilated area. Hand mixing is recommended to minimize entrainment of air bubbles. Entrainment of air in the high-viscosity resin will result in a porous casting that will fail after only a few parts have been molded. Casting the epoxy around the pattern should be done in a way that minimizes air encapsulation. This implies that the epoxy should be forced around the low end of the pattern, thus pushing the air ahead of the liquid front. Once the epoxy is cast, it should be placed in a convective oven at 90 to 110°C overnight to ensure first cure. When all sections of the mold have been so prepared and cured, the mold should be assembled, the parting lines filed, ejector pin locations drilled (obviously with care so as to avoid chipping the plastic material), gate, runner, and sprue areas added (machined into the epoxy mold if desired, although it is recommended that the epoxy materials be used only as mold inserts and that sprues, runners, and even gates be cut into the steel die block), and filing or machining in flash vents. The completed epoxy mold should then be returned to the oven for a second overnight cure at 150 to 175°C.

In economic considerations other than decreased cycle time, it should be pointed out that while material costs are higher than other mold material costs, the time and expense of machining are minimized. Furthermore, repairs to the mold are relatively easy. Only sanding or grinding of the damaged area and insertion of an epoxy patch is required. And replacement of a broken mold insert can be accomplished by simply casting a new mold over the plaster or ceramic pattern.

10.7.2 Zinc Alloys

Of the many types of castable zinc alloys that are candidate materials, the alloy known as Kirksite A in the United States and Zamak in Europe (Zamak 2 has essentially the same composition as Kirksite A) is the prime mold material. Castings are relatively hard ($25 kp/mm^2 = 40,000$ psi tensile strength) and can be cast against ceramic patterns at relatively low temperatures (380 to 400°C) without pressure, and it is one of the cheapest of the zinc alloys. Nevertheless, this material has not found acceptance for production tooling because it is relatively soft and, therefore, susceptible to long-term distortion under pressure, brittle failure due to accidental overpressuring, and surface deterioration due to operator abuse or erosive materials (such as glass fibers) in the plastic melt. It is also considerably heavier than other materials (6.7 sp. gr.) and has rather poor thermal properties: specific heat (25 to 100°C) = 0.1 cal/g·°C, thermal conductivity (70 to 140°C) = 0.25 cal/cm·sec·°C, and coefficient of thermal expansion (20 to 100°C) = 27.4 × 10^{-6} (°C)$^{-1}$. Cast against a steel pattern, the shrinkage is 0.5 to 0.7%; against a ceramic pattern, 1.0 to 1.4% shrinkage can be expected. Normal chemical composition includes 3.5 to 4.5% Al; 0.03 to 0.08% Mg; 0.1% (max.) Cu; 0.1% Fe; trace amounts of Pb, Cd, and Sn; and the balance a special high-grade zinc.

Because of the relatively low casting temperature of the zinc alloy, water lines of hard copper or Admiralty brass tubing can be cast in place. This helps to overcome the poor thermal conductivity of the zinc alloy. Standard steel ejector pins, sprue bushings, and runner-gate systems can be used with the zinc alloy casting, although it should be noted that for long runs the casting will wear and allow flash around ejector pins.

The properties of zinc alloys, when compared with aluminum, are very favorable for short-run tooling (the mold at the end of its useful life can be melted into an ingot for recasting into another shape), but surface deterioration, wear, poor cooling capability (resulting in longer mold cycle times), and the ever-present problem of overpressurization make it unattractive for production runs in excess of 10^4 parts.

10.7.3 Copper Alloys

High-copper-content alloys such as bronze (90% Cu, 10% Zn, trace amounts of Pb and Fe) and brass (red brass is about 85% Cu, 15% Zn, with trace amounts of Pb and Fe, whereas cartridge brass is 70% Cu, 30% Zn, with trace amounts of Pb and Fe) are sometimes used as mold inserts for small conventional injection mold parts (less than 450 g or so). The tensile strengths are in the range of 25 to 30 kp/mm^2 (40,000 to 50,000 psi). These materials can be cast against ceramic and have pouring temperatures 915 to 1025°C. The thermal properties for bronzes and brasses are excellent (owing to the copper) but decrease with decreasing copper content. For bronze, with a specific gravity of 8.80 (g/cm^3), the thermal conductivity (20°C) is 0.45 (cal/cm·sec·°C), the heat capacity (25°C) is 0.09 (cal/g·°C), and the coefficient of thermal expansion (20 to 100°C) is 18 × 10^{-6} (°C)$^{-1}$. For red brass, the values are 8.75 (sp. gr.), 0.38, 0.09, and 18.7 × 10^{-6}. For cartridge brass, the values are 8.53 (sp. gr.), 0.29, 0.09, and 20 × 10^{-6}.

Most of these high-copper-content alloys can be cold pressure-hobbed using a standard H-12 or H-13 tool steel hob. This eliminates some of the porosity that occurs in atmospheric casting. The normal shrinkage for casting brass or bronze against a ceramic pattern is about 1.2 to 1.5%. Against steel, it is 0.4 to 0.5%. There is no measurable shrinkage during hobbing. The cast surfaces will yield very finely detailed parts in CIM; this is not always the case with SFM parts, however, owing to the lower pressures available to push plastic into the detailings.

Machining of details as well as vents, gates, and ejector pin locations is very easy, and any flaws in the castings can be corrected by silver soldering and remachining. Thus, worn or cracked castings can be easily and quickly repaired.

Certain SFM materials cannot be processed in high-copper-content alloys because either their decomposition products or the blowing agent residues (such as NH_3) are corrosive to the alloys. As examples, the residue from Luvapor AZ6600 BA (commonly used with GPS) is NH_3 and is alkaline, and one of the residues from Genitron AC (commonly used with HDPE and PP) is isocyanic acid and thus is acid. Both these blowing agents attack high-copper-content alloys. Caution should be taken when selecting these alloys to ensure compatibility with appropriate SFM materials and their BAs.

In an effort to improve the mechanical strength of the high-copper-content alloys, the copper beryllium (CuBe) alloy was developed. This alloy is about 2% Be; 0.1 to 0.5% Ni, Co, and/or Fe; and nearly 98% copper. The tensile strength is nearly 130 kp/mm^2 (200,000 psi), and the pouring temperature is about 865°C. Pressure hobbing is difficult but can be done with shallow-draw hobs of S-7 tool steel. The shrinkage of CuBe against ceramic patterns is 1.5 to 1.8%. Against S-7 steel patterns, it is 0.5 to 0.7%. The thermal properties of CuBe are very good owing to the high copper content. The CuBe specific gravity is 8.25 (g/cm^3), the thermal conductivity (20°C) is 0.25 (cal/cm-sec·°C), and the specific heat is 0.10 (cal/g·°C). The coefficient of thermal expansion (20 to 100°C) is 16.6×10^{-6} (°C)$^{-1}$.

The very high strength of CuBe implies very long life of the mold, and wear of the mold by moving cores and ejector pins is relatively small. The very high thermal conductivity of the material yields the shortest possible cooling cycle time for SFM machines, and the detail in the mold is the finest that can be obtained for any cast mold process and almost equals that for electroformed nickel. Despite these ideal material properties, CuBe is rarely used for SFM. It is susceptible to etching and attack by the corrosive thermal decomposition products of certain resins (PVC) and residues of SFM blowing agents. It is very expensive (on the order of $4/lb for virgin ingots and $1.50/lb for processed material), and thus mold inserts are required. Molds can be repaired by silver soldering or brazing, but vaporization of beryllium can lead to respiratory problems. Thus, repair of damaged mold inserts should not be attempted; damaged molds should be scrapped.

10.7.4 Aluminum Alloys

Aluminum alloys are extensively used for low-pressure SFM production tools and for CIM short-run or prototype tools. The standard recommended alloys are 2024-T4 and 6061-T651. These materials contain about 1 to 2% Cu, 0.5 to 1% Mg, 0.5% Mn, 4 to 8% Si, 1% Ni, 1% (max.) Fe, small amounts of Ti and Zn, and the rest Al. The pouring temperature is around 550 to 600°C, the tensile strength is about 25 kp/mm^2 (40,000 psi), and the specific gravity is 2.68 (g/cm^3). The thermal conductivity of Al alloys is about 0.3 (cal/cm·sec·°C), the heat capacity is 0.23 (cal/g·°C), and the coefficient of thermal expansion (20 to 100°C) is 19×10^{-6} (°C)$^{-1}$. Three forms of aluminum tooling are used:

1. Tools that have been machined from ingots through conventional metal hogging, coarse and finish machining and polishing. Here shrinkage is essentially nil.
2. Tools that are "roughed out" using the Shaw Unicast Process, where the aluminum is poured over dried plaster or fired ceramic patterns. Machining and polishing are required to meet specifications. Shrinkage is typically 1.1 to 1.3%.
3. Tools that have been "roughed out" using a pressure casting process. After molten aluminum is poured into a frame or sand pack, a steel hob is forced into the frame and held there during cooling of the aluminum. Finish machining is rarely necessary, but polishing is required. The shrinkage of 0.5 to 0.7% is less than that for ceramic cast tools. Although detailing is better with steel hobbing, the ceramic pattern is preferred from an economic viewpoint.

In the casting operation the primary problem is porosity in the mold. This problem is aggravated by increasing processing pressure and thermal cycling and can lead to localized mechanical fatigue and early mold failure. Porosity is not so serious a problem in low-pressure SFM as it is with CIM foam molding or high-pressure SFM. Nevertheless, some mold life can be lost if porosity is excessive.

Corrosiveness of the resin, thermal decomposition products, and blowing agent residue are normally not problems with aluminum. Erosion near gates when processing glass-filled resins and wear around ejector pins or moving cores can pose problems, but these can be overcome through steel bushing inserts.

Some surface hardening of aluminum tools can be achieved by oil annealing treatment, but, in general, the maximum Brinell hardness numbers are on the order of 100 to 130. Tool steel values are in excess of 400. Thus, care must be taken by operators when removing parts stuck in aluminum molds.

Despite these shortcomings, aluminum is an ideal prototype and short- to medium-production-run tool material. As a result, most U.S. low-pressure SFM molders are using aluminum tools.

10.7.5 Electroformed Nickel

Riverside Plastics, Hicksville, New York, one of the first licensees of the Union Carbide low-pressure SFM process, extensively used electroformed

nickel for tool materials. As a result, many major SFM mold innovations
have evolved (and have been patented) from their electroplating technology.
For example, the first large castings (for foam parts of 50 lb or more) were
of electroformed nickel. Very fine details, such as those required in period furniture parts, have been reproduced on electroformed nickel.

Normally a very pure nickel (99.95% with trace cobalt) is used for
plating. The specific gravity of the material is 8.9 (g/cm^3), and the thermal
properties are fair: thermal conductivity (25°C) = 0.22 (cal/cm·sec·°C),
heat capacity (100°C) = 0.112 (cal/g·°C), and coefficient of thermal expansion (0 to 100°C) = 13.3 × 10^{-6} (°C)$^{-1}$. The tensile strength is about 25
kp/mm^2 (38,000 psi).

The keys to plating are in pattern surface preparation so that the
plating reproduces it accurately and in plating with a relatively cold bath.
The pattern surface must be conductive. If the pattern is metal (aluminum,
Wood's metal, zinc), care must be taken to eliminate all grease or oil from
the surface. Fingerprints on the surface will reproduce in the electroplated
mold. Degreasing is mandatory. Some platers will apply a very fine coat
of graphite on the metal surface to help eliminate "cold" spots. For non-
conductive patterns (wood, epoxy, plaster, ceramic), a thin urethane coating
is first applied to the surface to seal it. The urethane is then oven-cured.
A second coating is applied to the first, and, while this is still tacky, a very
fine coating of powdered graphite is applied. Because the graphite powder
is so fine, any detailing, such as wood grain or filigree, will be accurately
reproduced.

Once a suitably conductive coating has been applied to the part, it is
immersed in a "cold" plating bath, and nickel is laid down at about 0.1 mm/
day until 1.5-3 mm have been deposited on the pattern. Frequently, at this
point copper is then plated against the nickel to a total thickness of 10 to 12
mm to strengthen the material and improve the heat transfer characteristics
of the mold. At this point, water lines can be soldered in, and zinc, aluminum, or copper can be cast into the cavity to finish the mold back. Ejector
pins, sprue bushings, vents, and the like can be machined into the composite
mold. More likely for the larger parts, however, the Ni-Cu composite is
machined to fit a special aluminum or brass die block containing water lines,
sprue bushing, runners, gates, ejector pin placements, core pulls, and so on.

It should be pointed out that "cold" nickel plating is a very slow process requiring much lead time, and thus it is not an attractive process to the
prototype tool maker. For production, however, properly prepared electroformed nickel molds can yield 10^6 parts or more. Attempts have been made
to "hot"-plate nickel (at rates of 1 mm/day or more). These molds show
poor mold thickness uniformity, many areas of low density, and high porosity.
Hot plating will produce a very grainy surface that cannot be polished to a
mirror finish without exposing additional defects. Molds also show very short
lifetimes owing to fatigue cracks. Normal shrinkage of electroformed
nickel is on the order of 0.1 to 0.3%, but allowances must be made for the

thickness of the urethane and graphite coatings that are required to make
the pattern conductive.

10.7.6 Steels

Many types of steels have been used for conventional injection mold
tooling. A list of more than 20 AISI types has been developed by the committee of Tool Steel Producers on behalf of the American Tool Steel Industry.
The characteristics of these steels are given in Table 10.7-1. Of these,
the most popular for high-pressure injection molding are the AISI S-7 and
H-13. These have good dimensional stability during hardening and excellent
toughness. S-7 is desired in many applications because of its good machinability after annealing. This allows for machining of runners, gates, and
vents after annealing. Although these materials are normally machined with
deep cavities "hogged" from billets, satisfactory shallow cavities can be
hot-hobbed (300 to 500° C). The typical hob for S-7 or H-13 is H-13 with
either copper cladding or a very thin coating of MoS_2. These coatings improve material flow, reduce scoring, and enhance hobbing details. Note
also that S-7 and H-13 are recommended hobbing steels for softer steels
such as P-1 or P-2.

For foam molding, P-1 can be used. This enables the tool maker
to use cold hobbing techniques and/or softer hobbing steels. If hobbing is
not needed or desired, a water-hardening low-carbon steel such as an AISI
"W" steel can be used. This type of tool steel is easily machined but is
usually hardened after machinig if abrasion or corrosion resistance is needed.

Casting against a ceramic pattern should also be considered. Molders using the Shaw Process of investment casting have cast steel tools
weighing 1 1/2 tons or more. The casting steel is frequently of the water-
hardening category. According to Sors [45], a casting ceramic can be made
from a mixture of ethyl silicate and quartz powder. After casting a slurry
of this mixture around the pattern, the mold halves are dried for 24 to 48
hr in dehumidifed room-temperature air and then heated slowly (3 to 4°C/hr)
to 800°C. The halves are held at this temperature for 3 to 4 hr before slowly cooling to room temperature. Slow pouring of casting steel at 1100°C is
required to prevent air entrainment and thermal stressing of the ceramic.
Suitable surface finishing is required to improve surface gloss and to determine the extent of surface or subsurface defects such as porosity or micro-
cracks. After surface finishing, the casting steel can be heat-treated to
improve abrasion resistance.

As mentioned earlier, most common chemical blowing agents have
residues that are slightly acidic. Thus, there exists a potential for corrosion of the mild steel mold. For molds that are to be used for 10^5 parts or
more, some corrosion protection is recommended. Chromium or nickel
plating is a relatively easy way of protecting the mold surface and improving
the surface finish at the same time. Normally, a 1- to 5-mm nickel plating
is applied over a thin (0.5 mm or less) base plate of copper, and a "brush"

plate of chromium (0.5 mm or less) is then applied over the nickel. This plating technology is identical with that of conventional injection mold plating technology.

One of the most widely used blowing agents, 1,1'-azobisformamide (ABFA), has biureac acid and ammonia as its decomposition residues. Thus, for molds that are to be used with this BA and a glass-filled resin such as nylon or polypropylene, stainless steels in the 400 series are sometimes used. These materials offer excellent abrasion and corrosion resistance. Countering these excellent properties are the reduced thermal conductivity of stainless steel, much higher billet costs, decreased machinability, and increased repair costs. As a comparison of the thermal properties, for a low-carbon water-hardening steel of the "W" category, the specific gravity is 7.86 (g/cm^3), the thermal conductivity (20°C) is 0.088 (cal/cm·sec·°C), the heat capacity (25°C) is 0.11 (cal/g·°C), and the coefficient of thermal expansion (20 to 100°C) is 11×10^{-6} (°C)$^{-1}$. The comparable values for 400 series stainless steel are 7.75 (g/cm^3), 0.055 (cal/cm·sec·°C), 0.11 (cal/g·°C), and 5.5×10^{-6} (°C)$^{-1}$. The tensile strengths for the two are about 110 and 70 kp/mm^2, respectively (and depend on the type of annealing). The hardness of the two materials depends on the types of steels and the method of hardening but are on the order of 230 Brinell hardness number.

Again, although tool steel is not required for SFM molding, it is used extensively because many tool and die operators are more familiar with the processing and machining characteristics of steel than with those of, say, aluminum.

10.8 Orientation, Shrinkage, and Other Process-Related Problems

As seen earlier, injection molding is, in essence, the high-pressure squeezing of a highly complex compressible fluid through a very small hole (gate) into a cold cavity and then the packing of additional material into the cavity to allow for shrinkage as the material cools. Ultimately, the degree of orientation in the material upon cooling to room temperature and the amount of distortion, warpage, and shrinkage are important. Consider first orientation for both amorphous and crystalline materials.

As Koda [46] points out, there are two areas of work in relationship to molding conditions on the mechanical properties of the part. The first deals with the effects of molding conditions on the mechanical strengths, and the second, on the residual strains. The primary method of analysis of frozen orientation in glassy polymers is birefringence. Probably the earliest work on birefringence in glassy materials was by Bailey [47] in 1948. The method was extensively exploited by Spencer and Gilmore [48] in attempting to find a relationship between molded-in strains and the

Table 10.7-1

Characteristics of Plastic Molding Die Steels

AISI type	Annealed hobability	Machinability annealed condition	Heat-treated core strength	Wear resistance	Toughness
Cold hobbing steels					
P-1	Best	Poor	Poor	Good, carb.[a]	Good
P-2	Best	Good	Fair	Good, carb.[a]	Good
P-3	Good	Fair	Fair	Good, carb.[a]	Good
P-4	Good	Good	Good	Good, carb.[a]	Good
P-6	Poor	Fair	Good	Good, carb.[a]	Good
Prehardened steels					
P-20	Poor	Good	Fair	Fair	Good
P-21	Poor	Good	Fair	Fair	Good
Heat hardening steels					
O-1	Poor	Good	Good	Good	Good
A-2	Nil	Fair	Best	Best	Fair
A-4	Nil	Good	Best	Best	Fair
A-6	Nil	Good	Best	Good	Fair
D-2	Nil	Fair	Best	Best	Poor
S-7	Nil	Good	Best	Good	Best
H-13	Nil	Fair	Best	Fair	Best
Special-purpose steels, corrosion resistant					
420	Nil	Fair	Good	Fair	Good

Characteristics of master hob steels, cold and hot hobbing

AISI type	Annealed machinability	Wear resistance	Compressive strength	Toughness	Dimensional stability in hardening
O-1	Best	Good	Good	Good	Good
A-2	Fair	Very good	Very good	Fair	Best
S-1	Fair	Good, carb.[a]	Very good	Best	Good
S-7	Best	Good, carb.[a]	Very good	Best	Best
D-2	Fair	Best	Best	Poor	Best
H-13	Fair	Fair	Very good	Best	Good
H-23	Poor	Fair	Good	Poor	Good

[a]Carburized case.

Dimensional stability in hardening	Comments (Note: all steels in group either carburized or nitrided)
Poor	Maximum ease in hobbing, satisfactory for low pressure
Good	Suitable for medium-high molding pressure
Good	Suitable for medium-high molding pressure
Best	For shallow impressions, minimum size change, for maximum pressure
Good	Difficult to hob, normally machined, has maximum strength
—	Suitable for all types of injection molds; popular for holder blocks; suitable for compression molds when carburized
—	Satisfactory for injection molds
Good	Used for small injection or compression molds requiring abrasive resistance
Best	Used for small, long-run molds requiring abrasive resistance or optical finish
Best	Used for small, long-run molds requiring abrasive resistance or optical finish
Best	Used for small, long-run molds requiring abrasive resistance or optical finish
Best	Used only in compression molds requiring ultimate in abrasive resistance
Best	Best combination of strength, toughness, and stability for injection and compression molding
Good	Good combination of hardenability and size change at intermediate hardness level
Fair	Select for resistance to corrosion

Red hardness and resistance to heat checking	
Poor	Satisfactory compressive strength and ease of machining
Poor	High compressive strength, wear resistance, and stability
Fair	Tough with good compressive strength and wear resistance when carburized
Good	Ideal combination of strength, toughness, stability, machinability, and wear resistance when carburized; good red hardness
Poor	Best wear resistance, poor toughness
Very good	Good red hardness and dimensional stability
Best	Excellent for long-run hot hobbing duty; difficult to machine

Table 10.7-1 (continued)

		Typical chemical analysis			
C	Mn	Ni	Cr	Mo	Other
0.10	—	—	—	—	—
0.07	—	0.50	2.00	0.20	—
0.10	—	1.25	0.60	—	—
0.07	—	—	5.00	0.75	—
0.10	—	3.50	1.50	—	—
0.35	—	—	1.25	0.40	—
0.20	—	4.00	—	—	Al 1.20
0.90	1.00	—	0.50	—	W 0.50
1.00	—	—	5.00	1.00	—
1.00	2.00	—	1.00	1.00	Pb may be added
0.70	2.00	—	1.00	1.00	S may be added
1.50	—	—	12.00	1.00	—
0.50	—	—	3.25	1.50	—
0.35	—	—	5.00	1.50	V 1.00
0.30	0.25	—	13.00	—	—
0.90	1.00	—	0.50	—	W 0.50
1.00	—	—	5.00	1.00	—
0.50	—	—	1.50	—	W 2.50
0.50	—	—	3.25	1.50	—
1.50	—	—	12.00	1.00	—
0.35	—	—	5.00	1.50	V 1.00
0.30	—	—	12.00	—	W 12.00

AISI type	Applications and recommendations	Method of fabrication	Annealed hardness (BHN)	Heat treatment	Usual working hardness Rc
Cold hobbing steels					
P-1	Injection and compression molds	Hobbed	90	Water	60/64°
P-2	Injection and compression molds	Hobbed	103	Oil	60/64
P-3	Injection and compression molds	Hobbed	116	Oil	60/64
P-4	Injection and com pression molds	Hobbed	121	Air	60/64
P-6	Injection and compression molds	Machined	207	Oil	59/61
Prehardened steels					
P-20	Injection and compression molds when carburized	Machined	—	—	30/36
P-21	Injection and compression molds when carburized	Machined	—	—	36/39
Deep hardening steels					
O-1	Injection, compression, and transfer molds	Machined	202	Oil	59/61
A-2	Injection, compression, and transfer molds	Machined	212	Air	59/62
A-4	Injection, compression, and transfer molds	Machined	229	Air	59/62
A-6	Injection, compression, and transfer molds	Machined	225	Air	58/60
D-2	Injection, compression and transfer molds	Machined	223	Air	58/61
S-7	Injection, compression, and transfer molds	Machined	197	Air	56/58
H-13	Injection, compression, and transfer molds	Machined	223	Air	50/54
Special-purpose steels, corrosion resistant					
420	Injection, compression, and transfer molds	Machined	183	Oil-air	50/54
Master hob steels, cold and hot hobbing					
O-1	Cold hobbing	Machined	202	Oil	58/61
A-2	Cold hobbing	Machined	212	Air	59/62
S-1	Cold hobbing	Machined	212	Oil	58/60
S-7	Hot or cold hobbing	Machined	197	Air	58/60
D-2	Cold hobbing	Machined	223	Air	59/62
H-13	Hot hobbing	Machined	223	Air	50/52
H-23	Hot hobbing	Machined	241	Salt	40/42

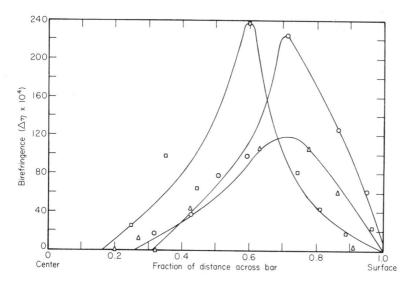

Figure 10.8-1   GPS stress birefringence in 0.085-in.-thick molded bar.
Circles and squares, short shots; triangles, full shot, not flashed [49]

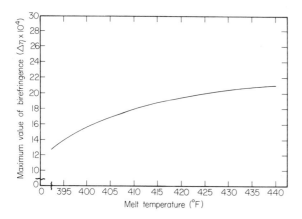

Figure 10.8-2   Effect of melt temperature on orientation (as measured by
maximum birefringence value) for 0.075-in.-thick mold and narrow gate.
Injection pressure, 900 psi; $T_{mold} = 80°F$; GPS [49].

pressure is of no use. The rate of heat removal, of course, determines
the point where the material strains are fully frozen, and thus raising the
mold temperature increases the time required to freeze in any birefringence.
See Fig. 10.8-7. Changing gate dimensions also influences the extent of

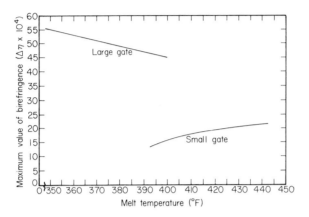

Figure 10.8-0 Effect of melt temperature on orientation (as measured by maximum birefringence value) for 0.075-in.-thick plaque mold and two gate sizes. Injection pressure, 900 psi; $T_{mold} = 80°F$; GPS [49].

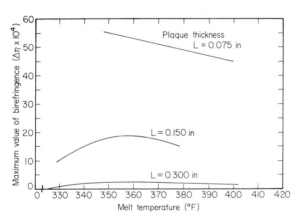

Figure 10.8-4   Effect of melt temperature on orientation (as measured by maximum birefringence value) for three thickness of plaque mold; wide gate. Injection pressure, 900 psi; $T_{mold} = 80°F$; GPS [49].

birefringence in the sample as a function of time of packing pressure, as shown in Fig. 10.8-8.

Before considering crystalline orientation factors, consider the effects of orientation in amorphous materials on the desired physical properties. Jackson and Ballman [50] extended their birefringence study on polystyrenes to consider the mechanical properties of their specimens. They measured tensile strength, elongation at break, elastic modulus, and notched Izod as functions of birefringence [as defined in Eq. (10.8-1)] in both the

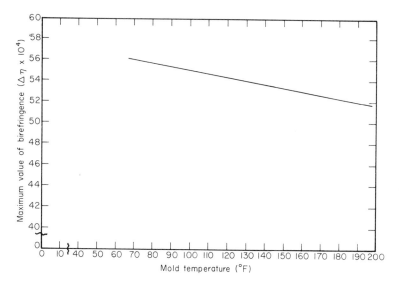

Figure 10.8-5 Effect of mold temperature on orientation (as measured by maximum birefringence value) for 0.075-in.-thick plaque mold; wide gate. Injection pressure, 900 psi; melt temperature, 350°F; GPS [49].

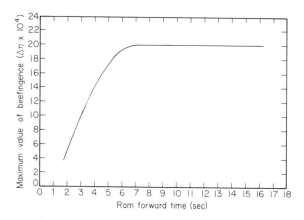

Figure 10.8-6 Effect of ram forward time on orientation (as measured by maximum birefringence value) for 0.075-in.-thick plaque mold; wide gate. Injection pressure, 900 psi; melt temperature, 425°F; mold temperature, 80°F; GPS [49].

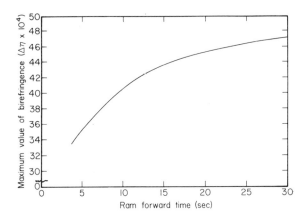

Figure 10.8-7   Same as Fig. 10.8-6 except $T_{mold} = 130°F$ [49].

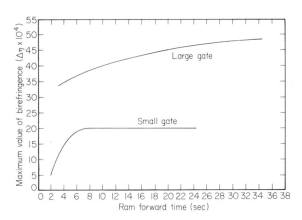

Figure 10.8-8   Effect of gate size on orientation (as measured by maximum birefringence value) for 0.075-in.-thick plaque mold.   Melt temperature, 425°F; mold temperature, 80°F; injection pressure, 900 psi; GPS [49].

cross-machine and in-line machine direction. Their results are shown in Figs. 10.8-9 to 1.8-14. These figures show rather graphically that orientation normally leads to increased mechanical properties. Note that in Fig. 10.8-9 the machine direction tensile strength increases nearly threefold with an increase in orientation birefringence from 10 to 60×10^{-4} units, whereas the cross-machine direction material properties seem to decrease somewhat. The same is apparent with percent elongation to break and notched Izod impact strength. The elastic modulus seems to be independent

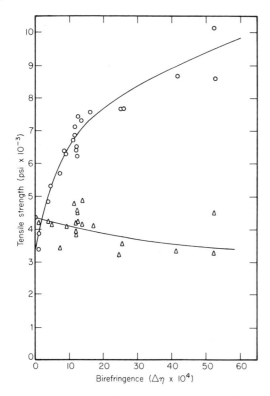

Figure 10.8-9   Effect of injection conditions on tensile strength of GPS. Circles, parallel to injection direction; triangles, transverse to injection direction; 0.100-in.-thick plaque [50].

of the degree or direction of orientation, as shown in Fig. 10.8-4, where the stress-strain curve indicates linearity.

Thomas and Hagen [51] attempted to relate some of the orientation effects of amorphous polystyrene with the major processing parameters, such as molecular-weight distribution. As shown in Fig. 10.8-15, there is little difference in the monodisperse polystyrene (MWD = $\overline{M}_w/\overline{M}_n$ = 1.06) and the polydisperse material (MWD = 2.6). The molding area diagram, shown as Fig. 10.8-16, shows that the materials process in about the same manner, the circles being the polydisperse material and the triangles the monodisperse material. As shown in Table 10.8-1, however, the monodisperse material has 10 to 15% greater tensile strength and as much as 60% greater elongation at break than the polydisperse material. Recalling the work by Ballman and coworkers, this indicates that the monodisperse material is more oriented than the polydisperse material. This was substantiated

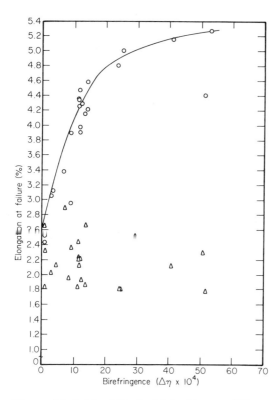

Figure 10.8-10   Effect of injection conditions on elongation at failure of GPS. Circles, parallel to injection direction; triangles, transverse to injection direction; 0.100-in.-thick plaque [50].

by measuring tensile strengths of microtensile specimens milled from the two materials, and the data also indicated that the polydisperse material had a much higher degree of anisotropy (nonuniformity in material properties) than the monodisperse material. Their attempt to relate this anisotropy to die swell from capillaries and then to the plasticizing capability of the lower-molecular-weight materials is inconclusive. Thus, for amorphous materials, orientation leads to improved material properties in the machine direction.

Before considering shrinkage in amorphous materials, consider Hubbauer's unique work on the effect of packing pressure, as determined by density measurements, on surface material properties [52]. In many products, it is necessary to achieve a high-gloss surface. ABS can yield a very high-gloss surface provided processing conditions are proper. As shown in Fig. 10.8-17, gloss (ASTM D-523) is a strong function of molding

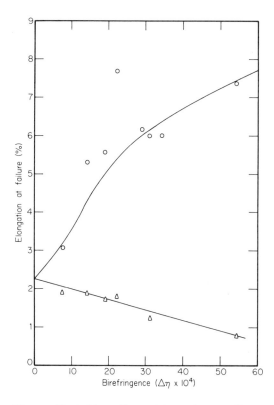

Figure 10.8-11   Effect of injection conditions on elongation at failure for GPS. Circles, parallel to injection direction; triangles, transverse to injection direction; 0.020-in.-thick plaque. Compare results with Figure 10.8-10 [50].

variables. In Fig. 10.8-18, Hubbauer shows that gloss decreases with increasing injection time, is nearly independent of melt temperature, and increases rapidly with increasing mold temperature and thus cooling time. In Table 10.8-2, packing pressure is necessary to achieve relatively high part density—as indicated by the sample thickness—and good gloss. This is also shown in Fig. 10.8-19.

Consider now the shrinkage of amorphous materials and the molding parameters that seem to affect it. Rubin [42] culled shrinkage data from several sources, and while these should be used only as guidelines, they do give a good indication of the effect of molding parameters. As shown in Figs. 10.8-20 to 10.8-26, for amorphous materials represented by polymethyl methacrylate, mold temperature seems to have little effect on overall shrinkage, increasing injection pressure decreases overall shrinkage a

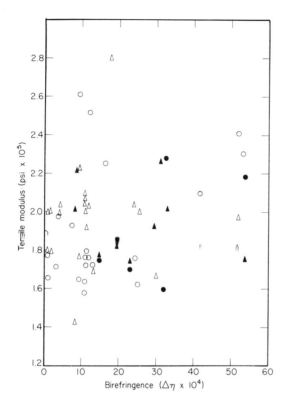

Figure 10.8-12   Effect of orientation on the tensile modulus of GPS. Circles, parallel to orientation; triangles, transverse to injection direction; closed points, 0.100-in.-thick plaque; open points, 0.020-in.-thick plaque [50].

little as does rate of injection, and increasing melt temperature seems to decrease shrinkage. This examination is supported by Ebnuth and Lube [53]. It can thus be concluded that shrinkage in amorphous materials is not so important as orientation to the final part quality. One additional characteristic of amorphous materials is crazing. It is fairly well documented that increasing orientation can lead to increased tendency to craze. Curtis has studied the fracture surface energy of PMMA as a function of its orientation as measured with birefringence [54] and has found an abrupt decrease in fracture surface energy in the machine direction to the extent that attempts to break a sample in the cross-machine direction always resulted in fibrillation of the material in the machine direction instead. This is shown in Fig. 10.8-27. Menges and Wubken [56] recently reexamined shrinkage data from amorphous styrene in an effort to determine shrinkage in machine and cross-machine directions as functions of the distance from

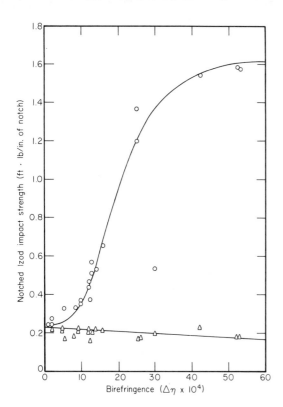

Figure 10.8-13 Effect of injection molding conditions on impact strength of GPS. Circles, parallel to injection direction; triangles, transverse to injection direction; plaque thickness: 0.100 in. [50].

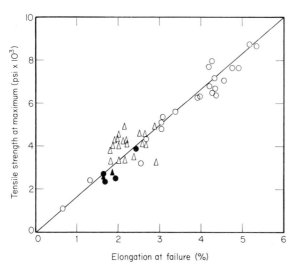

Figure 10.8-14 Cross-plot of tensile strength at ultimate to elongation at break for GPS showing no apparent correlation with degree of orientation. Circles, data parallel to direction of flow orientation; triangles, data perpendicular to direction of flow orientation; solid shapes, 0.020-in.-thick plaque; open shapes, 0.100-in.-thick plaque [50].

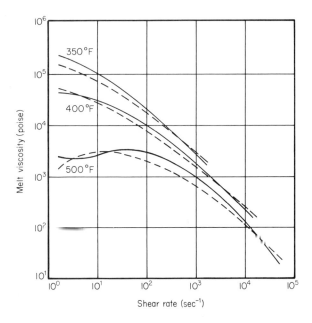

Figure 10.8-15   Viscosity curves for monodisperse (solid line) and poly-disperse (dashed line) GPS of same weight-average molecular weight $\overline{M}_W$. Three temperatures are shown [51].

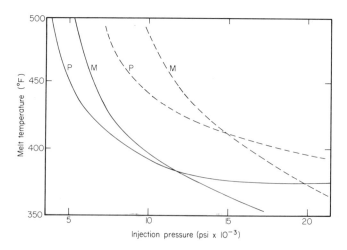

Figure 10.8-16   Molding area diagrams of monodisperse (M) and polydisperse (P) GPS of same weight-average molecular weight $\overline{M}_W$. Solid line, short shot; dashed line, flashing [51].

Table 10.8-1

Tensile Properties of Monodisperse and Polydisperse Polystyrenes of
Similar Weight-Average Molecular Weight [51]

$$QX4451: \quad \overline{M}_w = 190,000, \quad \overline{M}_w/\overline{M}_n = 2.6$$

$$S109: \quad \overline{M}_w = 193,000, \quad \overline{M}_w/\overline{M}_n = 1.06$$

Melt temp. and material	Injection press. (psi)	Tensile strength (psi)	Yield behavior	Rupture elongation (%)
375°F				
QX4451	17,500	6650	Very brittle; failure; no yield	3.3
S109	16,400	7720	Almost yields	5.2
425-430°F				
QX4451	9,600	6320	Very brittle; failure; no yield	3.4
S109	10,200	7020	Went through a yield point	5.9
460-465°F				
QX4451	7,700	5870	Very brittle; no yield	3.3
S109	8,650	6500	Almost yields	4.2
495°F				
QX4451	5,600	6160	Very brittle	3.4
	9,000	6030	Failure	3.1
	10,600	5610	No yield	3.5

the surface of the sample. Again machine parameters such as injection
pressure, melt temperature, and injection rates were examined carefully.
Their results basically support the orientation data of Ballman and coworkers, as shown in Figs. 10.8-28 to 10.8-32. Extensive shrinkage in the
highly oriented material near the surface is expected if the material is reheated to 120°C for 5 min. Their results again support the early birefringence work of Ballman and coworkers and many others since.

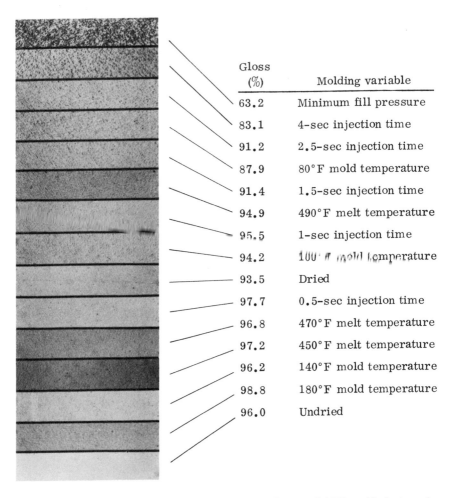

Gloss (%)	Molding variable
63.2	Minimum fill pressure
83.1	4-sec injection time
91.2	2.5-sec injection time
87.9	80°F mold temperature
91.4	1.5-sec injection time
94.9	490°F melt temperature
95.5	1-sec injection time
94.2	100°F mold temperature
93.5	Dried
97.7	0.5-sec injection time
96.8	470°F melt temperature
97.2	450°F melt temperature
96.2	140°F mold temperature
98.8	180°F mold temperature
96.0	Undried

Figure 10.8-17   Photomicrographs of the surfaces of ABS molded at various processing conditions.   Note in the accompanying table the measured gloss of the surface [52].

Extensive effort has been brought to bear on the problem of orientation of crystalline materials. Clark [57] and Collier and Neal [58] have examined the morphological aspects of crystallization under nonisothermal high-shear environs. Clark shows that during the growth of crystallities of Delrin acetal two or possibly three layers of material are formed. Each layer has mechanical characteristics unique unto itself, as we shall point out later. Let us consider the microscopic process of crystal growth. As

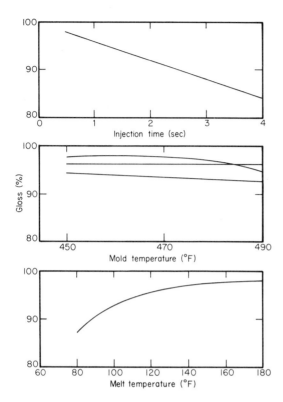

Figure 10.8-18   Effect of several operating parameters on surface gloss of three types of Marbon ABS as measured with the Gardner glossmeter. Injection time and mold-temperature effects for all three resins appear similar [52].

Table 10.8-2

Effect of Maximum and Minimum Fill Pressures on the Thickness and Gloss of Several Compounds [52]

Compound	Pressure (psi)	Thickness (mils)[a]	Gloss (%)
A	7150	99.9	93.1
	1430	96.7	63.4
B	7150	99.8	96.0
	2600	97.0	80.8
C	7150	99.8	96.5
	3250	97.3	78.3

[a]Average thickness of dead end of plaque.

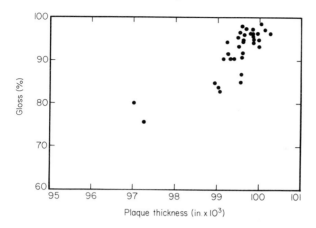

Figure 10.8-19 Effect of thickness of nominal 100-mil plaque at plaque dead end on measured percent gloss for multiple grade ABS [52].

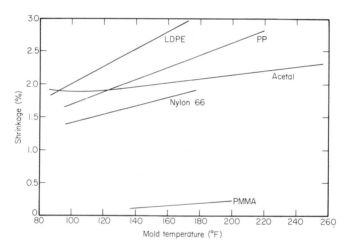

Figure 10.8-20   Effect of mold temperature on shrinkage of several plastics [42].

Statton and Geil [59] and Reneker [60] have pointed out, a polymer crystal grows by laying the molecule back and forth across the end of an already formed crystal. Clark shows that the orientation of these folded chains depends on surface nucleation, temperature gradients, and induction time for spontaneous nucleation.   Crystallization begins at the mold wall and proceeds into the interior as is shown in Fig. 10.8-33.   The polymer at the mold surface is subjected to the high-shear flow of melt filling the remaining

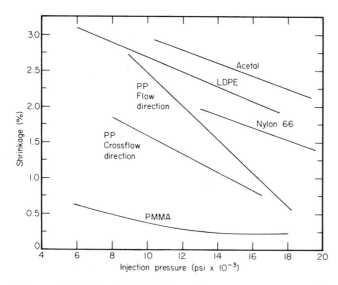

Figure 10.8-21   Effect of injection pressure on shrinkage of several plastics [42].

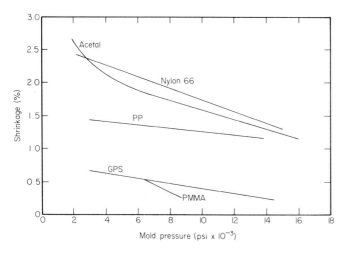

Figure 10.8-22   Effect of in-mold injection pressure on shrinkage of several plastics [42].

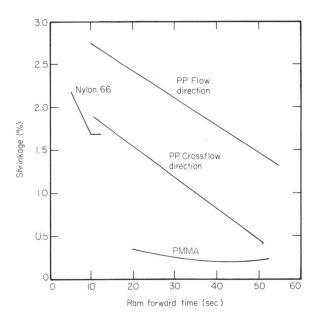

Figure 10.8-23   Effect of ram forward time on shrinkage of several plastics [42].

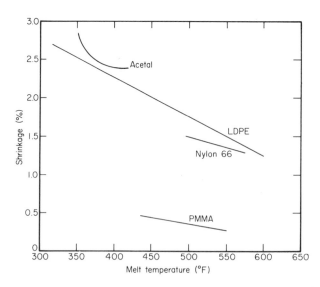

Figure 10.8-24   Effect of melt temperature on shrinkage of several plastics [42].

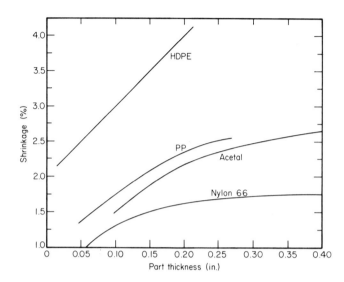

Figure 10.8-25   Effect of part thickness on shrinkage of several plastics [42].

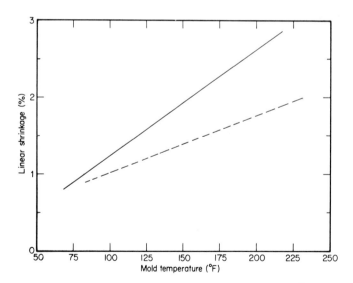

Figure 10.8-26   Effect of mold temperature on shrinkage of Avisun (Amoco) PP. Solid line, in flow direction; dashed line, perpendicular to flow direction [52].

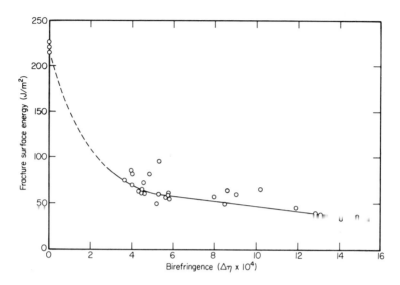

Figure 10.8-27 Effect of orientation (as measured by birefringence value) on fracture surface energy for PMMA [55].

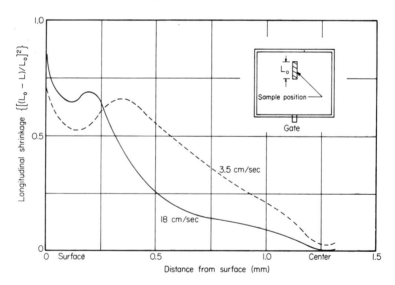

Figure 10.8-28 Effect of injection speed on longitudinal shrinkage as a function of depth into wafer; measured before and after annealing.

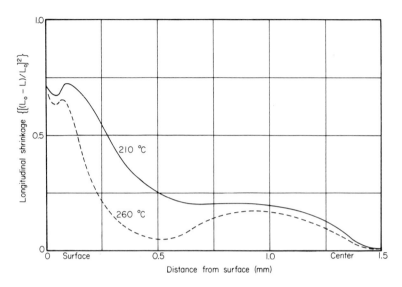

Figure 10.8-29   Effect of melt temperature on longitudinal shrinkage as a function of depth into wafer; measured before and after annealing; same sample position as in Fig. 10.8-28 [56]

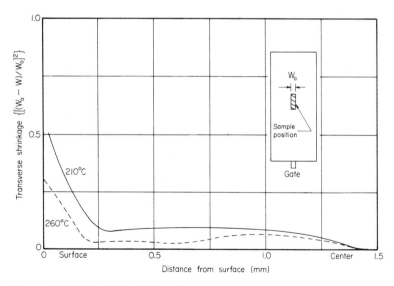

Figure 10.8-30   Effect of melt temperature on transverse shrinkage as a function of depth into wafer; measured before and after annealing [56].

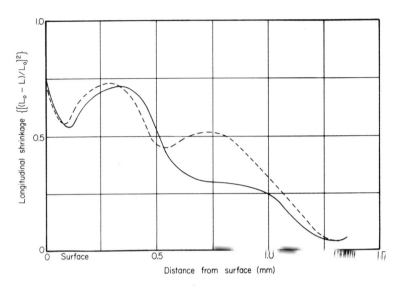

Figure 10.8-31   Effect of proximity to gate on longitudinal shrinkage as a function of depth into wafer. Solid line, wafer far from gate; dashed line, near gate; measured before and after annealing [56].

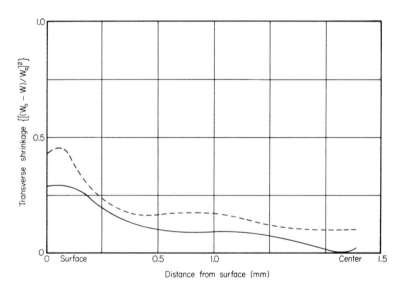

Figure 10.8-32   Effect of proximity to gate on transverse shrinkage as a function of depth into wafer. Solid line, wafer far from gate; dashed line, near gate; measured before and after annealing [56].

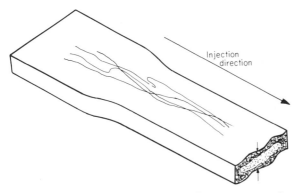

Figure 10.8-33   Clark's proposed mechanism of crystallization of homopolymer polyacetal during injection. See accompanying photomicrograph, Figure 10.8-34 [57].

portion of the parts. This creates a line of nuclei in which the molecular chains are oriented parallel to the flow direction. Rapid flow of heat from the melt toward the cold-mold-wall establishes a thermal gradient, and since lamella nuclei grow preferentially in the direction of the thermal gradient, these molecular chains are oriented perpendicular to the melt flow direction. If cooling is sufficiently rapid, the core material will drop below the temperature at which nuclei are spontaneously formed, and thus the spherulites formed have no overall preferred orientation of lamellae. As shown in Fig. 10.8-34, these two or three layers, depending on how the surface layer is categorized, are clearly observed and can be measured. Clark observed that for his 0.125-in.-bar, the lamellae oriented by both thermal gradients and surface nucleation form a layer about 0.2 mm in thickness. The lamellae oriented only by thermal gradients are 0.2 to 0.9 mm below the surface, and the unoriented lamellae are found from 0.9 mm to the centerline. Furthermore, Clark [62] shows that increasing the temperature difference between the melt and the mold increases the penetration of the molecules having preferred orientation. Increasing injection pressure increases the freezing point (about 1°F/1000 psi or so), and thus an increase in the depth of penetration of lamellae having preferred orientation with increasing pressure and constant mold wall temperature is seen. Collier and others have subsequently found that the time-temperature-pressure-shear history of the material while in the melt state can also influence the depth of penetration of the lamellae with preferred orientation.

This molecular interpretation of the phenomena that occur while the material is flowing into a cold mold at high pressures is necessary to the understanding of shrinkage of crystalline materials. It is certain that to some extent the model described by Clark is applicable to most highly

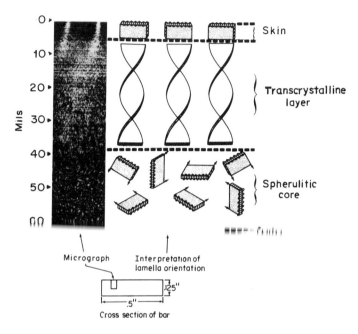

Figure 10.8-34   Photomicrograph of polyacetal homopolymer cross-section showing skin, where lamellae are perpendicular to surface and extrusion direction; transcrystalline region, where lamellae are perpendicular to surface only; and the spherulitic core, where no preferred orientation is seen [57, 61].

crystalline plastics, such as nylon, acetal, chlorinated polyethers, PP, and HDPE. To a lesser extent, the semicrystalline materials such as PVC and LDPE are also experiencing crystallization while under nonisothermal and anisotropic stress fields. Scherpereel [63] has examined these parameters for polypropylene not only in the center of an end-gated tensile test bar mold but also in the gate region. In Fig. 10.8-35, Clark's hypothesis of two- or three-layer lamellar growth holds for PP as well. More importantly, the type of crystallization, as given in Table 10.8-3, and the balance between the two main types do affect the tensile test results. His results are not entirely consistent with the Clark hypothesis in that increasing mold temperature always seems to increase tensile properties, but increasing melt temperature, and thus increasing thermal driving force, seems to decrease tensile properties. Nevertheless, they do indicate that the mold-to-melt temperature balance must be controlled in order to control material properties. Scherpereel has also observed another type of material which he refers to as structureless material that seems to predominate in the gate

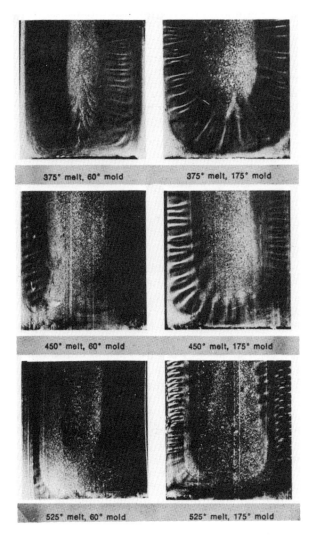

Figure 10.8-35   Optical photomicrographs of injection-molded polypropylene tensile bars. Samples taken at midpoint of bars, perpendicular to flow direction. Sample thickness, 10 μm; column, change in melt temperature at constant mold temperature; row, change in mold temperature at constant melt temperature [63].

Table 10.8-3

Mechanical Properties of Four ABS Materials [63]

Material	Yield (psi)		Break (psi)		Elongation at break (%)	
	Nonweld	Weld	Nonweld	Weld	Nonweld	Weld
A	7120	—	6145	4610	94	8
B	7550	—	6590	3710	30	6
C	7170	7010	5630	6400	40	20
D	8320	—	7040	2820	46	5

region at low melt and mold temperatures. This is shown in Fig. 10.8-36. Ahead of this structureless material is a rather uniform spherulitic crystal growth, and thus Scherpereel believes that this structureless material is frozen before it can nucleate and grow into lamellae of either type. Thus, it would seem that gate regions may have entirely different material properties from those of the bulk part. This is an important fact if it is subsequently found that material failures are beginning in the vicinity of a gate in a crystalline material. Brydson reproduces some data that show that in addition to the lamellar orientation espoused by Clark crystalline materials such as PP are also affected by orientation. As shown in Fig. 10.8-37, the gate is moved to various positions on the rectangular plaque. At various locations on the plaque, samples are cut in the machine direction and in the cross-machine direction and subjected to flexural load to break. As can be seen, changes in gate geometry and location while holding temperatures and pressures constant can lead to very nonuniform material properties.

As shown in Figs. 10.8-20 to 10.8-26, average shrinkage values for crystalline materials are 10 times those for amorphous materials. They are also much more sensitive to machine parameters. In Fig. 10.8-10, for example, increasing the mold temperature on PP from 100 to 220°F doubles the overall shrinkage from about 1.5% to nearly 3%. As expected, increasing the pressure on the material decreases the shrinkage, since the cavity is being packed, as shown in Fig. 10.8-11. This packing, according to Clark, increases the freezing temperature of the material, and thus a greater amount of oriented lamellae is expected. This, combined with Fig. 10.8-11 data, indicates that orientation of the lamellae results in higher-density parts and thus lower shrinkage. This is substantiated in Figs. 10.8-12 and 10.8-13. Conversely, increasing the thickness of the part

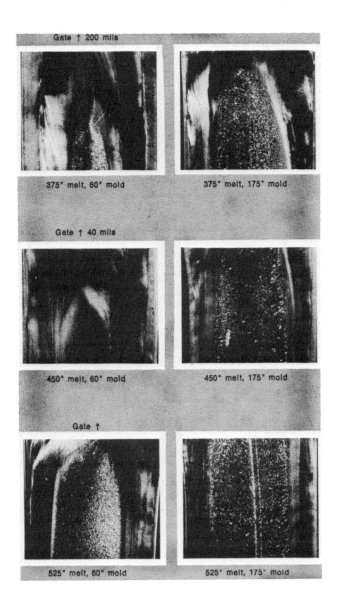

Figure 10.8-36 Optical photomicrographs of injection-molded polypropylene tensile bars. Samples taken in the gate region showing minimization of crystalline organization when compared with Figure 10.8-35. Column, constant mold temperature, varying melt temperature; row, constant melt temperature, varying mold temperature [63].

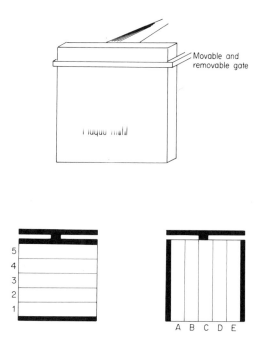

Figure 10.8-37 Schematic of test plaque mold cavity used to obtain data given in accompanying table. Note that gate section can be altered and positioned to obtain any gate configuration, including all shown in the accompanying table. Note also the numbering and lettering scheme used to identify flexural bars. Shaded areas indicate discarded plaque material [64].

Table for Figure 10.8-37

Effect of Gate Type, Length, and Position on Orientation Effects in Plaque Moldings [64]

Gate type and position

		Sample									
		M	N	O	P	Q	R	S	T	U	V
Gate length (in.)		0.125	0.06	0.125	0.06	0.125	0.06	0.125	0.06	0.125	0.06
Flexural load to break (kg)	1	2.55	2.55	4.5	4.7	2.2	2.1	4.6	4.65	4.35	4.45
	2	2.4	2.3	4.75	4.75	2.1	1.95	4.65	4.75	4.4	4.25
	3	2.4	2.4	4.9	4.95	2.05	2.15	4.9	4.9	4.50	4.65
	4	2.4	2.4	5.25	5.3	2.10	1.95	5.0	5.25	4.85	4.90
	5	2.65	2.05	5.4	5.6	2.2	2.0	5.5	5.40	5.25	5.3
	A	6.1	6.1	4.65	4.65	5.15	5.25	4.6	4.6	4.3	4.4
	B	6.3	6.3	4.75	4.75	5.45	5.5	4.95	4.85	4.4	4.65
	C	6.3	6.2	5.0	5.0	5.3	5.5	5.1	5.0	4.5	4.7
	D	6.25	6.1	5.3	5.35	5.25	5.35	5.3	5.3	4.9	5.1
	E	5.9	5.8	5.3	5.35	5.05	5.15	5.3	5.5	5.1	5.2

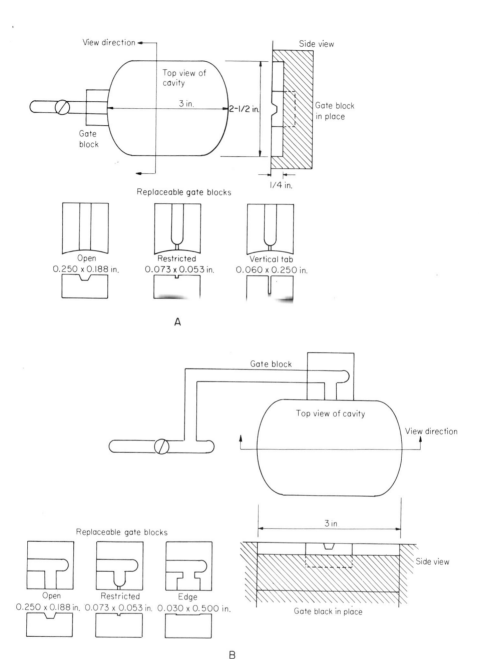

Figure 10.8-38   A: Schematic of mold design used to generate data given in Figs. 10.8-39 and 10.8-40. Note the removeable blocks. B: Schematic of mold design used to generate data given in Figs. 10.8-39 and 10.8-40. Again note removeable gate blocks [65].

increases the proportion of material that is unoriented, and an increase in shrinkage should occur. This is supported in Fig. 10.8-15. Again, location of gates should play an important role in the extent of shrinkage. Williams and Pancoast [65] examined the sizes and locations of gates on the overall shrinkage of PP in a small oval disk. Their experimental setup is shown in Fig. 10.8-38, where both end and edge gating configurations are available. The effect of gate configurations on volumetric shrinkage is shown in Fig. 10.8-39. The relationship of gate configuration to volumetric shrinkage is shown in Fig. 10.8-40 to be dependent on length of packing pressure application as well as mold temperature. Woebcken [66] has examined the shrinkage data of many crystalline plastics in an attempt to categorize their extent of shrinkage. He concludes that no satisfactory theory is available to date that explains all types of shrinkage effects in crystalline materials and believes that semicrystalline materials such as LDPE are even more unpredictable in terms of knowing suitable shrinkage values a priori.

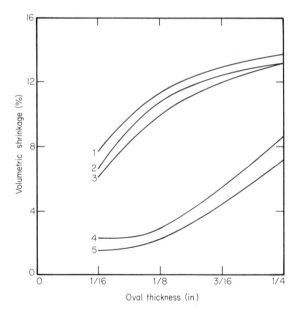

Figure 10.8-39   Effect of PP oval part thickness on volumetric shrinkage for various gating schemes of Fig. 10.8-38. 1, edge, side gate; 2, restricted, side gate; 3, restricted, end gate; 4, open, side gate; 5, open, end gate [65].

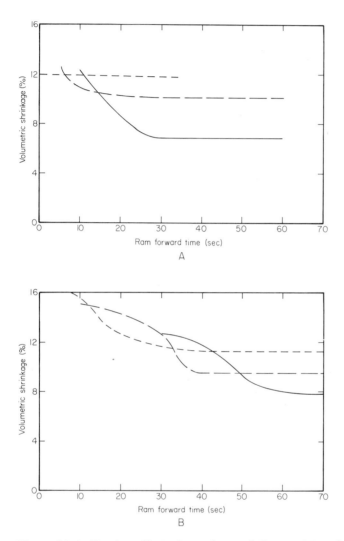

Figure 10.8-40   A: Effect of ram forward time on PP volumetric shrink-age for three gating schemes. Solid line, open gate, 0.041-in.^{2} area; dashed line, vertical tab gate, 0.015-in.^{2} area; dotted line, restricted gate, 0.004-in.^{2} area; mold temperature, 55°F. B: Same as A except mold temperature is 190°F [65].

The strength of weld lines in molded parts remains a very difficult property to predict. However, here too, some headway is being made. Hubbauer [52] examined weld lines in ABS and PVC to determine the types of molding conditions that affect their strength. As shown in Fig. 10.8-41 the weld line of PVC is much weaker than that for ABS, relative to the tensile strength of the bulk material. Part of this is certainly attributable to the absence of crystallinity in ABS and the presence of some crystallinity in the PVC. This result is not substantiated at this time for these two materials, however. Hagerman [67] examined the weld line area of ABS using a scanning electron microscope and found that the weld area shows little "cross-over" flow. Thus, when a weld line is broken, there is no fibril drawing from the craze field as there is with a break in the bulk material. His micrograph is shown as Fig. 10.8-42. The material strengths are also abnormally low, as shown in Table 10.8-4. Thus, at least for amorphous materials, weld lines are zones of weakness, and one might speculate that without proper knitting across the weld line crystalline materials might exhibit even poorer weld line performance.

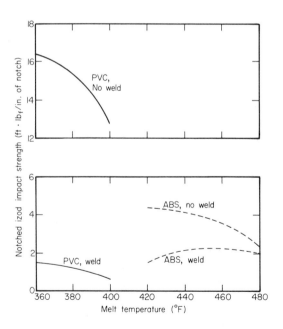

Figure 10.8-41   Effect of weld line on notched Izod impact strength of PVC and ABS [52].

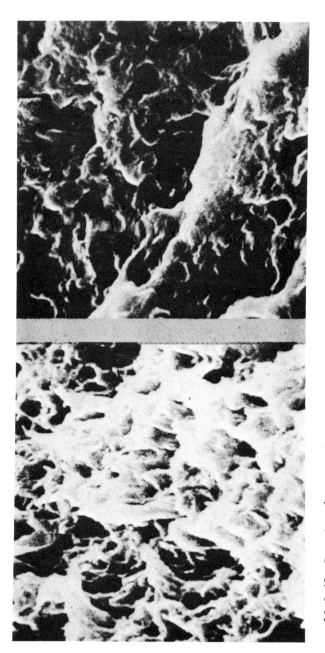

Figure 10.8-42 Scanning electron micrograph of the broken edge of an ABS tensile test bar. The left photo shows characteristic drawing at the fracture of a solid unwelded flow. The right photo shows little drawing or knitting across the weld line. A characteristic failure surface for all cold-knit materials, including thermosets [67].

Table 10.8-4

Effect of Molding Conditions on Weld Strength of ABS [51]

Injection pressure (psi)	Injection speed (sec)	Izod impact strength (ft·lb/in.)
1,300	2	2.0
15,600	2	2.1
6,500	4	1.9
6,500	0.5	2.0
1,300[a]	2	1.8

[a]Using silicone mold release.

10.9 Conclusion

We are just beginning to understand the complexities of injection molding. We are still not sure how to predict flow patterns in complex molds and flows in parts with variable wall thickness, warpage, differential shrinkage, and so on. Certainly there are many engineers working on these problems, but the area is so large and frought with false starts that many workers feel that true prediction of flow into injection molds (regardless of mold geometry) will not be a reality until the 1980s and that adequate prediction of final material properties from set molding conditions will not occur before the next century. Nevertheless, good guidelines are available. Benchmarks and/or starting points for most injection molding operations can be obtained using known engineering methods.

Problems

P10.A Following the arguments of Ballman et al. [6-8], show that Eq. (10.2-4) results from integration of Eq. (10.2-1). Note that I is shown to be the mean value of the rheological constant across the stream. Physically, what is another meaning for I? In the arguments leading to Eq. (10.2-9), the point is made that the polymer is incompressible. In Sec. 10.4, considerable discussion centers around the Spencer-Gilmore form for the compressibility of a polymer. Resolve this apparent dichotomy.

P10.B MIPS has a power-law coefficient and index value of $K = 1.7$ and $n = 0.35$ at 400°F. The viscosity at 10^2 sec^{-1} is 0.095 lb·sec/in.^{2} at this temperature. The Arrhenius energy of activation is 22 kcal/mol. The mold wall temperature is $100°F$. The temperature profile across the material filling the runner is as follows: $T = 400°F$ for $0 < r < 0.45d$. T decreases linearly from 400 to $100°F$ over $0.45d < r < 0.5d$. Obtain an approximate value for I, the mean value of the rheological coefficient, from Eq. (10.2-5). Show that if the region of influence of wall temperature does not change over the interval $0 < x < X$, $I' = I$ in the equation. Then show that for isothermal flow, $I' = A$. What are the units on $R(t)$ in Eq. (10.2-9)?

P10.C Experimentally it has been found that the resistance to flow in a long cylindrical channel increases exponentially with time according to

$$R(t) = R_0 \exp(at) \qquad\qquad (10.C-1)$$

$R_0 = 150 \ lb_f \cdot sec^{1/m}/in.^{3+1/m}$. The reservoir pressure is 10,000 psi, and $a = 0.5 \ sec^{-1}$. Make a plot of the injection velocity as a function of time t in seconds. The runner diameter is now reduced 10%. What is the new value for R_0? Make another plot of the injection velocity as a function of time. All things being constant, replot this new injection velocity as a function of time if the temperature of the resin has increased $15°F$. Use the data given in P10.B for this calculation.

P10.D Show that Eq. (10.2-19) is correct from Eq. (10.2-18). What is the functional meaning of B in terms of resistance, viscosity, Arrhenius dependency, etc.?

P10.E Defend the heuristic argument leading to Eq. (10.2-22). Support these results by extracting the necessary information from Fig. 10.2-3.

P10.F From the Frizelle-Paulson data of Fig. 10.2-7, extract the terms X_∞/R and b using the exponential expression (10.2-23). Plot your calculated results, and compare them with the experimental data of Fig. 10.2-7.

P10.G Carry through the argument that doubling the cavity depth in an isothermal runner decreases the pressure loss by 3.4 times, assuming $n = 0.37$. Then show that if $n = 0.535$, doubling the cavity depth decreases pressure loss by 4.2 times, in accordance with the Frizelle-Paulson data. Describe qualitatively the effect a decreasing mean temperature would have on these results.

P10.H Barrie states in Eq. (10.2-25) that the hoop stress is dependent on a Newtonian form of tensile viscosity. In Chap. 5 it was argued that extensional viscosity is probably related to Troutonian viscosity, which in turn is three times the shear viscosity. That being the case, rewrite Eq. (10.2-25) for a power-law Troutonian viscosity. Then consider injection molding of a 0.93-sp. gr. LDPE having a viscosity at $374°F$ of 0.2 $lb_f \cdot sec/in.^2$ at $10^2 \ sec^{-1}$ and 0.014 at $10^4 \ sec^{-1}$. The shear stress $4Q/\pi R_0^3$ in the cavity is $10^3 \ sec^{-1}$. The sprue-gate radius R_0 is 0.25 in. Plot the pressure loss in shear [Eq. (10.2-24)] and in tension [Eq. (10.2-25)] for an 0.100-in.-thick disk as a function of radius R. Give a rational argument for the shape of the curve of pressure loss in tension.

P10.I   Take the Harry-Parrott data of Figs. 10.2-10B and C, and replot
the velocity data against distance on semilogarithmic coordinates. Obtain
the terms needed in the Ballman et al. empirical equation (10.2-19). What
conclusions can you draw about the Harry-Parrott data?

P10.J   Given the following data on ABS, obtain the Kamal-Kenig dimensionless groups, the Reynolds number, the Prandtl number, and the Brinkman
number:

$$P_f = 15,000 \text{ psi}$$

$$\rho_m = 1.07 \text{ g/cc}$$

$$R_0 = 0.05 \text{ in.}$$

$$h = 0.0625 \text{ in.}$$

$$\text{viscosity} = 0.1 \text{ lb}_f \cdot \text{sec/in.}^2 \text{ at } 200 \text{ sec}^{-1} \ (450° \text{F})$$

$$n = 0.5$$

$$T_1 = 450° \text{F}$$

$$T_0 = 70° \text{F}$$

$$c_p = 0.5 \text{ Btu/lb} \cdot ° \text{F}$$

$$k = 0.14 \text{ Btu/ft} \cdot \text{hr} \cdot ° \text{F}$$

Combine these values to determine the coefficient in Eq. (10.2-29). What
must be said about the orders of magnitude of the two terms on the right of
the equal sign?

P10.K   Consider the flow of an ABS having the properties given in P10.J
in a half-round runner with a radius of 1/8 in. Two cavities split off the
runner at each junction point. Obtain the hydraulic radius R_h for this type
of runner. Then determine the pressure drop down the main runner before
the junction point if the shear rate based on the hydraulic radius is 1000
sec^{-1}. If the main objective of the runner is to split the flow at the junction
such that one-eighth of the original material goes to each cavity and the rest
continues along the main runner, what must be the radius of the half-round
runner connecting the main runner to the cavity?

P10.L   As an extension of P10.K, consider the second junction where two
more cavities split off the runner. Again, one-eighth of the original material must go to each cavity. What is the radius of each of these half-round
runners? Extend this analysis to include the third junction point and the
last junction point. Remember, the objective is to get an equal amount of
material to each cavity by changing the runner dimensions. Neglect gate
resistance.

P10.M   Gates between runners and cavities have frequently been referred
to as valves. Why is this the case? Consider the form for gate resistance,
Eq. (10.3-8).

P10.N Flow in a manifold runner system can be compared under certain conditions to the water hammer problem of a hydraulic surge tank. Consider a large reservoir with a pressure P_1 to which is connected a pipe of diameter D and length L. Consider a single cavity at point L, where the pressure is P_2. The cavity has a diameter d, and the melt level in the cavity at time t is h. Show that the material balance at the instant when the melt reaches the end of the runner system is

$$(\pi d^2/4) \, dy/dt = (\pi D^2/4)V \tag{10.N-1}$$

where V is the velocity of the material in the runner. Then show that a momentum balance gives

$$-\rho L(\pi d^2/4) \, dV/dt = (P_1 - p_2)(\pi D^2/4) + \tau(\pi DL) \tag{10.N-2}$$

Let the shear stress at the wall be given by

$$\tau = K\{(3n + 1)/nR\}^n |V|^{n-1}V \tag{10.N-3}$$

where R = D/2. Select boundary conditions such that $V = V_0$ and y = Y when t = 0. What is an appropriate value for Y in terms of V_0? Hint: Set all time derivatives to zero and solve the equations. Then let $\delta = (D/d)^2$, $P = (P_1 - p_2)/\rho V_0^2$, $v = V/V_0$, $y' = y/Y$, and $\theta = V_0 t/L$. Show that these equations become

$$dy'/d\theta = Bv \tag{10.N-4}$$

$$dv/d\theta = (P + A|v|^{n-1}v) \tag{10.N-5}$$

Define A and B. Then show that P can be written in terms of y' such that

$$dv/d\theta = (Cy' + A|v|^{n-1}v) \tag{10.N-6}$$

Note that these equations must be solved simultaneously subject to $v(0) = 1$ and $y'(0) = 1$. Show that if A is zero, one solution to $y'(\theta)$ is a sinusoidal function. Make a plot of y' vs. θ and y' vs. v for various values of A, B, and C. What is the physical significance of A = 0? Note that even when A is not zero the rate of flow into the cavity is nonlinear. Plot $dy'/d\theta$ for several values of A, B, and C to demonstrate this. Hypothesize what would happen if (1) the fluid were nonisothermal, (2) if the gate resistance were nonzero, and (3) if the gate resistance were highly exponential with regard to V_0. Write appropriate modifications to the equations given to take into account nonzero gate resistance. Propose a series of experiments to determine whether fluid pumped into a manifold runner can result in oscillatory flow.

P10.O Obtain Eqs. (10.3-31) and (10.3-32). What are the critical assumptions that lead to these equations? How can these be relaxed?

P10.P Make a plot of $P'_{i0}/X'G_0$ and $P'_{gk}/X'G_0$ against X/X' for various values of $1 + r^4x'$ from Eqs. (10.3-33) and (10.3-34). Change r values by 10% for each plot.

P10.Q Determine the time to freeze the gate, the centerline temperature at that time, and the location of the freezing line in the part for the cooling of polycarbonate initially at 550°F. The freezing temperature is approximately 300°F, and the mold temperature is 105°F. $k = 0.14$ Btu/ft·hr·°F, $c_p = 0.5$ Btu/lb·°F, $\rho = 80$ lb/ft^3, the gate half-width is 0.020 in., and the cavity radius is 0.25 in. The configuration of the gate is a rectangle, 0.040 × 0.200 in. wide. The cavity can be approximated by a cylinder.

P10.R When two interfaces meet too quickly at a weld line area, the plastic can scorch. In all but vacuum molds, a layer of air is pushed ahead of the advancing fluid front. When two fluid fronts are advancing, the air is compressed adiabatically from atmospheric pressure to a pressure approximating that of the injection pressure. This means that the temperature can rise very rapidly. The adiabatic compression equation is given as

$$T_2/T_1 = (P_2/P_1)^{(k-1)/k} \qquad (10.R\text{-}1)$$

Calculate the theoretical increase in temperature if $P_2 = 1500$ psi, $P_1 = 15$ psi, and $T_1 = 70°F = 530°R$. $k = 1.4$ for air.

P10.S Consider cooling a mold block 12 in. on a side with water at 50°F. There are four gun-bored holes through the mold block, each with a 7/16-in. diameter. Determine the velocity required to achieve a Reynolds number, $Re = DV\rho/\mu$, of 10,000. If the lines are serpentined with 7/16-in. hose 12 in. long, calculate the pressure drop between the inlet and exit lines at the mold block. If the lines are manifolded with 7/16-in. hose 6 in. long on each side of the block, calculate the pressure drop. For a water velocity of 2 ft/sec in the 7/16-in. holes, calculate the individual convective film heat transfer coefficient from Eq. (10.6-4). Assuming that the Kamal-Kenig conduction resistance coefficient is infinite and the fouling factor is negligible, determine the overall heat transfer coefficient for an aluminum mold. The water lines are spaced 2 in. apart and have centerlines 1 in. from the mold surface.

P10.T The circulating pump on the chiller has a rated output pressure of 60 psig. A 50:50 mixture of water and ethylene glycol is the refrigerant exiting the chiller at 20°F. The coolant must pass through 20 equivalent feet of 5/16-in.-diameter channel before exiting into a recirculating reservoir at 0 psig. The overall heat transfer coefficient per unit length of channel is 20 Btu/ft^2·hr·°F. Eight feet of channel actively transfer heat. Determine the exit temperature of the coolant and its velocity through the channel.

P10.U (Following Acrivos et al. [31]) Consider a manifolded mold where coolant flows into the bottom of the entrance manifold and out of the bottom of the exit manifold. Assume turbulent water flow in each manifold chamber and also in the connecting tubes. Set up the differential equations and the boundary conditions that determine the variation of velocity V_1 in the entrance manifold and V_2 in the exiting manifold as a function of the distance from the bottom of the manifold block. Find a solution to the equations when the friction effect in the manifolds is negligible. Express the result in the form of the difference in the maximum connecting tube velocities as a function of the pertinent dimensionless quantities. Hints: Make a momentum balance on the entrance manifold, assuming incompressible fluid, no pressure drop in the block, all momentum destroyed at each hole in the manifold, a large number of holes per unit length of manifold tube, a constant cross-section in the manifolds, and evenly spaced holes in the block, all the same size. Your equation should look like

$$-k(2\rho SV \; dV) = S \; dP + \tau p \; dx \qquad (10.U\text{-}1)$$

where K is the empirical constant introduced to account for nonideality in the destruction of x-momentum, S is the cross-sectional area of the manifold, P is the pressure, p is the perimeter, τ is the shear stress at the wall, V is the velocity up the manifold, and x is the distance up the manifold. Then make a material balance showing that

$$- S\rho dV = \alpha p \; dx \; C\rho[2(P - P_0)/\rho]^{1/2} \qquad (10.U\text{-}2)$$

where C is an orifice coefficient and α is the fraction of intersurface that is open. Make the following changes in variables: $v = V/V_0$, where V_0 is the velocity at the inlet; $\pi = (P - P_0)/2k\rho V_0^2$, where P_0 is the pressure in the mold block tubes; and $y = 4\alpha C(2K)^{1/2}x/D$, where D is the pipe diameter. Let the shear stress at the wall be given in terms of the velocity as

$$\tau = (f/2)|V|V \qquad (10.U\text{-}3)$$

where f is the friction factor, given as

$$f = f_0 V^{-1/4} \qquad (10.U\text{-}4)$$

Show that the momentum balance and mass balance become

$$v \; dv/dy + d\pi/dy + F_0 v^{7/4} = 0 \qquad (10.U\text{-}5)$$

$$dv/dy = -(2\pi)^{1/2} \qquad (10.U\text{-}6)$$

where $F_0 = f_0/2^{5/2}\alpha Ck^{3/2}$. Now make similar balances on the exit manifold, and than set $F_0 = 0$ for both sets of equations. Use subscripting to distinguish the two sets of equations. Then show that the following equation satisfies the velocity equation:

$$v_1 = b \cosh ay_1 - bw \sinh ay_1 \qquad\qquad (10.U-7)$$

where $a = [d^2(1 + b^2)/(d^2 - b^2)]^{1/2}$, w is the initial mass flow rate of cool-ant, $d = (p_1\alpha_1C_1S_2k_1^{1/2}/p_2\alpha_2C_2S_1k_2^{1/2})$, and $b = S_2/S_1$. Then show that $v_2 = (1/b)v_1$. Plot v_1 vs. v_2 for $w = 1$, $b = 1$. Discuss the expected effect F_{01}, F_{02} will have on the solution. Explain how the solution of these cou-pled equations can be used to size manifold lines and to balance lines running to mold blocks to improve balance of flow.

P10.V Repeat P10.U with the exit manifold emptying from the top rather than from the bottom. What changes in flow distribution throughout the mold block is to be expected?

P10.W The individual convective film heat transfer coefficient flro water flowing through a 1/2-in. mold block water line is calculated to be 100 Btu/$ft^2 \cdot hr \cdot °F$. The water lines are spaced 1 1/2 in. apart and have centerlines 1 in. from the mold surface. The coolant passage length is 6 ft. Calculate the total increase in coolant temperature if the mold material is CuBe, if it is Kirksite, and if it is aluminum-filled epoxy with a thermal conductivity of 1 Btu/ft$\cdot$hr$\cdot$°F. $Q = 6000$ Btu/hr.

References

1. Mod. Plast. Encyclopedia, 49(10A):604 (1972).
2. J. F. Stevenson and P. R. Patel, Experimental and Simulation Studies of Pressure Distribution in Mold Cavities During Filling, First Inter-Conference on Polymer Processing, M.I.T., Proceedings, M.I.T. Press, 1978.
3. J. F. Stevenson, A. Galskoy, K. K. Wang, I. Chen, and D. H. Reber, Polym. Eng. Sci., 17:706 (1977).
4. J. F. Stevenson and R. A. Hauptfleisch, Tech. Report #7, Cornell University, Ithaca, N.Y., 1976.
5. D. C. Paulson, SPE ANTEC Tech. Pap., 13:1009 (1967).
6. R. L. Ballman, L. Shusman, and H. L. Toor, Ind. Eng. Chem., 51: 847 (1959).
7. R. L. Ballman, L. Shusman, and H. L. Toor, Mod. Plast., 36(9):105 (1959).
8. R. L. Ballman, L. Shusman, and H. L. Toor, Mod. Plast., 36(10): 115 (1959).
9. R. S. Spencer and G. D. Gilmore, J. Colloid Sci., 6:118 (1951).
10. J. Eveland, H. J. Karam, and C. E. Beyer, Soc. Plast. Eng. J., 12(5):30 (1956).
11. W. G. Frizelle and D. C. Paulson, SPE ANTEC Tech. Pap., 14:405 (1968).
12. I. T. Barrie, Plast. Polym., 37:463 (1969)
13. I. T. Barrie, Plast. Polym., 38:47 (1970).

14. I. T. Barrie, Soc. Plast. Eng. J., 27(8):64 (1971).
15. F. N. Cogswell and P. Lamb, Plast. Polym., 38:331 (1970).
16. D. H. Harry and R. G. Parrot, Polym. Eng. Sci., 10:209 (1970).
17. J. R. A. Pearson, Mechanical Principles of Polymer Melt Processing, Pergamon, Elmsford, N.Y., 1966, p. 128.
18. M. R. Kamal and S. Kenig, Polym. Eng. Sci., 12:294 (1972).
19. M. R. Kamal and S. Kenig, Polym. Eng. Sci., 12:302 (1972).
20. J. L. Berger and C. G. Gogos, SPE ANTEC Tech. Pap., 17:1 (1971).
21. P. C. Wu, C. F. Huang, and C. G. Gogos, SPE ANTEC Tech. Pap., 19:197 (1973).
22. G. Williams and H. A. Lord, Polym. Eng. Sci., 15:533 (1975).
23. H. A. Lord and G. Williams, Polym. Eng. Sci., 15:569 (1975).
24. J. L. Throne, Principles of Thermoplastic Structural Foam Molding: A Review, First International Conference on Polymer Processing, M.I.T., Proceedings, M.I.T. Press, Cambridge, Mass., 1978.
25. J. Byine, paper presented at Injection Mold Die Design Meeting, University of Wisconsin Extension, Milwaukee, WI, April 1973.
26. J. F. Carley, Polym. Eng. Sci., 6:158 (1966).
27. J. M. McKelvey, Polymer Processing, Wiley, New York, 1962, p. 67.
28. G. M. B. Pye, Injection Mould Design, Iliffe, London, 1971, p. 142.
29. J. L. Throne, Polym. Eng. Sci., accepted for publication, 1978.
30. J. D. Keller, J. Appl. Mech., 23(3):77 (1949).
31. A. Acrivos, B. D. Babcock, and R. L. Pigford, Chem. Eng. Sci., 10:112 (1959).
32. R. S. Spencer and G. D. Gilmore, J. Appl. Phys., 21:513 (1950).
33. S. Kenig and M. R. Kamal, Soc. Plast. Eng. J., 26(7):50 (1970).
34. W. L. Sifleet, N. Dinos, and J. R. Collier, Polym. Eng. Sci., 13:10 (1973).
35. Y. T. Tam, N. Dinos, and J. R. Collier, SPE ANTEC Tech Pap., 19:601 (1973).
36. R. C. Progelhof and J. L. Throne, SPE ANTEC Tech Pap., 21:455 (1975).
37. L. Temesvary, Mod. Plast., 43(12):125 (1966).
38. A. Prasad, paper presented at SPE NATEC, 1974, Detroit. Preprint p. 254.
39. J. L. Throne, paper presented to Milwaukee Section, SPE, March 1974.
40. F. Kreith, Principles of Heat Transfer, 2nd ed., International Textbook Co., Scranton, Pa., 1965, Sec. 3.4.
41. V. Paschkis, Analog Methods, In: Handbook of Heat Transfer (W. M. Rosenow and J. P. Hartnett, eds.), McGraw-Hill, New York, 1973, Sec. 5.
42. I. I. Rubin, Injection Molding: Theory and Practice, Wiley-Interscience, New York, 1972, Chap. 2.
43. J. H. DuBois and W. I. Pribble, eds., Plastics Mold Engineering, Van Nostrand Reinhold, New York, 1965, Chap. 9.
44. W. P. Benjamin, Plastic Tooling: Techniques and Applications, McGraw-Hill, New York, 1972.

45. L. Sors, Plastic Mould Engineering, Pergamon, Elmsford, N.Y., 1967, p. 108.
46. H. Koda, J. Appl. Polym. Sci., 12:2257 (1968).
47. J. Bailey, India Rubber World, 118:225 (1948).
48. R. S. Spencer and G. D. Gilmore, Mod. Plast., 28(4):97 (1950).
49. R. L. Ballman and H. L. Toor, Mod. Plast., 38(10:113 (1960).
50. G. B. Jackson and R. L. Ballman, Soc. Plast. Eng. J., 16(11):1147 (1960).
51. D. P. Thomas and R. S. Hagen, SPE ANTEC Tech. Pap., 14:49 (1968). (1968).
52. P. Hubbauer, Plast. Eng., 29(8):37 (1973).
53. H. Ebnuth and G. Lube, Plastverarbeiter, 23:171 (1972).
54. J. W. Curtis, J. Phys. D, 10:1413 (1970).
56. G. Menges and G. Wubken, SPE ANTEC Tech. Pap., 19:519 (1973).
57. E. S. Clark, Soc. Plast. Eng. J., 23(7):46 (1967).
58. J. R. Collier and L. Neal, SPE ANTEC Tech. Pap., 14:63 (1968).
59. W. O. Statten and P. H. Geil, J. Appl. Polym. Sci., 3:357 (1960).
60. D. H. Reneker, J. Polym. Sci., 59:S39 (1962).
61. E. S. Clark, Appl. Polym. Symp., 20:325 (1973).
62. E. S. Clark, Soc. Plast. Eng. J., 23(7):46 (1967).
63. D. E. Scherpereel, Plast. Eng., 29(12):46 (1973).
64. J. A. Brydson, Flow Properties of Polymer Melts, Van Nostrand Reinhold, New York, 1970.
65. R. F. Williams, Jr., and L. H. Pancoast, Jr., Mod. Plast., 44(9): 185 (1967).
66. W. Woebcken, Kunststoffe, 63:632 (1973).
67. E. M. Hagerman, Plast. Eng., 29(10:67 (1973).

11.

Rotational Molding

11.1 Introduction

Rotational molding, as a method of making hollow plastic articles, had its beginnings in "slush" molding of PVC plastisols in the late 1930s or early 1940s. A series of symposiums offered by USI Chemicals in the early 1960s alerted the plastics engineer to the potentials of rotational-molding LDPE powders [1,2]. Since that time extensive experimentation on materials including nylons, polycarbonates, HDPE, HIPS, CAB, and ABS has been reported in the literature [3]. However, very little engineering information has been available until recently [4-7]. The reason for this is apparent when one begins to contrast the complexity of the fluid flow–heat transfer problem with the potential growth of the market. Undoubtedly, rotational molding, like thermoforming (discussed in the next chapter), will remain a rather specialized method of producing objects that could not be produced economically any other way.

To simplify some of the discussion concerning rotational molding from a process viewpoint, it is necessary to immediately define the main elements of the process and describe the major steps required to convert a

plastic resin into a useful hollow plastic part. As shown in schematic form
in Fig. 11.1-1, a typical machine consists of a multiple-arm device with
each arm containing two concentric driven spindles. Attached to the ends
of the arms is a mold holder or "spider," and fastened to the spider are
one or more hollow two- or three-piece thin-wall metal molds. The operation of the machine from a material viewpoint is as follows: The molds are
opened, cleaned, and charged with a preweighed amount of plastic powder
(or liquid). The molds are then closed and secured with latch-type mold
closures [4]. The molds are rotated in the polar and equatorial directions
simultaneously through planetary gears in the concentric spindles. Since
each directional speed can be independently controlled, it is the objective
of the molder to determine the appropriate overall speed of rotation and the
ratio of the speeds of rotation of the spindles necessary to produce a part
with uniform wall thickness. Once the spindles are revolving, the molds
are inserted into a forced convection oven held at a constant preset temperature. As the mold heats, the tumbling powder contacts the hot surface and
begins to stick. With sufficient time in the oven, all the material will stick
to the mold surface. The material must then sinter-melt into a homogeneous
liquid adhering to the mold wall. The operator must determine the length
of time required for this process. If he allows the material to remain in

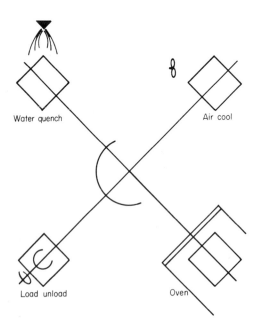

Figure 11.1-1   Schematic for four-station rotational molding process [4].

the oven for an excessive period of time, it may degrade [4]. If he takes
the material from the oven before the part is fully melted, it will be friable
or porous, and this constitutes a "short shot."

Once the part is fully melted, it must be cooled. If the material is
quenched by flooding the mold surface with cold water, the partial vacuum
drawn in the still-soft part can cause it to collapse. Furthermore, crystalline materials such as HDPE are quite sensitive to cooling through the region
of maximum crystal growth rate, and uneven part walls, dimensional variations, and severe warpage can result with improper cooling methods [5].
Once the part is at or near room temperature, it can be removed from the
mold, and the mold is ready for another cycle. Thus, the process is cyclic
as regards the part but semicontinuous as regards the machine operation.

In this chapter, the temperature problems faced in transiently heating
a plastic powder in a rotating system and in the process of sinter melting
will be analyzed. As will be seen, the ability of a plastic powder to form a
useful hollow part depends entirely on its ability to sinter-melt into a homogeneous liquid. And the rather simple approach to cooling that can be used
to determine the extent of part distortion will be detailed. Furthermore,
some of the problems inherent in rotational molding of liquids will be discussed briefly. It is interesting to note that although PVC plastisol was the
first material rotationally molded and although it passes through a gel and
cure stage in a manner not unlike that of the cross-linking reactions of
reactive liquids such as catalyzed unsaturated polyesters or epoxies, basic
studies on the flow behavior of reactive liquids during rotation are just beginning [8,9].

The questions of fluid flow of the plastic melt (after sinter melting
and before quenching) will not be considered here. This has been discussed
at length elsewhere [4,10]. Although some evidence indicates that flow can
occur across the mold surface, the effect is controllable and is apparently
rather small as long as the aspect ratio of the mold is rather near unity and
the positioning of the mold on the spider is not peculiar when compared with
the geometry of the part or the neighboring molds.

11.2 Heat Transfer to the Mold Materials

When a thin sheet of metal is introduced to a step change in environmental
temperature, it can be shown that the internal resistance to heat transfer is
small when compared with the surface resistance [11]. Thus, if at time
$t = 0$, the mold at a uniform temperature T_0, is moved into an air environment at temperature $T = T_\infty$, we can show that

the change in internal   the net heat flow into the mold
  energy of the mold   =   from the environment
    during time dt   during time dt

$$\rho c_p L \, dT = h(T_\infty - T) \, dt \qquad (11.2\text{-}1)$$

where ρ is the density of the metal, c_p is its specific heat, L is the thickness (or characteristic dimension) of the sheet, h is the conventional convection heat transfer coefficient, t is time, and T is the instantaneous temperature of the sheet. This rather simple differential equation can be integrated to obtain a typical first-order heat transfer response to a step change in surface temperature:

$$(T_\infty - T)/(T_\infty - T_0) = \exp(-h\alpha/Lk)t \qquad (11.2\text{-}2)$$

where α is the thermal diffusivity of the metal, $= k/pc_p$, and k is its thermal conductivity. By curve-fitting measured data, it has been found that $h \doteq 5$ Btu/ft$^2 \cdot$hr$\cdot$°F for a conventional forced convection oven. The experimental data for a 0.125-in.-thick aluminum mold are shown in Fig. 11.2-1 along with the curve of Eq. (11.2-2) for h = 5. In some instances, it is desirable to bring the mold surface temperature up to T_∞ as rapidly as possible. This implies that h must be increased by baffling the heating chamber

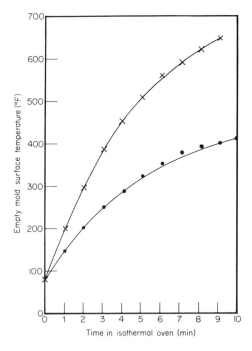

Figure 11.2-1   Experimental and theoretical temperature response for an empty 0.125-in.-thick aluminum disk mold in an isothermal McNeil convection oven. X's are for an isothermal oven temperature of 750°F; solid circles for 450°F [4].

or increasing the circulation blower capacity, L must be decreased by reducing the thickness of the metal used in the mold, or α/k must be increased
by changing materials. Given in Table 11.2-1 are several common rotational
molding mold materials. It is apparent that aluminum offers the highest
α/k value. The criterion that determines the lower limit on L is the hot
strength of the metal at oven temperatures of 500 to 1000°F. From experience, even steel molds distort and warp under repeated thermal cycling if
L is below 0.050 in. Changing the heat transfer coefficient by choosing
other modes of heating has been tried and in some cases is commercially
successful. For example, hot oil or salt can be used as a heating medium
and the arm simply immersed in the bath. The oil material losses and the
potential contamination of the plastic make the process marginal. Some
operators are using jacketed molds and piping the hot oil through the jacket.
This is effective if many parts are to be made from a single mold, but otherwise the expense of the double-wall mold can be prohibitive.

    Note that the heating equation (11.2-2) also applies for the cooling of
a mold. Here, the heat transfer coefficients are normally significantly
larger, with water quench, for example, and thus the thermal lag in the
mold can be ignored. The surface of the plastic can thus be considered approximately equal to the quench temperature. This point will be amplified
later on.

    Note that the rate of heating of the empty mold, as given by Eq.
(11.2-2), represents the maximum rate of heating. Any thermal sink in
contact with the mold will cause it to heat at a slower rate [7]. Nevertheless,
for the following discussion, we can assume that Eq. (11.2-2) represents a
good first approximation to the temperature profile during the early stages
of rotational molding.

Table 11.2-1

Thermal Properties of Rotational Mold Materials

Material	α, Thermal diffusivity (ft^2/hr) (500° F)	k, Thermal conductivity (Btu/ft·hr·°F) (572°F)	α/k (appropriate units)
Aluminum	3.33	133	0.0250
Copper	4.42	212	0.0208
Nickel	0.60	32	0.0188
Steel	0.49	25	0.0196

11.3 Melting Plastic in a Rotating System

The surface that the plastic contacts is seen to be heating in a transient, albeit rational, way. Now it is necessary to determine how to characterize the heating of powder in this system. Consider (1) the shape of the mold; (2) the rate of revolution of the mold; (3) the ratio of the rates of revolution of the two spindles; (4) the effect of the set-point temperature of the oven; (5) the physical characteristics of the powder, such as maximum particle size, particle size distribution, dynamic repose angle, and so on; (6) the physical characteristics of the sinter-melt, such as thermal diffusivity, melting temperature, latent heat of fusion, if any; and (7) time in the oven, or oven cycle time, abbreviated OCT, among others. Since economics are based on the controlling step in any process, it has been found [3, 4] that the oven cycle time is frequently but not always the controlling step in the rotational molding process. Two important exceptions to this assumption are the cooling step in rotationally molded foam and the materials handling step when many small parts are being made on a very large machine.

For the purpose of this analysis, consider the rotational molding of a cylindrical object rotating about its horizontal axis. This geometry is applicable to the rotational molding of garbage cans, liquid containers, fertilizer drums, piggy banks, coffins, manikins, dress forms, and spherical objects such as lamp globes and beach balls. It can be used to model rotational molding of surf boards and lamp poles but certainly not toilet seats or lids.

Consider first the physical characteristics of the powder used in rotational molding. Available to the molder today are powders having physical characteristics ranging from perfect spheres (Cosden HIPS) to cubes with rounded corners (USI Chemicals Microthene LDPE) to particles of arbitrary shape, often with long filamentous tails (regrind). Some of the characteristics of powder that are desirable in hopper design were discussed in Chap. 8. Most of these characteristics are desirable in rotational molding powder as well. Note that the thermal characteristics of the process will influence the physical characteristics of the powder, since heat is transferred to a particle through contact with other particles and the mold surface and contact with the surrounding air. The latter is a notoriously poor mode of heat transfer, as seen in Chap. 6. Therefore, maximum contact of particles for maximum conduction heat transfer is desired. For a flat sheet of thickness L, the surface-to-volume ratio (a measure of the efficiency by which heat is transferred through the particle surface and into the particle) is $1/L$. For a cube of side L, it is $6/L$, and for a sphere of radius L, it is $3/L$. However, for contact conduction, if only one portion of the particle is in contact with a heated surface (such as the mold), the surface-to-volume ratio for a flat sheet becomes $1/2L$, for a cube, $1/L$, and for a sphere, zero. Thus, the more spherical the powder becomes, the easier it flows, but the poorer it conducts heat. Furthermore, the

smaller the powder particle becomes, the larger the surface-to-volume
ratio becomes, and the more efficient contact heat transfer becomes. Thus,
powder is used for rotational molding rather than, say, extruder pellets or
beads.

There is a practical lower limit on particle size. First, grinding of
powders to, say, 100 to 150 mesh leads to excessive material losses owing
to the high degree of fines greater than 200 mesh produced and the overheating of the grinding equipment [4]. Furthermore, this overheating owing
to excessive shear can degrade the plastic and lead to losses in mechanical
strength and undesirable product characteristics such as discoloration
and streaking. A second point is that the tumbling of very fine powders
within a metal mold will build up high static charges that aid in agglomerating
the powders. This leads to uneven melting and nonuniformity of part wall
thickness. And keep in mind that finer powders imply greater surface area
per unit weight and thus greatly enhanced surface adsorption tendencies for
moisture. Moisture is a characteristic source of pinhole defects in nylons
and polycarbonates [3].

This does not imply that fines should be excluded from rotational
molding powders. Inclusion of up to 10% fines greater than 200 mesh aids
in powder flow during tumbling, perhaps acting as a lubricating agent for
the larger particles. This provides for smoother powder flow across the
mold surface. Furthermore, fines aid in densifying the sinter-melt in that
they fill the interstices between the larger particles and thus promote capillary action. The proper proportion of fines to coarser particles has not
been established for any molding powder. Experiments [4] have shown that
the materials that rotationally mold best have the approximate makeup
shown in Table 11.3-1, as determined by rototapping the materials up to 8
hr on a multiple-sieve shaker table.

It seems, however, that significant deviations of these percentages
are possible without significant changes in the physical or mechanical characteristics of the final part. For example, an 80 mesh material (Rexene
324C8 HIPS) has been molded with apparently the same ease and mechanical
characteristics as a nominal −35 mesh material [4]. Thus, it is recommended that from heat transfer and powder flow processing viewpoints the
most desirable particle size is −35 to +75 mesh and the most desirable particle shape is cubic with very generous corner radii.

Table 11.3-1

Mesh Size of MIPS-Rotomolding Powder [4,12]

Material— rounded cubes:	+30 mesh	−30 +50 mesh	−50 +70 mesh	−70 +100 mesh	−100 mesh
	1%	50%	30%	12%	7%

In addition to these data, the thermal diffusivity of the powder is needed. Diffusivities of powders, in general, depend on the particle sizes, size distribution, packing density, and particle shapes. From laboratory tests, it has been determined that the thermal diffusivity of a -35 mesh USI Chemical Microthene LDPE with a bulk density of about 20 to 25 lb/ft^3 is approximately 1/40 of the value for the solid material. These values agree with the orders of magnitude difference between loose sand and silica [13]. This point is discussed later and is certainly challengable. Unfortunately, even though these data are needed for extrusion and injection molding as well as rotational molding, very little effort has been made in determining representative values, particularly as functions of temperature and pressure [14].

Two other pieces of information also not available are the temperature at which a plastic first becomes tacky and the retardation time of the plastic if it exhibits viscoelasticity. The tack temperature is needed to determine the temperature at which the plastic first begins to stick to the mold and to itself. This is important, because until the mold reaches this temperature, the powder remains in a "powder pool" near the bottom of the mold as shown in Fig. 11.3-1. Once the mold surface has exceeded this temperature, the powder sticks to the wall and depletes the powder pool of material. This depletion continues until all the powder is adhering to the mold wall. Since no data are available for this temperature, the Vicat softening temperature or a heat distortion temperature extrapolated to zero pressure can be used in place of a more appropriate tack temperature. As will be shown later, a precise value for this temperature is not critical; a 50°C variation can yield results within engineering accuracy.

The importance of the viscoelastic retardation time for polymer melts will be seen when the mechanism of sinter-melting of the porous three-dimensional structure into a homogeneous void-free plastic melt is considered later. There is no way of replacing the actual value of the retardation time with a similar approximate property, and thus the data available here are very limited. Frankly, at times, an appropriate value must be estimated.

The geometry of powder flow in a horizontal rotating cylinder is shown in Fig. 11.3-1. It is apparent that the friction between the loose powder in the pool and the rotating mold surface serves to move the powder up the rotating wall until the effect of gravity is sufficient to allow the powder to break away and fall across the top of the stagnant pool of powder to the opposite mold wall. The property that characterizes the point at which the powder is released from the mold wall is the dynamic angle of repose [15]. For most powders, plastic powders included, the dynamic angle of repose is between 25° and 50° from the horizontal. The more spherical or regular the shape of the powder particle, the lower the dynamic angle of repose and consequently the more slowly the powder will fall from the point of release to the point of recontact with the opposite mold surface. As is

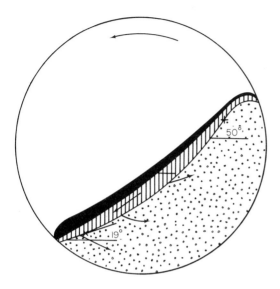

Figure 11.3-1   Schematic showing circulation of powder in rotating sand
bed in horizontal cylinder; 6-in. diameter, 5.6 rpm [15].

apparent from Fig. 11.3-1, the falling powder is mixing and is exchanging
material and latent heat with the powder that is statically in contact with the
rotating mold wall.

Kurihara and Kuno [16] have proposed a circulation model for mixing
in the powder pool, and Suzuki and Tanaka [17] have calculated a viscosity
for falling powder based on the decrease in velocity of a falling powder layer
owing to shear with a solid inclined plane. Goodman and Cowin [18] have
considered a more sophisticated approach using continuum mechanics to
describe the flow process down an incline. They conclude that the simple
approach of Suzuki and Tanaka is not entirely valid and that the circulation
model of Kurihara and Kuno is probably useful only in a material balance
sense. They find that the velocity of the falling powder is not necessarily
maximum at the granular surface and that a viscosity of sorts can be defined
as long as the shear surface is static. If the shear surface can exchange
material with the flowing surface as in the case of vertical shear, the flowing fluid can either decrease or increase in mass, depending on the effect
of external factors such as gravity field and boundary conditions.

Thus, there are two aspects of powder flow. One is the release of
the material at the mold surface, which from a continuum viewpoint takes
place uniformly but from actual observations is seen to be a very dynamic,
cyclic buildup and release of material [15]. The second is the flow of powder across the static powder surface, mixing and exchanging mass and energy

with the granular shear surface. It should be apparent from this brief discussion and the earlier consideration of hopper design in extruders that concentrated experimental and theoretical effort is needed before the powder motion can be modeled in even relatively simple geometries. Therefore, consider here the simple and expedient solution of assuming that the powder in contact with the rotating mold is static relative to the mold wall; that the powder being released from the mold surface falls across the static powder without exchange of material, momentum, or energy (e.g., is freely falling); and that during free fall the powder is intimately mixed thermally so that upon contact with the mold surface it has attained a uniform temperature.

Thus, the model is simplified, as shown in Fig. 11.3-2. Here the static powder is subject to transient heat conduction from the mold wall, and the falling powder adiabatically mixes during free fall. The equation that describes the transient heat conduction into the powder is the conventional equation, with the thermal diffusivity being that of the loose powder:

$$\partial T/\partial t = \alpha_{effective}(\partial^2 T/\partial x^2) \qquad (11.3-1)$$

Since the powder contacts the mold surface for a relatively short time, it can be considered to be infinitely deep relative to the thermal source. As

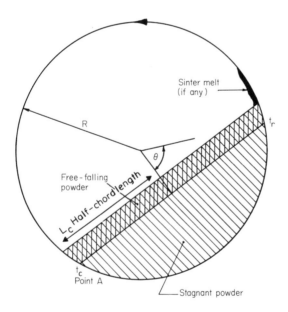

Figure 11.3-2 Idealized flow configuration for flow in a horizontal cylinder [4].

seen in Eq. (11.2-2), the mold is heating in an exponential fashion. Thus, the surface boundary condition can be written as

$$T_s = T(x = 0, t) = T_\infty (1 - e^{-\beta t}) + T^* \tag{11.3-2}$$

where $\beta = h\alpha/Lk$ from Eq. (11.2-2) and T^* is an offset temperature, to be discussed shortly. Also $T(x = \infty, t) = T_0$, where T_0 is the average temperature of the powder recontacting the mold after mixing and free fall. An exact solution for Eq. (11.3-1) subject to similar prescribed boundary conditions has been given by others [19,20]. Here, however, an approximate cubic temperature profile is used to solve the equation [4,21]:

$$T = T_s(1 - x/\delta)^3 \tag{11.3-3}$$

where δ is a thermal penetration thickness. After substitution into Eq. (11.3-1), the thermal penetration thickness δ, an artificial concept much like the fluid mechanical boundary layer thickness, is given as

$$\delta = \frac{2\sqrt{6}}{T_s(t)} \left[\int_0^t \alpha_{effective} T_s(t)^2 \, dt \right]^{1/2} \tag{11.3-4}$$

where T_s is given by Eq. (11.3-2) or any other surface-temperature profile so desired. As a quick check of Eq. (11.3-4), the thermal penetration thickness for a simple step change in surface temperature can be found:

$$\delta = (24\alpha_{effective} t)^{1/2} \tag{11.3-5}$$

This agrees exactly with theoretical and experimental values for both heat and mass transfer [22,23]. As further use of this method, it can be used to determine the effective thermal diffusivity of powder or melt simply by measuring the temperature of the material at a known distance L above the heated surface [4]. For the case of a linear heated surface temperature ($T_s = et$), δ is given from Eq. (11.3-4) as

$$\delta = (8\alpha_{effective} t)^{1/2} \tag{11.3-6}$$

and thus the temperature at a known distance L from the heated surface is given from Eq. (11.3-3) as

$$T = et[1 - L/(8\alpha_{effective} t)^{1/2}]^3 \tag{11.3-7}$$

Given in Fig. 11.3-3 is the experimental linear temperature profile for a laboratory hot plate and in Fig. 11.3-4 is the experimental temperature profile for HIPS powder at a distance of 0.040 in. above the plate. From Eq. (11.3-7), an effective thermal diffusivity $\alpha_{effective} = 8.8 \times 10^{-8}$ ft^2/hr

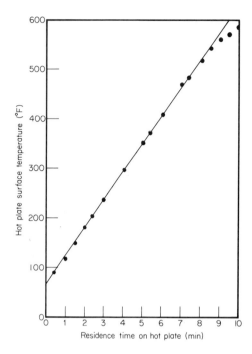

Figure 11.3-3 Experimental measurement of surface temperature of
450-W Fisher Scientific Co. Autemp Hot Plate [4].

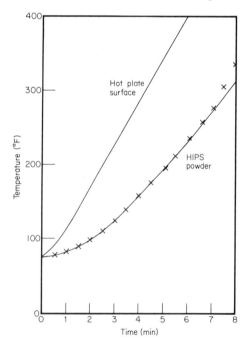

Figure 11.3-4 Experimental verification (X's) of temperature penetration
theory into HIPS on Fisher Hot Plate; measured at 0.040 in. above hot plate
surface; solid line from Fig. 11.3-3. Effective diffusivity of HIPS powder
determined to be 7.62×10^{-4} in.^{2}/min [4].

is found. This value is consistent with measured values of others [13]. The form for the thermal penetration thickness when the surface-temperature profile is given by Eq. (11.3-2) is

$$\delta = \frac{2\sqrt{6\alpha_s}}{T_\infty (1 - e^{-t_2}) + T^*} \left\{ t_c [T_\infty^2 + 2T_\infty T^* + T^{*2}] + \left(\frac{2T_\infty^2}{\beta} \right. \right.$$

$$\left. + \frac{2T_\infty T^*}{\beta} \right)(e^{\beta t_2} - e^{-\beta t_1}) - \frac{T_\infty^2}{2\beta}(e^{-2\beta t_2} - e^{-2\beta t_1}) \right\}^{1/2} \qquad (11.3\text{-}8)$$

Here t_1 is the time of initial contact of powder with mold surface, t_2 is the time of powder release from mold surface ($t_c = t_2 - t_1$), and T^* is the initial temperature difference between the mold surface and the powder recontacting the mold surface. Thus, the temperature profile for the powder at the time of release from the mold surface is given by Eq. (11.3-3) with Eq. (11.3-2) defining the surface temperature and Eq. (11.3-3) defining the penetration thickness. The average temperature is given as the integral of this temperature over the depth of the powder. If the depth is assumed to be equal to the half-chord length L_c of the static powder, as shown in Fig. 11.3-2, an average temperature is obtained:

$$T_{av} = T_0(x/L_c) + T_s(t)(x/L_c) \qquad (11.3\text{-}9)$$

where T_0 is the average bulk powder temperature at t_1.

If the powder between $x = 0$ and $x = x_1$, say, is above the tack temperature at the point of release from the mold surface, this material is assumed to stick to the mold or itself. Thus, the average temperature is given as

$$T_{av} = T_0(\delta - x_1)/(L_c - x_1) + [\delta T_s(t)/(L_c - x_1)](1 - x_1/\delta)^4 \qquad (11.3\text{-}10)$$

The thickness of the material that is at or above the tack temperature at the point of release is given as

$$x_1 = \delta\{1 - [T_m/T_s(t)]^{1/3}\} \qquad (11.3\text{-}11)$$

where T_m is the tack temperature and T_s and δ are calculated from Eqs. (11.3-2) and (11.3-8), respectively.

The time of free fall of powder between the release point and the recontact point can be calculated from [15]

$$t_f = (2L_c/g \cos \beta')^{1/2} \qquad (11.3\text{-}12)$$

where, again, L_c is the half-chord length and β' is the dynamic angle of repose. For a cylinder, $L_c = R \sin \theta/2$, where R is the cylinder radius and θ is the chord angle. The time of contact with the mold surface t_c is directly proportional to the chord angle θ, the radius of the cylinder, and inversely proportional to the speed of rotation.

The basic method of analysis requires bookkeeping. A clump of powder is followed as it makes the circuit from the static heat conduction zone to the dynamic adiabatic mixing zone, to, again, the static heat conduction zone. The total time for one circuit will depend on the degree of filling of the mold with powder (e.g., the chord angle), the geometry of the mold (e.g., R), and the speed of rotation (rpm). Of course, fixing the initial chord angle and the geometry of the mold establishes the ultimate part thickness. The area of the segment of powder is given as $A_S = \theta R - R^2 \sin \theta/2$. Thus, the part thickness is

$$tk = (A_s/2\pi R)(\rho_{powder}/\rho_{plastic}) = (\theta - R \sin \theta/2)(2\pi)$$
$$\cdot (\rho_{powder}/\rho_{plastic}) \qquad (11.3\text{-}13)$$

where $\rho_{powder}/\rho_{plastic}$ is the ratio of the bulk densities of the powder and homogeneous plastic. Usually this ratio is on the order of 0.5 to 0.6 and should be no less than 0.3 to 0.4. For ratios below these values, there may be insufficient room in the mold for the powder needed to produce a part with a useful wall thickness.

Now as the mold turns, it is apparent that the time for one mold revolution is simply the reciprocal of the rpm. For the plastic, however, by following a clump of plastic it is found that it takes $t_c + t_f$ time units. This time is normally not the same as that for the speed of the mold. As a matter of fact, as the material collects on the mold surface, $t_c + t_f$ approaches zero. Thus, T* must be adjusted for the appropriate time based on an analysis that follows the powder revolution. The moment the mold surface has reached the tack temperature of the plastic, the thickness of plastic that is at or above this temperature can be calculated. Through a material balance based on the half-chord length, this material can be removed from the mass of powder falling to recontact the mold surface. Then the material balance can be corrected for the missing material by shortening the half-chord length and the calculation continued. Consider the material adhering to the mold wall as a three-dimensional matrix with the thermal properties of the unmelted or unsoftened powder in the pool. Therefore, the calculation can safely neglect the heat of melting, thermal transport through a material with thermal properties different from those of the powder, and so on. The effect of oven set-point temperature on times for initial and final powder pickup by the mold mass is shown in Fig. 11.3-5 as a function of mold rpm for a material tack temperature of 200°F and a cylinder radius of 6 in. The internally calculated temperatures for both the powder in the pool and that on the mold wall are shown in Figs. 11.3-6. The effect of rpm, oven set-point temperature, and tack temperature on the extent of material attached to the mold wall as functions of temperature are shown in Fig. 11.3-7. Some experimental data for HIPS are also shown.

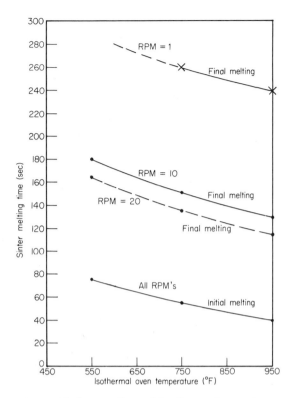

Figure 11.3-5 Effect of isothermal oven temperature on time for initial and final sinter-melting assuming tack temperature $T_m = 200°F$. Parameter is speed of rotation [4].

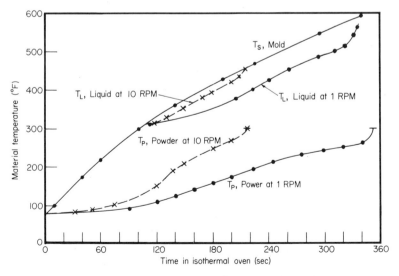

Figure 11.3-6 Calculated sinter-melt (T_L) and powder (T_p) temperatures for two values of rpm. Tack temperature, $T_m = 300°F$; isothermal oven temperature $T_\infty = 750°F$.

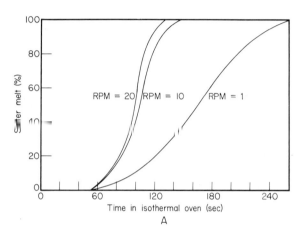

A

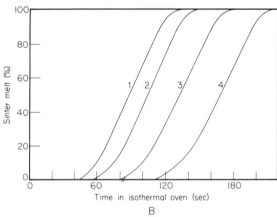

B

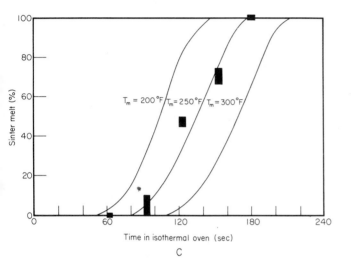

Figure 11.3-7 A: Effect of mold rpm on sinter-melting rate for isothermal oven temperature of 750°F and tack temperature of 200°F. B: Effect of isothermal oven temperature on extent of sinter-melting. 1: Tack temperature T_m = 200°F, isothermal oven temperature T_∞ = 950°F; 2: T_m = 200°F, T_∞ = 750°F; 3: T_m = 200°F, T_∞ = 550°F; 4: T_m = 300°F, T_∞ = 750°F. C: Effect of tack temperature T_m on extent of sinter-melting. Isothermal oven temperature is 750°F. Blocks represent experimental data on microthene LDPE in 1/8-in.-thick aluminum disk mold at T_∞ = 673°F [4].

As mentioned above, the use of the empty mold response to a step change in environmental temperature is a first approximation to the heating of plastic in contact with the metal mold. Recently [7], an extensive analysis of the static heat transfer problem has been presented. To understand how static heat transfer fits the above model, consider the hypothetical case where the material, by some means or other, uniformly coats the mold (e.g., there is no rotating pool). In this case, we can apply simple transient heat conduction solutions to the material response to a step change in environmental temperature. Recall, however, that as the material reaches the melt temperature it rapidly changes from a powder to a liquid. This implies that the thermal diffusivity increases 40-fold. The characteristic transient heat transfer equation of Chap. 6 is therefore applied to both powder and liquid layers, with thermal coupling occurring at the moving interface.

$$T_m(x = x_0) = T_p(x = x_0) \tag{11.3-14}$$

$$-k_m (\partial T_m / \partial x)_{x=x_0} = -k_p (\partial T_p / \partial x)_{x=x_0} \tag{11.3-15}$$

where the moving interface is $0 \leq x_0 \leq L$. The first coupling condition ensures equal temperatures at the interface; the second ensures equal heat fluxes at the interface. Note also that there is a significant difference in the densities of powder and liquid. This can be included in the corresponding one-dimensional transient heat conduction equations. The solution to the set of equations coupled through the boundary conditions was by finite difference using a continuous system simulation. In the example shown in Fig. 11.3-8 for a five-element plastic and one-element mold model, the mold surface temperature follows rather carefully the predicted curve for the empty mold until element 1 begins to liquify. Subsequent liquifications of the elements result in greater and greater deviations of mold temperature from that predicted by the empty mold. Once all elements have liquified, the mold surface temperature again begins to climb. The wiggles in the curve are due to the discrete nature of the numerical method. The materials modeled are 0.100-in. (liquid) MIPS in a 0.060-in. steel mold. The bulk density of the powder was half that of the liquid, whereas its thermal conductivity was 1/40 that of the liquid. Several other aspects of this static model have been shown [7]. For example, it was found that the time to full liquification was insensitive to the melting temperature range. It was also found that prediction of the time to full sinter melt (or liquification) compared quite well with previous experimental data [4]. And finally, it was found that there was excellent agreement between the predicted position of the moving interface x_0 and that predicted using the simple Goodman penetration theory model, Eq. (11.3-5).

The model enables the engineer to predict fully the entire heating and cooling curves for any combination of mold and plastic materials and any heating and cooling profiles. Several examples are given in the references.

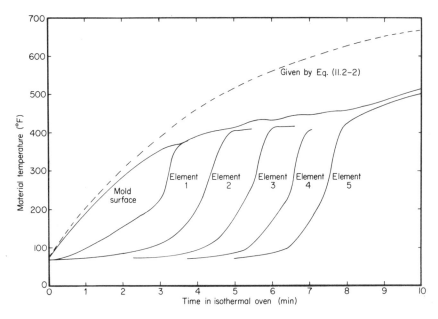

Figure 11.3-8   Static heat conduction modeling of the transient heating of a densifying powder in a rotational molding system.   Shown are temperature profiles for the mold and five elements. $T_\infty = 750°F$; final plastic thickness, 0.200 in., mold, 0.060-in. steel; lower melting temperature, 200°F; upper melting temperature, 250°F [7].

11.4 Melt Sintering and the Densification of a Part

Once the plastic is entirely on the mold wall, it is necessary to begin "expressing" the voids from the part. If the mold were quenched and opened at this time (e.g., the part "short-shot"), it would be found that the particles at the free surface could be broken away from the rest of the part with little effort. The particles are still identifiable and separable. The particles in contact with the surface of the mold, on the other hand, would have formed a rather strong web and could be separated only with much effort. Nevertheless, the structure is a three-dimensional network of plastic, with air being the primary continuous phase. The objective in heating the part for an additional period of time is to consolidate the voids and move them toward the free surface, thus obtaining a homogeneous plastic part.

The technology of melt-sintering is well developed in glass and metal processing, but only two major efforts have been attempted for plastics. The first, by Kuczynski and Neuville [24], extends Frenkel's glass sintering model [25] to polymethyl methacrylate. Frenkel shows that from thermodynamics if two glass spheres touch, the system acts to lower the free

energy by decreasing the total surface area. This is done by increasing
the thickness of the "web" that forms between the two spheres. Thus, he
found that

$$x = (3a\gamma t/2\eta)^{1/2}$$
                                                                (11.4-1)

where x is the thickness of the web formed between the two spheres, a is
the sphere radius, γ is the surface tension of the molten glass, η is its viscosity, and t is time. Thus, the growth of the web is proportional to the
square root of time. Kuczynski and Neuville used empirical relationships
to find a curve fit for their materials. Their resulting equation appears as

$$x = a^{3/5}\{K(T)t\}^{1/n}$$
                                                                (11.4-2)

where n decreased in value from 5 to 0.5 as the isothermal sintering temperature increased, and K(T) likewise was very temperature dependent.
This empirical approach can be useful for determining the sintering time
for glassy plastic materials but requires curve fitting of the isothermal sintering data for each material to obtain the empirical relationships for n and
K(T). A more important effort by Lontz [26] shows that for rubbery materials the Frenkel model and the Kuczynski-Neuville empirical model are not
correct, since there is no way of accounting for the material elastic resistance to surface tension. As a result, Lontz offered the following modified
Frenkel equation:

$$x = (3a\gamma t/2\eta)^{1/2}\{1 - \exp(-t/\tau)\}^{-1/2}$$
                                                                (11.4-3)

The second term on the right contains a correction to the Newtonian viscosity
to include the viscoelasticity of the polymer melt. The term τ is a retardation time characteristic of a relaxing viscoelastic material of the Maxwell
model. As can be seen, if τ is very small, as is the case for a glassy
polymer, the Lontz model reduces to the Frenkel model. In essence, then,
the web growth for glassy polymers should continue at a rate proportional
to $t^{1/2}$ for isothermal cases. If a material such as ABS is rotationally
molded, the very large relaxation time of the material should dramatically
alter the web growth rate. Thus, for very large values of τ, $1 - \exp(-t/\tau)$
can be approximated by t/τ. Thus, for a highly viscoelastic material, the
web thickness at an isothermal condition would approach a constant, given by

$$x/a = 3a\gamma\tau/2\eta$$
                                                                (11.4-4)

In other words, for viscoelastic materials, the isothermal sintering would
stop when x reached the value given by Eq. (11.4-4). Increasing the temperature in an attempt to increase the web thickness and thus reduce the
extent of porosity of the final part may be detrimental, since τ and γ decrease with increasing temperature. This may counteract the decreasing
value of viscosity with temperature. Proper determination of the value for
the retardation time now shows that the curves are in agreement with data
for HIPS (Fig. 11.4-1). The proper choice for retardation time, of course,

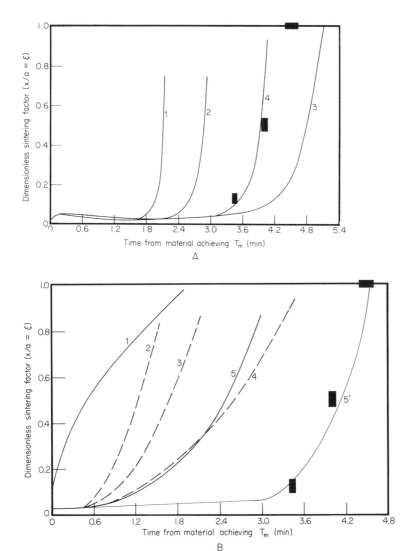

Figure 11.4-1 A: Extent of sintering based on Kuczynski-Neuville empirical model for sinter-melt temperature $T_m = 200°F$. 1: Isothermal oven temperature, $T_\infty = 950°F$; 2: $T_\infty = 750°F$; 3: $T_\infty = 550°F$; 4: transient hot plate temperature response (Fig. 11.3-3). Blocks: experimental data for -35-mesh Rexene 324C8 MIPS on hot plate. B: Extent of sintering based on Lontz viscoelastic model for sinter-melt temperature $T_m = 200°F$. 1: Isothermal surface temperature, $400°F$; 2: isothermal oven temperature $T_\infty = 950°F$; 3: $T_\infty = 750°F$; 4: $T_\infty = 550°F$; 5: transient hot plate temperature response (Fig. 11.3-3); 5': transient hot plate temperature resonse with Lontz data corrected for larger retardation time. Blocks: experimental data for -35-mesh Rexene 324C8 MIPS on hot plate [4].

depends on the method of evaluation [27], and a more complex model for viscosity should be chosen. Much effort is needed here to determine why certain materials such as ABS can be molded only if the rubber content is kept very low.

The important point to keep in mind here is that for glassy plastics such as GPS or PMMA the sintering can be accomplished without significant processing problems. With high-rubber-content materials, such as ABS or HIPS, experience shows that sintering is never fully complete. This is pragmatic support for a Lontz-type modification to the viscosity in the Frenkel sintering model. Lontz also discusses the effect of particle shape on sintering times. The Frenkel model assumes spherical particles with point contacts. As seen above, the optimum shape for rotational molding plastic particles is a rounded cube. Lontz feels that sintering times will increase with increasing irregularity of particle shape but gives no support for this. One possible reason is that the force required to form a new surface (e.g., the principal force being surface tension or surface traction force) is increased with increasing surface-to-volume ratios. More importantly, however, it would appear that irregular surfaces may encapsulate voids, thus leading to mechanical weakness without significantly increasing sintering times. Again additional work in thermodynamic rheology is needed, with the important groundwork having been founded in allied fields of metal and glass sintering processes.

Once the web-to-particle radius ratio exceeds about 0.5 or so [24,26], the material character is changed dramatically. The continuous three-dimensional structure begins to collapse toward the mold surface, and the interstitial voids between the lattice structure become more like spherical bubbles. It is important that the melting rate of the plastic structure not progress too rapidly (e.g., the melt temperature not increase too rapidly), or the bubble will become encapsulated. Once a bubble is encapsulated, it will either coalesce with others to form a long bubble parallel to the mold surface or will remain identifiable in the part forever. The latter is most probable. Although one can easily calculate the rate of the bubble rise in a horizontal casting (see Chap. 7), the calculation here is meaningless since the gravity field revolves.

The filling of voids by capillary action can be calculated again using the Frenkel theory for the densification of glass [25]. It can be shown that the time required to fill a void Z units in depth and R units in radius can be found from the differential form for the Frenkel equation:

$$Z^2 = (r/2) \int_{t_1}^{t_2} \{\gamma(t)/\eta(t)\} \, dt \qquad\qquad (11.4-5)$$

For most plastics, the surface tension is linear with respect to temperature, and the viscosity is exponential with respect to temperature. Thus,

$$\gamma(T) = \gamma_0 - (\partial\gamma/\partial T)(T - T_0) \tag{11.4-6}$$

and

$$\eta = \eta_0 \exp[-E/R(T - T_0)] \tag{11.4-7}$$

where E is the activation energy (kcal/g·mol). In Table 11.4-1 are given representative values for γ_0, T_0, E, η_0, and $\partial\gamma/\partial T$ for LDPE and GPS. In Fig. 11.4-2 is given the rate of flow into a void of radius 0.01 cm for these two materials while exposed to the exponential temperature profile of Eq. (11.2-2), for various oven set-point temperatures and a constant oven cycle time of 5.5 min. To refer this to filling rates of voids during densification, note that at an oven temperature of 600°F for GPS a void of 0.01 cm in length (Z) can be filled by Frenkel forces in 5.5 min. At 950°F, the void length over the same period of time is increased to 0.1 cm. In other words, the Frenkel filling or densification rate increases from about 2×10^{-3} to 2×10^{-1} cm/min with increasing oven temperature from 600 to 950°F. Thus, one can envision the welling up of melt along the formed sintered material from the hotter melt plane below. If the sintered material does not remain strong enough to allow the Frenkel forces to fill the void area and expel the air ahead of a planar melt front, the sintered material can collapse above the melt front and encapsulate air voids. Thus, a plane of encapsulated bubbles would appear that is parallel to the melt surface in the interior of the final part. The obvious solution to this problem is to densify at a lower oven set-point temperature, thus reducing the rate of melting and the rate of softening of the three-dimensional structure above the planar melt front.

Note also that the Frenkel forces can also be affected by the viscoelasticity of the melt, since Eq. (11.4-5) contains viscosity in the same form as the equation modified by Lontz. Thus, the rate of filling of voids can be retarded by viscoelastic forces in much the same way as sintering is retarded. This means that even if a rubbery material could reach a fully sintered state, it might not be able to densify to a homogeneous part.

Table 11.4-1

Thermodynamic Properties for LDPE and GPS

Material	T_0 (°C)	η_0 (poise)	E (kcal·mol)	γ_0 (dyn/cm^2)	$\partial\gamma/\partial T$ (dyn/cm^2·°C)
LDPE	105	5.22×10^6	14.4	31	0.058
GPS	177	4.92×10^5	22.6	30	0.062

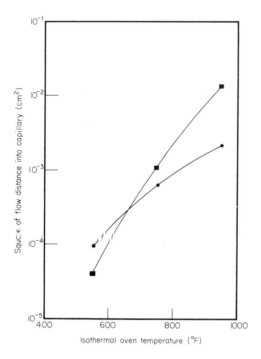

Figure 11.4-2   Square of capillary flow distance into capillary radius of
0.01 cm. Total time in isothermal oven = 5.5 min. Solid squares, MIPS;
solid circles, HDPE [4].

     Note that for GPS approximately 5.0 min are required to sinter the
material at an oven temperature of 550°F. The melt front at this temperature and time has theoretically progressed 0.01 cm into the plastic, and
thus no significant encapsulation is expected. On the other hand, at 950°F,
complete sintering is accomplished in less than 2 min, but the melt front
has moved only 0.08 cm into the material in this period of time. This means
that at the higher temperature there is greater tendency for melt collapse
of the three-dimensional structure and resulting encapsulation of air. From
a practical viewpoint, then, to produce homogeneous parts of most glassy
and low-rubber-content plastics, lower oven set-point temperatures and
longer cycles are necessary. While parts of LDPE can be made at 1000°F
oven temperatures [12], the mechanical properties of these parts are probably lower than those made at lower oven temperatures of 600°F or so.
This of course ignores any consideration of thermal or oxidative degradation and considers only the mechanism of melt-sintering and densification.
Thus, contrary to previous work [4], it is necessary to balance the rate of

sintering with the Frenkel rate of densification to avoid the encapsulation of bubbles.

As mentioned, degradation is a further complication. It is known that while polyethylene increases in tensile strength with increasing residence time in the oven, thermally sensitive materials such as GPS, HIPS, ABS, acrylics, nylons, PC, cellulosics, and acetals can be thermally and chemically degraded. Thus, one must balance the oven time and temperature with an allowable degree of deterioration in mechanical properties that result from degradation. Apparently, the only rotationally molded plastic that has mechanical properties comparable with injection-molded sample is LDPE. As shown in Fig. 11.4-3, after 3 min of oven exposure to 550°F set-point temperature, a rotational molding grade of MIPS loses about one-third of its tensile strength when compared to the injection-molded specimen. At an oven set-point temperature of 750°F for 3 min, it has lost nearly half its strength [4]. Some antioxidants can be used to inhibit oxidative degradation, and inert gas (N_2 and CO_2) is also useful. But above 500°F or so, thermal degradation of most rotational molding materials is inevitable. Thus, to achieve optimum rotational molding properties characterized by high densification, minimum void density, high mechanical strength, etc., oven temperatures no higher than 600 to 650°F are recommended. Thus, long oven cycle times can make fabrication of many plastic parts economically marginal.

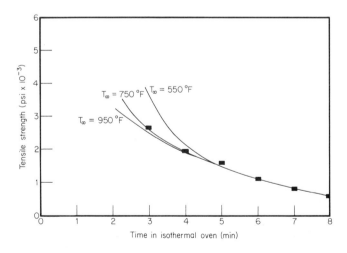

Figure 11.4-3   Comparison of computed and measured tensile strength for MIPS Rexene 314C8 as a function of time in isothermal oven. Solid blocks are experimental data points at T_∞ = 673°F. Oven temperature is parameter [4].

11.5 The Cooling Process

As mentioned earlier, there are two steps to the cooling process. The
first is air cooling, where the mold is allowed to rotate in an ambient or
forced air stream for several minutes. The second involves external
quenching of the rotating mold with a water stream, a fog, or mist. An
additional technique of internal quenching with cold CO_2 or N_2 is used when
the mold is well vented and the concentric shafts of the machine have provision for introduction of cold gas into the molded part. Contrary to earlier
thought [5], the air cooling step is not used primarily to cool the part through
a crystalline freezing or glass transition temperature at a rate sufficiently
slow to allow for uniform crystallization and minimization of part shrinkage
and/or distortion. More recent work indicates that the air cooling step
serves primarily to average the temperature profile and invert it. Recall
that during the oven cycle time the material in contact with the mold surface
has a temperature that is significantly higher than the material at the free
surface, simply because conduction of heat into the melt is controlled by the
poor thermal conductivity of the melt. It can be assumed that an approximate
temperature profile in the melt during the latter stages of the oven cycle
time is a linear one. If q is the surface heat flux (Btu/ft^2·hr), then the
appropriate temperature profile, assuming that at x = L the melt surface is
insulated, would be

$$T = T_s - (T_s - T_a)(x/L) = T_s - (qx/k) \qquad (11.5\text{-}1)$$

where $T_a = T_s - (qL/k)$ is the temperature at the free surface and T_s is the
surface temperature given by Eq. (11.2-2). As long as q is supplied at a
known rate [$q = h(T_\infty - T_s)$ according to the work done earlier], T_a will continue to increase but will lag T_s because of the small value of the melt thermal conductivity k. Consider now the removal of the heating supply. Assume that T_s drops slowly with temperature and that the value of q becomes
very small or even negative. Note that $T_\infty - T_s$ approaches zero or becomes
negative if $T_\infty \ll T_s$, which is the case if the mold surface is hotter than
the surrounding environment. Now the mold surface temperature drops,
and the free surface temperature approaches the mold surface temperature
in magnitude. The temperature profile thus flattens, since there is no
energy input and relatively little energy is being expended to the environment. As cooling continues in this fashion, the surface temperature begins
falling since the surface is cooled by convection in nearly the same way as
it was heated. But the rate of cooling is much slower owing to the much
lower heat transfer coefficient for natural convection. Furthermore, the
heat being removed is from the hottest part of the plastic, and the time required to invert the temperature profile, when calculated using standard
tables [20], is on the order of 2 to 4 min. The progression of the maximum
temperature through the plastic part is determined in part by the thermal
conductivity of the melt. When finally the temperature profile is again

linear, with the minimum temperature at the mold interface and the maximum temperature at the free surface, the part can be suitably quenched.

This heuristic argument is also related in a way to part distortion, which was the primary argument originally given for the two-step cooling process. Suppose the surface of a part is to be quenched from 600 to 50°F instantaneously, while the free surface is nearly isothermal at say 450°F. The surface would "freeze" in much the same way as it does during injection molding into a cold cavity. Without pressure behind the frozen front, however, the very large volume shrinkage would cause the surface to either crack or pull away from the mold surface. If the latter happens, heat transfer to the quenched material stops, and the surface of the plastic is reheated immediately by the hot melt directly behind the surface. Now, however, the material has no mode of cooling and thus remains soft and at a nearly uniform temperature. Other areas that have not pulled from the surface continue to cool and shrink and now begin to draw material from the still-soft hot spot. In this way, shrinkage and distortion can occur, regardless of the glass-transition temperature, crystallization temperature, or any other characteristic temperature. Thus, it can be concluded that if sufficient time is allowed for inversion of the temperature profile, quenching can take place, and this "freezing-thawing" effect should be minimized. Note that the time for this first step of cooling is approximately proportional to the part thickness, whereas the time for the second step, quenching, is proportional to the square of the part thickness. Internal quenching of the part with cold gas during the first step will accentuate shrinkage. Internal cooling is probably most beneficial in the latter stages of cooling. Incidentally, there is sufficient knowledge available to support the hypothesis that quenching with warm or hot water (100 to 150°F) will cause much less distortion and warpage than quenching with tap or refrigerated water. Again, the cooling of plastic materials in injection molds supports this also. Gentle cooling rates allow conduction to cool the part more uniformly, albeit at a slower overall rate. This is particularly true for materials that pass through a zone of crystallization. This point has been commented on in the literature [5].

11.6 Liquid Rotational Molding

The earliest type of rotational molding, as mentioned earlier, was "slush" molding of PVC plastisol. With the advent of LDPE powder by USI Chemicals in the early 1960s the engineering effort to improve "slush" molding stopped. This is unfortunate, for at least three characteristics of the process make liquid rotational molding attractive. First, most liquid plastics cure or react with relatively little external heat. Second, liquid plastics are not plagued with sinter-melting and densification problems. And third, most foam moldings of large parts require uniform coating of a mold

surface to produce a uniform skin backed with the foam core, and this is
attainable through rotation of the mold during foaming. Also note that many
otherwise suitable plastic parts rotationally molded from powder have unac-
ceptable surfaces because of microvoids in the surface caused by the sinter-
melting process of the powder. These "pores" are intrinsic to the processing
of powders and are completely absent when rotationally molding liquids. In
this section, therefore, consideration is given to the engineering problems
that are unique to rotational molding of liquids.

It is well known that epoxies, unsaturated polyesters, rubber latexes,
and liquid PVCs exhibit long induction times once they are catalyzed or reaction is initiated with moderate heating of the mass. The reaction characteristics of these materials are completely unlike urethane reaction characteristics. Once urethane material components are mixed, exothermicity occurs,
and the material heats and cross-links very rapidly. For the nonurethane
liquids, however, early cross-linking reaction does not heat the liquid or
increase its viscosity significantly. During this induction period, the isothermal viscosity remains relatively low (10 to 100 P) and nearly constant
[28]. At a critical time, known as the "gel time," sufficient cross-linking
sites have been activated to cause the viscosity to rise very rapidly in a
very short time until the material is solid-like. Thus, the objective in rotational molding of liquids, very simply put, is to get a uniform coating of
liquid on the walls and keep it there until gelation occurs. Owing to the long
induction times of 5 to 10 min and longer, rotational speeds early in the
process must be carefully controlled to ensure the uniform coating of the
mold surface. If the rotational speed is too fast, the mold moves through
the liquid pool near the bottom of the mold too rapidly to allow buildup of a
uniformly thick layer on the mold surface. If the speed is too slow, the portion of the mold first in contact with the liquid will collect nearly all the
liquid from the pool. Thus, the material will move with the mold surface
and gradually spread in an uncontrolled fashion over the rest of the mold
surface. The latter problem has been seen in the in situ casting of Nylon 11
[3] and by those working with urethanes.

The key to the proper processing of these materials, then, is the balancing of rotational speed with time and temperature. To begin, consider
that the machine is stopped with a catalyzed liquid polymer in a pool in the
bottom of the mold. As the mold begins to move, liquid is drawn from the
pool by the moving mold surface. This is known as withdrawal [29,30,35].
There are basically four forces that describe the fluid flow in the region where
film is formed on a moving surface from a pool. These forces are viscous
($\eta V/\delta$), gravitational ($\rho g \delta$), capillary (γ/δ), and inertial (ρV^2). Here η is
the local viscosity, V is the velocity of the moving surface (or rpm of the
rotating mold), δ is the thickness of the film, ρ is the liquid density, γ is
the surface tension, and g is the gravitational constant. These forces result
in three dimensionless groups, given in this form:

$$T_D = f(Re, Ca) \tag{11.6-1}$$

where $T_D = \delta(\rho g/\eta V)^{1/2}$, a dimensionless film thickness; $Re = $ V $/$, a film Reynolds number; and $Ca = \eta V/\gamma$, a capillary number. For all liquid polymers of interest, $Re \ll 1$ and $Ca \gg 1$. Thus, the relationship between the dimensionless film thickness and the Reynolds and capillary numbers is [31]

$$T_D = 0.66 \text{ and } Q = 0.56 \tag{11.6-2}$$

where $Q = (q/V)(\rho g/\eta V)^{1/2}$ and q is the flux of liquid per unit width and unit time. Examination of this reveals that through rearrangement of T_D the thickness of the liquid film is proportional to the square root of the viscosity and the velocity of the moving surface. The amount of liquid contained in the film is proportional, therefore, to the velocity of the moving surface to the 3/2 power and to the square root of the viscosity.

Now consider an isothermal process with a very gradually increasing (or constant) viscosity, as would be the case for a curing PVC plastisol. The thickness can be controlled by beginning the rotational speed at a value that allows uniform coating of the wall, as calculated from Eq. (11.6-2). Assume, for example, that the surface velocity of the mold for $\delta = 1$ cm, $\rho = 1$ g/cm^3, and $\eta = 10$ P is desired. From $T_D = 0.66$, V = 225 cm/sec. Thus, if the mold is 30 cm in diameter, its rotational speed is 15 rpm. If the rate at which viscosity is increasing with time is known, the machine can be programmed so that the speed can be decreased in proportion to the increase in viscosity owing to curing of the PVC.

Heating a reactive fluid results in a dramatic decrease in the viscosity prior to gelation, although a greatly shortened induction time. The shorter induction time appears attractive because of the increased productivity of the machine, and therefore this aspect of liquid rotational molding is considered here. To maintain part thickness as the viscosity decreases prior to gelation, the speed of rotation must be increased dramatically. Also remember that the material is actively reacting and generating heat and that this heat if accumulated in the pool will act to reduce viscosity and shorten reaction time. Therefore, if a proper speed of rotation is not programmed, the fluid will remain in the pool or will return to the bottom of the mold to form a new pool. Now follow through the consequences of this pooling. From earlier work the heat of a cross-linking reaction is generated volumetrically and removed geometrically. The walls of the mold exhibit a volume-to-surface ratio proportional to the thickness of the liquid attached to the wall. If the pool in the bottom of the mold is cylindrical with a radius R, the volume to surface ratio is proportional to the radius. But the cylinder is in contact with a heat sink at only one line along the cylinder. Thus, the effective surface-to-volume ratio for heat transfer is nearly zero. This means that the pool will heat at a much faster rate than the material adhering to the walls. More importantly, the pool will reach the gel point before the material on the walls. Since gels are in general quite sticky, the pool will mechanically wipe the material from the wall as

it passes through the pool and add that material to the pool. The result
will be that early in the rotational programming most of the material was
on the mold walls and a little was in the pool. Now nearly all is forming
into a cylinder in the bottom of the mold, and when the mold is opened, the
result will be disastrous.

The obvious solution to the problem is to program the mold rotation
so that the material adheres initially in a uniform layer on the mold wall and
then remains there as the temperature of the mold and resin increases and
the resin viscosity decreases [28]. Unfortunately, because of the typical
Arrhenius exponential dependency of viscosity on temperature, it is necessary to increase the speed of the machine very rapidly when the mold is
subjected to a transient temperature similar to that of Eq. (11.2-2). Normally, conventional machines are not capable of achieving and maintaining
these increased speeds. An alternate method uses viscosity modifiers, such
as $CaCO_3$, aluminum powder, or talc, which are added directly to the liquid
resin. These materials make the normally power-law resins Bingham-type
plastics and also act as heat sinks. In a recent study [28], as much as 50
wt% $CaCO_3$ was added to liquid polyester to stiffen it and thus prevent re-
pooling of the fluid once it is on the mold walls. The major fault with this
technique is that the mechanical properties of the final part are decreased
significantly through the addition of certain fillers such as $CaCO_3$, and other
undesirable properties such as electrical conductivity are forced upon the
material through the use of aluminum or copper powder. As recommended
in [28], the programming of rotational molds with liquid resins must incor-
porate new machine designs to achieve higher speeds of rotation; the liquid
resins must have higher viscosities, more gradual viscosity increases with
cross-linking reaction progression, and some Bingham plastic character
through the addition of stiffeners; and the heating technique must be more
carefully controlled so that isothermal temperatures of 200°F or so can be
maintained until gelation and then raised to cure the resin into a thermoset
part [8,9]. (See also Sec. 7.3 for additional discussion.) The technology
of rotational-molding thermosetting resins other than urethanes has long
been neglected and offers some exciting alternatives to the present thermoplastic molding techniques.

11.7 Other Processes

Potentially the most important market available to rotational molders is
that of large piece structural foam moldings. While most of the structural
foam technology is in its infancy, serious attempts to make useful structural
foam articles through rotational molding are underway in many laboratories.
One of the keys to the utility of this method is the capability of sizes in excess of 1000 lb or more, with McNeil Femco advertising a 5000-lb machine.
Thus, articles such as all-terrain vehicle cabs, entire camper tops, and

15- to 20-ft boat hulls are possible with modifications in present machinery. It is well known that structural foam parts are characterized by surface densities of near that of the compact material and core densities of 20 to 30 lb/ft^3. These are achieved by tumbling a blowing agent such as 1,1'-azobisformamide (ABFA) in concentrations of less than 1% onto the surface of the resin powder or pellet. From injection foam molding, it is known that powder gives a more uniform cellular product than does pellet resin [32]. It is a happy coincidence, then, that powders of -35 mesh are needed for rotational molding. All molds must be reinforced to withstand 150- to 200-lb/in.^{2} pressure generated by the blowing agent. Tumbling the powder in the rotating mold during blowing ensures a rather uniform and unoriented bubble pattern in the molded part. This is not easily achieved by conventional injection foam molding. The key with foam rotational molding is to rapidly achieve the decomposition temperature of ABFA (about 200°C) while rapidly rotating the mold. Hot molds will yield thick smooth skins with few, if any, splay marks and fine cells within the part. Quenching the mold is done without the air cooling cycle, since the skin is held against the mold surface by the still-blowing ABFA. Because the molds are normally much thicker than those for conventional rotational molding, the oven cycle time on heating is much longer, but the higher temperatures required to sinter and densify the material are apparently not needed since the voids naturally fill with blowing agent. The molds should be well vented, as is the case with injection foam molds.

Another way of making a structural foam part using rotational molding is to use a liquid resin and a foamable solid resin. An example is PVC organisol with polystyrene powder presoaked in heptane or some other suitable physical blowing agent. The idea here is to allow the plastisol to adhere to the mold surface and the styrene to remain the interior of the part. Now as the temperature of the mold increases, the plastisol begins to cure. Since the plastisol acts somewhat like a thermal insulator, it will reach curing temperature before the volatile physical blowing agent can begin to swell the heat-softened styrene. Once the styrene reaches its softening point, it expands to a core density of 15 to 20 lb/ft^3. In concept, then, the process is quite simple. In actual operation, however, there is difficulty in uniformly coating the mold surface with liquid, as discussed earlier. This coating process is further complicated by the presence of the styrene powder, which is somewhat compatible with the PVC organisol and thus sticks and partially dissolves in it. Attempting to coat a metal mold with this sticky mess is not unlike trying to process hot caramel popcorn. The process should be examined in more detail, however, by considering much finer styrene powders, 100 μm or less in dimension, and handling the mixture as a filled liquid resin, as was discussed for talc-filled polyester and epoxies earlier in this chapter.

A novel way of simultaneously rotational-molding at least two semi-miscible plastics having melting or softening points of at least 10°C difference has been patented [33]. This is done quite simply. All plastic powders are loaded into the mold simultaneously. As the mold heats, the lowest melting material sticks to the mold surface preferentially, and the higher melting materials remain in the powder pool. As the "free" surface of the lowest-melting material reaches the melting point of the second-lowest-melting material, it is preferentially removed from the powder pool, and so on until the several materials are layered onto the mold surface in order of increasing melting temperature. In theory, these layers should be completely separate. In practice, however, the sticky nature of the melting powder on the mold surface will tend to accumulate some higher-melting material as a solid. The interfaces, for slightly miscible materials, at least, are somewhat indistinct, thus indicating that tear strength at the interface should be very high. This idea will not work well for incompatible materials such as LDPE and GPS, and one should expect delamination at the interface. The standard problems of porosity of the surface, incomplete sinter-melting, and so on are present here in essentially the same way as in one component molding. Nevertheless, it should be realized that the technique could be useful if a project called for molding a container with a barrier material inside (a polyester, for example) and an impact material on the outside (HIPS, for example). Also HDPE/LDPE and HDPE/PP combinations could be made this way. The important facets of this development are that more than two layers (barrier material, impact material, UV-absorbing material) can be molded as long as the temperatures are compatible and that this can be done on a machine that has not been modified in any way.

11.8 Conclusions

In this chapter, the reader has been given a viewpoint of the technical level of work being carried out in rotational molding. At present, much emphasis is being placed on the molding of powders to the near-neglect of liquid resins. The potential in liquid resins has been spelled out. Newer ideas that lend themselves to rotational molding, such as structural foam molding and two-component molding, have been discussed briefly. The field is potentially very large, although at present it seems to lack imagination and suitable engineering know-how.

Problems

P11.A A mold of 0.125-in. steel is placed in an oven at $T_\infty = 550°F$. Room temperature is 70°F. If the convection heat transfer coefficient is 5 Btu/ $ft^2 \cdot hr \cdot °F$, how long does it take the mold to heat to 500°F? If the mold is cooled from 500°F with 50°F water, how long does it take the mold to cool to 100°F if the convection heat transfer coefficient is 150 Btu/$ft^2 \cdot hr \cdot °F$?

P11.B A 0.100-in. electroformed nickel mold is replaced with a 0.250-in. aluminum mold. Does the time to reach a specified temperature after insertion into the isothermal oven increase or decrease? Show all work.

P11.C Work through the substitution of Eq. (11.3-3) into Eq. (11.3-1) to obtain the penetration thickness δ. Then assume that the surface temperature T_s can be represented by a polynomial, $T_s = a + bt + et^2$. Obtain a form for the penetration thickness as a function of time. Show that when $a = c = 0$, δ reduces to Eq. (11.3-6).

P11.D Obtain Eq. (11.3-8).

P11.E Form a different temperature profile than the one given as Eq. (11.3-3), and obtain a new form for the penetration thickness for constant T^*.

P11.F Following the bookkeeping suggestions following Eq. (11.3-12), develop a computer logic that enables a temperature profile as a function of time to be produced for both powder and melt. Make sure to include the testing procedure for melted material and the subsequent redefinition of the amount of material remaining as a powder.

P11.G In [4], it was found that for the data given in Lontz [26] the following equation fit the Kuczynski-Neuville empirical relationship [24]:

$$s/a = (b't)^{1/n} \tag{11.G-1}$$

where

$$b' = b_0' \, \exp(-b_1' T_c) \tag{11.G-2}$$

$$n = n_0 \, \exp(-n_1 T_c + n_1 T_{mc}) \tag{11.G-3}$$

where $b_0' = 1.24 \times 10^5$, $b_1' = 2.01 \times 10^3$, $n_0 = 5$, and $n_1 = 0.023$. T_c is the local temperature in °C, and T_{mc} is the melt temperature in °C. Make a plot of x/a for $T_{mc} = 100$ and $150°C$. Then let $T = 10(1 + 5t)$, where t is in seconds. Make a plot of x/a against time with $T_{mc} = 100°C$. Can a nonisothermal expression be reconciled with the isothermal form for the Frankel equation? If not, why not?

P11.H Given particles of LDPE 500 μm in diameter, with a surface tension $\gamma = 30$ dyn/cm^2 and a viscosity of 10 P, make a plot of x/a vs. time for γ various values of τ. Show that for large values of τ the results are approximated by Eq. (11.4-4). Show that for very small values of τ the results are approximated by Eq. (11.4-1). Expand the exponential in Eq. (11.4-3) in a series so that higher-order terms can be plotted. Can the results of Eq. (11.4-1) and (11.4-4) be approximated over their regions using higher-order terms?

P11.I The viscous energy of activation of a polymer is 15 kcal/mol, and its viscosity is 10^6 P at T = 150°C. The surface tension at this temperature is 30 dyn/cm^2, and the rate of change of surface tension with temperature is given as 0.06 dyn/cm$^2 \cdot °$C. Rao and Throne [4] show that densification also follows the Frankel equation according to

$$1 - (r/r_0) = \gamma t / 2\eta r_0 \qquad (11.I-1)$$

where r is the initial void radius in cm and r is the instantaneous radius in cm. Make a plot of r/r_0 for 150 and 200°C. Then assume that the temperature increases linearly from 150 to 200°C in 5 min. By differentiation, find the rate of densification dr/dt over this temperature range and the size of the void after 5 min.

P11.J Consider the flow of a polymer material on a vertical surface. Assume at the low shear stresses induced by gravity that the fluid is Newtonian. Show that the maximum shear stress is given by

$$\tau_{max} = \rho g d \qquad (11.J-1)$$

where g is the gravitational constant, d is the thickness of the melt layer, and ρ is the density of the melt. Then calculate the shear rate for a 0.25-cm-thick melt layer for a 1.05-sp. gr. GPS with a 1.7×10^5 P viscosity at 186°C.

P11.K Assume that the rotationally molded sample is held in ambient air until it equilibrates at 500°F. The material is GPS with a thermal conductivity of 0.14 Btu/ft$\cdot$hr$\cdot°$F, a specific heat of 0.45 Btu/lb$\cdot°$F, and a specific gravity of 1.05. The mold is 0.050-in. aluminum and should not lead to any resistance in cooling. The mold is then quenched with 50°F water. Assuming that the inner or free surface is insulated, determine the time required to cool the material to an average temperature of 100°F. Then determine the inner or free surface temperature at this time. Carefully list all assumptions.

P11.L (Following McKelvey [34]) The temperature dependence of the rate constant for the expansion of GPS is given approximately by

$$k = A/(B - T) \qquad (11.L-1)$$

where A and B are constants. For B = 400 and A = 50, obtain values of k for the temperature range T = 350 to 400°K. One form for the rate of change of volume of a polymer with temperature is

$$dv/dT = b' + (k/r)(a + bT - v) \qquad (11.L-2)$$

For GPS, $b' = 2.5 \times 10^{-4}$ cm^3/g$\cdot°$C, a = 0.845 cm^3/g, b = 3.75 $\times 10^{-4}$ cm^3/g$\cdot°$C, and r = 0.01°C/sec. Introduce Eq. (11.L-1) into this expression, and integrate it to obtain an explicit expression for v as a function of T. Compute v for GPS heated from 300 to 400°K. Explain how this expression can be used to determine differential volume changes in a rotationally molded part.

References

1. USI Chemicals Symposium on Rotational Molding, Chicago, Nov. 1963.
2. USI Chemicals Symposium II on Rotational Molding, Chicago, Nov. 1964.
3. P. F. Bruins, ed., Basic Principles of Rotational Molding, Gordon & Breach, New York, 1971.
4. M. A. Rao and J. L. Throne, Polym. Eng. Sci., 12:237 (1972).
5. J. L. Throne, Polym. Eng. Sci., 12:335 (1972).
6. Anon., Engineering Design Handbook: Rotational Molding of Plastic Powders, Army Materiel Command Pamphlet, AMCP 706-312, ADA-013178, Apr. 1975.
7. J. L. Throne, Polym. Eng. Sci., 16:257 (1976).
8. R. C. Progelhof and J. L. Throne, Polym. Eng. Sci., 16:680 (1976).
9. J. A. Deiber and R. L. Cerro, Ind. Eng. Chem. Fund., 15:102 (1976).
10. G. Ong, personal communication on PC, Oct. 1973.
11. F. Kreith, Principles of Heat Transfer, International Textbook Company, Scranton, Pa., 1965.
12. Phillips Petroleum Co., Bulletin 17, Rotational Molding, Bartlesville, Okla., undated.
13. F. G. Troppe and R. W. Roberts, paper presented at 63rd Annual Meeting, AIChE, New York, N.Y., Dec. 1970.
14. R. C. Progelhof, J. L. Throne, and R. R. Ruetsch, SPE Eng. Prop. Structure DIVTEC, Oct. 7, 1975, Hudson, OH, Proceedings, p. 221.
15. R. L. Brown and J. C. Richards, Principles of Powder Mechanics, Pergamon Press, Elmsford, N.Y., 1966, p. 24.
16. K. Kurihara and H. Kuno, J. Phys. Soc. Jpn., 7:727 (1965).
17. A. Suzuki and T. Tanaka, Ind. Eng. Chem. Fund., 10:84 (1971).
18. M. A. Goodwin and S. C. Cowin, J. Fluid Mech., 45:321 (1971).
19. J. G. Knudsen and D. L. Katz, Fluid Dynamics and Heat Transfer, McGraw-Hill, New York, 1958, p. 363.
20. P. J. Schneider, Conduction, In: Handbook of Heat Transfer (W. M. Rohsenow and J. P. Hartnett, eds.), McGraw-Hill, New York, 1973, Sec. 3.
21. T. R. Goodman, Integral Methods in Nonlinear Heat Transfer, Advances in Heat Transfer, Vol. 1 (T. F. Irvine, Jr., and J. P. Hartnett, eds.), Academic Press, New York, 1964, Chap. 2.
22. C. O. Bennett and J. E. Myers, Momentum, Heat and Mass Transfer, McGraw-Hill, New York, 1962, p. 489.
23. P. V. Danckwerts, Gas-Liquid Reactions, McGraw-Hill, New York, 1970, p. 143.
24. G. C. Kuczynski and B. Neuville, paper presented at Notre Dame Conference on Sintering and Related Phenomena, Lafayette, IN, June 1950.
25. Y. I. Frenkel, J. Phys. Moscow, 9:385 (1945).

26. J. F. Lontz, Sintering of Polymer Materials, In: Fundamental Phenomena in the Material Sciences, Vol. 1 (L. J. Bonis and H. H. Hausner, eds.), Plenum, New York, 1964, p. 214.

27. R. S. Lenk, Plastics Rheology, Wiley-Interscience, New York, 1968, p. 125.

28. J. L. Throne, SPE ANTEC Tech. Pap., 20:367 (1974).

29. J. A. Tallmadge and C. Gutfinger, Ind. Eng. Chem., 59:18 (1967); errata, 60:74 (1968).

30. P. Groenveld, Chem. Eng. Sci., 25:1571 (1970).

31. P. Groeveld, Chem. Eng. Sci., 25:33 (1970).

32. J. L. Throne, Thermoplastic Structural Foams, SPE Educational Seminar, Norwalk, Conn., Oct. 1974.

33. F. J. Reilly, U.S. Patent 3,542,912 (Nov. 24, 1970).

34. J. M. McKelvey, Polymer Processing, Wiley, New York, 1962, p. 132.

35. P. Groenveld, Chem. Eng. Sci., 25:1267 (1970).

12.

Thermoforming

12.1 Introduction

Thermoforming is a generic name for shaping a sheet of plastic that has been heated until soft. While some processes may use a softened pool of plastic rather than a sheet, it is generally agreed that the use of sheet plastic distinguishes thermoforming from other types of processing such as blow molding or compression molding. Some of the more familiar but more specialized subgroups of thermoforming are vacuum forming, drape forming, matched mold forming (of sheet stock), pressure bubble-vacuum forming, plug assist vacuum forming, vacuum snap-back forming, and trapped sheet pressure forming [1]. Some of these are discussed in more detail below.

The thermoforming business, like the rotational molding business, is in a period of rapid growth but remains a relatively minor plastic processing area. The major areas of growth in thermoforming within the last few years have been in packaging and disposable containers. Basically, these areas deal with high-volume, very thin sheet stock products such as fresh- and frozen-food trays, blister packs, utility trays, hot drink containers and attendant snap-on or press-on lids, cold drink containers, and containers suitable for both quick-freezing and heating. The low-volume areas are not growing so rapidly but are probably more profitable than the high-volume areas because of the uniqueness of thermoforming to produce high-quality plastic parts with relatively little setup time and tooling costs. Some of the areas in which low-volume thermoforming is paying off include exterior signs, turnpike signs, heavy-duty pallets, tote boxes, refrigerator door liners, camper tops, canoes, sailboats, snowmobile shrouds, room humidifier cabinets, seating of all types including children seats and airport lounge seats, equipment cases, parts trays, and bathtubs. Most architectural designers use thermoformed prototypes for esthetic evaluation. Many parts that will be made by other means (rotational molding, structural foam injection molding, and even blow molding) may be developed as proto-types using thermoformed sheet. Scale models of furniture and camper styles (to determine aerodynamic suitability) have been made of thermoformed plastic sheet.

In this chapter we shall concentrate on a rather abstract concept of thermoforming. Three components of thermoforming will be detailed: the heating source, the sheet stock, and the mold. Our purpose will be to establish some engineering relationships between these three components. Note, however, that the engineering of the basic thermoforming process is very pragmatic. Overall extrapolations or rules of thumb at this point in the process development seem premature.

12.2 Thermoforming Techniques—Variations

Shown in Fig. 12.2-1 is the simplest type of thermoforming. Here a sheet of plastic is heated in an oven or with radiant heaters until soft and then draped over a mold. The hot sheet is then smoothed down to minimize the wrinkles and entrapped air bubbles and then is held in place by hand until cold. This very old technique is known as drape forming. Vacuum forming, shown in Fig. 12.2-2, uses external atmospheric air pressure to force the sheet onto the mold rather than hand forming. Vacuum forming uses either a male or female mold as shown. Simple drape forming and vacuum forming are restricted to rather shallow draw depths (H/W of 2:10 or so, where H is the depth of the draw and W is the width of the narrowest mold dimension), simply because deeper draws require "prestretching" of the sheet. This is particularly true if the sheet being drawn is somewhat rubbery when soft, such as ABS. Furthermore, deep draws of sheet that has not been prestretched will yield parts that are very thin in the corners, as will be seen. Therefore, two major variations and many minor variations have been devised. The first is a plug-assisted draw. As shown in Fig. 12.2-3, a plug stretches the sheet before it contacts the mold surface. This helps thin the sheet in

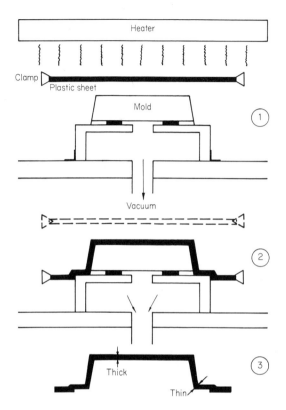

Figure 12.2-1 Schematic example of drape forming. Step 1: plastic clamped and heated; step 2: formable plastic draped onto cold mold, vacuum applied to pull sheet into intimate contact with mold; step 3: final wall thickness distribution, thick in center, thin at corners [1].

the center and wall area so that additional stretching of the sheet into the corner will take place from relatively thick plastic. A second technique uses pressure to expand the sheet into a bubble prior to contact with the mold. This blown vacuum forming technique is shown in Fig. 12.2-4. In some cases, both techniques can be used effectively, as shown in Fig. 12.2-5. Other variations include combining air pressure and vacuum during drawing (Fig. 12.2-6), using just air pressure and venting the mold during drawing (Fig. 12.2-7), and simply clamping the plastic sheet between two halves of a matched mold (Fig. 12.2-8). All of these variations are useful in some aspect of the thermoforming industry. The most popular and most used technique is the plug-assisted vacuum forming technique. The

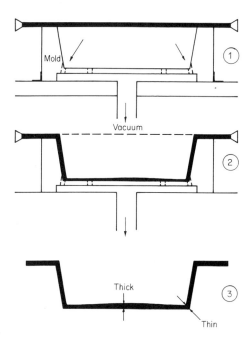

Figure 12.2-2   Schematic example of conventional vacuum thermoforming into female mold.   Step 1: formable, heated sheet clamped and sealed against mold surface; step 2: vacuum applied, drawing sheet into intimate contact with cold mold; step 3: formed part, thick in center, thin in corners, edges [1].

ramification of this process is considered in some detail in this chapter. Also some of the materials that can be used successfully as patterns for thermoforming will be discussed.

12.3 Heating Sources

As is now known, there are three methods of heating plastic materials: · conduction, convection, and radiation. Direct contact between the heating source and the plastic sheet is the most efficient way of heating. The attractiveness of conduction, therefore, has not been overlooked by inventors, and many attempts have been made to exploit it. Most have not worked. Most of the attempts have been made on moving sheet such as that used in high-speed thermoforming machines because the heating cycle controls the overall machine cycle. The problem is that the sheet cannot be allowed to

slide over the heating surface, for very fine asperities in the heating surface will score lines into the plastic sheet surface. Therefore, the heater
must move with the sheet. Most of the problems center around the practical
problems of clamping and unclamping the sheet from the heater surface.
The very large thermal expansion of the sheet will cause it to sag or bulge,
and clamping or constraining will cause wrinkles in the sheet. If the sheet
is heated too rapidly, it can become sticky or tacky, and thus it will stick
to the heater surface. The controlling phenomenon for any heating method
is not the rate of energy brought to the surface of the sheet but the rate of
conduction of that energy into the sheet. It would seem, then, that conduction heat transfer is not acceptable owing to problems other than the heating
process itself.

Convection ovens have been extensively used, either as a primary
source of heat or in conjunction with radiant heat, since the early days of
thermoforming. It is, nevertheless, a rather inefficient method of heating,
since the heat transfer coefficients for even the most efficient air convection
oven do not exceed 10 Btu/ft^2·hr·°F. Thus, in addition to the thermal resistance of the plastic to conduction of energy within the sheet, there is
added resistance at the surface of the sheet owing to convection heat transfer. Recall the discussion of Chap. 6; the dimensionless group that describes
this ratio of resistances is the Biot number Bi = hL/k, where h is the heat
transfer coefficient, L is the half-thickness of the sheet assuming convection
heating on both sides of the sheet, and k is the sheet thermal conductivity.
For sheets with half-thickness on the order of 1/4 in. or less the Biot number is less than unity. This means that the rate of heating is substantially
reduced, as can be seen from Fig. 6.2-5. This reduced rate of heating can
be useful if the sheet is very thick or the material is thermally sensitive,
such as opaque PMMA or PVC. In general the time for heating controls the
thermoforming process, the productivity of the machine, and ultimately the
profits of the corporation. As will be seen, convection cooling, in combination with radiant heating, can be an effective way of controlling excess surface temperatures.

At present, the most popular way of heating thermoplastic sheet stock
is with electric heaters. Wound Nichrome wire on glass rods, steel "calrod"
rods, flat steel sheet either heated by conduction from embedded cartridge
heaters or radiation from remote "calrod" heaters, incandescent (tungsten
wire) or carbon arc heaters have been used. Rod heaters, backed with
reflectors to provide more uniform heat to flat surfaces, remain the simplest
and most versatile of the electric heaters, although not always the most
efficient. Whenever opaque plastic sheet is thermoformed, remember that
all the energy incoming to the sheet is absorbed on the surface of the sheet,
regardless of the mode of heat transfer (conduction, convection, or radiation). For radiation, the "radiant interchange" equation yields the heat flux
delivered to the plastic surface from the heater bank [2,3]:

$$Q = FA\sigma(T_1^4 - T_2^4)$$

$$(12.3-1)$$

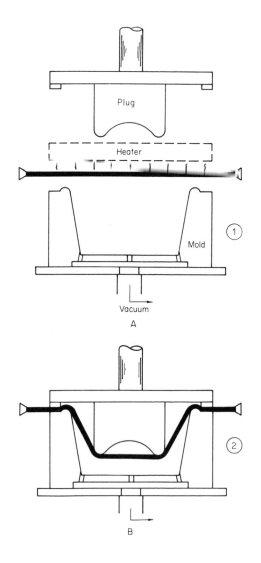

Plug

Heater

Mold

①

Vacuum

A

②

B

620

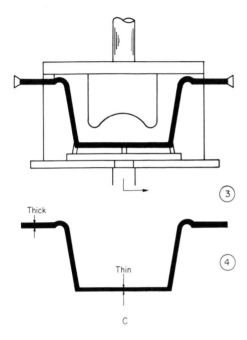

Thick

Thin

C

Figure 12.2-3 A: Schematic example of plug-assisted vacuum forming.
Step 1: after sheet heater is removed from vicinity, plug assist positioned
over open mold cavity. B: Step 2: plug then forced against formable
softened plastic, sealing the edges against the mold surface; plug assist
may be heated. C: Step 3: vacuum then applied, drawing softened plastic
against cold mold surface. Step 4: characteristic cross-section of formed
part, thin in center, thick in clamp area [1].

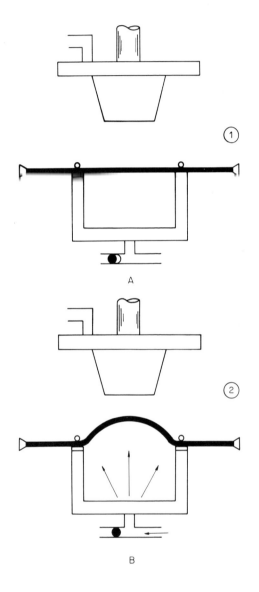

1

A

2

B

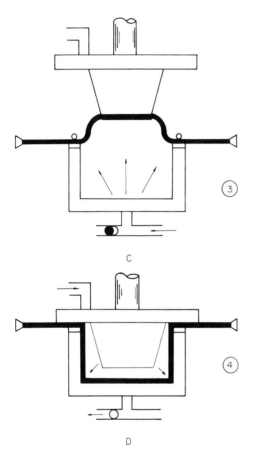

C

D

Figure 12.2-4 A: Schematic example of pressure bubble–plug assist
vacuum forming. Step 1: formable sheet clamped, sealed against the mold
surface. Note extra clamp ring on sheet top. B: Step 2: air pressure (10
to 30 psig) inflates formable sheet. C: Step 3: plug assist now actuated
into bubble center; air pressure still maintained during plug action. D:
Step 4: vacuum in mold, air pressure through plug (optional), now applied,
forcing sheet into intimate contact with mold surface [1].

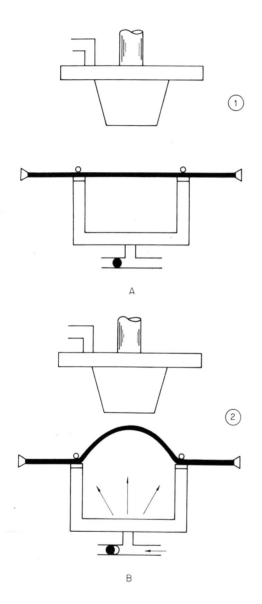

A

B

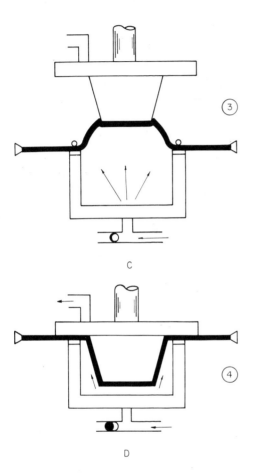

C

D

Figure 12.2-5   A: Schematic example of pressure bubble-vacuum snap-back thermoforming.   Step 1: formable, softened sheet clamped against mold surface.   B: Step 2: air pressure applied, forcing sheet to deform into bubble; air pressure initially 10 to 50 psig.   C: Step 3: plug assist now forced into bubble, initiating eversion; air pressure within mold cavity maintained.   D: Step 4: vacuum now applied to sheet through ports in plug, forcing the sheet into intimate contact with the plug; therefore, mold cavity acts only as reservoir for plug [1].

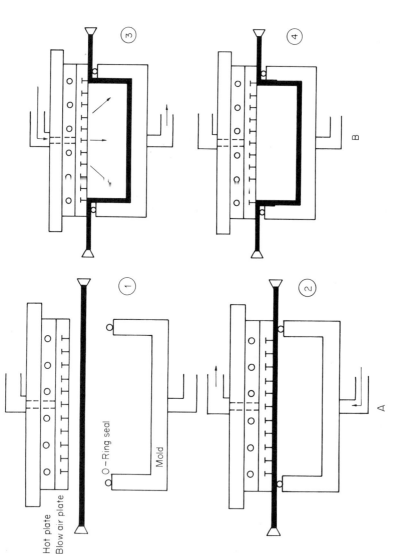

Figure 12.2-6 A: Pressure forming of softened plastic sheet, a schematic example. Note in this example that contact heat transfer is used to heat the sheet to the softening point. Step 1: components of pressure forming. Step 2: sheet heated by contact with hot plate. Note: Air pressure in mold greater than atmospheric to ensure intimate contact of plastic with hot plate. B: Step 3: vacuum then applied to mold box, pressure applied through ports on contact hot plate, forcing sheet into intimate contact with mold cavity walls. Step 4: optional cutting step can be carried out at this point by increasing force on hot plate–blow air plate [1].

where A is the surface area of the plastic (ft^2), T_1 is the temperature of
the heater (°R), T_2 is the temperature of the plastic sheet (°R), σ is the
Stefan-Boltzmann constant (= 0.173×10^{-8} Btu/ft$^2 \cdot$hr$\cdot$°R^4), and F is the
view factor or configuration factor. The determination of F, the view fac-
tor, can itself be the subject of a textbook, as discussed in Chap. 6. If the
system is considered to be ideally made of two parallel planes with the
reflectors behind round rod heating sources acting to uniformly distribute
the radiant energy so that the source appears planar, the following definition
for F is valid:

$$F = (1/e_1 + 1/e_2 - 1)^{-1} \tag{12.3-2}$$

where e_1 is the emissivity of the heater source and e_2 is that of the plastic
sheet. Normally, the emissivity of roughened steel is about 0.9 to 0.95,
and plastic is on the order of 0.85 to 0.9 (with the exception of a mirrored
plastic). Thus, F has the range of about 0.78 to 0.86 or so. If both surfaces were perfect black bodies, $e_1 = e_2 = 1$, and thus F = 1. And, as discussed in Chap. 6, a Biot number for radiant heat transfer can be defined
in much the same way as one is defined for convection heat transfer. The
radiant heat transfer coefficient is defined as

$$Q = h_r A(T_1 - T_2) \tag{12.3-3}$$

where

$$h_r = F[\sigma(T_1^4 - T_2^4)/(T_1 - T_2)] = FF_T \tag{12.3-4}$$

where F_T is the temperature factor, as defined by the bracket in the second
equation. A plot of F_T against heater temperature and plastic surface temperature is given in Fig. 12.3-1. Remember here that F_T is a function of
plastic sheet temperature and that therefore so is h_r. Thus, h_r is also time
dependent and increases with increasing sheet temperature.

Of course, consideration of a radiant heat transfer coefficient can be
completely circumvented by discussing energy input simply in terms of the
energy flux to the surface. This is a convenient way of considering heat
transfer for numerical calculations. Thus, fictitious heat transfer coefficients and the difficult problems of secondary reflectivities from the sheet
surface and from the heater source reflector surface can be neatly avoided.
The limitations of the method come when the results are applied to real
processing problems.

There are other approaches to determination of the view factor that
enable consideration of other types of geometries for heater sources. A
powerful numerical technique, known as Monte Carlo, has been used extensively to determine view factors for unusual shapes of sources and sinks
[4, 5]. One calculation, as an example, enables consideration of the consequences of removing parabolic reflectors and adding a second row of electric

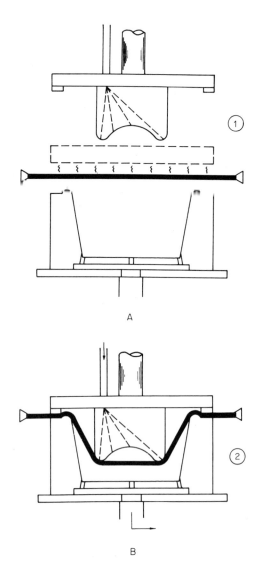

A

B

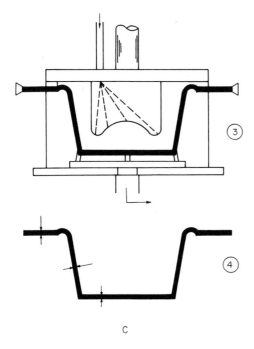

C

Figure 12.2-7 A: Plug assist pressure forming, a schematic example.
Step 1: after sheet is brought to formable temperature, heater shuttled out
of forming zone. B: Step 2: plug assist now driven into formable sheet,
clamping it against the mold rim. C: Step 3: air pressure now forces sheet
into intimate contact with mold wall. Note that no vacuum is used; the mold
cavity is allowed to vent. Step 4: uniformly thin walls, center, thick rim
material [1].

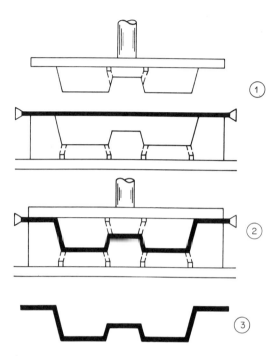

Figure 12.2-8   Schematic example of matched die molding. Molds can be
of wood, plaster, plastic, aluminum, or steel. Step 1: formable sheet
clamped, positioned over mold form; step 2: plug driven into formable
sheet, vents provided for exhausting residual air from plug, cavity; step 3:
formed part characterized by uniform wall thickness [1].

heaters and a planar nonradiating wall. A typical curve is shown in Fig.
12.3-2.

     Note that neither the standard rod heaters with reflectors nor stan-
dard opaque plastic sheet stock are black-body radiant absorbers (where
e = 1) or gray-body radiant absorbers (where e is a constant, independent
of wavelength). A radiant energy spectrum of a white GPS sheet, for example, might show a relatively low emissivity (e = 0.85 or so) in the visible
range of radiant wavelength and a high value (e = 0.95 or more) elsewhere.
For a white ABS sheet, however, the emissivity may be relatively uniform
(at, say, 0.90) over the entire spectrum. This means that the simple average value of emissivity may be improper when attempting to do careful
temperature predictions of these materials. Usually, however, this variation is sufficiently small that it can be neglected in actual machine operation.

     Platzer [7] has shown that the time-temperature profile of an 80-mil
HIPS sheet will depend on the type of heater used, the distance between the

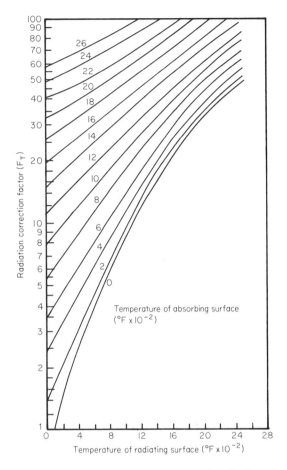

Figure 12.3-1   Radiation correction factor F_T as a function of the temperature of the radiating and absorbing surfaces (in °F) [2].

heater and the sheet, and the heater temperature. See Fig. 12.3-3. An increase in the heater temperature is expected to cause an increase in the sheet temperature from examination of Eq. (12.3-1). Note that as the sheet temperature approaches the heater temperature, the heat flux (Q/A) decreases dramatically. This is the apparent cause of the sigmoidal or S-shaped curves seen in Fig. 12.3-3. As can be shown, an increase in source temperature from 700 to 1000°F increases the black-body heat flux (with sheet temperature at 140°F) from about 3000 Btu/ft^2·hr to more than 7500 Btu/ft^2·hr. The effect of distance between sheet and heater blank is normally not a significant design factor. Nevertheless, two points should be

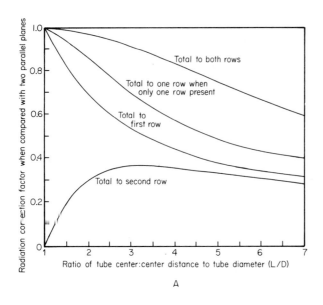

A

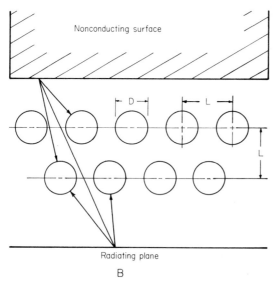

B

Figure 12.3-2   A: Effect of rows of tubes between and parallel to radiating plane. Curve is compared with the case of an infinitely deep row of
tubes or a parallel absorbing plane. B: Schematic for A [6].

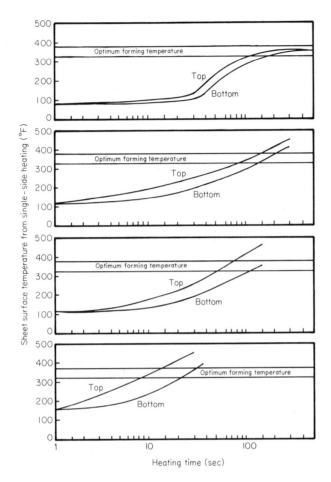

Figure 12.3-3 Measured sheet temperatures for various infrared radiant heater configurations. Material, 0.080-in. white HIPS; top curves, Pyrex glass heater, 8 in. from sheet; second set of curves, Sill glass-fiber heater, 700°F, 4 1/2 in. from sheet; third set of curves, Chromalox strip heater, 700°F, 3 1/2 in. from sheet; fourth set of curves, Chromalox strip heater, 1000°F, 3 1/2 in. from sheet [7].

made. Recognize that there is an air layer between the heater bank and the sheet and that this air is normally transparent to radiant energy. It is not transparent to convection heating, however, and heat is given off from both the sheet and the heater bank to the surrounding air by convection. This has been called "convection cooling" in the literature [8] and can be included in the calculations of the efficiency of the heating system in a standard method outlined in Chap. 6. Basically, the air is heated by natural convection for

which the heat transfer coefficient is no more than 2 Btu/ft^2·hr·°F. As
long as the radiant heaters are relatively close to the sheet surface, this
column of air cannot circulate with sufficient velocity to decrease the effectiveness of the radiant heat transfer process. At heater-to-heater distances
on the order of 1 ft or so, the vortices can be rather rapidly established
and quite effective in removing heat from both the heater bank and the sheet.
The effectiveness of the natural convection cooling is increased as the sheet
width and/or length decreases relative to the heater-to-sheet distance.
There is an added factor that is secondary to this convection cooling. Radiant energy can be absorbed by water vapor and CO_2 that might be present in
the air between the sheet and the heater bank. It is well known that these
materials are excellent radiant energy absorbers [2], and their effectiveness
increases with increasing distance between heaters and sheet. This factor
is probably not important with conventional strip or rod heaters. With the
new heaters that use natural gas combustion and reflectors for radiant energy, the engineer should realize that H_2O and CO_2 are combustion by-products
and design the heater bank efficiencies to include this additional radiant energy sink.

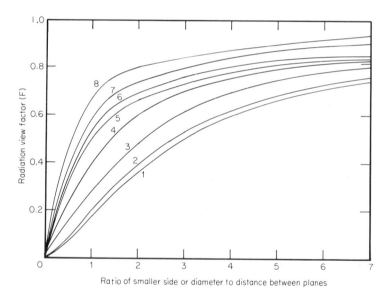

Figure 12.3-4   Radiation view factor corrections for radiant transfer between parallel planes, directly opposed, with various geometries. Curves
1-4: direct radiation between planes; curves 5-8: planes connected by nonconducting but reradiating walls; curves 1 and 5: discs; curves 2 and 6:
squares; curves 3 and 7: 2:1 rectangles; curves 4 and 8: long, narrow rectangles [9]

Also note that as the heater-to-sheet distance increases, some of the energy emitted by the heater bank will be lost to the surroundings beyond the edges of the sheet. Effectively, this reduces the value used for F, the view factor, as shown in Fig. 12.3-4. The sheet can also radiate to its surroundings. This energy loss can be calculated if the surrounding area temperature is known (normally room temperature), and the sheet temperature can be approximated from solution of the heat conduction equation with appropriate constraints or measured using an optical pyrometer or surface thermocouple. A plot of the view factor for various dimensional ratios can be made as in Fig. 12.3-5. Again, there may be minor losses (less than 5% or so) for most sheets, but these can lead to increased cycle times and decreased production rates.

As seen in a later section, there are advantages to be gained in cycling the heater banks. Normally, for opaque materials, however, the heater banks are kept on continuously, and the heater temperature is controlled. On-off cycling of the heaters can lead to shorter heater life, and thus this technique should be used only when the materials are thermally sensitive or transparent (technically, "semitransparent") to thermal radiation.

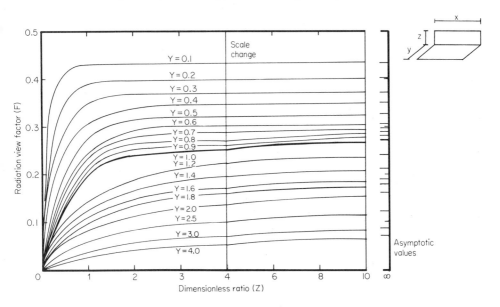

Figure 12.3-5 Radiation view factors for radiant interchange between adjacent rectangles in perpendicular planes. Y is y/x in sketch, defined as the length of unique side of that rectangle on whose area the heat transfer equation is based to the length of the common side; Z is z/x in sketch, defined as the length of the unique side of the other rectangle to the length of the common side [9].

12.4 Mold Materials

Before considering the process of drawing a softened sheet into the mold
cavity, consider the mold materials. There are several types of materials
that have been used very successfully as molds for thermoforming. The
mold material application is classified according to the number of pieces
to be taken from each mold. If the parts being made are prototype parts
(less than 1000 pieces, say), a rather inexpensive mold that can be reworked
(or replaced) relatively easily and at rather low cost is desired. If the
parts are for a long production run (for butter tubs, for example), production
tooling should be considered. Long mold life, good heat transfer (to improve
the cycle), and faithfulness of detail should be considered as primary reasons for material selection.

   The cheapest mold material is wood. A seasoned, kiln-dried hard-
wood such as ash or hickory should be chosen. Grain can be accentuated
with sand or grit blasting prior to sealing. It should be remembered that
the process uses rather low pressures (less than 50 psi) against the sheet,
and thus details such as wood grains or "leather" require more accentuation
in the mold than would be the case if the mold were being used for injection
molding, say. The wood should be sealed with a penetrating wood sealer.
After a final light sanding, a rather heavy coat of high-gloss urethane varnish should be applied with a minimum of brushing to minimize bubble entrainment. To ensure the toughness of the urethane varnish, the mold is
placed in a convection oven at 100 to 150°C for 2 to 4 hr after the urethane
has set. This finishes the "cure" on the urethane. At this point, very
small (1/16 or 3/32 in. in diameter) holes should be drilled through the primary surface to the unvarnished portion of the mold. These holes should
be placed in corners and at places where air pockets might be formed during
thermoforming of the sheet. Certainly working with wood enables the designer to change part dimension and venting with relatively little effort.
Good wood molds will last for thousands of pieces. A normal lifetime of
wood is on the order of 500 pieces. The predominant mode of mold failure
is drying and cracking because of the heat of the plastic sheet.

   Many designers prefer to work with plaster rather than wood because
of its dimensional stability and lack of grain. Some seal plaster in much
the same way as they seal wood. This helps minimize dusting or chalking
of the plaster and gives the surface some added strength. Because of the
brittle character of plaster, it is difficult to drill vent holes in a final casting.
As a result, vent tubes (small plastic straws or steel wires in many cases)
must be cast in during the development of the plaster mold. Plugging of
the small tubes during final mold surface preparation is a continuing problem.
Recent developments indicate that incorporation of an organic sintering agent
that can be burned out by firing the ceramic will yield a porous mold, and
thus vent holes are unnecessary. The process requires elevated temperatures, however, and does not yield a very smooth surface. Any surface

preparation plugs the pores. Nevertheless, for a few pieces in either male or female molds, plaster remains the workhorse of the prototype tool and die maker.

One of the more interesting tool and die developments for thermo-formers in the last few years has been metal-filled epoxy. There are sev-eral brands available on the market, and all use the same basic ingredients: a reactant, a catalyst, and a metal filler such as aluminum or steel (to 70 wt%). Normally a positive pattern of wood or plaster is prepared as described above for mold preparation. Then a very thin coating of unfilled epoxy (less than 5 mils) is applied over the pattern. This is allowed to cure to tackiness, and a heavy coating (5 to 10 mils or more) of metal-filled epoxy is applied with a squeegee. This coating is allowed to cure to tackiness. If this is a short-life prototype, additional coats of metal-filled epoxy are applied until the mold wall thickness is on the order of 100 to 150 mils. The inside of the cavity (assuming it to be a male mold) is then filled with reinforcement, such as woodflour mixed with low-viscosity epoxy or plaster. Vent holes are then drilled through the finish surface, and any chipping of the surface is repaired with unfilled epoxy. This type of prototype mold is good for thousands of parts and can be readily repaired with unfilled epoxy if damaged during molding. A more important advantage of the liquid mold material is the ability to include heating or cooling passages behind the finish surface during mold preparation. For example, copper tubing can be fastened into the tacky epoxy coating. The next coat of metal-filled epoxy is applied directly over the tubing. Although prototype tool makers are rarely con-cerned with cycle efficiency, the ability to heat a mold enables the molder to thermoform thermally sensitive materials such as PP or PC and to mini-mize cracking or crazing of highly oriented material. This is discussed in more detail later.

Aluminum has been used extensively as a mold material. Most of the thermoform tooling is made from machined stock of the 2024-T4 types. Some tooling has been made by cold-hobbing the aluminum, but the ductility of aluminum is notoriously poor and cold-hobbed molds are normally shallow drawn. The ease of machining of aluminum makes it very attractive to the production tool maker. Cooling lines, when necessary, can be relatively far apart because of the high thermal conductivity of the metal. The sur-face can take a very high polish but will oxidize over many thermal cycles. Thus, for long-lived high-gloss surfaces, the aluminum is normally nickel or chromium plated.

Electroplated nickel can be used quite effectively for thermoformed molds. Here the pattern, either wood or plaster, is made conductive by applying a low-viscosity urethane coating to seal the surface. While this coat is still tacky, a thin powder coating of carbon black or graphite is ap-plied. If the finish surface is to be highly polished, the graphite coating can be buffed prior to plating. The conductive pattern is then placed in the electroplating bath, and nickel is plated against the carbon at the rate of

5 mils/day. Hot plating (50 mils/day) normally yields a surface with many pinholes and imperfections [10]. However, these are less serious to thermoforming molders than to injection molders. When the nickel has built to 100 to 120 mils, the plater can add water lines, if necessary, and continue plating with either nickel or copper. Or he can lay the water lines in place and cast Kirksite or aluminum around them and against the back of the nickel mold surface. One tool maker has patented a method of plating nickel in a manner that leaves very small vent holes at discrete locations across the entire mold surface [11]. This eliminates the obvious problem of drilling vent holes in very hard nickel plate. For high production rates, rapid thermal cycling, and excellent surface detail, electroplated nickel is preferred as a mold material.

Steel is also used extensively by the production tool maker. Steel is much cheaper than either aluminum or nickel, and more tool makers know how to work with it than with aluminum or nickel. A good low-carbon water hardening steel of the AISI "W" series is easily machined, polished, and drilled for vent holes and/or water lines and can be plated if necessary. Tool steels such as AISI S-7 or H-13 are not needed since there is relatively little pressure applied to the mold and the temperatures are, in general, quite moderate. Nevertheless, good tool steels will yield excellent mold surfaces and very long life and thus are frequently called for by the molder.

Stainless steels, particularly 403 and 404, have been used for thermoform mold inserts in areas of very high wear such as at the lips or edges of a mold having a deep draw. They are not common tool materials in production, however, because of their high cost and relatively poor thermal conductivities when compared with low-carbon steel, aluminum, or nickel.

Throughout this section, the heat transfer aspects of the mold have been discussed. Before considering the effect of mold temperature on the drawing ability of thermoformed sheet, consider the dual role played by the mold. Its primary function is that of the form into or onto which the sheet is drawn. For that purpose, it must be rigid, as the earlier discussion on mold materials illustrated. It has a secondary function of heat removal from the sheet while the sheet is in contact with it. The objective of mold cooling is not necessarily to reduce the drawn sheet stock to room temperature. Normally the time required to do that cannot be afforded. The mold must remove enough heat from the sheet to allow the sheet to retain the mold form when it is removed. The sheet should be somewhat soft, however, so that it can be removed from slight undercuts in the mold surface. In general, the thermal capacity of the mold is very large when compared with that of the sheet, and thus the mold remains nearly isothermal during the cooling stage. For wood and plaster molds, however, the mold temperature when drawing a rather thick sheet can increase rather substantially over several molding cycles. This increase in temperature will increase the cooling and forming cycle time and may change the wall thickness dimensions across the part as well.

The more important aspect of mold temperature is its effect on stock temperature during drawing of the sheet. The heat transfer aspects of this problem can be attacked by assuming that the stock and the mold are isothermal at T_S and T_m, respectively, at the time of contact. Since the free surface of the stock is in ambient air, there is essentially no heat transfer from this surface relative to that transferred to the mold surface. This means that the problem is essentially an unsteady-state heat conduction problem. From Fig. 12.4-1, the free-surface time-temperature profile is obtained. In Fig. 12.4-2, the internal time-temperature profiles of the sheet are found. Of course, the times and temperatures are given in dimensionless units. Knowing the thermal diffusivity of the sheet and its thickness, one can obtain the real time as a function of the Fourier number. The actual temperature can be calculated from the known initial temperatures of the sheet stock and the mold. These time-temperature profiles are used later to determine the drawing ability of a sheet. At this point, it is apparent that increasing the mold temperature, assuming that it remains nearly isothermal during sheet stock cooling, will slow the cooling rate of the sheet. If the drawing rate of the sheet is directly proportional to its temperature, the drawing rate will not slow as rapidly. Incidentally, control of blown bubble air temperature is also critical to the surface temperature of the sheet stock for much the same reason. However, note that heat transfer from the sheet stock to air is very poor and that the high thermal resistance at the sheet stock-air interface will result in a much reduced effect on sheet cooling rate.

Two additional points are worth discussing briefly. First, a method for calculating the efficiency of cooling in a mold has been outlined in Chap. 10. The method of shape factors can be applied directly here to determine

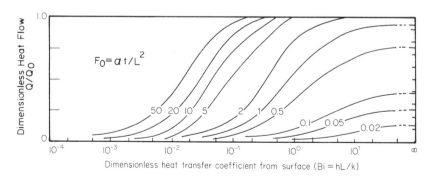

Figure 12.4-1 Fraction of heat flow from plastic slab when exposed to a step change in environmental temperature. Note: Parameter is dimensionless time, $F_0 = \alpha t/L^2$ (where L is the half-thickness of the slab) [2].

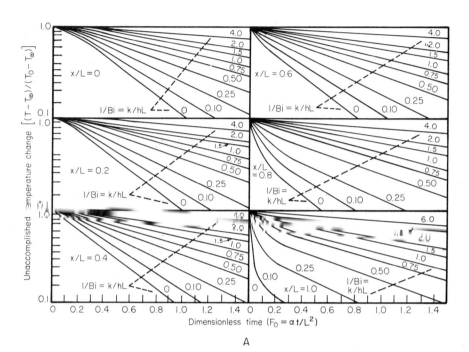

A

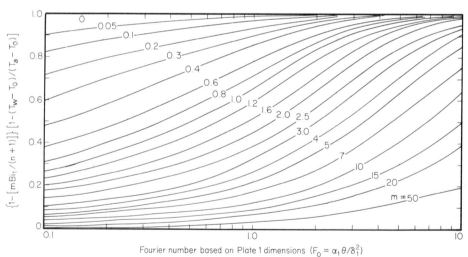

B

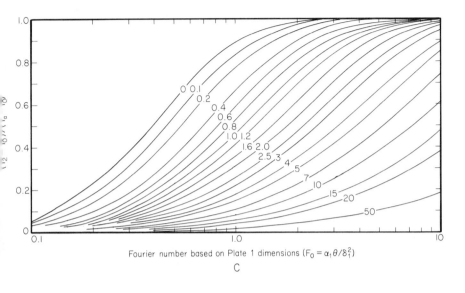

Fourier number based on Plate 1 dimensions ($F_0 = \alpha_1\theta/\delta_1^2$)

C

Figure 12.4-2   A: Dimensionless temperature as a function of dimension-less time, $F_0 = \alpha t/L^2$, for various thicknesses into slab. Note: Parameter is dimensionless heat transfer coefficient from sheet surface, $Bi = hL/k$. See Chap. 6 for more details regarding graph interpretation [2]. B: Two-plate response to a change in surface heating conditions, with plate 1 convectively heated and the rear of plate 2 perfectly insulated. This curve represents plate 1 surface response $x_1/\delta_1 = 0$. The curve below represents the interfacial temperature response. The parameter $m = n + (n + 1)/Bi$, where $n = \rho_2 C_2 \delta_2 / \rho_1 C_1 \delta_1$. ρ is the slab density, C is its specific heat, and δ is its thickness. $Bi_1 = n\delta_1/k_1$, the Biot number, where h is the heat transfer coefficient for convection, k_1 is the thermal conductivity of plate 1 material [3]. C: Same conditions as in B, except temperature response at $x_1/\delta_1 = 1$, the interface between the two plates [3].

641

the efficiency of cooling for steady heat transfer. This, coupled with the rather simplistic unsteady heat transfer calculations given in Chap. 10, can be used directly to determine the areas for increased cooling. And finally, the importance of holding sections of the mold at different temperatures should be noted. Frequently, it is necessary to keep the center and the middle of the wall of a deep drawn female mold quite warm relative to the corners and the lip, because the drawing bubble of plastic touches these portions first and it is from these portions that the material for the corner must come. The corners are cold, of course, to help speed the cycle. Varying the temperature across the mold surface is sometimes easier than attempting to vary the sheet stock temperature. Yet, as will be seen, the latter is the frequent way of controlling part wall thickness.

12.5 Characteristics of Sheet During the Thermoforming Process

As can be anticipated, thermoforming is another combined unsteady-state flow and heat transfer process. The effects can be separated somewhat by assuming that during the preliminary stages of heating (until the sheet has reached its drawing temperature) the process is heat transfer controlling. During the early stages of drawing (before the drawn sheet touches the mold surface at any point), a form of extensional flow can be assumed. Once the sheet touches the mold surface at some point other than the rim, the combined heat transfer-extensional flow problem must be considered. Again, as is typical with the "minor" areas of plastics processing, the technology is a curious mixture of theory and experiments that are primarily academic but nevertheless give some insight into the real process taking place and into some very pragmatic rules of thumb which only give a very distorted glimpse as to the proper engineering approach to be taken in thermoforming.

12.5.1 Heat Transfer

Consider first the problem of heat transfer to sheet stock below its drawing or forming temperature. If the sheet is opaque, all energy (radiant and/or convection) is transmitted to the surface of the sheet, and the interior of the sheet is then heated by conduction. The basic equation for heat transfer is that given in Chap. 6:

$$\rho c_p \, \partial T / \partial t = k(\partial^2 T / \partial x^2) \tag{12.5-1}$$

where ρ, c_p, and k are the density, specific heat, and thermal conductivity of the plastic, respectively; T is the sheet temperature; and t and x are the time and perpendicular distance into the sheet, respectively. The boundary and initial conditions can be given as

$$T(x,0) = T_0 \tag{12.5-2}$$

$$(\partial T/\partial x)_{x=L,t} = 0 \tag{12.5-3}$$

$$-k(\partial T/\partial x)_{x=0,t} = -(1 - r)Q_s + h[T(x = 0,t) - T_0] \tag{12.5-4}$$

The first condition is, of course, the initial temperature distribution in the sheet. T_0 is room temperature in most cases. The second condition is a symmetry condition at the center of the sheet that is being heated uniformly on both sides. If the sheet is being heated on only one side, this term represents effective insulation of the unheated side of the sheet, a reasonable first approximation. The third condition is at the primary heating surface. The term on the left represents the rate of heat conduction into the sheet. The first term on the right represents the amount of incident radiant energy that is absorbed by the sheet. Here, of the energy Q_s that falls on the surface, rQ_s is reflected, where r is the reflectivity of the sheet surface, and the remainder must be absorbed. The second term on the right represents the extent of heating or cooling owing to convection between the ambient air and the sheet surface. h can again represent the sum of the convective and effective radiative heat transfer coefficients, as discussed earlier. For the rather simple case where the sheet stock is entirely opaque to incident radiation, a tabulated solution can be used [3,12]. The solution to this equation is given in Chap. 6 for the case of radiative heating alone. It should be apparent, however, from inspection of boundary condition (12.5-4) that if Q_s is increased by increasing the heater bank temperature and/or decreasing the distance between the heater bank and the sheet surface, the temperature gradient is increased at the sheet surface. Since k is in general very poor for plastics, severe thermal gradients in the sheet stock can occur rather quickly. One measure of the temperature uniformity in sheet stock during heating is the "evenness index." Basically, this is a ratio of extreme temperature difference to the maximum sheet temperature. For opaque materials the maximum sheet temperature always occurs on the surface; for transparent materials such as PMMA or transparent ABS, the maximum sheet temperature may occur within the sheet owing to volume absorption of radiant energy. It is apparent that the evenness index is always maximum initially. What is desired is a rapid drop of the index toward zero. The more rapid the decrease in the index, the more uniform the sheet stock temperature. As shown in Fig. 12.3-3, for various heating sources, the Pyrex glass heater yields the most rapid decrease in the evenness index. The evenness index is also a function of the extent of heat loss from the surface, as shown in Fig. 12.5-1. Here the Biot number represents the dimensionless rate of heat removal from the surface, $Bi = hL/k$, where L is the half-thickness of a sheet heated on both sides. The source temperature in this case is 2000°F (e.g., the radiant heater bank is emitting energy equivalent to a black-body radiator operating at 2000°F). As the heat transfer coefficient increases from

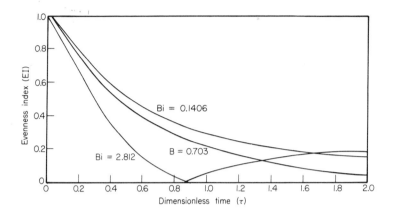

Evenness index for symmetric heating of softened sheet with surface cooling. Note: Parameter is dimensionless heat transfer coefficient, Bi = hL/k; radiant source temperature T_s = 2000°F [8].

approximately 1 to 20 Btu/ft$^2 \cdot$hr$\cdot$°F, the evenness index decreases very rapidly with increasing dimensionless temperature (τ = Fo = $\alpha t / L^2$). Near a Biot number of about 2, the maximum temperature occurs inside the sheet. These calculations were carried out for PMMA, a transparent plastic. This temperature inversion cannot happen for an opaque sheet, but the decrease in evenness index with increasing surface cooling is similar in effect. Note, of course, that increased surface cooling will automatically lead to increased time to reach forming temperature, regardless of the sheet material. Thus, a more uniform thermal profile must be economically balanced against a much lower heating rate.

Studies of on-off pulsing of the heater banks have found that temperature profiles can be made remarkably uniform. Although most of the work has been done with transparent sheet, it is expected that the results are at least qualitatively applicable to opaque sheet [12]. As shown in Fig. 12.5-2, decreasing the total period (for 50% on, 50% off cycle) of the heater banks does not affect the maximum temperature of the sheet but does decrease the evenness index dramatically. If the ratio of on-to-off time of the heater bank cycle is decreased while holding the total period constant, both the evenness index and the maximum cycle can be decreased, as shown in Fig. 12.5-3. The temperature profiles for symmetric and unsymmetric heating are shown in Fig. 12.5-4 for the case of pulsed "on-off" heating and for two dimensionless times. The uniformity of the temperature profile for symmetric heating is quite good. It should be noted here that the Biot number

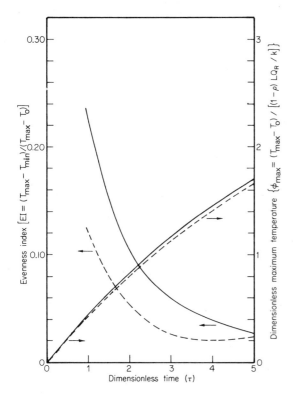

Figure 12.5-2   Effect of frequency of period for "on-off" pulsed symmetric heating of softened plastic sheet. "On" period 50% of total period; solid line, total period T = 0.05; dashed line, total period T = 0.20 [13].

corresponds to a natural convection heat transfer coefficient of 1 Btu/ $ft^2 \cdot hr \cdot °F$.

Note that it is possible to heat even the most thermally sensitive materials by cycling the heater banks, albeit at the cost of a longer heating cycle.

12.5.2 Stretching the Sheet

The importance of a uniform temperature profile in the sheet is apparent when the next step in thermoforming is considered. Regardless of

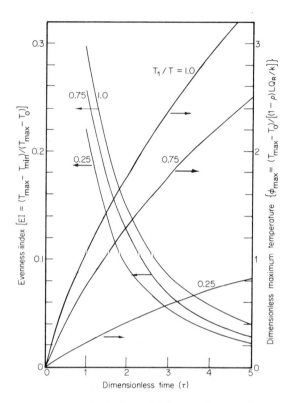

Figure 12.5-3   Effect of heating time on temperature and evenness index for thermoplastic sheet being heated in an "on-off" fashion. The total heating time T = 0.05. The parameter T_1/T is the ratio of "on" time to total time of period [13].

whether a bubble is blown prior to snap-back molding or a drape is formed over a male plug or the sheet is drawn into a cavity in conventional vacuum forming, the heat-softened sheet is being biaxially stretched. Recently, much interest has been centered on biaxial deformation [14,15]. While most of the work has centered around the blowing of a bubble, recent investigators have also looked at a uniaxial extension and shear extension in cylinders of heat-softened sheet. Since the deformation of a bubble seems to be more applicable to the early stages of deformation, this mode of stretching is considered here.

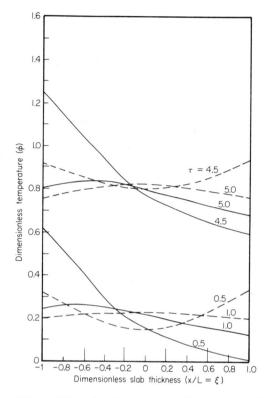

Figure 12.5-4   Comparison of time-dependent temperature profiles across a thermoplastic sheet being heated radiantly and cooled convectively. Solid lines, unsymmetric heating; dashed lines, symmetric heating. Note: values for symmetric case divided by 2 to correspond with same total rate of incident radiation as symmetric case. Total time period T = 1. "On" period 50% of total time period. T_s = 1500°F; dimensionless heat transfer coefficient Bi = hL/k = 0.1406; material, 1.0-cm-thick acrylic sheet. Note: τ is dimensionless time [13].

Schmidt and Carley consider a modified Treloar [16] form of the strain-energy density equation (relationship between the work of deformation of the bubble and its strain invariants, I_1 and I_2):

$$W = C_{10}(I_1 - 3) + C_{02}(I_2 - 3)^2 \qquad (12.5-5)$$

The first term on the right represents the standard "Hookean" stress-strain term. The second represents the non-Hookean effects of deformation. That the term should be squared is a matter of conjecture. Mooney [17], who originally modified the Treloar equation by adding the term as a linear function, was not able to get a good fit to his experimental data for deformation of rubber sheet when the elongation exceeded about 1000%. The squared term is used by Schmidt and Carley as a tool in explaining their experimental elongation data. They express their principal extension ratios in terms of cylindrical coordinates: λ_1 is the extension ratio in the meridional direction ($= d\xi/d\rho$), λ_2 is the extension ratio perpendicular to λ_1 and to r (the blowing bubble radius) in the latitudinal direction ($= 2\pi r/2\pi\rho$), and λ_3 is the extension ratio perpendicular to the others in the bubble thickness direction (e.g., normal to the bubble surface) ($= H/H_0$, where H is the thickness of the blowing bubble and H_0 is the initial thickness of the sheet). ρ is the radius of a given circle on the original sheet, and ξ is the arc length measured along a meridian from the pole of the deformed middle surface to a point (r, θ, z). λ_1 is obtained from the continuity equation:

$$\lambda_1 = 1/(\lambda_2 \lambda_3)$$

$$(12.5\text{-}6)$$

λ_2 was estimated from knowing the bubble radius as a function of time during blowing. λ_3 was measured directly from slicing through the cold bubble after forming was completed. Thus, λ_1 could be determined from Eq. (12.5-6). An example of the experimental data obtained (for $\lambda_1, \lambda_2, \lambda_3$) is shown in Table 12.5-1 for Dow Styron 666. The extension ratio is plotted in Fig. 12.5-5 and compared with the best curve fit of the Hookean and "Mooney" forms of the strain-energy density equation. The agreement is poor, particularly when the data are compared with the Schmidt-Carley modification of the equation, as shown in Fig. 12.5-6. The maximum areal elongations observed in this work were more than 10,000% for the styrenes and 1000% for CAB. These values are near the limit for conventional thermoforming. Note that localized cooling of the blowing sheet leads to unsteady growing of the bubble. Schmidt and Carley attempted to make the sheet as uniform in temperature as possible but still found that across the 1.72-in.-diameter disk the temperature initially varied by more than 30°F. They monitored temperatures during blowing and their data, shown as Fig. 12.5-7, show nearly isothermal local temperatures during blowing until the last 1.5 sec of expansion. At that point the temperature at the center of the sheet and thus at the top of the bubble dropped precipitously. Since the sheet did not contact a cold mold surface during this time, Schmidt and Carley attribute this to a reduction in the extent of viscous dissipation, which prior to this point had been making up the energy loss to the surroundings. The authors admit, however, that other effects might be taking place or that the temperature measurement might be erroneous. The ultimate effect of sheet cooling during blowing is a reduction (local or overall) in the extent of deformation possible with a given air pressure and/or a given period of time for inflation. Irrespective of the thermal effects, however, it seems that isothermal biaxial stretching of the sheet can be qualitatively predicted from the experiments and theory of Schmidt and Carley.

Table 12.5-1

Experimental Data for Free-Blown Softened Polystyrene Sheet

Material: crystal polystyrene (Styron 666)
Initial sheet thickness: 0.060 in.
Clamp radius: 0.870 in.

Circle no.	Left half			Right half			Average		
	Radius (in.)	Angle (deg)	Thickness (in.)	Radius (in.)	Angle (deg)	Thickness (in.)	λ_1	λ_2	λ_3
0	0.000	0.0	0.00056	0.000	0.0	0.00056	10.351	10.351	0.0093
"H"	0.585	14.0	0.00032	0.585	22.5	0.00076	13.666	9.474	0.0090
1	1.150	29.0	0.00044	1.010	33.9	0.00100	11.126	8.639	0.0170
2	1.900	48.5	0.00082	1.590	59.5	0.00145	7.908	7.121	0.0190
3	2.315	83.0	0.00113	1.795	89.5	0.00212	7.163	5.552	0.0271
4	2.085	57.3	0.0020	1.405	62.5	0.0036	6.302	3.523	0.0467
5	1.550	33.8	0.0039	1.235	45.0	0.0082	4.841	2.280	0.1009
6	1.045	26.7	0.012	0.935	36.5	0.020	2.950	1.341	0.2667
7	0.870[a]	—	0.032	0.870[a]	—	0.032	1.853	1.012	0.5333

[a] Radius to middle surface.

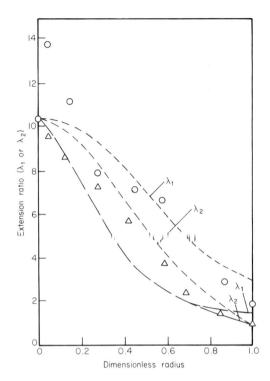

Figure 12.5-5   Extension ratios for crystal polystyrene inflated bubble.
Circles, λ_1, observed extension ratio; triangles, λ_2, observed extension
ratio; dotted lines, computed extension ratios from neo-Hookean behavior;
dashed lines, computed extension ratios from Mooney strain-energy function [14].

        Lai and Holt [18,19] have continued the Carley-Schmidt experiment,
finding that PMMA and high-impact polystyrene have time-dependent stress-
strain behaviors at elevated temperatures (165 and 122°C, respectively).
Their data are correlated by

$$\sigma = Kt^m \epsilon \qquad\qquad\qquad (12.5\text{-}6a)$$

where K is a temperature-dependent Hookean constant, t is time, and m is
on the order of -0.05 for PMMA and -0.33 for HIPS. From their analysis,
they point out that an increase in flow stress with strain depends on the inherent rate of strain hardening of the material being stressed and on the
rate of flow stress relaxation. For increasing rates of strain hardening at
low rates of stress relaxation, thickness uniformity should improve. Conversely, the poorest thickness uniformity in biaxial stretching occurs when
the relaxation time is large and there is no strain hardening (m = 0).

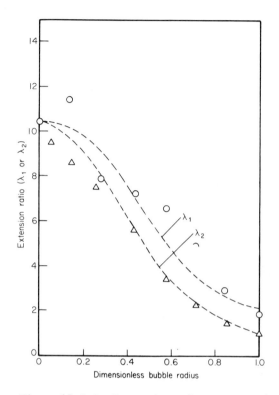

Figure 12.5-6 Comparison of experimental and computed extension ratios for inflated bubble of crystal polystyrene sheet. Dashed lines, computed from neo-Hookean model; circles, λ_1, extension ratio, experimentally determined [14].

Certainly, additional experiments will follow that will refine this basic work to include such effects as sheet cooling and will develop methods of predicting to coefficients C_{10} and C_{02} for various sheet materials.

12.5.3 Forming

Now consider the complex problem of drawing the sheet from the point where it touches a cold mold surface. Platzer has given some clues as to the effect of temperature on the drawing of a sheet into a rectangular form, as shown in Fig. 12.5-8. Note first in the top curve that the sheet is nonisothermal across its thickness and probably not entirely isothermal across its area as well. As the sheet is drawn into the mold, it contacts the mold first at the rim and then at the center bottom of the mold. If the sheet were stopped from drawing as this instant ("short-shot" the thermoforming), the sheet wall thickness would be very similar to that given by

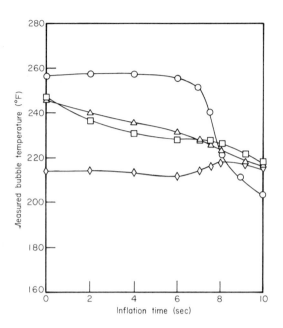

Figure 12.5-7 Measured temperature profiles during bubble inflation of crystal polystyrene sheet. Circles, infrared polar thermometer measurement at dimensionless radius of zero, radius made dimensionless by dividing by effective clamping radius of specimen; squares, embedded thermocouple measurement at dimensionless radius of 0.36; triangles, same, at dimensionless radius of 0.64; diamonds, same, at dimensionless radius of 0.93 [14].

Schmidt and Carley for the appropriate areal elongation shown in Table 12.5-1. If the sheet is drawn, the material in contact with the cold rim and center bottom of the mold will decrease in temperature, thus causing a relatively rapid increase in resistance to drawing. Therefore, the sheet is stretched outward from the point at the center bottom of the mold and downward from the rim. Very little additional thinning takes place at the rim or in the region around the bubble where it first touched the bottom of the mold. As a result, the total amount of material available for drawing decreases, and the bubble thins. As more material touches the walls and bottom of the container, it cools, and the already-thin hot sheet that remains must be extended even more in order to fill in the rest of the mold. For the relatively shallow draw shown (3/14 or 0.2:1), the 82-mil-thick sheet was drawn to 31 mils in the center of the part where the bubble first touched, 14 to 19 mils in the wall, and only 4 mils in the corner. Remember also that regardless of the uniformity of the sheet temperature either through

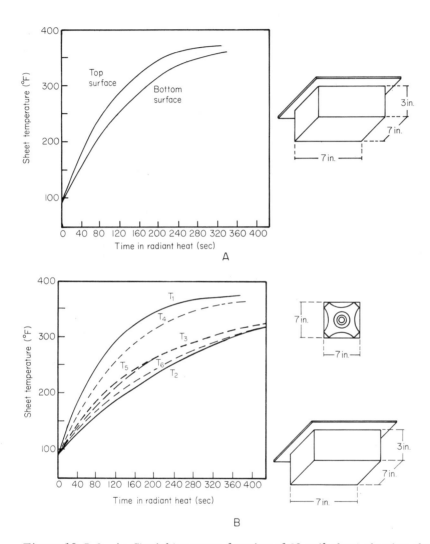

Figure 12.5-8 A: Straight vacuum forming of 82-mil sheet showing characteristic nonuniform thinning in drawn article. Note: Formed part corner thickness, 4 mils; edge, 14 to 19 mils; center, 31 mils. B: Vacuum forming of 82-mil sheet, with surface shading, using four layers of tissue paper, as shown in schematic on upper right. Sheet temperature locations: T_1, top center; T_2, top side; T_3, top corner; T_4, bottom center; T_5, bottom side; T_6, bottom corner. Note: sheet thickness nearly uniform throughout part at 24 to 26 mils [7].

the sheet or across its surface the cooling of the sheet begins immediately upon contact with the cold mold surface. Fortunately, if the process is operating at a very rapid drawing cycle, the effect of chilling of the sheet is minimized, and thus material can be drawn from the bulk of the sheet in contact with a cold mold. But, in doing this, the sheet material is being oriented. Denson has referred to this form of orientation as "shear elongation." This has been discussed in some detail in Chap. 5. What is probably more important to the molder is that this orientation is "frozen" into the sheet as it cools to the mold temperature. For most materials, orientation implies increase in strength. Thus, the container is strong in the direction of draw but weak in hoop strength, and this weakness is further enhanced by a very nonuniform wall thickness.

Certain materials, such as PP, exhibit excessive thinning during processing. Funt [20] has considered a stress-strain-rate of strain model for PP based on a Noll simple fluid with memory. Under isothermal conditions, he shows that the general constitutive equation of state can be reduced to

$$\sigma_i \lambda_i = 2 \left[\left(\frac{\partial W}{\partial II} + II \frac{\partial W}{\partial I} \right) \lambda_i^2 - \frac{\partial W}{\partial I} \lambda_i^4 + III \frac{\partial W}{\partial III} \right] \qquad (12.5\text{-}7)$$

where σ_i is the stress term in the i-direction, λ_i is the elongational term in the i-direction, W is the strain-energy function, and I, II, and III are the Cauchy strain invariants. I is the trace of the strain tensor, II is the sum of the cofactors of the matrix formed by the strain tensor, and III is its determinant. For an incompressible fluid, III = 1. For equal biaxial extension, the following simplifications can be made:

$$\lambda_1 = \lambda_2 = \lambda \qquad (12.5\text{-}8)$$

$$\lambda_3 = 1/\lambda^2 \qquad (12.5\text{-}9)$$

$$\sigma \lambda = \sigma_1 \lambda_1 - \sigma_3 \lambda_3 = \sigma_2 \lambda_2 - \sigma_3 \lambda_3 \qquad (12.5\text{-}10)$$

and the constitutive equation becomes

$$\sigma \lambda (\lambda^2 - 1/\lambda^4)^{-1} = 2 \frac{\partial W}{\partial II} + 2 \frac{\partial W}{\partial I} \lambda^2 \qquad (12.5\text{-}11)$$

Funt finds that for PP the deformation to strain behavior can be described by

$$\lambda = \exp(\epsilon) \qquad (12.5\text{-}12)$$

and that

$$(\partial W/\partial I) = 0, \qquad (\partial W/\partial II) = f(T, t_1) \qquad (12.5\text{-}13)$$

The functional relationship for $f(T, t_1)$ depends on the material and an appropriate elastic stress-relaxation time t_1. For PP, Funt finds that

$$f(T, t_1) = a'(165 - T)[1 - \exp(-\epsilon/t_1)] \tag{12.5-14}$$

where T is in °C (with 165°C being the melting point) and t_1 is the temperature-dependent stress relaxation time, obtained from 100-sec isochronous creep data.

Williams [21] has made an extensive study of biaxial stretching of sheet, obtaining relationships between applied force and draw-down ratios, in a manner similar to that of Funt. His analysis, however, allows for determination of wall thickness variation as a function of depth of draw for several classical types of stretching modes. For conventional vacuum thermoforming into a cylindrical mold, for example, he shows that the wall thickness varies with the depth of draw, according to

$$t/t_0 = \exp(-2s/a) \tag{12.5-15}$$

where t_0 is the initial sheet thickness, s is the distance down the wall, and a is the cylinder radius. This model has been verified experimentally for PET and GPS forming onto cold molds. To determine the maximum draw-down ratio, the following theoretical pressure relationship can be used:

$$P = (2N_0/a)[1 - 0.125 \exp(-6s/a)] \tag{12.5-16}$$

where $N_0 = \phi t_0$ and ϕ is a material property obtained from the stress-strain data:

$$\sigma\lambda = \phi(\lambda^2 - 1/\lambda^4) \tag{12.5-17}$$

Funt's analysis yields the proper form for ϕ from Eq. (12.5-14). As an example of the functionality of the Williams-Funt approach, for a 10-mil sheet being vacuum-formed (P $\doteq$ 14 psi), draw ratios of s/a in excess of 4 can be achieved with a sheet temperature of 130°C, s/a of about 2 at 120°C, and s/a of 0.6 at 110°C. Similarly, a sheet at 150°C can be vacuum-drawn only to an s/a = 2.6. Note that although we could achieve s/a in excess of 2 with proper sheet temperature control, t/t_0 at s/a = 2 is 0.018. This means that for a 10-mil sheet the wall thickness at the bottom will be only 0.18 mil.

It is apparent that to achieve uniform wall thickness with uniformly heated sheet alternative means are required. Williams teaches us that stretching of a sheet with direct force can yield dividends. If the cylindrical mold of radius a is again considered and the sheet is being stretched a distance H using a cylindrical plug of radius b, we can show that the relationship between the distance from the rim s and the local radius of the sheet r is

$$s = H(a - r)/(a - b) \qquad (0 \leq s \leq H) \tag{12.5-18}$$

The reduced thickness of the stretched sheet is given as

$$t/t_0 = \{1 + [H/r \ln(a/b)]^2\}^{-1/2} \tag{12.5-19}$$

Williams also shows that the ratio of deflection to plug height H is given as

$$w/H = \ln(a/r)/\ln(a/b) \qquad (12.5\text{-}20)$$

For plugs with diameters approaching those of the mold (a $\simeq$ b), this ratio reduces to a linear form.

To couple plug assist with vacuum forming, we can make a gross assumption that the additional surface generated by the plugging action reduces the ultimate s/a. One simplistic way of doing this is to replace t_0 in Eq. (12.5-15) with the t calculated from Eq. (12.5-19). Then a, the cylinder radius, can be replaced with an effective cylinder radius which includes the new surface area. Consider the surface area of the plastic on the plug as that of a frustrum of a right cone:

$$A = \pi[H^2 - (a - b)^2]^{1/2}(a + b) + \pi b^2/4 \qquad (12.5\text{-}21)$$

Now a can be replaced with

$$a_{eff} = (4A/\pi)^{1/2} \qquad (12.5\text{-}22)$$

and in keeping with the grossness of the assumptions, the appropriate value for s at the bottom of the mold is half the distance between the base of the plug and the bottom of the mold. Application of this technique to plug-assisted thermoforming of PP on a warm mold has shown that the minimum wall thickness of 6.2 mils is predicted to occur at 1.2 in. from the rim of a 30-mil sheet being formed into a cup 2 in. deep $\times$ 2 1/2 in. in bottom diameter. The measured wall thickness of 7.9 mils occurs at 1.125 in. from the rim. Thus, it would appear that the sequential application of the Williams' stretching and drawing theories yields a suitable starting point for the prediction of plug-assisted thermoforming.

12.5.4 Other Tricks

As mentioned earlier, one way to control the uniformity of the draw is to keep the mold warmer in the regions of first contact with the drawing bubble such as the rim and the center bottom of the mold. An alternative to this is shown in the second curve of Fig. 12.5-8. Here several layers of tissue paper have been placed on the surface of the sheet in a way that allows the center and outside edges of the sheet to heat faster than the walls and corners. Thus, a very uneven heat distribution occurs throughout the sheet. Now as the center of the drawing bubble touches the bottom of the mold, it remains sufficiently warm to allow material to be drawn into the corners. As a result, the drawn sheet has a nearly uniform part wall thickness of 24 to 26 mils. This technique requires much trial and error on the part of the molder but yields parts that can meet rather stringent wall thickness specifications. Cycle time and temperature uniformity have been sacrified, however. The drawing temperature of the sheet was considered to be 300°F

(minimum). With the unscreened sheet, this temperature was achieved at
the bottom of the sheet in about 4.5 min. With the screened sheet, the portions of the sheet beneath the screening took more than 7 min to reach
forming temperature.

Incidentally, tissue paper is not normally used for screening in production runs. Wire screen or mesh is used. To analyze the radiant interchange among the heater bank, the wire screen, and the plastic surface, the
screen wire can be considered to be a series of cylinders that are parallel
to the heater bank. It is obvious that wire is made of a series of cylinders
that are parallel in two directions, not one. But the simplistic consideration of a single row of parallel cylinders will be used here to represent a
single screen and two rows of parallel cylinders to represent two screens.
Earlier, the effect of removing the parabolic reflectors from a single bank
of rod heaters was mentioned. Note that the curve of the "view factor" for
various geometric ratios is applicable here as well. Here assume that the
screens are made of cylinders d in diameter and that the two screens are
spaced L units apart. If the diameter of the wire of the screen and the distance between the wires are known, L/d can be determined. From Fig.
12.3-1, the F value that is needed to determine the amount of energy absorbed
by the wire screen is obtained. Suppose, for example, that both the screen
and heater bank are assumed to be black bodies (e.g., emissivities equal to
unity). Then

$$Q = F'A\sigma(T_1^4 - T_s^4) \qquad\qquad (12.5\text{-}23)$$

where F' is the value obtained from Fig. 12.3-2. If, in addition, the plastic can be considered a black body, the total amount of energy that is interchanged between the heater and the wire and the heater and the plastic is
given by

$$Q = A(1 - F')\sigma(T_1^4 - T_2^4) \qquad\qquad (12.5\text{-}24)$$

Here T_1 is the temperature of the heater bank (absolute degrees), T_s is
that of the wire, and T_2 is that of the plastic surface. In addition, the screen
also radiates energy to the plastic. This energy is given as

$$Q'' = A''F'\sigma(T_s^4 - T_2^4) \qquad\qquad (12.5\text{-}25)$$

Note that A is the area of the sheet but that A'' is the area of the wire. Thus,
the total amount of energy received by the plastic sheet is the sum of the
energy transmitted through the wire according to Eq. (12.5-23) and that
emitted by the wire according to Eq. (12.5-25). Think of this in the following way. The sheet is seeing a surface with a grid radiating at T_s and the
holes in the grid radiating at T_1. The smaller the holes become, the smaller the amount of energy absorbed from the source at temperature T_1.

Since T_1 is normally larger than T_s, the surface is cooled as the hole size decreases. Of course, this calculation can be made using a synthetic "radiant" heat transfer coefficient, but it is more cumbersome than with the conventional radiation equation (12.5-23). Incidentally, it is advantageous to put the screens as close to the heater surface as possible to minimize "spotlighting" or localized heating of the surface of the plastic directly beneath the "window" in the screen. It is obvious that this can also be minimized by using screen with rather fine mesh. The use of screens is much more practical than layers of tissue paper, since the screens can be attached directly to the heater bank.

12.5.5 Transparent Sheets

Earlier, the problems of heating transparent sheets such as PVC, PMMA, PC, or GPS were discussed briefly. When these materials must be heated, it is found that some of the incident radiant energy is absorbed on the surface of the sheet. Depending on the absorption or transmission spectrum of the material and the thickness of the sheet, the remainder is either absorbed within the volume of the sheet or transmitted through the sheet. For the most part, emphasis on the heating of transparent sheet has been placed on volume absorption of radiant energy is perpendicularly incident with the surface. If the radiant energy source is specular rather than diffuse, then specular reflection from the back surface of the sheet will polarize the reflected energy. It is well known that there is a critical internal angle of reflection that prevents the reflected energy from ever leaving the sheet. The energy is simply "piped" along the surface of the sheet until it is completely absorbed. For most plastics, the critical angle between the incident radiant wave and the surface of the plastic is small, and this "stray" amount of energy can be neglected in subsequent calculations. For PMMA, however, the critical angle is about 42°, and thus a relatively large portion of the incident radiant energy can be trapped within the sheet and be absorbed. In effect, then, the sheet behaves as if it were volumetrically opaque.

It should be apparent that the more energy that is transmitted through the sheet, the slower the sheet will heat at the same radiant energy flux. Furthermore, the more energy that is absorbed within the sheet relative to that absorbed on the surface, the more uniformly the sheet will heat at the same fraction of radiant energy transmission. For very thin sheets such as PVC for blister packs, the extent of transmitted energy is very high, depending, of course, on the sheet thickness and the transmission spectrum of the stock. What is frequently done to heat these materials is to use either two heater banks with the sheet between or one heater bank and a highly reflective mirrored surface with an insulated back. Thus, an energy interchange between two heater banks of approximately the same temperature with a lightly absorbing layer between must be considered. Cess describes pure radiation heat transfer in an absorbing medium with primary radiation from parallel black plates [22]. Although his work is concerned primarily with absorbing gases, the work is directly applicable here. In his notation, for a nonabsorbing medium,

$$q_r = e_1 - e_2 \qquad (12.5\text{-}26)$$

where e_1 is the emissive power of plate 1 and e_2 is that of plate 2. Now, according to Cess, "if the space between the plates is instead occupied by an absorbing gas (read "absorbing transmitting plastic sheet"), the effect upon the net heat transfer, q_r, will be similar to that of placing a radiation shield between the plates. One would thus expect the heat transfer to be decreased as the absorption ability (or optical thickness) of the intervening medium is increased." If ϕ is the dimensionless emissive power of the intervening layer (e.g., the plastic),

$$\phi = (e - e_1)/(e_1 - e_2) \qquad (12.5\text{-}27)$$

and if the optical thickness is defined as

$$\tau_0 = \int_0^L a\ dy \qquad (12.5\text{-}28)$$

where a is the absorptivity of the intervening layer and L is the total thickness of the layer, then $q_r/(e_1 - e_2)$ can be plotted against τ_0. For very thin absorbing layers or layers with very low average absorption coefficients (e.g., small values of τ_0), it is an optically thin plate, and

$$q_r/(e_1 - e_2) \doteq 1 - \tau_0 \qquad (12.5\text{-}29)$$

If a very large average volumetric absorption coefficient or a relatively thick layer is present, then the layer is considered to be optically thick:

$$q_r/(e_1 - e_2) \doteq (4/3\tau_0) \qquad (12.5\text{-}30)$$

The entire curve of $q_r/(e_1 - e_2)$ as a function of τ_0 is shown in Fig. 12.5-9. Here it has been assumed that a is a function only of temperature and not of wavelength (the so-called gray approximation). If a is independent of temperature, $\tau_0 = aL$. If a is a function of wavelength, as it is for most transparent plastics, the wavelength spectrum can be sectioned into narrow bandwidths. Then a is assumed to be constant in each of these bandwidths (a = a_λ), the corresponding $\tau_{0\lambda} = a_\lambda L$ is obtained, and the $[q_r/(e_1 - e_2)]$ is calculated for that bandwidth, which incidentally represents the amount of energy absorbed by the intervening layer in that bandwidth. Then all these values are added in proper fashion to obtain the average amount of energy absorbed by the transparent plastic. For relatively thin sheets of plastic, it can be assumed that uniformly heating both sides of the sheet will yield a nearly uniform temperature across the sheet. Thus, the temperature of the sheet can be easily determined. This gives a first attempt at the determination of temperatures in transparent sheet stock.

Recently an extension of early work on transparent sheets [23] has
been made to coextruded sheets where one of the sheets is transparent and
the other opaque [24]. Typical examples of this type of sheet are PMMA
(for UV barrier) and ABS (for impact strength), acrylonitrile (for barrier
properties) and HIPS (for impact strength), and HDPE and LDPE (for lower
costs). In early attempts at laminating, it was found that delamination during
thermoforming left small bubbles at the interface between the sheets. With
the newer hot extrusion techniques, this is minimized. Nevertheless, a
gassy plastic such as ABS can evolve gaseous degradation products if overheated, and these gases will accumulate at the interface. Sufficient differential thermal expansions of the sheets will lead to delamination. To avoid
this, it is necessary to heat the sheet in a rather uniform manner. As
shown in Fig. 12.5-10, changing the extent of heating of the two surfaces
changes the shape of the temperature profile so that the maximum temperature moves away from the interface. Controlling the cycle time of the
process should control the evenness index in much the same way as for a
single sheet (opaque or transparent). There is apparently little reason to
consider coextruded sheets impossible to thermoform using conventional
radiant heater banks without delaminating the sheets. Care and the understanding of the phenomena occurring during heating are requisite to a successful processing.

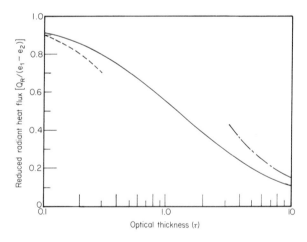

Figure 12.5-9   Heat transfer between radiant parallel black plates separated by absorbing medium. Solid line, theory for all thicknesses of absorbing media; dashed line, optically thin approximation; dash-dot line,
optically thick approximation [22].

12.6 Conclusions

As discussed earlier in the chapter, the engineering aspects of thermoforming have not been fully explored owing to the rather minor economic
role played by the process to date. Only recently have computations and
experiments been carried out to determine engineering parameter interactions. It is obvious that this type of work is necessary to the growth and
optimization of the process. For too long, thermoforming has been considered capable of making parts using relatively simple and, by implication,
crude molds and heating systems. New developments, such as the injection
thermoformer, called a "monoformer," of Hayssen Manufacturing Co. [25],
give insight into the potential of the process. Here, a disk of molten or
softened plastic is injected into a cavity with a conventional reciprocating
screw injection molding machine. The disk is clamped on the rim, and a
plug is used to begin stretching the disk into a two-piece cavity. Once the
plastic has been sufficiently stretched into the cavity, air pressure and/or
vacuum is applied in the mold and the plastic is pulled or pushed to the mold
surface, where it cools. The part is then ejected by opening the two-piece
mold. The process uses a combination of injection molding, thermoforming,
and blow molding to form a part that has no scrap. Other processes, such
as the Dow scrapless thermoforming process [26], are similar in concept
but seem to lack the flexibility of the Hayssen process.

Another technology that is growing rapidly is thermoforming of foam
sheet. GPS foam sheet can be made using sodium bicarbonate/citric acid
nucleator and isopentane or butane physical blowing agent. Commonly, a
two-stage system is used wherein the polymer is melted in the first stage
screw and then transferred to a second, where the blowing agent is introduced. The foamed product is extruded either through a fishtail die or a
pipe die, slit, and flattened. The foam sheet density is nominally 0.05 to
0.1 g/cm^3. The only precaution taken in forming foamed sheet is to minimize expansion of the sheet by overheating. Although quartz lamps are
recommended to enable deep penetration of radiation into the cellular structure, most fabricators continue to use the more reliable resistance heaters.
Benning [27] has detailed the processing aspects of GPS foamed sheet.

Of course, the ideal process for any thermoform molder is to begin
with either powder, pellets, or regrind and in one step produce a thermoformed process, thus eliminating the energy expended in heating and cooling
the plastic. This approach still has not been fully exploited by the engineer,
however, and the pragmatic thermoformer must rely on optimization of the
three basic components of his process: the heater bank, with its peculiar
thermal characteristics; the mold, made of one of many materials, including
wood and stainless steel; and the sheet stock itself, with its unique processing
problems, including poor thermal conductivity and biaxial orientation.

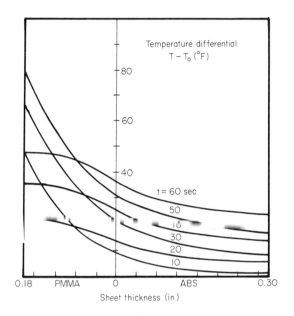

A

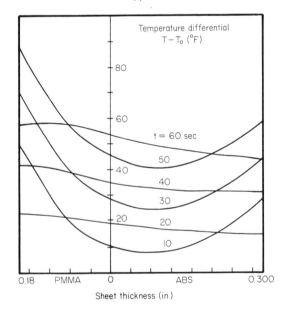

B

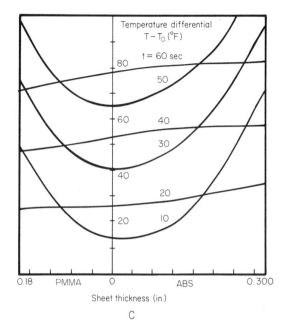

Temperature differential
$T - T_0$ (°F)

t = 60 sec

80

50

60   40

30

40

20

20   10

0.18   PMMA   0   ABS   0.300

Sheet thickness (in.)

C

Figure 12.5-10   A: Temperature profiles in pulsed radiant heating of co-extruded sheet. Total cycle time period T = 20 sec; T_1, "on" period, 50%; sheet material, 0.18-in. PMMA, 0.30-in. ABS (ABS opaque); Q_{PMMA} = 2000 Btu/ft^2·hr; Q_{ABS} = 0. Parameter is time (in sec). B: Temperature profiles in pulsed radiant heating of coextruded sheet. Same conditions as A except Q_{PMMA} = 2000 Btu/ft^2·hr = Q_{ABS}. C: Temperature profiles in pulsed radiant heating of coextruded sheet. Same conditions as A except Q_{PMMA} = 2000 Btu/ft^2·hr and Q_{ABS} = 5000 Btu/ft^2·hr [24].

In this chapter, then, these three components of the process have been put in an engineering perspective, so that the thermoform molder can understand how they interact and can effect some degree of optimization of his (her) process.

Problems

P12.A A sheet of plastic at 200°F is held in a large room with quiescent air at 80°F. The heat transfer coefficient from convection between the sheet and the air is 1 Btu/ft^2·hr·°F. Estimate the rate of heat transfer per foot of sheet (a) if the sheet is mirrored with an emissivity of 0.1 and (b) if the sheet is dull gray with an emissivity of 0.85. Assume that the sheet may be considered completely surrounded by the air. Thus, the configuration factor F = Ae, where A is the area of the sheet and e is the emissivity of the sheet. Note that the combined convection-radiation heat transfer coefficient is given as

$$h = h_r + h_c \qquad\qquad (12.A-1)$$

P12.B Assume that the radiant heater source is at 2000°F and that the source is planar, with an emissivity of 0.1. Calculate the effective radiant heat transfer coefficient if the plastic sheet initially is 80°F with an emissivity of 0.9. Calculate the effective radiant heat transfer coefficient when the sheet has reached a forming temperature of 450°F. Then assume an average effective radiant heat transfer coefficient and determine the approximate time required to heat the sheet 0.100 in. thick with a specific gravity of 1.05 and a specific heat of 0.5 Btu/lb·°F from 80 to 450°F (average temperature).

P12.C Show that in Eq. (12.3-4) h_r increases with increasing sheet temperature.

P12.D A sheet of opaque GPS with an emissivity of 0.85 is being heated with a single row of rod heaters. The surface behind the heaters is nonconducting. The rod heaters are 1 in. in diameter and are spaced 4 in. apart. If the emissivity of the rod heaters is 0.9 and the temperature is 1500°F, what is the amount of radiant energy absorbed by the sheet initially at 80°F? A second row of rod heaters is placed behind the first, also at 4-in. intervals. Now what is the amount of energy transmitted?

P12.E Examine the data given in Fig. 12.3-3 to determine if an appropriate configuration factor can be determined for each of the types of heaters. If not, what additional information is necessary? From appropriate handbooks, supply the missing data.

P12.F A 0.125-in. sheet at the average forming temperature of 400°F is placed in contact with a water-cooled aluminum mold at 50°F. Assume that the mold temperature remains constant. How long does it take to cool the free surface of the sheet to 125°F? The thermal diffusivity of the sheet stock is constant at 5×10^{-3} ft^2/hr.

P12.G A 0.050-in. sheet of PMMA is being heated in a convection oven. The walls of the oven are at 500°F, and the sheet stock temperature initially is 100°F. If the sheet is 4 ft square and the oven is 4 ft on a side, what is the difference in radiant energy transmitted to the edges as compared to that transmitted to the center? The sheet is centered in the oven. The emissivities of the sheet and the oven are 0.8.

P12.H Assume that in P12.G the walls of the oven are removed and the sheet radiates to ambient air at 80°F. Now compute the energy absorption by the edges, and compare with that in the center of the sheet.

P12.I A very fine metal shield is used to block radiation to one portion of a very thick GPS sheet. Develop an expression for the reduction in radiant heat transfer between the planar incandescent source and the sheet. Assume that the emissivity of the fine metal shield is 0.4, that for the incandescent source, 0.7, and that for the sheet, 0.9. Ans.: Energy is reduced to 20% of original energy.

P12.J Extend P12.I to a general form for n such metal shields. Edge effects are to be neglected.

P12.K In the text, it was pointed out that for transparent sheet stock the maximum temperature occurs within the sheet. Justify this mathematically. Further, it was stated that the maximum temperature in an opaque sheet always occurs at the primary surface of the sheet. Justify this also.

P12.L According to Mooney, $\lambda_1 = 1/\lambda_2 \lambda_3$. If this empirical, or can it be shown mathematically? What continuity equation must be used to justify this?

P12.M A wire mesh of 0.062-in. strands 0.125 in. apart on a square pattern is used to mask a portion of a Nylon 66 sheet during heating. What is the reduction in radiant energy transmitted to the sheet if the wire is suspended 0.125 in. from the sheet? The wire is nonradiating. The planar heating source is 0.250 in. from the sheet surface at a temperature of 600°F. The sheet initial temperature is 60°F. The emissivities of the wire, sheet, and heating source are 0.75.

P12.N Progelhof et al. [13] have shown that the absorption spectrum for Lucite acrylic can be approximated by the following step-function approximation:

Wavelength, λ (μm)	Absorption coefficient α (cm^{-1})
0-0.4	∞
0.4-0.9	0.02
0.9-1.65	0.45
1.65-2.2	2.0
2.2-∞	∞

A 0.060-in.-thick sheet of acrylic is applied to an opaque ABS sheet. Assume that the acrylic acts as if it were an absorbing transmitting layer (equivalent to Cess' absorbing gas). Calculate the optical thickness τ_0, and determine the reduction in radiant energy received by the ABS sheet owing to the presence of the acrylic sheet. Repeat the calculation assuming that the sheet is 0.005 in. thick. Can a very thin sheet of acrylic appreciably reduce the amount of UV received by the ABS sheet?

References

1. E. Jones, Mod. Plast. Encyclopedia, 49(10A):548 (1972).
2. F. Kreith, Principles of Heat Transfer, International Textbook Company, Scranton, Pa., 1965, Chap. 5.
3. E. R. G. Eckert, Radiation: Relations and Properties, In: Handbook of Heat Transfer (W. M. Rohsenow and J. P. Hartnett, eds.), McGraw-Hill, New York, 1973, Sec. 15A.
4. R. Siegel and J. Howell, Thermal Radiation Heat Transfer, McGraw-Hill, New York, 1972.
5. R. C. Progelhof and J. L. Throne, J. Spacecr. Rockets, 7:1365 (1970).
6. A. I. Brown and S. M. Marco, Introduction to Heat Transfer, McGraw-Hill, New York, 1951.
7. N. Platzer, Sheet Forming, In: Processing of Thermoplastic Materials (E. C. Bernhardt, ed.), Van Nostrand Reinhold, New York, 1959, Chap. 8.
8. R. C. Progelhof, J. Quintiere, and J. L. Throne, SPE ANTEC Tech. Pap., 17:112 (1971).
9. H. C. Hottel, Radiant-Heat Transmission, In: Chemical Engineer's Handbook (J. H. Perry, ed.), McGraw-Hill, New York, 1950, p. 483.
10. Personal communication, Akron Standard Mold Co., Akron, Ohio, 1973.
11. Personal communication, R. Vodra, United Nickel Co., Wooster, Ohio, 1971.
12. H. Lunka, Soc. Plast. Eng. J., 26:48 (1970).
13. R. C. Progelhof, J. Quintiere, and J. L. Throne, J. Appl. Polym. Sci., 17:1227 (1973).
14. L. R. Schmidt and J. F. Carley, SPE ANTEC Tech. Pap., 19:284 (1973).
15. C. D. Denson and R. J. Gallo, Polym. Eng. Sci., 11:174 (1971).

16. L. R. G. Treloar, The Physics of Rubber Elasticity, Oxford University Press, New York, 1958, p. 155.
17. B. D. Mooney, J. Appl. Phys., 11:582 (1940).
18. M. O. Lai and D. L. Holt, J. Appl. Polym. Sci., 19:1209 (1975).
19. M. O. Lai and D. L. Holt, J. Appl. Polym. Sci., 19:1805 (1975).
20. J. M. Funt, Polym. Eng. Sci., 15:817 (1975).
21. J. G. Williams, J. Strain Anal., 5:49 (1970).
22. R. D. Cess, The Interaction of Thermal Radiation with Conduction and Convection Heat Transfer, In: Advances in Heat Transfer, Vol. 1 (T. F. Irvine, Jr. and J. P. Hartnett, eds.), Academic Press, New York, 1964, p. 1.
23. J. G. Quintiere, J. L. Throne, and R. C. Progelhof, Soc. Plast. Eng. J., 29(1):35 (1973).
24. R. C. Progelhof and J. L. Throne, Polym. Eng. Sci., 14:810 (1974).
25. Hayssen Manufacturing Co., Division of Bemis Manufacturing Co., Sheboygan, Wisc.
26. Dow Chemical Co., Midland, Michigan, 1973.
27. C. J. Benning, Plastic Foams, Dekker, New York, 1968.

13.

The Technology of Blow Molding

13.1 Introduction

Like thermoforming, blow molding is a generic term that describes an expanding class of plastics processing. The central theme is expansion or inflation of a "sleeve" of molten polymer against a cold mold. The primary objective of blow molding is the formation of a hollow object such as a liquid container, although nonfunctional objects such as signs have been made by blow molding. As Kovach [1] points out, the technology of conventional blow molding has been adapted from the glass blowing industry. There are several forms of blow molding, as illustrated in Figs. 13.1-1 to 13.1-3. In conventional parison blow molding, by far the largest form of blow molding, a cylinder of molten plastic is extruded through an annular die in a vertical position (Fig. 13.1-1). The cylinder, called a parison, is pinched off at both the top and bottom, forming a hollow bag of molten plastic. Normally, contained in one of the pinchers is a mechanism for inflation of the hollow bag outwards against the mold walls. The mold walls are well-vented, and sometimes a vacuum is applied to certain portions of the mold to ensure good contact between the drawn molten plastic and the cold mold. When the hollow plastic object is sufficiently cool to retain the shape of the mold, the mold halves are opened and the part ejected.

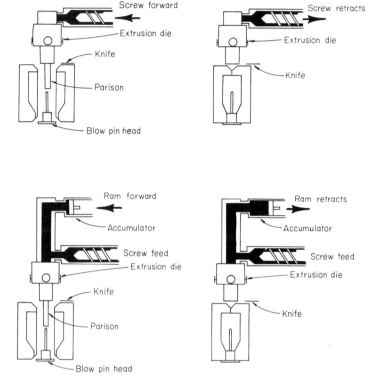

Figure 13.1-1 Schematics of two types of intermittent blow molding concepts. Top, extrusion using reciprocating screw; top left, extrusion of parison; top right, blowing of bottles; bottom, extrusion using ram accumulator; bottom left, parison extrusion; bottom right, blowing of container. (Courtesy of <u>Modern Plastics Magazine</u> [1].)

Another technique that is growing in interest is injection blow molding. Here a cylinder of molten plastic is injected around a "blowing mandrel," as shown in Fig. 13.1-2. The cylinder and mandrel are then rotated into a split blowing mold, where the cylinder is inflated through the mandrel. As is seen later, one of the problems with conventional blow molding is that it is difficult to control the thickness or material distribution of the parison. Injection blow molding has helped minimize this problem. However, most successful injection blow molding operations are restricted to rather small containers (less than 1 liter in volume).

As mentioned in the last chapter, the Hayssen process and similar processes use injection molding to provide a softened plastic disk suitable for "monoforming." Monoforming is, in reality, a form of injection blow molding or injection thermoforming.

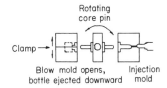

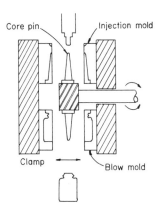

Figure 13.1-2 Schematics of two types of injection blow molding systems, both being two-station systems. Top, Piotrowski system; bottom, Schjeldahl system [1].

Dip blow molding uses a warm mandrel that is inserted into a pool of molten plastic. The plastic adheres to the mandrel as it is withdrawn, and this plastic cylinder and mandrel are then rotated into a split blowing mold, and blowing takes place in much the same fashion as injection blow molding. There is a difference, incidentally, in the characteristics of the polymer melt used in injection blow molding and dip blow molding. In injection blow molding, considerable orientation in the melt can be achieved during injection. If the material has sufficient memory in the melt, this orientation can be retained during blowing and frozen in during quenching against the cold mold. In dip molding, however, the melt remains nearly unoriented.

Since conventional parison blow molding uses a conventional extruder to produce the parison obviously with some start-stop control on the flow of melt from the extruder, it would seem natural that the development of a cold parison would be logical. Here, the extruder delivers (either horizontally or vertically) a continuous tube or pipe of plastic. This tube is cooled by conventional means and sliced into lengths. At some later time, these "cold" parisons, now called "preforms," are reheated either in a furnace or with a hot mandrel, clamped into a split blowing mold, and inflated in the conventional blow molding sense. This is now called "reheat blow molding." This is shown in Fig. 13.1-3.

All of these processes are variations on the basic theme of expansion of a cylinder of softened plastic by internal air pressure and/or external vacuum until the cylinder conforms to the shape of the cold split blowing mold. The method to be chosen for a given application will depend on the use of the object, the environmental conditions that the plastic will be subjected to, the characteristics of the plastic (amorphous or crystalline), the thermal sensitivity of the plastic (PVC or LDPE, for example), and the rheological character of the material since increasing die swell will lead to increasing part weight. Certainly, economics will play an important role as well. Injection blow molding requires a reciprocating injection molding machine, and normally this type of machine is more expensive than a conventional extruder. Therefore, we would expect that for the same bottle weight, for example, the machine cost for injection blow molding would be higher than that for extrusion blow molding. Certainly, if this cost can be made up in better parison wall control, thus reducing the amount of material

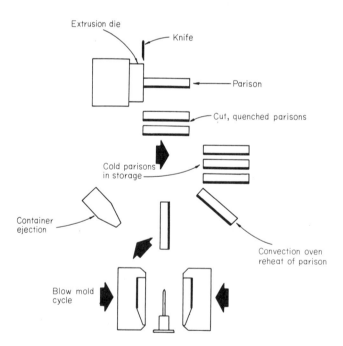

Figure 13.1-3   One type of cold parison blowing process, similar to the Beloit Orbet process. Parisons are cut from pipe die, reheated in oven to forming temperature, and then blown into bottles. (Courtesy of <u>Modern Plastics Magazine</u> [1].)

required to make a bottle of a given volume, say, or higher production rates, then injection blow molding would be preferred from an economic viewpoint. At the present time, parison blow molding is still economically preferred for high-volume low-profit margin items such as bleach bottles, cosmetic bottles, and the like. Reheat blow molding is the preferred method for highly oriented bottles of PP and PET, for primary use as pressurized or oil containers.

The various machine types and applications are summarized in Table 13.1-1.

In this chapter, some of the problems of dropping a parison with a reasonably uniform wall dimension are considered. The problems of cooling are similar to those in thermoforming except that the materials may be highly crystalline, such as PP, PET, or HDPE, and thus the effect of morphology on the final part must be involved. The cooling problem is also considered briefly. The mold design from a thermal viewpoint is not unlike that for injection molding, rotational molding, or thermoforming. Mold temperature control has not been considered an important aspect of blow molding technology to date. With attempts to blow more sophisticated resins being made, it will undoubtedly become important in the very near future. The emphasis, then, will be on the problem of die swell of the extrudate issuing from the annular die.

13.2 Rheological Properties of Blow Molding Resins

As mentioned, the objective during the fluid flow stage of blow molding is to rapidly extrude a parison through an annular die that is held in a vertical position. Since this process is cyclic and not continuous like conventional extrusion, the extrusion technology developed in an earlier chapter cannot be easily applied. Instead effort must be concentrated on the unsteady behavior of an extruding parison that is exhibiting the effects of gravity and normal stress relaxation resulting in die swell. Several excellent isolated attempts to study the variables that affect parison die swell and drape have been made in an effort to separate the two effects and to then relate the effects to the rheological characteristics of the melt. First, however, it is necessary to itemize the various physical and material parameters that can affect the parison. Again, the first objective of parison blow molding is to deliver a parison of minimum weight with, if possible, controlled wall thickness variations. The variations in parison wall thickness will allow drawing from thicker sections during blowing, thus producing a part of rather uniform wall thickness. Working against this idealized concept of a parison are the unsteady nature of extrusion, the lack of a drawing force (other than gravity), the time delay between the end of the extrusion of the parison and the clamping and inflating of the parison bag (thus allowing stress relaxation within the material), the relatively high viscosity of the melt required to minimize the effect of draw down, the extrusion of a hot melt into a rather cool environment,

Table 13.1-1

Applications for Various Blow Molding Machines [1]

Machine type	Applications
Continuous extrusion transfer and injection transfer	Very small to small containers; aspirin bottles; containers for nose spray, deodorant, sunburn cream, glue, hair coloring; small toys, doll parts, etc.; materials: PVC, polystyrene, PE, PP, acrylics
Continuous extrusion preform transfer	Small parts (9-oz and 1-qt sizes up to 1 gal); containers for eye lotion, aerosols, nondairy cream, bath oil, shampoos, motor oil, mustard, milk, pharmaceuticals; hospital syringes; labware; special liners; water pitchers; urinals; gas tank floats; no scrap recycle; process extends limits of injection transfer
Continuous extrusion ferris wheel and rotary table	Containers to 5 gal, including all previously listed; not generally used for parts other than bottles; special machine design required for bottles with offset handles; generally most popular bottle-making process
Continuous extrusion cold tube	Oriented containers with very accurate walls; generally similar products to ferris wheel; process allows additional steps and control not available in the direct expansion of parison from extruder
Intermittent extrusion reciprocating screw	1/2- to 5-gal containers; acid bottles; labware; special liners; bulk milk bottles; auto horns; urinals; wax dispensers; electric coffee pots; special disposable containers, hospital syringes, etc.; best used with stable compounds
Intermittent extrusion ram accumulator	5-gal containers to 300-gal drums; riding toys; street light globes; water coolers; displays; lawn ornaments; vaporizers; waste baskets; seal backs and bottoms; boats; under fender skirts; auto gas tanks; large figurines; must use stable compounds

and probably other effects such as improper alignment of the die mandrel (thus allowing for an uneven wall extrudate) or mandrel spider legs that leave orientation marks and possibly material property changes in the extrudate. Probably the most concentrated effort has been applied to the interrelationship between the die swell and the normal stresses of the plastic. Several very recent works have concentrated efforts in this area with some very interesting results. These efforts are considered in depth, and other parametric studies are discussed briefly.

Die swell in emerging plastic streams has been a concern not only of the blow molder but of the filament extruder as well. Much work is described in standard texts on the rheological properties of plastics [2-5,20] and will not be detailed here. Middleman approaches the subject from a viewpoint that is perhaps useful here to the early understanding of the problem. He notes that fluid, as it issues from a die, exerts an axial thrust on the die. This thrust is defined as

$$F = \int_0^R \rho v^2 2\pi r \, dr - \int_0^R T_{zz} 2\pi r \, dr \tag{13.2-1}$$

where F is the force, r is the radius of the circular die or the characteristic width of the annular die, v is the velocity of the fluid at any r, and T_{zz} is the normal tension in the axial direction. The first term on the right represents the momentum of the exiting melt, and the second term represents the axial tension in the fluid. If this equation is inverted to solve for the axial tension,

$$(T_{zz})_R = \rho V^2 \left[(3n + 1/n) - 2 \int_0^1 (v/V)^2 (r/R) \, d(r/R) \right]$$

$$- (F/\pi R^2)[1 + (1/2n)(d \ln F/d \ln 8V/D)] \tag{13.2-2}$$

which looks very much like a Rabinowitsch-Mooney equation. If F is replaced with $\rho \pi R_j^2 V_j^2$, where R_j is the radius of the jet and V_j is its velocity, Eq. (13.2-2) becomes

$$(T_{zz})_R = \rho V^2 \left[(3n + 1/n) - 2 \int_0^1 (v/V)^2 (r/R) \, d(r/R) \right]$$

$$- (1/nX^2)[1 + n - d \ln X/d \ln(8V/D)] \tag{13.2-3}$$

where $X = R_j/R$. As Middleman points out, this ignores surface tension and gravity. Hopefully, the effects of gravity can be extracted from rheological work on the relationship between draw down and elongational viscosity.

Middleman then shows that $(T_{zz})_R$ is related to the normal stresses in the fluid in the following way:

$$(T_{zz})_R = -p(0, L) + (\tau_{11} - \tau_{22})_R - \int_0^R (\tau_{22} - \tau_{33}) \, dr/r \qquad (13.2\text{-}4)$$

where $\tau_{22} = \tau_{rr}$, the radial normal stress; $\tau_{33} = \tau_{\theta\theta}$, the angular normal stress; and $\tau_{11} = \tau_{zz}$, the axial normal stress. If the Weissenberg normal stress difference ($\tau_{22} - \tau_{33}$) is zero, a reasonable first assumption, the third term in Eq. (13.2-4) is zero everywhere.

Now if $X = R_j/R = D_j/D = L_j/L$ is measured, where L_j is the thickness of the parison and L is the thickness of the die land, as a function of the shear rate at the exit of the die land, which can be obtained using the Rabinowitsch-Mooney equation in the die land of the blow molding die,

$$8V/D = \dot{\gamma}_w = (1/\pi R^2)[3Q + \Delta P(dQ/d \Delta P)] \qquad (13.2\text{-}5)$$

all the terms on the right-hand side of Eq. (13.2-3) are known. This gives the axial tension of the fluid directly. A more exact method, of course, is to measure the normal stress difference ($\tau_{11} - \tau_{22}$) using a rheometer and to calculate $(T_{zz})_R$ directly from Eq. (13.2-4). Examining Eq. (13.2-3) for a moment reveals that the second term contains the effects of axial tension of the fluid on die swell. As $(T_{zz})_R$ becomes larger or equivalently the normal stress difference becomes larger, $X = L_j/L$ must also become larger. Now while Middleman's theory cannot be applied directly to the unsteady-state extrusion of a parison under gravitational draw down, some rheological consequences of the melt on the thickness of the parison wall can be seen. Proving this is very difficult, but Macosko and Lorntson [4] have looked at two commercial blow-molding-grade HDPEs. As can be seen from Table 13.2-1, the molecular-weight distributions and the relative viscosities of the resin are nearly identical. The resin labeled M-2 shows a 42-wt% die swell, however, and the M-1 resin only a 28-wt% die swell.

Although the data indicate that M-2 has a higher melt strength than M-1, a test made in accordance with standard practice, a comparison of the elongational curves does not support this, as seen by comparing Figs. 13.2-1 and 13.2-2. Furthermore, the viscosity-shear rate curves are nearly identical, as shown in Figs. 13.2-3. It is only when the normal stress difference $\tau_{11} - \tau_{22}$ is plotted against shear rate that any appreciable difference in material behavior is seen. This is given in Fig. 13.2-4. The normal stresses for M-2 are 27 to 45% greater than those for M-1 over the 2 decades of shear rate, and this compares well with the 15% higher die swell observed for M-2. This is qualitatively what is expected from the analysis of Eq. (13.2-3).

Continuing this analysis, consider the very important experiments of Wilson et al. [6] and Wu and coworkers [7,8]. The idea of extruding a

Table 13.2-1

Physical Properties of Two Blow Molding Grade HDPEs [4]

	M-1	M-2
Melt index	0.6	0.5
Gel permeation chromatography:		
$\overline{M}_n$	8,500	9,500
$\overline{M}_w$	131,500	133,000
$\overline{M}_z$	907,000	1,035,000
$\overline{M}_w/\overline{M}_n$	15.5	14.0
Branching	Little	Some long chain
Standard 100-g bottle-blown bottle wt (g)	128	142
Die swell ($\dot{\gamma} = 10^3$ sec^{-1})	65	75
Melt strength (g):		
$\dot{\gamma} = 92$ sec^{-1}	6.2	10.5
$\dot{\gamma} = 1.85$ sec^{-1}	7.5	11.8

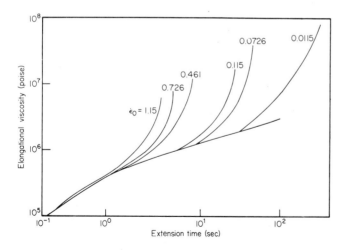

Figure 13.2-1   Elongational viscosity as a function of extension time and rate ($\dot{\epsilon}_0$, sec^{-1}) for one type of blow molding HDPE (M-1) at 180°C [4].

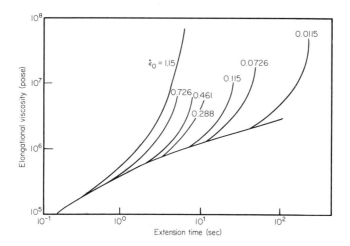

Figure 13.2-2   Elongational viscosity as a function of extension time and rate ($\dot{\epsilon}_0$, sec^{-1}) for one type of blow molding HDPE (M-2) at 180°C [4].

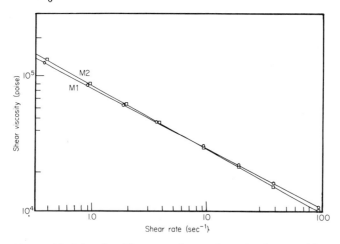

Figure 13.2-3   Capillary or shear viscosity for two blow molding HDPEs (M-1 and M-2) at 180°C [4].

parison into a pinch-off mold that is sectioned so that the weight of the parison at various distances from the bottom of the die lips was proposed by Sheptak and Beyer [9], but it wasn't until recently that methods of uncoupling the effects of parison die swell and draw down were accomplished experimentally. In the work of Wilson et al., the concern was the determination of the diameter swell and weight swell as functions of the extrusion shear rate of the parison. For their work, they defined diameter swell as follows:

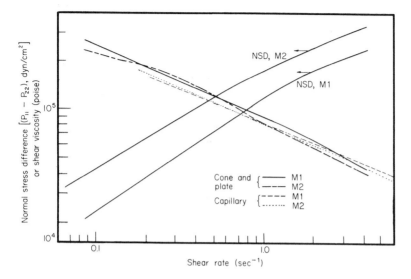

Figure 13.2-4 Comparison of shear viscosities and normal stresses of two blow molding HDPE's (M-1 and M-2) at 180°C [4].

$$S_d = (D_p/D_d - 1) \qquad (13.2\text{-}6)$$

where D_p is the parison diameter (external) and D_d is the effective bushing diameter (the external dimension on the die). The weight swell is given as

$$S_w = (W_a/W_t - 1) \qquad (13.2\text{-}7)$$

where W_a is the actual weight (in g/in. of parison length) and W_t is the theoretical weight (e.g., 100 g in Table 13.2-1). Since the purpose of Wilson et al.'s work was the determination of shear rate influence, they did not determine the die swell as a function of the distance below the die. To obtain W_a, they averaged the weight of the pillows cut from the pinch plate molding. As shown in Fig. 13.2-5, die swell either measured as diameter swell or the more accurate weight swell shows a marked increase with increasing extrusion shear rate. This, of course, is supported by the rheological work of Macosko in that his normal stresses increase very rapidly with increasing shear rate. It is difficult to extract from Wilson et al.'s work the importance of the melt index. Certainly, if Macosko's hypotheses are correct, increased branching should lead to increased die swell (in wt%, say). Therefore, for polyethylene, the lower-density material (0.950 sp. gr.) should show more die swell than the higher-density material (0.956 and 0.960 sp. gr.) at the same melt index, assuming, of course, that the power-law index remains relatively the same over the shear rate ranges considered here. This seems to be the case for the 0.3 melt index polyethylenes of Fig. 13.2-6 for weight swell.

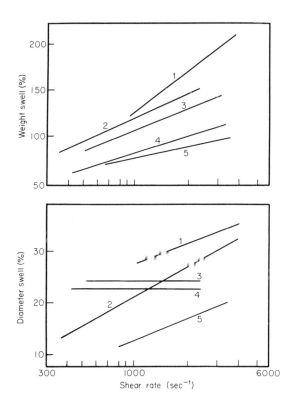

Figure 13.2-5 Characteristic weight and diameter swells of blow-molding-grade HDPE resins: 1: 0.9 MI, 0.96 sp. gr.; 2: 0.3 MI, 0.95 sp. gr.; 3: 0.3 MI, 0.956 sp. gr.; 4: 0.3 MI, 0.96 sp. gr.; 5: 2 MI, 0.925 sp. gr. [6].

Wu and his coworkers have been concentrating on determining die swell from the bottom of the parison to the top. They obtain the weights for each of the pillows formed in the pinch-off mold, and by varying the shear rate, mold-close delay time, or melt temperature can determine these effects on the individual parison pillow weight. As shown in Fig. 13.2-7, they corroborate the work by Wilson, showing that die swell, on the average, increases with increasing shear rate. They also show that increasing melt temperature, on the average, decreases die swell but probably increases draw-down effect. This is shown in Fig. 13.2-8. The effect of draw down is shown in Fig. 13.2-9, where the die swell for the bottom pillow (actually pillow 11 of 12) is seen to be smaller than the average die swell of Fig. 13.2-7 but in general is about as sensitive to shear as the average. The effect of pendant weight on die swell is shown for pillow 4 for two different shear rates in Fig. 13.2-10. The overall effect is one of decreasing die swell,

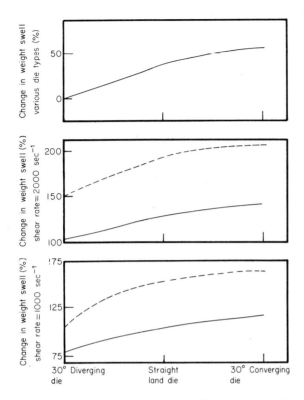

Figure 13.2-6 Effect of material change and shear rate on parison weight swell for three types of dies. Dashed line, 0.3 MI, 0.95 sp. gr.; solid line, 0.3 MI, 0.96 sp. gr. [6].

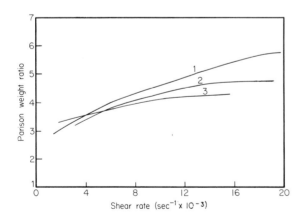

Figure 13.2-7 Effect of shear rate on overall parison weight increase for three characteristic HDPE blow molding resins [7].

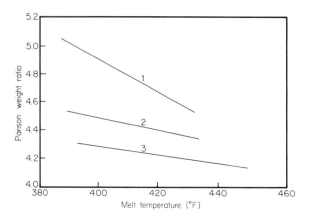

Figure 13.2-8   Effect of melt temperature on overall parison weight increase for three characteristic HDPE blow molding resins; measurement after 2 sec; shear rate, 10,000 sec^{-1} [8].

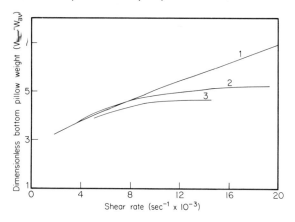

Figure 13.2-9   Effect of shear rate on weight of bottom pillow (divided by average pillow weight) of parison for three characteristic HDPE blow molding resins; measurement after 2 sec [8].

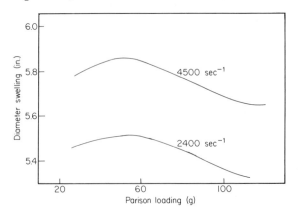

Figure 13.2-10   Effect of parison weight and extrusion shear rate on pillow 4 diameter swell for one blow molding HDPE resin; data points not shown [8].

probably corresponding to increasing elongation, with increasing loading and decreasing shear rate. The latter, of course, agrees with the qualitative experiments of Wilson et al. and the normal stress measurements of Macosko and coworkers.

Wu and his coworkers then consider separation of die swell and draw-down time in a somewhat heuristic manner. They assume that the increase in weight is rate dependent on draw-down time and length and on pure elastic swelling to an equilibrium value. This can be expressed as

$$dS_L/dt = k_L L \qquad (13.2\text{-}8a)$$

$$dS_s/dt = k_s(S_\infty - S_s) \qquad (13.2\text{-}8b)$$

The first equation gives the weight swell contribution to draw down, and they assume it to be directly proportional to the length of the parison hanging beneath the point of interest. k_L is a proportionality constant. The second equation gives the weight swell contribution to pure elastic swelling. It was first proposed in this form by Clegg [10] and Arai [19] for conventional extrusion of filaments. S_∞ is the weight swell at equilibrium. To integrate this equation, S_{s0}, the weight swell at zero time, is needed. Again, k_s is a proportionality constant. Integration of these equations yields

$$S_L = k_L Lt \qquad (13.2\text{-}9)$$

$$S_s = S_\infty - (S_\infty - S_{s0})e^{-k_s t} \qquad (13.2\text{-}10)$$

If the total increase in material weight is the sum of the effect of draw down and elasticity, then

$$S_w = S_L + S_s = k_L Lt + S_\infty - (S_\infty - S_{s0})e^{-k_s t} \qquad (13.2\text{-}11)$$

Bagley et al. [11] have also studied polyethylene die swelling but for a continuous extrusion. Nevertheless, the work of Wu and coworkers and Bagley et al. offers many similar results with regard to the explanation for the various constants involved in Eq. (13.2-11). For example, it is apparent that the strength of the melt should be reflected in the value for k_L. The higher the value for melt strength, the smaller the effect of S_L and thus the smaller the value for k_L. According to Fig. 13.2-11, resin E has a much higher melt strength than resin C. Similarly, for Macosko's resin, as shown in Table 13.2-1, resin M-2 has a higher melt strength than M-1. In both cases cited here, the total weight swell is higher, although it is apparent that viscosity per se is not important in Macosko's work. Bagley et al. and Wu and coworkers show that S_∞, S_{s0}, and k_s are process dependent since the former were working with a capillary die, whereas Wu and coworkers show a linear dependence of S_∞ with shear rate and no dependency of S_{s0} on shear rate. Bagley et al.'s work is shown in Fig. 13.2-12, and that of Wu for S_∞ in

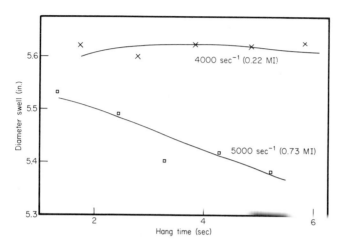

Figure 13.2-11   Effect of parison hang time on pillow 4 diameter swell for two blow molding HDPE resins [8].

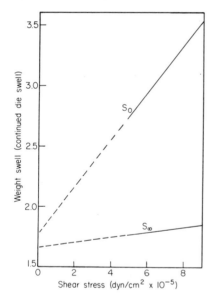

Figure 13.2-12   Change in zero shear and equilibrium shear swell values with shear stress; continuous die swell measurement on HDPE [11].

Fig. 13.2-13. Huseby and Gogos [13] show linear relationships between
S_∞ and shear rate for HDPE (Marlex 6009) in support of Wu's work. Bagley
et al. also show that k_S is linearly shear rate dependent, as shown in Fig.
13.2-14. Thus, an increase in shear rate should lead to an increase in the
equilibrium value for pure elastic die swell, and that equilibrium value
should be reached more rapidly. Wu also proposes a linear relationship
between k_L and k_S and an inverse relationship between k_S and zero-shear
viscosity. Macosko's work seems to indicate that the latter relationship is
not correct, and there seems to be little reason for assuming that the former relationship is relevant. There is certainly sufficient ground for assuming that continuous extrusion die swell from capillaries can be discussed
together with the discontinuous parison die swell from a blow molding die.
Certainly the melt must undergo large-scale deformation in the region of
the die land prior to exiting as a parison or a filament. It is fairly well
established now that many of the postextrusion problems are attributable to
the configuration of the die. In particular, the die length-to-diameter ratio
is critical to filament processing, and this ratio should play an important
role in the extrusion of the blow molding parison as well.

    Recently, Pritchatt and coworkers [12] carried out an extensive
study of die head configuration effects on parison swell. They found, for
example, that the parison weight swell ratio was approximately proportional
to the square root of the die outlet angle, whereas the diameter swell ratio
was approximately proportional to the square of the die outlet angle. They
then analyzed the flow of polymer through a tapered die, relying on the

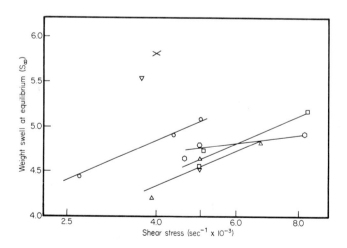

Figure 13.2-13   Effect of shear rate on equilibrium weight swell for several PEs [12].

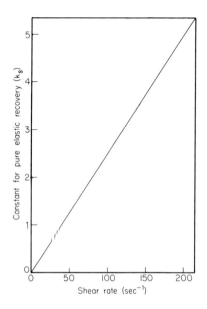

Figure 13.2-14   Effect of shear rate on proportionality constant for pure elastic recovery [11].   Data points not shown.

Frederickson-Bird analysis for flow through an annulus [14].   For flow in a converging die of angle β, the pressure drop equation for a power-law fluid is given as

$$\Delta P = (Q/2\pi)^n 2\eta_0 (1 - Y^{3n})/(3n \tan \beta R_0^{3n}) \qquad (13.2\text{-}12)$$

where Q is the volumetric flow rate; $Y = y_0/y_1$ is the dimensionless axial coordinate; η_0 is the power-law viscosity at $\dot{\gamma}_0 = 1 \ \sec^{-1}$; R_0 is the outside radius of the annulus; $\tan \beta = R_0/y_0$; and n is the power-law index.   Pritchatt et al. also give equations for the pressure force and drag force.   This viscous-only pressure drop must be modified to include contributions from the axial normal stress gradient.   Although more sophisticated examples of flow in diverging channels are available, Pritchatt shows that for dies with small taper angles an averaged pressure drop over a small element in the annulus due to extensional flow can be approximated by

$$(\Delta P_e/2)[(R_0^2 - R_1^2)_{in} + (R_0^2 - R_1^2)_{out}] = \sigma_T[(R_0^2 - R_1^2)_{in} - (R_0^2 - R_1^2)_{out}]$$

$$(13.2\text{-}13)$$

where σ_T is the local tensile stress, defined as the product of the extensional rate $\dot{\epsilon}$ and the uniaxial extensional viscosity η_e.   They recommend approximating the extensional rate by dividing the bulk polymer velocity by the flow length. As mentioned in Chap. 5, there are more sophisticated ways of determining rates of deformation. Nevertheless, the effect of extensional flow on the pressure drop can be approximated using Eq. (13.2-13). Pritchatt et al. also point out that an approximate value of the pressure drop due to normal stress gradient can be obtained by taking the difference in normal stresses:

$$\sigma_{zz,in} - \sigma_{zz,out} \doteq \Delta P_n \qquad (13.2\text{-}14)$$

In certain cases, as mentioned earlier, the normal stress data are available for this calculation. If not, Worth and Parnaby [15] point out that the swell ratio can be obtained from

$$S/S_0 = 2(R_0^2 - R_1^2)^{-1} \int_{R_1}^{R_0} (\sigma_{zz}/E + 1)r \; dr \qquad (13.2\text{-}15)$$

Or, knowing S/S_0, an approximate value for σ_{zz} can be obtained. E is Young's modulus for the polymer. Other, more reliable methods have been discussed above. Pritchatt et al. point out that Eq. (13.2-13) and (13.2-15) for ΔP_e and ΔP_n should be used only to determine if these pressures are significant when compared with that obtained from Eq. (13.2-12). If they are, more sophisticated analyses are required.

Now it can be seen how to approach the practical problem of timing the parison extrusion rate with the mold-close delay time. Consider a material with a very high melt strength such that k_L is small. To minimize postextrusion swelling then, the parison should be dropped rather slowly through a die with a rather long L/D ratio. Thus k_s, S_∞, and S_{s0} remain rather small, and swelling to the equilibrium value is slowed. Higher melt temperature can help minimize the pure elastic swell also, since apparently S_∞ and S_{s0} are reduced with increasing temperature. Watch, however, since the melt strength probably decreases in an Arrhenius fashion with increasing melt temperature. Furthermore, the mold should probably be closing during the latter stages of parison extrusion to minimize the mold-close delay time. If the material has a low melt strength, such that k_L is very large, the parison should be dropped as quickly as possible in an effort to minimize draw-down weight increase. Since increasing the shear rate is the result of a faster parison extrusion, k_s, S_∞, and S_{s0} are all going to be quite large. A very fast extrusion reduces the effect of the third term on the right in Eq. (13.2-11), but nevertheless the pure elastic die swell should approach the equilibrium value rather rapidly. Decreasing the melt temperature might help reduce the effect of k_s on the elastic die swell. Wu and coworkers have given a plot of the effect of mold-close delay time on the total increase in material weight, and this is shown in Fig. 13.2-15.

Wu develops an additional correlation from his experimental data that may be of interest here. Recall in the chapter on thermoforming that elongation

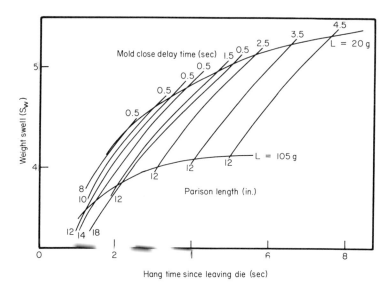

Figure 13.2-15 Composite graph showing effect of parison length, parison weight, and mold-close delay time on weight swell as a function of time since material left die; HDPE [8].

of a material in one direction resulted in a decrease in the thickness of that material in proportion to the square root of the extent of elongation. Cogswell [16] shows that for continuous extrusion of pipe the swelling thickness is proportional to the square of the swell diameter ratio. If t is the thickness of the parison, Cogswell's theory yields

$$2t/(D_d - D_m) = (D_p/D_d)^2 \qquad (13.2-16)$$

The actual area is given as

$$(4A/\pi)_A = D_p^2 - D_i^2 \qquad (13.2-17)$$

where D_p is the outside diameter of the parison and D_i is its inside diameter. The theoretical cross-section of the parison is given as

$$(4A/\pi)_T = D_d^2 - D_m^2 \qquad (13.2-18)$$

where D_d is the external diameter of the die and D_m is the diameter of the mandrel (e.g., the internal diameter of the die). Now $D_p - D_i = 2t$. Using this, D_i can be eliminated from Eq. (13.2-17), obtaining

$$(4A/\pi)_A = 4t(D_p - t) \qquad (13.2-19)$$

Now $S_w = W_A/W_T$ according to Wu's work, and $W_A = A_A L\rho$ and $W_T = A_T L\rho$, where ρ is the melt density. This can be rewritten then in terms of the ratios of the actual to theoretical areas assuming the lengths to be equal:

$$S_w = (4A/\pi)_A/(4A/\pi)_T \qquad (13.2-20)$$

Substitution into this equation yields

$$S_w = 4t(D_p - t)/(D_d^2 - D_m^2) \qquad (13.2-21)$$

or

$$S_w = (2D_d S_d^3 - 2S_d^4)/(D_d + D_m) \qquad (13.2-22)$$

In other words, it appears that S_w is proportional to S_d^3 for most cases, with a fourth-power effect added in. Wu finds an excellent, albeit empirical, correlation between the swell diameter and the fourth root of the swell weight ratio:

$$S^w = \alpha(S_d)^4 \qquad (13.2-23)$$

His experimental agreement is shown in Fig. 13.2-16. The theoretical equation, Eq. (13.2-22), seems to agree with the data of Wilson et al. more closely. Under any conditions, the flow is certainly not isotropic.

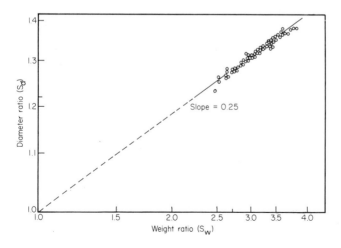

Figure 13.2-16 Correlation of individual pillow diameter swell ratio and individual pillow weight ratio showing slope of 1/4 [8].

13.3 Stretching the Parison

Once the parison has been dropped, it must be readied for inflation. In
some smaller bottle blowing operations, the parison is pinched off at the
bottom from the previous blowing step. This parison is allowed to remain
free in the mold while the top of the parison is pinched off around a blowing
pipe. The blowing pipe closure may include bushings necessary for the formation of external threads or other types of closures on the blown bottle.
For larger containers and rather high-speed operations, the pinched-off
section of the parison is allowed to drop below the halves of the split mold,
and thus pinch-off of the parison occurs at both the top and the bottom of the
parison.
    Inflation of the parison against the mold walls is accomplished by a
combination of air pressure from the blowing pipe (50 to 500 psig or more)
and venting at various places on the split mold surface. Since most molds
are cored for cooling, the regions around the pinch off freeze first and thus
are very thick in the final part in comparison with the side walls of the container. In some cases where an integral handle is required, inflation of the
parison begins before the mold closes fully. This enables inflation of the
plastic in the rather narrow handle section. Otherwise, partial collapse of
the material in this section can occur. Here, too, freezing of the material
occurs much earlier than in the body of the bottle, and thus the handle section can be rather thick when compared to side walls of the bottle. There is
relatively little phenomenological difference between inflation of the parison
and deep drawing of the softened sheet in thermoforming. Some points are
worth repeating, however. Inflation of the parison again depends on the material resistance to elongation. In the early stages of elongation, the flow is
probably best described as pure shear elongation, as defined in Chap. 5. At
very rapid rates of elongation, the fluid resistance increases dramatically.
Since the supply pressure is constant, the greatest elongation, here defined
as the differential change of diameter of the parison per unit diameter,
should occur immediately upon initiation of the inflation pressure. Uneven
parison wall thickness at this point could lead to a very rapid thinning of the
parison in a local spot or line along the length of the parison. This effect
probably would not be seen as a thin spot in the final part wall thickness but
might be a source of local environmental stress cracking or embrittlement.
In other words, uneven parison wall thickness might not lead to physical defects but probably leads to material defects. As mentioned earlier, uneven
parison wall thicknesses or even unusual melt orientation may be caused by
the die or spider holding the mandrel or even the method of filling the accumulator ahead of the die. So far, most of the studies have been made on continuous profile extrusions, but one might anticipate the problems to be similar in
the extrusion of a parison. Of course, uneven melt temperature in the parison can also dramatically affect the uniformity of inflation of the parison.
    Once the parison bag has contacted the mold wall, the elongation prob-
lem is nearly the same as that of draw down of plastic sheet in thermoforming,

and thus the thinning problems in the corners and edges of a square bottle
will not be considered here in detail. There are some problems with blown
objects that cannot be coped with in the same fashion as with thermoforming.
For example, recall that by screening the plastic sheet in thermoforming,
the center bottom of the sheet can be kept softer than the edges, and thus
material is drawn from the center to achieve a uniform wall thickness.
This is not possible with a parison, since the center bottom of a bottle, say,
is held in the cold pinch-off. Thus, the drawing for the corner must come
from the bulk of the parison and not from the center bottom. The standard
bottle formed by parison blow molding will thus show a very heavy pinch-off
cross-section and adjacent to this a very thin corner. There have been
many ideas patented to attempt to circumvent this problem, but most require addition of hardware to the conventional parison blow molding machine
or the mold. For example, parison wall thickness programming is now
standard on many of the more expensive machines. One method of programming the amount of material extruded through the die at any time is to
keep the die opening constant and vary the material flow rate. In this way,
for example, the parison extrusion can begin at a rather high shear rate,
thus causing a rapid diameter or weight swell early in the parison length;
then the flow rate can be slowed, thus reducing weight swell in the center
of the parison; and finally the parison extrusion can be finished at a very
high shear rate again, causing a high weight swell at the top of the parison.
Now when blowing occurs, there is sufficient material in the top and bottom
of the parison from which the corner material can be drawn. Rate programming requires much trial and error for a new mold, is difficult to control for materials with rather poor melt strengths, and leads to problems
in molds with integral handles. It seems to work best with cylindrical "wasp-
shaped" containers.

Another programming technique controls the die opening and holds
the extrusion speed constant. While this type of parison programming requires a rather slow extrusion speed for the parison, it is probably easier
to build and control and thus has become popular than the variable-speed
programming process. There are many ways of changing the die opening.
One technique uses a variable land length. Here the mandrel vertical position is changed relative to the stationary bushing. This variation changes
the effective land length at the tip of the parison die as well as the back pressure applied to the melt in the reservoir behind the die. With a tapered
land, the die opening varies the thickness of the parison irrespective of the
weight swell. Recall, however, that variation in the land length, L/D in the
case of simple capillary extrusion, leads to variation not only in the shear
rate but also in the extent of weight swell or die swell in the case of capillary extrusion. An example of this is shown in Fig. 13.3-1. In essence,
then, the molder is faced with a programmable way of controlling the thickness of the extruding parison but cannot account for the land length effect on
weight swell in a predictable manner. Again, the parison program must be
developed by trial and error for each mold.

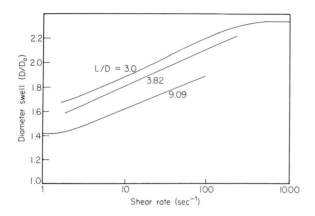

Figure 13.3-1   Effect of capillary L/D ratios on 0.9-MI, 0.96-sp. gr.
HDPE as a function of shear rate; data points not shown [13].

The newer orientation blow molding methods deal with a preform or
sack of softened plastic. The parison, in this case, has a formed bottom,
and thus pinch-off during mold closing is of no interest. More importantly,
a stretch rod is used to stretch the parison to the bottom of the mold during
blowing. In this way, very high degrees of orientation are achieved (2.5 to
4×). Furthermore, this simultaneous stretch-and-blow allows much more
uniform container wall thickness, particularly in corners. This area is developing rapidly for containers of surfaces of revolution; however, the formation of odd-shaped containers of uniform wall thickness from odd-shaped
preforms is entirely trial and error. High-performance containers of the
future will certainly utilize the stretch-and-blow technology.

In effect, then, the technology of this phase of blow molding is similar to that of thermoforming. Additional experimentation and new ideas in
the stretching of this sack of molten plastic must be presented before this
aspect of blow molding can be brought to a predictable level of engineering.

13.4 Cooling the Blown Parison

As with thermoforming, once the hot plastic touches the cold mold, the layer adjacent to the mold surface rapidly drops in temperature to that of the
mold. The rate of cooling of a blow-molded part can be calculated using the
standard transient heat conduction curves, provided some account is given
for free-surface cooling. Normally, the inflating air is cold, although experimentation in this area has not been thorough enough to determine if this
is desirable from a rheological viewpoint during the stretching of the parison.
Thus, cooling occurs from the inside or free surface of the plastic skin as

well. To account for this, a rear surface resistance can be included in the form of a Biot number, $= hL/k$, where L is the half-thickness of the plastic in contact with the mold wall, h is the heat transfer coefficient between the plastic and the inflating air, and k is the thermal conductivity of the plastic. Normally, h is on the order of 1 to 10 $Btu/ft^2 \cdot hr \cdot °F$ for air. Thus, despite a rather poor thermal conductivity, the Biot number remains rather small for thin-walled containers. Hunkar [17] has proposed using other means for increasing the heat transfer between the free surface and the environment in an effort to improve the cycle times. Depending on the process, the cooling time can represent 50 to 95% of the total machine time. The primary effort here should be placed on very high-speed multiple-cavity multiple-station molding of a crystalline material such as HDPE or on the blow molding of thick-walled containers, such as 200-liter barrels. Hunkar's process, called "internal surface cooling" or ISC, utilizes a mixture of water and very high-pressure air (1000 psi or more). The mixture, carefully metered, is passed through a let-down valve after initial inflation. Owing to the very great expansion of the air, the moisture freezes into very fine ice crystals, which coat the inside of the container. The very hot plastic is rapidly cooled by the ice and in turn melts the ice to water droplets and then evaporates the water. The water vapor is removed with the air as it is exhausted from the mold. The result is a dry container that has been cooled by melting and evaporating water. The key here is that more than 1200 Btu can be removed per pound of water injected, and the thermal resistance to this form of heat removal is very small. Thus, while the amount of moisture in the air is small (less than 10 wt%), the presence of this heat sink acts as if another mold wall, albeit a mold wall with an increasing surface temperature, were placed against the "free" surface. This means, in effect, that L, the thickness of the plastic layer between the "two" mold faces, is one-half that for a layer with one surface against a mold and the other a free, nearly insulated surface. Since the Fourier number is proportional to the square of the thickness L, from the standard transient conduction heat transfer charts with zero surface resistance it is apparent that to achieve the same degree of cooling only one-fourth the amount of time is needed, assuming that both surfaces remain at essentially the same temperature. Since the inner "mold surface" is increasing in temperature as the moisture passes from ice to water droplets to vapor, the one-fourth rule cannot be considered absolute. Nevertheless, as shown in Fig. 13.4-1, the cooling time for a 45-mil wall part is reduced from about 8 sec to 4.5 sec and for a 30-mil wall part from 2.5 to about 1.5 sec. If the one-fourth rule were correct here, the cooling time for the 45-mil wall part should be reduced to about 2 sec and for the 30-mil wall part, about 0.6 sec. Undoubtedly, some of the difference can be attributed to the increasing temperature of the internal surface coolant over this period of time, and some can be attributed to the time delay in metering the coolant into and out of the mold cavity. Nevertheless, the effect is in the proper direction with regard to the rules of heat transfer.

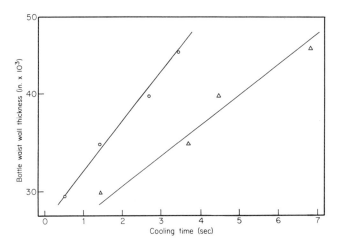

Figure 13.4-1   Comparison of patented internal surface cooling (ISC; circles) and conventional 100-psi air (triangles) for blow-molded HDPE. Mold
temperature, 38°F; melt temperature, 390°F; part cooled to 114°F in each
case [17].

       Incidentally, one would expect less part distortion with this ice-to-
steam method than with a thermal cooling method using, for example, CO_2.
This can be attributed to the fact that the temperature on the inner surface
gradually increases with time, and this warming would tend to reduce differential shrinkage, particularly in a thick-walled part made of a crystalline
material such as HDPE. There is no reason, furthermore, why this technique is not directly applicable to the internal cooling of rotational molds as
well. Certainly it seems that some of the effort spent on programming a
parison would be better spent on developments in the stretching and cooling
of the parison. Hunkar's ISC method shows one rather simple concept that
can yield reductions in cooling cycles and thus production improvements of
up to 55% with polyethylene.

13.5 Additional Considerations

As mentioned earlier, there have been many modifications of parison blow
molding in recent years. Most important economically is injection of the
parison around the blowing pipe. If the injection cavity is separated from
the blowing cavity, as shown in Fig. 13.1-2, the parison can be shaped and
its surface temperature can be controlled during injection. Until recently
injection blow molding has been restricted to containers less than 250 ml or
so. The principal restriction has been the size of the mandrel for extrusion

and the problem of keeping it hot and clear of molten plastic residue. Newer techniques of injection indicate that this problem can be solved with some
effort, and thus injection blow-molded containers of 1 to 5 liters in volume
are expected to be produced in volume in the next few years. It is also
necessary to point out that in injection blow molding a rotational or translational staging of the parison is required to bring the injected parison from
the injection mold to the blowing mold. Some recent work on extrusion of
the parison around a blowing mandrel followed by pinch-off and inflation
indicates that for small containers (50 ml or so) a single machine can handle
dozens of blowing molds. These molds are in line with the horizontal cross-
head die manifold of a continuous extruder, and the entire mold assembly
and mandrel shuttle to and fro at a speed matched to the extrusion rate of
the extruder. The blown bottles are ejected in a horizontal attitude from
the multiple molds. And one additional development, by Nissei [18], shows
that laminated injection blow molding of bottles such as cosmetic containers
is entirely feasible. After injection of the first and internal resin around
the blow pipe, the blow pipe holding the tube of resin rotates from the first
injection machine to a second injection mold on a second injection machine,
where the second or external resin is injected around the first. The blow
pipe, now holding a laminate of two resins, is rotated to the blow mold where
inflation occurs. It is obvious that the rheological implications of stretching
a laminate sheet are very complex and certainly have not been studied in
any depth at this point. After a rather modest inflation, the bottle is allowed
to cool and is then rotated to a fourth station for ejection. While the idea of
two-stage injection blow molding is quite interesting, it would seem that a
similar idea could easily be applied to coextrusion of a two- or three-component parison, with pinch-off and blowing taking place in a conventional split
blowing mold. Certainly the extrusion of multiple layers of parison is diff-
cult, but the opportunities for plastic containers with oil-resistant inner
layers and impact-resistant outer layers (oil cans, for example) or impact-
resistant inner layers and UV barrier outer layers (milk containers, for
example) are huge. The development of the dual head die and mandrel system for such a system is logically based on some of the information given
above, extended no doubt with common sense and extensive trial and error.

13.6 Conclusions

Blow molding, in each of its many forms, is simply described as the extension of a hollow sleeve of molten plastic (a parison) that is subsequently
expanded by internal air pressure against a cold-mold wall. The parallels
to glass blowing are generous and well acknowledged, as Kovach has amply
illustrated. The uniqueness of plastic lies in its elastic memory that causes
the parison walls to swell. Much effort is directed toward the understanding
and control of this effect, often to the neglect of the cooling and materials

handling problems. Application of the simple concepts of fluid mechanics and transient heat transfer lend credence to some of the more recent discoveries, such as the ISC process of Hunkar. At present most of the material blown into bottles are polyolefins and PVC, and this is certainly correct considering the tremendous market in disposable, "squeezable" containers. New developments needed to meet the rigid, high-pressure container market of soft drinks and beers and the barrier problems of oil and gasoline containers are presently being commercialized, primarily with reheat blow molding of an injection-molded parison.

Problems

P13.A From experiments with HDPE parisons, an empirical relationship between the thickness of the parison and the thickness of the die has been found as a function of the shear rate:

$$L_j/L = 1 + (\dot{\gamma})^{1/2} \tag{13.A-1}$$

The power-law index is 0.3, and the velocity across the jet at the point of issuing from the die is

$$v = V_0[1 - (2r/H)^{(n+1)/n}] \tag{13.A-2}$$

where

$$V_0 = [(n + 1)/(2n + 1)]V \tag{13.A-3}$$

If the melt has a 0.9 sp. gr., determine $(T_{zz})_R$, according to Eq. (13.2-3). Then obtain the force exerted by the sheet as it issues from the die with 0.100 in. = H thickness.

P13.B Using the experimental data for Nylon 66 given at 565°F in Fig. 5.2-4, obtain the form for the Rabinowitsch-Mooney equation (13.2-5). Then repeat P13.A for flow in a slit die. Discuss how the normal tension in the axial direction might be used to determine the extent of increase in material needed in extrusion of a parison.

P13.C Obtain Eqs. 13.2-10 and 13.2-11. Then, following Wu's work, where S_∞ is a linear function of shear rate and S_{s0} is independent of shear rate, obtain a plot of S_w as a function of shear rate. Bagley shows that k_s is also linearly dependent on shear rate.

P13.D Given a bushing diameter of 2.502 in., a mandrel diameter of 2.445 in., and a diameter swell of $S_d = 1.44$, determine S_w from Eq. (13.2-11) using Wu's correlation of Eq. (13.2-19). Compare your results with Wu's experimental value of $S_w = 2.72$. Ans.: $S_w = 4.3$.

P13.E A parison 0.060 in. thick along the sides and 0.100 in. thick at the
bottom contacts an aluminum mold held isothermally at 70°F. How long
does it take to cool the parison side to a free-surface temperature of 150°F
if the melt temperature is uniform initially at 450°F? At this time, what
is the free-surface temperature and the average temperature of the material
at the bottom? The thermal diffusivity of the material is constant at 5 ×
10^{-3} ft^2/hr. The free surface is considered to be insulated.

P13.F Hunkar [17] proposes using high-pressure air containing water to
cool the inside of the parison. The water in the air freezes as it enters the
parison, forming ice on the inner surface. The melting of this ice increases
the rate of cooling, according to Hunkar. First, verify that if the parison
is initially at 390°F and contacts an isothermal mold at 38°F, it takes approximately 3 sec to cool a 0.030-in.-thick part to an average temperature
of 114°F and approximately 8 sec to cool a 0.048-in.-thick part to the same
temperature. Is the shape of the second curve given in Fig. 13.4-1 correct?
If not, what is the proper shape? The thermal diffusivity for HDPE is approximately 4 × 10^{-3} ft^2/hr. Assume an internal heat transfer coefficient
of 5 Btu/ft^2·hr·°F for low-pressure air. Now repeat the calculation assuming
that the effective heat transfer coefficient for melting ice is 500 Btu/ft^2·hr·°F.
Do your results agree qualitatively with Hunkar's observations? If not,
what is an appropriate effective heat transfer coefficient? Suppose that the
38°F low-pressure air is replaced with -140°F CO_2. Can the new cooling
times match those of Hunkar? If not, why not?

References

1. G. P. Kovach, Forming of Hollow Articles, In: Processing of Thermoplastic Materials (E. C. Bernhardt, ed.), Van Nostrand Reinhold, New
   York, 1959, Chap. 9.
2. J. A. Brydson, Flow Properties of Polymer Melts, Van Nostrand Reinhold, New York, 1970, p. 63.
3. S. Middleman, The Flow of High Polymers, Wiley-Interscience, New
   York, 1968, p. 41.
4. C. W. Macosko and J. M. Lorntson, SPE ANTEC Tech. Pap., 19:461
   (1973).
5. R. I. Tanner, J. Polym. Sci. Part A, 8(2):2067 (1970).
6. N. R. Wilson, M. E. Bently, and B. T. Morgan, Soc. Plast. Eng. J.,
   26(2):34 (1970).
7. K. C. Chao and W. C. L. Wu, Soc. Plast. Eng. J., 27(7):57 (1971).
8. E. D. Henze and W. C. L. Wu, SPE ANTEC Tech. Pap., 18:735 (1972).
9. N. Sheptak and C. E. Beyer, Soc. Plast. Eng. J., 21(2):190 (1965).
10. P. L. Clegg, In: The Rheology of Elastomers (P. Mason and N. Wookey,
    eds.), Pergamon Press, Elmsford, N.Y., 1957, p. 174.

11. E. B. Bagley, S. H. Storey, and D. C. West, J. Appl. Polym. Sci., 7:1661 (1963).

12. R. J. Pritchatt, J. Parnaby, and R. A. Worth, Plast. Polym., 43:55 (1975).

13. T. W. Huseby and C. G. Gogos, Polym. Eng. Sci., 5:130 (1965).

14. A. G. Frederickson and R. B. Bird, Ind. Eng. Chem., 50:347 (1958).

15. R. A. Worth and J. Parnaby, Trans. J. Chem. Eng., 52:368 (1974).

16. F. N. Cogswell, Plast. Polym., 38:391 (1970).

17. D. B. Hunkar, SPE ANTEC Tech. Pap., 19:448 (1973).

18. Nissei Plastic Industrial Co., Ltd., Kyoto, Japan, 1977.

19. T. Arai, Nippon Gumu Kyokaishi, 30:993 (1957).

20. J. M. McKelvey, Polymer Processing, John Wiley and Sons, New York, N.Y., 1962, p. 93.

where V is the actual volume of the system. The total derivative of the volume is formed as follows:

$$dV = (\partial V/\partial T)_P \, dT + (\partial V/\partial P)_T \, dP \qquad (14.2\text{-}4)$$

where $(\partial V/\partial X)_Y$ represents the rate of change of volume with the variable X while the variable Y is held constant. Rewriting Eq. (14.2-4),

$$dV/V = V^{-1}(\partial V/\partial T)_P \, dT + V^{-1}(\partial V/\partial P)_T \, dP \qquad (14.2\text{-}5)$$

Now define the first bracketed term as the volume expansivity β:

$$\beta = V^{-1}(\partial V/\partial T)_P \qquad (14.2\text{-}6)$$

Note that the units of volume expansivity are degrees^{-1}. The negative of the second bracketed term of Eq. (14.2-5) is the isothermal compressibility.

$$\kappa = (-V^{-1})(\partial V/\partial P)_T \qquad (14.2\text{-}7)$$

κ has the units of pressure^{-1}. Sometimes the reciprocal of κ is given in the literature. B, the bulk modulus, is defined as

$$B = (-V)(\partial P/\partial V)_T \qquad (14.2\text{-}8)$$

Therefore, the percentage increase in volume (percent volume expansion) is written in terms of differential temperature and pressure as

$$dV/V = \beta \, dT - \kappa \, dP \qquad (14.2\text{-}9)$$

As will be seen shortly, Griskey and his coworkers have examined the thermodynamics of polymers using these two physical properties of the materials. They found that κ and β are nearly independent of pressure and temperature, but when working with very accurate dilatometric measurements this assumption should be checked occasionally.

Occasionally, it is necessary to know how pressure changes with temperature when the volume of the system is held constant. This is found from this theorem:

$$(\partial P/\partial V)_T(\partial V/\partial T)_P = -(\partial P/\partial T)_V \qquad (14.2\text{-}10)$$

Rearranging and using the physical properties of volume expansivity and isothermal compressibility,

$$[(-V)(\partial P/\partial V)_T][V^{-1}(\partial V/\partial T)_P] = -(\partial P/\partial T)_V \qquad (14.2\text{-}11)$$

or

$$(\partial P/\partial T)_V = \beta/\kappa \qquad (14.2\text{-}12)$$

This important equation is considered later.

As mentioned, the van der Waals equation of state relates pressure, volume, and temperature. However, the equation that relates stress with rate of deformation for a viscoelastic fluid is also an equation of state. Hooke's law, for example, is one form of an equation of state that relates tension F to elongation $L - L_0$ at constant temperature:

$$F = C_1(L - L_0) \tag{14.2-13}$$

This equation can be differentiated much as was done for the volume equation:

$$dL = (\partial L/\partial T)_F \, dT + (\partial L/\partial F)_T \, dF \tag{14.2-14}$$

In an analogous fashion, the linear expansivity is obtained:

$$\alpha = L^{-1}(\partial L/\partial T)_F \tag{14.2-15}$$

and the isothermal Young's modulus:

$$Y = (L/A)(\partial F/\partial L)_T \tag{14.2-16}$$

And applying the analogous theorem of Eq. (14.2-10),

$$(\partial F/\partial T)_L = -\alpha A Y \tag{14.2-17}$$

As is anticipated, this type of equation of state has been extensively investigated by elastomer researchers. Surface tension can also be defined as a constitutive equation of state. For most rather simple membranes, the surface tension is only related to the temperature:

$$\sigma = \sigma_0(1 - T/T_0)^n \tag{14.2-18}$$

and therefore the analogous theorem of Eq. (14.2-10) is not needed.

Recently, however, there are indications that surface tension for polymers is a function of the rate of elongation of the surface $\dot{e}$ as well as the temperature. This can be written as

$$\sigma = f(\dot{e}, T) \tag{14.2-19}$$

And thus we obtain an isothermal extensibility ϵ, defined as

$$\epsilon = \sigma_0^{-1}(\partial \sigma/\partial \dot{e})_T \tag{14.2-20}$$

and a thermal thinning coefficient τ, defined as

$$\tau = -\sigma_0^{-1}(\partial \sigma/\partial T)_{\dot{e}} \tag{14.2-21}$$

Note that there are extensive interrelationships between the "gas" equations of state and those for other effects such as surface tension and elongation.

To continue this analogy, consider the internal energy equation in an effort to obtain the heat capacities. The internal energy equation is written as

$$dU = dQ - dW \qquad (14.2\text{-}22)$$

where dQ is the change in heat added to the system and dW is the change in work done by the system. For the "gas" system, work is given as

$$dW = P \ dV \qquad (14.2\text{-}23)$$

For the tension-elongation system, work is

$$dW = -F \ dL \qquad (14.2\text{-}24)$$

and for surface tension.

$$dW = -\sigma \ dA \qquad (14.2\text{-}25)$$

where A is the surface area of the film. Now heat capacities can be defined in terms of differential changes in heat added to the system as given in Table 14.2-1. In general, for most polymeric systems, the heat capacities are highly temperature dependent and slightly pressure dependent. However, note that the pressure drops experienced by materials as they flow from extruder dies or into injection molds are on the order of thousands of atmospheres, and thus a relatively weak dependence on pressure can mean 10 to 20% change in heat capacity during processing due to pressure changes. Kamal and Levan have explicitly pointed this out [3]. As an example of the use of the internal energy equation, consider the tension form for the equation of state:

$$dU = dQ + F \ dL \qquad (14.2\text{-}26)$$

Table 14.2-1

Heat Capacities

"Gas" equation:	$C_v = (\partial Q/\partial T)_V$	(constant volume)
	$C_p = (\partial Q/\partial T)_P$	(constant pressure)
Tension equation:	$C_F = (\partial Q/\partial T)_F$	(constant force)
	$C_L = (\partial Q/\partial T)_L$	(constant extension)
Surface tension equation:	$C_\sigma = (\partial Q/\partial T)_\sigma$	(constant surface tension)
	$C_A = (\partial Q/\partial T)_A$	(constant surface area)

Now, differentiating with respect to temperature at constant length yields

$$(\partial U/\partial T)_L = (\partial Q/\partial T)_L + F(\partial L/\partial T)_L \qquad (14.2\text{-}27)$$

The second term on the right is identically zero. Thus,

$$(\partial U/\partial T)_L = (\partial Q/\partial T)_L = C_L \qquad (14.2\text{-}28)$$

If Eq. (14.2-26) is differentiated with respect to temperature at constant force,

$$(\partial U/\partial T)_F = (\partial Q/\partial T)_F + F(\partial L/\partial T)_F \qquad (14.2\text{-}29)$$

But from Eq. (14.2-15) the second term on the right can be written in terms of the linear expansivity:

$$(\partial L/\partial T)_F = -L\alpha \qquad (14.2\text{-}30)$$

or

$$(\partial U/\partial T)_F = C_F - L\alpha \qquad (14.2\text{-}31)$$

Probably the most use of internal energy is made in the definition of enthalpy. In the most classic definition, enthalpy H is given as

$$H = U + PV \qquad (14.2\text{-}32)$$

By differentiation,

$$dH = dU + P\,dV + V\,dP \qquad (14.2\text{-}33)$$

If a substitution for dU is now made from Eq. (14.2-22) using the gas form for dW,

$$dH = dQ + V\,dP \qquad (14.2\text{-}34)$$

Note that if Eq. (14.2-34) is differentiated with respect to temperature at constant pressure,

$$(\partial H/\partial T)_P = (\partial Q/\partial T)_P + V(\partial P/\partial T)_P \qquad (14.2\text{-}35)$$

The second term on the right is zero, and the first term is the heat capacity at constant pressure from Table 14.2-1:

$$(\partial H/\partial T)_P = (\partial Q/\partial T)_P = C_P \qquad (14.2\text{-}36)$$

Thus, the heat capacity at constant volume is identical to the rate of change of internal energy with respect to temperature at constant volume, and the heat capacity at constant pressure is identical to the rate of change of enthalpy with respect to temperature at constant pressure. Thus, it is possible to evolve enthalpy charts and tables for the various polymers when concepts such as heat capacity and compressibility factors are considered.

Much effort is made in discussing entropy. For this discussion, entropy represents a measure of disorder of the system. Increasing entropy implies overall increasing disorder in the system. It is expected, for example, that by changing water from ice to liquid water to steam, the entropy of the system is increased because the disorder of the system is increased. It can be shown that changes in entropy are related to changes in heat added to the system. For example, using Maxwell's equations,

$$(\partial Q/\partial T)_V = C_V = T(\partial S/\partial T)_V \qquad (14.2\text{-}37)$$

$$(\partial Q/\partial T)_P = C_P = T(\partial S/\partial T)_P \qquad (14.2\text{-}38)$$

Obviously, there are equivalent equations for the other equations of state. At near-equilibrium conditions, the changes in heat in the system can be approximated by the entropy term

$$dQ \doteq T \, dS \qquad (14.2\text{-}39)$$

Combining the internal energy equation and the expressions for entropy yields the following two equations:

$$T \, dS = C_V \, dT + T(\partial P/\partial T)_V \, dV \qquad (14.2\text{-}40)$$

$$T \, dS = C_P \, dT - T(\partial V/\partial T)_P \, dP \qquad (14.2\text{-}41)$$

Note that these equations can be written in terms of the physical properties of the material. For example, Eq. (14.2-41) can be written in terms of the volume expansivity as

$$T \, dS = C_P \, dT - TV\beta \, dP \qquad (14.2\text{-}42)$$

Thus, entropy can be related to measurable quantities such as compressibility, Young's modulus, heat capacities, and so on. Perhaps it is useful to pursue the concept of entropy one step further. Consider an isothermal polymer system. Now $dT = 0$. Now $T \, dS$ can be replaced by dQ as in Eq. (14.2-39) if near equilibrium. Thus,

$$Q = -T \int V\beta \, dP \doteq -T\beta \int V \, dP \qquad (14.2\text{-}43)$$

The second part of Eq. (14.2-43) is an approximation, assuming that the volume expansivity is nearly independent of pressure. For most liquids over rather narrow pressure ranges $V = f(P)$ is a rather weak function. Therefore, for the purposes of illustration, it can be removed from the integral sign and replaced with an average volume:

$$Q \doteq -T\beta V_{av}P \qquad (14.2\text{-}44)$$

This equation represents the heat liberated during isothermal compression of a pure plastic. Another application of Eq. (14.2-42) is the effect of adiabatic changes in pressure temperature. This condition would occur with no heat liberation through the system walls. Thus, $dQ = 0 = T\,dS$. Hence,

$$C_p\,dT = TV\beta\,dP \tag{14.2-45}$$

Or, rearranging,

$$dT = (TV\beta/C_p)\,dP \tag{14.2-46}$$

For relatively small changes in conditions, the rate of change of temperature of the system with change in pressure is directly proportional to the volume expansivity and inversely proportional to the heat capacity. This equation and earlier ones are used to check the equations of state that have been proposed for polymers.

14.3 Equations of State for Polymers

One of the first, and certainly the most widely used, equations of state for polymer systems is the Spencer-Gilmore equation [4]:

$$(P + \pi_i)(V - \omega) = R'T \tag{14.3-1}$$

This equation was patterned after the successful two-constant van der Waals equation and contains three constants: π_i, ω, and R'. Sometimes, the expression for R' is written as $R' = R/M$, where R is the standard gas constant and M is the empirical coefficient. According to a recent survey by Kamal and Levan, many workers have assigned M a value corresponding to the molecular weight of the monomer unit, thus reducing the Spencer-Gilmore equation to two constants.

In a search for suitable equations of state, remember that the empirical equation must be at once as simple as possible (e.g., containing the fewest number of adjustable constants) and as general as possible (thus being accurate over wide ranges of temperature and pressure). Thus, Ku [5] considers the isothermal Murnaghan equation of state,

$$P = (B_0/S)[(V_0/V)^S - 1] \tag{14.3-2}$$

where B_0 is the initial bulk modulus at V_0 and S is a constant, and the isothermal Tait equation as applied to polymers,

$$V/V_0 = 1 - A\,\ln[(B + P)/(B + 1)] \tag{14.3-3}$$

where A and B are constants for the system. As shown in Tables 14.3-1 and 14.3-2, values for B_0 and S for the Murnaghan equation and A and B for the

Table 14.3-1

Parameters in the Murnaghan Equation of State for High Polymers [5]

Polymer	B_0 (atm)	S	v_0	T (°K)	Static pressure range (atm)
Low-density polyethylene	30063.2	9.4902	1.0895	293.2	10,000
	25138.4	8.4305	1.1269	343.2	10,000
High-density polyethylene	40148.4[a]	16.9992	1.028	292.6	2,000[a]
	35034.6[a]	12.2738	1.042	332.8	2,000[a]
Polymethyl methacrylate	39712.3[a]	9.0004	0.842	293.2	2,000[a]
	28191.3[a]	10.9414	0.862	373.2	2,000[a]
Polystyrene	36388.4[a]	8.5179	0.955	293.2	2,000[a]
	14073.2[a]	15.7109	0.994	399.?	? 000[a]
Polychloro trifluoroethylene	....18.0	10.4331	0.4662	293.2	10,000
	33993.5	9.8502	0.4745	343.2	10,000
Polytetrafluoro- ethylene	29267.0	14.9218	0.4484	293.2	1 to 5,000
	21040.0	12.8603	0.4484	293.2	6,000 to 10,000
Polyvinyl fluoride	41687.7	7.8864	0.7956	293.2	10,000
	31102.5	9.3506	0.8135	343.2	10,000
Polyvinylidene fluoride	34757.8	9.5880	0.6439	293.2	10,000
	32208.2	9.3077	0.6582	343.2	10,000
Polyvinyl alcohol	58233.0	10.7629	0.7720	293.2	10,000
	46258.8	11.0647	0.7861	353.2	10,000
Nylon 66	56688.2[a]	7.7507	0.858[b]	300.0	40,000[a]
Nylon 610	55470.7[a]	7.8852	0.917	300.0	40,000[a]

[a]The unit is in kg/cm^2.
[b]Based on average density of amorphous and crystalline samples.

modified Tait equation have been determined at given temperatures for several polymers. Rehage and Breuer [6] have subsequently modified the Tait equation to include temperature in the following form:

$$V = V_0 + \phi_0 T - (k_0/a)(1 + bT) \ln(1 + aP) \qquad (14.3\text{-}3a)$$

where a and b are the constants for the system and V_0, ϕ_0, and k_0 are the volume, slope of the V-T curve, and the derivative $(\partial V/\partial P)_T$, respectively, all evaluated at 0°C and zero pressure.

Table 14.3-2

Parameters in the Tait Equation of State for High Polymers [5]

Polymer	A	B (atm)	Temperature (°K)
Low-density polyethylene	0.086812	2510.1	293.2
	0.097213	2383.0	343.2
High-density polyethylene	0.055221	2310.9[a]	292.6
	0.054046	1809.7	332.8
Polymethyl methacrylate	0.098701	3922.7[a]	293.2
	0.080562	2253.7	373.2
Polystyrene	0.090000	3200.0[a]	293.2
Polychlorotrifluoroethylene	0.086480	3430.0	293.2
	0.094714	2777.4	353.2
Polytetrafluoroethylene	0.05587	1523.5	293.2 (1 to 3,000 atm)
	0.06367	1216.8	293.2 (6,000 to 10,000 atm)
Polyvinyl fluoride	0.10650	4400.0	293.2
	0.088401	2656.8	353.2
Polyvinylidene fluoride	0.088353	3019.4	293.2
	0.090014	2834.2	343.2
Polyvinyl alcohol	0.081121	4676.8	293.2
	0.077595	3513.8	353.2

[a]The unit of B is in kg/cm^2.

Griskey and Whitaker have used the reduced form for an equation of state, following traditional chemical engineering principles [7]. They found from correlation of large amounts of experimental data that

$$V = 0.01205RP^{n-1}(T/T_g)^{m+1}/\rho_0^{0.9421} \qquad (14.3\text{-}4)$$

where n and m are universal parameters that depend on pressure, R is the gas constant, T_g is the glass-transition temperature for the polymer, and ρ_0 is the density of the material at 25°C and 1 atm.

Kamal and Levan, in their analysis of the various equations of state, developed an inverse-volume equation of state, which appears to be quite similar to the theoretical virial equation of state:

$$1/V = \rho = (1/V_\infty + aP + bP^2) + T[(\partial\rho/\partial T)_{P=0} + cP + dP^2/2] \quad (14.3\text{-}5)$$

Here a, b, c, and d are adjustable parameters, V_∞ is the specific volume at $0°K$ and 0-atm pressure, and $(\partial\rho/\partial T)_{P=0}$ is the density derivative with temperature at zero pressure.

As can be seen, the proposed equations of state range in number of adjustable parameters from two for the Spencer-Gilmore equation (with M a known quantity) and the Griskey equation to four for the inverse-volume equation of Kamal and Levan assuming that V_∞ and $(\partial\rho/\partial T)_{P=0}$ can be readily found. Kamal and Levan have tested several of these equations by evaluating their predicted results for volume expansivity and isothermal compressibility. As can be seen from Table 14.3-3, the two forms of the Spencer-Gilmore equation with and without floating values of M (SGC and SGF), the Rehage-Breuer-Tait equation (RBT), the inverse-volume equation (IV), and the Whitaker-Griskey equation (WG) have been differentiated using the above definitions for volume expansivity, Eq. (14.2-6), and isothermal compressibility, Eq. (14.2-7). The variation in values as compared with experimental values for four of the five equations is shown i ~~ ~~ ~~ 14.3-4 for five homopolymers at temperat~~ ~~ ~~ 200 C. According to Kamal and Leva~~ ~~ ~~itaker-Griskey equation gives standard deviations that are approximately one order of magnitude higher than the values shown in Table 14.3-4. An example of the variation of compressibility with pressure is shown for polystyrene at 145.2°C in Fig. 14.3-1 and of the volume expansivity with temperature for polystyrene in Fig. 14.3-2. For four of the

Table 14.3-3

Expressions for β and κ for Different Equations [3]

	β	κ
SGC	$\dfrac{1}{V}\dfrac{R}{M(P+\pi)}$	$\dfrac{1}{V}\dfrac{RT}{M(P+\pi)^2}$
SGF	$\dfrac{1}{V}\dfrac{R'}{P+\pi}$	$\dfrac{1}{V}\dfrac{R'T}{(P+\pi)^2}$
RBT	$\dfrac{1}{V}\left[\phi_0-\dfrac{k_0 b}{a}\ln(1+aP)\right]$	$\dfrac{1}{V}\left[\dfrac{k_0(1+bT)}{1+aP}\right]$
IV	$-V\left[\left(\dfrac{\partial P}{\partial T}\right)_{P=0}+cP+\dfrac{dP^2}{2}\right]$	$V[a+2bP+T(C+dP)]$
WG	$\dfrac{m+1}{T}$	$-\left[\ln\dfrac{T}{T_g}\cdot\dfrac{dm}{dp}+\dfrac{n-1}{P}+\ln P\cdot\dfrac{dn}{dp}\right]$

Table 14.3-4

Average Absolute Percent Deviation for Thermal Expansion Coefficient (β) and Compressibility Coefficient (κ) [3]

Model ID	Polyethylene, medium density		Polyethylene, low pressure		Polyethylene, high pressure		Polyisobutylene		Polystyrene	
	β	κ	β	κ	β	κ	β	κ	β	κ
SGC	8.49	17.5	87.1	7.48	19.8	5.91	72.3	37.8	26.6	11.0
SGF	7.61	16.8	33.1	6.4	10.9	6.12	45.9	17.2	8.5	6.24
RB	6.0	7.5	17.9	8.2	10.75	9.3	39.0	18.0	8.3	5.4
IV	9.15	2.78	19.0	5.8	8.8	5.4	29.0	12.1	5.6	3.03

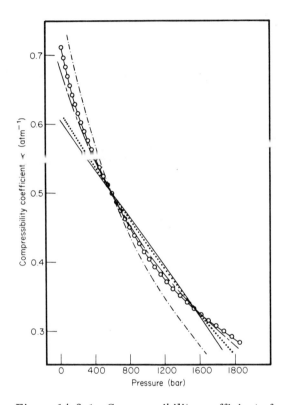

Figure 14.3-1 Compressibility coefficients for GP styrene at 145.2°C.
Solid line, experimental; solid line with circles, Rehage-Breuer-Tait equation; dash-dot line, two-constant Spencer-Gilmore equation; dash-dash line, three-constant Spencer-Gilmore equation; dotted line, Kamal-Levan four-constant equation [3].

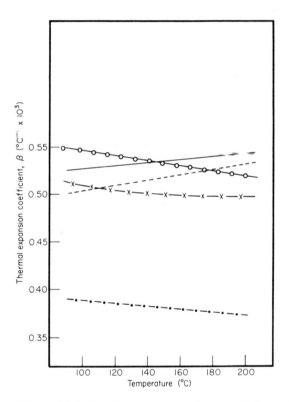

Figure 14.3-2   Thermal expansion coefficients for GP styrene. Solid line, experimental; solid line with circles, Rehage-Breuer-Tait equation; dash-dot line, two-constant Spencer-Gilmore equation; solid line with X's, three-constant Spencer-Gilmore equation; dotted line, Kamal-Levan four-constant equation [3].

five polymer systems examined, the accuracy of the equations is quite good.
The notable exception is butadiene. Examination of the dependence of β, the
volume expansivity, with temperature, $\partial\beta/\partial T$, in Fig. 14.3-2 indicates
that the standard equations, such as the Spencer-Gilmore equation, predict
a decrease in volume expansivity with increasing temperature, whereas the
experimental data are indicating an increase. The multiple-constant inverse-
volume equation can take this fact into account in the evaluation of its coefficients and thus is the only empirical equation that shows the proper sign
for $\partial\beta/\partial T$. Nevertheless, Kamal and Levan concede that although the four-
or six-constant inverse-volume equation is most accurate, the Spencer-Gilmore equation with M as an adjustable parameter (e.g., R') yields useful
results and is certainly the equation most similar to the gas form of the
equation of state. The Whitaker-Griskey equation, which takes the fewest
parameter evaluations, yields poor volume expansivity and isothermal compressibility predictions. For the analysis of the thermodynamic properties
of polymerically associated bulk follow, then, we shall use the following Spencer-
Gilmore definitions for the volume expansivity and isothermal compressibility:

$$\beta = V^{-1}(\partial V/\partial T)_P = R'/V(P + \pi) \tag{14.3-6}$$

$$\kappa = (-V^{-1})(\partial V/\partial P)_T = R'T/V(P + \pi)^2 \tag{14.3-7}$$

In Table 14.3-5 are given some values for this form of the Spencer-Gilmore
equation of state. As will be seen, other values can be obtained from experimental P-V-T data.

The ultimate goal of this analysis, however, is not to determine β
and κ but to determine, for example, the amount of energy required to heat
a polymer from one temperature to another at constant pressure or to
determine the amount of heating that takes place as a polymer is dropped
in pressure from, say, the conditions in an injection molding accumulator
to those in the mold cavity. To obtain this information, enthalpic and
entropic tables and/or graphs should be constructed. Fortunately, these
have been done by Griskey and his coworkers for many of the common polymers. The tables are given as Tables 14.3-6 through 14.3-13 [9-20].

(Text continues on p. 736)

Table 14.3-5

Some Values for Constants in the Spencer-Gilmore Equation [8]

Material	R'	ω (ml/g)	π (psi)
PS	11.6	0.822	27,000
PMMA	12.05	0.734	31,300
LDPE	43.0	0.875	47,600

Table 14.3-6

Thermodynamic Properties of Polyethylene—14.7 to 28,000 psia [9]

Pressure (psia)		Temperature – °F) 80	120	140	160	211	248	30	322	342	360	400	440
14.7	H[a]	22.40	42.00	52.15	62.30	92.91	118.98	166.?6	181.56	200.16	216.90	254.10	278.40
	S[b]	0.0436	0.0789	0.0962	0.1136	0.1606	0.1993	0.2?37	0.2899	0.3144	0.3365	0.3855	0.4120
1000	H	25.23	44.80	54.73	65.01	94.55	121.30	168.?7	184.27	202.87	218.76	256.26	281.11
	S	0.0428	0.0780	0.0955	0.1126	0.1584	0.1940	0.2?74	0.2885	0.3129	0.3341	0.3834	0.4105
2000	H	28.10	47.58	57.31	67.71	96.19	123.61	171.?8	186.98	205.58	220.63	258.62	283.93
	S	0.0421	0.0772	0.0948	0.1116	0.1562	0.1887	0.2?51	0.2871	0.3115	0.3317	0.3815	0.4092
3000	H	30.97	50.34	59.89	70.38	97.99	125.92	174.?5	189.76	208.36	222.88	261.11	286.83
	S	0.0414	0.0763	0.0941	0.1106	0.1542	0.1843	0.2?50	0.2859	0.3103	0.3298	0.3799	0.4080
4000	H	33.82	53.09	62.47	73.05	99.79	128.22	177.?5	192.55	211.15	225.14	263.74	289.90
	S	0.0407	0.0754	0.0935	0.1095	0.1522	0.1800	0.2?39	0.2847	0.3091	0.3279	0.3784	0.4070
5000	H	36.66	55.83	65.05	75.73	101.70	130.53	180.?2	195.42	214.02	227.64	266.37	293.05
	S	0.0399	0.0745	0.0929	0.1085	0.1504	0.1764	0.2?75	0.2837	0.3080	0.3264	0.3770	0.4062
6000	H	39.53	58.55	67.63	78.40	103.63	132.88	182.?9	198.29	216.89	230.14	269.01	296.25
	S	0.0391	0.0735	0.0923	0.1075	0.1486	0.1727	0.2?10	0.2827	0.3070	0.3250	0.3756	0.4054
7000	H	42.23	61.25	70.23	81.07	105.63	135.25	185.?4	201.24	219.84	232.84	271.66	299.50
	S	0.0381	0.0726	0.0917	0.1064	0.1469	0.1697	0.2?97	0.2818	0.3061	0.3238	0.3742	0.4047
8000	H	44.88	63.94	72.83	83.75	107.64	137.64	188.?0	204.20	222.80	235.54	274.32	302.76
	S	0.0371	0.0716	0.0911	0.1054	0.1453	0.1667	0.2?84	0.2810	0.3052	0.3237	0.3728	0.4041
9000	H	47.49	66.63	75.45	86.45	109.71	140.04	191.91	207.21	225.81	238.37	276.99	306.01
	S	0.0359	0.0706	0.0905	0.1045	0.1438	0.1641	0.2?62	0.2797	0.3044	0.3218	0.3711	0.4035

714

10,000	H^a	78.07	111.79	142.42	19_.93	210.23	228.83	241.20
	S^b	0.0900	0.1424	0.1616	0.?040	0.2783	0.3036	0.3209
12,000	H	83.85	116.11	145.33	19?98	213.28	231.88	247.08
	S	0.0890	0.1398	0.1571	0.1?86	0.2710	0.3029	0.3195
14,000	H	88.67	120.51	148.59	201.04	216.34	234.94	253.08
	S	0.0880	0.1374	0.1530	0.1?35	0.2636	0.3022	0.3184
16,000	H	93.99	124.95	152.13				259.26
	S	0.0871	0.1352	0.1493				0.3175
18,000	H	99.35	129.44	155.87				265.57
	S	0.0862	0.1331	0.1459				0.3168
20,000	H	104.75	134.00	159.76				271.97
	S	0.0853	0.1311	0.1427				0.3163
22,000	H	110.19	138.64	163.78				
	S	0.0845	0.1292	0.1398				
24,000	H	115.63	143.35	167.88				
	S	0.0837	0.1274	0.1371				
26,000	H	121.11	148.13	172.06				
	S	0.0829	0.1257	0.1345				
28,000	H	126.63	152.96	176.38				
	S	0.0822	0.1241	0.1321				

[a]H = enthalpy in Btu/lb.
[b]S = entropy in Btu/lb.
°R datum: enthalpy and entropy as zero at 14.7 psia and 32°F.

Table 14.3-7

Thermodynamic Properties of Polypropylene [10]

Pressure (psia)		80	120	160	200	240	280	320	400	440
14.7	H[a]	22.40	42.00	62.30	85.60	112.20	146.10	179.70	254.10	278.40
	S[b]	0.0436	0.0789	0.1136	0.1499	0.1890	0.2405	0.2875	0.3855	0.4120
1000	H	25.23	44.80	65.01	88.24	114.52	147.41	179.91	256.26	281.11
	S	0.0428	0.0780	0.1126	0.1488	0.1874	0.2389	0.2831	0.3834	0.4105
2000	H	28.10	47.58	67.71	90.86	116.82	149.76	180.56	258.62	283.93
	S	0.0421	0.0772	0.1116	0.1471	0.1858	0.2374	0.2793	0.3815	0.4092
3000	H	30.97	50.34	70.38	93.47	119.12	153.13	181.45	261.11	286.83
	S	0.0414	0.0763	0.1106	0.1465	0.1842	0.2359	0.2759	0.3799	0.4080
4000	H	33.82	53.09	73.05	96.08	121.38	155.58	182.70	263.74	289.90
	S	0.0407	0.0754	0.1095	0.1459	0.1926	0.2345	0.2729	0.3784	0.4070
5000	H	36.66	55.83	75.73	98.72	123.65	158.05	184.01	266.37	293.05
	S	0.0399	0.0745	0.1085	0.1443	0.1810	0.2332	0.2701	0.3770	0.4062
6000	H	39.53	58.55	78.40	101.36	125.96	160.56	185.36	269.01	296.25
	S	0.0391	0.0735	0.1075	0.1432	0.1794	0.2319	0.2672	0.3756	0.4054
7000	H	42.23	61.25	81.07	104.01	128.28	163.11	186.73	271.66	299.50
	S	0.0381	0.0726	0.1064	0.1422	0.1779	0.2306	0.2645	0.3742	0.4047
8000	H	44.88	63.94	83.75	106.67	130.62	165.70	188.16	274.32	302.76
	S	0.0371	0.0716	0.1054	0.1411	0.1764	0.2295	0.2618	0.3728	0.4041
9000	H	47.49	66.63	86.45	109.34	132.97	168.32	189.65	276.99	306.01
	S	0.0359	0.0706	0.1045	0.1401	0.1750	0.2284	0.2592	0.3711	0.4035

Temperature (°F)

[a] H = enthalpy in Btu/lb.
[b] S = entropy in Btu/lb.
°R datum: enthalpy and entropy as zero at 14.7 psia and 32°F.

Table 14.3-8

Thermodynamic Properties of Ethylene–Propylene Copolymer [12]

Pressure (psia)		Temperature (°F)								
		140	180	220	300	340	380	420	460
14.7	H^a	75.00	98.00	130.00	200.00	226.00	248.00	273.30	324.00
	S^b	0.1379	0.1738	0.2223	0.3125	0.3568	0.3795	0.4190	0.4880
1000	H	77.36	100.11	131.88	202.87	228.90	249.97	276.05	326.71
	S	0.1364	0.1719	0.2200	0.3113	0.3553	0.3772	0.4176	0.4866
2000	H	79.71	102.24	133.81	205.75	231.90	252.07	278.87	329.51
	S	0.1348	0.1700	0.2178	0.3101	0.3543	0.3751	0.4164	0.4853
3000	H	82.09	104.41	135.75	208.60	235.05	254.23	281.71	332.34
	S	0.1334	0.1682	0.2157	0.3089	0.5535	0.3731	0.4152	0.4841
4000	H	84.47	106.58	137.63	211.46	238.33	256.45	284.98	335.27
	S	0.1319	0.1664	0.2134	0.3077	0.3529	0.3712	0.4140	0.4831
5000	H	86.85	108.72	139.52	214.31	241.58	258.70	287.29	338.25
	S	0.1305	0.1646	0.2113	0.3065	0.3522	0.3694	0.4128	0.4821
6000	H	89.24	110.86	141.44	217.08	244.83	260.96	290.12	341.29
	S	0.1291	0.1628	0.2091	0.3053	0.3516	0.3676	0.4117	0.4812
7000	H	91.63	113.11	143.37	219.74	248.16	263.27	292.90	344.42
	S	0.1277	0.1612	0.2071	0.3040	0.3511	0.3659	0.4106	0.4804
8000	H	94.01	115.42	145.32	222.37	251.52	265.62	295.64	347.58
	S	0.1263	0.1597	0.2050	0.3026	0.3506	0.3642	0.4094	0.4797
9000	H	96.37	117.70	147.30	225.02	254.84	267.98	298.38	350.81
	S	0.1249	0.1581	0.2030	0.3012	0.3502	0.3626	0.4082	0.4790

[a] H = enthalpy in Btu/lb.
[b] S = entropy in Btu/lb.
° R datum: enthalpy and entropy as zero at 14.7 psia and 32°F.

Table 14.3-9

Thermodynamic Properties of Polymethyl Methacrylate [13]

Pressure (psia)		Temperature (°F)						
		70	140	212	230.2	248	266	282
14.7	H^a	12.73	51.84	91.80	107.08	123.12	139.23	155.00
	S^b	0.0249	0.0953	0.1590	0.1828	0.2075	0.2315	0.2547
500	H	13.87	52.95	92.59	108.06	124.09	140.31	156.00
	S	0.0248	0.0951	0.1583	0.1823	0.2071	0.2313	0.2544
1,000	H	15.05	54.09	93.40	109.14	125.10	141.42	157.02
	S	0.0246	0.0949	0.1576	0.1822	0.2067	0.2310	0.2540
1,500	H	16.22	55.23	94.22	110.24	126.12	142.52	158.10
	S	0.0245	0.0947	0.1570	0.1819	0.2063	0.2307	0.2537
2,000	H	17.40	56.37	95.05	111.34	127.14	143.62	159.18
	S	0.0244	0.0945	0.1563	0.1817	0.2060	0.2305	0.2534
3,000	H	19.74	58.66	96.71	113.59	129.22	145.84	161.35
	S	0.0241	0.0941	0.1550	0.1812	0.2052	0.2300	0.2529
4,000	H	22.08	60.94	98.36	115.89	131.31	148.03	163.51
	S	0.0238	0.0937	0.1536	0.1809	0.2046	0.2294	0.2523
5,000	H	24.41	63.22	100.01	118.25	133.44	150.22	165.67
	S	0.0235	0.0933	0.1523	0.1806	0.2040	0.2289	0.2518
6,000	H	26.75	65.51	101.66	120.66	135.61	152.40	167.84
	S	0.0232	0.0930	0.1510	0.1804	0.2035	0.2284	0.2513
7,000	H	29.08	67.80	103.24	123.07	137.80	154.60	170.11
	S	0.0230	0.0926	0.1496	0.1803	0.2030	0.2279	0.2509
8,000	H	31.40	70.09	104.82	125.48	140.04	156.83	172.37
	S	0.0227	0.0923	0.1482	0.1801	0.2026	0.2275	0.2505

9,000	H	33.72	72.39	106.46	127.?0	142.33	159.11	174.63
	S	0.0224	0.0920	0.1496	0.1? 0	0.2023	0.2272	0.2502
10,000	H	36.04	74.69	108.13	130.?4	144.67	161.43	176.88
	S	0.0222	0.0917	0.1457	0.17?	0.2020	0.2269	0.2498
11,000	H	38.36	76.98	109.86	132.?	147.03	163.78	179.12
	S	0.0219	0.0914	0.1445	0.17?	0.2018	0.2267	0.2494
12,000	H	40.67	79.29	111.65	135.?	149.42	166.13	181.35
	S	0.0216	0.0911	0.1435	0.179?	0.2017	0.2265	0.2491
13,000	H	42.98	81.59	113.46	137.8?	151.80	168.49	183.58
	S	0.0213	0.0908	0.1424	0.179?	0.2015	0.2263	0.2487
14,000	H	45.28	83.90	115.28	140.2?	154.18	170.84	185.81
	S	0.0210	0.0906	0.1415	0.1798	0.2014	0.2261	0.2484
15,000	H	47.58	86.20	117.12	142.77	156.57	173.21	188.04
	S	0.0208	0.0903	0.1405	0.1798	0.2012	0.2260	0.2480
16,000	H	49.87	88.51	118.97	145.25	158.95	175.57	190.27
	S	0.0205	0.0900	0.1396	0.1798	0.2011	0.2257	0.2477
17,000	H	52.21	90.82	120.83	147.72	161.35	177.95	192.51
	S	0.0203	0.0898	0.1386	0.1798	0.2010	0.2256	0.2473
18,000	H	54.55	93.12	122.69	150.21	163.75	180.32	194.74
	S	0.0201	0.0895	0.1377	0.1798	0.2009	0.2255	0.2470
19,000	H	56.89	95.43	124.55	152.73	166.18	182.68	197.06
	S	0.0199	0.0893	0.1368	0.1798	0.2008	0.2253	0.2468
20,000	H	59.22	97.73	126.41	155.25	168.64	185.05	199.51
	S	0.0197	0.0891	0.1359	0.1798	0.2008	0.2252	0.2468
21,000	H	61.55	100.03	128.28	157.77	171.09	187.43	201.97
	S	0.0196	0.0889	0.1351		0.2008	0.2251	0.2468
22,000	H	63.87	102.34	130.18	160.29	173.55	189.81	204.43
	S	0.0194	0.0886	0.1342		0.2008	0.2250	0.2468

Table 14.3-9 (continued)

Pressure (psia)		70	140	212	230.2	248	266	282
23,000	H	66.20	104.64	132.11	162.7?	176.00	192.20	206.88
	S	0.0192	0.0884	0.1334		0.2008	0.2249	0.2468
24,000	H	68.51	106.95	134.00	165.3?	178.44	194.58	209.34
	S	0.0190	0.0882	0.1326		0.2008	0.2248	0.2468
25,000	H	70.83	109.24	135.82	167.?5	180.88	196.95	211.79
	S	0.0188	0.0880	0.1317		0.2008	0.2247	0.2468
26,000	H	73.18	111.53	137.61	170.?3	183.32	199.31	214.36
	S	0.0187	0.0878	0.1307		0.2008	0.2246	
27,000	H	75.54	113.82	139.40	173.01	185.75	201.66	216.94
	S	0.0186	0.0876	0.1298		0.2008	0.2245	
28,000	H	77.90	116.11	141.21	17?.59	188.18	204.02	219.43
	S	0.0185	0.0874	0.1288		0.2008	0.2244	

Temperature (°F)

[a] H = enthalpy in Btu/lb$_m$.
[b] S = entropy in Btu/lb$_m$·°R.

Table 14.3-10

Thermodynamic Properties of Polyvinyl Chloride [11]

Pressure (psia)		Temperature (°F)				
		70	123.8	179.	195	206
14.7	H[a]	9.31	23.12	56.85	69.28	74.47
	S[b]	0.0182	0.0431	0.101	0.1216	0.1296
500	H	10.28	24.08	57.51	70.17	75.33
	S	0.0181	0.0430	0.104	0.1214	0.1293
1,000	H	11.29	25.07	58.25	71.07	76.21
	S	0.0180	0.0428	0.099	0.1211	0.1290
1,500	H	12.26	26.06	59.0	71.96	77.08
	S	0.0178	0.0427	0.0924	0.1028	0.1287
2,000	H	13.25	27.04	59.77	72.84	77.96
	S	0.0177	0.0425	0.098	0.1205	0.1284
3,000	H	15.27	29.03	61.29	74.65	79.74
	S	0.0175	0.0423	0.097	0.1199	0.1278
4,000	H	17.30	31.01	62.80	76.52	81.68
	S	0.0173	0.0420	0.096	0.1195	0.1274
5,000	H	19.32	32.98	64.25	78.44	83.63
	S	0.0171	0.0417	0.0958	0.1191	0.1271
6,000	H	21.34	34.94	65.65	80.42	85.58
	S	0.0169	0.0414	0.0945	0.1188	0.1268

Table 14.3-10　(continued)

Pressure (psia)		Temperature (°F)				
		70	123.8	179.5	195	206
7,000	H	23.35	36.91	67.15	82.38	87.52
	S	0.0167	0.0412	0.0936	0.1186	0.1265
8,000	H	25.37	38.90	68.75	84.31	89.45
	S	0.0166	0.0410	0.0928	0.1182	0.1262
9,000	H	27.44	40.90	70.45	86.23	91.37
	S	0.0165	0.0408	0.0921	0.1179	0.1259
10,000	H	29.44	42.91	72.53	88.17	93.29
	S	0.0163	0.0406	0.0916	0.1176	0.1256
11,000	H	31.45	44.93	74.05	90.12	95.20
	S	0.0161	0.0404	0.0911	0.1174	0.1252
12,000	H	33.45	46.95	75.02	92.11	97.11
	S	0.0159	0.0403	0.0907	0.1172	0.1249
13,000	H	35.45	48.97	77.00	94.10	99.01
	S	0.0158	0.0401	0.0904	0.1170	0.1246
14,000	H	37.44	51.00	79.1	96.08	100.91
	S	0.0156	0.0400	0.0901	0.1168	0.1243

15,000	H	39.43	53.02	■.63	98.05	102.81
	S	0.0154	0.0399	0.0898	0.1166	0.1240
16,000	H	41.42	55.04	8■.56	100.01	104.70
	S	0.0152	0.0397	0.0895	0.1164	0.1237
17,000	H	43.41	57.05	8 .47	101.99	106.60
	S	0.0150	0.0396	0.0892	0.1162	0.1234
18,000	H	45.30	59.06	8 .36	103.95	108.50
	S	0.0148	0.0395	0.0889	0.1160	0.1231
19,000	H	47.37	61.08	8■.23	105.91	110.43
	S	0.0147	0.0394	0.0886	0.1158	0.1228
20,000	H	49.35	63.10	91 12	107.87	112.52
	S	0.0145	0.0393	0.882	0.1156	0.1228
21,000	H	51.33	65.12	93 02	109.83	114.61
	S	0.0143	0.0392	0.■80	0.1154	0.1228
22,000	H	53.30	67.14	94 ■4	111.82	116.70
	S	0.0141	0.0391	0.■77	0.1152	0.1228
23,000	H	55.27	69.15	96.■9	113.82	118.79
	S	0.0139	0.0390	0.0■75	0.1151	0.1228
24,000	H	57.24	71.16	98. 4	115.85	120.86
	S	0.0137	0.0389	0.0■73	0.1150	0.1228

Table 14.3-10 (continued)

Pressure (psia)		Temperature (°F)					
		70	123.8	170.6	195	206
25,000	H	59.20	73.17	107.79	117.89	122.93
	S	0.0136	0.0388	0.0871	0.1150	0.1228
26,000	H	61.17	75.19	109.72	119.91	125.00
	S	0.0134	0.0387	0.0869	0.1149	0.1228
27,000	H	63.13	77.21	111.66	121.90	127.07
	S	0.0132	0.0386	0.0867	0.1148	0.1228
28,000	H	65.09	79.23	113.59	123.87	129.14
	S	0.0130	0.0385	0.0865	0.1146	0.1228

[a]H = enthalpy in Btu/lb_m.
[b]S = entropy in Btu/$lb_m \cdot °R$.

Table 14.3-11

Thermodynamic Properties of Polystyrene [15]

Pressure (psia)		70	141	205	278	293	324	354	397
14.7	H^a	10.77	35.01	76.28	115.89	1?5.71	147.24	171.28	202.60
	S^b	0.0211	0.0643	0.1329	0.1910	0.?049	0.2349	0.2679	0.3080
500	H	11.96	36.26	76.35	117.01	1?.88	148.34	172.38	203.72
	S	0.0207	0.0641	0.1308	0.1906	0.?046	0.2345	0.2674	0.3076
1,000	H	13.19	37.55	76.48	118.16	12?.07	149.48	173.52	204.89
	S	0.0204	0.0638	0.1289	0.1901	0.?042	0.2340	0.2670	0.3072
1,500	H	14.55	38.84	76.67	119.31	12? 27	150.62	174.66	206.07
	S	0.0203	0.0636	0.1270	0.1897	0.?38	0.2336	0.2665	0.3068
2,000	H	15.76	40.14	76.91	120.47	130.47	151.77	175.81	207.25
	S	0.0199	0.0634	0.1252	0.1893	0.2?34	0.2331	0.2661	0.3064
3,000	H	18.20	42.73	77.57	122.79	132.?1	154.08	178.13	209.61
	S	0.0192	0.0630	0.1218	0.1884	0.2?28	0.2323	0.2653	0.3056
4,000	H	20.74	45.34	78.41	125.14	135.?8	156.39	180.43	212.00
	S	0.0187	0.0626	0.1188	0.1877	0.20?2	0.2315	0.2644	0.3049
5,000	H	23.28	47.95	79.44	127.48	137.?	158.72	182.72	214.41
	S	0.0182	0.0622	0.1161	0.1870	0.20?5	0.2307	0.2636	0.3042
6,000	H	25.74	50.58	80.66	129.83	140.??	161.06	185.04	216.83
	S	0.0175	0.0619	0.1136	0.1861	0.20?	0.2300	0.2628	0.3036

Table 14.3-11 (continued)

Pressure (psia)		70	141	205	278	293	324	354	397
					Temperature (°F)				
7,000	H	28.06	53.21	82.05	132.18	147.85	163.42	187.37	219.24
	S	0.0166	0.0616	0.1114	0.1854	0.2005	0.2292	0.2621	0.3029
8,000	H	30.29	55.84	83.53	134.54	145.35	165.80	189.72	221.62
	S	0.0156	0.0612	0.1094	0.1847	0.2000	0.2286	0.2614	0.3022
9,000	H	32.70	58.47	85.06	136.90	147.86	168.17	192.07	223.98
	S	0.0148	0.0610	0.1074	0.1840	0.1995	0.2279	0.2607	0.3016
10,000	H	35.22	61.10	86.64	139.27	150.38	170.55	194.42	226.33
	S	0.0143	0.0607	0.1056	0.1834	0.1991	0.2273	0.2600	0.3009
11,00	H	37.79	63.73	88.29	141.63	152.90	172.94	196.79	228.71
	S	0.0139	0.0604	0.1038	0.1827	0.1986	0.2267	0.2593	0.3003
12,000	H	40.26	66.37	90.03	144.01	155.42	175.33	199.15	231.18
	S	0.0134	0.0601	0.1022	0.1821	0.1982	0.2261	0.2587	0.2997
13,000	H	42.75	69.01	91.82	146.39	157.94	177.72	201.52	233.70
	S	0.0128	0.0599	0.1007	0.1815	0.1978	0.2255	0.2581	0.2993
14,000	H	45.24	71.64	93.64	148.78	160.47	180.12	203.89	236.20
	S	0.0123	0.0596	0.0992	0.1809	0.1974	0.2249	0.2575	0.2989

15,000	H	47.84	74.27	95.47	151.18	163.00	182.52	206.27	238.65
	S	0.0120	0.0594	0.0977	0.1803	.1970	0.2243	0.2569	0.2984
16,000	H	50.45	76.90	97.29	153.60	55.56	184.93	208.67	241.07
	S	0.0117	0.0591	0.0963	0.1798	.1966	0.2238	0.2564	0.2978
17,000	H	53.02	79.53	99.10	156.04	58.13	187.34	211.07	243.50
	S	0.0114	0.0589	0.0948	0.1793	1963	0.2233	0.2558	0.2973
18,000	H	55.52	82.17	100.98	158.49	70.71	189.76	213.46	246.00
	S	0.0109	0.0586	0.0935	0.1788	1960	0.2228	0.2553	0.2966
19,000	H	58.08	84.80	102.87	160.97	73.29	192.17	215.85	248.52
	S	0.0106	0.0584	0.0921	0.1784	958	0.2223	0.2548	0.2964
20,000	H	60.55	87.42	104.73	163.49	75.86	194.58	218.24	251.02
	S	0.0100	0.0582	0.0908	0.1781	955	0.2218	0.2542	0.2962
21,000	H	62.97	90.05	106.56	166.06	78.42	197.00	220.65	253.49
	S	0.0094	0.0580	0.0894	0.1778	952	0.2213	0.2537	0.2958
22,000	H	65.38	92.67	108.38	168.66	80.98	199.43	223.06	255.94
	S	0.0088	0.0579	0.0879	0.1776	949	0.2209	0.2533	0.2953
23,000	H	67.81	95.30	110.23	171.28	83.53	201.86	225.46	258.43
	S	0.0083	0.0576	0.0866	0.1774	46	0.2205	0.2528	0.2950
24,000	H	70.37	97.92	112.16	173.91	86.07	204.30	227.87	260.95
	S	0.0079	0.0574	0.0853	0.1772	44	0.2200	0.2523	0.2946

Table 14.3-11 (continued)

Pressure (psia)		70	141	205	278	293	324	354	397
					Temperature (°F)				
25,000	H	72.94	100.53	114.15	176.58	183.63	206.73	230.27	263.47
	S	0.0077	0.0572	0.0842	0.1771	0.1941	0.2196	0.2519	0.2943
26,000	H	75.48	103.15	116.14	179.28	181.23	209.17	232.68	265.98
	S	0.0073	0.0570	0.0830	0.1771	0.1939	0.2192	0.2514	0.2940
27,000	H	77.87	105.76	118.12	182.02	183.88	211.60	235.09	268.48
	S	0.0067	0.0568	0.0819	0.1771	0.1938	0.2188	0.2510	0.2937
28,000	H	80.23	108.37	120.05	184.79	186.58	214.04	237.52	270.97
	S	0.0060	0.0566	0.0807	0.1771	0.1938	0.2184	0.2506	0.2933

[a]H = enthalpy in Btu/lb$_m$.
[b]S = entropy in Btu/lb$_m$·°R.

Table 14.3-12

Thermodynamic Properties of Nylon 610 [16]

Pressure (psia)		Temperature (°F)									
		77	151	199	250	300	351	390	399	410	421
14.7	H[a]	17.10	59.40	86.71	122.19	152.34	187.97	241.07	305.51	419.58	563.76
	S[b]	0.0333	0.1081	0.1517	0.2056	0.2471	0.2946	0.3681	0.4639	0.6327	0.8444
500	H	18.27	60.65	88.00	123.49	153.65	189.29	242.31	306.69	420.60	564.14
	S	0.0330	0.1080	0.1516	0.2055	0.2471	0.2945	0.3680	0.4637	0.6323	0.8433
1,000	H	19.48	61.94	89.32	124.81	154.99	190.64	243.58	307.91	421.64	564.52
	S	0.0327	0.1079	0.1515	0.2055	0.2470	0.2945	0.3679	0.4635	0.6319	0.8421
1,500	H	20.69	63.23	90.63	126.12	156.33	191.99	244.85	309.13	422.68	564.91
	S	0.0324	0.1077	0.1514	0.2053	0.2470	0.2944	0.3677	0.4633	0.6315	0.8410
2,000	H	21.89	64.52	91.93	127.41	157.65	193.34	246.13	310.37	423.72	565.29
	S	0.0321	0.1076	0.1513	0.2052	0.2469	0.2944	0.3676	0.4631	0.6311	0.8398
3,000	H	24.30	67.12	94.53	129.95	160.26	196.04	248.68	312.84	425.83	566.08
	S	0.0315	0.1074	0.1511	0.2049	0.2467	0.2943	0.3673	0.4627	0.6303	0.8375
4,000	H	26.70	69.73	97.13	132.43	162.81	198.75	251.24	315.32	427.96	566.89
	S	0.0309	0.1072	0.1509	0.2045	0.2464	0.2942	0.3670	0.4624	0.6296	0.8353
5,000	H	29.09	72.35	99.71	134.85	165.33	201.45	253.81	317.83	430.11	567.76
	S	0.0303	0.1070	0.1507	0.2041	0.2460	0.2941	0.3668	0.4620	0.6288	0.8331
6,000	H	31.48	74.97	102.28	137.22	167.81	204.15	256.38	320.35	432.28	568.73
	S	0.0297	0.1069	0.1504	0.2036	0.2456	0.2940	0.3665	0.4617	0.6281	0.8310

Table 14.3-12 (continued)

Pressure (psia)		Temperature (°F)									
		77	151	199	250	300	351	390	399	410	421
7,000	H	33.87	77.59	104.84	139.56	170.25	206.85	258.95	322.87	434.52	569.81
	S	0.0292	0.1067	0.1502	0.2030	0.2452	0.2939	0.3663	0.4614	0.6275	0.8291
8,000	H	36.25	80.20	107.39	141.86	172.66	209.54	261.48	325.41	436.81	570.99
	S	0.0286	0.1065	0.1499	0.2024	0.2447	0.2938	0.3660	0.4612	0.6269	0.8272
9,000	H	38.62	82.78	109.92	144.14	175.04	212.22	264.02	327.96	439.14	572.28
	S	0.0280	0.1064	0.1497	0.2018	0.2442	0.2937	0.3657	0.4609	0.6264	0.8255
10,000	H	40.98	85.34	112.46	146.40	177.37	214.87	266.59	330.53	441.50	573.64
	S	0.0274	0.1061	0.1494	0.2011	0.2437	0.2935	0.3655	0.4606	0.6259	0.8239
11,000	H	43.07	87.88	114.98	148.65	179.66	217.45	269.18	333.11	443.91	575.09
	S	0.0263	0.1059	0.1492	0.2005	0.2430	0.2934	0.3653	0.4604	0.6255	0.8224
12,000	H	45.15	90.39	117.50	150.92	181.91	220.04	271.82	335.70	446.34	576.62
	S	0.0252	0.1056	0.1489	0.1999	0.2424	0.2933	0.3651	0.4602	0.6251	0.8210
13,000	H	47.49	92.89	120.02	153.19	184.13	222.54	274.47	338.32	448.81	578.25
	S	0.0247	0.1053	0.1487	0.1993	0.2417	0.2923	0.3650	0.4600	0.6247	0.8197
14,000	H	49.83	95.37	122.54	155.49	186.31	224.96	277.16	340.93	451.32	579.97
	S	0.0241	0.1050	0.1484	0.1987	0.2409	0.2925	0.3649	0.4598	0.6244	0.8185

15,000	H	52.16	97.83	125.06	157.80	188.48	2?7.37	279.84	343.55	453.86	581.80
	S	0.0235	0.1047	0.1482	0.1982	0.2402	0.?919	0.3648	0.4597	0.6242	0.8174
16,000	H	54.48	100.30	127.57	160.13	190.64	22?.72	282.48	346.17	456.42	583.73
	S	0.0229	0.1044	0.1479	0.1977	0.2394	0.?914	0.3646	0.4595	0.6239	0.8164
17,000	H	56.81	102.76	130.07	162.47	192.79	23?.04	285.13	348.77	458.98	585.76
	S	0.0223	0.1040	0.1477	0.1972	0.2387	0.?08	0.3645	0.4593	0.6237	0.8156
18,000	H	59.13	105.22	132.58	164.83	194.94	234.35	287.74	351.36	461.57	587.89
	S	0.0217	0.1037	0.1474	0.1967	0.2379	0.2?03	0.3643	0.4591	0.6235	0.8149
19,000	H	61.45	107.69	135.08	167.20	197.09	236.?4	290.33	353.93	464.15	590.08
	S	0.0211	0.1034	0.1471	0.1963	0.2372	0.2??7	0.3641	0.4589	0.6233	0.8142
20,000	H	63.77	110.15	137.59	169.58	199.26	238.?3	292.89	356.52	466.75	592.31
	S	0.0205	0.1031	0.1469	0.1959	0.2365	0.28?1	0.3639	0.4587	0.6231	0.8136
21,000	H	66.09	112.62	140.09	171.97	201.44	241.??	295.44	359.11	469.35	594.58
	S	0.0199	0.1028	0.1466	0.1955	0.2358	0.288?	0.3636	0.4585	0.6229	0.8131
22,000	H	68.41	115.09	142.60	174.38	203.63	243.5?	298.00	361.78	471.77	596.87
	S	0.0194	0.1025	0.1464	0.1951	0.2351	0.288?	0.3634	0.4581	0.6225	0.8125
23,000	H	69.69	117.56	145.10	176.81	205.84	245.7?	300.66	363.83	473.42	600.98
	S	0.0168	0.1022	0.1461	0.1947	0.2344	0.2874	0.3633	0.4576	0.6212	0.8121
24,000	H	72.00	120.03	147.60	179.24	208.07	248.09	303.24	366.20	476.05	603.66
	S	0.0162	0.1019	0.1459	0.1944	0.2338	0.2869	0.3631	0.4572	0.6211	0.8120

Table 14.3-12 (continued)

Pressure (psia)		70	151	199	250	300	351	390	399	410	421
						Temperature (°F)					
25,000	H	74.32	122.49	150.10	181.69	210.32	250.40	305.83	368.73	478.67	605.97
	S	0.0156	0.1016	0.1456	0.1941	0.2332	0.2865	0.3629	0.4569	0.6210	0.8115
26,000	H	76.64	124.95	152.60	184.15	212.59	252.72	308.42	371.24	481.31	608.29
	S	0.0151	0.1013	0.1454	0.1938	0.2327	0.2828	0.3628	0.4567	0.6210	0.8115
27,000	H	78.93	127.41	155.10	186.62	214.89	255.06	311.03	373.72	483.95	610.60
	S	0.0144	0.1009	0.1451	0.1935	0.2321	0.2854	0.3626	0.4564	0.6207	0.8115
28,000	H	81.22	129.86	157.61	189.11	217.22	257.42	313.65	376.17	486.59	612.87
	S	0.0138	0.1006	0.1449	0.1932	0.2317	0.2349	0.3625	0.4560	0.6206	0.8110

[a] H = enthalpy in Btu/lb$_m$.
[b] S = entropy in Btu/lb$_m \cdot$ °R.

Table 14.3-13

Thermodynamic Properties of Polytetrafluoroethylene [17]

Pressure (psia)		Temperature (°F)							
		50	75	100	125	150	175	200	225
14.7	H[a]	4.11	12.65	19.15	25.09	31.07	37.15	43.30	49.55
	S[b]	0.008193	0.02452	0.03641	0.04677	0.05830	0.06954	0.07905	0.08834
500	H	4.35	13.20	19.74	25.68	31.61	37.74	43.90	50.14
	S	0.007410	0.02433	0.03630	0.04666	0.05819	0.06943	0.07894	0.08823
1,000	H	4.87	13.72	20.37	26.30	32.29	38.36	44.52	50.73
	S	0.006994	0.02409	0.03619	0.04655	0.05808	0.06932	0.07883	0.08812
2,000	H	5.69	14.77	21.57	27.51	33.44	39.56	45.72	51.96
	S	0.006163	0.02353	0.03597	0.04633	0.05786	0.06910	0.07860	0.08789
3,000	H	6.65	15.83	22.81	28.75	34.73	40.80	46.96	53.16
	S	0.005334	0.02303	0.03577	0.04623	0.05765	0.06888	0.07850	0.08767
5,000	H	8.50	17.92	25.26	31.19	37.17	43.24	49.38	55.62
	S	0.003696	0.02199	0.03538	0.04573	0.05724	0.06847	0.07797	0.08724
10,000	H	13.70	23.25	31.42	37.34	43.31	49.36	55.50	61.72
	S		0.01997	0.03448	0.04481	0.05632	0.06752	0.07700	0.08626
25,000	H	31.19	40.12	49.10	55.16	61.63	67.78	74.03	80.37
	S		0.01480	0.03115	0.04166	0.05400	0.06537	0.07503	0.08444
50,000	H	60.51	69.42	80.33	86.48	87.44	94.20	104.94	111.72
	S		0.01051	0.03050	0.04135	0.04454	0.05663	0.07333	0.08342
100,000	H	117.69	126.59	142.11	148.04			163.46	160.68
	S		0.00484			0.02672	0.03959	0.06958	0.08033
150,000	H	173.20	182.09	197.61	203.53			218.96	226.19
	S		0.00075			0.02252	0.03540	0.06538	0.07613

[a] H = enthalpy in Btu/lb$_m$.
[b] S = entropy in Btu/lb$_m$·°R.
Datum: enthalpy and entropy as zero at 14.7 psia and 32°F.

Table 14.3-14

Thermodynamic Properties of Nylon 66 [18]

Pressure (psia)		77	122	167	212	257	302	347	392
						Temperature (°F)			
14.7	H[a]	13.75	32.60	51.25	74.20	9_.88	121.25	146.25	173.20
	S[b]	0.0157	0.0640	0.0970	0.1290	0.1620	0.1940	0.2270	0.2590
500	H	17.2	38.4	57.0	78.0	1_3.0	125.5	148	182
	S	0.0	0.0584	0.0955	0.1285	0.1615	0.1934	0.2255	0.2600
1,000	H	18.4	40.6	58.3	79.2	1_6	129	149.5	184
	S		0.0576	0.0944	0.1280	0.1600	0.1918	0.2242	0.2586
2,000	H	20.1	42.4	60.9	81.6	1_8.6	130.2	150.1	185.8
	S		0.0562	0.0938	0.1270	0.1594	0.1912	0.2240	0.2573
3,000	H	21.9	44.3	62.8	82.9	1_1	134	153	188
	S		0.0558	0.0933	0.1265	0.1590	0.1904	0.2235	0.2570
4,000	H	23.5	45.4	64.0	84.2	1_2	136	154	189
	S		0.05455	0.0930	0.1261	0.1584	0.1900	0.2228	0.2560
6,000	H	27.8	48.8	67.5	88.2	_14	137	160	190
	S		0.05450	0.0923	0.1255	0.1580	0.1898	0.2225	0.2552
8,000	H	31.5	53.6	72.2	94.6	_18	142	167	194
	S		0.0541	0.0920	0.1255	0.1578	0.1890	0.2224	0.2548

10,000	H	36	58	76.8	100	124	148	174	201
	S		0.0536	0.0918	0.1254	.1577	0.1890	0.2224	0.2545
15,000	H	44.4	68.1	89.0	113	35.4	161	180	213
	S		0.0535	0.0912	0.1250	.1555	0.1880	0.2223	0.2544
20,000	H	55.2	78.0	99.55	123.2	46.18	170.25	195.25	222.40
	S	0.0033	0.0529	0.0912	0.1243	.1548	0.1878	0.2223	0.2543
30,000	H	74.0	98.5	124	149	73	196	224	248
	S		0.0500	0.0910	0.1230	1545	0.1876	0.2210	0.2538
40,000	H	97.4	123.29	146.95	171.60	4.68	218.65	243.55	271.10
	S		0.0447	0.0863	0.1213	1543	0.1863	0.2193	0.251
60,000	H	138	164	192	218		270	295	322
	S		0.0415	0.0845	0.1204	540	0.1854	0.2186	0.2503
80,000	H	184	214.10	239.98	266.20	2 .48	313.25	338.15	366.10
	S		0.0340	0.0795	0.1176	506	0.1826	0.2156	0.2476
100,000	H	225.6	254.6	284.0	307.0	33.5	355.4	376.0	405.0
	S		0.0318	0.0780	0.117	496	0.1808	0.2145	0.463
120,000	H	273.9	305.1	331.6	358.3	38 7	405.6	430.5	458.6
	S		0.0287	0.0757	0.1150	79	0.1799	0.2129	0.2449

^{a}H = enthalpy in Btu/lb$_m$.

^{b}S = entropy in Btu/lb$_m$·°R.

Datum: enthalpy and entropy as zero at 14.7 psia and 32°F.

These include the properties of high-density polyethylene, polypropylene, ethylene-propylene copolymer, PMMA, rigid PVC, GPS, Nylon 66, Nylon 610, PTFE, and PCTFE. Recall that enthalpy is defined as

$$(H_T - H_{T_0})_{P_0} = \int_{T_0}^{T} C_P \, dT \tag{14.3-8}$$

from the integration of Eq. (14.2-36) and that for pressure changes in enthalpy at constant temperature

$$(H_P - H_{P_0})_T = \int_{P_0}^{P} [V - T(\partial V/\partial T)_P] \, dP \tag{14.3-9}$$

or, using the definition for volume expansivity,

$$(H_P - H_{P_0})_T = \int_{P_0}^{P} (V - TV\beta) \, dP \tag{14.3-10}$$

For a Spencer-Gilmore-type polymer, this can be rewritten as

$$(H_P - H_{P_0})_T = \int_{P_0}^{P} [V - R'T/(P + \pi)] \, dP \tag{14.3-11}$$

But, substituting the expression for V from the Spencer-Gilmore equation (14.3-1),

$$(H_P - H_{P_0})_T = \omega(P - P_0) \tag{14.3-12}$$

where ω is obtained either from Table 14.3-5 or the tabulated data of Griskey et al. As can be seen, this is one way of obtaining the experimental values for the constants of the Spencer-Gilmore equation of state.

Furthermore, the entropy dependence on pressure and temperature is given by

$$(S_T - S_{T_0})_{P_0} = \int_{T_0}^{T} C_P \, dT/T \tag{14.3-13}$$

$$(S_P - S_{P_0})_T = \int_{P_0}^{P} - (\partial V/\partial T)_P \, dP \tag{14.3-14}$$

As mentioned previously, the isentropic case (no heat transfer, dS = 0) is probably more important than the changes in entropy with temperature and pressure. Equation (14.2-46) can be used to obtain the isentropic changes for the Spencer-Gilmore form of the polymeric system:

$$dT = (TV\beta/C_P) \, dP = [R'T/C_P(P + \omega)] \, dP \qquad (14.3-15)$$

Upon integrating this equation between the conditions of T_0 and T, P_0 and P,

$$T/T_0 = [(P + \omega)/(P_0 + \omega)]^{R'/C_P} \qquad (14.3-16)$$

This expression is remarkable in its similarity to the isentropic compression equation for compressible gases.

There are many examples where isentropic expansion or compression can occur in processing of plastic materials. For example, flow of a plastic between nip rolls of a calender indicates that the time of compression and expansion of the material is very small when compared with the rate of change of energy in the material owing to heat transfer to the calender rolls. Likewise, flow through a screen pack or from a nozzle into a sheeting die is essentially isentropic in that the changes in pressure are rapid when compared with the changes in the enthalpy of the material owing to heat transfer from the die walls. Recently, injection speeds achieved in thermoplastic structural foam molding are so high (10 to 50 lb/sec) that this aspect of injection molding might be considered, as a first approximation, as isentropic flow.

14.4 Heat Capacity and Related Thermal Properties

Although heat capacity can be defined in terms of the rate of change of enthalpic energy or of internal energy with respect to temperature, as given in Eqs. (14.2-37) and (14.2-38), normally the total energy curves at constant volume or pressure are not available as functions of temperature. In this case, it becomes necessary to either measure or estimate values for the heat capacity of a given material. Many ways of experimentally obtaining the entire enthalpic energy curve were discussed in Chap. 4. Here, effort will be concentrated on methods for estimating these values, based primarily on very accurate predictions that have been made for simple organic liquids. Sakiadis and Coates [21] have presented a very accurate method which has been shown to be accurate to ±2% for more than 100 organic liquids. Griskey and coworkers [22,23] have shown that with proper assumptions about the unit transmitting sensible heat the Sakiadis-Coates method can be directly applied to polymers with little additional error.

Sakiadis and Coates assume that energy is transmitted by translational motion of the molecules or molecular segments. Thus, the segment acts as a liquid harmonic oscillator, and the liquid behaves as a semicrystalline solid. The specific heat that is obtained through this type of analysis is the heat capacity at constant volume. The total molecular energy is then considered to be the sum of its component parts: translational, external rotational, internal rotational, vibrational, and electronic. Of these the last is considered to be negligible. Furthermore, the motion can be considered to be either as an acoustical branch, where the vibrations in each molecule cause it to move as a whole, or an optical branch, where atoms or groups of atoms are moving relative to one another. For the acoustical branch, theoretical molecular mechanics show that the translational and external rotational motions can be written as

$$C_{vt} = 3R'f_D \qquad (14.4\text{-}1)$$

$$C_{vr} = 3R'f_E \qquad (14.4\text{-}2)$$

where R' is the gas constant. f_E and f_D are the Einstein and Debye functions, respectively. It can be shown that $f_D \cdot f_E$ for most practical situations, and thus the total molecular energy of the acoustical branch is given as

$$C_{vt} = C_{vr} = 6R'f_E \qquad (14.4\text{-}3)$$

where f_E is given as a function of the wave number ω and the absolute temperature (°R):

$$f_E/R' = u^2 \exp(u)/[\exp(u) - 1]^2 \qquad (14.4\text{-}4)$$

where $u = 1.4384\omega/T$. In general, for liquids at 77°F or higher, $f_E \doteq 1$. For other conditions, the values for ω can be obtained from Table 14.4-1, and f_E is given in Fig. 14.4-1.

To obtain proper values for energy in an optical branch, a summation of all interacting bonds must be carried out. For internal rotation, all are considered to be harmonic oscillators. This assumes that double bond interaction is negligible. Thus, the internal rotation contribution is given as

$$C_{vr'} = n_{r'}R'f_D \qquad (14.4\text{-}5)$$

where $n_{r'}$ represents the number of single bonds about which internal rotation of groups can take place, such as C——C. Stretching and bending motions are also harmonic oscillations, and the total molecular energy contribution can best be represented in terms of Einstein functions as

$$C_{v(\nu+\delta)} = \sum_i q_i f_{E_{\nu_i}} + \sum_i q_i f_{E_{\delta_i}} \qquad (14.4\text{-}6)$$

Table 14.4-1

Molecular Vibration Contribution to Total Heat Capacity [24]

Bond	Stretching, ω_ν (cm^{-1})	Bending, ω_δ (cm^{-1})
$>$C—H	2960	—
—C$\overset{H}{\underset{H}{\diagup}}$H	—	1450
$\cdot\cdot$C$\overset{H}{\underset{H}{\diagup}}$H	—	1000
$>$C$\overset{H}{\underset{H}{\diagdown}}$	—	1450
$\overset{\diagdown}{\diagup}$C—H	3020	—
$=$C$\overset{H}{\underset{H}{\diagup}}$	—	1100
$\equiv$C—H	3300	700
C—C	900	370
C$=$C	1650	—
C$\equiv$C	2050	—
C—C$=$C	—	600
C—C$\equiv$C	—	300
C$=$C$=$C	—	350
C—O	1030	205
C$=$O	1700	390
C—N	900	370

Table 14.4-1   (continued)

Bond	Stretching, ω_v (cm^{-1})	Bending, ω_δ (cm^{-1})
C$=$N	1620	845
—C$\equiv$N	2250	—
C—S	650	330
C$=$S	1550	530
$\geq$C—F	1100	530
$\geq$C—Cl	660	330
$\geq$C—I	500	260
O—H	3400	1150
$\geq$N—H	3350	1320
N—O	1030	205
N$=$O	1700	390
N—N	990	390
S—H	2570	1050
S—S	500	260

where $f_{E_{v_i}}$ and $f_{E_{\delta_i}}$ represent the contribution to the heat capacity owing to stretching (v) and bending (δ) vibrations of the ith bond. q_i is the total number of bonds in the molecule. As Sakiadis and Coates point out, there are additional contributions to the degrees of freedom. These are given as

$$3n - 6 - n_{r'} - 2 \sum_i q_i \qquad\qquad (14.4\text{-}6a)$$

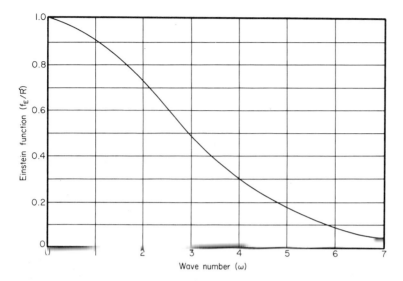

Figure 14.4-1 Correction in specific heat calculations for Einstein function as function of wave number [24].

These contributions are assumed to be of the same type as that of the bending motion. Thus,

$$C_{v(\nu+\delta)} = n_{r'}R'f_D + \sum_i q_i f_{E_{\nu_i}} + \left[\frac{3n - 6 - n_{r'} - \sum_i q_i}{\sum_i q_i}\right] \sum_i q_i f_{E_{\delta_i}}$$

(14.4-7)

is the total contribution of the optical branches. Obviously, then, the heat capacity is the sum of the acoustical branches, Eq. (14.4-5), and the optical branches, Eq. (14.4-7).

Experimental agreement with the computed values is usually excellent with the higher-molecular-weight hydrocarbons but shows errors with low-molecular-weight hydrocarbons and acids, alcohols, and ethers. Thus, Sakiadis and Coates add an adjustable constant that depends on molecular weight and type of liquid. Values of this constant S are shown for four series of materials in Figs. 14.4-2 and 14.4-3. This constant is a multiplier to the last term on the right of Eq. (14.4-7), as shown below:

$$C_{v(\nu+\delta)} = n_{r'}R'f_D + \sum_i q_i f_{E_{\nu_i}} + \left[\frac{3n - 6 - n_{r'} - \sum_i q_i}{\sum_i q_i}\right] S \sum_i q_i f_{E_{\delta_i}}$$

(14.4-7a)

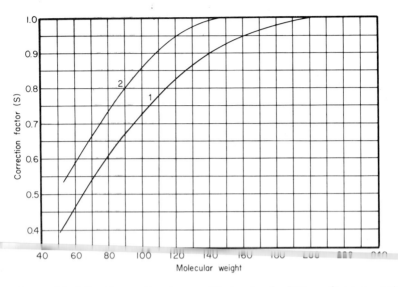

Figure 14.4-2 Correction to specific heat calculations for increasing molecular weight of various compounds. Curve 1: aliphatic hydrocarbons, naphthenes, esters; curve 2: aromatic hydrocarbons, halides, esters, ketones [21].

Note that the heat capacity thus calculated is for constant volume. The thermodynamic principles of Sec. 14.3 can be used to obtain a value for heat capacity at constant pressure. This form is much more useful in computations for enthalpy, for example. It can be shown that the isothermal compressibility of a liquid, as given in Eq. (14.3-7), can be written in terms of the velocity of sound in the liquid:

$$\kappa = (C_p/C_v)/U^2\rho \qquad\qquad (14.4\text{-}8)$$

where U is the velocity of sound and ρ is the liquid density. Furthermore, the difference in heat capacities is given as

$$C_p - C_v = T\beta^2/\kappa\rho \qquad\qquad (14.4\text{-}9)$$

where β is the volume expansivity, given in Eq. (14.3-6). Proper substitution yields

$$C_p/C_v = 1 + (T\beta^2 U^2/C_p) \qquad\qquad (14.4\text{-}10)$$

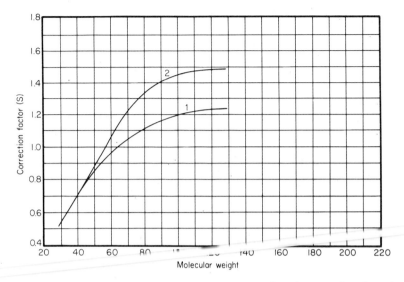

Figure 14.4-3   Correction for increasing molecular weight of various compounds to specific heat calculations. Curve 1: aliphatic acids, alcohols, ethers; curve 2: aromatic alcohols, amines, mercaptans [21].

Thus, to obtain C_p, C_v is calculated from Eqs. (14.4-5) and (14.4-7), and the value is substituted into Eq. (14.4-10) with the appropriate values of volume expansivity and velocity of sound. C_p is an implicit solution of this equation.

According to Griskey [20,22], the heat capacity of a representative unit of the polymer will represent the heat capacity of the entire molecule. For example, for HDPE, the representative unit is the mer

$$
\begin{array}{cc}
\text{H} & \text{H} \\
| & | \\
-\text{C}-\text{C}- \\
| & | \\
\text{H} & \text{H}
\end{array}
$$

The molecular weight needed to obtain S, for example, is that of the mer. The velocity of sound that is needed in Eq. (14.4-10) is that for the mer, which is very similar to the value for the monomer, ethylene. For polymers with side groups, such as GPS, the mer

$$
\begin{array}{cc}
\text{H} & \text{H} \\
| & | \\
-\text{C}-\text{C}- \\
| & | \\
\text{H} & \phi
\end{array}
$$

should be chosen. If the polymer has side chain branching, the intersection can be included, but unless there are unusually many side chain branches and they are short, their contribution to the heat capacity would be very small. Heat capacities for co- and terpolymers can also be determined once the percentages of the various components are known. For polymer blends and extensive amounts of additives or reinforcements, the weight fractions should be used to determine an additive heat capacity. Representative values are given in Table 14.4-1A.

Representative values for thermal conductivity can also be obtained using a molecular approach [25]. Nearly all the work has been done on pure simple organic materials, however, and it again becomes necessary to extrapolate these data to polymeric materials having similar molecular characteristics. In Table 14.4-2 are the thermal conductivity values at 100°F for more than 50 common organic liquids. In addition, the differential change in the value with temperature is given over the normal operating range of the materials. Sakiadis and Coates point out that a simple model can be obtained for thermal conductivity if it can be assumed that isothermal planes of liquid molecules having the same mean energy can be formed. In this way, molecules vibrate about a fixed position, and the energy is transferred from one layer to another in much the same way as mechanical energy is transferred from point to point along an interconnecting network of springs. L is the distance between these molecular surfaces, and d is the molecular diameter. Thus, $x = d + L$ in the notation to follow.

An energy drop per molecule along the row of molecules in the direction of heat flow is given as

$$Q = -[(dQ/N)/dx]L = -L[(dQ/N)/dT](dT/dx) \qquad (14.4-11)$$

Assume that the thermal vibrations are transmitted at the velocity of sound U. Thus, the total energy passing a fixed point in the row is

$$-L[(dQ/N)/dT](dT/dx)(U/x) \qquad (14.4-12)$$

Now is yz is the cross-sectional area corresponding to x, the heat flux is

$$-q/A = -L[(dQ/N)/dT](dT/dx)(U/x)(1/yz) \qquad (14.4-13)$$

As seen in Chap. 6, the equation for heat conduction related the heat flux to the thermal gradient as

$$-q/A = -k(dT/dx) \qquad (14.4-14)$$

Now is the molecular volume $xyz = M/\rho N$, where M is the molecular weight, ρ is the density, and N is Avogadro's number, Eq. (14.4-13) becomes

$$k = [(dQ/N)/dT](LU\rho N/M) \qquad (14.4-15)$$

From Table 14.2-1, it is seen that

$$[(dQ/N)/dT] = MC_p/N \qquad (14.4-16)$$

and thus

$$k = C_p U\rho L \qquad (14.4-17)$$

Table 14.4-1A

Polymer Specific Heats [22]

Polymer	Specific heats a temperature (°C) (cal/g·°C)							
	120	150	180	210	240	270	300
Polymethyl methacrylate	0.66	0.75	0.80	0.89	1.12	1.54	2.52
Polydimethylaminoethyl methacrylate	0.40	0.42	0.53	0.72	0.93	1.47	2.36
Polycyclohexyl methacrylate	0.31	0.33	0.44	0.59	0.76	1.14	2.21
Polyallyl methacrylate	0.19	0.28	0.41	0.56	0.83	1.14	1.58
Polyethyl acrylate	0.17	0.21	0.32	0.41	0.48	0.54	0.63

Now according to theoretical physics [2], the velocity of sound in liquids can be approximated by

$$U = (R\rho/M)^3 \qquad (14.4\text{-}18)$$

Thus, Eq. (14.4-17) becomes

$$k = C_p R^3 \rho^4 L/M^3 \qquad (14.4\text{-}19)$$

Small corrections for special groups are found in Table 14.4-3. Sakiadis and Coates show that for common organic liquids the thermal conductivity decreases to an asymptote with increasing number of carbons for most materials, as shown in Fig. 14.4-4. Values for the intermolecular distance L are given in Table 14.4-4 for several linear materials. Note that as the chain length increases, the intermolecular distance L decreases, albeit rather slowly. An arithmetical form for the intermolecular distance L is

$$L = x[1 - (v_0 \rho N/M)^{1/2}] \qquad (14.4\text{-}20)$$

where v_0 is the minimum volume possible when the molecules are touching on a rectilinear pattern. One form for v_0 is based on the volume at the critical point and the conditions at the freezing point:

$$v_c/v_0 = \rho_0/\rho_c = 2\left[1 + \left(\frac{\rho_f/2 - \rho_c}{T_c - T_f}\right)(T_c/\rho_c)\right] \qquad (14.4\text{-}21)$$

According to Gambill [24], this ratio is approximately 4 for most small-molecule organic materials. The extension of this equation to polymeric materials does not seem practical, however, since many materials have no defined freezing point and none have critical conditions. As a first approximation, however, the values for the monomeric unit might be used to get an estimate of the intermolecular distance. The value of the thermal conductivity depends directly on this distance, however, and any error in calculation should be directly reflected in the error for the value of the thermal conductivity. Caution is thus advised.

Representative values for the constant R are given in Table 14.4-5. These values can be used to obtain the velocity of sound values needed in both the heat capacity and thermal conductivity equations. The heat capacity is essentially constant, as is expected from the work discussed above. Since the molecular weight is increasing, the thermal conductivity should decrease, as is the case in Fig. 14.4-4. It would appear, then, for polymeric materials that to obtain a representative thermal conductivity the representative unit should contain at least 8 to 10 carbons in the backbone. This is contrary to the recommendations given for heat capacity. Although the accuracy for simple molecules is excellent, interstitial voids in polymers should cause the calculated values to deviate considerably from experimental values.

Table 14.4-2

Observed Values of Thermal Conductivity [21]

Liquid	k at 100°F $(Btu/hr \cdot ft^2 \cdot °F/ft)$	$dk/dT \times 10^4$ $(Btu/hr \cdot ft^2 \cdot °F/ft \cdot °F)$	Temp. range (°F)
n–Hexane	0.0714	-1.40	91–135
n–Heptane	0.0725	-1.30	91–170
n–Octane	0.0751	-1.15	93–170
n–Nonane	0.0777	-1.30	92–171
n–Decane	0.0770	-1.40	106–169
n–Hexene–2	0.0715	-1.60	100–131
n–Heptene–2	0.0747	-1.30	91–170
n–Octene–2	0.0769	-1.40	92–171
n–Octene–1	0.0740	-1.40	103–172
2–Methyl pentane	0.0617	-1.00	87–121
3–Methyl pentane	0.0628	-1.00	89–128
2,2–Dimethyl butane	0.0560	-0.93	85–110
2,3–Dimethyl butane	0.0592	-0.80	86–122
2,2,4–Trimethyl pentane	0.0560	-1.05	101–171
2,2,5–Trimethyl hexane	0.0623	-1.40	94–171
Methyl alcohol	0.1187	1.77	95–138

Table 14.4-2   (continued)

Liquid	k at 100°F (Btu/hr ft² °F/ft)	dL/dT × 10⁴ (Btu/hr·ft²·°F/ft·°F)	Temp. range (°F)
Ethyl alcohol	0.0981	-1.20	95–167
n–Propyl alcohol	0.0912	-0.95	95–168
n–Butyl alcohol	0.0885	-1.02	94–170
n–Amyl alcohol	0.0863	-0.91	94–170
n–Hexyl alcohol	0.0878	-0.93	94–170
n–Heptyl alcohol	0.0903	-0.81	95–169
n–Octyl alcohol	0.0927	-1.00	95–170
n–Decyl alcohol	0.0947	-1.18	95–170
i–Propyl alcohol	0.0814	-0.95	94–171
i–Butyl alcohol	0.0803	-0.81	94–170
t–Butyl alcohol	0.0670	-0.75	93–171
Ethylene glycol	0.1510	+0.36	97–169
Propylene glycol	0.1215	-0.35	96–169
Glycerol	0.1789	+0.53	96–171
Dipropylene glycol	0.1007	-1.08	96–169
Methyl acetate	0.0931	-1.92	96–120
Ethyl acetate	0.0826	-1.62	106–145

n-Propyl acetate	0.0796	-1.40	99-169
n-Butyl acetate	0.0795	-1.26	98-170
n-Amyl acetate	0.0782	-1.25	96-170
n-Octyl acetate	0.0815	-1.45	104-170
Methyl propionate	0.0849	-1.55	108-145
Ethyl propionate	0.0800	-1.55	107-168
n-Propyl propionate	0.0795	-1.34	108-170
n-Amyl propionate	0.0790	-1.39	106-169
n-Ethyl butyrate	0.0781	-1.38	107-169
n-Amyl chloride	0.0676	-0.85	108-169
1-Chlorodecane	0.0754	-1.23	106-169
n-Propyl bromide	0.0571	-1.19	96-136
n-Butyl bromide	0.0581	-1.13	103-170
n-Amyl bromide	0.0599	-1.06	102-170
n-Hexyl bromide	0.0614	-1.02	96-169
n-Propyl iodide	0.0503	-0.90	105-170
n-Heptyl iodide	0.0573	-0.98	102-169
Nitromethane	0.1170	-1.97	110-168
Nitroethane	0.0962	-1.55	108-168
1-Nitropropane	0.0873	-1.44	110-169

Table 14.4-3

Structural Contribution to the Thermal Conductivity
of Liquids at $T_r = 0.6$ [21]

		dk (Btu/ft·hr·°F)
Aliphatic alcohols	—OH	-0.0070
Esters	—C(=O)O—	+0.0070
Alkyl halides	—Cl	-0.0168
	—Br	-0.0248
	—I	-0.0310
Nitrated alkanes	—NO$_2$	0
Isomerization	For 1 —CH$_3$ group	-0.0060
	For 2 —CH$_3$ groups	-0.0104
	For 3 —CH$_3$ groups	-0.0142
Bonding	One double bond	+0.0010

		Effective number of carbon atoms, x
Aliphatic hydrocarbons	$C_n H_{2n+2}$	$x = n$
Aliphatic alcohols	$C_n H_{2n+1} OH$	$x = n$
Esters	$C_n H_{2n} O_2$	$x = n - 2$
Alkyl halides	$C_n H_{2n+1} Cl$	$x = n$
	$C_n H_{2n+1} Br$	$x = n + 1$
	$C_n H_{2n+1} I$	$x = n + 2$
Nitrated alkanes	$C_n H_{2n+1} NO_2$	$x = n$

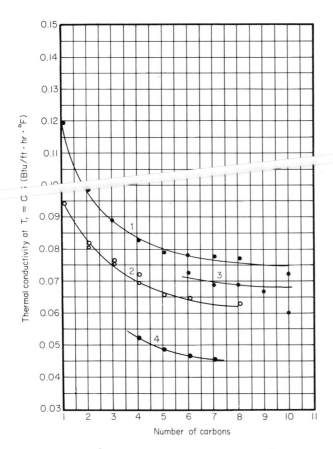

Figure 14.4-4   Effect of increasing molecular weight on thermal conductivity of several organics. Curve 1: alcohols; curve 2: esters; curve 3: hydrocarbons; curve 4: alkyl bromides [21].

Table 14.4-4

Specific Heat, Density, and Intermolecular Distance L for
Several Linear Materials [21]

Liquid	C_p (Btu/lb·°F)	ρ_c (lb/ft^3)	$L \times 10^{-9}$ (ft)	R/M
n-Hexane	0.534	14.49	0.268	0.369
n-Heptane	0.526	14.59	0.251	0.362
n-Octane	0.522	14.59	0.245	0.358
n-Nonane	0.521	14.59	0.230	—
Ethyl alcohol	0.575	17.20	0.259	0.318
n-Propyl alcohol	0.563	17.02	0.230	0.318
n-Butyl alcohol	0.563	16.95	0.217	0.319
n-Amyl alcohol	0.560	16.90	0.206	—
Methyl acetate	—	—	—	0.273
Ethyl acetate	—	—	—	0.280
n-Propyl acetate	—	—	—	0.285

Table 14.4-5

Structural Contributions to the Velocity of Sound [24]

	R
Basic radicals	
Methyl	9.50
Benzene	23.25
Cyclohexane	27.50
Naphthalene	33.67
Additional radicals	
Carbon with zero to three hydrogens	4.47
Carbon, two oxygens	6.25
Carbon, oxygen, hydrogen	2.30
Ketone	4.47
$-NH-$	3.27
$-NH_2$	2.45
$-COOH$	4.83
$-CN$	4.20
$-O-$	1.40
$-O-$	0.70
$-Cl$	3.13
$-Br$	3.55
$-I, -NO_2$	4.58
$-S, =S$	2.82
Bonding	
Double	-1.30
Triple	-2.60
Position	
Ortho	0
Meta	0.30
Para	0.60

14.5 Stretching Rubbery Materials

As has been seen earlier, film stock is frequently stretched biaxially in order to achieve thinner films and to orient the film to achieve greater strength.
Fine filaments are formed by rapid elongation of the plastic as it leaves the
spinnerette. The amount of force required to elongate a rubbery plastic
can be calculated quite accurately from thermodynamics. If the elongational
process can be considered isothermal and thermodynamically reversible,
the following equation holds:

$$dU = T \, dS + dW \tag{14.5-1}$$

where dW represents the amount of work that is done on the rubber during
elongation. It can be shown that for isothermal systems this work is the
Helmholtz free energy dA. For uniaxial elongation, assuming that there is
little volume change,

$$dA = dW = f \, dL = D \, dV \cdot f \, dT \tag{14.5-2}$$

or, in another form,

$$f = (\partial W/\partial L)_T = (\partial A/\partial L)_T \tag{14.5-3}$$

f is the force required to elongate the rubber, and L is the length of the
rubber. Therefore, the force required for elongation is directly proportional to the change in Helmholtz free energy per unit extension. For small
strains, Eq. (14.5-3) can be written as

$$f = (\partial A/\partial L)_T = (\partial^2 A/\partial L^2)_T (L - L_0) \tag{14.5-4}$$

Thus, $(\partial^2 A/\partial L^2)_T$ represents a constant of proportionality, Young's modulus. For large deformation, however, the force-elongation curve is not
linear. Note that the Helmholtz free energy can be written as

$$A = U - TS \tag{14.5-5}$$

or for isothermal conditions,

$$dA = dU - T \, dS \tag{14.5-6}$$

By combining Eqs. (14.5-2) and (14.5-6), the tension can be written as

$$f = (\partial U/\partial L)_T - T(\partial S/\partial L)_T \tag{14.5-7}$$

The first term on the right represents the change in internal energy with
extension and the second the change in entropy with extension. This term
is related to the temperature coefficient of extension. Note that for any
change

$$dA = dU - d(TS) \tag{14.5-8}$$

and since $dW = f\, dL$, it can be shown that

$$dA = f\, dL - S\, dT \tag{14.5-9}$$

It is apparent then that for isothermal conditions the second term drops out and the equation reduces to Eq. (14.5-6). By differentiation, the following results:

$$(\partial A/\partial L)_T = f \tag{14.5-10a}$$

$$(\partial A/\partial T)_L = -S \tag{14.5-10b}$$

Now since

$$(\partial A/\partial T)_L/\partial L = \partial(\partial A/\partial L)_T/\partial T \tag{14.5-11}$$

the following results:

$$\cdots _T = (\partial / \partial T)_L \tag{14.5-11a}$$

When this is substituted into Eq. (14.5-7), the change in internal energy with respect to measurable quantities results:

$$(\partial U/\partial L)_T = f - T(\partial f/\partial T)_L \tag{14.5-12}$$

As a result, for isothermal conditions, the rate of change of internal energy and entropy can be determined by inspection of the force applied on the specimen at various temperatures. This can be seen in Fig. 14.5-1.

It can be shown from molecular considerations that stored energy is a function of the elongational ratios (a_1, a_2, a_3) in three dimensions [27, 28],

$$W = (G/1)(a_1^2 + a_2^2 + a_3^2 - 3) \tag{14.5-13}$$

and the stress-strain relationships are given as

$$e_1 - e_2 = G(a_1^2 - a_2^2) \tag{14.5-14a}$$

$$e_2 - e_3 = G(a_2^2 - a_3^2) \tag{14.5-14b}$$

$$e_3 - e_1 = G(a_3^2 - a_1^2) \tag{14.5-14c}$$

Here G is the Hookean modulus of rigidity. For uniaxial extension, $a_1 = a$, $a_2 = a_3 = a^{-1/2}$. Thus, the stored energy term is given as

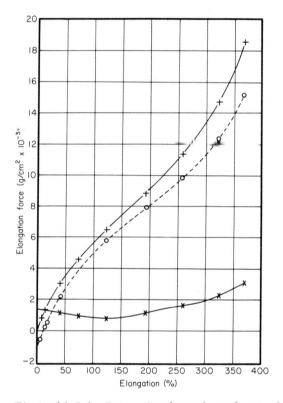

Figure 14.5-1   Retraction force for vulcanized natural rubber at 20°C (+'s).
Also plotted are $(\partial U/\partial L)_T$ (circles) and $-T(\partial S/\partial L)_T$ (X's) [26].

$$W = (G/2)(a^2 + 2/a - 3) \tag{14.5-15}$$

If f is the only force acting in the direction of extension a, its value per unit cross-sectional area is given as the rate of change of the stored energy as a function of elongation:

$$f = dW/da = G(a - 1/a^2) \tag{14.5-16}$$

As can be seen in Fig. 14.5-2, this equation predicts the force required to uniaxially compress or elongate a specimen for values of a less than 1.5 or so. Other experimenters have found agreement to values of a of 3.5 to 4. Above this, deviation is probably caused by shear-induced crystallization, nonisothermal conditions, and volume change in the specimen. For biaxial stretching, $a_1 = a_2 = a$, $a_3 = a^{-2}$. If the sheet is of initial thickness d_0, the force required for stretching is given by

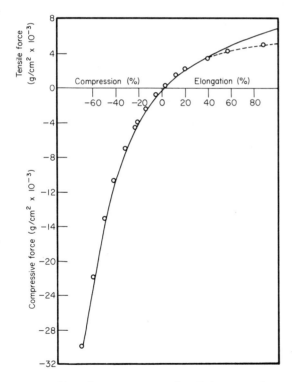

Figure 14.5-2 Experimental and theoretical stress-strain curve for vulcanized natural rubber at 20°C [26].

$$f = e_1 d_0/a = G d_0 (1 - 1/a^6) \qquad\qquad (14.5\text{-}17)$$

This obtains since $e_3 = e_1 = e_2$, and

$$e_1 = G(a^2 - 1/a^4) \qquad\qquad (14.5\text{-}18)$$

Stein [29] has carried out extensive thermodynamic analyses for anisotropic sheets in biaxial orientation to obtain internal pressures required to inflate sheet biaxially. For example, he finds that if the pressure is inversely proportional to the negative normal stress perpendicular to the skin, the isothermal compressibility is given as

$$\beta = -(L_3)^{-1}(\partial L_3/\partial P)_{T, L_1, L_2} = (G_{33})^{-1} \qquad\qquad (14.5\text{-}19)$$

where G_{33} is the Young's modulus normal to the film. Analyses for other variables are given also. Note the immediate relevance of this thermodynamic analysis to the stress-strain analysis on formable sheets below their melting point in Chap. 12. The reader would do well not to neglect these thermodynamic aspects when considering stretching and flow of polymeric materials in all types of processing.

14.6 Polymer-Polymer Systems

In many physically blended systems, the materials are sufficiently homologous to allow for dispersion of one polymer system in another down to a molecular level. This microscopic dispersion allows for the blending of HDPE and LDPE, or HDPE and PP, without appreciable separation. On the other hand, some rubber-glassy polymer systems cannot be formed because the affinity of one material to another is so poor that even if the system seems initially homogeneous, processing shear will separate the phases. If the affinity is marginal, the phases may appear homogeneous during processing, but the final part may be cheesey or friable. As an example of the last type of system, Paul and coworkers [30, 31, 71] have extensively examined the mechanical properties of incompatible physical blends of scrap polyolefins, styrenics, and vinyls. As shown in Fig. 14.6-1, the energy to break, defined as the area under the stress-strain curve, is lowest at an equal weight mixture of the three materials. For any binary pair, the energy to break is usually the lowest at an equal weight mixture. Thus, according to Rosen [32], these pairs have little interfacial bonding. To achieve a mutually beneficial two-phase systems, spherical particles must be rubbery and the continuous matrix must be glassy. The rubber inclusions should act

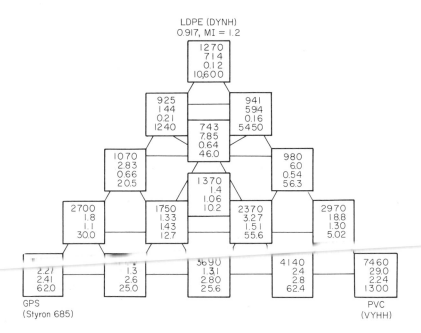

LDPE (DYNH)
0.917, MI = 1.2

Figure 14.6-1 Ternary diagram of mechanical properties of mixtures of LDPE, GPS, and PVC (VA copolymer). The key to the data is as follows: top number, tensile strength, psi; second number, elongation at break, percent; third number, tensile modulus, 100,000 psi; last line, energy to break (area under stress-strain curve), psi [30].

as stress concentrators and crack stoppers, and they must be large enough to exhibit macroscopic rubber characteristics. If there is insufficient bonding at the interface, separation and void formation can occur preferentially there, thereby weakening the glassy material.

Consider the process of mixing two pure polymers at constant pressure and temperature to form a true molecular solution. If dG is the change in Gibbs free energy, defined as G = H - TS, where H is the enthalpy of the system, then

$$dG = dH - T\,dS \tag{14.6-1}$$

Accordingly, if dG is negative, the system should tend toward a true solution. If dG is positive, the tendency is toward a two-phase system. The change in entropy dS is always positive since the randomness of the system is increasing. Thus, an entropy contribution always favors true solutions. However, the order of magnitude of the entropy contribution is very small primarily because there are very few molecules to move about during mixing.

Thus, the change in enthalpy must be approximately zero or negative for true mixing to occur. As will be detailed later, one measure of the compatibility of the two plastics is the compatibility of their solubility parameters. If both plastics have the same solubility parameter, one will swell the other, and thus the system can become, in time, a true solution. For solubility parameters that are not equal,

$$dE \doteq dH = \phi_1 \phi_2 (\delta_1 - \delta_2)^2 \qquad (14.6\text{-}2)$$

where δ_1, δ_2 are the solubility parameters of the two materials and ϕ_1, ϕ_2 are the volume fractions of the components. $dE = dH$ assumes that the volume change upon mixing is negligible. Several solubility parameters are given in Table 14.6-2, and δ^2, the cohesive energy density, for several other materials is given in Table 14.6-1. Note that $dH \geq 0$. Hence, to achieve a nearly zero change in enthalpy, and thus a true solution, the solubility parameters must be nearly identical. Therefore, most of the polymer-polymer solutions are two-phase, with hydrogen bonding being the primary influence on compatibility at the interface. Rosen points out that the only true polymer-polymer solution known is PVC-butadiene-acrylonitrile (NBR

Table 14.6-1

Cohesive Energy Densities of Linear Polymers [72]

Polymer	Repeat unit	Cohesive energy density, δ^2 (cal/cm^3)
Polyethylene	$-CH_2CH_2-$	62
Polyisobutylene	$-CH_2C(CH_3)_2-$	65
Polyisoprene	$-CH_2C(CH_3)=CHCH_2-$	67
Polystyrene	$-CH_2CH(C_6H_5)-$	74
Polymethyl methacrylate	$-CH_2C(CH_3)(COOCH_3)-$	83
Polyvinyl acetate	$-CH_2CH(OCOCH_3)-$	88
Polyvinyl chloride	$-CH_2CHCl-$	91
Polyethylene terephthalate	$-CH_2CH_2OCOC_6H_4COO-$	114
Polyhexamethylene adipamide	$-NH(CH_2)_6NHCO(CH_2)_4CO-$	185
Polyacrylonitrile	$-CH_2CHCN-$	237

Table 14.6-2

Typical Values of the Solubility Parameter δ [72]

Solvent	δ_1 [(cal/cm^3)$^{1/2}$]	Polymer	δ_2 [(cal/cm^3)$^{1/2}$]
n–Hexane	7.24	Polyethylene	7.9
Carbon tetrachloride	8.58	Polystyrene	8.6
2–Butanone	9.04	Polymethyl methacylate	9.1
Benzene	9.15	Polyvinyl chloride	9.5
Chloroform	9.24	Polyethylene terephalate	10.7
Acetone	9.71	Nylon 66	13.6
Methanol	14.5	Polyacrylonitrile	15.4

rubber), and even with this system, the range of true solubility is restricted
to narrow molecular-weight ranges and specific ratios of the materials.
The strong affinity of the chloride and nitrile groups apparently make this
system a solution. The strong bonding characteristics of the chloride
groups form the basis for Locke and Paul's methods of improvement of
scrap polymer blends, but the authors warn that while CPVC-PVC system
solubility can be improved, CPVC-PS and CPVC-PE systems are nearly
unchanged, since there is no group on either PS or PE with which the chloride
group can form an affinity. It now appears that polyphenylene oxide-poly-
styrene blends also form true polymer solutions.

14.7 Polymer-Solvent Interaction

As discussed earlier, polymers are frequently precipitated from the polymerizing medium through careful selection of the solvents used for the
polymerization. Dissolution of polymers with solvents is necessary when
certain adhesives are employed during fastening steps. Certain linoxus
require that the medium be somewhat soluble in the polymer in order to activate the polymer interfaces. Frequently, solvent resistance is required for
a specific end use, as with gasketing materials, hoses, and containers. To
understand the polymer-solvent interaction, it is necessary to consider
ideal and nonideal liquid solutions.

As seen above, solution thermodynamics indicates that if the change
in Gibbs free energy upon mixing is negative, a true solution can be formed
[34]. If it is positive, two phases are preferred. Equation (14.6-1) shows
that the change is the sum of the changes in enthalpy and entropy upon mixing. It can be shown that for ideal mixing the change in entropy is given as

$$S = -R(n_1 \ln \phi_1 + n_2 \ln \phi_2) \qquad (14.7-1)$$

where n_1, n_2 are the number of moles of solvent and polymer, respectively,
and ϕ_1, ϕ_2 are the volume fractions of solvent and polymer, respectively.
ϕ_1 and ϕ_2 are given as

$$\phi_1 = n_1/(n_1 + mn_2), \qquad \phi_2 = mn_2/(n_1 + mn_2) \qquad (14.7-2)$$

m is the ratio of molar volumes of polymer and solvent.

Scatchard [35] found that the heat of mixing ΔH could be approximated
by

$$\Delta H \doteq wn_1\phi_2 \qquad (14.7-3)$$

where for low-molecular-weight mixtures

$$w = V_1(\sigma_1 - \sigma_2)^2 \qquad (14.7-4)$$

for which V_1 is the molar volume of solvent and σ_1, σ_2 are related to the energy required to evaporate the materials:

$$\sigma_1 = (\Delta E_{1,vap}/V_1)^{1/2}, \qquad \sigma_2 = (\Delta E_{2,vap}/V_2)^{1/2} \qquad (14.7\text{-}5)$$

Therefore, the free energy of mixing two liquid-like materials can be written as

$$G = RT(n_1 \ln \phi_1 + n_2 \ln \phi_2) + wn_2\phi_2 \qquad (14.7\text{-}6)$$

Flory [27] and Huggins [36] found that Eq. (14.7-6) was suitably accurate at very dilute solutions of polymer in solvent. However, at concentrations of polymer in excess of 0.1%, significant error was seen. The last term on the right of Eq. (14.7-6) is an excess free-energy term. Investigation showed that although the first two terms on the right were correct, the last term had to be modified as follows:

$$\Delta G/RT = n_1 \ln \phi_1 + n_2 \ln \phi_2 + \chi \phi_1 \phi_2 \qquad (14.7\text{-}7)$$

χ is a dimensionless quantity that is a function of the interaction energy characteristic of any solvent-polymer pair. This parameter, called the interaction parameter or the Flory interaction parameter, is an experimentally determined correction factor for the nonideality of the solvent-polymer system. The chemical potential μ_1 of the solvent in solution relative to its potential μ_1° in the free state is given as the rate of change of the free energy with respect to n_1:

$$\partial(\Delta G/RT)/\partial n_1 = (\mu_1 - \mu_1^\circ)/RT$$
$$= \ln(1 - \phi_2) + (1 - 1/m)\phi_2 + \chi\phi_2^2 \qquad (14.7\text{-}8)$$

According to thermodynamic stability, the conditions for incipient phase separation are

$$(\partial\mu_1/\partial\phi_2)_{T,P} = (\partial^2\mu_1/\partial\phi_2^2)_{T,P} = 0 \qquad (14.7\text{-}9)$$

Applying these criteria to Eq. (14.7-8) and reducing, we obtain

$$\phi_{2c} = (1 + m^{1/2})^{-1}, \qquad \chi_c = (1 + m^{1/2})^2/2m \doteq (1 + m^{1/2})/2 \qquad (14.7\text{-}10)$$

Thus, the polymer and solvent could be miscible below these critical values over the entire range of ϕ_1. Above these values, two phases could form. Note that m in general is a very large number, on the order of 10^4. Thus, ϕ_{2c}, should be less than 0.01, and the value of χ_c approaches 1/2. Subsequent work has shown that these criteria are valid for most nonpolar systems but that additional terms must be added to account for polarity in both solvent and polymers.

One of the useful guides that can be obtained from this early work is the concept of a theta temperature or theta solvent. Flory [27] shows that with complete generality the excess chemical potential from Eq. (14.7-8) can be written as

$$(\mu_1 - \mu_1^\circ)^E/RT = (k - y)\phi_2^2 \tag{14.7-11}$$

where k and y are enthalpy and entropy terms represented by

$$\Delta H^E = RTk\phi_2^2, \qquad \Delta S^E = RTy\phi_2^2 \tag{14.7-12}$$

Comparing Eqs. (14.7-11) and (14.7-8), it can be shown that

$$k - y = \chi - 1/2 \tag{14.7-13}$$

If an ideal temperature for solution is defined as θ, it can be shown that

$$\theta = kT/y \tag{14.7-14}$$

such that Eq. (14.7-12) becomes

$$(\mu_1 - \mu_1^\circ)^E/RT = -y(1 - \theta/T)\phi_2^2 \tag{14.7-15}$$

One way of looking at the meaning of θ is to consider that for a poor solvent k and k/y are positive. Thus, θ is also positive. At a temperature where $\theta = T$, the right-hand side of Eq. (14.7-15) is zero and thus so is the excess molar free energy. Thus, the system should be an ideal solution. Note also that this implies that χ is an approximate function of $1/T$. It can be shown also that Eq. (14.7-10) can be used to obtain

$$1/T_c = (1/\theta)[1 + (1/y)(\chi_c - 1/2 + 1/2m)] \tag{14.7-16}$$

Again, as m becomes large, $T_c = \theta$. Thus, the critical temperature is called the theta temperature and represents the temperature at which two phases form. Several sources give extensive lists of θ-solvents [37, 73]. In Table 14.7-1 are representative values of θ. The primary uses of θ-solvents are in selective precipitation of a given material from a solution and in separation of one material from another for the purposes of chemical analysis.

As mentioned, the Flory-Huggins form for the Gibbs free energy of mixing is not as accurate as one might want. Several investigators have modified the theory, either through empirical means or through molecular thermodynamic considerations, to arrive at analyses that are more accurate. Unfortunately, the resulting equations are considerably more complex than the Flory-Huggins equation and as a result require input data that are not readily available. For example, Heil and Prausnitz [38] review all the

Table 14.7-1

Thermodynamically Ideal Polymer-Solvent Systems [37]

Polymer	Solvent	θ (°C)
Cellulose nitrate (12.6% N)	Ethanol	.28
Cellulose tricaprylate	Dimethyl formamide	140
	γ-Phenylpropanol	48
Gutta-percha	n-Propyl acetate	60
Polyacrylic acid	1-4-Dioxane	30
Polybutadiene	Diethyl ketone	a
	Diisobutyl ketone	a
	Methyl amyl ketone	a
	Methyl-n-propyl ketone	a
Poly-n-butyl methacrylate	Isopropanol	21.5
Polychlorotrifluoroethylene	2-5-Dichlorobenzotrifluoride	130
Polydimethylsiloxane	Methyl ethyl ketone	20
	Phenetol	83
Poly-n-hexyl methacrylate	Isopropanol	32.6
Polyisobutylene	Anisole	105.5
	Benzene	24.4
	Diisobutyl ketone	58.0
	Diphenyl ether	148
	Ethyl benzene	-24
	Ethyl caproate	57
	Ethyl-n-heptanoate	34
	Phenetol	86
	Toluene	-13
Polyisoprene	Diethyl ketone	a
	Methyl amyl ketone	a
	Methyl isobutyl ketone	a

Table 14.7-1   (continued)

Polymer	Solvent	θ (°C)
Polymethyl methacrylate	Di-n-propyl ketone	32
	m-Xylol	27
Polystyrene	Cyclohexane	34.1
	Ethyl cyclohexane	70
	Methyl cyclohexane	70.5
Rubber	Methyl-n-propyl ketone	14.4

[a]Depends on polymerization conditions.

existing theories and propose replacing the global volume fractions ϕ with local volume fractions z:

$$\Delta g/RT = x_1 \ln z_{11} + s_2 x_2 \ln z_{22} + (1 - s_2)x_2 \ln \phi_2$$

$$+ x_1 z_{21}(g_{12} - g_{11})/RT + s_2 x_2 z_{12}(g_{12} - g_{22})/RT \qquad (14.7\text{-}17)$$

where Δg is the Gibbs free energy per mole of mixture; s is the number of segments in a polysegmented molecule; x_1, x_2 are the mole fractions of solvent and polymer, respectively; and $g_{12} - g_{11}$ and $g_{12} - g_{22}$ are adjustable parameters representing molecular interaction energies. z's are local volume fractions that obey conservation equations:

$$z_{12} + z_{22} = 1, \qquad z_{21} = z_{11} = 1 \qquad (14.7\text{-}18)$$

The z's are found to be functions of the g's such that only $g_{12} - g_{11}$ and $g_{21} - g_{22}$ are unknown parameters. Nevertheless, there are now two parameters rather than the one Flory interaction parameter. Heil and Prausnitz state that any description of phase equilibrium must give a reasonably accurate representation of the variation of chemical potential with polymer concentration. It must contain only a small number of adjustable parameters that can be uniquely defined and can be somehow related to physical processes. The form should predict equilibrium conditions for mixed solvent-polymer systems from solvent-polymer binary data. And the expression should be thermodynamically consistent. While their approach satisfies all these criteria, it nevertheless requires interpretation of interactions in local volumetric regions around the segments of the molecules. But the Heil-Prausnitz model should be considered as an extension of the Flory-Huggins model. In the limit, it reduces to the Flory-Huggins model.

There has been one notable attempt to characterize polymer solutions from traditional nonstatistical activity coefficient considerations. Maron and coworkers [39-45] consider the behavior of various types of molecules in solution in terms of the effective volume that they occupy. They include also an interaction parameter that is dependent on temperature and concentration. The effective volume or packing factor e is related to the volume fraction of polymer ϕ_2 by

$$1/e = 1/e_0 + S\phi_2 \qquad (14.7\text{-}19)$$

where e_0 is the value of e at $\phi_2 = 0$. S is a constant. This is rewritten in terms of e_∞, the smallest effective volume factor the solute may have in solution:

$$e/e_0 = 1/[1 + (e_0 - e_\infty)\phi_2] \qquad (14.7\text{-}20)$$

If e_0 and e_∞ are chosen correctly, this expression should be ~~~~ binary systems, whether polar or ~~~ ~~ coworkers assume that e is independent ~~ ~~perature in the range of 20 to 50°C. They obtain the following expressions for the Gibbs free energy of mixing, the enthalpy of mixing, and the entropy of mixing:

$$\Delta G/RT = \chi° + \mu n_1 \phi_2 \qquad (14.7\text{-}21)$$

$$\Delta H/RT = T\lambda n_1 \phi_2 \qquad (14.7\text{-}22)$$

$$\Delta S/R = \chi° - (\mu + T\lambda)n_1 \phi_2 \qquad (14.7\text{-}23)$$

$\chi°$, μ, and λ are forms of interaction parameters:

$$\chi° = n_1 \ln \phi_1 + n_2 \ln \phi_2 + (n_1 + n_2) \ln(V°/V) + n_2 \ln(e/e°) \qquad (14.7\text{-}24)$$

$$\mu = \mu_{12} + \mu_{11} + \mu_{22} \qquad (14.7\text{-}25)$$

$$\lambda = \lambda_{12} + \lambda_{11} + \lambda_{22} \qquad (14.7\text{-}26)$$

μ is an interaction parameter, in much the same form as the g_{ij} of Heil and Prausnitz. The interaction between solvent molecules is given as λ_{11}, between polymer molecules as λ_{22}, and between solvent and polymer molecules as λ_{12}. The superscript is the pure state. The interaction parameters are defined as the rate of change of the μ-interaction parameters with temperature:

$$\lambda_{ij} = (\partial \mu_{ij}/\partial T)_{\phi_2, P} \qquad (14.7\text{-}27)$$

Application of this theory shows that for a rubber-benzene system the heat of mixing depends on the derivative of the interaction parameter with temperature λ rather than directly on the interaction parameter μ, as predicted by the Flory-Huggins theory. This is shown in Fig. 14.7-1. Therefore, the heat of mixing for this system yielded an expression for μ as a function of ϕ_2 and temperature with an accuracy of 4%. The theory was applied to osmotic pressure behavior of polystyrene and polyisobutylene in various solvents. They found that for each polymer-solvent system at a given temperature the interaction parameter μ, being linearly dependent on ϕ_2, was sufficient to account for the behavior of solutions as concentrated as $\phi_2 = 0.90$. The linear dependence is

$$\mu = \mu^\circ + \sigma^\circ \phi_2 \qquad (14.7\text{-}28)$$

where μ° and σ° are constants that can be determined from osmotic pressure data at polymer concentrations necessarily higher than those needed for molecular-weight determinations. In Fig. 14.7-2 is the dependence of μ° and σ° on polymer molecular weight. Note also in Fig. 14.7-3 for these plastics in several solvents that $\mu - \sigma \phi_1$ is linearly dependent on ϕ_2 also. Extrapolation to zero values of ϕ_2 yields the constants. They also show that the Flory interaction parameter does not measure explicitly the interactions in either dilute or concentrated solutions. The relationship between μ and χ is given as

$$\mu - \sigma \phi_1 = \chi \qquad (14.7\text{-}29)$$

A graphical method for obtaining μ and σ is outlined in detail.

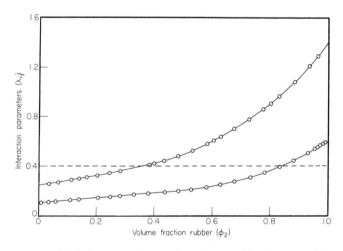

Figure 14.7-1   Experimental and theoretical observations of Flory interaction parameters for rubber in benzene at 25°C. Dashed line represents Flory-Huggins theory [39].

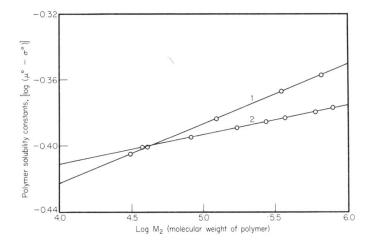

Figure 14.7-2   Effect of molecular weight of polymer on osmotic solubility constants for two polymer-solvent systems. Curve 1: polystyrene-toluene; curve 2: polyisobutylene-cyclohexane [43].

Extending the theory to critical miscibility, Maron and coworkers found that for the limit of infinite molecular weight, where $\chi_c = 1/2$, $\mu_c = \mu_c^\circ = 2/3$, $\sigma_c^\circ = 1/6$, and $\mu_c^\circ - \sigma_c^\circ = 1/2$. Note that $\sigma = (\partial\mu/\partial\phi_2)_{T,P}$ in Eq. (14.7-29) and in Fig. 14.7-2. These results were confirmed in the specific case of polystyrene-cyclohexane, and partial miscibility diagrams for this system agreed much better with the Maron theory than with the Flory-Huggins theory. The experimental and computed results for this system are shown in Fig. 14.7-4. Note also in Fig. 14.7-5 that the critical volume fraction of polymer decreases strongly with increasing molecular weight.

These semiempirical methods have been generalized to multicomponent solutions without further assumptions being necessary. Heil and Prausnitz demonstrate a rapid graphical method that allows prediction of solubility limits in ternary systems containing one polymer and a mixed solvent. Extensive verification of these theories are limited owing to the lack of reliable thermodynamic data. As shown in Fig. 14.7-6, phase diagrams among solvents, nonsolvents, and polymers can be established using the Gibbs free energy of mixing as a base. Note the effect of molecular weight, as represented by increasing values of m, on the solubility envelope. The solubility threshold is found by extending a line from the 100% polymer apex to a tangent with the solubility curve. A typical phase diagram for a polymer-polymer-solvent system is shown in Fig. 14.7-7. The polymer-polymer interaction parameter is the factor that positions the solubility curve. A typical experimental phase diagram for rubber-PMMA-benzene is shown in Fig. 14.7-8.

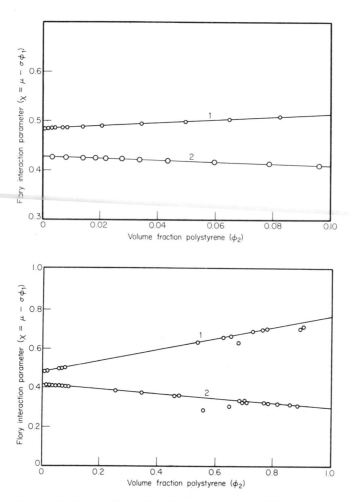

Figure 14.7-3   Effect of solvent type on the Flory interaction parameter
$\chi = \mu - \sigma\phi_1$ for polystyrene and polyisobutylene.   Curve 1: methyl ethyl ketone (27°C); curve 2: toluene (27°C); curve 3: benzene (24.5°C); curve 4:
cyclohexane (30°C); curve 5: ethyl acetate (27°C); curve 6: methyl ethyl
ketone (49.05°C); curve 7: toluene (69.20°C) [40].

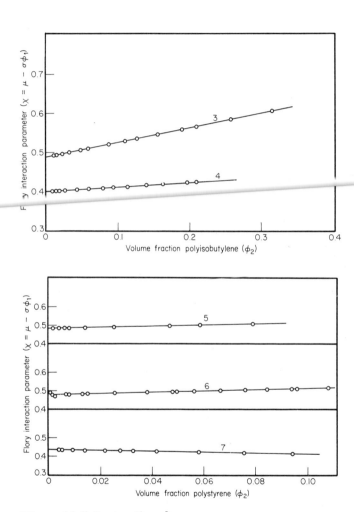

Figure 14.7-3 (continued)

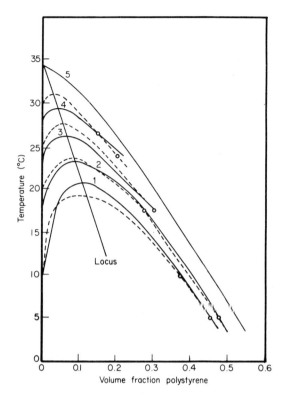

Figure 14.7-4 Solubility of polystyrene in cyclohexane as a function of temperature. Parameter is molecular weight. Curve 1: $M_n = 43,600$; curve 2: $M_n = 89,000$; curve 3: $M_n = 250,000$; curve 4: $M_n = 1,270,000$; curve 5: $M_n = \infty$. Dotted lines, observed data; dashed lines, predicted from Maron theory. Locus line represents critical volume fraction [44].

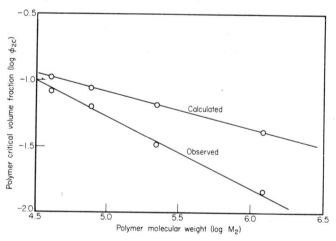

Figure 14.7-5 Comparison of calculated and observed dependency of critical polymer volume fraction (ϕ_{2c}) on polymer molecular weight (M_2) for GPS in cyclohexane [44].

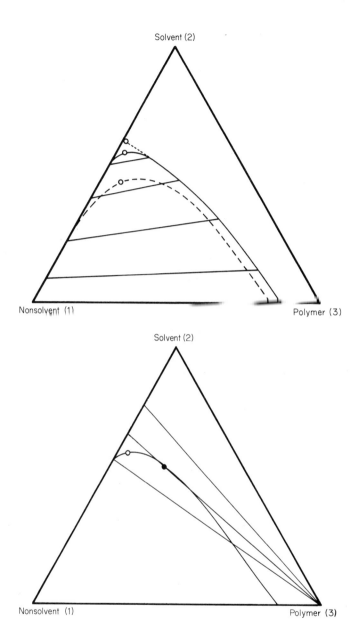

Figure 14.7-6   Three-component phase diagram schematics, $x_1 = x_2 = 1$.
On top, $x_3 = 10$ (dashed line), $= 100$ (solid line), $= \infty$ (dotted line). Flory
interaction parameters, $\chi_{12} = \chi_{13} = 1.5$, $\chi_{23} = 0$. Two-phase tie lines are
shown for $x_3 = 100$. Critical points are open circles. On bottom diagram,
precipitation threshold indicated by solid circle for $x_3 = 100$. Three solvent
ratio lines are shown [27].

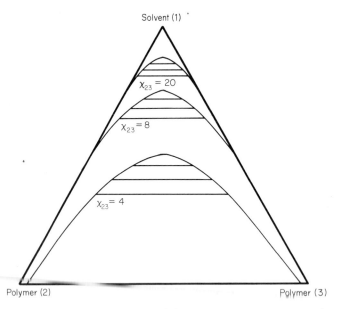

Figure 14.7-7   Schematic of three-component system where polymers (2) and (3) interact equally with solvent (1). Polymer-polymer Flory interaction parameters are shown. $x_2 = x_3 = 1000$ [27].

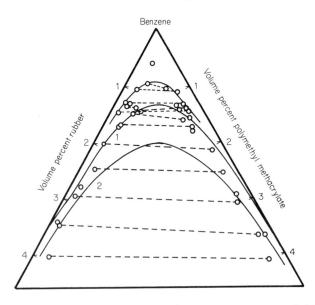

Figure 14.7-8   Experimental three-component solubility diagram for natural rubber-PMMA-benzene.   Dotted lines are tie lines.   Curve 1: binodial for $m\chi_{12} = 154$; curve 2: binodial for $m\chi_{12} = 100$ [46].

Two molecular weights of PMMA were used: $\overline{M}_n = 1.6 \times 10^6$ and $= 3 \times 10^5$. From this diagram, values for m and x_{12}, x_{23}, and x_{13} can be obtained. Flory states that the tie lines should be horizontal for nonpolar groups. The data in Fig. 14.7-8 show nonhorizontal tie lines, indicating that the difference in polarity and possibly in chain length may also be a factor in determining the solubility of these systems [46,47].

Paxton [48] has examined polybutadiene-PS-solvent in a similar fashion and has applied the Maron theory to his experimental data. As shown in Fig. 14.7-9, the solubility envelope is quite dependent on the choice of solvent. Further, Paxton points out that the packing factor at infinite dilution e_0 is directly related to the intrinsic viscosity:

$$e_0 = \eta_{int,v}/2 \tag{14.7-30}$$

e_∞ for polystyrene solutions is approximately 4, and the effective volume of a bulk polymer is similar to the value of close-packed spheres $e^\circ = 1.35$. Suitable ～ to allow for multiple phases yields an expression for x°:

$$-x^\circ/n_2\phi_3 = (n_1\phi_2/n_2\phi_3)[x_{12} + (\phi_3/\phi_2)x_{13}] + x_{23} \tag{14.7-31}$$

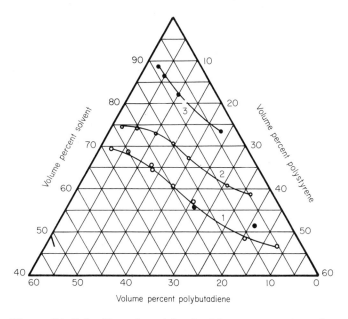

Volume percent polybutadiene

Figure 14.7-9 Experimental miscible concentrations for GPS and polybutadiene for various solvents at 27°C. Curve 1: benzene; curve 2: toluene; curve 3: CCl_4 [48].

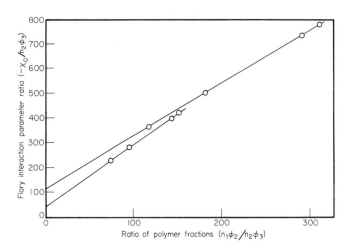

Figure 14.7-10   Experimental verification of Eq. (14.7-31) for obtaining Flory interaction parameters. Data for two types of polybutadienes shown [48].

As shown in Fig. 14.7-10, Eq. (14.7-31) predicts a straight line, in agreement with experimental data. Thus, with suitable modifications, the Maron theory can also be adapted for prediction of the thermodynamic properties of multicomponent polymer-solvent systems.

14.8   Gas-Polymer Systems

It is frequently necessary to know the rate of diffusion of gases through polymer systems. Films used in food wrap can have great sensitivity to permeation of O_2 and thus must be selected carefully. The formation of structural foams depends on the solubility and diffusion of gases through the polymer melt system. The rate of diffusion of water vapor through plastic pellets must be known in order to design proper dehydration systems. It has been widely demonstrated that Henry's law holds for the relationship between the concentration of sorbed gas and the partial pressure of the gas at an interface. This can be written as

$$c = Hp \qquad\qquad\qquad (14.8\text{-}1)$$

where c is the gas concentration [cc (STP)/g plastic], p is partial pressure (atm), and H is Henry's law constant for the system [cc (STP)/g atm]. As shown in Fig. 14.8-1, for permanent or noncondensible gases in glassy polymers, Henry's law seems applicable to gas-polymer systems with pressures of more than 20 atm [52, 56, 74, 75]. Some experimental values for H are given in Table 14.8-1.

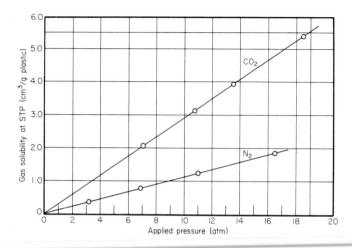

Figure 14.8-1   Solubility of N_2 and CO_2 into polyethylene at 372°F [56].

As expected, the Henry's law constant is Arrhenius dependent on temperature, as

$$H = H_0 \exp[-E_s/R(T - T_0)] \tag{14.8-2}$$

Veith and Sladek [49] consider inclusion of voids in the glassy materials. For many materials examined, it appeared that there was a decrease in solubility accompanying crystallization. This decrease was smaller than the corresponding reduction in the amorphous volume fraction, however. Thus, it was postulated that because crystallization was a volume reduction phenomenon, small microvoids were formed in the amorphous region. They thus added a Langmuir sorption isotherm to the right-hand side of Eq. (14.8-1), yielding

$$C = C_H + C_D = C_{H'}bp/(1 + bp) + Hpa \tag{14.8-3}$$

$C_{H'}$ is a hole saturation constant, b is a hole affinity constant (atm^{-1}), and a is the amorphous volume fraction. As is seen for the semicrystalline material, polyethylene terephthalate, in Fig. 14.8-2, the solubility curve is not linear, as would be predicted by Eq. (14.8-1). Thus, it appears that for any semicrystalline material the presence of microvoids might explain the nonlinear behavior of the solubility of a gas in the amorphous regions of a plastic.

The solubility of small molecules of water vapor has been investigated by Klute [50, 51]. He found that Henry's law consistently underestimated the solubility of water in LDPE and attributed the error to the crystalline nature of the polymer. If the diffusion coefficient is assumed to be

Table 14.8-1

Experimental Values of Henry's Law Constant at 371°F [56]

Polymer	Gas	Henry's law constant [cc (STP)/g atm]
Polyethylene	Nitrogen	0.111
	Carbon dioxide	0.275
	Monochlorodifluoromethane	0.435
	Argon	0.133
	Helium	0.038
Polypropylene	Nitrogen	0.133
	Carbon dioxide	0.228
	Monochlorodifluoromethane	0.499
	Argon	0.176
	Helium	0.086
Polyisobutylene	Nitrogen	0.057
	Carbon dioxide	0.210
	Krypton	0.114
	Argon	0.102
	Helium	0.043
Polystyrene	Nitrogen	0.049
	Carbon dioxide	0.220
	Monochlorodifluoromethane	0.388
	Argon	0.093
	Helium	0.029
Polymethyl methacrylate	Nitrogen	0.045
	Carbon dioxide	0.260
	Krypton	0.122
	Argon	0.105
	Neon	0.126
	Helium	0.066

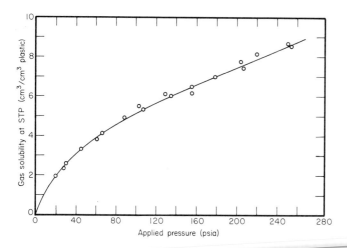

Figure ....y of CO_2 in one-mil Mylar film at 40°C [49].

independent of concentration, Fick's law predicts the rate at which the non-condensible materials pass a given point under the applied concentration gradient:

$$J = -D(dC/dx) \qquad (14.8-4)$$

where J is the flux [cc (STP)/time area], D is the diffusion coefficient or diffusivity, and x is the distance into the polymer, measured from the surface. If steady state across a membrane L units thick can be assumed, $C = C_1$ at $x = 0$ and $C = C_2$ at $x = L$. Thus,

$$J = D(C_1 - C_2)/L \qquad (14.8-5)$$

If H is a function only of temperature and the system is isothermal, Eq. (14.8-1) can be substituted for concentration, yielding

$$P = DH = JL/(p_1 - p_2) \qquad (14.8-6)$$

Thus, the product of the diffusivity and the Henry's law constant is the permeability P. Accordingly, for a semicrystalline material, the permeability is only a function of the amount of amorphous material present. Klute writes this as

$$P' = PaY \qquad (14.8-7)$$

where a is the volume fraction of amorphous material and Y is a parameter that describes the distortion of lines of diffusion around the spherulitic crystals dispersed throughout the material. Y is seen to be a nearly linearly decreasing function with the extent of crystallization in Fig. 14.8-3. The simplest form for this function is

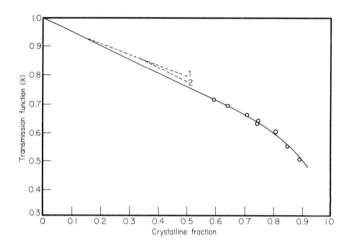

Figure 14.8-3   Effect of extent of crystallinity on transmission function Y
for gas permeability into plastics. Data are for mixed spheres. Curve 1:
theoretically predicted from Maxwell's equation; curve 2: theoretically
predicted from Runge's equation [50].

$$Y = 2/(3 - a) \tag{14.8-8}$$

which has been derived for identical spheres randomly embedded in a matrix.
Y is frequently called the transmission function. Note that this form of the
equation is nearly identical with the microvoid theory discussed above.

The temperature dependency of the diffusivity has been analyzed by
several investigators [52, 56]. It is important to establish the proper form
for the diffusivity, however, before proper analysis of the data can be accomplished. As mentioned above, if the diffusivity is assumed to be independent of concentration, Fick's law holds. If the process for determining
concentration changes is time dependent, then

$$\partial C / \partial t = D(\partial^2 C / \partial x^2) \tag{14.8-9}$$

A typical experimental program will use a fresh sample of plastic that is
exposed to the selected diluent instantaneously at $t = 0$. The pressure drop
or concentration is monitored as a function of time. Since the Henry's law
constant is independent of concentration, measurement of pressure is directly
related to measurement of gas concentration. Since the diffusional process
into a polymer is very slow, the polymer can be assumed to be infinitely
deep over normal periods of time. Therefore, the solution to Eq. (14.8-9)
is analogous to the solution of the transient one-dimensional heat conduction
problem when the surface has been exposed to a step change in surface temperature. This solution has been detailed in Chap. 6. For the infinitely
deep polymer, the following describes the concentration profile:

$$C'' = C_0'' \, \text{erf}[x/2(Dt)^{1/2}] \tag{14.8-10}$$

where C_0'' is the initial excess concentration and the excess concentration at the surface is assumed to be zero after $t = 0$. The change in C_0''/C_0'' as a function of $x/2(Dt)^{1/2}$ is shown in Fig. 14.8-4 [53]. Normally, for liquid diluents, the concentration profile can be monitored for all x, using birefringence or interferometric methods [54,55]. Thus, exact values of D can be determined as functions of time, position can be superimposed, and as a result the concentration at one position at one time should be relatable to that at another position at another time. If this is not the case, the diffusivity may be concentration dependent. An experimental concentration profile for the diffusion of dimethylformamide into polyacrylonitrile is shown in Fig. 14.8-5. For gaseous diluents, the sample is frequently saturated with the gas at an elevated pressure. In this manner, $C = HP^* = C^*$ ($0 \le x \le L$, $t \le 0$). At $t = 0$, the sample is introduced into an environment where $C = HP^{**} = C^{**}$ ($x = 0$, $t \ge 0$). The base of the specimen is covered with a non-permeable membrane. The solution to Eq. (14.8-10) using these boundary conditions is

$$(C - C^*)/(C^{**} - C^*) = 1 - (4/\pi) \sum_{n=0}^{\infty} (-1)^n/(2n + 1) \, \exp[-D(2n + 1)^2 \pi^2 t/4L^2]$$

$$\cdot \cos[(2n + 1)\pi x/2L] \tag{14.8-11}$$

The experimental drop in pressure and that calculated through best fit for a given value of diffusivity for argon-polyethylene is shown in Fig. 14.8-6. Experimental values of the diffusivity are given in Table 14.8-2. Since a

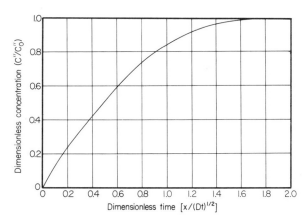

Figure 14.8-4 Concentration as a function of dimensionless time into a semiinfinite slab [53].

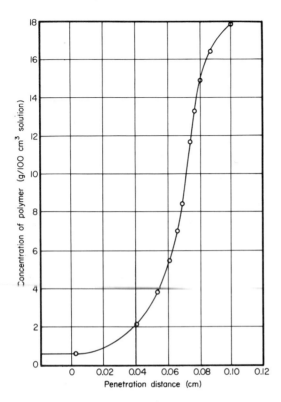

Figure 14.8-5   Concentration profile of dimethylformamide diffusing into polyacrylonitrile after 60 sec; measured using interferometric methods [54].

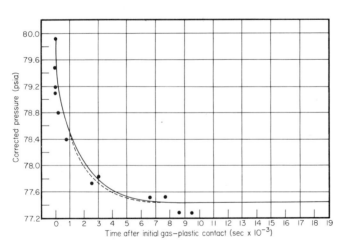

Figure 14.8-6   Pressure profile for argon-polyethylene at 77.4 psia and 368°F. Solid line, $D = 9.19 \times 10^{-5}$; dashed line, $D = 9.61 \times 10^{-5}$ [56].

Table 14.8-2

Experimental Values of Diffusion Coefficient at 371°F [56]

Polymer	Gas	Diffusion coefficient $(cm^2/sec \times 10^5)$
Polyethylene	Nitrogen	6.04
	Carbon dioxide	5.69
	Monochlorodifluoromethane	4.16
	Argon	9.19
	Helium	17.09
Polypropylene	Nitrogen	-
	Carbon dioxide	4.25
	Monochlorodifluoromethane	4.02
	Argon	7.40
	Helium	10.51
Polyisobutylene	Nitrogen	2.04
	Carbon dioxide	3.37
	Krypton	7.30
	Argon	5.18
	Helium	12.96

proper temperature dependency for diffusivity is the Arrhenius equation, the corresponding energies of activation for diffusion E_d are given in Table 14.8-3 for selected systems. Note from Eq. (14.8-6) that the permeability for gases in glassy polymers is the product of the diffusivity and Henry's law constant. Therefore, the permeability should also be Arrhenius dependent on temperature. This is shown in Table 14.8-4 for the permeability of water vapor in polyethylene.

Note that the above analysis is based on a diffusion coefficient that is independent of diluent concentration. Several polymer-diluent systems show abnormal behavior, however, with diffusivities that are based on the Fickian form of the diffusion and mass transfer equation showing high concentration dependence. Diffusion with concentration-dependent coefficients is called non-Fickian diffusion. There are two general approaches to the

Table 14.8-3

Heats of Solution and Energies of Activation [52]

Gas	Polymer	Temperature range (°C)	E_s (kcal/mol)	E_d (kcal/mol)
CO_2	Molten polyethylene	188–224	-0.80	4.4
N_2	Molten polyethylene	125–188	0.95	2.0
CO_2	Molten polyethylene	188–224	-1.7	3.0
N_2	Thermally softened polystyrene	120–188	—	10.1
H_2	Thermally softened polystyrene	119–188	—	9.6
N_2	Thermally softened polystyrene	136–188	-1.7	—
H_2	Thermally softened polystyrene	126–188	-1.9	—
CH_4	Thermally softened polystyrene	125–188	1.05	3.6

Table 14.8-4

Activation Energies for the Permeation of Water Vapor in LDPE
Calculated from Data Reported in the Literature [51]

P^a	T (°C)	

A. Data for a cast film

0.032	15.0	E_p = 6590 cal/mol·°C
0.045	25.0	$\log P_0$ = +3.50
0.068	35.0	$\log P_{(25°)}$ = -1.35

B. Data for a cast film

0.045	25.0	E_p = 7900 cal/mol·°C
0.086	40.0	$\log P_0$ = +4.44
		$\log P_{(25°)}$ = -1.35

C. Data for molded bottles[b]

0.038	25.0	E_p = 8140 cal/mol·°C
0.067	37.8	$\log P_0$ = +4.55
0.134	54.5	$\log P_{(25°)}$ = -1.42
0.263	73.9	

D. Data for a calendered film

0.005	0.0	E_p = 8390 cal/mol·°C
0.042	25.0	$\log P_0$ = +4.77
0.060	32.0	$\log P_{(25°)}$ = -1.37
0.077	38.0	

E. Data for a cast film

0.017	15.0	E_p = 9380 cal/mol·°C
0.022	20.0	$\log P_0$ = +5.35
0.025	22.0	$\log P_{(25°)}$ = -1.54
0.055	40.0	
0.154	60.0	
0.347	80.0	

F. Data for a blown film

0.017	25.0	E_p = 14,540 cal/mol·°C
0.025	30.0	$\log P_0$ = +8.90
		$\log P_{(25°)}$ = -1.78

[a] $P = P_0 \exp(-E_p/RT)$, where P is expressed in g·mm/m^2·day·cm Hg.
[b] Calculated from permeability to liquid water; published value at 21.1°C rejected as being in error.

understanding of non-Fickian diffusion. Rewriting the Fickian diffusion
equation in terms of activities rather than concentrations or partial pressures to allow the introduction of nonideality, as discussed in Sec. 14.7 is
a viable suggestion. An alternative is the direct introduction of the concentration-dependent diffusion coefficient $D*(C)$. These are explored briefly
below.

Jost [57] shows that the most general form for the diffusion equation
relates the mass flux J to the spatial gradient of the chemical potential μ.
Fick assumed that it was related to the spatial gradient of the concentration,
as shown in Eq. (14.8-4). Therefore,

$$J = -(Cu/N)(\partial\mu/\partial x) \tag{14.8-12}$$

where C is the molar concentration, u is the mobility of the diffusing particle for a unit force, and N is Avogadro's number. The chemical potential
for any system is given in terms of its activity a as

$$\mu - \mu^0 = NkT \ln a \tag{14.8-13}$$

For an ideal solution, a is equal to the mole fraction x of the diluent. For
a nonideal solution, $a = C\gamma$, where γ is commonly called an activity coefficient. It represents the deviation from nonideality [1] and is directly de-
rivable from the work for polymer-solvent systems given in Sec. 14.7. Now
Eq. (14.8-12) can be rewritten in terms of the activity and its coefficient:

$$J = -CukT(\partial \ln a/\partial x) = -D°(\partial \ln a/\partial \ln C)(\partial C/\partial x)$$

$$= -D°[1 + (\partial \ln \gamma/\partial \ln C)](\partial C/\partial x) \tag{14.8-14}$$

In comparing Eq. (14.8-14) with Eq. (14.8-4), it is apparent that

$$D = D°[1 + (\partial \ln \gamma/\partial \ln C)] \tag{14.8-15}$$

For many simple systems, activities and their activity coefficients are known
over the full range of concentration. As was seen in Sec. 14.7, however,
the data for polymer-solvent systems are available only for very low concentrations. Morrison [33], for example, compares the diffusivity based on
concentration with that based on chemical potential for three diluent vapors
in polyvinyl acetate. An example is shown in Fig. 14.8-7 for acetone in
polyvinyl acetate to a concentration of 9% weight. Although the diffusivity
based on chemical potential should be less sensitive to concentration than
that based on concentration, it is not independent of concentration as many
have assumed. And since the former diffusivity requires accurate activity-
concentration data, it is more difficult to display. Therefore, most experimenters report the concentration-based diffusivity and note that it is highly
concentration dependent.

The concentration dependency of the diffusivity based on concentration
has been examined in detail by many investigators [58,59], and mathematical
and empirical attempts have been made to describe this dependency. At

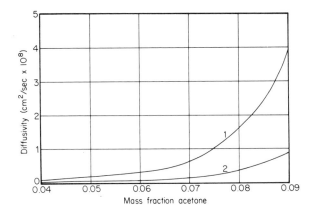

Figure 14.8-7 Concentration-dependent diffusivity of acetone in polyvinyl acetate. Curve 1: diffusivity based on mass fraction ~~~~~~~~~~~~~~~~~~~~~~~ ~~~~~~~~~~~~~~ thermodynamic activity [33].

present, however, no method is fully reliable. Kokes and Long propose an exponential concentration dependence of the form

$$D(C) = D* \exp(dC) \qquad (14.8\text{-}16)$$

where d is a characteristic constant that has some association with the Flory interaction parameter [59, 60]. As is seen in representative data in Figs. 14.8-8 and 14.8-9 for polymethyl methacrylate and ethyl acetate and polymethyl methacrylate and water vapor, there is no clear choice for d. Furthermore, hysteresis can also take place upon absorption and desorption, thus making the establishment of rules governing d very obscure. Nevertheless, the choice of an exponentially dependent form for the diffusivity is analytically satisfying to analysts solving the transient one-dimensional diffusion equation:

$$C/\partial t = D*\{\partial[\partial \exp(dC)C/\partial x]/\partial x\} \qquad (14.8\text{-}17)$$

Crank and Park [61] list many solutions to this form of the non-Fickian equation. Others warn that care must be taken in selecting the form for the diffusion equation (14.8-16) for certain forms violate the requirement of material objectivity [62]. At best, it would appear that Eq. (14.8-16) should be applied only over a very limited concentration range. Note that through suitable manipulation the concentration activity coefficient can be related to the fictitious characteristic constant d:

$$\gamma(C) = (1/dC)[\exp(dC) - 1] \qquad (14.8\text{-}18)$$

Remember, however, that this equation simply relates two artifical parameters that have been introduced to explain the nonideality of the diffusional process into polymers.

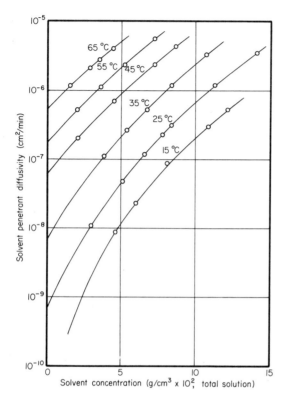

Figure 14.8-8 Diffusivity values for the diffusion of ethyl acetate into poly-methyl acrylate at various temperatures. Note that the diffusivity is plotted as a function of the penetrant concentration [58].

While analytical solutions to Eq. (14.8-17) are available [63], nu-merical methods are preferred. An example of the concentration profile as a function of the dimensionless diffusional distance into a semiinfinite slab is shown in Fig. 14.8-10 for various values of d. The curve for d = 0 is the graphical form of Eq. (14.8-10). Other solutions for various boundary conditions are also given [63].

As mentioned, the primary interest in gas-polymer systems has al-ways been in determining the barrier properties of finished articles. There are two processing areas in which this technology is now being applied. In thermoplastic structural foam molding, a gas, usually nitrogen, is dissolved under relatively high pressures (500 atm) in the plastic melt. When this mixture is injected into the mold cavity as a short shot, expansion takes place as the gas comes from solution. The result is a finished part having

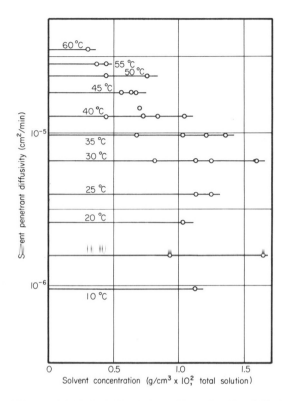

Figure 14.8-9   Diffusivity values for the diffusion of water into polymethyl acrylate at various temperatures. Note that the diffusivity is plotted as a function of the penetrant concentration. Compare the concentration dependency with that in Fig. 14.8-8 [58].

a relatively dense skin and a relatively low-density foam core. Although this technology received its impetus by Union Carbide Corporation in the late 1960s, projections indicate that more than 1 billion lb will be used by the mid-1980s [64].

An older area that uses gas-polymer diffusion and solubility is expanded polystyrene bead production. Products from this operation, commercialized in the late 1950s, include flotation devices, insulated picnic coolers, wig forms, and hot drink cups. Extensive experimental studies [65, 66] show that when a small bead of polystyrene is soaked in a poor solvent, such as a hydrocarbon-based liquid (methylene chloride, pentane, butane), and then exposed to 100°C steam, the bead expands and softens. Expansion rates are reported to be twice those predicted by simple vaporization of the solvent. As a result very low-density foams are achieved (less than 0.5 lb/ft^3).

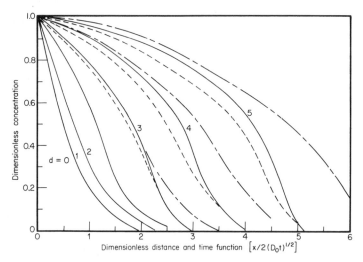

Figure 14.8-10   Computed dimensionless concentration as a function of dimensionless distance and time for diffusion into a semiinfinite slab, assuming an exponential dependency of diffusivity on concentration: $D = D^* \exp(dC)$. Solid lines, computed using the Boltzmann equation; dashed lines, computed by solving the partial differential equations for transient response; dash-dot lines, correction to the analog computer model for simple approximations. Parameter is the exponential factor d [63].

Table 14.8-5

Gas Permeability in Thermoplastic Films [65]

| Film material | Permeability of film to[a] | | | |
	N_2	O_2	CO_2	H_2O
Polyvinylidene chloride[b]	0.0094	0.053	0.29	14
Polychlorotrifluoroethylene	0.03	0.10	0.72	2.9
Polyethylene terephthalate[c]	0.05	0.22	1.53	1,300
Polyamide (Nylon 6)	0.10	0.38	1.6	7,000
Rigid PVC	0.40	1.20	10	1,560
Cellulose acetate	2.8	7.8	68	75,000
HDPE (0.954-0.960)	2.7	10.6	35	130
LDPE (0.922)	19	55	352	800
GP polystyrene	2.9	11	88	12,000
PP (0.91)	—	23	92	680
Ethylcellulose (plasticized)	94	265	2000	130,000

[a] Permeability units are $(cm^3/cm^2/mm/sec/cm\ Hg) \times 10^{10}$; N_2, O_2, and CO_2 data at 30°C; H_2O data at 25°C and 90% relative humidity.
[b] Saran.
[c] Mylar A.

Skinner et al. [65] attribute this to the very rapid diffusion of water vapor into the bead and the very slow diffusion rate of solvent vapor out of the bead. As shown in Table 14.8-5, the permeability data for several resins substantiate this hypothesis.

14.9 Diffusional Mass Transfer

In spite of the extremely high viscosities of polymer-solvent systems, it is frequently necessary to remove the solvent to concentrate the system. Stripping in the conventional sense of distillation cannot be practiced successfully owing to the very high viscosity. The terminal velocities of bubbles rising in these materials are extremely small, as was seen in Chap. 7. Special columns have therefore been built that allow the polymer-solvent mixture to flow down a vertical wall in very thin films. These thin films are contacted either with stripping steam or with zero-humidity air, and placo at the interiace. Thin film evaporation is used for dehydration of condensation polymers as well. In another application, metal surfaces are dip-coated with a vinyl latex, which cures when contacted with air. The volatile solvents are driven from the latex, and O_2 permeates the matrix, hastening the cure. The curing of catalyzed thermoset polyester, particularly when mixed with styrene, requires evaporation of the styrene from the surface in order to achieve optimum cure. High humidity or high pressure translates into low styrene partial pressure, and this inhibits cure. Dryer design depends on the optimum diffusion and convection from the surface of pellets in order to achieve optimum drying times. Differential shrinkage of water-extended polyesters and compression-molded α-cellulose-filled melamine can be enhanced by differential water removal from the panels.

Two resistances to mass removal from polymeric materials act in series. Within the bulk of the material, molecular diffusion controls. At the interface, convective mass transfer controls. In many instances, one or the other can be considered dominant. For example, in the drying of bulk resin, the relative mass of the material and the very low values for diffusivity dictate that diffusion to the free surface will control the rate of moisture removal. For very porous materials, such as water-extended polyesters, the rate of movement of water to the surface is so rapid that it does not control. Rather, the rate of evaporation of water from the surface controls. The rate of removal of solvents from the interface of a polymer-solvent thin film also controls mass transfer, provided that the film is very thin and the system is stirred occasionally.

Gas diffusion in plastics matrices is frequently characterized as the diffusion of two gases, one into the matrix and the other out of the matrix. Such is the case, for example, with the diffusion of freon out of and O_2 or N_2 into a urethane foam. Thus, equimolar diffusion is written as

$$J_A = -D_{PA}(dC_A/dx) = k_A(C_{A0} - C_A) \tag{14.9-1}$$

$$J_B = -D_{PB}(dC_B/dx) = k_B(C_{B0} - C_B) \tag{14.9-2}$$

Now $J_A = J_B$. Thus, $D_{PA} = D_{PB}$. k represents a proportionality constant, here called a mass transfer coefficient. It has a great similarity to the individual film heat transfer coefficient of Chap. 6. On the other hand, if the diluent is diffusing out of the polymer matrix with no replacing diluent, as would be the case if the diluent were evaporating at the surface, then the process is concerned with the diffusion through a stagnant system. The mass flux for this system, which in reality deals with a mass balance that must move with time owing to the depletion of diluent, can be written as

$$J_A = [-D_{PA}/(1 - C_A)](dC_A/dx) \tag{14.9-3}$$

If the system can be considered quasi-stationary, then $C_A(x = 0) = C_A(x = L) = C_{AL}$, and Eq. (14.9-3) can be integrated to give

$$J_A = [-D_{PA}/L][\ln(1 - C_{AL})/(1 - C_{A0})] \tag{14.9-4}$$

If a logarithmic form for concentration is defined,

$$C_{A,\ln} = (C_{A0} - C_{AL})/\ln(1 - C_{AL})/(1 - C_{A0}) \tag{14.9-5}$$

Eq. (14.9-4) can be written as

$$J_A = [D_{PA}/LC_{A,\ln}](C_{A0} - C_{AL}) \tag{14.9-6}$$

or

$$J_A = k_A(C_{A0} - C_{AL}) \tag{14.9-7}$$

where k_A is the previously discussed mass transfer coefficient.

Again, these mass transfer coefficients are usually applied only when convection takes place, although they can be defined anywhere mass transfer takes place. For the combined resistance problem, the standard solution of the transient one-dimensional molecular diffusion equation should be coupled with convective diffusion through the stagnant layers of diluent adjacent to the interface. For this case, an overall mass transfer coefficient can be formed as

$$1/K = 1/k_g + 1/k_p \tag{14.9-8}$$

where k_g is the mass transfer coefficient in the layer of gas adjacent to the polymer interface and k_p is the diffusional form of the mass transfer coefficient within the polymer itself. In general, however, the value of K, the overall mass transfer coefficient, cannot be calculated from first principles for most cases. It can be shown from a quasi-stationary problem, however, that the mass flux is the same on each side of the interface. If the concentration at the interface is given as C_i,

$$J = k_g(C_{0g} - C_{ig}) = k_p(C_{iL} - C_{0L}) \qquad (14.9-9)$$

Here C_g is the concentration of the diluent in the gas phase and C_L is the value in the glassy polymer phase. It has been shown that the concentration at any interface is related to Henry's law such that

$$C = Hp \qquad (14.8-1)$$

where p is the partial pressure of the diluent in the gas phase. Therefore, Eq. (14.9-9) can be rewritten in terms of other coefficients as

$$J = K_g(p_0 - p_e) = K_p(C_e - C_0) \qquad (14.9-10)$$

Rearranging yields

$$1/K_g = (1/k_g)[(p_e - p_0)/(C_{ig} - C_{0g})] \qquad (14.9-11)$$

and

$$1/K_g = (1/k_g) + (1/k_p)[(p_e - C_{ig})/(C_{ig} - p_0)][(C_{ig} - p_0)/(C_{0L} - C_{iL}0] \qquad (14.9-12)$$

Now, applying Eq. (14.8-1) across the interface,

$$1/K_g = 1/k_g + H/k_p \qquad (14.9-13)$$

This illustrates the combined resistance through the polymer and the stagnant film at the interface very well. Many empirical forms for k_g based on convection from an interface have been found. Two forms seem applicable for most polymer diluent interfaces, however. Higbie [67] found that by considering that the gas surface is alternately stagnant and highly mixed, the penetration theory model could be used. In this way, the solution to Eq. (14.8-9), for partial pressure over very short periods, yields the following mass flux:

$$J = 2(p_i - p_0)(D_A t/\pi)^{1/2} \qquad (14.9-14)$$

In comparing this solution with Eq. (14.9-9), it is apparent that the mass transfer coefficient in the gas phase can be represented as

$$k_g(t) = 2(D_A t/\pi)^{1/2} \qquad (14.9-15)$$

If this is averaged over the contact time t_c, it becomes

$$k_g = 2(D_A/\pi t)^{1/2} \qquad (14.9-16)$$

Higbie's model shows that a short contact time yields the maximum mass transfer coefficient in the gas phase, thus implying that the rate of dissolution is enhanced with increasing frequency of mixing and settling. This is why surface renewal in the scraped-film thin film evaporator is necessary for a high rate of mass transfer from the polymer film. Danckwerts [68] replaced Higbie's model with another more realistic model in which each element of the surface is exposed to the environment in a random period of time. In this way the average desorption rate is given as

$$\int_{t=0}^{t=\infty} f \exp(-ft) \, dt \tag{14.9-17}$$

where f is the fraction of surface renewed. Thus, the mass flux is given as

$$J = (p_i - p_0)(D_A f)^{1/2} = k_g(p_i - p_0) \tag{14.9-18}$$

Again, k_g is defined as the proportionality constant in Eq. (14.9-18). Both theories yield similar results.

For more convective systems, such as high-velocity dryer designs, the convective gas-phase mass transfer coefficient is found to be a strong function of the velocity of the environment. If the surface can be assumed to be planar or continuous, the Gilliland-Sherwood correlation appears applicable [69]:

$$Sh = (k_g d/D_A) = 0.023(dU\rho/\eta)^{0.8}(\eta/\rho D_A)^{0.44} \tag{14.9-19}$$

The term on the left is the Sherwood number and is similar to the Nusselt number of heat transfer. The first term on the right is the characteristic Reynolds number and the second is the Schmidt number Sc, a ratio of the molecular viscous motion to the molecular diffusive motion. It is analogous to the Prandtl number of heat transfer. If the gas is flowing around discrete particles such as fluidized beads or pellets, the proper correlation is based on the diameter of the particle:

$$Sh = (k_g d_p/D_A) = 2 + (d_p U\rho/\eta)^{1/2}(\eta/\rho D_A)^{1/3} \tag{14.9-20}$$

This equation should be used for determining drying rates in convective driers, provided that diffusion of the moisture to the pellet surface is sufficiently rapid to allow neglect of the second term on the right of Eq. (14.9-8). On the other hand, the stagnant model for gas-phase diffusion should be used to calculate the rate of absorption of water vapor into the pellets.

14.10 Conclusions

Other aspects of polymer dilution are covered elsewhere. For example, the technique of solvent bonding depends on different interpretations of the thermodynamics presented above. The process of reverse osmosis is dependent on the compatibility of a polymer membrane with one type of molecule and its incompatibility with others. The dispersion of catalyst within a micelle or an atomized drop of sprayed polyester depends on diffusion with chemical reaction. In liquids, microcirculation should be considered as an important aid to diffusion. Nevertheless, the material presented in this chapter should allow the engineer to understand the general terminology and approaches to the thermodynamics of polymers.

Problems

P14.A Make a plot of the specific gravity of GPS as a function of pressure for T = 200°C and T = 240°C. Use the Spencer-Gilmore equation of state and the constants given in Table 14.3-5.

P14.B Calculate th al   GPS using the data in Table 14.3-5.

P14.C In a plunger injection molder, GPS is compressed isothermally at 200°C from 1 to 1500 atm. Using an average density calculated from P14.A, determine the heat liberated during compression.

P14.D LDPE is being foamed by dissolving N_2 into melt at 190°C and 2000 atm and then allowing the melt to be forced at high speed into a mold at atmospheric pressure. If the system is considered to be adiabatic, what is the temperature change if the specific heat of LDPE is 0.4 Btu/lb·°F?

P14.E Obtain the increase in enthalpy at 266°F for PMMA when the pressure is raised from atmospheric to 20,000 psia. Use the data given in Table 14.3-9 for your computations. Then determine the value for ω, and compare this value with that given in Table 14.3-5.

P14.F Calculate the heat capacity for PMMA using the Sakiadis-Coates method of calculation. Compare your results with those shown in Table 14.4-1A.

P14.G Calculate the thermal conductivity of linear polyethylene, and compare your value with long-chain linear hydrocarbons. Discuss the differences in values, if any.

P14.H Following the analysis presented, arrive at Eq. (14.5-16). Then show that the same result can be obtained by letting $e_1 = af$ and $e_2 = 0$ in Eq. (14.5-14). What is the physical significance of this assumption? Then follow through the argument on biaxial stretching leading to Eq. (14.5-18). Give a physical meaning to Stein's argument regarding the pressure being inversely proportional to the negative normal stress perpendicular to the skin. Then arrive at Eq. (14.5-19), and explain how this isothermal compressibility compares with that obtained using a Spencer-Gilmore form of an equation of state.

P14.I   Obtain Eqs. (14.7-13) and (14.7-16).

P14.J   (Following Rosen [70]) Discuss qualitatively how a good solvent can be effectively used in the design of a stirred solution polymerization reactor. Then determine how a θ-solvent can be used to precipitate a polymer of a specific chain length from the solution.

P14.K   Diisooctyl phthalate is frequently used as a plasticizer in PVC. For a plasticizer to be effective, must it be a good solvent or a poor solvent in the PVC? Diisooctyl phthalate has a lower partial pressure and a lower diffusivity than dioctyl phthalate. Which would dissolve in PVC more rapidly at the same temperature? Which would lend better long-term stability to PVC?

P14.L   From Fig. 14.7-8, compute the fraction of PMMA in benzene when the initial mixture was equal parts rubber, PMMA, and benzene. Do the same for the fraction of GPS in benzene with equal parts of GPS, benzene, and polybutadiene in Fig. 14.7-9.

P14.M   A scrap mixture of LDPE and GPS is treated with benzene to strip the GPS from the scrap. Every 100 lb of scrap contains 25 lb of GPS and 75 lb of LDPE. The solubility of GPS in benzene is 0.5%. That is, for every 100 lb of benzene used, 0.5 lb of GPS can be dissolved. Consider the process to be a countercurrent stripping operation. For the process to be economical, only 0.5 lb of GPS can remain in every 100 lb of LDPE after stripping. Tests have shown that the scrap retains 0.5 lb of benzene per pound of dry, GPS-free LDPE as it moves from stage to stage. How many leaching stages are required? Hint: Consider the problem on the basis of 100 lb of benzene- and GPS-free LDPE. Since the ratio of benzene to LDPE is constant, flow rates should be based on pounds of benzene. This problem is very similar to conventional chemical engineering leaching problems.

P14.N   LDPE-coated cardboard is being recycled in a countercurrent stripping operation. The cardboard contains 20% LDPE which is being stripped with refluxing hexane. Ninety percent of the LDPE is recovered in a solution containing 50% by weight of LDPE. The paper is contacted initially with hot fresh hexane, and 1 lb of solution is removed in the underflow in association with every 2 lb of cardboard pulp. How many ideal stages are required?

P14.O   A laboratory test to determine the amount of glass contained in glass-reinforced Noryl (modified PPO) uses chloroform as a stripping agent. To establish proper testing procedures a known sample containing 18% glass with a specific gravity of 2.7 is used. To achieve the desired 98% extraction, how many stages are needed? Each stage requires 200 volumes of liquid per 100 volumes of plastic in order for the mixture to be pumped to the next stage. The strong solution should have a concentration of 100 g/liter.

P14.P   A thin film evaporator is being used to strip styrene monomer from GPS polymer at 371°F. The diffusion coefficient for the system is 2×10^{-5}

cm^2/sec. The initial monomer concentration is 0.01 g/100 g of polymer. The film is exposed to zero concentration for 10 sec. What is the average concentration after this time if the film thickness is 1 cm? 0.1 cm? 10 cm? Assume no resistance to mass transfer at the gas-liquid interface.

P14.Q Examine Morrison's data of Fig. 14.8-7 to obtain an analytical relationship between the diffusivity based on activity and that based on concentration. Fit an appropriate curve to the data to obtain a form for the activity coefficient. Then relate this to the Maron thermodynamic coefficient. Will Eq. (14.8-18) work? If not, what form can be used?

P14.R Calculate the permeability constant for N_2 in polyethylene at 371 and 440°F. Similarly, calculate the permeability constant for N_2 in GPS at 371 and 150°F. Is the last value valid? If not, why not?

P14.S Polycarbonate containing 0.5% (wt) H_2O is to be dried in a forced convection dryer. The humidity of the dry air is 15% RH at 70°F, and it is heated to 250°F prior to drying. After passing through the bed, the air is refrigerated to remove the moisture and recirculated. The average velocity through the bed is 1 ft/sec based on free volume. The pellets are 1/8 in. on a side. Calculate the rate of moisture removal if the diffusivity is 6 × 10^{-5} cm^2/sec at 250°F. Assume that the pellets are sufficiently large to permit use of the penetration theory for calculation of the mass transfer coefficient through the polymer. Then use the Sherwood correlation to determine the rate of diffusion through the film adjacent to the pellet surface. Note that the rate of mass transfer is dependent on time. Determine the time required to remove 90% of the moisture initially present. List all assumptions.

P14.T One method of making a thermosetting structural foam is to cast resin containing a large amount of water. When the resin has cured, the water is removed by diffusion and drying from the surface. In this way, if a foam of density 60% that of the polymer is needed, 40% water is emulsified into the casting resin. Water-extended polyester is a typical example of this type of material. A slab of water-extended polyester is being dried in a batch dryer under constant drying conditions. Eleven hours are required to reduce the moisture content from 40 to 10%. The critical moisture content, where the drying rate changes from a constant rate to a declining rate, is found to be 15% and the equilibrium moisture is 2%. Assuming that the rate of drying during the declining rate period is proportional to the free-moisture content, how long will it take to dry the polymer from 40% water to 5%? From 60 to 5%? All moisture contents are on a dry basis.

References

1. J. M. Smith and H. C. Van Ness, Introduction to Chemical Engineering Thermodynamics, McGraw-Hill, New York, 1959, p. 90.

2. J. O. Hirshfelder, C. F. Curtiss, and R. B. Bird, Molecular Theory of Gases and Liquids, Wiley, New York, 1954, p. 634.
3. M. R. Kamal and N. T. Levan, SPE ANTEC Tech. Pap., 18:367 (1972).
4. R. S. Spencer and G. D. Gilmore, J. Appl. Phys., 20:502 (1949).
5. P. S. Ku, Equations of State of Organic High-Polymers, G.E., Philadelphia, U.S. Dept. Commerce, AD 678 887, Jan. 1, 1968.
6. H. Breuer and G. Rehage, Kolloid Z., 216-217:166 (1967).
7. H. L. Whitaker and R. G. Griskey, J. Appl. Polym. Sci., 11:1001 (1967).
8. J. M. McKelvey, Polymer Processing, Wiley, New York, 1962, p. 130.
9. R. G. Griskey and N. Waldman, Mod. Plast., 43(3):119 (1966).
10. R. G. Griskey and N. Waldman, Mod. Plast., 43(3):121 (1966).
11. R. G. Griskey and N. Waldman, Mod. Plast., 43(4):160 (1966).
12. G. N. Foster III, N. Waldman, and R. G. Griskey, Mod. Plast., 43(5):245 (1966).
13. R. G. Griskey, N. W. Din, C. A. Gellner, and N. Waldman, Mod. Plast., 43(6):103 (1966).
14. R. G. Griskey, C. A. Gellner, and M. W. Din, Mod. Plast., 43(7):119 (1966).
15. R. G. Griskey, M. W. Din, and C. A. Gellner, Mod. Plast., 43(9):165 (1966).
16. R. G. Griskey, M. W. Din, and C. A. Gellner, Mod. Plast., 43(11): 129 (1966).
17. W. H. Wagner and R. G. Griskey, Mod. Plast., 44(4):134 (1967).
18. R. G. Griskey, G. Van Riper, and W. H. Wagner, Mod. Plast., 44(10): 144 (1967).
19. R. G. Griskey and J. K. P. Shou, Mod. Plast., 45(6):138 (1968).
20. R. G. Griskey, Mod. Plast., 45(9):215 (1968).
21. B. C. Sakiadis and J. Coates, AIChE J., 2:88 (1956).
22. R. G. Griskey and D. O. Hubbell, J. Appl. Polym. Sci., 12:853 (1968).
23. R. G. Griskey and G. N. Foster, SPE ANTEC Tech. Pap., 14:235 (1968).
24. W. R. Gambill, Chem. Eng., 64(3):237 (1957).
25. B. C. Sakiadis and J. Coates, AIChE J., 1:275 (1955).
26. P. J. Flory, J. Chem. Phys., 10:51 (1942).
27. P. J. Flory, Principles of Polymer Chemistry, Cornell University Press, Ithaca, N.Y., 1953, Chap. 11.
28. L. R. G. Treloar, The Physics of Rubber Elasticity, Oxford University Press, New York, 1958, Chap. 2.
29. R. S. Stein, Soc. Plast. Eng. Trans., 1:164 (1961).
30. D. R. Paul, C. E. Vinson, and C. E. Locke, Polym. Eng. Sci., 12:157 (1972).
31. C. E. Locke and D. R. Paul, Polym. Eng. Sci., 13:380 (1973).
32. S. L. Rosen, Polym. Eng. Sci., 7:115 (1967).
33. M. E. Morrison, AIChE J., 13:815 (1967).

34. H. C. Van Ness, Classical Thermodynamics of Non-Electrolyte Solutions, Macmillan, New York, 1964.
35. G. Scatchard, Chem. Rev., 8:321 (1931).
36. M. L. Huggins, J. Am. Chem. Soc., 64:1712 (1942).
37. R. L. Baldwin and K. E. Van Holde, Fortschr. Hochpolym. Forsch., 1:451 (1960).
38. J. F. Heil and J. M. Prausnitz, AIChE J., 12:678 (1966).
39. S. H. Maron, J. Polym. Sci., 38:329 (1959).
40. S. H. Maron and N. Nakajima, J. Polym. Sci., 40:59 (1959).
41. S. H. Maron and N. Nakajima, J. Polym. Sci., 42:327 (1960).
42. S. H. Maron and N. Nakajima, J. Polym. Sci., 47:157 (1960).
43. S. H. Maron and N. Nakajima, J. Polym. Sci., 47:169 (1960).
44. S. H. Maron and N. Nakajima, J. Polym. Sci., 54:587 (1961).
45. S. H. Maron, N. Nakajima, and I. M. Krieger, J. Polym. Sci., 37:1 (1959).
46. G. M. Bristow, J. Appl. Polym. Sci., 2:120 (1959).
47. H. Tompa, Polymer Solutions, Butterworth's, London, 1956, p. 174.
48. T. R. Paxton, J. Appl. Polym. Sci., 7:1499 (1963).
49. W. R. Veith and K. J. Sladek, J. Colloid Sci., 20:1014 (1965).
50. C. H. Klute, J. Appl. Polym. Sci., 1:340 (1959).
51. C. H. Klute and P. J. Franklin, J. Polym. Sci., 32:161 (1958).
52. P. L. Durrill and R. G. Griskey, AIChE J., 15:106 (1969).
53. F. Kreith, Principles of Heat Transfer, International Textbook Company, Scranton, Pa., 1965, p. 157.
54. R. M. Secor, AIChE J., 11:452 (1965).
55. H. Matsuo, K. Iino, and M. Kondo, J. Appl. Polym. Sci., 7:1833 (1963).
56. P. L. Durrill and R. G. Griskey, AIChE J., 12:1147 (1966).
57. W. Jost, Diffusion in Solids, Liquids, Gases, Academic Press, New York, 1960, Chap. 3.
58. H. Fujita, Fortschr. Hochpolym. Forsch., 3:1 (1961).
59. R. J. Kokes and F. A. Long, J. Am. Chem. Soc., 75:6142 (1953).
60. J. R. Kuppers and C. E. Reid, J. Appl. Polym. Sci., 4:124 (1960).
61. J. Crank and G. S. Park, Diffusion in Polymers, Academic Press, New York, 1968.
62. P. F. Lesse, J. Polym. Sci. Part A, 9(2):755 (1971).
63. R. F. Riek, T. J. McAvoy, and D. C. Chappelear, J. Polym. Sci. Part A, 6(2):1863 (1968).
64. Bruce C. Wendle, ed., Engineering Guide to Structural Foam, Technomic, Westport, Conn., 1976.
65. S. J. Skinner, S. Baxter, S. D. Eagleton, and P. J. Grey, Mod. Plast., 42(1):171 (1965).
66. G. A. Pogany, Br. Plast., 37:506 (1964).
68. R. Higbie, Trans. AIChE, 31:365 (1935).
69. P. V. Danckwerts, Ind. Eng. Chem., 43:1460 (1951).

70. S. L. Rosen, Fundamental Principles of Polymeric Materials for Practicing Engineers, Barnes & Noble, New York, 1971.

71. D. R. Paul, C. E. Locke, and C. E. Vinson, Polym. Eng. Sci., 13:202 (1973).

72. F. W. Billmeyer, Jr., Textbook of Polymer Science, Interscience, New York, 1966, Part 2, pp. 21-26.

73. M. Kurata and W. H. Stockmayer, Fortschr. Hochpolym. Forsch., 3:196 (1963).

74. N. N. Li and E. J. Henley, AIChE J., 10:666 (1964).

75. A. W. Myers, C. E. Rogers, V. Stannett, and M. Swarc, Mod. Plast., 34(10):157 (1957).

15.

Assembly Techniques

15.1 Introduction

It is frequently necessary to fasten plastics either to themselves or to non-plastic materials in order to complete a product. Many of the useful methods rely upon the unique properties of the plastics materials themselves. For example, most plastics have very poor thermal conductivities. Therefore, frictional heat generated at a surface is only very slowly transmitted to the interior. If the frictional heat buildup is rapid enough, the surface can melt without softening the interior. Thus, two plastic surfaces can be rubbed together very rapidly until the interface is melted and then held together until the heat is dissipated into the interior. Friction welding is thus a rather unique way of fastening plastics that utilizes the inherent properties of the materials. There are, of course, adhesives and solvents and mechanical fasteners that work very well in certain instances. In this chapter, several of the more popular techniques will be reviewed, and the engineering implications behind them will be disclosed.

15.2 Adhesive Bonding

Whenever a plastic sheet is to be fastened to a foreign substrate, adhesive
bonding is recommended. Here an adhesive is applied either to the plastic
or the substrate, and the two surfaces are pressed together and held until
the adhesive cures. Typical operations using adhesive bonding include
laminating of "Formica" melamine sheet to plywood and fastening vinyl tile
to a suitable flooring substrate. What is needed is excellent compatibility
between the adhesive and the substrate and the adhesive and the plastic. In
an engineering sense, the plastic must be sufficiently chemically attacked
by the adhesive in order to render a good bond between the plastic and the
adhesive. A physical bond is insufficient here, regardless of the roughness
of the plastic surface. Normally the adhesive contains a vehicle, such as a
ketone, that swells and activates the top layers of the plastic, thus providing
sufficient active sites for the cross-linking of the adhesive to attack. Certain plastics, such as the olefinic series (HDPE, LDPE, PP), have notoriously poor bonding characteristics, and thus no adhesive, not even epoxy,
can attack the surface sufficiently to afford a good bond. The reason for
this is that the olefins are relatively chemically inert, exhibiting well-
protected backbones, and thus there is no site for the reactive agent in the
adhesive to attack and react with.
    Recent developments in solvent etching solutions [1] have led to very
powerful formulations that activate the olefins for a short time, thus allowing painting and staining bases to adhere. No suitable solutions have been
found, however, that allow for adhesive bonding. It has been known for some
time, however, that the olefins have active surfaces for a short time after
removal from the molding process (injection molding, blow molding, injection molding), and in this short period the surface can be attacked directly
with suitable paints and stencils. To enhance this effect, a corona discharge
can be used to deliberately etch the surface [2]. Note that bonding energies
are thermodynamic properties. Failure of bonds between plastics, adhesives,
and substrates should in general be failures in a cohesive mode through the
homogeneous material rather than failure at the bond interface. If a sheet
of rather inert material, such as LDPE, is passed under an electric discharge or flame spray, a fair amount of formation of polar groups occurs on
the surface of the plastic. Porter [3] has considered corona discharge in a
quantitative way. The rate of activation of the surface is proportional to the
product of the average rate at which the surface is bombarded with charged
particles and the probability that a charged particle will strike a certain
reactive site, such as an exposed portion of the backbone of the molecular
chain. Porter finds that this can be expressed as

$$dy/dt = K(Gf)(p) \qquad\qquad\qquad (15.2\text{-}1)$$

where dy/dt is the rate of activation of the surface, K is a proportionality
constant, G is the number of particles swept against the surface of the film
in a single exposure to the corona discharge, f is the frequency of exposure

to the discharge, and p is the probability of a particle striking an active site. Porter then shows that

$$p = aN_0(1 - y) \qquad (15.2-2)$$

where a is the area of a single reactive site, N_0 is the number of active sites per unit area of the surface, and $1 - y$ represents the fraction of sites available for reaction. Integrating this equation, and assuming that the sheet passes through the discharge at a linear speed of U, where A_d is the area of discharge, yields

$$y = 1 - \exp(-aN_0 KGfA_d/U) \qquad (15.2-3)$$

To correlate the extent of activation, Porter derived a relationship between the extent of activation and the peel strength of a pressure-sensitive tape adhered to the LDPE. The experimental data are tabulated in Table 15.2-1. To relate peel strength to the extent of activation, Porter proposed that

$$y = (S - S_0)/(S_1 - S_0) = (F - F_0)/(F_1 - F_0) \qquad (15.2-4)$$

where F_1 and F_0 represent the forces required to separate the tape from the sheet when the sheet is completely activated and completely unactivated, respectively. The S's represent the peel strengths (in g/in.) for the same conditions. Porter assumes that the differential changes in the F's are directly proportional to the differential changes in the S's. He then calculates $k = (aN_0 KGf)A_d/A$ from the data given in Table 15.2-1 and obtains y as a function of kt, as shown in Fig. 15.2-1. Further, he shows that $k = k_0(I/I_0)$, where I is the current. For the example given, $k = 0.0019I$, where I is in the microamperes.

Note that this activation does not remain constant. Within minutes of activation, the sites will begin deactivating. Thus, if the postoperation is one of printing, stenciling, painting, or metal plating, it must occur immediately after corona discharge. If deactivation occurs as a first-order reaction, one form for the rate of deactivation might be

$$dy/dt = -k'y \qquad (15.2-5)$$

where k' is a rate constant that is probably related to a chemical reaction between the deactivating species and the site. The solution for this equation is

$$y = y_0 \exp(-k't) \qquad (15.2-6)$$

It is apparent that the more aggressive the deactivating species or the higher the temperature (since k' is probably Arrhenius dependent), the more rapidly the surface becomes deactivated.

One of the processes that require a highly activated surface is hot foil stamping. This method of decorating is gaining wide acceptance as a way of covering swirls in structural foam surfaces and as a way of adding

Table 15.2-1

Kinetic Data for the Surface Activation of Polyethylene [3]

Time (sec)	Peel strength (g/in.)
4	330
10	530
20	710
25	840
30	910
40	990
50	1100
60	1200
70	1170
80	1190
90	1200
100	1220
300	1270
500	1310
900	1370
1000	1380
1100	1370

$S_0 = 160$ g/in. $\quad$ U $= 63$ ft/min

$S_1 = 1260$ g/in. $\quad$ I $= 65$ μA

$\quad$ f $= 60$ cycles/sec $\quad$ V $= 8650$ V

Gap $= 50$ mils $\quad$ T $= 25°$ C

decorations to the otherwise sterile plastic surface. Early work by Kraus and Manson [4] showed that the adhesive strengths of LDPE and GPS when applied to steel were very close to the tensile strength of the plastics. For example, for LDPE the adhesion at 22°C was 194 kg/cm^2 and the tensile

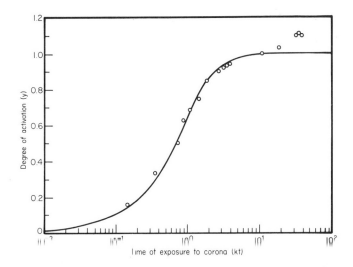

Figure 15.2-1 Degree of activation of polyethylene surface via corona discharge as function of the kinetic time kt. Solid line represents theory [3].

strength was 141 kg/cm^2. For GPE, at 25°C the values were 412 and 377, respectively. This implies that at separation failure occurs cohesively in the plastic interior rather than adhesively at the interface. Rutzler and his coworkers [5, 6] worked through the energies required to theoretically separate these materials when all the adhesive energy was attributable to dispersion forces. The values obtained were 1 to 20 times greater than the actual adhesive strengths, thus indicating that only a small fraction of the surface is used for actual adhesion. They estimate that this fraction could be as low as 1%. Thus, it is apparent that to achieve a bond that fails cohesively, it is necessary to involve as much of the surface in the bonding process as possible.

Many other postmolding operations depend on activation of the polymer surface. For example, the dyeing of plastic fabrics requires good adhesion between the dye and the polymer surface. Most dyes are in aqueous, or disperse, media. As a result, the polymers that have the greatest dyeability are the ester-linked resins, such as polyesters, polycarbonates, acrylics, cellulosics, and nylons. Polystyrene will not dye with conventional textile techniques, and although the olefins and vinyls seem to accept the surface dye, long-term stability is very poor. Surface treatment of olefins can improve this, however. One operation uses a combination of a graft initiator and a silver salt to create reactive sites on the olefin backbone [7]. This treatment is then followed by a conventional textile dye, which attacks and reacts with the active site.

Electroplating requires good adhesion between the plating and the plastic substrate. One of the first resins to be electroplated was polypropylene. However, many of its applications have been taken over with plating-grade ABS and modified polyphenylene oxide. Optimum conditions for plating of many resins are found at high melt temperatures, high mold temperatures, slow fill rates, and gates properly positioned to minimize residual stresses. The typical electroplating preplating process steps include cleaning, etching, neutralizing, catalysis, acceleration, and electroless nickel or copper brush plating, with rinses at each intermediate point. The extent of plating required is dependent on application of the final part, as shown in Table 15.2-2 [8].

Table 15.2-2

Recommended Electroplate Thickness [8]

| Service conditions | Plating parameters | |
	Batch sequence	Recommended thickness (in.)
Mild exposure indoors in normally warm, dry atmospheres	Electrolytic strike	Adequate to cover
	Bright acid copper	0.0006–0.0008
	Bright nickel	0.0002–0.0003
	Conventional chromium	0.000010–0.000015
Moderate exposure to high humidity and mildly corrosive atmosphere	Electrolytic strike	Adequate to cover
	Bright acid copper	0.0006–0.0008
	Semibright nickel	0.0003–0.0004
	Bright nickel	0.0002–0.0003
	Conventional chromium	0.000010–0.000015
Severe exposure to high humidity, with temperature variations and severe corrosive atmosphere	Electrolytic strike	Adequate to cover
	Bright acid copper	0.0006–0.0008
	Semibright nickel	0.0004–0.0006
	Bright nickel	0.0003–0.0004
	Special chromium[a]	0.0001
	Conventional chromium	0.000010–0.000015

[a]Required for reducing microporosity or microcracking.

Vacuum metallization requires surface conditions that are more stringent than those required for electroplating. The primary reason for this is that resins that out-gas will not plate evenly. In general, large-diameter vacuum cylinders are loaded with parts to be metallized. Vacuum of 0.1 to 0.5-μm is achieved in a three-stage vacuum system in 5 to 20 min (depending on the size of the vacuum cylinder). The aluminum-coated tungsten filament is then heated very rapidly to 1800 to 2000°F. This causes the aluminum to vaporize and deposit on the cooler plastic surfaces. The vaporization plating takes 10 to 30 sec, and a coating of 2 to 5 μin. is deposited. This plating is then immediately coated with a lacquer or polyurethane coating, which is air dried and oven cured. One of the largest single uses for vacuum metallized parts is in plastic model kits. However, the second-surface optical depth of more recent techniques has led to extensive use in automotive and appliance fields. In fact, techniques are now available to continuously metallize film at feed rates of 500 ft/min and widths to 60 in.

Hot stamping is frequently used to apply a trademark or decoration, although it has been used on occasion to completely cover a plastic substrate. In general, the hot stamp foil is a series of layers, including carrier, high-temperature wax, pigment, grained coat, wax, toner coat, and heat-sensitive adhesive coating. The adhesive coating is normally activated at 100 to 200°C. The foil thickness is 0.003 to 0.010 in. but can be considerably thicker, depending on the stamping stresses involved. Note that the standard method of applying the foil is to heat the foil with direct contact of a laminating platen and then press the softened foil against the room-temperature substrate. Good adhesion thus depends first on the softening of the heat-sensitive coating and second on the partial softening or activation of the substrate surface. The principles of transient heat conduction (Chap. 6) are definitely applicable here in order to determine the time required to achieve optimum bond.

In-mold decoration has been the primary method of decorating thermosets. The foil, usually a melamine-impregnated paper, is inserted into the transfer or compression mold after the resin is partially cured. The mold is then reclosed and pressurized to 2000 to 4000 psi at 300 to 350°F, until fusion takes place. This is one way of "coloring" thermosets that are characterized by black, brown, or amber natural resin colors. Some experimentation with in-mold decorating of thermoplastics has been going on for years, with some partial successes. The greatest challenge, surface improvement of the characteristic swirled surface of low-pressure thermoplastic structural foam, has not been met, owing to excess gas generation that is frequently trapped between the foil and the structural foam surface. In other, nonfoam, applications, however, plastic-backed paper foil has been used to attach labels to square food containers, and metallic foil has been used to achieve a highly reflective decorator jar or cap. In general, the foils are 0.005 to 0.020 in. thick. The primary processing problem is lack of adhesion with the plastic. As has been shown elsewhere [9], the foil must act as an insulator at the mold surface. The greater the amount of insulation, the slower the plastic cools at its interface with the foil, and thus the more effective the adhesion at the interface becomes. This effect, as again analyzed using the transient heat conduction arithmetic of Chap. 6, is shown in Fig. 15.2-2. It is interesting to note that the factors that make an economic process in hot stamping, e.g., rapid heating of the foil owing to its thinness, are diametrically opposed to the factors that make an economic process in in-mold decoration, e.g., slow cooling of the plastic-foil interface owing to its thickness.

If the plastic is to be painted, in order to achieve improved appearance, environmental resistance, or multicolored appearance, surface treatment is normally required. For many materials, such as ABS and GPS, simple hydrocarbon solvent wiping is sufficient to prepare the plastic as a substrate. For others, the above-mentioned ritual of chemical, solvent, electron beam, or corona etching is required. For some difficult materials, such as nylon or PP, combinations of etches, with ancillary rinses and intermediate sandings and buffings, are required. The reader is advised to carefully select the paint and surface preparation as a system. Also important is the method of applying paint. With emphasis on safety and environment

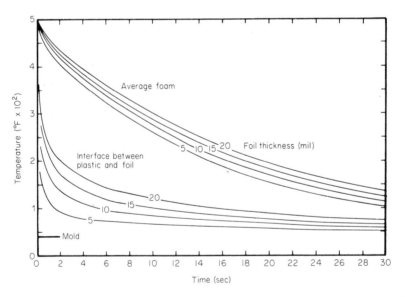

Figure 15.2-2 Effect of foil layer thickness on the temperature profile between the plastic and foil and on the average temperature profile in a 1/4-in. structural foam slab (PS at 0.7 sp. gr.). Mold temperature assumed to be 60°F, melt temperature to be 500°F [9].

control, oil-based and solvent-based paints are giving way to either aqueous paints or 100% solids paints. The latter are frequently much more expensive than the aqueous paints, but there is superior adhesion compatibility between the 100% solids paints and their substrates. Paint application then also depends on the type of paint being applied. Among the dozens of choices, dip coating is used only when a very heavy coat is required. The thickness of the resulting layer is frequently quite uneven. Spray coating is preferred for automation, with airless spraying being phased out as the newer techniques of electrostatic attraction and electrostatic atomization are developed. It appears that charging the plastic part is frequently more difficult than charging the paint drop, particularly with aqueous paints, and as a result, considerable research is being spent on electrostatic atomization coating. One of the newer techniques that is being considered as a way of achieving heavy coats of filled paint, or really resin, is reaction painting. A reactive resin is applied in some manner (roller-coated if the resin has high viscosity, electrostatically atomized if it has a water like viscosity) to the substrate and then rapidly reacted with either UV, infrared, atomic rays, X-rays, ozone, or electron beam irradiation. UV-activated resins are the most popular, even to the point of being sold in ready to use form in hobby stores. This technology offers a "way out" for the fabricator and/or designer working with off-color resins, varying amounts of regrind, thermoplastic structural foam, or phenolic-based thermosets.

Printing requires substrate conditions that are similar to those for any decorating technique in that oily surfaces, rough surfaces, and surfaces with poor adhesion will all defeat most printing operations. Newer techniques in electrostatic silk screen printing will enable the decorator to work around some of the surface limitations. This has been, until recently, restricted to single color printing per pass.

Incidentally, excellent bonds between certain plastics and metals can be achieved without treating the plastic surface. For example, in fluidized bed coating, the objects to be coated, such as plier handles, wire racks, and so on, are heated to temperatures far above the melting temperature of the plastic. The objects are then dipped into a bed of fine plastic powder of 50 to 200 mesh. The bed is fluidized by blowing air through it via a porous plate in the bottom. The fine plastic powder sticks to the surface of the metal and sinters and densifies into a homogeneous coating. The arithmetic of sintering and densifying is nearly identical to that encountered in the latter stages of rotational molding and thus will not be repeated here. Another technique for coating metal with plastic is powder spray coating. A hot air gun equipped with a powder dispenser is used for application. The metal surface is normally heated, either prior to application or with contact with the heated air. The powder is blown from the gun and mixes with high-velocity air having temperatures of 500 to 1000°F. This softens the powder surface prior to impingement on the solid surface. The powder particles are sticky when they touch the surface and each other and thus stick together.

Continued application of heat sinters them. Normally the coating obtained this way is very porous, but the surface is quite a bit smoother than that obtained in fluid bed coating. There are many applications for powder spray coating such as nonstick frying pan coatings and corrosion protection. In general, it seems that this technique is most useful for applications of plastics that are very difficult to process in the melt state. Teflon is the best example of this type of plastic.

As understood from the work in adhesion, the reason this process works without adhesive interlayer is that the high temperature of the metal and/or air activates the plastic upon contact. Thus, a clean, active metal surface and the newly activated plastic surface bond well together.

After the surface has been activated, many common adhesives can be used to bond plastics to substrates. Stokes [10] lists many adhesive types, given here in Tables 15.2-3 to 15.2-5. Thus, for example, it is possible to find the types of adhesives that can be used to bond nylon to ceramics, fabrics, leather, metal, paper, rubber, or wood. Note that the characteristic behavior is that the adhesive acts as an intermediary between the plastic and nonplastic or dissimilar plastic.

Note that plastics can be glued to plastics as well as to other substrates. For example, nylon can be glued to acetal using epoxy. Failure of the union should occur cohesively within one or the other plastic and not adhesively at the epoxy-plastic interface. As a result, the cohesive strength of the adhesive is usually greater than that of either of the plastics. This point will be touched on later. Although the tables list several adhesives for joining two pieces of the same plastic, there seems to be little need for this type of adhesive. Solvent bonding or ultrasonic welding is preferred. Nevertheless, for quick repair, it is handy to know that PVC can be bonded with polyvinyl acetate and that cellulosics cannot be bonded with epoxy but can be bonded with a urethane-based adhesive. In fact, it is more important at times to know what materials cannot be bonded together with specific adhesives. This information can eliminate frustrations of, say, trying to bond HDPE with epoxy.

Note that most of the adhesives listed are unsaturated plastics. These either cure on contact with air or by mixing catalyst with resin. As detailed in Chap. 4, thermosetting resins have superior strength to most thermoplastic materials. This ensures that bonding failures will occur within the plastics and not cohesively within the adhesive. Therefore, the tensile strength of the bonding material need not be excessively greater than that for either of the adherents. The use of an expensive adhesive with extreme strength is unwarranted when both adherents are quite weak. Resorcinol adhesives between layers of wood to form plywood are not needed, therefore, for their very high cohesive strength but for their water resistance.

Table 15.2-3

Adhesive Types for Bonding Nonplastics to Thermoplastics [10][a]

Surfaces	Acetal	Cellulosics[b]	Ethyl cellulose	Nylon	Polycarbonates	Polyethylenes	Polypropylenes[c]	Polystyrene	Polyurethane	Polyvinyl chloride	Tetrafluoro-ethylene	Polymethyl methacrylate
Ceramics	23	4	14	4, 23	23, 36	3, 41	1, 41	42	4	4, 5	23	3, 4
Fabrics	23, 4	4, 5, 42	14	3, 4	23, 36	3, 41	1, 41	5	5, 36	4, 5, 42	22	4
Leathers	23, 4	4, 5, 42	14	3, 4	23, 36	3, 41	1, 41	36, 5	4, 5	4, 5, 41, 42	22	3, 4, 42
Metal	23, 4	3, 4	14	3, 23	23	3, 41, 31	1, 2	3	5, 4	3, 4, 15, 36	22, 23	3, 4
Paper	23, 4	42	14	4, 41	36	41	1, 2, 41	36, 5	5, 36	42	22, 23	42
Rubber	4	1-5	14	2, 3	36, 5	3, 41	1, 2, 41	2-5	5, 36	4, 5, 15	23	1-5
Wood	23	4	14	3, 4	23, 36	3, 41	1, 2, 41	36	36	4, 42, 36	23	3, 4, 42

[a]See Adhesive Number Code Key for Tables 15.2-3, 15.2-4, and 15.2-5.
[b]Cellulose acetate, cellulose acetate butyrate, cellulose nitrate.
[c]Special surface treatment recommended.

Table 15.2-4

Adhesive Types for Bonding Nonplastics to Thermosets [10][a]

Surfaces	Diallyl phthalate	Epoxies	Melamine	Phenolics	Polyesters	Polyethylene terephthalate	Urea
Ceramics	5, 24	23, 31	3	3	3	36	4
Fabrics	36	4	4	4	4	5, 36	4, 42
Leather	31, 36	4	3, 4	3, 4	5	5, 36	3, 4
Metal	31	23, 31	4	3	5	36	3, 4
Paper	31, 36	4	41, 42	42	41	5, 36	42
Rubber	3	4	3, 4, 2	3, 4	1-5	36, 13	1-5
Wood	31, 36	23, 31	3	3, 42	3	36	3, 42

[a]See Adhesive Number Code Key for Tables 15.2-3, 15.2-4, and 15.2-5.

Table 15.2-5

Adhesive Types for Bonding Plastics to Plastics [10][a]

Surfaces	Adhesive type(s)
Acetal	23, 4
Cellulosics[b]	14, 4, 36, 5
Ethyl cellulose	14
Nylon	23, 22, 3, 36
Polycarbonates[c]	
Polyethylenes	23, 41, 31
Polymethyl methacrylate	13, 6, 2, 36, 5, 31
Polypropylenes	23, 31, 41
Polystyrene	23, 13, 2, 6, 5, 31, 36
Polyurethane	5, 23, 4, 36
Polyvinyl chloride	4, 11, 5, 42, 36
Tetrafluoroethylene	23, 22
Diallyl phthalate	4, 23, 31, 3, 36
Epoxies	4, 23, 31, 3, 36
Melamine	3, 4, 23, 31, 36
Phenolics	4, 23, 31, 3, 36, 5
Polyester-fiberglass	4, 23, 31, 36
Polyethylene terephthalate	5, 36
Urea	4, 23, 31, 3

[a]See Adhesive Number Code Key for Tables 15.2-3, 15.2-4, and 15.2-5.
[b]Cellulose acetate, cellulose acetate butyrate, cellulose nitrate.
[c]No adhesive type recommended; solvent cement preferred method.

Adhesive Number Code Key to Tables 15.2-3, 15.2-4, and 15.2-5

Elastomeric

1. Natural rubber
2. Reclaim
3. Neoprene

Key to Tables 15.2-3, 15.2-4, and 15.2-5 (continued)

---

4. Nitrile
5. Urethane
6. Styrene-butadiene

Thermoplastic

11. Polyvinyl acetate
12. Polyvinyl alcohol
13. Acrylic
14. Cellulose nitrate
15. Polyamide

Thermosetting

21. Phenol formaldehyde (phenolic)
22. Resorcinol, phenol-resorcinol
23. Epoxy
24. Urea formaldehyde

Resin

31. Phenolic-polyvinyl butyral
32. Phenolic-polyvinyl formal
33. Phenolic-nylon
36. Polyester

Miscellaneous

41. Rubber latexes (water based) (natural or synthetic)
42. Resin emulsions (water based)

15.3 Thermal Bonding

A common way of bonding thin sheet stock is by simply clamping the two
pieces of material and locally heating the region to be bonded. This is done
commercially on a continuous basis by running the sheet stock between pre-
heaters and then pressing the sheets together at the nip of heated rollers.
The primary objective is to get the interior surfaces at or above a tacky
temperature while preventing the sheets from shrinking or wrinkling under
the applied heat. The most important physical properties for heat sealing
are thermal conductivity, specific heat, and heat distortion temperature [11].
This is a classic transient heat conduction problem, as discussed in detail
in Chap. 6. Briefly, heat must be transmitted through the sheet to a nearly

insulated interior surface. As discussed, the rate of heat transmission is dependent on the thermal diffusivity of the material (thermal conductivity divided by density and specific heat) and the thickness of the sheet. As shown in Fig. 15.3-1, the exposed surface of the sheet is always hotter than the interior surface. Thus, the goal is to prevent completely melting the exposed surface before the interior surface has reached the heat distortion temperature. It should be apparent that the thicker the sheet, the more gentle the heating process must be in order to prevent the exposed surface from sticking to the heating source.

In addition to heat, of course, pressure is required to force the surfaces together. Knight and Funk [13] have shown that for heat sealing 1.5-mil LDPE at a heat source temperature of 400°F, increasing the clamping pressure from 40 to 240 psi reduces the clamping time from 0.42 to 0.25 sec. Their graphs are shown in Fig. 15.3-2. Apparently, increased pressure _____ ___ air from between the interior surfaces. Any air present acts as a thermal insulator or a reduc____ _____ ____ ____ _____ the inter-face, thus reducing the rate of activation of the surfaces at the interface. And even the smoothest sheet has microscopic asperities. Zero-pressure contact of two sheets results in touching only at the highest asperities.

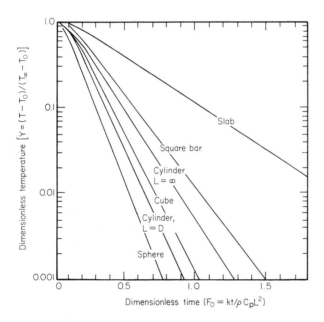

Figure 15.3-1   Dimensionless centerline temperature of various solids as function of dimensionless time, $F_0 = \alpha t/L^2$ [12].

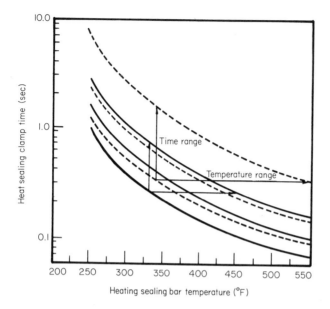

Figure 15.3-2   Effect of heat sealing temperature on clamp time for 1.5-mil polyethylene film at 40- and 240-psi sealing pressures. Curves represent analysis of experimental data. Dashed lines, 40 psi; solid lines, 240 psi [13].

Increasing pressure forces more asperities to contact. The softer the material, the lower the pressure needed to involve the same fraction of surface. It is apparent from earlier discussions that as much of the surface as possible must be involved in contact in order to achieve an acceptable seal. One might speculate that very high pressures might be needed to effect a good seal if the sheet stock were very thick or the material were very stiff.

Very heavy plastic stock can be welded. If the plastic is sufficiently thermally stable and does not degrade or form microscopic bubbles due to included air, water, or monomer, heating the surfaces to be joined with a hot air gun can be an effective means of softening the interfaces. The hot air gun should have a directional tip and be capable of achieving 750°F. The objective is to heat the interfaces between the two pieces as rapidly and uniformly as possible while the pieces are near but not touching each other. When the surfaces appear shiny and are tacky, they must be forced together and clamped until cool. According to Neubert and Mack [14], the purpose of the clamping pressure should be to maintain complete control over the amount of deformation of material in the melt zones. The quality of the final weld depends on the relative amount of molten material displaced from the interface at the moment the parts are contacted. The displacement generates a small amount of shear, which helps mixing in the interfacial region. It is well documented that molecular mixing or diffusion is a very slow process [15]. The force also helps dispel oxidized or degraded material from the weld area. Obviously, discretion must be used, since excessive force displaces all the melted plastic and leaves the weld region with large amount of solid material having little if any adhesive qualities.

When dealing with very thick stock or pipe, hot gas welding is economically unattractive and requires very careful attention to proper temperatures. It is therefore not suitable for in-line forming of seams. As a result, direct contact with the heating source and hot-rod welding are preferred. In the latter, a heating source, such as a directed tip hot air gun, is directed to the weld area between the two pieces. As the surfaces are heated, a rod containing a similar material is brought into the heated field. The rod material melts and flows into the weld area. The process is very similar to conventional gas welding of metals in that the rod is consumed and a bead is formed which must be "puddled" along the weld area. It is necessary to ensure that the region on either side of the weld is heated sufficiently to achieve a strong bond. Care must be taken in this type of welding since the region around the weld can be a source of internal stresses. Frequently, a very heavy weld is annealed in much the same way as a metal weld. This welding process is very slow and quite time-consuming.

Neubert and Mack discuss in detail the continuous direct contact weld process. The heating element must be controlled very carefully to ensure a strong consistent weld. The element is normally a planar electric resistance heater. They point out that both the heating and cooling of the weld region is critical to a water-tight weld. In Fig. 15.3-3 are experimental material temperatures measured as functions of time for various element surface temperatures. Note the characteristic exponential decay of the temperature profile in the material. This profile can be described mathematically [16]. Consider a heat source moving at a constant velocity on a sufficiently large body. A quasi-steady state can be found, since if the source were considered to be stationary and the material were removed, the temperature profiles around the source would be considered stationary. Assume that the source emits heat at the rate Q and moves at a velocity v. Assume further that the source is planar. The basic transient heat conduction equation is

$$\alpha[\partial^2 T/\partial x^2] = \partial T/\partial t \qquad\qquad (15.3\text{-}1)$$

where x is in the direction of the moving source. Now it can be shown that w = x - vt, where w is a new coordinate attached to the source. Thus, Eq. (15.3-1) becomes

$$\alpha[\partial^2 T/\partial w^2] = -v(\partial T/\partial w) + (\partial T/\partial t) \qquad\qquad (15.3\text{-}2)$$

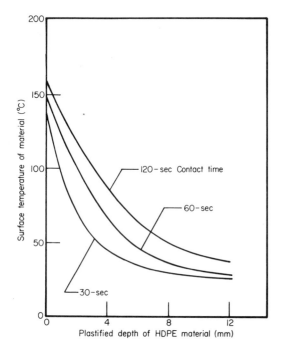

Figure 15.3-3   Surface and penetration temperatures of HDPE pipe for welding studies. Note material is plasticized and weldable at 120°C. Heating element surface temperature is 195°C. Three contact times are shown. Curves represent experimental data [14].

Now since it has been assumed that the system is quasi-steady, the second term on the right is zero. If T_0 is the initial temperature, then the following equation describes the temperature increase:

$$\theta = T - T_0 = \exp(-vw/2\alpha)\,\phi(w) \tag{15.3-3}$$

where $\phi(w)$ is given as the solution to this equation:

$$(d^2\phi/dw^2) - (v/2\alpha)^2\phi = 0 \tag{15.3-4}$$

Now the boundary conditions are $dT/dw = 0$ when $W = \pm\infty$ and

$$-k(dT/dw) = q'' \qquad \text{when } w = 0 \tag{15.3-5}$$

The first condition assumes that the surfaces an infinite distance from the source remain at the initial temperature, and the second assumes that the rate of heat conduction into the material at the source is directly proportional to the rate of heat produced per unit area by the planar source at $w = 0$. The general solution to Eq. (15.3-4) is

$$\phi = M \exp(-vw/2\alpha) + N \exp(vw/2\alpha) \tag{15.3-6}$$

It can be shown that after inclusion of the appropriate boundary conditions, the following obtains:

$$T - T_0 = (q''\alpha/kv) = \theta_{max} \qquad (w < 0) \tag{15.3-7a}$$

$$T - T_0 = (q''\alpha/kv) \exp(-vw/\alpha) \qquad (w \geq 0) \tag{15.3-7b}$$

Note that the temperature remains at θ_{max} after the source has moved on, because no heat loss term was included in Eq. (15.3-1). A temperature profile is shown in Fig. 15.3-4. If heat loss is added to Eq. (15.3-1) so that it becomes

$$\partial T/\partial t = \alpha[\partial^2 T/\partial x^2 - m^2(T - T_0)] \tag{15.3-8}$$

the appropriate general solution becomes

$$\theta = M\exp\{-[(v/2\alpha)^2 + m^2]^{1/2} + (v/2\alpha)\}w + N\exp\{-[(v/2\alpha)^2 + m^2]^{1/2} - (v/2\alpha)\}w \tag{15.3-9}$$

Using the appropriate boundary conditions, it is found that

$$\theta = \theta_{max} \exp\{[(v/2\alpha)^2 + m^2]^{1/2} + (v/2\alpha)\}w \qquad (w < 0) \tag{15.3-10}$$

$$\theta = \theta_{max} \exp\{[(v/2\alpha)^2 + m^2]^{1/2} - (v/2\alpha)\}w \qquad (w \geq 0) \tag{15.3-11}$$

where

$$\theta_{max} = q''/\{2k[(v/2\alpha)^2 + m^2]^{1/2}\} \tag{15.3-12}$$

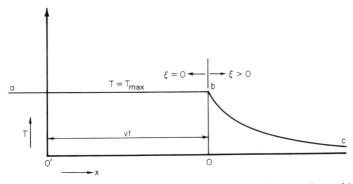

Figure 15.3-4 Temperature distribution for linear flow of heat, originating from a moving heat source, assuming no heat losses. v is the velocity of the moving heat source [16].

A schematic of the one-dimensional problem with heat losses is given in Fig. 15.3-5. Jacob [16] also shows how to analyze the problem of a two-dimensional heat flow without losses ($m = 0$). Here the basic equation in $\phi(w, y)$ is

$$(\partial^2 \phi / \partial w^2) + (\partial^2 \phi / \partial y^2) - (v/2\alpha)^2 \phi = 0 \tag{15.3-13}$$

Then the geometry around the source is conducive to replacing w and y by a radial coordinate such that $r^2 = w^2 + y^2$. As a result, Eq. (15.3-13) is reduced to cylindrical form as

$$d^2 \phi / dr^2 + r^{-1} \, d\phi/dr - (v/2\alpha)^2 \phi = 0 \tag{15.3-14}$$

The solution of this equation is in the conventional Bessel function form

$$\phi = MI_0(vr/2\alpha) + NK_0(vr/2\alpha) \tag{15.3-15}$$

where I_0 and K_0 are the modified zero-order Bessel functions of the first and second kind, respectively. These functions are tabulated in convenient form elsewhere [17]. After substitution of the boundary conditions, in cylindrical form, the following solution is obtained:

$$\theta = T - T_0 = (q'/2\pi k) \exp(-vw/2\alpha)K_0(vr/2\alpha) \tag{15.3-16}$$

where q' is the heat output per unit time and unit length.

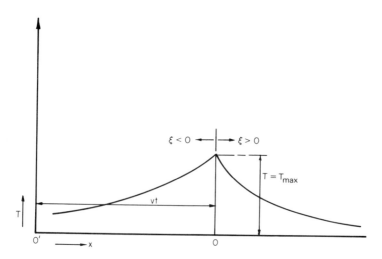

Figure 15.3-5   Temperature distribution for linear flow of heat, originating from a moving heat source, assuming heat losses. v is the velocity of the moving heat source [16].

In the heuristic analysis of the heat conduction equations, it is apparent that if m^2 represents a convection heat loss from the surface, the larger it becomes, according to Eqs. (15.3-10) and (15.3-11), the slower the plate heats and the more rapidly it cools. Likewise, note from Eq. (15.3-12) that the maximum excess temperature θ_{max} decreases with increasing heat removal from the plate, and thus to achieve a welding temperature dependent on, say, the heat distortion temperature, more unit energy must be expended. In Eq. (15.3-15), note that the modified Bessel function depends on a radial dimension around the heating source. As a result, to obtain the temperature profile at various distances into the slabs, it is necessary first to plot the temperature at various radial distances from the source and to cross-plot this to obtain the desired values. Nevertheless, it can be seen from Eq. (15.3-15) that as y (and as result r) increases away from the source at a constant value of w, it can be shown that $K_0(ax)$ approaches ln x as x approaches 0 and approaches $(\pi/2a)^{1/2}$ exp(-ax) as x approaches infinity. Hence, as y becomes large, K_0 becomes exponentially small, in good agreement with the data of Neubert and Mack. Note further in Fig. 15.3-6 that the characteristic heating and cooling curves show the exponential increase and decrease in temperature predicted by Eqs. (15.3-10) to (15.3-12). Thus, it can be concluded that the mechanism of welding

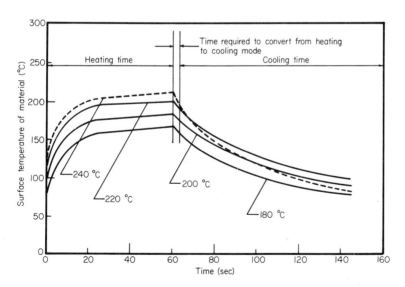

Figure 15.3-6   Surface temperature of PVC pipe with 3/8-in. walls as a function of time for heating element surface temperature of 180, 200, 220, and 240°C for pipe welding. Welding surface area is 2 cm^2. Dashed line represents material degraded by excessive heat [14].

large sections of plastic can be predicted according to standard heat transfer rules.

Frictional welding has been extensively used to assemble similar
plastic parts, such as halves of bottles. If the objects to be welded are
cylindrical, the process is known as spin welding. The process relies on
friction between the plastic parts and the very low thermal conductivity of
plastics in general to effect the weld. One-half of the container is held in a
fixed chuck, and the other is a spinning chuck that is equipped with a clutch.
The parts are formed such that there is an interference fit between them.
The spinning chuck is set in motion and the spinning half of the container
brought against the stationary half. In less than a dozen revolutions, the
friction between the two halves will have heated the interface to temperatures
above the melting point. Since most plastics are very tacky when soft, the
torque required to keep the halves separated will increase rapidly. When
the torque required to keep the spinning half moving exceeds the torque setting of the clutch, the clutch will disengage and the two halves will remain
bound very tightly. Spin welding machines have been built that will yield
1800 welds/hr on 1/2-pint containers.

It is apparent that the key to spin welding is good frictional coefficients between the two plastic surfaces. The shearing force applied to the
surfaces has been found to be nearly in proportion to the normal force applied to the surfaces [18]:

$$F = uN \qquad\qquad\qquad (15.3\text{-}17)$$

where u is the dynamic coefficient of friction between the two surfaces. As
shown in Table 15.3-1, the coefficients of friction for some materials can
increase with increasing speed, assuming the same normal forces [19].
Chang and Daane show that for certain materials the coefficients decrease
with increasing speed to a constant value and that there is a trend toward
increasing coefficients with increasing temperature. Furthermore, since it
can be assumed that all the frictional energy is transformed into heat, the
total amount of heat generated per unit time can be written as

$$Q = JNV \qquad\qquad\qquad (15.3\text{-}18)$$

where V is the velocity of the moving surface; N is the normal force, given
above; and J is the conversion from ft·lb to Btu. One way of writing the
normal force is in terms of the pressure applied over the contacting surface.
Therefore, Eq. (15.3-18) can be rewritten as

$$Q = JPAV = (JA)PV \qquad\qquad\qquad (15.3\text{-}19)$$

In the design of conventional sliding surfaces, the PV factor is a criterion
for loading of the wear surfaces. Here it is a measure of the rate of heat
that is generated at the sliding surface interface. Note that this expression
can be combined with Eq. (15.3-17) to obtain an expression of the total generation as a function of the shear force at the fluid interface. It should be

Table 15.3-1

Coefficients of Friction [19]

Specimen		Sliding speed (mm/sec)					
Slider	Plate	0.03	0.1	.4	0.8	3.0	10.6
Unlubricated tests							
PP (as molded)	PP (as molded)	0.54	0.65	.71	0.77	0.77	0.71
Nylon (as molded)	Nylon (as molded)	0.63	—	.69	0.70	0.70	0.65
PP (as abraded)	PP (as abraded)	0.26	0.29	.22	0.21	0.31	0.27
Nylon (machined)	Nylon (machined)	0.42	—	.44	0.46	0.46	0.47
Mild steel	PP (abraded)	0.24	0.26	.27	0.29	0.30	0.31
Mild steel	Nylon (machined)	0.33	—	.33	0.33	0.30	0.30
PP (abraded)	Mild steel	0.33	0.34	.37	0.37	0.38	0.38
Nylon (machined)	Mild steel	0.39	—	.41	0.41	0.40	0.40
Water-lubricated tests							
PP (abraded)	PP (abraded)	0.25	0.26	.29	0.30	0.28	0.31
Nylon (machined)	Nylon (machined)	0.27	—	.24	0.22	0.21	0.19
Mild steel	PP (abraded)	0.23	0.25	.26	0.26	0.26	0.22
Mild steel	Nylon (machined)	0.23	—	.20	0.20	0.19	0.17
PP (abraded)	Mild steel	0.25	0.25	.26	0.26	0.25	0.25
Nylon (machined)	Mild steel	0.20	—	.23	0.23	0.22	0.18
Liquid-paraffin-lubricated tests							
PP (abraded)	PP (abraded)	0.29	0.26	.24	0.25	0.22	0.21
Nylon (machined)	Nylon (machined)	0.22	—	.15	0.13	0.11	0.08
Mild steel	PP (abraded)	0.17	0.17	.16	0.16	0.14	0.14
Mild steel	Nylon (machined)	0.16	—	.11	0.09	0.08	0.08
PP (abraded)	Mild steel	0.31	0.30	0.30	0.29	0.27	0.25
Nylon (machined)	Mild steel	0.26	—	.15	0.12	0.07	0.04

apparent that increasing the frictional coefficient, either through increased pressure or temperature, increases the instantaneous amount of heat generated at the interface. In the former case, a friction or interference fit is called for at the mating surfaces. In the latter case, as frictional energy is translated into heat, the surface temperatures rise dramatically, thus increasing the coefficient of friction, which in turn increases the rate of frictional energy. No satisfactory form for the transient process of conversion of frictional energy into heat and the subsequent melting and dissipation into the body of the plastic has been presented.

Another technique that deserves attention here is electromagnetic induction bonding. This process is finding extensive use in in-field welding of plastic pipe and fittings. Basically a resin containing a high concentration of magnetic materials, such as iron filings, is placed between the two non-magnetic materials to be joined. An induction generation in the 2- to 25-mHz range with 1- to 10-kW power supplies a copper induction coil which is wrapped around the area to be bonded. The induction source heats the magnetic-filled resin, which in turn melts the nonmagnetic materials, affecting the polymer-to-polymer bond. Welding times are normally 0.5 to 5 sec, with welds capable of withstanding loads up to the yield point of the nonmagnetic materials [20].

15.4 Mechanical Fastening

Whenever plastics are to be mechanically fastened to other plastics, the design should allow for snap-fit assembly. This type of fastening can be most useful in forming hinges or swivels. Ideally, one part should be of soft, supple plastic and the other of stiff, rigid material. Ears, nipples, or tabs can be molded into the stiffer material and undersized holes or slots provided in the softer material. In this way, the stiffer material can be simply snapped into the softer part. For similar soft materials, the tabs or nipples can be barbed. Once the tab is pushed into the slot, the barbs prevent it from working out.

Commercially there are many forms of metal inserts. The primary objection to their use has been the extensive labor required to set them into the mold prior to molding around them. Not only is the labor expensive, but rejects require removal of the metal inserts before the scrap can be reground. The obvious advantage of metal inserts is that a tapped or threaded metal member is already in place at the time of assembly and thus no additional postmolding operations are needed. Metal inserts also offer superior holding power over a self-tapped hole and thus are usually reserved for heavy-duty applications such as commercial hand tools and furniture. As can be seen in Fig. 15.4-1 [21], most inserts are knurled or have inverse taper barbs that provide additional surface area for bonding with the plastic. Even with these, however, overtightening of the mating piece can break the insert loose and cause it to either drop out, crack the plastic around the insert, or cause the insert to turn freely in the plastic, thus preventing complete subsequent tightening of the bolt or nut. Other problems with inserts include inadvertent filling of the insert cavity or threads with plastic owing to improper sealing of the insert to the mold wall and changes in the physical properties of the plastic owing to its presence as a heat sink in the plastic during processing. It should be apparent that these obstructions can cause dramatic weld lines and anisotropic stresses which can result in long-term part failure in the vicinity of the insert.

Direct tapping into molded holes is useful for medium-duty applications. The primary caution here is that the material should not be highly notch sensitive. Self-cutting screws have one or two slots along the threads which help relieve materials and remove cuttings as they are driven home. Self-cutting screws are recommended if the flexural modulus of the plastic is greater than 100,000 psi. The higher the flexural modulus, the finer the screw pitch should be. For materials with lower moduli, there are thread-forming screws, with large open threads much like sheet metal screws. They have nearly flat cut threads, however, so that the softer plastic will push back against the screw root, thus securing it and preventing back out. This type of screw looks much like a helical barb. In some very soft plastics, such as LDPE, screws of this type can be driven without screwing, much the way wood screws can be hammered into soft wood. See Fig. 15.4-1.

A technique used frequently with thermosets and with some of the tougher thermoplastics employs a molded capstan. The article to be assembled has a molded-in hole through which the capstan is inserted. Then a Tinnerman nut or one of similar design is press-fit over the surface. Speed clips can also be used. This technique is particularly useful in fastening decorative articles and those exposed to low levels of stress or if the part does not need to be disassembled.

Staples and rivets can also be used to hold certain types of plastic together. Many structural foam furniture pieces have fabric stapled directly to the plastic part. Special staples that flair outward when driven into the plastic are used in this application. Some of the staples have barbs to offer greater pullout resistance. Rivets can be either metal or plastic. The metal rivet is peened over after it is pressed through the plastic. The plastic rivet is either spread by cutting apart a webbing between the two prongs of the rivet root or by heating the tip of a solid rivet with a hot iron. Both types have found applications, but remember that significant stresses can occur around the hole and that tearing or distortion of the hole can cause the rivet to pull out.

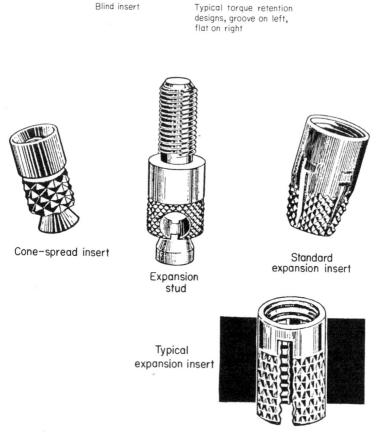

Figure 15.4-1 Many examples of molded-in inserts, driven inserts, and self-tapping screws [21].

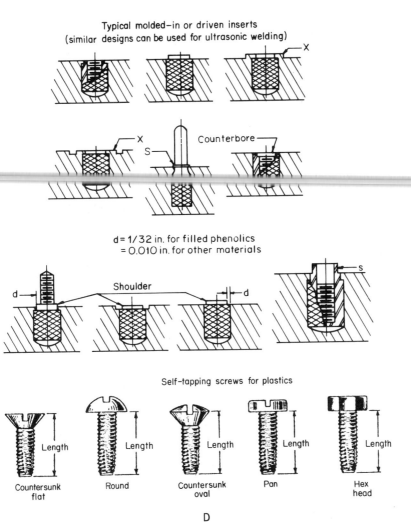

D

Figure 15.4-1 (continued)

15.5 Ultrasonic Welding

Ultrasonic welding has gained a great measure of acceptance in recent years as a controlled method of locally welding plastics together and of putting inserts into plastic parts. The standard device uses high-frequency energy on the order of 20 to 200 kHz with rather low amplitude. It is generated either with a piezoelectric transducer or a magnetostrictive transducer. The energy is directed with a specially shaped horn to a target or director that has been shaped to concentrate the energy in a very small area of the plastic interface. In concentrated doses, the energy forces the plastic to generate heat owing to the frictional forces that occur when the excited molecules in the plastic vibrate in a longitudinal mode. The frictional forces generate sufficient heat to bring the target area temperature very quickly above the welding temperature of the plastic. Because of the insulating quality of the plastic, very little heat is conducted from the target area, and with proper pressure, a suitable weld can be effected in a matter of seconds. A properly designed part has an energy director molded into one portion of the plastic and a receiver in the other. The receiver should be designed to direct the energy back toward the director, thus allowing energy intensification at the interface.

Aloisio et al. [22] have developed methods for analyzing ultrasonic bonding in terms of the thermoviscoelastic character of the plastic materials being welded. Experimentally they found that the temperature profile of the weld area, as measured by placing a thermocouple at the weld point between a lap joint, shows that the significant temperature rise occurs after the impulse from the ultrasonic device has been removed. This temperature profile is shown in Fig. 15.5-1, and the experimental setup is shown in Fig. 15.5-2. Apparently, to form a good part, the plastic must first be heated to the point where the melt at the interface will flow. Then pressure must be applied. For many materials, the viscosity of the material may be quite high in the areas peripheral to the concentrator. As a result, high pressures must be used even though the center material may be quite fluid. Most of the resistance will be felt in the rim of fluid around the director. Shaping the director so that the hotter fluid will flow outward away from the director is thus very important. Furthermore, once the plastic is molten and has flowed sufficiently far to provide a suitable bond, the pressure must be retained on the weld until it cools by conduction into the surrounding material. Note the close similarity in the flow process with that described in detail for compression molding in the section in Chap. 7.

Cooling rates also depend on the amount of material heated and the amount of energy discharged into the plastic. The more the plastic is heated, the longer it takes to cool to the rigid condition. Furthermore, flowing plastic beyond the target area can yield a poor bond, since the bond strengths between the melt and solid plastic are normally not so strong as between melt and melt.

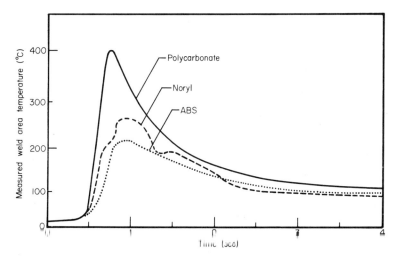

Figure 15.5-1 Measured temperatures for three materials during ultrasonic welding weld time, 0.6 sec; hold time, 1.0 sec; applied static pressure, 40 psi; 1:1 coupler [22].

If a linear viscoelastic solid is subjected to a sinusoidal strain input of frequency f and amplitude A, the strain behavior of the plastic is written as

$$e = A \sin ft \tag{15.5-1}$$

From linear viscoelasticity (see Chap. 4), the corresponding uniaxial stress is

$$s = E^*A \sin(ft + \phi) \tag{15.5-2}$$

where $E^{*2} = E'^2 + E''^2$. E' and E'' are the storage and loss moduli, respectively. The stress is shifted by an angle ϕ, where the shift angle is related to the storage and loss moduli by

$$\tan \phi = E''/E' \tag{15.5-3}$$

$\tan \phi$ is referred to as the loss tangent. If a viscoelastic solid is thought of as a combination of springs and dashpots, the force and displacement for a spring are in phase and the force leads the displacement for a dashpot by 90°. Thus, as seen in Chap. 4, the in-phase modulus E' is the pure elastic modulus, and the out-of-phase modulus E'' is the dissipative modulus. In other words, the storage modulus is a measure of the energy stored and recovered during each cycle of deformation, and the loss modulus is a measure of the energy dissipated as heat during this same cycle. $E''/E' = \tan \phi$ is on the order of 0.001 for metals but can range from 0.001 to 1 for

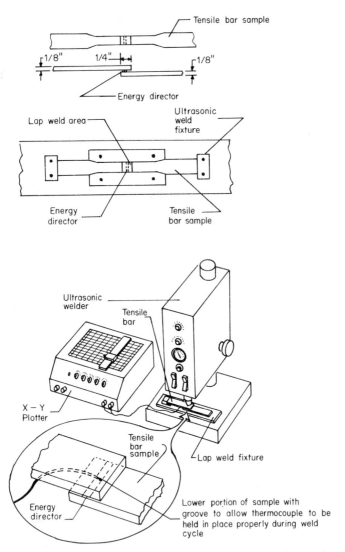

Figure 15.5-2 Schematic of apparatus used to obtain ultrasonic heating data of Fig. 15.5-1 [22].

polymers, depending on the frequency of deformation. Aloisio et al. then characterize the transient heating of the plastic at the interface as transient one-dimensional heat conduction, according to

$$k(\partial^2 \theta / \partial x^2) = \rho c_p (\partial \theta / \partial t) - D \qquad (15.5\text{-}4)$$

where D is the dissipation per unit time. D can be adequately replaced by a function of the dissipative modulus as

$$D = fA^2 E''/2 \qquad (15.5-2)$$

To simplify the work, the x dependency of θ, the temperature, is assumed to be small compared with the time dependency. Thus, Eq. (15.5-4) with these assumptions becomes

$$d\theta/dt = fA^2 E''/2\rho c_p \qquad (15.5-6)$$

A sinusoidal deformation in amplitude dL is applied to the system at the interface between the part and the horn. The force of reaction F at each interface must be equal for the system to be in equilibrium. If there is a linear stress variation through the thickness of the E_1 material, and if s_1 and A_1 are the stress and cross-section defined midway through the E_1 slab, then

$$s_1 A_1 = s_2 A_2 = F \qquad (15.5-7)$$

The displacement amplitude dL is

$$dL = 2e_1 L_1 + e_2 L_2 \qquad (15.5-8)$$

where e_1 is the strain midway through the E_1 material. The notation carries over for the second material with "2" subscripts. Now $s_1 = e_1 E_1$ and $s_2 = e_2 E_2$. Hence,

$$dL = (dL)_2 [1 + 2(L_1 E_2 A_2/L_2 E_1 A_a)] \qquad (15.5-9)$$

where $(dL)_2 = FL_2/A_2 E_2$, the energy director displacement. It is apparent that the smaller the second term in the brackets on the right-hand side of Eq. (15.5-9) becomes, the greater the proportion of the horn displacement is applied to deforming the energy director. As an example, Aloisio et al. used the data in Table 15.5-1 for the ultrasonic bonding of polycarbonate. Thus, half of the horn displacement is applied to the director. They point

Table 15.5-1

Data for Ultrasonic Bonding of Polycarbonate [22]

$A_1 = 0.125$ in.^{2}	$E_2/E_1 = 1$; $A_2/A_1 = 0.08$
$A_2 = 0.010$ in.^{2}	
$L_1 = 0.125$ in.	$2L_1/L_2 = 12.5$; $(dL)_2 = L/2$
$L_2 = 0.020$ in.	

out that it the magnitude of L_1 is such that these regions flex under applied deformation, as is the case with crystalline materials such as nylons, very small amounts of horn displacement are actually applied to the target. In Table 15.5-2 are given the physical properties for three weldable materials. The E" data at 20 kHz was extrapolated from experimental data at 110 Hz using standard time-temperature-frequency superposition methods [23]. Also given in the table are the calculated and experimental values for the rate of energy generation $d\theta/dt$. There is rather good agreement for all materials considering the extensive assumptions that have been made.

Again, referring to Fig. 15.5-1, note that peaking of the thermal curve occurs 0.5 to 1 sec after initiation of welding. Note that the total weld time was 0.6 sec and that the part was held between the weld fixture and the welder at a pressure of 20 psi for an additional 1 sec. In that period of time, the director heated the plastic lap at the rate calculated until flow began. Flow caused collapse of the director, and since the horn could no longer direct the energy, the rate of heating dropped rapidly. The temperature at the interface continued to rise, however, as the stored energy was released. As seen, the Lexan polycarbonate peaked at about 400°C, about 125°C above its recommended minimum processing temperature. Noryl-modified PPO and ABS are also 20 to 50°C above their recommended minimum processing temperatures. Thus, the plastics at the interface are indeed melted in 0.6 sec of directed ultrasonic energy. In Table 15.5-3 are given parametric effects of weld times and static pressures on the shear strengths of the lap joints for the three plastics considered by Aloisio et al.

Certainly additional work is needed on the time-temperature-pressure profiles for other ultrasonically weldable materials. More must be known about the shaping of directors, methods of clamping and holding pressure, and data for E" over the 20- to 200-kHz frequency range of the ultrasonic welders. Nevertheless, it is apparent that ultrasonic welding offers a unique way of directing energy to the interfaces to be joined without changing the thermal characteristics of exterior surfaces. The ultrasonic energy is transformed into internal mechanical dissipation, characterized here by E",

Table 15.5-2

Physical Property Data for Three Ultrasonically Weldable Plastics [22]

Material	C (cal/g·°C)	ρ (g/cm^3)	$\dot{\theta}$ calc. (°C/sec)	$\dot{\theta}$ exp. (°C/sec)	E" (dyn/cm^2)
Polycarbonate	0.30	1.2	1165	1650	1.38×10^8
ABS	0.36	1.03	1805	1100	2.09×10^8
Noryl	0.32	1.08	1125	1400	2.00×10^8

Table 15.5-3

Mold Shear Strength as Function of Weld Time and Applied
Pressure for Three Materials; Hold Time, 2 sec [22]

Weld time (sec)	Bond area (in.^{2})	Shear strength (psi)	Static pressure (psi)
		ABS	
	Energy director: $0.019 \times 0.020 \times 0.50$ in.		
	Shear yield stress: 1920 psi		
0.2	0.0582	910	20
0.3	0.0886	1550	20
0.4	0.100	1120	20
0.5	0.144	>1420	20
0.6	0.148	1380	20
0.7	0.148	1260	20
0.8	0.158	1280	20
0.9	0.174	>1275[a]	20
0.2	0.0722	1590	40
0.3	0.127	905	40
0.4	0.138	1250	40
0.5	0.167	>1200[a]	40
0.6	0.173	>1130[a]	40
0.7	0.169	>1200[a]	40
0.8	0.173	>1220[a]	40
0.9	0.162	>1360[a]	40
		Polycarbonate	
	Energy director: $0.020 \times 0.020 \times 0.50$ in.		
	Shear yield stress: 2777 psi		
0.2	0.0456	900	20
0.3	0.0487	3180	20
0.4	0.0976	1900	20
0.5	0.0786	2670	20

(continued)

Table 15.5-3   (continued)

Weld time (sec)	Bond area (in.^{2})	Shear strength (psi)	Static pressure (psi)
0.6	0.0770	>2980[a]	20
0.7	0.0560	>4800[a]	20
0.8	0.0608	>3740[a]	20
0.9	0.123	>2730[a]	20
0.2	0.0652	1200	40
0.3	0.054	>3350[a]	40
0.4	0.096	2120	40
0.5	0.106	>2920[a]	40
0.6	0.067	>1340[a]	40
0.7	0.136	>2640[a]	40
0.8	0.142	>1740[a]	40
0.9	0.160	>1920[a]	40

Noryl
Energy director: 0.025 × 0.50 in.
Shear yield stress: 2642 psi

Weld time (sec)	Bond area (in.^{2})	Shear strength (psi)	Static pressure (psi)
0.2	0.0185	973	20
0.3	0.0513	1793	20
0.4	0.0918	2200	20
0.5	0.1302	2004	20
0.6	0.1296	1790	20
0.2	0.0237	84	40
0.3	0.1040	1827	40
0.4	0.0796	2977	40
0.5	0.0372	2016	40
0.6	0.1340	1798	40

[a]When the shear strength is indicated as "greater than," the specimen failed in tension prior to bond failure. Bond areas were determined after subsequent separation of bonds unless uniform flashing indicated complete bond area coverage.

in order to raise the temperature at the target below the energy director to a point where the viscosity of the material is sufficiently low for rapid flow and wetting of the interfaces. Once flow occurs, the energy director collapses, and the rate of energy dissipation decreases rapidly. Note that the shaping and collapse of the concentrated energy source acts as an effective internal temperature-limiting mechanism that can reduce the incidence of thermal degradation during bonding.

15.6 Solvent Bonding

Unlike adhesive bonding, where a foreign substance is used that remains as a layer between the two articles being bonded, solvent bonding uses a low-molecular-weight chemical that normally partially dissolves into and swells the plastic. This makes the interface tacky. Under pressure, the interfaces adhere, and with some shearing action, sufficient comingling takes place to ensure a suitable bond. Solvent bonding is normally restricted to plastic bonding and is frequently reserved for bonding of similar plastics. For some plastics, such as GPS, there are many solvents that can be used. Some of the aldehydes, such as formaldehyde, and most of the ketones, such as acetone and methyl ethyl ketone, are excellent bonding solvents. Many plastics have very few solvents, however, as seen in Chap. 14. Nylon, for example, requires hot formic acid for bonding, and thus it is better to bond nylon to itself with an adhesive.

Theta solvents were discussed in detail in Chap. 14, as was the close relationship between the free energy of the plastic and that of the solvent. Ideally, a suitable solvent for bonding should soften and swell the plastic rather than attacking and dissolving it. Further, the solvent should be sufficiently gentle in its attack so as to prevent solvent stress cracking. Furthermore, it should be noted that solvent that is absorbed into the structure of the plastic, thus swelling it, will remain concentrated at the weld after processing. The rate of diffusion of the solvent through the plastic and migration to the free surfaces for evaporation must also be considered when selecting a suitable solvent. Normally, solvents will decrease the physical properties of the plastic, and frequently even small residual amounts can lead to part failure under long-term loads much earlier than predicted from standard failure tables.

Ueberreiter [24] graphically illustrates some of the solution processes that can take place. As shown in Fig. 15.6-1, the normal dissolution process has several layers of material in various stages of solution. Above the solid material there is an infiltration layer, where molecular diffusion of solvent occurs. Above this is the solid swollen layer, followed by the gel layer. Then occurs the liquid solution layer, where the polymer is in complete solution with the solvent. And above this is the pure solvent. The liquid solution layer has a large concentration gradient, with a high solvent concentration near the solvent interface and a large polymer concentration layer

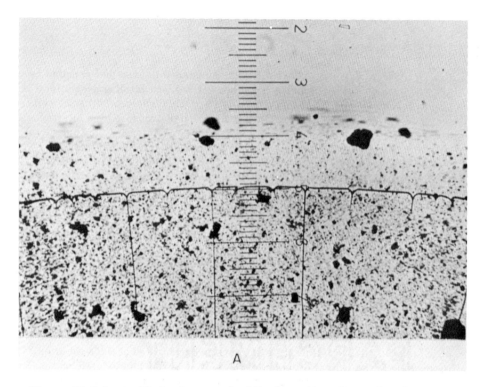

Figure 15.6-1   A: Photomicrograph of the formation of a surface layer at
the beginning of a normal dissolution process. Compare the development of
the gel layer with that in Fig. 15.6-2. B: Photomicrographic interferometric development of the normal dissolution process. The solid polymer is on
the left, solvent on the right, and developing surface or gel layer in the middle [24].

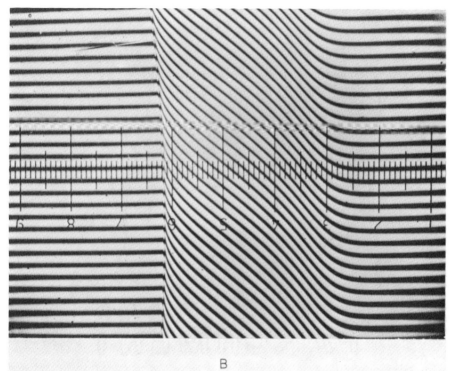

B

near the gel layer. The gel layer is characterized by the untangling of the long-chain polymer molecules. The swollen layer is characterized by the stretching of the polymer network by the infiltrating solvent molecules. In the infiltration layer, the polymer-solvent system is in a glassy state. Here single molecules of solvent are penetrating the great number of voids of molecular dimensions that are common in glassy polymers. No noticeable swelling is occurring in this layer, although there is solvent infiltration. Note the dramatic change in the index of refraction during dissolution (Fig. 15.6-1B).

Some materials exhibit normal dissolution down to very low temperatures. Polystyrene is typical of this type of material. Even at -40°C, toluene solvent attack on polystyrene shows very distinct bands of solvent, liquid layer, gel layer, swollen layer, and infiltration layer. Some materials, such as polymethyl methacrylate, exhibit gradual thinning of the gel layer with decreasing temperature, until at a temperature known as the gel temperature the gel layer disappears entirely. Note in Fig. 15.6-2 the

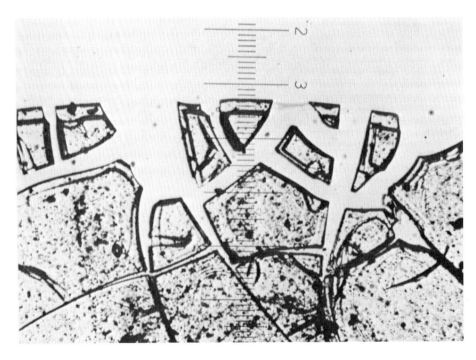

Figure 15.6-2 Photomicrograph of dissolution process without formation of gel layer. Characteristic stress cracking below the gel forming temperature. Compare this photomicrograph with that in Fig. 15.6-1 [24].

complete absence of any gel layer. The solvent drives into the plastic and cracks large pieces from the interface.

Ueberreiter defines the velocity of penetration $\dot{s}$ as the rate at which the solvent penetrates the polymer. For normal dissolution this rate is constant at a given temperature. It is strongly dependent on temperature, however, exhibiting an Arrhenius dependence:

$$\dot{s} = \dot{s}_0 \exp(-A/T) \qquad\qquad (15.6\text{-}1)$$

where $\dot{s}_0$ is some reference velocity of penetration and A is a form of solvent dissolution activation energy. At the point where the gel layer disappears, the slope of $\dot{s}$ vs. $1/T$ changes dramatically. This is shown in Fig. 15.6-3 for PMMA having a gel temperature of 84°C. An enormous increase in A is observed, indicating that a different dissolution mechanism is taking place. The gel layer has disappeared, and cracking is taking place. Fur-

as is typical with normal dissolution. Ueberreiter has shown that dissolution by stress cracking mechanism for a given polymer-solvent system can be changed into the normal dissolution process by raising the solvent-polymer interfacial temperature above the gel temperature.

As shown for polystyrene in Table 15.6-1, the rate of solution and the thermodynamic definition of a good solvent given in Chap. 14 are not the same. Note that only toluene and perhaps benzene are both suitable thermodynamic and kinetic solvents in that they are thermodynamically suitable

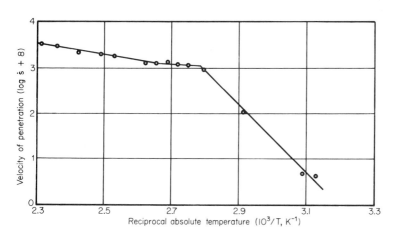

Figure 15.6-3 Velocity of solvent penetration into polymer as function of reciprocal temperature. Note that the sharp break in the curve represents the gel forming temperature. Polymer, 105,000-molecular-weight PMMA; solvent, dimethylphthalate [24].

Table 15.6-1

Solvent-Polystyrene Systems [24]

Solvent	Thermodynamic suitability	Kinetic suitability
Toluene	+	+
Carbon tetrachloride	+	−
Ethylacetate	−	+
Methyl ethyl ketone	−	+
Amylacetate	−	−

for dissolution and that the dissolution forms a normal layered system. Furthermore, as shown in Fig. 15.6-4, the diffusional rate of the solvent depends on the viscosity of the solvent as well as being exponentially dependent on the temperature. Thus, a decrease in the rate of solution implies that the volume of material suitably prepared for bonding decreases. Thus, it will take longer to soften the material to a bonding or gel consistency as the temperature decreases and the solvent viscosity increases.

Some materials are so susceptible to stress cracking dissolution that no suitable solvent can be found to minimize detrimental product appearance and performance. PMMA is one material that demonstrates these characteristics at room temperature. Thus, small amounts of PMMA are dissolved in a suitable solvent, thus making a syrup. In this way, the material seems to have a ready-made gel layer protecting the surface, and the increased viscosity slows the speed of penetration of the solvent layer. GPS and rigid PVC can also be bonded at a slower rate using this type of solvent-polymer solution. Noryl, attacked strongly by chloroform, can also be bonded this way, although a 60:40 mixture of chloroform and carbon tetrachloride make a near-theta solvent at room temperature, thus permitting normal dissolution.

Again, the object of solvent bonding is to soften and swell the plastic, not dissolve it. A sufficiently thick gel layer containing solvent must be developed on each surface so that when these are pressed together and clamped the tackiness of the gel will allow bonding. Excess solvent will dissolve other areas of the plastic, and thus solvents should be used with care. The clamping pressure is normally quite light and used primarily to express air and excess solvent from the bonding area. Further, excess solvent trapped within the gel layer and swollen solid layer will only slowly diffuse to the solid surface and evaporate. Thus, most solvent-welded parts will exhibit lower physical properties in the solvent-bonded regions.

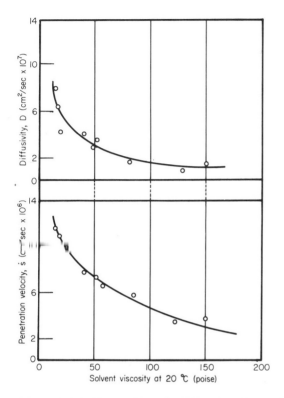

Figure 15.6-4   Kinetic solubility of various solvents in PVC-PVA (87:13) copolymer at 120°C as function of solvent viscosity at 20°C. Top set of curves represents diffusivity; bottom set of curves represents velocity of penetration ṡ in cm/sec [24].

15.7 Comments on Part Design for Easy Assembly

When mating two pieces of plastic, the location of the mating line can be as important esthetically as the location of a parting line. Frequently, melt, adhesive, or solvent-swollen polymer is expressed from the weld region to form an esthetically unattractive bead. Several ways of over-coming this bead have been discussed elsewhere [21]. For example, the bead region can be embedded in a series of parallel beads deliberately molded into the part. Or stepped lines can be used, where one part is off-set 10 to 20 mils from the other. This allows for welling of the bead and also possible registry problems and seems to yield the best results.

Molded-in internal threads require unscrewing mechanisms in the mold. However, with materials having a Shore A hardness of 60 or less,

strippable molded-in threads can be used. Normally, threads having gen-
erously rounded or $\vee$ profiles with 12 to 32 threads/in. can be stripped
without significant damage to the thread undercut surfaces.

To handle small metallic inserts, the inserts are partially punched
from a continuous strip. The edges of the strip are retained so that the inserts remain tied together in an endless belt. This belt is then fed semi-
automatically into the mold. After the plastic has been molded around the
insert, the edges are broken away. This minimizes tedius hand labor.

And certainly the use of the integral hinge bears mention. With some
materials, the hinges need not be used more than a dozen times. As an
example, the plastic cigarette box closure is only used 20 times. Thus,
the hinge area can be allowed to stress-crack under flexure. For thousands
of flexes, however, a material having living hinge qualities, such as PP,
must be selected. Then the hinge must be worked when the part remains
hot to minimize excessive crystallization in the flexing area.

In general, the key to assembly is the selection of a method that is
compatible with the materials being assembled. Solvent resistance, notch
sensitivity, flexibility, and mechanical strengths must all be considered
when selecting an assembly method.

Problems

P15.A An adhesive requiring at least 250°F temperatures to cure is placed
between two pieces of plastic each 0.25 in. thick. A hot plate at 350°F is
placed against each surface. The adhesive layer can be considered to be
negligible in thickness. How long does it take to heat the adhesive to reacting
temperature? The thermal diffusivity of the plastic is 5×10^{-3} ft^2/hr, and
the initial temperature is 65°F.

P15.B Given the kinetic data in Table 15.2-1, show that the rate constant
k defined as $k = aN_0 KGfA_d/A$ is approximately 0.037 sec^{-1}. Then translate
the equation into activation of a sheet moving linearly at a velocity of U to
show that

$$U = -\pi Dk/\ln(1 - y) \qquad\qquad (15.B-1)$$

where D is the diameter of the drum used to obtain the experimental data
tabulated in Table 15.2-1. If D = 6 in., show that to achieve y = 0.9 the
linear speed should be less than 1.5 ft/min. How much activation is achieved
if the linear speed is increased to 100 ft/min? If k is directly proportional
to the frequency, what increase in frequency is necessary to achieve 90%
activation at 100 ft/min?

P15.C In P15.B, the applied current was held constant. If the rate constant
is directly proportional to the applied current, what increase in current is
necessary to achieve 90% activation at 100 ft/min? Make a plot of Eq.
(15.B-1) for the range of $0.2 < y < 1$.

P15.D Porter [3] has found experimentally that the gap between the electrical discharge and the sheet also affects the rate constant. For example, at 50 mils the rate constant for the data given in Table 15.2-1 for gap distance is about 0.11 sec^{-1} at 65 μA and 8650 V and 0.08 sec^{-1} at 200 mils of gap distance. If the gap distance can be assumed to affect the rate constant exponentially, determine the gap distance necessary to achieve 90% activation at 100 ft/min if the frequency and applied current are the same values as those given in P15.B.

P15.E The rate constant for deactivation of PP film has been found to be 0.15 sec^{-1}. The rate constant for activation is approximately 0.2 sec^{-1}. These values were obtained on a 12-in.-diameter rotating drum. Translate these values into linear velocity, according to Eq. (15.B-1). Then determine the speed necessary to have 50% of the surface active when the printing process is 24 in. from the corona discharge.

welding. The planar source is at $450°$F, and the sheet stock is tacky at $250°$F. Approximately 0.1 in. of sheet stock must be melted in order to achieve a strong weld. Using the proper figure in Chap. 6, determine the time required to achieve this condition. What is the temperature 0.2 in. into the sheet stock?

P15.G Work through the arithmetic to obtain Eq. (15.3-7).

P15.H A heat source of 1000 Btu/ft^2 is used to weld plastic pipe, having a thermal diffusivity of 2.5×10^{-3} ft^2/hr and a thermal conductivity of 0.15 Btu/ft·hr·°F. To heat the pipe from 70°F to a welding temperature of 350°F, what is the maximum velocity at which the heating source can move? Assume no heat losses.

P15.I In P15.H, the parts being welded are adjacent to a stainless heat sink, so that the heat loss to it from the plastic region is approximately 100 Btu/ft^2·hr·°F. Can the part now be welded? Now what is the welding heat source speed? Through repositioning, the heat sink is moved some distance from the weld area such that the heat loss is reduced to 10 Btu/ft^2·hr·°F. What is the increase in welding speed?

P15.J Work out the analysis for the velocity of a moving heat source that approaches a heat sink positioned along the weld path.

P15.K Teflon has an approximate value of the dissipative modulus E'' of 6×10^8 dyn/cm^2. Using a horn of 20 kHz with an amplitude of 0.002 cm, determine the temperature rise per unit time in the material. It has a density of 0.95 g/cm^3 and a specific heat of 0.4 cal/g·°C. If the weld must be 50°C above the melting temperature of 265°C, how long must the bonding energy be applied?

P15.L Reconsider Aloisio et al.'s equation (15.5-4). This equation is a special form for the nonhomogeneous diffusion equation which can be written in shorthand as

$$\theta_t = a\theta_{xx} + D(x,t) \tag{15.L-1}$$

Since the conditions on x are infinite in that the time interval is very short, multiply both sides of this equation by $(2\pi)^{-1/2} \exp(i\xi x)$ and integrate from $-\infty < x < \infty$ to obtain

$$\theta'(\xi,t) = (2\pi)^{-1/2} \int_{-\infty}^{\infty} \theta(x,t) \exp(i\xi x)\, dx \tag{15.L-2}$$

Show that this satisfies the following differential equation:

$$d\theta'/dt + a\xi^2\theta = D'(\xi,t) \tag{15.L-3}$$

where $D'(\xi,t)$ represents the Fourier transform of $D(x,t)$. Now since $\theta'(\xi,0) = 0$, show that

$$\theta'(\xi,t) = \int_0^t \exp[-a\xi^2(t - t')]D'(\xi,t')\, dt' \tag{15.L-4}$$

If the Fourier integral transform is defined as

$$\theta(x,t) = (2\pi)^{-1/2} \int_{-\infty}^{\infty} \theta'(\xi,t) \exp(-i\xi x)\, d\xi \tag{15.L-5}$$

show that through substitution and interchanging the order of integration,

$$\theta(x,t) = (2\pi)^{-1/2} \int_0^t dt' \int_{-\infty}^{\infty} \exp[-i\xi x - a\xi^2(t - t')]D'(\xi,t')\, d\xi \tag{15.L-6}$$

If $F(\xi) = \exp[-a\xi^2(t - t')]$ is the Fourier transform of the function

$$f(x) = 2a(t - t')^{-1/2} \exp[-x^2/4a(t - t')] \tag{15.L-7}$$

show that the final solution to Aloisio et al.'s equation is

$$\theta(x/t) = (4\pi a)^{-1/2} \int_0^t dt'/(t - t')^{1/2} \int_{-\infty}^{\infty} \exp[-(x - \eta)^2/4a(t - t')]D(\eta,t')\, d\eta \tag{15.L-8}$$

Now work out the expression if D is a constant. Compare your result with Eq. (15.5-6) for $d\theta/dt$, and comment on the assumption of the independency of θ on x.

References

1. Personal communication, Red Spot Paint and Varnish Co., Evansville, Ind., 1974.
2. J. M. McKelvey, Polymer Processing, Wiley, New York, 1962, Chap. 6.
3. J. C. Porter, doctoral thesis, Washington University, St. Louis, 1960.
4. G. Kraus and J. E. Manson, J. Polym. Sci., 6:625 (1952).
5. D. Taylor and J. E. Rutzler, Ind. Eng. Chem., 50:928 (1958).
6. F. J. Bockoff, E. T. McDonel, and J. E. Rutzler, Ind. Eng. Chem., 50:904 (1958).
7. Personal communication, Plastics Research Corporation of America, Inc., Brooklyn, 1976.
8. W. L. Cottrell and S. W. Nolan, Mod. Plast. Encyclopedia, 53(10A): 432 (1976).
9. R. C. Progelhof and J. L. Throne, SPE ANTEC Tech. Pap., 22:532 (1976).
10. E. P. Stokes, Mod. Plast. Encyclopedia, 49(10A):680 (1972).
11. B. P. Rouse, Jr., and T. M. Hearst, Sealing and Welding of Thermoplastics, In: Processing of Thermoplastic Materials (E. C. Bernhardt, ed.), Van Nostrand Reinhold, New York, 1959, Chap. 10.
12. W. H. McAdams, Heat Transmission, McGraw-Hill, New York, 1954.
13. R. M. Knight and W. U. Funk, Mod. Plast., 34(12):133 (1957).
14. W. Neubert and W. A. Mack, Plast. Eng., 29(8):40 (1973).
15. R. B. Bird, W. E. Stewart, and E. N. Lightfoot, Transport Phenomena, Wiley, New York, 1960, Chap. 16.
16. M. Jacob, Heat Transfer, Vol. 1, Wiley, New York, 1949, Sec. 17.
17. F. W. J. Olver, Bessel Functions of Integer Order, In: Handbook of Mathematical Functions (M. Abramowitz and I. A. Stegun, eds.), Dover, New York, 1965, Sec. 9.
18. H. Chang and R. A. Daane, SPE ANTEC Tech. Pap., 20:335 (1974).
19. G. Ross and A. W. Birley, In: Thermoplastics: Properties and Design (R. M. Ogorkiewicz, ed.), Wiley, New York, 1974.
20. Personal communication, M. Chookazian, EMABond, Inc., Englewood, N.J.
21. J. H. DuBois, Plastics Product Design, In: Plastics Mold Engineering (J. H. DuBois and W. I. Pribble, eds.), Van Nostrand Reinhold, New York, 1965, Chap. 2.
22. C. J. Aloisio, D. G. Wahl, and E. E. Whetsel, SPE ANTEC Tech. Pap., 18:445 (1972).
23. J. D. Ferry, Viscoelastic Properties of Polymers, Wiley, New York, 1970.
24. J. D. Ueberreiter, The Solution Process, In: Diffusion in Polymers (D. Crank and D. P. Park, eds.), Academic Press, New York, 1968, Chap. 7.

16.

Platics Economics

16.1 Introduction

There are several viewpoints to consider when pricing plastics processes
and parts. The process can be viewed from the custom molder's viewpoint.
He (she) must have sufficient experience in estimating his internal costs
when bidding on a job. He must know how to accurately determine his overhead, direct labor, machine utility costs, depreciation, and so on. From
the purchaser's viewpoint, the unique properties of the materials to be
used and the total number of pieces to be made are important. The designer
must know how to design the necessary mechanical strengths and esthetic
features of plastics into the part and how to design around the obvious weaknesses. The mold maker must determine the proper placing of water lines,
types and location of gating, the degree of shrinkage or growth of the materials and the mold, the placement of ejector pins, the methods of clamping
the mold into the machine, and the mold material.

  The most important consideration must lie with the process engineer,
who must carry through several levels of thinking regarding the fabrication
of the final plastic product. He must be able to justify economically the
use of plastic rather than more conventional materials such as die-cast
zinc, brass, wood, ceramic, sheet metal, or cast iron. He must know the
ultimate use of the final part, so that he can determine what plastic material,

847

if any, will serve well throughout the expected life of the part. He must
know how to correlate accelerated test results with actual use, at least on
a qualitative basis. He must be capable of meeting many of the esthetics
demanded by the customer and yet must have a part that remains service-
able at a low cost. He must know how the part will change during molding
and postmolding treatment, including shrinkage, warpage, and sink marks,
and must be able to convince the designer that these things do happen to
plastic materials. He must work carefully with the mold designer to prevent gating in areas that can leave blemishes, gate blush, sink marks, or
weld lines exposed, thus damaging the esthetics of the part. He must insist
that cooling lines be placed in reasonable proximity to the plastic part that
is being cooled and that the size of the water lines be large enough to accom-
modate adequate cooling rates. Coolant Reynolds numbers in excess of
2000 are desirable so that advantage can be made of turbulent heat transfer
coefficient. He must be aware of the alternative processing methods for
making the part and their inherent strengths and weaknesses. And he must
always guard against a miracle resin or process. Most radically new
developments do not work much better than known developments, and since
the new developments have not had the exposure of many conditions and
many applications, they may be susceptible to off-spec conditions, poor
quality control on the materials, dirt, off-color, and poor compatibility
with color concentrates, additives, and reinforcements. For new machinery
developments, inability to get replacement parts, poor maintenance records,
poorly made critical areas such as screws, toggle clamps, control circuits,
and drive train gears, and inconsistency from machine to machine are common.

Dearle [1] discusses in great detail many preliminary considerations
to establishing a working molding shop. He includes plant layout, production planning, estimating, and management and control of profits. He has
work sheets and forms for several small operations. In this chapter, the
emphasis will be placed primarily on how to compare different processes
from an economic viewpoint, weakness/strength considerations, how to
make a quick estimate as to the manufacturing cost of a plastic part so that
a comparison with existing parts can be made, and how to cost out to a
return on investment (ROI) on plastic parts. Since most molders have their
own methods of accounting, only one method will be used as an example of
this form of complete detailed cost analysis.

16.2 Comparative Plastics Processes

In many cases, the choice of the plastic process to make a given item is
dictated by the shape or use of the item itself. For example, to replace
copper tubing with vinyl, the extrusion process is the clear-cut choice. If
a part is very small with many convolutions and details and is presently die
cast or machined, the injection molding process is competitive. If a large
part, such as a compressor housing, is needed and if the part must have

varying wall thicknesses and must withstand soldering or high motor temperatures, compression or transfer molding of a thermoset would be considered competition to die casting or forging. If a very large part such as
a camper top is required, thermoforming should be compared with sheet
metal assembly and painting. These decisions seem clear-cut. Nevertheless, there are certain new processes that might also be included in an
efficient cost analysis. For example, thermoplastic structural foam molding
can be used to produce parts that can compete with compression molding of
thermosets or with large-part thermoforming. New processes should be
included as a way of learning more about the state of the art for that process
and can be excluded in some cases on the basis of insufficient engineering
knowledge and limited materials at the time the analysis is made even if the
price seems competitive.

There are areas, however, where there are competing processes.
Forming a hollow object or container is one area. Frequently, blow molding,
injection molding, rotational molding, and thermoform molding are all economical under certain conditions. The various characteristics of these
processes are listed in Table 16.2-1. A. D. Little [2] produced a rather
extensive study of the economics of fabricating a $12 \times 12 \times 12$-in. box with
1/8-in. walls using these four processes. The results are shown in Table
16.2-2 for 1966 dollars. Although the value of the dollar has changed, the
comparative analysis seems valid today. These analyses contain capital
investment, tooling cost, material cost to make the box of LDPE, the manufacturing cost, and the conversion cost for 10^5, 3×10^5, and 10×10^5 pieces
annually. Inclusion of the Hayssen process, described in Chap. 12, is probably justified as a new process. In another comparison, they fabricated a
20-gal garbage pail using blow molding, injection molding, and rotational
molding, and their comparative analyses are given in Table 16.2-3. Crater
[3] has compared these processes for the fabrication of 1-, 5-, and 55-gal
containers as unit cost per annual production. As shown in Fig. 16.2-1,
blow molding is preferred in the 1-gal size when the annual production rate
exceeds about 15,000 units. Rotational molding is preferred in the 5-gal
size, regardless of the production rate, and injection molding is preferred
for large production rates of 55-gal containers. It should be noted, however,
that there is probably little demand for 1 million 55-gal containers a year
in any operation. Thus, rotational molding is preferred for 1000 to 10,000
units of these containers. In sizes above 55 gal, of course, rotational molding is in competition with blow-molded or glass-wound thermoset polyester
containers and not with injection-molded products.

Frequently, a comparison of the cost of plastic processes must include the cost of fabrication by conventional means. As shown in Table
16.2-4, a 1966 survey [2] compared blow molding and rotational molding
with stamped and welded steel to form an 18-gal gas tank, $38 \times 30 \times 7 \times
0.100$ in. The recommended plastic is HDPE. Volkswagenwerk AG [4] be-
gan work in 1963 on a gas tank that at present is used in the VW Passat and
Golf. Their initial process was powder rotational molding, and for various

Table 16.2-1

General Characteristics of Several Plastics Processes

Characteristics	Blow	Injection	Rotational	Thermoform
Resin form	Pellets	Pellets	Powder	Sheet
Availability of material	Very good	Excellent	Fair	Good
Material cost	Standard	Standard	Price includes grinding	Price includes extrusion of sheet
Types of materials	Somewhat restricted	All	Very restricted	Low rubber content
Scrap reuse	Immediate	Immediate	No thermally sensitive materials	Economically required
Color	Colored pellets or liq. conc.	Same	Dry blend	Colored sheet
Processibility of thermally sensitive materials	Difficult	Moderate	Very difficult	Controllable
Thermoset materials	Not tried	Difficult	Possible	SMC
Variety of mold materials	Very limited	Same	Many	Very many
Cost of molds	High	Highest	Moderate to high	Lowest
Mold maker reliability	Fair	Excellent	Poor to fair	Fair
Mold closure	Butt	Butt	Tongue/groove	Rim clamp

Nonferrous tooling	Limited	Limited	Good	Excellent
Method of holding mold	Mechanical	Hydraulic/mechanical	Mechanical	Mechanical/vacuum
Thermal cycling	Moderate	Moderate	Severe	Mild
Major trial/error problems	Pinch-off/wall uniformity	Gating/weld lines	Ratio of speeds	Corner draw down
Cooling method	Mold core/air	Mold core	Water quench	Air jets
Part release	Push pins/air	Ejector pins	Manual	Air
Life of molds	10^5–10^6	10^6–10^8	10^4	10^3–10^4
Operating pressure	Moderate	High	Low	Lowest
Operating temperature	High	High	Highest	Moderate
Controlling part of cycle	Blowing	Cooling	Heating	Heating
Skill of operator	High	Moderate	Moderate	Low
Man/machine interaction	Low	Nil	Very high	Normally high
Filling methods	Automatic	Automatic	Manual	Manual/semi-automatic
Part removal	Automatic	Automatic	Manual	Manual/semi-automatic
Part wall uniformity	Good, controllable	Excellent	Fair	Good, some control
Flash	Moderate to high	Moderate	Low	Highest
Inserts	Questionable	Feasible, costly	Feasible	Questionable

Table 16.2-1   (continued)

Characteristics	Blow	Injection	Rotational	Thermoform
Material orientation	Uniaxially oriented	Oriented	Unoriented	Biaxially oriented
Stress retention	High	Highest	Little	High
Method of control of distortion, warp	Cooling	Pressure	Air/water cool	Heating
Method of forming closed part	NA	Welding	NA	Welding/2-sheet forming
Primary mechanical part failure	Thin side walls	Weld line	Poor tensile strength	Thin corners
Surface finish	Very good	Excellent	Good	Excellent
Parting line shape				

Table 16.2-2

Comparison of Fabrication Costs for $12 \times 12 \times 12 \times 1/8$-in. LDPE Box (1966 dollars) [2]

	Blow	Injection	Rotational	Thermoform
Scrap (flash)	15%	8%	5%	25%
Cap. invest.				
1×10^5	$200,000	$180,000	$100,000	$100,000
3×10^5	280,000	230,000	100,000	260,000
10×10^5	560,000	600,000	220,000	780,000
Tooling cost				
1×10^5	7,200	7,000	4,000	5,000
3×10^5	7,200	7,000	7,000	7,200
10×10^5	12,000	19,800	21,800	13,000
Material cost ($/part)				
1×10^5	$0.80	$0.50	$0.65	$0.90
3×10^5	0.50	0.25	0.50	0.80
10×10^5	0.25	0.20	0.50	0.65
Manufacturing cost (10×10^5)	$0.817	$0.673	0.793	$2.646
Conversion cost (10×10^5)	$0.246	$0.187	$0.125	$0.333

Table 16.2-3

Comparison of Fabrication Costs for 20-Gallon HDPE Garbage Pail
(1966 Dollars) [2]

	Blow	Injection	Rotational
Machine cost	$230,000	$75,000	$28,000
Mold cost	5,600	30,000	10,500
Power cost/yr	5,950	4,500	3,360
Labor cost/hr	1.75	1.75	3.50
Mat'l cost/lb	0.18	0.16	0.21
Mat'l cost/part	0.90	0.80	1.05
Rate/parts per hr	80	60	60
Cost/part			
10^4 parts	$1.63	$3.95	$2.20
10×10^4 parts	1.12	1.25	1.25
100×10^4 parts	1.07	0.98	1.16

Table 16.2-4

Costs of Comparative Processes for 18-Gallon Gasoline Tank [2]

	Blow	Rotational	Stamped steel
Items/hr	102	88	126
Material cost	$3.230	$3.220	$2.693
Amortization	0.136	0.073	0.262
Direct labor	0.230	0.266	0.460
Utilities	0.137	0.029	0.080
Overhead (factory only)	0.147	0.106	0.196
Manufacturing cost	3.880	3.694	3.691

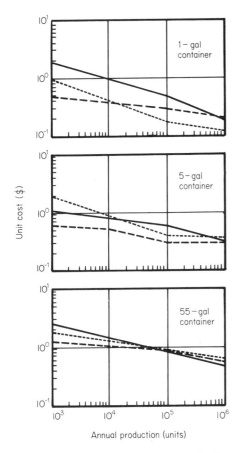

Figure 16.2-1   Comparative costs for various size containers made by three different processes as a function of the annual production required. Solid line, injection molding; dotted line, blow molding; dashed line, rotational molding [3].

reasons they experimented with Nylon 6 in situ reaction molding, injection molding, and thermoforming before they developed a process in 1975 for blow molding Nylon 6.   The cost of this tank is competitive with welded steel.

Note that the difference in material prices used in certain economic analyses can come about owing to scrap and specialized resin preparation. Rotational molding requires powder, and thus a grinding charge must be included.   Thermoformed materials are in sheet form, meaning that the prior processing and the web regrind charges must be passed on.   Blow

molding resins usually cost slightly more than injection molding resins
since the process requires higher-molecular-weight resins with narrower
molecular-weight distributions. Thus, the peculiarities of the process
must be included in any realistic economic analysis.

16.3 The Simple Cost Estimate

Before becoming embedded in a log sheet of details of inventory and plant
layout, it is instructive to determine whether the plastic part is competitive
to within ±25% or so of a part made in a more conventional process. Thus,
a very simple cost estimate should be carried out. This is particularly
true of decorator parts, parts presently made of wood with little hand work,
parts of sheet metal that have been simply painted, extruded metal such as
that used in window frames, folded and shaped metal sheet such as aluminum
siding or gutters, parts that are presently made of plastic, particularly
thermoset parts such as phenolic electrical boxes, and so on. The part
that is being developed in plastic should have realistic manufacturer's cost
associated with it in its present form. Never make a part of plastic because
the competition is making it of plastic. Always be critical of the constraints
on the part, such as durability, fatigue, creep, erosion, corrosion, colorfastness, flammability, and electrical or arc resistance, and list carefully
all the codes that the part must pass. Note that codes exist on all levels of
government and cover many facets of organized society. Representative
codes to examine include Underwriters, Factory Mutual, BOCA, FHA,
National Sanitation Foundation, local fire ordinances, state consumer acts,
FDA, and Department of Commerce. If the part must be tested and evaluated in order to gain certain acceptable labels from state, local, or federal
agencies, the test must be paid for, and sufficient time must be allotted for
necessary approval. In some cases with federal standards, 24 to 60 months
are required for such approval as FHA external application certification and
12 to 24 months for internal application for plastics in critical areas such
as drain pipes and siding. Food contact applications require 24-month rat
feeding studies. If the time seems inordinately long, return to conventional
approved materials and wait until someone else obtained the approval in that
application.
      Remember that there is nothing sacred about plastic. If a suitable
way of recycling glass bottles is found, there will be little need for a plastic
pressure-resistant bottle. Glass has superior strength and good impact
and excellent resistance to diffusion of CO_2 under pressure. No plastic,
regardless of the material source, will reproduce the thread produced on
steel by a threading machine. If clean sharp threads are needed, metal is
superior to plastic. And wood is a natural structural foam of unusual
strength. No plastics engineer should be embarrassed to state that he feels
that the part should not be made of plastic. In all probability, the proper

material is not yet commercial or the optimum or economically feasible process is not sufficiently reproducible.

Using an experimental-grade plastic to evaluate the feasibility of a new part or a redesign can lead to severe problems. The company supplying that plastic may decide that the market is not large enough and stop production; it may combine this experimental formula with another, thus leading to poorer mechanical properties; or it may go financially broke trying to market it. A backup supplier of a material of a similar grade of resin should always be sought. One recommended way of operating is to purchase truckload quantities from one supplier and drum-load quantities from another. Accurate records are required using this system, however, since processing behavior may vary from lot to lot. The logs should contain proper machine conditions, quality performance of the part, scrap generation, extent of regrind, and so on. In this way, if the major supplier has trouble meeting the quota or if the required number of pieces increases, the minor supplier can be contacted. This is also a commonsense hedge against strikes, slowdowns, and fickle energy redistribution programs. The smaller the molding house, the more effort the engineer must expend in keeping records of production performance, particularly with regard to the engineering aspects. As mentioned, material performance, including quality performance of incoming and final product materials; scheduling of various materials in various machines; performance of a given material in a given machine; machine performance, such as cycle time, mold water temperature, and barrel zone temperatures; problems in start-up and shutdown; and problems of scorching and excess warpage of parts are usually the most important items. Every material and every lot of every material will process slightly differently from the last batch owing to subtle changes in molecular-weight distribution, number-average molecular weight, changes in color concentration, condition of filler, moisture content, grinding conditions, and so on. And certainly every machine will have its idiosyncrasies. Common problems include surging for no apparent reason, sympathetic cycling of heater banks, uneven closure of molds, wearing of screws, slippage in drive mechanisms, burned-out heaters, and overshoot of temperature controllers, among many. The log should be kept at all times. The information obtained when things are sliding away from the established norm are as important in troubleshooting reoccurring troubles as that obtained when things are finally going right.

Frequently, marketing will require a plastic part that must compete with a part presently made of metal, glass, wood, or ceramic. It is necessary to determine in a few minutes how much that part would cost. Consider here that the part is larger than 15 g and smaller than 2000 g. Following the procedure below, a quick estimate can be made in a short time.

1. Determine how many parts are required per year. If only a few are needed, then the process is for prototype fabrication, and a small prototype shop could easily produce all that is necessary.

The unit cost, of course, would be quite high. If 10^6 parts/year are requested, determine if the number is inflated or is a realistic portion of the achievable market. Normally less than half the requested number should be produced the first year, until the market stabilizes.

2. Determine the material properties that are required, not desired. Most marketing evaluations associate weight with strength. If that were the case, mercury would be several times stronger than aluminum. Plastics are very light. Is it necessary to make the plastic part as thick or thicker than that presently designed? Flexural strength, for example, is proportional to the cube of the part thickness. Doubling the wall thickness on a part will double its weight but increase its strength by a factor of 8. Also, doubling the wall thickness will increase the molding time by a factor of 4.

3. Once the material properties based on the actual part performance requirements have been selected, a quick decision must be made as to whether any plastic can meet these properties. If not even glass-reinforced thermosetting plastics are strong enough or have enough temperature resistance, turn to more conventional materials. If several materials meet or exceed the required material properties, however, select the cheapest material that exhibits those properties. Use minimum physical properties selected from handbooks or company literature. As an example, if a part requires a notched Izod impact strength of 0.4 ft·lb/in. on a 1/8-in. specimen, medium-impact PS with a value of 1.0 and a cost of about $0.30/lb will perform satisfactorily, whereas high-impact PS with a value of 2.0 and a cost of $0.34/lb and ABS with a 7.0 value and a cost of $0.42/lb would be too expensive. Saving a little here can mean more engineering input in the production processing and tool cost area.

4. Now determine the amount of material required for the part. If the part is presently of die-cast metal, weight it, normalize it using the proper density of the die-cast material, and then include the specific gravity of the material selected. This yields the final part weight. If radical changes are being made in the part, material should be added or subtracted. At best guess, the weight will be within ±10%, and at worst, ±25%. Once the amount of material is known, add a small percentage to the total for lost scrap.

5. Determine an equitable cost for the material, in $/lb. Do not use truckload quantities unless the production projection will use 30,000 lb in less than 4 months. If the part weighs 0.1 lb, scrap included, 300,000 parts can be made from one truckload. In

general, a 2-month inventory is recommended, and thus 0.9 to 1.8 million parts/year is a valid breakpoint for 0.1 lb/part and truckload quantities. Many suppliers provide price breaks at 5000 lb, 2000 lb, 500 lb (drums), and 50 lb (bags). The resin may cost as much as 100% more for bag quantities than for truckload quantities. Thus, if the material costs $0.30/lb on truckload, it may cost $0.60/lb in bags. Improper cost analysis of this will directly affect the profit per item.

6. Determine the materials cost per part ($/item) assuming 100% good parts.

7. If the plastic is inexpensive (LDPE, HDPE, GPS, PVC, PP), multiply the material cost by a factor of 2.0 to 2.25 to get the direct labor per unit ($/unit). If the material is expensive, such as nylon, acetal, or polycarbonate, multiply the material cost by 0.8 to 1.5 to get the direct labor cost per unit ($/unit). It is assumed here that there is not a significant amount of hand labor involved in the part. Thus, several finishing steps, many assembly steps, hand lay-up of fiberglass-reinforced polyester, casting, or resin, and the like must have direct labor determined at a much higher rate. If excessive labor is involved in the process, a long-form cost analysis with detailed processing and labor steps identified and costed must be carried out.

8. Add the direct labor and the materials cost per unit. This value is a good first approximaton to the fixed and variable burden. If this value is doubled, a first approximation of the manufacturing cost can be obtained.

9. If selling price is the objective, doubling the manufacturing cost will yield a value that should include shipping costs, packaging, distribution, warehousing, dealer's overhead, and advertising.

In Table 16.3-1 is a representative cost analysis for the fabrication of a 450-g Nylon 6 blower wheel; 50,000 parts were required. The cost analysis shows that a manufacturing cost of just under $3/unit could be achieved. The final manufacturing accounting showed that on 20,000 the unit manufacturing cost was 3.60. The short cost analysis was 20% too low. In Table 16.3-2 is another analysis for a 1.5-oz LDPE dust cover; 100,000 parts were required, with no assembly. The projected manufacturing cost of $59/1000 was less than 1% below the actual manufacturing cost.

These are obviously excellent examples of the accuracy of this method of estimating. Nevertheless, experience shows that projected costs come to within 20% of the actual costs on more than 90% of the estimating jobs. Most important to the success of this method is the familiarization with the hard materials costs. This method is most useful in knowing beforehand whether custom molders are being fair in their quotations.

Table 16.3-1

Short Cost Analysis Example (1975 Dollars)

16-oz Nylon 6 blower wheel, 50,000 parts		
Materials cost ($0.82/lb in LTL quantities)	=	$0.82
Direct labor (expensive material × 1)	=	0.82
		$1.64
F&V burden (double the materials cost)	=	1.64
Class "C" estimate of mfg. cost:		$3.28
ACTUAL cost (with projected F&V burden)		
Materials cost ($0.78/lb and total part weight slightly more than 16 oz)	=	$0.793
Direct labor (injection molding)	=	0.472
		$1.265
F&V burden on first 10,000 parts (based on anticipated 120,000 parts total)	=	1.634
PROJECTED manufacturing cost:		$2.899/unit
ACTUAL cost—FINAL cost (only 20,000 parts made)		
Materials and direct labor costs	=	1.265
Actual F&V burden	=	2.347
FINAL manufacturing costs:		$3.612/unit

16.4 The Long-Form Cost Estimate

When the process has been crystallized, the materials have been selected,
and the part is considered feasible and profitable, a long-form cost estimate
is needed to determine the fair market value and the return on the investment (ROI). The estimate here included staffing, warehousing, taxes,
machine capacity, and inventory control. One of the prime requisites for
this estimate is the confidence of a company accountant familiar with the
various forms of overhead used in the company for similar operations.
This is particularly true if the operation is captive. It is essential to know
whether the part will make a real profit for the company, based on the assumption that the captive operation is a profit center. Only in this way can
it be determined if the process should go captive or custom. Local custom

Table 16.3-2

Short Cost Analysis Example (1973 Dollars)

1.5-oz LDPE dust cover, 100,000 parts projected		
Materials cost (at $0.105/lb)	=	$0.00985
Direct labor (2 cavity plunger injection molding machine) (material × 2)	=	0.01970
		$0.02955/unit
F&V burden (times 1 of the above line)	=	0.02955
Projected class "C" mfg cost (or $59.10/1000):		$0.05910/unit
ACTUAL cost: LDPE at 1.47 oz (with regrind; 250,000 parts made)		
Materials cost (at $0.116/lb)	=	$0.01066
Direct labor (4-cavity screw injection)	=	0.01822
		$0.02888
F&V burden (actual)	=	0.03040
ACTUAL manufacturing cost (or $59.28/1000):		$0.05928/unit

molders very often have more efficient operations, better automation, better quality control, and more interaction with mold makers in order to make small changes in mold cavity dimensions, gating locations, flash areas, and venting. Since the custom molder adds a return on his quotation, it normally will be slightly higher than an efficiently run captive operation, but the added cost may be beneficial in higher efficiency, reliable delivery, good quality control, guarantees, and commonsense engineering. If the captive operation cannot meet the product of the custom molder with equal efficiency, quality, and engineering, the custom molder should get the job.

To determine the cost of a captive operation it is necessary, then, to cost out the part on the basis that yields a return on investment. The following procedure should be followed:

1. Assuming that the part has been justified per se, the material properties that are required have been determined, at least one acceptable material meeting all or most of these properties has been found, and the results of the short estimate appear exciting, the design must be carefully considered. Try to make as many changes in the design as possible in order to simplify it and speed up the processing cycle. Working with the designer, determine the minimum amount of detail and/or decoration that

is required. In-mold decoration requires additional tooling costs; postmold decoration requires labor. The following engineering changes can help improve the processing of any plastic part:

1. Large radii everywhere
2. Uniform wall thickness wherever possible
3. Minimum of detail and matte finish surfaces
4. Curved rather than flat surfaces to hide warpage and sink marks
5. No plating if possible
6. No inserts if possible
7. Potential sink marks hidden under assemblies or with masking ribs

2. Try to combine as many pieces of material into one part as possible. Polyfunctionality will minimize assembly time and reduce the amount of ingenuity required in fastening methods.

3. Once the material of choice is known and the part has been fully engineered, the method of processing must be selected. Again this is usually dictated to a great degree by the part function. As has been seen, however, there are alternative ways of making the part. If more than one alternative process looks feasible, carry all through the calculation until a clear judgment as to the relative costs can be made.

4. An equitable market price should be agreed upon. Plastics are not cheap. Although the plastic part may be replacing a die-cast part, there is little justification for pricing it below the cost of the die-cast part. Consider a break-even cost first. The return on the investment should look good on this basis. If it does not, the change to plastics should be immediately questioned.

5. Working with marketing, carefully map out the number of pieces to be produced in the first, second, and third years. Determine the marketing areas and the approximate distances from the processing location to the end use. In this way, it is possible to determine whether a central processing location is satisfactory or whether regional locations are needed. Also, try to determine the firmness of the prospective design. If frequent design changes are to be made, tooling costs must reflect this.

6. The availability and capacity of in-house machines should be carefully documented. If the machines available are large and the part to be made is small, multiple cavity molds or the purchase of a small machine must be considered. If the available machines have small capacities and the part to be made is large, a large machine or outside custom molding must be considered. The shot capacity, clamp pressure, barrel pressure, platen dimensions, maximum and minimum daylight, and sequencing controls must be known. Remember, the mold must fit between tie rods, and the injection and clamp pressures must be sufficient to allow processing of the chosen material.

7. Determine cycle times. In general, this is a difficult step, requiring experience and input from several molders who have formed parts similar to that being proposed. Materials suppliers will have additional information. In injection molding, for example, most of the cycle time is due to cooling. Thus, the time required for cooling the thickest section of the part to a satisfactory level will give a good measure of the mold cycle time. To cool Delrin, duPont recommends multiplying the total time required to cool the centerline to the freezing temperature by 1.2 to get the total cycle time. Thus, if it takes 60 sec to cool Delrin to a centerline freezing temperature, an additional 12 sec should be allowed for opening and closing the mold and for injection and packing. It is better to be conservative at this point. Once approval has been given on the project based on the conservative cycle times, work can begin on reducing these times. Thus, the improvement here means increased ROI over and above that acceptable in the original cost estimate. Note that if the cost looks very close to a go/no-go decision, make several estimates of the cycle time in order to determine how critical a 10 to 15% variation in this time is in the final cost analysis. In management, this is called a "sensitivity analysis."

8. The number of parts to be made at one time must now be determined. In blow molding and injection molding, the number of cavities in a family must be determined. In rotational molding and thermoforming, the number of individual molds on a single frame must be determined. The number of parts to be made at one time is frequently constrained by the size of the part and the available machine capacity. Often, however, this number will not be obvious a priori. Therefore a reasonable estimate should be made. Remember that mold maintenance costs are approximately proportional to the number of cavities to the 0.5 to 0.6 power. Downtime on a mold with eight cavities is considerably more expensive than downtime on four molds with two cavities each. The more competent the mold repair shop is, the lower the cost of mold repair. Likewise, the simpler the mold, the lower the maintenance cost. Furthermore, if the part does not fit in any machine, consider custom molding. This is particularly true if the part to be made is precision and the in-house experience is predominantly in disposables.

9. Then determine approximate mold costs. Sanada [5] gives a general equation for mold costs:

$$\log_{10}[\text{mold cost/cavity (\$)}] = A \log_{10}[\text{part weight of each cavity (lb)}] + B$$

$$(16.4\text{-}1)$$

where A and B depend on the various processes, as shown in Table 16.4-1. These values are based on mid-1973 dollars. Note that for most processes the mold cost is approximately proportional to the part weight to the 0.6 power. One would expect $\pm 10\%$ accuracy on standard parts and as much as $\pm 50\%$ deviation on detailed parts or disposables. Certainly the cost of the mold should depend on the extent of detailing, number and complexity of

Table 16.4-1

Estimation of Mold Tooling Costs (1973 dollars) [5]

log [mold cost/cavity ($)] = A × log [part weight of each cavity (lb)] + B

Process	A	B
Injection molding of rigid plastics	0.57	4.26
Injection molding of foamed plastics	0.57	3.43
Injection molding of thermosets	0.57	4.26
Blow molding	0.34	3.62
Injection blow molding	0.40	4.15
Thermoforming	0.27	3.27
Rotational molding	0.27	3.27
Compression molding	0.57	4.26

side cores, precision required, and mold material. These values are for machined tool steel and thus must be corrected to allow for special materials and textures.

Another guide based on the machined surface area can be used. For simple shapes and single cavities, a machined surface would cost $20,000 to $30,000/ft^2. For detailed, deep sections, side cores, and single cavities, the cost could be $25,000 to $45,000/ft^2. This (1975) cost includes the die block, the insert, the cooling channels, and the finished machining and lapping.

For multiple cavities, the cost is usually proportional to the cavity number to the 0.6 to 0.7 power, depending on the matching and registry required. Thus,

$$\text{cost (\$/mold)} = [\text{single cavity cost (\$)}] \times [\text{no. cavities}]^{0.6-0.7}$$

$$(16.4-2)$$

Thus, if the single cavity costs $10,000, two cavities cost $16,200, four cost $26,400, and eight cost $43,000. Although an eight-cavity mold costs more than four times that of a single-cavity mold, its unit cost is about half. Further, if a single-cavity mold can produce a year's supply of parts in 4000 hr, an eight-cavity mold can do it in 500 hr. Part of the subjective decision that must accompany the choice of the number of cavities deals with the scheduling of the machine. If a small machine is available for 4000 uninterrupted hours, it should be considered. However, if frequent mold changes are required to meet other inventory demands, the setup time of the mold must be taken into consideration, and the downtime on the machine and manpower required for mold change must be included in the projected mold cost.

   10. Machine efficiency must now be determined. Normally a company has established a policy concerning average machine utilization. No machine runs at 100% efficiency. Materials must be changed, repairs must be made, molds must be replaced or repaired, materials may not process correctly, and the machine must be started and stopped. Manual operation of the machine normally requires more time than automatic operation. Normally second-shift machine efficiency is lower than first shift and third shift is lower than second. Many machine repairs and mold changeovers are made during second and third shifts. One recommended machine utilization rate for two shifts is 85%. For three shifts, it is 80%. Frequently, for a three-shift operation, the machine utilization rate is further broken down by shift at 90% on first shift, 80% on second, and 70% on third. This detailed breakdown is probably not needed, however, in this cost analysis.

   For most two-shift operations, a shift should include 48 min for wash-up, lunch, and morning and afternoon breaks. Furthermore, the shift should include weekend days (104), 11 paid holidays, and 3-weekday weeks of vacation or personal absence days. Thus, a normalized shift should have 235 days/year of 432 min each day. Thus, the most machine time possible is

$$432 \times 2 \times 235 = 203,000 \text{ min/year} \qquad (16.4\text{-}3)$$

At 85% two-shift efficiency, this is 172,500 min/year/machine and includes mold changeover time, maintenance, purging, start-up, shutdown, material changes, and so on.

   11. Machine capacity can now be determined. This uses the cycle time and the number of cavities in a given mold as

$$172,500 \text{ min/year} \times (\text{no. cavities})/(\text{cycle time}) = \text{total no. parts/year/machine}$$

$$(16.4\text{-}4)$$

Note that this does not specify whether the parts are good. This is just the machine capacity. If the number of parts required is a small fraction of machine capacity, a smaller machine is needed, fewer cavities can be used, or other jobs can be scheduled on that machine. If the number of parts required is significantly larger than the machine capacity, a larger machine is needed or more cavities must be provided.

   12. The number of good parts expected should now be obtained. This is not the number of good parts obtained per pound of a thermoplastic material, since undoubtedly rejects, trimmings, flash, scrap, runners, and purge will be reground and used again. For a thermoset, however, this number is quite close to the actual number of good parts per pound of resin since there usually is no way of recycling the rejects. If the final product demands very close tolerance, and if the material is thermally

sensitive or shows excessive gate blush and color segregation, the quality control might allow 85% of the total number of parts to be marked good. If the part is subcritical and made of a proven material, the acceptable quality limit could allow as much as 95% of the total to pass. Thus, the number of good parts that can be made by a given machine is

(no. good parts/year/machine) = (total no. parts/year/machine) × (fraction good)

$$(16.4-5)$$

Since only the good parts can be sold, the total sales on a given machine are

($/part) × (no. good parts/year/machine) = ($/year/machine)(16.4-6)

And all expenses are based on the total number of parts, good or bad, made on that machine in the same period of time. Note that machines make money, not parts.

13. To determine the number of people required on a direct labor basis, outline the entire operation in detail. Then estimate the amount of time required for each person to do each assigned task. For example, a time estimated must be made for each step in an operator's cycle, consisting of removing the part from the machine, degating it, deflashing it, inspecting it, assembling it to the other parts, labeling it, and placing it in a shipping container. Can a single operator operate more than one machine? What happens if the cycles are slightly different and both machines demand his attention at the same time? Does he (she) need to place inserts into the mold? Can a less skilled laborer degate and deflash the parts, or can it be done mechanically? Must he get his own material? Must he weigh, measure, and mix colorants, additives, and blowing agents into his material? Must he move his own stacks of finished parts? Is a secondary operation such as assembly, necessary, or can the customer assemble the part? Must he maintain his own machine? Does he have responsibility for setting machine operating conditions, such as cycle times, operating temperatures, screw speeds, and oven temperatures? If so, his working wage will be higher than average. If he accepts responsibility for meeting acceptable quality levels, his working wage will also be higher than average. Can automatic monitoring equipment be used to remove the tedious jobs? Once the process has been outlined in detail, the extent of direct labor involvement can be determined. Direct labor does not include supervisory personnel, materials handlers, inspection, packaging, shipping, warehousing, or traffic control, unless the operator does some of these jobs. The maintenance crew, engineering staff, secretarial pool, inventory control, and trucking are also not included. The direct labor should include all people who have intimate contact with the process or as-molded finished part.

Sanada [5] has shown, using another point of departure, that direct labor can be calculated according to the specific process, as shown in Table 16.4-2. His source for the labor costs is not referenced. They appear low

Table 16.4-2

Estimation of Labor Costs (1973 dollars) [5]

Process[a]	Operator	Inspector/packer	Supervisor	Material handler	Utility man
A	1/2	1/2	1/8	1/8	1/8
B	1	1/2	1/8	1/8	1/8
C	2	1/2	1/8	1/8	1/4
Labor cost ($/hr)	3.00	2.50	3.75	2.50	3.25

[a]A = injection molding, blow molding, injection blow molding, compression
    molding,
 B = thermoforming,
 C = rotational molding.

for northern operations and high for the south. In general, in urban areas,
skilled injection molding operators make $4.00 to $5.50/hr, and in rural
areas and certain portions of the south, the rate is $3.50 to $4.00/hr. Time
should also be allowed for shift relief. Some operations recommend a
floating direct laborer and charge an additional 0.15 to 0.25 man-hour/hr
of direct labor for this relief. Note that for two shifts a second shift differential of $0.20 to $0.50/hr is necessary. Try not to include overtime in
early calculations. If the machine must run 100 hr/week to meet inventory
quotas, schedule a third shift or increase the number of cavities so that the
machine can be run during a normal 864-min two-shift day.

 The direct labor cost can be determined as

$$(DL\$/hr) \times (40 \text{ hr/week}) \times (52 \text{ weeks/yr}) \times (2 \text{ shifts})$$

$$\times (1.05, \text{ shift differential}) = (DL\$/yr) \qquad (16.4\text{-}7)$$

Assume one man operating one machine. Then

$$(DL\$/\text{year/machine})/(\text{total no. parts/year/machine}) = (DL\$/\text{part})$$

$$(16.4\text{-}8)$$

This assumes 100% man utilization. In the northern tier of states, an 85%
man utilization can be assumed. In the southeast and south-central states,
70% man utilization should be used. Thus,

$$(DL\$/\text{part})/(\text{fraction man util.}) = \text{actual } (DL\$/\text{part}) \qquad (16.4\text{-}9)$$

To get the cost of direct labor per good part, use

$$\text{Actual (DL\$/part)/(fraction good parts)} = \text{(DL\$/good part)} \quad (16.4\text{-}10)$$

Equation (16.4-10) is used to determine labor costs, and Eq. (16.4-7) is used to determine fringe benefits, insurance, overhead if tied to direct labor charges, and so on.

14. More accurate material costs must now be determined. Some preliminary work should be used here. It is recommended that prototype testing of new molds and materials be done on the resin suppliers' machines. In this way accurate cycle times, temperatures, pressures, cooling rates, and the like can be found at little cost to the overall program. Material utilization should be approximately 90% of the purchase order. In this way, additional material will be available for quality control, spillage, purging, short shots, and display pieces. If, for example, a run uses 4500 lb of material, 5000 lb should be ordered. If only 4500 lb are ordered, additional material might be needed to meet these extraordinary losses, and the savings accrued on a price break at 5000 lb may be lost in ordering additional drums or bags. Once a material is working satisfactorily, avoid shopping for cheaper material, in spite of creeping inflation. The substitute material may not be acceptable to the customer and may not have the quality control desired.

Assume 95% utilization of material when determining material price. Assume short-term resin warehousing, a suitable price break, and a competitive market. This assumes that all rejects, scrap, and so on are reprocessable. If the material is thermally sensitive, a lower utilization rate must be assumed. If the material is a thermoset resin, assume zero scrap value. For tight molds and easy-flow materials, the scrap and rejects should be no more than 10% of the total process. For very difficult molding operations, it can be as high as 40% of the total process.

Do not base calculations on one type of material and then mold with another. The unit material costs may change dramatically. For example, if 10% glass-filled resin was specified and 30% glass-filled resin was required to meet certain material conditions, the total weight of the part increases by 10% or so. Furthermore, reinforcing additives normally increase the price of the material. Thus,

$$\text{(part wt)} \times \text{(mat'l \$/lb)/(fraction mat'l utilization)} = \text{(mat'l \$/part)}$$

$$(16.4\text{-}11)$$

15. Plant overhead must be determined accurately. It is imperative that normal operating costs for a plant similar to that being proposed are known. Dearle [1] gives several guides to the layout of a molding operation. Basically, a throughput scheme such as the one given below should be followed:

Truck dock—materials receiving—materials warehousing (old is used first)—transportation to bins or machine hoppers—materials loading into machines—machines—part rejection from machine—conveying of finished parts—regrind of scrap—finishing and assembly of plastic parts—packaging—cartoning and crating—warehousing of finished parts (old is first)—materials shipping—truck dock   (16.4-12)

With most molding operations, the floor space of the actual plants are small in relation to the volume of work produced. Rotational molding and thermoforming of large objects are exceptions to this, however, since large swing areas are required. Every machine should be accessible to overhead cranes to facilitate mold changes. As an example of a plant size, for the production of 12,000,000 lb/year of thermoformed bathtubs $5 \times 3 \times 3$ ft, the plant needed 135,000-ft^2 primary floor space and a final warehouse of 60,000 ft^2. In addition, 12,000 ft^2 were allowed for office space. In this particular operation, the (1972) DL\$/hr was fixed at \$3.60/hr. The unit selling price of the 40-lb units was \$30, with the actual selling price of \$31.58 reflecting 5% returns and allowances. A freight allowance of \$3/unit was allowed. For most of the comparative operations considered, the two-shift direct labor force was between 100 and 200 people. For this, an indirect labor force of 55 was estimated, as discussed below. The fringe on hourly employees was 25% and on salaried employees was 10%. The building cost was \$12/ft^2 (1969 rural values), and a depreciation of 25 years on the buildings and 15 years on the equipment was assumed. The tooling cost was written off on a yearly basis, with 5-year straight-line depreciation on the primary tools and 3 years on the smaller tooling. This implies that the primary tooling will last long enough to produce 1.5 million pieces. If this is not feasible, the tooling should be written off on the basis of a cost per unit. An accounts receivable rate of 12.5% on net sales, liabilities of 6% on net sales, inventories of 12.5% of net sales, and plant start-up expenses of \$300,000 on the first year and \$100,000 the second year were considered normal. A sound cost analysis should be carried through until all start-up expenses have been eliminated and the number of units demanded increases to a normal level. If the number of units to be made is fixed, the start-up cost should be added into the unit cost of the first 10 to 20% of the job.

Even if the job is based on a fixed number of pieces, building and equipment depreciation must be included in the cost estimate. Here the cost is prorated on the basis of the actual number of hours the machines are being used.

An example of an indirect labor breakdown is given in Table 16.4-3, and an example of plant overhead is shown in Table 16.4-4.

16. We now come to capital equipment estimation. At this point, the types and numbers of machines needed for the operation should be known.

Table 16.4-3

Example of an Indirect Labor Breakdown (1972 Dollars)

Category	Hourly	Salary	Annual salary	Indirect labor	Fringe	
					%	$
Supervision		1	$18,000	$18,000	10	1,800
Tech. and professional		11	12,000	132,000	10	13,200
Warehouse/shipping	8		7,000	56,000	25	14,000
Receiving/stores	2		7,000	14,000	25	3,500
Prop. equip. attend.	2		7,000	14,000	25	3,500
Mat'l handlers	2		7,000	14,000	25	3,500
Setup/tool const.	6		9,600	57,600	25	14,400
Janitor	2		7,000	14,000	25	3,500
Yard	1		7,000	7,000	25	1,750
Clerk/engrg		1	6,400	6,400	10	640
Quality control	6		7,000	42,000	25	10,500
Mat'ls mgmt		4	6,400	25,000	10	2,560
Accounting		4	6,400	25,600	10	2,560
Personnel		2	6,400	12,800	10	1,280
Mgr secretary		1	6,400	6,400	10	640
Shipping clerk	—	2	6,400	12,800	10	1,280
Totals	29	26		$458,200		$78,610

In addition, if the machines are to be purchased, decisions must be made in such areas as U.S. or foreign machines, semiautomatic or automatic, simple servo controls or solid-state controls, and manual sensing or feed-back/feed-forward control circuitry. Some blow molding operations, for example, have card-programmable parison extrusion. Some injection molding machines have computer memory in which the optimum running condition is established and recalled after running another job. Here

Table 16.4-4

Example of Plant Overhead Charges (1972 Dollars)

Account	$/Year
Indirect labor	458,200
Overtime Premium	15,000
Shift premium	4,000
Fringe benefits	
a. Indirect: $78,610	
b. Direct: 25% of DL$ [Eq. (16.4-8) times the number of machines]	
Operating supplies	50,000
Expense tools	15,000
Maintenance and repair tools	25,000
Utilities and heating	100,000
Purchased services	50,000
Tel. and tel. and travel	10,000
Taxes and insurance	60,000
Subtotal overhead (ex. fringe)	787,200

Depreciation:

  a. Building: 135,000 ft^2 at $12/ft^2 at 25 years = $64,800/year

  b. Equipment: 15 years, straight line = 6.67%

  c. Tool amortization: 5 years = 20%

                     3 years = 33.3%

memory storage includes proper barrel temperatures, injection pressures, clamping pressures, cycle times, injection speeds, and so on. It is apparent that the more controls that are needed on a given machine, the more expensive the initial investment will be on that machine. Furthermore, maintenance would be more expensive since greater skills are called for. And modifications and advancements in technology are taking place at rates that might make the very expensive machine obsolete before it is fully depreciated.

In general, low-maintenance features such as integral solenoid systems and solid-state controls should be sought. Foreign machines may not have adequate technical service representatives in the United States and furthermore are normally equipped in metric. Until the United States is fully committed to metric, in-house repair will necessitate two sets of tools, special nuts, bolts, springs, O-rings, fittings, and the like. Once a machine having the most equitable price has been found, a satisfactory negotiated price must be agreed upon. Normally, delivery times can be negotiated also. Frequently, if the cost warrants, a more expensive machine with a much shorter delivery time should be preferred, all other things being equal.

If the parts require deflashing, degating, preheating, drying, packaging, stamping, embossing, printing, welding, and so on, this cost must be included in the capital equipment cost as an ancillary equipment cost. If long overhead conveyors and bins are needed, these must be considered as ancillary also. Spray painting booths, spray guns, environmental protection equipment, color concentrate mixers, parts conveyors, drum tumblers, and grinders are included as ancillary.

For every new piece of equipment purchased, an installation charge is levied. Millwright charges can be as much as 30% of the cost of installation of a heavy press such as a 1000-ton injection molding machine. The larger the machine, the more site preparation is needed, and the higher is the percentage of the capital cost of the machine. Simply uncrating, moving a machine to a site, installing utilities, and leveling may amount of 10% of the purchase price of the machine. Normally, 20% of the total capital cost of both the primary machine and any ancillary equipment should be budgeted for installation charges.

Sanada [5] includes methods for calculating capital equipment cost, ancillary cost, installation charges, and so on. He finds that

$$\log_{10}[\text{primary mach. cost (\$)}] = A \times \log_{10}[\text{total part wt (lb)}] + B$$

$$(16.4\text{-}13)$$

where A and B are given in Table 16.4-5. Ancillary machine costs are determined as a percentage of the primary machine costs as follows:

1. All types of injection molding, 10%
2. Compression molding, 10%
3. Injection blow molding, 10%
4. Blow molding, 20%
5. Extrusion, 30%
6. Thermoforming, 40%
7. Rotational molding, 60%

Sanada recommends installation costs that are approximately 10% of the main machine costs. This is thought to be low, particularly on new installations and in areas where it is necessary to remove old equipment prior to installation of the new equipment. Sanada's dollar values should be considered as

Table 16.4-5

Estimation of Primary Machine Costs (1973) [5]

Process	A	B
Injection molding of rigid plastics	0.66	4.58
Injection molding of foamed plastics	0.64	4.48
Injection molding of thermosets	0.66	4.68
Blow molding[a]	0.49	4.72
Injection blow molding	0.50	5.02
Thermoforming	0.48	4.02
Rotational molding	0.48	4.02
Compression molding	0.61	4.40
Extrusion[b]	0.51	3.66

[a]Based on one molding station; main machine cost of N molding stations = N times main machine cost of 1 molding station.
[b]Based on output (lb/hr) instead of part weight.

mid-1973 values. The installation charge is a one-time charge and should therefore be added into the first-year capital cost only. It is not related to start-up charges, which are primarily due to costs involved in supplies, desk and office accessories, shop tools and machines, personnel training, and generally lower efficiencies owing to the new operation. The capital depreciation per good unit part is calculated as

$$(cap. \ equip. \ cost)/15\text{-year dep.} = (\$ \ cap. \ dep./year) \qquad (16.4\text{-}14)$$

$$(\$ \ cap. \ dep./year)/(no. \ good \ parts/year) = (\$ \ cap. \ dep./good \ part)$$

$$(16.4\text{-}15)$$

17. The mold life must now be estimated. This has been discussed earlier, and it is pertinent only if the mold life is to be longer than the total production run. For example, if the mold has an expected lifetime of 10 million shots and only 100,000 pieces are needed, the mold cost must be based on the 100,000 pieces. If the mold does not outlast the number of parts or if the order is a continuing one, the mold should be amortized over a 3- to 5-year period. The ownership of the tool should also be determined. In many cases, outside custom molders will include the amortization of the tool in their cost estimates. Determine whether the tool is completely

amortized at the completion of the first order. This prevents paying for the same tool order after order. If the mold is owned outright, outside molders may be reluctant to recommend changes in it in order to improve part quality. As a result, many molders will try to make quality parts on an inferior tool. Establish the ownership of the tool early in negotiations with outside molders.

Furthermore, a good mold maker will support his work with a letter of guarantee that the parts produced will meet certain dimensional specifications or surface finishes. At the very least, the mold maker should be willing to do minor rework on the tool if vents are found to be improperly placed, parting lines are too prominent, or blind gates must be opened. This work should be gratis in the interest of continuing business.

If the tool is now amortized over the number of good parts that constitute the first order, the result is

$$(\text{mold cost \$})/(\text{no. good parts}) = (\text{mold \$/part}) \qquad (16.4\text{-}16)$$

18. Additional part costs must also be included on a per-unit basis. This includes the costs of assembly parts such as nuts and bolts or inserts, solvents, cements, labels, package costs, shipping costs, crating, and so on. These should be figured on the unit basis and the 5% added for damages and returns.

19. The fixed and variable burden should now be calculated. This is the sum of the following:

a. Subtotal OVHD/no. good parts/year
   (Table 16.4-4)   =   OVHD \$/part
b. \$ cap. dep./part [Eq. (16.4-15)]   =   \$ cap. dep./part
c. Mold amrt./part [Eq. (16.4-16)]   =   mold \$/part
d. Fringe benefits:
   10% ind. labor/part
   25% DL/part   =   fringe/part
                                                _____

   Fixed and variable burden   F&V/part

20. To obtain manufacturing costs, add the direct labor charges, direct material costs, F&V burden, and additional part costs and packaging and freight:

a. Dir. mat'ls/part [Eq. (16.4-11)]
b. DL /part [Eq. (16.4-10)]
c. Pkg, extra parts (see Sec. 4.18)
d. Freight costs
e. F&V burden (see Sec. 4.19)
                                                _____

   Manufacturing cost/part   =   mfg \$/part

As an example of the analysis to this point, consider the thermoforming of a camper top from polymethyl methacrylate. As shown in Table 16.4-6, three sheet sizes are being considered. Because of the thinning in the corners of the sheet, the thermoformed top will be backed with a fiberglass-reinforced thermosetting polyester for added strength. The selling price to the dealers is to be $30 net. A fully automated plant is envisioned where the sheets are moved continuously through the heating section, into the thermoforming section where the sheet is formed into a camper top shape using a plug-assist pressure former, into a cooling zone, and into a trimming frame. The sheet is then manually placed on a fixture where it is manually sprayed with FRP and hand-rolled. Manual trimming, drilling, inspection, and packaging follow. An automatic forming machine can handle one 0.125-in. sheet every 4 min, one 0.080-in. sheet every 3 min, and one 0.060-in. sheet every 2 min, provided that a turntable device can be used to handle the sheet. Each machine can handle 41,000 units of 0.125-in. sheet every year, assuming 85% machine efficiency and 5% spoilage. The sales demand is for 300,000 units/year. Seven machines yield 287,000 good units/year, and so this is the number decided upon. For the FRP operation, each operator can handle 4-ft^2/min spray-up, and thus 2 men are needed for each machine. For each FRP operator, 2 helpers are needed to roll the FRP. At least 2 men are needed for trimming, packaging, inspecting, and repairing. Thus, 69 people are needed per shift to handle 0.125-in. sheet, 67 per shift for 0.080-in. sheet, and 65 per shift for 0.060-in. sheet. The total capital equipment cost for the seven machines needed for the 0.125-in. sheet is estimated at $944,000 from equipment manufacturers' quotations. Seven molds of aluminum are estimated to last for 10^6 formings, and each is estimated to cost $8000. The molds are amortized over a 3-year period. As is seen, all three show unit manufacturing costs of less than $30.

Recall, however, that this cost does not include the cost of liabilities, inventory, start-up costs, accounts receivable, and a viable return on investment. These must be added to the manufacturing cost in order to determine the feasibility of the project. In Table 16.4-7 is the first two-year analysis for each of the three sheet thicknesses. The return on the investment is figured beginning with the expected revenue from the sales of the camper tops. From this is subtracted the cost of the direct labor, direct materials, manufacturing overhead, depreciation and amortization charges on an annual basis, outbound freight, packaging, and start-up. The results are the pretax profits. Then the total annual investment is determined by adding the cost of the land, buildings, machinery, tooling, accounts receivable, and inventories in hand. This investment must be made in order to initiate the business. The return on gross assets is then the net pretax profit divided by the total annual investment. From the total annual investment must be subtracted the current liabilities and the accumulated depreciation. This yields the cumulative net investment. The return on net assets is the pretax profit divided by the total annual investment minus the accumulated depreciation.

Table 16.4-6

Cost Analysis for Three Thicknesses of Sheet for
Thermoformed Camper Top (1972 Dollars)

	0.125-in. Sheet	0.080-in. Sheet	0.060-in. Sheet
Units produced per year	302,000	288,000	259,000
Good units produced per year	287,000	273,000	246,000
Net sales @ $30/unit	$8,608,000	$8,198,000	$7,378,000
Materials costs			
Acrylic sheet	2,876,000	2,299,000	1,124,000
FRP	2,045,000	1,948,000	1,753,000
	$4,921,000	$4,247,000	$2,877,000
Direct labor			
No. people/2 shifts	138	134	130
$/year @ $3.60/hr	$1,036,000	$1,006,000	$ 976,000
Direct labor fringe (25%)	259,000	252,000	244,000
Capital equipment			
Building	$1,620,000	$1,620,000	$1,620,000
Depreciation @ 25 years	65,000	65,000	65,000
Equipment	944,000	880,000	616,000
Equipment dep. @ 15 years	63,000	59,000	32,000
Molds and tooling	56,000	40,000	32,000
Mold amort. @ 3 years	19,000	13,000	11,000
Subtotal overhead (ex. fringe)	787,000	787,000	787,000
Building depreciation	65,000	65,000	65,000
Equip. depreciation	63,000	59,000	41,000
Mold amortization	19,000	13,000	11,000
Fringe indirect	79,000	79,000	79,000
Fringe direct	259,000	252,000	244,000
F&V burden	$1,272,000	$1,255,000	$1,227,000
F&V burden/part	$4.43	$4.61	$4.98
Manufacturing cost/unit			
Materials	$17.15	$15.54	$11.70
Direct labor	3.28	3.32	3.58
Packaging	2.10	2.10	2.10
F&V burden	4.43	4.61	4.98
Total	$26.96	$25.57	$22.36

One-half of this can be considered the after-tax return on investment. As can be seen, none of the processes seem particularly attractive. A good operation should have an ROI of 20 to 25% after taxes. One way of improving the ROI is to increase the selling price. Another is to look for an alternative process.

Once an ROI has been determined, the final decision to proceed must be made. For the more established plastics processing areas such as extrusion, a 15% ROI is considered reasonable. For the untested processing areas such as liquid reaction molding or structural foam molding or where the materials or design have not been fully established, an ROI of 25 to 40% should be sought. Inflation must also be factored into the decision. If the present operation is functioning with a 5% ROI, increased labor and materials costs can quickly make this negative. Sanada shows that the manufacturing cost for all types of molding processes depends on the production rate of the machine and appears to be independent of the type of molding process. This is shown in Fig. 16.4-1. The cost declines dramatically with increasing production rate until production reaches 1 million lb/year. The manufacturing cost per unit resin cost asymptotically approaches a value of 2. Earlier, it was pointed out that for low-priced resin doubling the sum of the direct labor cost and resin materials cost would yield an approximate manufacturing cost. At very high rates of production, processes usually are highly automated, and thus the direct labor is a very small fraction of the unit sum, thus supporting the earlier arguments, even under extrapolation to very high production rates.

A complete analysis of a 1-gal plastic paint pail operation has been given by Neil and Waechter [6]. The decision to produce 5×10^6 pails/year was based on an eventual capture of 10% of the market. The engineering team considered injection molding, thermoforming, rotational molding, and blow molding as alternative processes for producing 1-gal and 5-gal pails. The choice of injection molding was based on large volume and a 13-month time constraint for entry in the market. The effort centered on injection molding 1-gal pails. Approximately 9×10^6 lb/year of white HDPE was required for the production. A 48-sec total cycle time for 0.090-in. pail walls was decided upon. Single and multiple cavities were considered, with a two-cavity bottom center-gated hot runner mold being most economical. The plant was to operate on a 5-day 24-hr/day week, 50 weeks/year. Six 1000-ton injection molding machines were needed for the pail bodies. In addition, 5×10^6 lids of PP were needed; 1.25×10^6 lb/year of PP were required in eight colors. A 28-sec cycle was needed, and single-cavity molds in nine machines were chosen for ease in changing color. A 2% scrap loss was assumed. The process was then layed out and the flow through the plant, beginning with materials storage bins and ending with palletized final products, was carefully planned. Storage of 20,000 lb of granulated scrap, 360,000 lb of PE (a 2-week supply), and 50,000 lb of eight colors of PP was provided for. Resin was purchased in 90,000-lb rail car quantities and stored in four 100,000-lb-capacity silos. Thus, a rail siding was also needed. Their unit

Table 16.4-7

Cost Analysis for Three Thicknesses of Sheet for Thermoformed Camper Top (1972) ($ in 1000s)

	0.125-in. Sheet		0.080-in. Sheet		0.060-in. Sheet	
	Year 1	Year 2	Year 1	Year 2	Year 1	Year 2
Net sales	$8608	$8608	$8198	$8198	$7378	$7378
Direct labor	1036	1036	1006	1006	976	976
Direct material	4921	4921	4247	4247	2877	2877
Manufacturing OVD	1125	1125	1118	1118	1110	1110
Deprec. & amort.	147	147	127	127	117	117
Outbound frt	861	861	820	820	738	738
Packaging	602	602	574	574	516	516
Start-up	300	—	300	—	300	—
Installation (assumed in equipment cost)						
Total costs	$8992	$8692	$8192	$7892	$6634	$6334
Pretax profits (PTP)	$(616)	$(316)	$ 6	$ 306	$ 744	$1044

Cash generation:						
50% PTP + deprec. and amort.	$(161)	$(11)	$130	$280	$489	$639
Cumulative cash generation	(161)	(172)	130	410	489	1128
Investment:						
Land	$200	$200	$200	$200	$200	$200
Buildings	1620	1620	1620	1620	1620	1620
Machinery	944	944	880	880	616	616
Tooling	56	56	40	40	32	32
Accts receivable	1076	1076	1025	1025	922	922
Inventories	1076	1076	1025	1025	922	922
Total annual invest (TAI)	4972	4972	4790	4790	4312	4312
Return on gross assets, PTP/TAI	Negative	Negative	0.1%	6.4%	17.2%	24.2%
Current liabilities	$516	$516	$492	$492	$443	$443
Accum. depreciation	147	294	127	254	117	234
Cumulative net investment	$4309	$4162	$4171	$4044	$3752	$3635
Return on net assets (less liabilities)	Negative	Negative	0.1%	7.6%	19.8%	28.5%
Return on net assets (after taxes)	Negative	Negative	Nil	3.8%	9.9%	14.3%

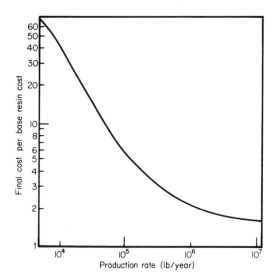

Figure 16.4-1   Ratio of final cost to base resin cost as function of production rate. Note that original graph includes data for the following processes: injection molding of rigid thermoplastics, injection molding of foamed plastics, injection molding of thermosets, blow molding, injection blow molding, thermoforming, rotational molding, and compression molding. All data lie along line shown within 20% (1973 dollars) [5].

cost is given in Table 16.4-8 at $0.734/unit. The total project cost is outlined in Table 16.4-9. No start-up costs are included. As shown in Table 16.4-10, the pails sell for $1.00/unit complete with handle and lid. The after-tax ROI is 15.3%.

This should complete the cost estimate, since all the information is now in a form accessible to marketing and management. According to Neil and Waechter [6],

> With this information, management can decide if the new venture is reasonable or not. If a larger net return on investment were required, the costs making up the total cost of the pail are well documented so that each could be analyzed to determine where particular costs could be reduced. Alternatively, additional market research could be performed to determine whether a selling price greater than $1/pail would significantly affect the market share.

Table 16.4-8

Unit Cost of 5 Million HDPE Pails (1973 dollars) [6]

Item	Annual cost ($)	Cost/pail (cents)
Direct material costs	1,864,000	37.3
PE for pails—14 cents/lb PP for lids—16 cents/lb Handles—6 cents each		
Direct labor	154,000	3.1
Operating supplies	1,500	0.03
Maintenance labor/materials	129,300	2.6
Finishing materials	295,000	5.9
Power—2522 kW	152,400	3.1
Heating	4,000	0.1
Total direct costs	2,600,200	52.1
Plant burden		
Salaries (administration)	231,000	4.6
Payroll, taxes, etc.	92,000	1.8
Miscellaneous	50,000	1.0
Total plant burden	373,200	7.4
General burden		
Depreciation, insurance, taxes	493,200	9.8
Total general burden	493,200	9.8
Total manufacturing cost	3,466,600	69.3
Overhead		
Sales expense	154,000	3.1
Administration	50,000	1.0
Total overhead	204,000	4.1
Cost of sales	3,670,600	73.4

Table 16.4-9

Cost Estimate Summary (1973 dollars) [6]

Item	Estimated cost ($)
Site preparation	51,900
Land clearance, excavation and filling, utility services, parking areas	
Foundations and building structure	487,000
45,000 ft^2, 180-ft clear building height, concrete block, aluminum sheet wall, 6-in. reinforced slab, lighting, sprinklers, building heat, engineering	
Process wiring and electrical equipment	113,400
Primary electrical distribution system (substation transformer by utility), motor and control wiring	
Process ventilation and air conditioning	57,300
Office air conditioning supply and exhaust air	
Process piping and instrumentation	22,500
Cooling water, gas and compressed air piping distribution systems, instrumentation, piping installations	
Subtotal	732,140
Equipment cost and installation	2,520,000
16 injection molders, scrap granulation and storage, polyethylene and polypropylene storage, pneumatic distribution system, and spare parts	
Engineering and construction management	145,000
Contingency (approx. 10%)	100,000
Capital cost	3,497,140
Land cost @ $4000/acre, 5 acres	20,000
Total project cost	3,517,140

Table 16.4-10

Return on Investment Analysis (1973 dollars) [6]

Average annual after-tax profits	
Sales (assume $1/pail)	5,000,000
Cost of sales	3,670,600
Annual profit before taxes	1,329,400
Taxes (assume 50% corporate taxes)	664,700
Annual profit after taxes	664,700
Total net investment	
Total project cost	3,517,140
Working capital requirements	823,000
Total net investment	4,340,140
[a]ROI = ($664,700)/($4,340,140) =	15.3%

[a]ROI = (annual average after-tax profits)/(total net investment).

16.5 Machine-Hour Costs

On short jobs, it is frequently necessary to use machine-hour costs. If the
number of pieces per hour can be determined, then the material cost per
unit is simply added to the machine-hour cost per unit to determine the
manufacturing cost. Hidden in the machine-hour cost is direct labor, fringe,
all overhead costs, depreciation and amortization, utilities, maintenance
costs, royalties, and installation. The use of this type of cost should be as
an alternative to the long-form cost estimate of Sec. 16.4. As an example
of the development of a machine-hour cost, consider the expenses of a 190-
ton, 14-lb single-nozzle structural foam machine. As shown in Table
16.5-1, the machine and its ancillary equipment cost $257,600 (1974 dollars).
Since the developments in structural foams are very rapid, a 5-year depreciation was used. Thus, based on 100% machine efficiency for a two-shift
operation, the cost of the machine per hour was $13.42. Based on the
physical dimensions of the machine, it was decided that 655 ft^2 of primary
floor space at $22/ft^2 and 3750 ft^2 of warehouse floor space at $12/ft^2 were
needed. The warehouse space was based on the 375-lb/hr plasticating
capacity. Using a 15-year straight-line depreciation, the floor space cost
$1.03/hr. Assuming 60% of the rated plasticating capacity, the utilities
were assumed to be $0.008/lb of resin or $1.79/machine-hour. This

Table 16.5-1

Machine Costs—190-ton, 14-lb Shot Low-Pressure
Structural Foam Machine (1974 Dollars) [7]

Base machine cost	$184,000
Ancillary machine cost (grinders, hopper loaders, powder blenders, etc.)	36,800
Installation	18,400
Nonproduction machine cost	18,400
Fixed capital investment	$257,600
Amortization, $/year (5-year SL depreciation)	51,520
Machine cost (100% eff.) (2-shift, 3840-hr/year operation)	13.42

included heater and chiller power, machine power, power to grinders, and makeup water. Maintenance and insurance were assumed annually to be 5 and 2.5% of the original fixed machine cost, or $2.40 and $1.20 per machine-hour, respectively. One and one-third men were needed at a DL cost of $4.50/hr. Indirect labor was assumed to be 25% of DL, administration was 20% of the total labor cost, fringe was 25% of the total labor, and overtime was 10% of the total labor cost. The total labor costs (100% man efficiency) were therefore $11.63, as shown in Table 16.5-2. Office expenses were assumed to be 25% of administration costs, adding another $0.37/machine-hour. Therefore, as shown in Tables 16.5-3A and B, the 100% machine efficiency, 100% man efficiency machine-hour cost for two shifts is $31.84; a 90% machine efficiency, 85% man efficiency machine-hour cost for two shifts is $35.87.

The way this machine-hour cost can be used is shown in Table 16.5-4, where a $0.30/lb material is processed on a 3-min total cycle using 80% of the maximum shot size of the machine; 5% scrap is included. Since this is a structural foam, blowing agent charges and royalties must be included. A single-cavity mold is used, and the mold cost is based on Sanada's formula, Eq. (16.4-2). The mold is amortized over a 3-year period. The process is expected to return 20% of the manufacturing cost, less materials, mold, packaging, and interest. The working capital was computed on the basis of 1-month equivalent of material, final product value (assumed to be $1/lb), wages, and salaries, and the interest on a 100% efficient operation was found to be $2.07/hr. Packaging and shipping charges were assumed to be $0.03/lb of finished material. Thus, in Table 16.5-5, the

Table 16.5-2

Labor Costs, Administrative Costs (1974 Dollars) [7]

Direct labor, 1 1/3 men ($4.50/hr, 100% eff.)	$ 6.00
Indirect labor, 25% DL	1.50
Administration, 20% total labor	1.50
Fringe, 25% total labor	1.88
Overtime, 10% total labor	0.75
Total labor, 100% eff.	$11.63
Office expense, 25% admin.	0.37
Office, labor, $/hr (100% eff.)	$12.00

Table 16.5-3A

Machine-Hour Costs, 100% Utilization (1974 Dollars) [7]

Machine cost, $/hr	$13.42
Floor space cost, $/hr	1.03
Maintenance and repair, $/hr	2.40
Insurance, $/hr	1.20
Utilities, $/hr	1.79
Office and labor, $/hr	12.00
Machine-hour cost (100% eff.)	$31.84

Table 16.5-3B

Machine-Hour Costs, 90% Machine Utilization, 85% Man Efficiency,
Two-Shift Operation (1974 Dollars) [7]

Machine cost, $/hr	$14.91
Floor space cost, $/hr	1.03
Maintenance and repair, $/hr	2.82
Insurance, $/hr	1.20
Utilities, $/hr	1.79
Office and labor, $/hr	14.12
Machine-hour cost	$35.87

Table 16.5-4

Custom Molding Costs for 190-ton, 14-lb Low-Pressure Structural
Foam Process—GPS Resin (1974 Dollars) [7]

Material cost, $/hr (100% utilization, $0.30/lb, 5% scrap included, 80% maximum shot, 3-min total cycle)	$67.20	
Blowing agent (1% of $1/lb), $/hr	2.24	
Royalty ($0.02/lb of resin)	4.48	
Mold cost (single cavity, A1 tool)		$12,800
Mold amort., 8/hr (3-year SL amort.)	1.11	
Minimum ROI (20% mfg cost, less materials, mold, packaging)	6.36	
Working capital (1-month equiv., mat'l, wages, final product value, salaries):		
Material		23,650
Final product, $/lb		71,680
Wages, salaries, $12/hr		3,840
		$99,170
Annual interest (8% of working capital)		7,930
Interest, $/hr	2.07	
Packaging, shipping, $/hr ($0.03/lb)	11.25	

Table 16.5-5

Manufacturing Costs—Two Shifts, 90% Machine Utilization,
85% Man Efficiency, ($/hr) (1974 Dollars) [7]

Machine cost	$ 35.87
Materials	73.92
Mold amort.	1.11
Minimum ROI	6.36
Interest	2.07
Pkg/shipping	11.25
Manufacturing cost	$130.58
Mfg cost/lb of resin	$ 0.583

manufacturing cost per hour, based on a two-shift operation with 90% machine utilization and 85% labor efficiency, is $130.58/hr. Based on the weight of the resin used, it is $0.583/lb of $0.30/lb of resin. Again, the manufacturing cost is approximately twice the resin cost. This approach yields sounder engineering information than that recommended by others [8,9].

16.6 Comments

There are many other ways of obtaining justification for the fabrication of plastics resins into finished parts. Each new development requires careful analysis of the process, the materials, the ancillary features, and the final part quality. Presented here are some proven ways of obtaining numbers upon which to base engineering decisions. Furthermore, these methods give management a clear picture of the various costs involved in developing a new product.

Problems

P16.A A customer requires 10,000 100-liter containers of HDPE with a threaded closure. The lids are to be of HDPE also. The container wall must be no less than 0.060 in. or no more than 0.125 in. Determine the approximate cost of this container if it is blow-molded or rotationally molded. The resin density is 0.96 g/cm^3, and costs are $0.32/lb in less-than-truckload quantities.

P16.B A swing seat is to be made of PP structural foam with an average density of 0.6 g/cm^2. The cost of PP is $0.305/lb in LTL quantities; 500,000 are required. The part dimensions are 18 × 8 × 1.25 in. The thermal diffusivity of PP structural foam is approximately 7 × 10^{-3} ft^2/hr. Determine the cycle time for this seat. Assume the mold to be eight cavity. Amortize the mold over a 2-year period of time. Use a 190-ton, 14-lb shot structural foam machine. Assume a two-shift operation with 85% man efficiency and 90% machine utilization. How many pieces will be manufactured per year?

P16.C A painted sheet metal guard for the end of a hot water baseboard radiator is to be replaced with an ABS guard. ABS costs $0.48/lb. The part wall thickness is 0.080 in., and the surface area is 95 in.^{2}. Estimate the cost of this guard in ABS having a specific gravity of 1.07 g/cm^3. Compare this cost with $0.22/guard presently paid for in sheet metal.

P16.D An order for an annual inventory of 10 × 10^6 bicycle grips has been received. The total material used in each grip is 45 g, including scrap. The material cost is $0.50/lb, and it has a specific gravity of 1.12 g/cm^3.

The total cycle is estimated to be 75 sec. The shop has the following work
schedule for the next year: The 100-ton, 4-oz (GPS) machine has 1800 hr
available, the 350-ton 25-oz (GPS) machine is available for the entire year
(3840 hr/year), the 750-ton, 150-oz (GPS) machine is available for 500 hr.
The 100-ton machine-hour cost, including 20% ROI, is $10/hr, the 350-ton
machine is $35/hr, and the 750-ton machine is $95/hr. The replacement
costs, including ancillary equipment, of each of these machines are $37,000
$122,000, and $175,000, respectively. How many molds are needed, and
what is the cavity count for each mold? Use no more than 80% of the rated
capacity of each machine is your calculations. Determine the schedule that
will allow the lowest unit cost on the grips. What is that unit cost?

P16.E A coffin manufacturer is considering polymethyl methacrylate
thermoforming. He will use $0.250 \times 7 \times 5$ft. stock to form the top and
bottom simultaneously. He wants 30,000 the first year and 40,000 every
year thereafter. One thermoformer can produce one coffin every 6 min
and costs $125,000, excluding ancillary equipment. Labor includes sheet
handlers, trimmers, and buffers. Building costs are $20/$ft^2$ including
warehousing. Lay out the entire plant, including traffic control. Determine
labor force, space requirements, and materials costs if the sheet stock
costs $1.25/$ft^2$. What should you charge the coffin manufacturer if you
expect a 40% pretax return on your investment? Use the long-form cost
estimate.

P16.F A cosmetic manufacturer needs 100 display containers and lids.
The container and lid must be made of hand-buffed vacuum-metallized ABS.
Each half weighs 50 g. The parts are to be assembled using two Rathbun
hinges each at $0.05/unit. The special ABS costs $0.65/lb. What process
would you use? What is the mold material? How much will you charge for
the finished product?

References

1. D. A. Dearle, Plastic Molding Technique, Chemical Publishing Co.,
   New York, 1970, Chap. 1.
2. A. D. Little, Inc., Boston, Mass., 1966.
3. W. deC. Crater, Rotational Molding of Nylon 11, In: Basic Principles
   of Rotational Molding (Paul F. Bruins, ed.), Gordon & Breach, 1971,
   p. 207.
4. H. Hablitzel, SPE ANTEC, Detroit, Mich., Nov. 1974, p. 96.
5. M. Sanada, SPE ANTEC Tech. Pap., 19:237 (1973).
6. W. M. Neil and C. J. Waechter, Soc. Plast. Eng. J., 29(3):27 (1973).
7. J. L. Throne, J. Cell. Plast., 12:264 (1976).
8. Anon., Mod. Plast., 44(4):96 (1967).
9. Anon., Bulletin TN4, Control Process, Inc., Plantsville, CN,
   March 1, 1972.

Index

A

Acetals, general properties, 16
Acrylics, general properties, 15
Addition polymerization, classification, 99
Adhesives (see also Adhesive bonding), 808
  classes of, 808
  thermosets, 808
Adhesive bonding, 800-805
  adhesive failure, 800, 803
  adhesive strength, 802
  applications, 800
  bond interface failure, 800, 803
  bonding to olefins, 800
  characteristics, 800
  cohesive failure, 800, 803
  dyeing, 803
  in-mold decoration, 805
  laminations, 800
  peel strength, 801
  physical bond, 800
  surface activation (see also Corona discharge), 803
  surface treatments, 803
Adiabatic conditions (see also Entropy), 706
Alternate processing methods (see Competing processes)
Amorphous polymers (see also Material properties, amorphous)
  glass transition temperature, 218

[Amorphous polymers]
  melting conditions, 218
  testing, 208
Analog methods, shape factor determination, 517
Assembly techniques, 799

B

Baekeland, Leo, 2
Batch processing
  definition, 353
  types, 353
Batch reactor (see also Batch reactor, homogeneous; Continuous stirred tank reactor)
  catalytic polymerization, 185
  control, 99
  definition, 98
  degree of polymerization, 98
  economics, 99
  heat transfer, 138, 139
  initiation reaction activity, 99
  mixing, 136
  molecular weight distribution, 98
  monomer concentration, 98
  scale-up, 99
Batch reactor, continuous, comparison with continuous stirred tank reactor (see also Continuous stirred tank reactor), 119, 120